53rd American Association of Swine Veterinarians Annual Meeting 2022

Defining Our Future

Indianapolis, Indiana, USA
26 February – 1 March 2022

ISBN: 978-1-7138-4500-3

Printed from e-media with permission by:

Curran Associates, Inc.
57 Morehouse Lane
Red Hook, NY 12571

Some format issues inherent in the e-media version may also appear in this print version.

Printed with permission by Curran Associates, Inc. (2022)

For permission requests, please contact American Association of Swine Veterinarians at the address below.

American Association of Swine Veterinarians
830 26th Street
Perry, IA 50220-2328
USA

Phone: 515-465-5255
Fax: 515-465-3832

aasv@aasv.org

Additional copies of this publication are available from:

Curran Associates, Inc.
57 Morehouse Lane
Red Hook, NY 12571 USA
Phone: 845-758-0400
Fax: 845-758-2633
Email: curran@proceedings.com
Web: www.proceedings.com

Table of Contents

Howard Dunne Memorial Lecture

Alex Hogg Memorial Lecture

Sunday: Research Topics

Sunday Concurrent Session #1: Student Seminar

Sunday Concurrent Session #2: Industrial Partners

Sunday Concurrent Session #3: Industrial Partners

Sunday Concurrent Session #4: Industrial Partners

Poster Session: Veterinary Students

Poster Session: Research Topics

Poster Session: Industrial Partners

Monday General Session: Defining Our Future

Monday Concurrent Session #1: PRRSV RSLP 1-4-4

Monday Concurrent Session #2: Sustainability and Animal Welfare

Monday Concurrent Session #3: Disease Prevention, Control, Elimination

Tuesday General Session: Foreign Animal Disease Preparedness & Response

Howard Dunne Memorial Lecture

Leaping into the future: Sit down, buckle up, and hang on

Angela Baysinger, DVM, MS, MSc

In remembrance and honor of
Howard W. Dunne

Dr. Howard W. Dunne received his veterinary degree from Iowa State University in 1941. He was in engaged in private practice for one year and industry for four years before entering graduate school at Michigan State University. After receiving his PhD in 1951, he spent two years as deputy chief of veterinary microbiology at the Chemical Corps Biological Laboratories in Camp Detrick, MD. In 1953, he joined the faculty of Pennsylvania State University, where he headed the veterinary science research program and the large animal diagnostic laboratory. Dr. Dunne was internationally known for his work on swine diseases, particularly hog cholera and the swine reproductive viruses then referred to as the SMEDI syndrome. Dr. Dunne authored more than 90 scientific publications, and is perhaps best known for initiating and editing the textbook *Diseases of Swine*, which is still widely considered the "bible" of swine disease.

In 1971, the American Veterinary Medical Association presented Dr. Dunne with the XII International Veterinary Congress Prize to recognize his contributions to the international understanding of veterinary medicine. At the 1974 Congress of the International Pig Veterinary Society (IPVS), Dr. Dunne was elected president of the 1976 IPVS Congress, to be held in Ames, Iowa. His sudden death at the age of 61 in September of 1974 saddened the swine veterinary community, and the 1976 IPVS Congress was dedicated to his memory.

At the suggestion of Dr. Allen Leman, a financial gift made by the 1976 IPVS Congress organizers to the fledgling American Association of Swine Practitioners was combined with additional association funds to endow this prestigious annual lecture in remembrance and honor of Dr. Howard W. Dunne.

HOWARD DUNNE MEMORIAL LECTURERS

1978	Michael Muirhead	1989	Steven Henry	2000	David Reeves	2011	Locke A. Karriker
1979	W. J. (Bill) Smith	1990	Stanley Curtis	2001	Max Rodibaugh	2012	R. Dean Boyd
1980	N.C. Nielsen	1991	Roy Schultz	2002	Tim Blackwell	2013	Craig Rowles
1981	Allen Leman	1992	Joseph Connor	2003	Lonnie King	2014	Daryl Olsen
1982	Michael Wilson	1993	Camille Moore	2004	Robert Desrosiers	2015	Gregory Stevenson
1983	Edward Bohl	1994	Robert Morrison	2005	Scott Dee	2016	John Harding
1984	Richard Ross	1995	Rodney Johnson	2006	Gregg BeVier	2017	Jeff Zimmerman
1985	Kent Kislingbury	1996	Paul Yeske	2007	Tom Burkgren	2018	William R. DuBois
1986	Ralph Vinson	1997	Barbara Straw	2008	Timothy J. Loula	2019	John T. Waddell
1987	Gary Dial	1998	Mike Daniel	2009	John U. Thomson	2020	Bret D. Marsh
1988	Thomas Alexander	1999	Larry Rueff	2010	James F. Lowe	2021	Jerome O. Geiger

https://doi.org/10.54846/am2022/1

Leaping into the future: Sit down, buckle up, and hang on

Angela Baysinger, DVM, MS, MSc
Bruning, Nebraska

Introduction

I am beyond humbled. How on earth did this Missouri hillbilly and now Nebraska Cornhusker get asked to present one of the most prestigious talks at the American Association of Swine Veterinarians (AASV) Conference? While researching and brainstorming the potential content of this presentation, I have questioned Dr. Mike Senn's choice and my qualifications for this task many times. Just take the time to review the previous individuals that have given the Howard Dunne Memorial Lecture. The first year I attended the AASV Conference (1993), Dr. Camille Moore strived for nursery/grow-finish performance. After finishing my master's program at the University of Nebraska (1997), one of my mentors and committee members, Dr. Barb Straw, examined veterinary practice as part of art, science, and politics. In all actuality, I can chronicle much of my career around the individuals who have given the Howard Dunne lecture and do not feel worthy. So, again I ask myself, how did I get here? However, if you know me, you will not be surprised by the title of this lecture nor the content therein. For those who do not know me, please do not take offense nor consider me an expert on the topics chosen. My desired outcome is to start the dialogue and fuel the discussions to take the AASV into the future.

AASV's mission

First, we must evaluate the purpose or mission of the AASV.[1]

It is the mission of the American Association of Swine Veterinarians to:

- increase the knowledge of swine veterinarians
- protect and promote the health and well-being of pigs
- advocate science-based approaches to veterinary, industry, and public health issues
- promote the development and availability of resources that enhance the effectiveness of professional activities
- create opportunities that inspire personal and professional growth and interaction
- mentor students, encouraging life-long careers as swine veterinarians

The AASV By-laws are somewhat broader reaching in their content.[2]

The objectives of this Association shall be:

- To endeavor to do all things necessary to promote the interests, to improve the public stature, and increase the knowledge of veterinarians in the field of swine practice.
- To elevate standards and ethics of swine veterinary medicine.
- To promote the relationship between the veterinary profession, swine practitioners, the swine industry, and the public.
- To promote understanding and goodwill among its members.
- To cooperate with veterinary and agricultural organizations and regulatory agencies.
- ALL POWER CLAUSE: To do everything and anything reasonable and lawfully necessary, proper, suitable, or convenient for the achievement of the purposes above stated, or for any of them, or for the furtherance of the said purposes.

Though intertwined in their direction, the mission statements are more introspective, and the Bylaw objectives are introspective with an eye toward the extension of our knowledge, leadership, and involvement globally. Therefore, after evaluating the mission and objectives, I reached out to AASV members and harvested their feedback and vision for the future of the AASV.

The survey

In preparation for this presentation, I surveyed two groups of AASV members. This survey was in no way scientifically conducted, and the results were not statistically evaluated. However, the outcome of the survey was a sampling of the "old" guard (> 40 years of age) and the "new" guard (< 40 years of age). The "old" guard was the previous Howard Dunne lecturers starting with the first year I attended the AASV conference (1993). The "new" guard was the members of the AASV's Early Career Committee. I will focus on the two questions that provided abundant material for further inquiry:

> What is the purpose of AASV?
> What are the top three opportunities for the AASV in the next ten years?

The AASV purpose question garnered the most extensive response that I distilled down to the following action list:

- Educate
- Advocate
- Empower
- Cultivate relationships
- Promote health & welfare of people and pigs
- Collaborate
- Adaptation
- Create opportunities
- Mentor
- Involve all members
- Lead the re-imagining of production systems
- Proactively provide industry leadership

The priorities that the AASV should focus on in the next ten years provided abundant material for my Howard Dunne lecture and future lectures or conference sessions.

- Animal welfare
- Foreign animal diseases (FADs)/transboundary diseases
- Adapt to changing business climate/structure nationally and globally
- Mental health
- Diversity, equity, inclusion (DEI)
- Veterinary education

- Move beyond just bugs and drugs (One Health, antimicrobial resistance, sustainability, policy, business acumen)
- Production technologies
- Training and sourcing of caretakers (employee deficits); staffing
- Social conversations
- Developing the future veterinarians/leaders of AASV

Since there is no way to fit all of the opportunities into this one lecture, I selfishly chose those I have a passion for and want to encourage the conversation within AASV. My intent is purely for the advancement of the AASV and our members.

Animal welfare

It goes without saying that animal welfare was going to be in this presentation regardless of the conference's theme or the initial focus of my presentation. But, with the AASV returning to an in-person conference - luckily, after the pandemic year of no personal contact and an ongoing infrastructure breakdown for the US, animal welfare is front and center.

If you have an eye on animal welfare, you will be intimately aware of the global initiatives that can, will, and currently are impacting the future US food animal industries. At the writing of this paper, some of the global policy issues are as follows:

EU Fitness check[3]: Evaluating the existing body of animal welfare laws within the EU for their effectiveness and relevance in light of scientific knowledge and public opinion. This process looks at the consistency with food, environmental and market rules, and sustainability goals of the EU's Green Deal and Farm to Fork Strategy.

European Green Deal[4]: Improve the well-being and health of citizens and future generations. For food animal agriculture, the Farm to Fork Strategy looks to make the EU food supply chain from farm to fork more sustainable; for a fair, healthy, and environmentally friendly food system.

"End the Cage Age" European Citizen's Initiative[5]: European Commission, on June 30, 2021, decided to revise current EU legislation with a commitment to:

- Put forth a proposal by the end of 2023 to phase out and prohibit the use of cages for hens, mother pigs, calves, rabbits, ducks, geese, and other farmed animals.
- Phase-out the use of cages for farmed animals across Europe by 2027.
- Ensure imported products in the EU comply with future cage-free standards

Within the US, all swine veterinarians and producers are aware of the recent ballot initiatives that have passed, been struck down, or recently voted into law:

California Prop 12[6]: The revised law requires that covered animals be housed in confinement systems that comply with specific standards for freedom of movement, cage-free design, and minimum floor space. Covered animals include veal calves, breeding pigs, and egg-laying hens. In addition, it covers the sale within the state of shell eggs, liquid eggs, whole pork meat, or whole veal meat.

Massachusetts Question 3 Initiative[7]: Delayed until January 1, 2023. Requires farmers to give chickens, pigs, and calves enough room to turn around, stand up, lie down, and fully extend their limbs. It will also prohibit the sale of eggs or meat from animals raised in conditions that did not meet these standards.

Colorado Initiative 16[8]: Failed. Protect Animals from Unnecessary Suffering and Exploitation. Livestock and fish could have been added to Colorado's animal cruelty law. The initiative defined a natural lifespan of 20 years for cows, 15 years for pigs and sheep, ten years for turkeys, eight years for chickens, and six years for ducks and rabbits. The proposed initiative only allowed the slaughter of an animal that had reached a quarter of its lifespan. The law also wanted to change what constitutes a "sexual act with an animal," which included practices used in breeding and animal husbandry.

Oregon Initiative 13[9]: The Oregon Changes to Animal Abuse, Neglect and Assault Law Initiative may appear on the ballot in Oregon as an initiated state statute on November 8, 2022. The initiative would make the following changes to Oregon law concerning animal abuse, neglect, and assault

- adds an exception for actions against an animal that is "necessary to defend him or herself against an apparent threat of immediate violence."
- removes the exception for "good animal husbandry," rodeo animals, fishing, hunting, trapping, scientific research using animals, and pest control.

Maine's Constitutional Amendment for a Right to Food[10]: Passed. This constitutional resolution declares that all individuals have a natural, inherent, and unalienable right to grow, raise, harvest, produce and consume the food they choose for their own nourishment, sustenance, bodily health, and well-being.

The AASV and all food animal industry participants should engage with their state legislature and participate in the national debate to represent the veterinary profession and the livestock industry. To determine if your state is a ballot initiative state, review the following website[11]:

https://www.multistate.us/insider/2017/11/6/direct-democracy-in-action-a-guide-to-state-ballot-measures

Looking back at 2020, the AASV and all terrestrial food animal veterinary groups were significantly impacted by the COVID-19 human infections that closed or slowed slaughter and processing starting in April of 2020. In addition, the cooperation and teamwork of all the "barnyard" veterinary groups (AASV, American Association of Bovine Practitioners, American Association of Avian Pathologists, American Association of Small Ruminant Practitioners) were exercised during the 2020 pandemic as the swine and poultry industries were forced to conduct mass depopulation due to having reduced or no access to slaughter for their animals for extended periods of time.

When dealing with unforeseen events of a human pandemic, the producers responded with measures to mobilize animal welfare strategies and put plans into action swiftly. The AVMA (American Veterinary Medical Association) Guidelines for Mass Depopulation of Animals (2019)[12] was the science-based document assembled by AVMA members experts from all species. In addition, it is the guidance document referenced by the US Department of Agriculture (USDA) for mass depopulation methods.[13] The AVMA Guidelines provide methods by species in the categories of "Preferred," "Permitted in Constrained Circumstances," and "Not Recommended."

Despite immense efforts to mitigate the need to depopulate animals, the reduction in slaughter plant throughput to 53.9 percent on April 29, 2020, forced producers to make difficult decisions.[14,15] Since the need for depopulation was not caused by an infectious transboundary disease, there was no governmental or regulatory support for the depopulation process. Depending on the availability of resources (eg, CO_2, ammunition, trained personnel), producers evaluated each depopulation method using an AVMA humane endings flowchart[16] and the guidelines[12] to determine feasibility and effective execution. A significant number of private practitioners, academics, and industry ally involvement occurred to determine the procedures and protocols to implement mass depopulation methods that had never been utilized in the US. There is a significant amount of National Pork Board (NBP), academic, and independent research on mass depopulation being conducted, reported, and published to build the science needed.[15-21]

The lessons learned from the infrastructure breakdown during the human pandemic will be a turning point for the food animal industry. All food animal groups (swine, cattle, poultry, small ruminants), henceforth known as the "barnyard," joined together to support and rally each other. As a result, there were lessons learned about communication and transparency with the organizations' membership, the barnyard, the AVMA, and the public. These lessons should be utilized to prepare for the next inevitable event that interrupts our food animal production world.

Diversity, equity, and inclusion

There is no way to adequately do justice to the DEI topic within the context of this lecture. Additionally, there is little to no data on the demographics of the AASV other than the Salary Survey[22] conducted every three years. Therefore, I choose to open the door to this discussion by reviewing the history of the AASV Salary Survey conducted seven times since 2002. The only clear-cut data extracted from the salary survey is gender, age, and practitioner vs. public/corporate job type.

So, what is the elephant in the room? The AVMA reported that as of December 31, 2020, 62.3 percent of private practice veterinarians are female.[23] The AVMA further broke down the information to show food animal exclusive practice to be only 24.7 percent female, food animal predominant was 28.4 percent female, and mixed animal practice was 46.6 percent female. In the AASV 2020 salary survey data, 55 percent of female and 45 percent of male respondents completed the salary survey (Figure 1).[22] Is the AASV predominantly female now, or are females more engaged with completing the salary survey? Why is the trend in veterinary medicine moving toward a more significant percentage of females? There are no apparent answers to these questions, but this does point the AASV and the veterinary profession toward needing a better understanding of the membership.

The second elephant in the room is the compensation differences between males and females. Yes, I said it. The first year the AASV salary survey was conducted, in 2002, the membership committee was the originator of the survey.[22] Being a committee member, there was significant discussion about reporting the gender differences, especially salary. Since only 12 percent of the survey respondents were female in 2002, there was concern that the data was skewed. The information published in 2002 and restated in the 2005 salary survey explained the gender differences as follows:

> Private Sector
> 1. No gender differences in primary area of focus, in primary function of practice or in the number of hours worked.
> 2. Significant differences were found among male and female veterinarians who earned an income from activities unrelated to veterinary medicine. Male veterinarians were more likely to report earning income from activities unrelated to veterinary medicine than were females. (Difference significant at the 0.001 alpha level).
> 3. Gender differences were found among those veterinarians who reported spending more than 50 percent of their time earning income from paid consulting. Only 7.1 percent of the female veterinarians report spending more than 50 percent of their time being paid for consulting. However, 23.7 percent of the male veterinarians reported spending 50 percent or more of their time on such activities (difference significant at the 0.05 alpha level).

It was not until the 2008 salary survey that comparison by gender was first published with 18 female private practitioners and 20 public/corporate female veterinarians. The apparent bias toward lower salaries for female veterinarians in the 2008 and 2011 survey, particularly among practitioners, was framed around the difference in median age between male (older) and female (younger) practitioners. The salary bias was not discussed in 2014 though the lower salary trend for female veterinarians continued. The salary bias was again evident in the 2017 survey, with the median age beginning to equalize in the Public/Corp, but the median age for practitioners remains skewed with an older median age in the male respondents. In 2020, the salary survey evaluated the gender and salary issue more extensively.

Several potential flaws in evaluating the salary survey data based on the respondents' reported information extend their income beyond a base salary alone (practice ownership, livestock ownership, other business ownership). This needs to be rectified in future surveys. Nonetheless, with the reported salary distribution by gender from the 2020 survey, there is a clear divide in gender equality based on salary:

	Median	Mean
Salary Male Practitioners (n = 148)	**$140,000**	**$219,186**
Salary Female Practitioners (n = 73)	**$98,000**	**$128,721**
Salary Male Public/Corp(n = 146)	**$155,000**	**$169,161**
Salary Female Public/Corp (n = 97)	**$123,850**	**$120,000**

What do we as an organization do with this information? First, we need to look at the distribution by gender and age that the salary survey provides (Figure 2). The most meaningful comparison is the apparent decline in female practitioners, particularly over the age of 40. You might argue that the female representation in the survey did not have a significant representation until the last five years (Figure 2) in the private practitioner's category. Conversely, if you reference the public/corp section of Figure 2, the distribution of females across age ranges increases, compared to males,

Figure 1: AASV salary survey responses by gender 2002 - 2020.

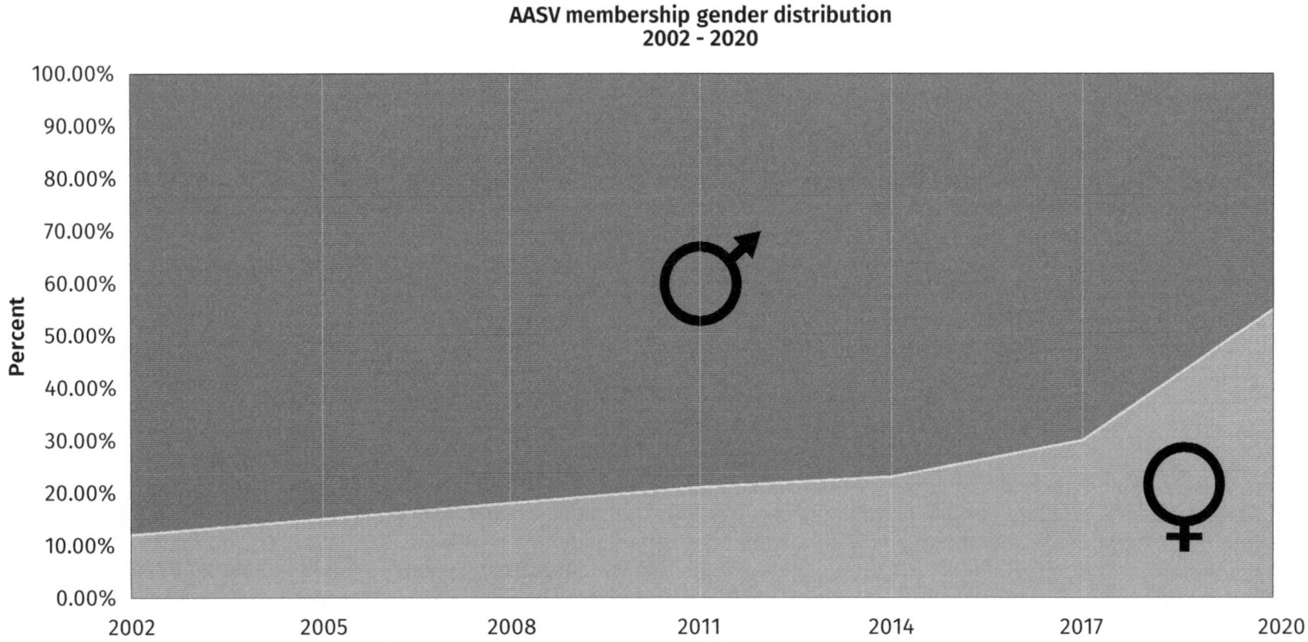

by age 30 and maintains. It would behoove the AASV to conduct further research to evaluate the apparent departure of females from private practice. With the veterinary profession becoming more female-dominated, what opportunities should swine practices utilize to attract and retain practitioners?

If we want to prioritize the retention of veterinarians in food animal practices, research in the US,[23] Germany,[24] and the UK[25], highlight potential opportunity areas to enhance. If we review the Villarroel et al[23] US study, the factors of high importance in the decision to leave rural veterinary practice (not ranked in order) were: (1) emergency duty, (2) time off, (3) salary, (4) family concerns, and (5) practice atmosphere. Additionally, the Kersebohm et al[24] study looked at work and life satisfaction for practitioners. German practitioners prioritized (1) a 'good' working atmosphere as most important, (2) a 'reasonable salary' was second, (3) 'holidays and leisure time' was third. Women prioritized the working atmosphere and family-friendly arrangements as statistically more significant than men, while men prioritized a 'reasonable' salary. The Adam et al[25] study focused on the management of the postgraduate veterinarian and their mentorship. The most recent publication focusing on recruitment and retention of food animal veterinarians, CAST (Council for Agricultural Science & Technology) Issue Paper,[26] the practice types, integrated activities, factors influencing recruitment and retention, and finally, strategies and tactics to increase recruitment and retention. The AASV should utilize this information to evaluate the salary discrepancy and the possible reasons for declining female private practitioners after 40 years of age. Is the current practice model adapting to the work and lifestyle needs of today's new graduates?

Summing it all up, with the next generation of swine veterinarians, how we train, mentor, and accommodate their professional and personal goals may determine retention in our industry and the AASV. Specific practice attributes such as caseload, facilities, practice atmosphere, the importance of personal life goals, and availability of mentorship are critical to retention. What else are we missing?

Finally, gender is one piece of the DEI puzzle. As society and graduates from colleges of veterinary medicine continue to look and act differently, the AASV also needs to explore member and industry sector data looking at race and ethnicity, societal backgrounds – more "city kids" becoming food animal veterinarians, sexual orientation, and more. In addition, as the global community becomes more diverse, inclusive, and equitable, we need to ensure that members and stakeholders see themselves in AASV.

Mental health

One of the most impactful events in the United States swine industry occurred during the human COVID-19 pandemic of 2020. The 2020 COVID-19 pandemic had a negative impact on swine farms throughout the United States as pork slaughter/processing plants closed and employee absenteeism delayed plant reopening. Plant closures led to a dramatic reduction in the number of pigs that could be slaughtered. Throughout the US swine industry, this lack of slaughter caused restrictions in animal movement and forced producers to make difficult decisions about the care of their pigs, including mass depopulation, due to the pigs' compromised welfare as a result of limited barn space availability, access to feed and water due to animal density, and potential aggression in overcrowded groups.[27-29] A significant responsibility of coordinating, managing, and application of this mass depopulation was placed on veterinarians. Though mass depopulation has been conducted, in the past, in response to foreign animal disease outbreaks, US veterinarians and swine producers had no previous experience performing mass depopulation for movement restrictions not caused by disease.

Veterinarians are at the center of the emergency mass depopulation process, providing guidance and animal welfare oversight. The responsibility of euthanizing animals, under any

Figure 2: 2020 AASV salary survey - Percentage comparison of age and gender for practitioners and public/corp veterinarians.

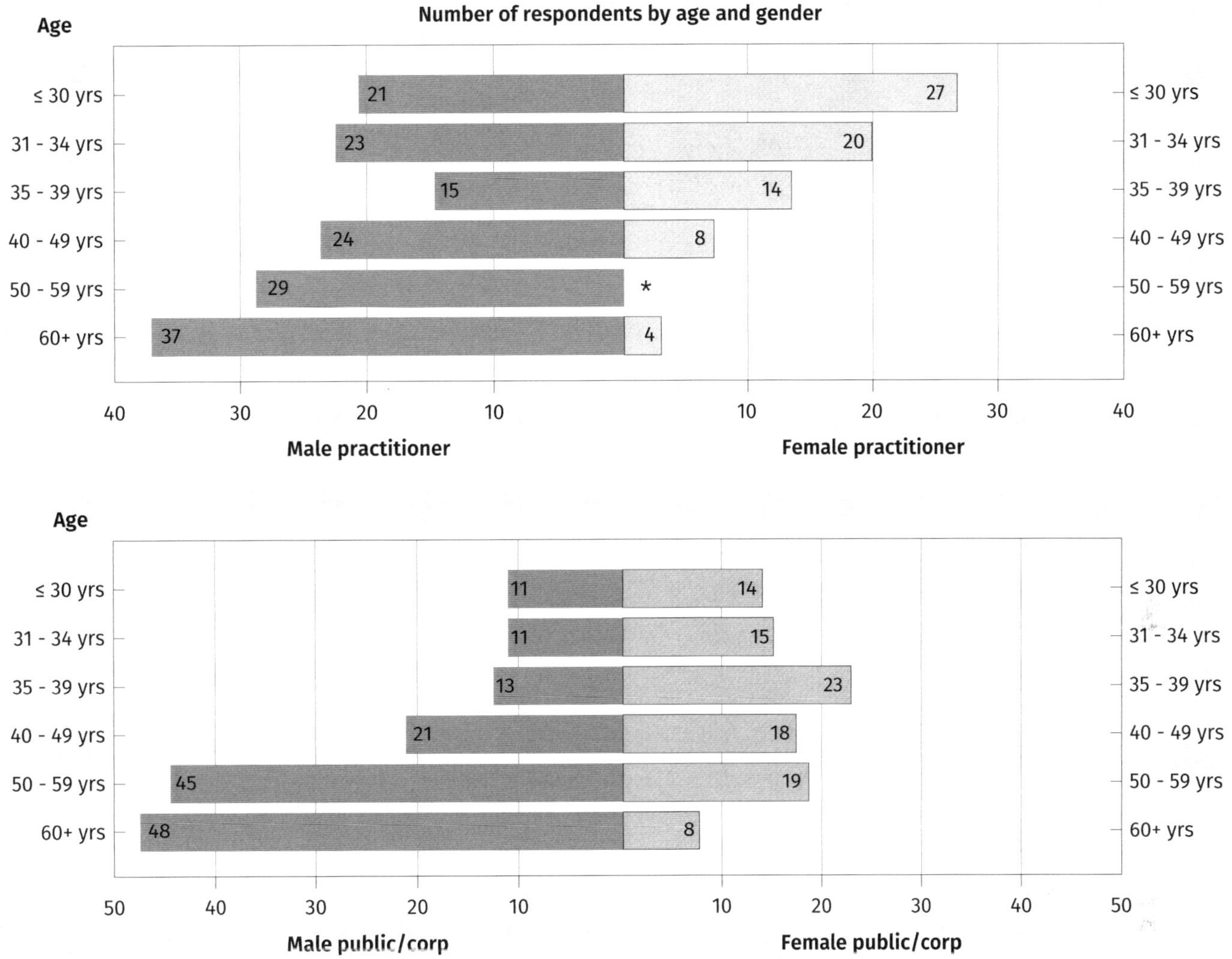

* numbers were not reported for less than 4 responses for purposes of anonymity.

Reprinted with permission from the American Association of Swine Veterinarians© 2020, page 20 **https://www.aasv.org/members/only/SalarySurvey2020.pdf**

circumstances, can create moral stress, often called the "caring-killing paradox," as veterinarians struggle to balance their love for animals with the necessity to euthanize.[30-32] Studies involving those whose work involves euthanizing animals have found that many of these individuals suffer from a form of post-traumatic stress disorder (PTSD) labeled perpetration-induced traumatic stress (PITS).[33,34] PITS differs from PTSD in the fact that PTSD follows a life-threatening or terrifying experience. Alternately, PITS does not directly threaten the individual; instead, it may threaten their ethical character or their identity as a veterinarian who cares for animals yet must implement euthanasia or the death of that animal.[35-37] Additionally, these adverse effects have been shown to be present in farm caretakers conducting routine applications[38,39] and farmers impacted by euthanizing animals during a foot-and-mouth disease outbreak.[40] A recent study by Hibi et al,[41] focusing on the mental distress of field veterinarians in the 2010 foot-and-mouth disease outbreak in Japan, found similar negative effects.

Yet, veterinarians who applied mass depopulation methods during COVID-19 faced an unprecedented challenge, and it is unknown how the depopulation events may have impacted their psychological and emotional well-being. Therefore, after consulting with Dr. Elizabeth Strand at the University of Tennessee-Knoxville, I engaged Dr. Lori Kogan, a psychologist at Colorado State University's Veterinary College, to study the mental health impact of the 2020 mass depopulation event on swine veterinarians.[42] The study's goals were four-fold:

1. Assess the mental health of swine veterinarians involved with mass depopulation events compared to swine veterinarians who were not involved in mass depopulation.
2. Measure the level of well-being, quality of life (QOL), psychological distress, burnout, and resilience in the veterinarians who conducted mass depopulation and the potential impact of depopulation methods on these factors.
3. Identify coping methods utilized by swine veterinarians for improved well-being.
4. Based on the study results, recommend intervention strategies and supportive services to assist veterinarians in future mass depopulation events.

Hopefully, by the time this information is presented, it will be accepted for publication. Nonetheless, the salient information that we, as members of the AASV, must acknowledge will be briefly reviewed. The survey asked all swine veterinarians surveyed about their mental health, medication use for mental health, outpatient care, and mental health counseling (Figure 3).

When asked their opinions about mental health services, between 50-65 percent of participants agreed with statements indicating a willingness to seek support and the importance of mental health services being available to those who need it. However, nearly half of respondents do not feel comfortable discussing mental health issues or feel that other veterinarians do not support veterinarians who need mental health services.

Furthermore, only a minority of veterinarians reported receiving mental health services. For example, 80 percent of respondents indicated they had not received outpatient mental health support, of which 16 percent reported feeling they needed it but did not obtain it.

We also asked all respondents about suicide ideation. Out of the 133 survey responses, 10 percent indicated that they had thought about suicide (Figure 4) at some point in their career. Three of those individuals had the ideation within the past 12 months, which coincided with the COVID-19 depopulation period. Compared to the Volk et al study [43] looking at veterinarians in the United States, 2.2 percent of swine veterinarians are better than the 7.5 percent of all veterinarians surveyed. This should, however, not negate the seriousness of the impact of mental health on veterinarians during agricultural emergencies such as depopulation[44-48] A more alarming statistic identified the prevalence of suicidal ideation for Canadian veterinarians in 2017 to be 26.2 percent.[49]

Suicide prevention requires both individuals and organizations to play a role, according to the experts. The Centers for Disease Control and Prevention (CDC)[50] provides the following recommendations:

- Strengthen economic supports.
- Strengthen access to and delivery of mental health and suicide care.
- Create protective environments, such as by reducing access to lethal means.
- Promote connectedness.
- Teach coping and problem-solving skills.
- Identify and support people at risk.
- Lessen the harms of suicide and prevent future risk.

The reasons swine veterinarians did not acquire mental health care fell primarily into three categories: concerns about counseling itself (ie, did not believe it would be effective), concerns about taking time off work, and financial concerns. To that end, only 42 percent of respondents reported their employee assistance program includes mental health; 30 percent did not know if their program included such benefits. Similarly, only 51 percent of respondents said their health insurance covers mental health, and 44 percent did not know. So how can swine veterinarians not know how to take care of themselves or their associates through the provision of health care coverage?

This study is the only one in the United States documenting the effect of mass depopulation of swine on veterinarians. However, it is well known that the veterinary profession has struggled with high incidences of mental health-related challenges. Despite the Perret et al[49] and Volk et al[43] studies of the general veterinary population in Canada and the United States, the Baysinger and Kogan study[42] spotlights the mental health of swine veterinarians during a disaster, mass depopulation.[51,52] Assuming the 2020 COVID-19 market disruption will not be the last time mass depopulation will be utilized in the US, the AASV and its members must take action now to do three things to prepare and support veterinarians, producers, and event participants:

1. Remove the stigma of mental health.
2. Educate veterinarians, producers, and the agriculture industry about mental health and suicide prevention.
3. Influence incorporation of mental health awareness and education into college curriculums.

With mental health, research shows that knowing or having contact with someone with mental illness is one of the best ways to reduce stigma.[53] What responsibility does the AASV have to its members? To remove the stigma associated with mental health, we need to create a culture of equality between physical and mental health. What are some of the ways this can be done?

Suicide prevention requires both individuals and organizations to play a role, according to the experts. Diving further into the CDC[50] recommendations:

- Strengthen economic supports.
  - This speaks to new graduates' student loan debt load and salary disparity between male and female veterinarians.
- Strengthen access to and delivery of mental health and suicide care.
  - Rural availability and/or telemedicine need to be promoted.
- Create protective environments, such as by reducing access to lethal means.
  - Is there an opportunity for the "barnyard veterinary" groups to manage a crisis hotline for veterinarians – a hotline with knowledgeable mental health professionals who understand the unique nature of being a swine veterinarian?
- Promote connectedness.
  - How we talk about mental health and provide the opportunity for discussion matters.
- Teach coping and problem-solving skills.
  - Are workshops or speakers at events hosted by the AASV inclusive of these skills?
- Identify and support people at risk.
  - Don't hesitate to ask people how they are doing and feeling.

According to the National Alliance on Mental Illness (NAMI),[54] stigmas around mental illness can be reduced by:

Talking openly about mental health
Educating yourself and others
Be conscious of language; mental health is equal to physical health
Show compassion
Be honest about treatment

We need to create a culture where discussing depression, anxiety, post-trauma, and other common mental illnesses become as mentionable as diabetes, hypertension, and migraines. The message is that mental illness is not the result of moral failing or limited willpower, but it is a legitimate illness with specific treatments. The AASV must dispel the negative stereotypes and help promote mental health within the membership, producers, and animal agriculture industries.

Figure 3: Responses to mental health questions from veterinarians involved in mass depopulation of swine during 2020.

Prescription Medications for mental health
- Yes: 7/134 (5%) in past 12 months
- Yes: 9/134 (7%) but not in past 12 months
- No: 118/134 (88%)

Outpatient care for mental health
- Yes: 7/134 (5%) in past 12 months
- Yes: 20/134 (15%) but not in past 12 months
- No: 107 (80%)

Needed mental health counseling but didn't get it
- Yes: 22/134 (16%) in past 12 months
- Yes: 9/134 (7%) but not in past 12 months
- No: 103 (77%)

Figure 4: Swine veterinarian's responses when asked if they had thought about suicide during 2020.[43]

Thoughts of suicide during 2020
Yes: 3 /133 (2%) in past 12 months
Yes: 11/133 (8%) but not in past 12 months
No: 119/133 (89%)

Developing our future swine veterinarians

Areas of additional expertise and focus that are all driving a more global and multi-species direction:

Sustainability
One Health
One Welfare

Within these three areas of focus, the AASV can and must develop our members to actively develop solutions for the swine industry in the US and the food supply chain globally. If we circle back to the priorities identified from the member survey, they can all be seen in Sustainability, One Health, and One Welfare. Reevaluation of the veterinary and animal science curriculums and training will be part of that endeavor to develop critical thinkers who can research, collaborate, and communicate the science for the future of the food animal industries.

Three quotes were gleaned from the survey from some of the "old" guard that applies to developing our future swine veterinarians, scientists, industry leaders, and the AASV:

Know and see the pigs. Solve problems through science. Serve producers & caretakers.
Jer Geiger, DVM, MS, Howard Dunne Memorial Lecture, 2021

Focus on "fighting the last war" – The industry has changed. We need to think differently, work differently, and act differently.
Jim Lowe, DVM, MS, Howard Dunne Memorial Lecture, 2010

Learn how to "be" with the changing structure of the swine industry.
Max Rodibaugh, DVM, Howard Dunne Memorial Lecture, 2000

Leaping into the future: Sit down, buckle up and hang on.
Angela Baysinger, DVM, MS, MSc, Howard Dunne Memorial Lecture, 2022

Conclusion

I am proud to be a swine veterinarian and member of the AASV. It is as much of who I am as being a farm kid from Missouri, a wife and mother, and an avid motorcyclist. As with every aspect of our lives, the world we are living in is changing. I have highlighted just a few of the changes I foresee affecting the AASV – animal welfare, DEI, mental health, and the future of our profession. While I have highlighted insights from my research and others and offered potential steps forward, I also have asked several questions. As I shared in my opening comments, I do not have all the answers, but I want to be part of the AASV that consistently seeks to improve while empowering its members and industry sector.

Will you join me?

I know the task ahead will require ongoing conversations and multiple attempts to "get it right." But, I am confident that, together, we can continuously achieve the AASV mission.

References

1. American Association of Swine Veterinarians. About the AASV. **Aasv.org. https://aasv.org/aasv/about.html** Accessed July 9, 2021.
2. American Association of Swine Veterinarians. Bylaws. **Aasv.org. https://aasv.org/aasv/bylaws.htm** Accessed July 9, 2021.
3. European Commission - E.U. Fitness Check. Initiative. **https://ec.europa.eu/info/law/better-regulation/have-your-say/initiatives/12400-Animal-welfare-evaluation-of-EU-rules-fitness-check-_en** Published 2020. Accessed October 7, 2021.
4. European Commission- Farm to Fork Strategy. Food Safety. **https://ec.europa.eu/food/horizontal-topics/farm-fork-strategy_en** Accessed September 7, 2021.
5. European Commission. Initiative. End the Cage Age. **https://ec.europa.eu/info/law/better-regulation/have-your-say/initiatives/12842-Animal-welfare-%E2%80%98End-the-Cage-Age%E2%80%99-European-citizens%E2%80%99-initiative_en** Published 2020. Accessed November 7, 2021.
6. California Department of Food and Agriculture (CDFA). Animal Health and Food Safety Services Division - Proposition 12 Implementation. **Cdfa.ca.gov. https://www.cdfa.ca.gov/ahfss/Prop12.html** Accessed September 8, 2021.
7. Massachusetts Question 3 Initiative. Session Law - Acts of 2016 Chapter 333. **Malegislature.gov. https://malegislature.gov/Laws/SessionLaws/Acts/2016/Chapter333** Published 2016. Accessed September 8, 2021.
8. State of Colorado. Colorado Initiative 16 - Protection of Animals from Unnecessary Suffering and Exploitation. **Sos.state.co.us. https://www.sos.state.co.us/pubs/elections/Initiatives/titleBoard/filings/2021-2022/16OriginalFinal.pdf** Accessed September 8, 2021.

9. Oregon Secretary of State. Initiative 13. **Egov.sos.state.or.us. http://egov.sos.state.or.us/elec/web_irr_search.record_detail?p_reference=20220013..LSCYYYANIMAL_ABUSE** Published 2021. Accessed September 8, 2021.

10. Maine - Right to Food. Right to Food Constitutional Amendment. **Legislature.maine.gov. https://legislature.maine.gov/legis/bills/getPDF.asp?paper=HP0061&item=1&snum=130** Published 2021. Accessed November 5, 2021.

11. Ballot Initiative Guide. Direct Democracy in Action: A Guide to State Ballot Measures. **Multistate.us. https://www.multistate.us/insider/2017/11/6/direct-democracy-in-action-a-guide-to-state-ballot-measures** Accessed November 8, 2021.

12. Leary S, Anthony R, Gwaltney-Brant S, et al. AVMA guidelines for the depopulation of animals: 2019 edition. Schaumburg, Ill: AVMA, 2019; 43-45.

13. USDA, APHIS. NAHEMS GUIDELINES: HEALTH AND SAFETY. **Aphis.usda.gov. https://www.aphis.usda.gov/animal_health/emergency_management/downloads/nahems_guidelines/fadprep-nahems-guidelines-health-safety.pdf** Published 2018. Accessed September 30, 2020.

14. USDA, Economics, Statistics, and Market Information System. Livestock, Dairy, and Poultry Outlook. **Downloads.usda.library.cornell.edu. https://downloads.usda.library.cornell.edu/usda-esmis/files/g445cd121/cc08j8026/vq280f939/LDP-M-320.pdf** Published 2021. Accessed March 6, 2021.

15. Baysinger A, Senn M, Gebhardt J, Rademacher C, Pairis-Garcia M. A case study of ventilation shutdown with the addition of high temperature and humidity for depopulation of pigs. *J Am Vet Med Assoc*. 2021;259(4):415-424. **doi:10.2460/javma.259.4.415**

16. American Veterinary Medical Association (AVMA). Humane Endings Flowchart. Avma.org. **https://www.avma.org/sites/default/files/2020-04/Humane-Endings-flowchart-2020.pdf** Published 2020. Accessed March 12, 2020.

17. National Pork Board (NPB). Application of Electrocution in Suckling Pigs for Mass Depopulation - Pork Checkoff. **Pork Checkoff. https://www.porkcheckoff.org/research/application-electrocution-suckling-pigs-mass-depopulation/** Published 2020. Accessed August 9, 2021.

18. National Pork Board (NPB). Determining the efficacy and safety of differing caliber/ammunition combinations for the humane euthanization and subsequent mass depopulation of market weight pigs - Pork Checkoff. Pork Checkoff. **https://www.porkcheckoff.org/research/determining-efficacy-safety-differing-caliber-ammunition-combinations-humane-euthanization-subsequent-mass-depopulation-market-weight-pigs/** Published 2021. Accessed September 9, 2021.

19. National Pork Board (NPB). Use of modified rendering trailers to achieve rapid depopulation of modern sow units - Pork Checkoff. Pork Checkoff. **https://www.porkcheckoff.org/research/use-of-modified-rendering-trailers-to-achieve-rapid-depopulation-of-modern-sow-units/** Published 2021. Accessed September 9, 2021.

20. National Pork Board (NPB). Validation of a mobile electrocution system for humane mass depopulation of swine - NPB#20-123. **Porkcheckoff.org. https://porkcheckoff.org/wp-content/uploads/2021/02/20-123-MOTE-final-rpt.pdf** Published 2020. Accessed September 9, 2021.

21. Koziel J, Lee M, Li Y. The prototype of a low-cost mobile CO2 vaporizer system for on-site humane swine depopulation, disposal and biosecure cleanup. In: *Agricultural And Biosystems Engineering Conference.* ; 2021. **https://lib.dr.iastate.edu/ abe_eng_conf/617** Accessed November 1, 2021.

22. American Association of Swine Veterinarians (AASV). AASV Publications - Resources - Salary Surveys. **Aasv.org. https://aasv.org/aasv/publications.htm** Published 2020. Accessed July 9, 2021.

23. Villarroel A, McDonald S, Walker W, Kaiser L, Dewell R, Dewell G. A survey of reasons why veterinarians leave rural veterinary practice in the United States. *J Am Vet Med Assoc*. 2010;236(8):859-867. **doi:10.2460/javma.236.8.859**

24. Kersebohm J, Lorenz T, Becher A, Doherr M. Factors related to work and life satisfaction of veterinary practitioners in Germany. *Vet Rec Open*. 2017;4(1). **doi:10.1136/vetreco-2017-000229**

25. Adam K, Baillie S, Rushton J. Retaining vets in farm animal practice: a cross-sectional study. *Veterinary Record*. 2015;176(25):655-655. **doi:10.1136/vr.103170**

26. CAST Issue Paper. Impact of recruitment and retention of food animal veterinarians on the US food supply. **Cast-science.org. https://www.cast-science.org/wp-content/uploads/2020/03/CAST_IP67_Vet-Students.pdf** Published 2021. Accessed September 9, 2021.

27. Kansas State University. Weekly Hog Slaughter | **AgManager.info. Agmanager.info. https://www.agmanager.info/livestock-meat/livestock-marketing-charts/weekly-hog-slaughter** Published 2020. Accessed November 6, 2020.

28. Gonyou H, Brumm M, Bush E et al. Application of broken-line analysis to assess floor space requirements of nursery and grower-finisher pigs expressed on an allometric basis. *J Anim Sci*. 2006;84(1):229-235. **doi:10.2527/2006.841229x**

29. Petherick J, Phillips C. Space allowances for confined livestock and their determination from allometric principles. *Appl Anim Behav Sci*. 2009;117(1-2):1-12. **doi:10.1016/j.applanim.2008.09.008**

30. Arluke, A. (1991). Coping with euthanasia: a case study of shelter culture. *JAVMA*. 1991; 198(7): 1176-1180.

31. Rollin, B. E. (1987). Euthanasia and moral stress. *Loss, Grief & Care, 1*(1-2), 115-126.

32. Pluhar, E. B. (1988). When is it morally acceptable to kill animals? *Journal of agricultural ethics, 1*(3), 211-224.

33. MacNair, R. (2002). *Perpetration-induced traumatic stress: The psychological consequences of killing*. Greenwood Publishing Group.

34. Bennett, P., & Rohlf, V. (2005). Perpetration-induced traumatic stress in persons who euthanize nonhuman animals in surgeries, animal shelters, and laboratories. *Society & Animals, 13*(3), 201-220.

35. Whiting, T. L., & Marion, C. R. (2011). Perpetration-induced traumatic stress—A risk for veterinarians involved in the destruction of healthy animals. *The Canadian Veterinary Journal, 52*(7), 794.

36. Manette, C. S. (2004). A reflection on the ways veterinarians cope with the death, euthanasia, and slaughter of animals. *Journal of the American Veterinary Medical Association, 225*(1), 34-38.

37. Sanders, C. R. (1995, June). Killing with kindness: Veterinary euthanasia and the social construction of personhood. In *Sociological Forum* (Vol. 10, No. 2, pp. 195-214). Kluwer Academic Publishers-Plenum Publishers.

38. Campler, M. R., Pairis-Garcia, M. D., Rault, J. L., Coleman, G., & Arruda, A. G. (2018). Caretaker attitudes toward swine euthanasia. *Translational animal science, 2*(3), 254-262.

39. Rault, J. L., Holyoake, T., & Coleman, G. (2017). Stockperson attitudes toward pig euthanasia. *Journal of animal science, 95*(2), 949-957.

40. Olff, M., Koeter, M. W., Van Haaften, E. H., Kersten, P. H., & Gersons, B. P. (2005). Impact of a foot and mouth disease crisis on post-traumatic stress symptoms in farmers. *The British Journal of Psychiatry, 186*(2), 165-166.

41. Hibi, J., Kurosawa, A., Watanabe, T., Kadowaki, H., Watari, M., & Makita, K. (2015). Post-traumatic stress disorder in participants of foot-and-mouth disease epidemic control in Miyazaki, Japan, in 2010. *Journal of Veterinary Medical Science*, 14-0512.

42. Baysinger A. and Kogan L. The mental health impact of mass depopulation of swine on veterinarians during the 2020 COVID-19 pandemic. In-process.

43. Volk J, Schimmack U, Strand E, Vasconcelos J, Siren C. Executive summary of the Merck Animal Health Veterinarian Wellbeing Study II. *J Am Vet Med Assoc*. 2020;256(11):1237-1244. **doi:10.2460/javma.256.11.1237.**

44. Hall M, NG A, Ursano R, Holloway H, Fullerton C, Casper J. Psychological Impact of the Animal-Human Bond in Disaster Preparedness and Response. *J Psychiatr Pract*. 2004;10(6):368-374. **doi:10.1097/00131746-200411000-00005**

45. Jenner M. The psychological impact of responding to agricultural emergencies. *The Australian Journal of Emergency Management*. 2007;22(2):25-31.

46. Mellanby RJ. Incidence of suicide in the veterinary profession in England and Wales. Vet Rec 2005; 157(14):415-417.

47. Platt B, Hawton K, Simkin S, et al. Suicidal behaviour and psychosocial problems in veterinary Surgeons: a systematic review. Soc Psychiatry Psychiatr Epidemiol 2012; 47(2):223-240.

48. Wasson E, Wieman A. Mental Health During Environmental Crisis and Mass Incident Disasters. *Veterinary Clinics of North America: Food Animal Practice*. 2018;34(2):375-388. **doi:10.1016/j.cvfa.2018.02.007.**

49. Perret J, Best C, Coe J, Greer A, Khosa D, Jones-Bitton A. Prevalence of mental health outcomes among Canadian veterinarians. *J Am Vet Med Assoc*. 2020;256(3):365-375. **doi:10.2460/javma.256.3.365**

50. Center for Disease Control (CDC). Suicide prevention strategies. **https://www.cdc.gov/suicide/prevention/index.html**. Accessed November 1, 2021.

51. Nett, R. J., Witte, T. K., Holzbauer, S. M., Elchos, B. L., Campagnolo, E. R., Musgrave, K. J., ... & Funk, R. H. (2015). Risk factors for suicide, attitudes toward mental illness, and practice-related stressors among U.S. veterinarians. *Journal of the American Veterinary Medical Association, 247*(8), 945-955.

52. Gyles, C. (2018). Surprising new findings on veterinarians' mental health and well-being. *The Canadian Veterinary Journal, 59*(10), 1041.

53. American Veterinary Medical Association (AVMA). **https://www.avma.org/javma-news/2021-11-01/education-communication-are-important-strategies-prevent-suicide-among**. Accessed November 1, 2021.

54. National Alliance on Mental Illness (NAMI). Pledge to Be StigmaFree Nami.org. **https://www.nami.org/Get-Involved/Pledge-to-Be-StigmaFree**. Accessed November 1, 2021.

Alex Hogg Memorial Lecture

Learning for the future

James A. Kober, DVM, MS, Diplomate ABVP SHM

In remembrance and honor of Alexander Hogg

Alexander Hogg
1920-2006

A native of Scotland, Alex (pronounced "Alec") Hogg immigrated to the United States with his family when he was six years old. After reaching adulthood, he served in the US Navy during World War II before entering veterinary school at Kansas State University. He received his DVM degree in 1950, and was a mixed animal practitioner in Iowa for twenty years. Dr. Hogg returned to college – this time, Iowa State University – to complete a Master's Degree in 1972. He spent the following years serving and educating swine veterinarians and producers in his roles as Nebraska Swine Extension veterinarian and professor at the University of Nebraska-Lincoln. Upon "retirement" from the university in 1990, Alex worked as a technical services consultant for MVP Laboratories. Ever an enthusiastic learner, he graduated from the University of Illinois Executive Veterinary Program in 1995.

An active and longstanding member of the AASV, Dr. Hogg joined what was then the American Association of Swine Practitioners shortly after it was formed. He served as secretary/treasurer of the association from 1974 to 1976, and was elected president in 1979.

The many awards and honors bestowed upon him over the years included several presented by the AASV: The Howard Dunne Memorial Award in 1985, the Meritorious Service Award in 1996, and in 2005, the first-ever AASV Foundation Heritage Award for a lifetime of achievement in veterinary medicine.

ALEX HOGG MEMORIAL LECTURERS

2011 Joseph F. Connor
2012 Lisa Tokach
2013 John T. Waddell
2014 Mark J. Engle
2015 C. Scanlon Daniels
2016 Peggy Anne Hawkins
2017 Matthew Turner
2018 Rodger Main
2019 Deborah Murray
2020 William L. Hollis
2021 Jeremy Pittman

https://doi.org/10.54846/am2022/2

Learning for the future

James A. Kober, DVM, MS, Diplomate ABVP SHM

I am deeply honored and humbled to be presenting the 16th Alex Hogg Memorial Lecture. It is truly a privilege to have been asked to do this - it is a great honor for me.

As I get going through this lecture, I certainly want it to be a tribute to Dr. Hogg. I am a bit of a history buff, so I am going to dip into some history as well.

The highlights from Alex's life are as follows:

Alex was born in Borgue, Scotland on April 15, 1920.

He immigrated to the US with his family in 1926 through Ellis Island and initially settled in Sand Springs, Oklahoma.

After high school, Alex completed two years at the Business College in Chillicothe, Missouri. He was married to his first wife, and they had one daughter, Janice.

In April 1941, Alex volunteered for military service in the US Navy where he started as a 3rd class Petty Officer. After the attack on Pearl Harbor, he volunteered for sea duty "so he would have something good to tell his grandchildren when they asked what he did during the great war."

After serving through World War II, Alex went to veterinary school at Kansas State University and graduated in 1950.

Alex was a mixed animal practitioner in Coin, Iowa for 20 years, then went to Iowa State to earn his master's degree in veterinary pathology in 1972 (at 52 years of age).

After receiving his master's degree, he became the Nebraska Swine Extension Veterinarian and a Professor at the University of Nebraska-Lincoln.

He retired from the university in 1990 and became the technical service veterinarian for MVP labs.

Alex was the author and editor of "Hogg Sense"... a monthly newsletter for 11 years (1994-2004), sponsored by MVP Laboratories, Inc, Ralston, Nebraska. The newsletters covered one side of a page and included 4-5 short topics, 1-2 sentences each, plus references.

> **PRRS:** Serum samples from cull sows are not a good indicator of the number of positive animals in a herd by IFA assay. Use of the IFA as a test to detect herd infection should be done from young pigs (late nursery-early grower) or breeding animals that have had recent signs. (Dr. Merwin Frey, NVSL). Hogg Sense, Volume 1, No. 1

Alex's AASP/AASV service and accomplishments:

- Secretary /treasurer 1974-1975
  - "The association experienced tremendous growth during his term because he and his wife worked tirelessly mailing invitations to practitioners across the country to join AASP." (History of American Association of Swine Practitioners 1969-1994, Meyer, Kenneth and Brandt, Wallace).
- Finance committee, IPVS, Ames, Iowa 1976
- AASP President 1979
- Howard Dunne Memorial Award 1985
- Meritorious Service Award 1996
- First Heritage Award 2005

Here are some quotes about Dr. Hogg when he received the first Heritage Award:

"Dr. Hogg exemplifies all that is good about the United States of America, immigrating to this country at a young age, and rising to the top of his chosen profession," said Dr. K.T. Wright, president of the AASV Foundation. While presenting the award, Dr. John Waddell, immediate past president of AASV, stated "Dr. Hogg epitomizes the ideal of increasing the knowledge of swine veterinarians."

Alex traveled extensively, both in the US and around the world giving pork production lectures. He loved teaching students, producers, and veterinarians. He authored many, many articles and newsletters but remained shy and humble.

My connection to Alex is the second Executive Veterinary Program (EVP) class at the University of Illinois (1993-1995). I was a relatively young veterinarian at the time, and I was amazed that a veterinarian of Alex's stature would sit down with me as a total equal, even though he was 40 years my senior. I found him to be smart, practical, and a genuinely nice and humble man. When we graduated in 1995, Alex was 75 years old and still learning new things! He remains the oldest graduate of the EVP program.

Alex was a tremendous lifelong learner and in turn, a lifelong teacher.

> **"Moving/Regrouping Pigs**: Every time a pig is moved or regrouped, days to market are increased by 7-10 days. (James Allison, DVM, Independence, KS.)" Hogg Sense, Volume 2, No. 3.

Mentorship

Most successful people have one or more significant mentors in their lifetime. The word mentor is both a noun and a verb. As a noun, a mentor is a trusted and experienced advisor. As a verb, a mentor is one who trains and advises someone, usually a younger colleague. Dr. Hogg took his mentorship responsibilities very seriously. I spoke to Mary Lou Chapek Hogg about people who may have been Alex's mentors, but she was not aware of who they were. Alex himself was undoubtedly a mentor to many students and veterinarians throughout his lifetime. Here is a comment from Dr. John Waddell about Dr. Hogg's mentorship:

"One of my most memorable quotes from Alex came from him 'recruiting' me to give a practice tip at AASV soon after I arrived (wet behind the ears) in Nebraska. I was, of course, hesitant but he reassured me with 'You can do it and I will help!' This would have been in 1981 or 1982. Another fond memory was when he was awarded the Heritage Award just weeks before he passed. The bronze weaned pig award was set before him and he exclaimed 'Is this my pig?!!!' He was so grateful and excited to add to his collection of pigs!"

If you will bear with me, I would like to introduce you to my three significant mentors.

First is Dr. Brad Thacker. Brad came to MSU as the Extension Swine Veterinarian when I started veterinary school (1983). Keep in mind that when I started vet school, I didn't necessarily know what type of practice I wanted after graduation. I saw an advertisement for student membership in an organization called AASP. I had no idea what this was, so I tracked Dr. Thacker down and asked him if AASP was legit. He assured me it was, then asked if I was interested in pigs. When I said I was, he immediately offered me a job in his lab. That started a great relationship. I literally spent at least one day a week going to farms and teaching herds with Dr. Thacker instead of attending class throughout veterinary school. I learned an incredible amount from him. This was a true case of "Don't let school interfere with your education."

Next is Dr. Gary Fedore. Dr. Fedore was my first boss after graduation. Dr. Fedore was a mixed animal practitioner, but his passion was horses. He was incredible with horses and with their owners (no easy task!). He was a non-boarded equine surgeon, and his success rate at colic and leg surgery was higher than most universities. I learned quickly that you don't need a whole room full of people to do a colic surgery - we did many with just three people, most in the middle of the night. He was very patient with me as I started practice and taught me an incredible amount about both medicine and surgery as well as client and employee communication.

Finally, the third significant mentor to push me onto a successful career was Dr. Max Rodibaugh. I went from my first job in a mixed animal practice to Dr. Rodibaugh's swine-only practice in 1989. I stayed with Dr. Rodibaugh for four and a half years and the number of things he taught me was incredible. This includes swine medicine as well as simply being a good person with uncompromising integrity. Many of you would agree that there are few people who would be a better mentor than Dr. Rodibaugh.

Many people direct our paths over the years of education and practice. Take a minute to think back about who your significant mentors are. Many of us don't do a great job keeping in touch with those people, even though they were an important person in our life. I encourage all of you take a minute to text or email your mentor(s). Better yet, call or send them a handwritten note. Try to stay in touch with them. Consider mentorship a lifelong endeavor, not just a point in time. That is what Alex did.

"**Secondary Bacterial Infections that cause Finisher Stall out**: The five most common bacterial isolates that cause chronic pneumonia and stall out in finisher swine are: *Pasteurella multocida* (61%), *Streptococcus suis* (45%), *Hemophilus parasuis* (41%), *Bordatella bronchiseptica* (33%), and *Actinomyces pyogenes* (22%)." (M. Yeager, DVM, PhD, ISU)." Hogg Sense, Volume 5, No. 2.

Opportunities for lifelong learning

I am going to go through some of the things I have done personally to increase my knowledge and be a lifelong learner over the years. You certainly don't have to take the same path as I did. I simply want to introduce you to some of the opportunities that are out there. And this is definitely NOT an all-inclusive list. Some of these examples are more suited to students or younger practitioners, but some can apply to more experienced practitioners as well.

Student externships: These are available to all veterinary students. Many swine and mixed animal practices have hosted externs for many years. It is up to the individual student to make connections and get these scheduled. Other than a small number of veterinary schools that still have an actual swine medicine course, most of your swine medicine and production knowledge will be on your own through externships.

Internships and residencies: Students will have to contact individual veterinary schools to see if these programs are available. Historically, internships are a way to get more experience with medicine and production. Residencies tend to lead to a master's or PhD degree. There are lots of these programs out there. Use your mentors to guide you if you choose this path.

Executive Veterinary Program (EVP) at the University of Illinois: This program was started in the early 1990s by Dr. Leroy Biehl and taken over by Dr. Larry Firkins. This program allows practitioners to get together every other month for 2-3 days and discuss specific topics such as nutrition, ventilation, epidemiology and statistics, and economics just to name a few. The program lasts for two years and culminates with everyone participating in and presenting a small group project. There have been six classes completed in swine health management (222 graduates) with a seventh class taking place now. The program has now been extended to beef practitioners, small animal practitioners, and even a pork producer EVP class.

As I mentioned, this is where I met Dr. Hogg and got to know him. **He was 75 years old when he completed EVP!**

See Table 1 for list of EVP graduates by class

- EVP classes:
    - Pioneer class, 1993
    - Pride of 95 class, 1995
    - Alliance class, 1998
    - Millennial class, 2000
    - Partners in progress, 2009
    - Horizons class, 2016
- There is a current class, which took a little break during the pandemic, but they are now back in session.
    - Contact: Dr. Larry Firkins or Dr. Jim Lowe for more information on EVP.

Practice-based master's degrees: Traditional, in-person MS degrees were mentioned above, but there are also practice-based master's. When I completed EVP in 1995, Dr. Thacker had moved to Iowa State. He developed an opportunity for EVP graduates to continue the same schedule (3 days, every other month) taking master's level courses at ISU. The EVP courses counted as credits. We did an on-farm research project and

Table 1: EVP graduates

Pioneer 1993	Pride of '95 1995	Alliance 1998
David Baum	Butch (Rodney) Baker	Keith Aljets
George Bergman	John Baker	Michael Beisbier
Robert Bidwell	Don Bend	Jeff Blythe
Terry Bolton	Pat Boe	Tamara Boettcher
Ron Brodersen	Wendell Bohmont	Danny Burns
Robert Bryant	Shawn Bolton	John Calhoun
J.R. Bush	Cary Christensen	Joseph Campbell
David Conrad	Jack Coleman	Sherrie Clark-Deener
Terry Cowan	Mike Dierenfeld	Rich Collins
Fred Cunningham	Alden Frischmeyer	Daniel Damman
G. Michael Daniel	Eric Goff	Patrick Drone
John Ehrhardt	Tyler Holck	Paul DuBois
Lawrence Firkins	Alex Hogg	Jerome Feichtner
Merlyn Holmbeck	Mike Huebert	Charles Francisco
Thomas Hughes	Clark Huinker	Randall Larson
Raymond Huston	Bob Ketter	Dennis Mefford
Carol Jakel	Jim Kober	Michelle Michalak
William Johnson	Darwin Kohler	Michael Mull
Kerry Keffaber	Bud Larson	Roger Pray
Michel Kuhn	Ken Lorenzen	Wade Price
Edward Lents	Tim Loula	Jerald Sprau
David Madsen	Mike Meade	Dave Striegel
Michael Male	Steve Menke	Jeff Waite
Robert Nutsch	Jim Moest	Ron White
Stephen Pate	Art Mueller	Robert Wills
Stephen Patterson	Bryan Myers	Sheldon Yoder
Joseph Rudolphi	Charles Schelkopf	
Michael Schelkopf	Brad Shive	
Jan Schuiteman	Dave Smidt	
Gary Schultz	Brad Thacker	
Roy Schultz	Dwight Van Genderen	
Ralph Vinson	Paul Van Roekel	
Alan Whitman	Jaime Velasquez	
Dennis Winter	Paul Von Behren	
Teddi Wolff	Tom Wetzell	
	Warren Wilson	
	Mac Wilt	

Table 1: Continued

Millennial 2000	Partners in Progress 2009	Horizons 2016
Sandra Amass San Miguel	Grant Allison	Gustavo Alvarez
William Armbruster	Keith Bretey	Natalie Baker
David Bane	Kristen Clements	Melissa Billing
Lisa Becton	Joe Connor	Corinne Bromfield
R.C. Ebert II	Todd Distad	Cesar Caballero
Mark Eisenhart	Tara Donovan	Jean Paul Cano
Patrick Fairbrother	Bill DuBois	Mitch Christensen
Steven Feuerbach	Noel Garbes	Dyneah Classen
Michael Hannon	Tom Gillespie	Jack Creel
Karla Cunningham	Jennifer Greiner Stevens	Kate Dion
Douglas King	Jeff Harker	Joseph Fent
John Korslund	Peggy Anne Hawkins	Kelly Greiner
Christopher Kuster	Alan Hertz	Abbey Harding
Kelly Lager	Derald Holtkamp	Catherine Harris
Jamie Lehman	Marlin Hoogland	Jennifer Hasty
Duane Long	Erin Lowe	Daniel Hendrickson
James Lowe	Randy Jones	Lynette Holman
Korb Maxwell	Locke Karriker	Micah Jansen
Dale Mechler	Lynn Keller	Clayton Johnson
Isabelle Moreau	Mike Lemmon	Troy Kaiser
Steven Moreland	Jason Miller	Seth Krantz
Darrell Neuberger	Bill Minton	Aaron Lower
Max Rodibaugh	Pablo Moreno	Jeff Luebbe
Brian Roggow	Jeff Okones	Darin Madson
Clifford Shipley	Daryl Olsen	Jana Morgan
Fred Sick	Tom Painter	Deborah Murray
Alan Snodgrass	Brian Payne	David Nolan
Dale Spindler	Mike Pierdon	Carissa Odland
Kurt Van Hulzen	Doug Powers	Arturo Oropeza Munoz
Alan Wildt	Sarah Probst Miller	Jeremy Pittman
Marvin Wilke	Craig Rowles	Maryn Ptaschinski
	Alan Scheidt	Nicolas Rippel
	Kent Schwartz	Jessica Risser
	Harry Snelson	Nathan Schaefer
	Matthew Turner	Jessica Seate
	John Waddell	Amanda Sponheim
	Doug Weiss	Angie Supple
	Todd Williams	Grant Weaver
	Paul Yeske	Brandon Whitt
		Todd Wolff

ended up with a "practice-based master's." There were six of us who took advantage of this opportunity: me, Drs. Teddi Wolff, Clark Huinker, Mike Dierenfeld, Butch Baker, and Warren Wilson. The master's degree was in Swine Health Management.

- University of Illinois has an online master's program designed for practicing veterinarians in livestock production that can be coupled with your EVP credits (contact Dr. Jim Lowe).

Online master's degrees: There are many online master's degree programs available now in multiple fields including food safety, business administration, and public health. Some of these can be incorporated into the DVM program for concurrent degrees. There are also concurrent DVM/PhD programs for the really motivated. For more swine- or disease-focused programs:

- ISU offers totally online MS and PhD degree programs (contact Dr. Daniel Linhares).
- University of Minnesota offers online programs as well, but some classes may require in-person participation, depending on the emphasis (contact Dr. Maria Pieters).

Board certification: You can become board certified in Swine Health Management through the American Board of Veterinary Practitioners (ABVP). This unique board certification organization certifies by species rather than system.

History and mission of ABVP.

"ABVP was founded in 1978 and its Diplomates have a common desire and willingness to deliver superior, comprehensive, multi-disciplinary veterinary service to the public. They are veterinarians who have demonstrated expertise in the broad range of clinical subjects relevant to their practice and display the ability to communicate medical observations and data in an organized and appropriate manner." (ABVP, 1978)

In 1985, a survey of AASP members confirmed the need to expand services to members. Three major plans were initiated:

Expand the AASP newsletter

The newsletter expanded its contents in 1989 and became the journal of the AASP and was called *Swine Health and Production.* It is now called the *Journal of Swine Health and Production.* The first editor was Dr. Bob Morrison.

Establish the AASP foundation

The foundation was chartered in 1989. We are all familiar with the foundation and the excellent work it does. Mary Lou Hogg has generously contributed to the foundation again and again. Thank you, Mary Lou!

Establishment of a board certification program

Several AASP members pushed this idea and put in a lot of work to see this through, including Drs. Al Leman, Al Scheidt and Mike Wilson. This was put in place specifically for the practicing veterinarian to become board certified. Here is the process to become board certified:

a. Entrance/qualifying exam - offered at the annual AASV meeting.
b. After the entrance exam, you go to work to complete credentialing.
c. Credentials:
    i. Diploma
    ii. License to practice
    iii. Synopsis of practice
    iv. Self-report-job experience
    v. Record of continuing education
    vi. Curriculum Vitae
    vii. Applicant evaluation form (3)
    viii. Case report
    ix. Publication
d. Pass certification exam - 3 components
    i. Practical
    ii. Essay
    iii. Oral
e. 16 practitioners took the entry/qualifying exam for the first time in 1992. The first certification exam was administered in 1993. The charter diplomates were Drs. Roy Schultz, Alan Scheidt, and Rodrick Tubbs.
f. The current Swine regent is Dr. Jeff Husa. (**www.abvp.com**)
g. Current Diplomates can be found in Table 2.

The first three certifications offered by ABVP in 1978 were Companion Animal, Food Animal, and Equine. ABVP currently awards certification in eleven recognized veterinary specialties (RVS):

Table 2: ABVP Diplomates

Suzanne Burlatschenko	John Harding	Sandra Amass San Miguel (emeritus)
Glen Cassar (emeritus)	Steve Henry (emeritus)	Anthony Scheiber
Matheus Costa	Jeff Husa	Gary Schultz
Robert Desrosiers (emeritus)	James Kober	Brad Thacker
Keith Erlandson	John Kolb	Lisa Tokach
Thomas Fangman	Eric Neumann	
Thomas Gillespie	Jeremy Pittman	

- Avian Practice
- Beef Cattle Practice
- Canine and Feline Practice
- Dairy Practice
- Equine Practice
- Exotic Companion Mammal Practice
- Feline Practice
- Food Animal Practice
- Reptile and Amphibian Practice
- Shelter Medicine Practice
- Swine Health Management

It is a rigorous pathway to become board certified. It is meant to be that way. Diplomates must recertify every 10 years by either retaking the exam or maintenance of certification pathway (credit for CE, lectures, publications, etc). It is not for everyone, but it is very gratifying to finally become certified.

Board certification, other: There are a total of 22 "Recognized Veterinary Specialties" through the AVMA. Examples are the American College of Animal Welfare (contact Dr. Meghann Pierdon or Dr. Tom Parsons), the American College of Veterinary Behaviorists, the American College of Poultry Veterinarians, and the American College of Veterinary Nutritionists, etc. Some of these colleges are more applicable to practitioners; others are more applicable to academicians or PhD veterinarians. The entire list of RVSs is on the AVMA website if you are interested in these colleges and board certifications.

"**The Future of Antibiotics:** Increasing concern on issues such as drug residues and antibiotic resistance could mean more and more drugs being withdrawn from use. Time is running out for pig producers to convince the consumer lobby that medicines are being used responsibly. Drugs can be saved if their use is controlled by the veterinarians and farmers do not experiment on their own". (International Pig Topics, Volume 13, No. 4 (1998) pg. 5). Hogg Sense, Volume 6, No. 9.

It's never too late! Examples of true lifelong learning:

Many of the examples of Lifelong Learning I talked about today are generally aimed toward the younger to middle career veterinarians. Older (wiser) veterinarians fear not! I am going to give you a few examples of accomplishments and degrees older veterinarians have earned.

The best example? Dr. Alex Hogg!

Alex went back to graduate school when he was 50 years old and completed his master's degree at age 52! Alex graduated from EVP at 75 years old! He is still the oldest graduate of that program.

My great friend Dr. Bob Seger, a solo beef practitioner from Montana (also known simply as "Beef")

Bob was older than 50 when he became board certified through ABVP. Bob sold his practice when he was 62 years old and promptly went back to graduate school and earned a master's degree in ruminant nutrition. Bob couldn't find a job (great recession of 2008), so he went back to school and earned a PhD in ruminant nutrition at the age of 67. He then worked for a large beef company in Russia for three years before retiring. Bob has seven kids, and they all are college educated by a one-man beef practice. What is even more amazing is that Bob was a world class hammer thrower. He competed and won many tournaments all over the globe.

My wife Donna is a lifelong learner

In her words, "she gets bored in a single job after a few years." I tell her that she has trouble keeping a job.

Donna earned her DVM in 1987 and spent time in a few small animal practices as we moved around in the first few years after graduation. She loves economics and she earned her MBA in the late nineties. She started working for the veterinary school at Michigan State in the mid-2000s teaching career development, practice management, and personal finance to veterinary students. She still does this at MSU and at St. Matthews University in the Cayman Islands. Donna began working on another (online) MS degree in 2016 and finished with a master's degree in "Strategic Foresight" (This degree would be right up Drs. Dale Polson's and Tom Stein's alley). I'm not entirely sure what this is, but she is energized and loves talking about it. Just so she doesn't get bored, she still does surgery every Monday at a non-profit spay neuter clinic. And on top of it all, she has done a fantastic job raising our three kids.

My blueberry farm!

I bought a commercial blueberry farm when I sold my practice. This wasn't a particularly steep learning curve because I grew up on a fruit and livestock farm, but I had to learn water requirements, trimming, marketing, etc.

Other things I have done:

Back up veterinarian for the Indianapolis Zoo, horse track veterinarian for a local thoroughbred track, back up veterinarian for a local outdoor discovery center (elk, birds of prey, non-returnable wildlife) and occasional inspection of poultry going to slaughter in Canada. You all passed the national board exam. Technically we can all work on any species with some quick learning.

Lesson:

Just because you are in your fifties, sixties, and even seventies, there is **no reason you should quit learning**.

"**Circoviral Postweaning Multisystemic Wasting Syndrome (PWMS):** There has been confirmation of two outbreaks of PWMS in the UK from 1986." (AAVLD, pg. 9). Hogg Sense Volume 8, No 11.

What we have yet to learn as an industry

"To be healthy, swine should have pure water and shade; **care in breeding and management is of much more importance than medicine**. Simple disorders, however, will exist and can be treated."
"A clean pen makes a fat hog and healthy pork...." Daniels, A.C., *Our Moneymakers*, p. 23, 1912.

Behavioral vices, specifically tail biting

This is an incredibly frustrating vice that has driven many of us crazy. Is this truly multifactorial? Genetics, nutrition, density, air quality, and boredom have all been blamed over the years, depending on what you are selling. This has contributed to considerable mortality in some cases. And the thing with this mortality is that it tends to be in the late-stage finisher when you already have 90% or more of the cost of production into the pig. We push today's pigs to eat, grow, and convert at an incredible rate. Are we pushing too hard? Do we truly know how to manage the pig we have created?

Bacterial disease, specifically upper respiratory disease

All of us have seen dirty, weepy eyes with moderate to severe conjunctivitis. Many of us have tried to culture these eyes and will find *Chlamydia, M hyorhinis, G parasuis, Streps,* and *Staphs* to name a few. Which is it? Something else? This syndrome typically does not kill pigs, but it certainly stunts pigs. In many barns, a high majority of the pigs left after the first and second cut will have eyes like these. We need to figure this out so it can be prevented.

Most genetic suppliers have gone away from routine slaughter checks, specifically, nasal turbinate scoring. Well guess what? Many of the pigs with eyes like these will have significant nasal turbinate atrophy, even though there is no twisted snout. I have personally swabbed many of these noses and never found toxigenic *Pasteurella type D,* so I am told it is not atrophic rhinitis. What is causing turbinate atrophy if not *PmD*? I urge you to have your producers check with their breeding stock company to see if they are conducting routine nasal turbinate inspections and ask to see the results.

Late-finishing mortality

In my case, much of this mortality is due to tail-bitten pigs that die or need to be euthanized. But what about dog sitting lameness? How many of these pigs die? What about good-looking finisher pigs that die and have no gross lesions at postmortem exam? Other than breeding stock, these are the most expensive animals on the farm. Why isn't more being done to investigate this? Is this the new normal?

Sow mortality. 10+%. Is this the new normal?

Water quality

Most agree quality water is important to maximize production, but there is not a good simple course of action when a water test comes back. Here are my suggestions:

a. Collect a water sample at the entry to the barn and at the farthest faucet from the entry. This will tell us if the water is bad coming into the barn or if gets bad once in the pipes in the barn.
b. Clean the water lines. There are good ways and not so good ways to do this. This act alone will improve the water quality if done properly. This may be all that is needed in many cases.
c. Water treatment. There are also good ways and not so good ways to do this. This takes more of a commitment, both financially and at a management level.
d. The poultry people are ahead of us in this field. A high percentage of their barns have some sort of water treatment.

New infectious agents:

Sapelovirus, Teschovirus, Seneca Valley Virus, etc.

Old agents that cost a lot:

PRRS, *M hyopneumoniae*, influenza, rotavirus, Salmonella, etc.

How to eliminate bacterial agents

What happened to Medicated Early Weaning? Is it possible or practical in today's large units?

Finishing pig density requirements

240 lbs. @ 7.5 sq. ft. vs 290 lbs. @ ?? sq. ft.

How are we going to be able to compete with manufactured proteins?

"Competing with manufactured protein: are dark days ahead for the food animal industry?" Bartlett, JAVMA, Vol 259, No. 2, 132-134.

"Disruption, Implications, and Choices; Rethinking Agriculture 2020-2030", Rethink X report. Tubb and Seba, September 2019.

The final Hogg Sense entry:

False Tails Fight Vice: While the search goes on to find the causes of tail biting, affected units usually try to curb the vice by giving the pigs something else to chew. This tends to be an object suspended over the pen, such as a length of chain or an old car tire hanging on a rope. (A new idea is) an inverted synthetic cone fixed to hang just above the pig, head high. Dangling down from the cone are four sticks of pliable plastic, about the same thickness and length as a pig's tail. The occupants of the pen can bite and bend these artificial tails without breaking them. A Danish producer was one of the first to try the plastic 'tails' and says that they have been far more effective than more conventional distractions, such as tires and pipes". (Pig International, Volume 31). Hogg Sense Volume 11, No. 12.

To summarize

I have given you a pretty deep look into the life of Dr. Alex Hogg. I hope you have gained a better appreciation of Dr. Hogg and his accomplishments. Mentorship of our up-and-coming swine veterinarians is an important part of our culture, so this organization stays strong. I gave you multiple (formal) ways to continue your path on lifelong learning, and there are likely many other options out there. Your age should not be a deterrent. Surely there is something out there you would like to learn more about, whether it is within the realm of veterinary medicine or not, it doesn't matter! Finally, we have learned an incredible amount about swine medicine and production over the last century plus, but there are still issues we need to learn more about. Never stop learning!

To conclude, I would like to tell you a little more about Alex.

Alex would have been 102 years old this year.

Alex's favorite song was "Scotland the Brave", specifically on the bagpipes.

Alex's favorite author was a poet by the name of Robert Burns, and his favorite poem was called "Ode to a Mouse"

But little mouse, you are not alone,
In proving foresight may be vain:
The best laid schemes of mice and men
Go often askew,
And leave us nothing but grief and pain, For promised joy!
Still, you are blessed, compared to me!
The present only touches you:
But oh! I backward cast my eye,
On prospects dreary!
And forward, though I cannot see,
I guess and fear!

Alex's favorite place to visit besides Scotland was Paris because that's where his daughter and grandkids lived. Professionally, his favorite place to visit was the Pasteur Institute in Bucharest, Romania.

Regarding attending AASV meetings, he enjoyed the scientific sessions and presenting lectures, but his greatest joy at the meetings were greeting old friends and meeting new people.

Finally, Alex's favorite saying for colleagues and family, especially his grandkids was "That's the stuff."

I would like to thank the following people:

- Mary Lou Hogg. Mary Lou and I had some delightful conversations and many emails as I gathered details on Alex's life. Thank you, Mary Lou for sharing Alex with us and for all you do for AASV and the AASV Foundation.
- My wife Donna who continually pushes me forward.
- My kids, Ben, Sarah, and Amy. Two physicians and a hospital administrator. I couldn't be prouder.
- Sue Schulteis - what more can you say about Sue.
- Dr. Mike Senn and the Program Committee. Thank you for letting me give this lecture.
- All my colleagues in this great group who influenced who I am today.

Thank You

https://doi.org/10.54846/am2022/3

Breed-to-finish risk factors associated with increased proportion of lightweight pigs marketed

E. S. Magalhães[1]; D. Rosero[2]; T. Donovan [2]; G. Trevisan[1]; D. J. Holtkamp[1]; G. S. Silva[1]; D. C. L. Linhares[1]
[1]Veterinary Diagnostic and Production Animal Medicine Department, Iowa State University, Ames, Iowa;
[2]The HANOR Company, Enid, Oklahoma

Introduction

Lightweight pigs are a consequence of poorer performance throughout the pigs` life-time[1,2] so understanding which factors are impacting it is important to decrease economic losses at marketing. This study consisted of analyzing 20 months of data from closeouts of pigs marketed in a US swine production system, with the objective of identifying the major whole-herd risk factors associated with the increased proportion of lightweight pigs. For this purpose, the dataset utilized in this study consisted of historical information for each group of marketed pigs, from breeding-to-market (BTM), containing data related to productivity parameters, closeout data of growing pig lots, and breeding-to-wean and growing pig health status, environment, and diagnostic information.

Methods

The dataset analyzed in this study consisted of closeout data for 1913 groups marketed between June 2018 to February 2020 in a large production system in the Midwestern USA. The outcome was the percentage of pigs marketed with a carcass weight lower than 200 pounds (# of pigs < 200 lbs. ÷ total # of pigs marketed in the group). Grow-finish performance data of each group was linked to the pre-weaning performance of the cohorts in the sow farm, creating a whole-herd dataset containing information on health and productivity throughout all stages of production. The health data consisted of weekly PRRSV breeding herd status, and a combination of PRRSV PCR results and diagnostic codes from tissues submitted to the VDL for the growing phase. SAS software was utilized for data management and analysis. Once the BTM dataset was built, continuous variables related to productivity were divided into quartiles, and then univariate analysis was conducted separately between each predictor collected in the dataset and the outcome percentage of lightweight pigs, using Linear Mixed Models on SAS (PROC GLIMMIX). The objective of this step was to screen prediction variables to determine which were associated with the outcome. A multivariable model was built utilizing the variables with a significant association ($P < 0.10$) in the previous step, and that remained significant in the multivariable model ($P < 0.05$). The effect of the predictors on the outcome using the multivariable model was reported, utilizing Tukey-test for pairwise comparison between the category means. Multicollinearity was investigated, and variables were excluded from the multivariable model as described by Cohen, et al.[3]

Results

The mean percentage of lightweight pigs in this study population was 16.7% (95% CI 16.3% - 17.1%). At the screening step, 15 of 21 variables had a significant association ($P < .10$) with the outcome. Four variables were excluded from the multivariable model due to multicollinearity, and one became not significant. The final set of risk factors in the model included: (a) average sow parity at weaning, (b) PRRS status at weaning, (c) weaning age, (d) season of the year when animals were marketed, (e) PRRS PCR detection in the growing phase, (f) pre-weaning mortality, and (g) average nursery mortality. The variables days on feed, interval to sell the group, and nursery stocking weight of feeder pigs were utilized as covariates. A summary of the differences in lightweight pigs of the cohorts are described in Table 1.

Discussion and conclusions

This study demonstrated the importance of analyzing lightweight carcass with the breeding-to-market retrospective performance of the closeouts to identify the impact of risk factors collected under the production system`s condition. Among the population analyzed, the major risk factors associated with a higher percentage of lightweight pigs were groups from sow farms with young average parity in the week of weaning, low weaning age, high pre-weaning mortality, unstable status for PRRSV, groups with positive PCR for PRRSV during grow-finish, and relatively high nursery mortality. These results can help guide future decisions regarding on-farm disease prevention and control efforts related to carcass performance by providing information that will support decision-makers to reduce the proportion of lightweight pigs and, thus, increased profitability.

References

1. De Grau et al. 2005. Can. J. Vet. Res. 69: 241–245.

2. Larriestra et al. 2006. Can. J. Vet. Res. 47: 560-566.

3. Cohen, J. 1988. *Statistical power analysis for the behavioral sciences* (2nd ed.). Hillsdale, NJ: Lawrence Earlbaum Associates.

Table 1: Average percentage of lightweight pigs for each category of W2F groups, based on different comparisons

Characteristic	Category A	Category B	Category C	Category D
(a) Average parity at wean - % lightweight	2.6 – 17.6%[a]	3.4 – 15.4%[b]	3.7 -14.8%[b]	4.6 – 14.5%[b]
(b) PRRS status at wean - % lightweight	IA – 17.7%[ab]	IB – 18.1%[a]	II-vx – 15.5%[b]	IV – 10.9%[b]
(c) Weaning age (days) - % lightweight	18.6 – 16.8%[a]	20.8 – 15.6%[ab]	21.9 -15.3%[ab]	23.5 – 14.3%[b]
(d) Season - % lightweight	Fall – 17.2%[a]	Winter – 15.6%[b]	Spring – 13.8%[c]	Summer - 15.6%[b]
(e) PRRS PCR result - % lightweight	Positive - 16.2%[a]	Negative - 14.8%[b]	–	–
(f) Pre-weaning mortality - % lightweight	13.1% – 15.3%[ab]	16.4% – 15.0%[a]	18.8% – 15.0%[a]	24.8% - 16.9%[b]
(g) Nursery mortality - % lightweight	1.4% - 15.4%[ab]	2.8% - 14.9%[a]	3.9% - 15.1%[a]	7.4% - 16.8%[b]

[abc] Different letters represent statistical difference of % of lightweight for the closeouts.

https://doi.org/10.54846/am2022/4

Analysis of a novel production model for nursing piglets from to birth to the end of nursery phase

Mark Schwartz[1,2]; Cipriano De Abreu[1]; Maria Pieters[1,3]
[1]Veterinary Population Medicine Department, College of Veterinary Medicine, University of Minnesota, St. Paul, Minnesota; [2]Schwartz Farms, Inc., Sleepy Eye, Minnesota; [3]Swine Disease Eradication Center, University of Minnesota, St. Paul, Minnesota

Introduction

Intense selection of swine genetics has resulted in an increase in litter size which has both improved the efficiency of swine producers as well as having created the unintended consequence of increased piglet pre-weaning mortality. The practices of cross fostering of piglets and the use of nurse sows to attempt to effectively manage large litters has led to the creation of multiple methods and theories, with varied effectiveness of such practices to reduce pre-weaning mortality. In addition to the evaluation of production parameters, the effect of large group rearing with a novel production method termed Multi-Suckle Common Creep (MSCC) evaluated the effect on the transmission of various virus and bacteria including *Mycoplasma hyorhinis.* Recent studies have been conducted to evaluate the socialization effect of large group production methods from birth to end of nursery phase on production parameter daily gain.

Materials and methods

The study was carried in a 5,000-sow commercial breed to wean farm. A total of 192 sows and litters were enrolled in the study with 96 sows and litters in the treatment group (MSCC) and 96 sow and litters in the control (conventional) group. All piglets (2,634) were individually weighed, tagged (LeeO Systems, Prairie Systems, Spencer, Iowa, USA), administration of injectable iron within 18 hours of birth, prior to cross fostering or comingling. In each of four rooms of 48 farrowing stalls, 24 litters were raised in individual stalls and three groups of eight litters were raised with the MSCC method, balanced by parity. From each of the four rooms, six sows from each treatment were randomly selected for sample collection, balanced by parity. At 24 to 48 hours prior to parturition, a vaginal swab, udder wipe and oropharyngeal swabs was collected from each of the 48 sows. Within 18 hours of birth and prior to cross fostering or comingling, two piglets were randomly selected and identified with a second ear tag and a tonsillar and nasal swab was collected from each of the 96 piglets. At mid-lactation an oropharyngeal swab and udder wipe was collected from each of the 48 sows and a tonsillar and nasal swab was collected from each of the 96 piglets. Within 24 hours of weaning an oropharyngeal swab and udder wipe was collected from each of the 48 sows and a tonsillar and nasal swab and serum sample was collected from each of the 96 piglets. At weaning, individual weaning weight was collected from each on-test piglet (2,334) and piglets were randomized into treatment group in the nursery. Individual piglet treatments and mortality data were collected from birth to the end of the nursery phase. All vaginal, oropharyngeal, tonsillar and nasal swabs were processed for testing using real-time PCR (QuantiFast Probe PCR kit, Qiagen Inc, Germantown, MD, USA) utilizing customized primers and probes for *M hyorhinis*. Statistical analysis was carried out with R (R Foundation for Statistical Computing, Vienna, Austria).

Results and discussion

No difference was observed between the treatment groups for pre-weaning mortality, growth rate from birth to weaning and weaning to end of nursery phase (P-value > 0.05). Fewer pigs were treated in the MSCC group during the nursing phase (P-value using Fisher's Exact Test < 0.05). No difference in nursery phase average daily gain was observed (P-value > 0.05). A lower nursery phase mortality was observed in the MSCC group (P-value using Fisher's Exact Test < 0.05). *M hyorhinis* genetic material was not detected in pre-farrow vaginal or sow oropharyngeal swabs. *M hyorhinis* genetic material was detected in 6.2% (3/48) sow pre-farrow and mid-lactation oropharyngeal swabs and 4.3% (2/46) weaning oropharyngeal swabs. *M hyorhinis* genetic material was detected in 16.6%, 11.1%, and 51.1% of MSCC piglet nasal swabs at birth, mid-lactation and weaning, respectively, and 2.0%, 4.4% and 45.4% of control piglet nasal swabs, respectively, and 0%, 11.1% and 60% of treatment group piglet tonsillar swabs at birth, mid-lactation and weaning, respectively, and 0%, 2.0% and 54% of control group piglet tonsillar swabs at birth, mid-lactation and weaning respectively.

The results of this study did not indicate a reduction in pre-weaning mortality nor an increased rate of gain from birth to weaning. A statistical difference in both proportion of piglets individually treated with antibiotic during the nursing phase, and the difference in nursery phase mortality was observed as described. Patterns of transmission from dam to piglet could not be determined. No difference in *M hyorhinis* prevalence between treatment groups was observed (P-value > 0.05). Additional study of the MSCC production method effect on production parameters and pathogen transmission between piglets prior to weaning.

https://doi.org/10.54846/am2022/5

Quantification of decontamination strategies for semi-truck cabs

C. Grace Elijah[1], DVM; Cassandra K. Jones[2], PhD; Caitlin Evans[3], PhD; Haley K. Wecker[3], BS; Charles R. Stark[3], PhD; Jianfa Bai[1], PhD; Elizabeth G. Poulsen Porter[1], MS; Allison K. Blomme[3], BS; Jason C. Woodworth[2], PhD; Chad B. Paulk[3], PhD; Jordan T. Gebhardt[1], DVM, PhD
[1]Department of Diagnostic Medicine/Pathobiology, College of Veterinary Medicine, Kansas State University;
[2]Department of Animal Sciences and Industry, College of Agriculture, Kansas State University;
[3]Department of Grain Science and Industry, College of Agriculture, Kansas State University

Introduction

The introduction of porcine epidemic diarrhea virus (PEDV) into North America led to a need to better understand the epidemiological link between potentially contaminated feed, pig delivery trucks, feed delivery trucks, feed mills, and production sites. A report by Greiner (2016)[1] suggested that other areas in the feed mill, like truck pedals and the feed mill office, could be a potential source of disease during PEDV or porcine delta coronavirus (PDCoV) outbreaks. While there is evidence to suggest that the inside of semi-truck cabs for the feed delivery chain may serve as a potential source for disease, there is a lack of standardized protocols or scientific evidence for methods of disinfection within the semi-truck cabs. Therefore, the objective of this project was to evaluate commercially available disinfectants and disinfection application methods against PEDV and porcine reproductive and respiratory syndrome virus (PRRSV) on various surfaces within semi-truck cabs.

Materials and methods

Three different surface types common in vehicle cabs (fabric, plastic, and rubber) were cut into 4 × 4 inch coupons and inoculated with either PEDV or PRRSV. Once inoculated, surfaces were placed in one of 3 semi-truck cabs (figure 1 and figure 2) and disinfectant treatment was applied. Disinfectant treatments were as follows: 1) no-disinfectant, 2) hurricane fumigation with 1:256 dilution of Synergize, 3) hurricane fumigation with 1:64 dilution of Intervention, 4) pump sprayer with 1:256 dilution of Synergize, 5) pump sprayer with 1:64 dilution of Intervention, 6) pump sprayer with 10% bleach, 7) no chemical with 10 hr downtime, and 8) gaseous fumigation over a 10 hr period with water based chlorine dioxide. Once disinfectant treatment was applied, the coupons were environmentally swabbed as previously described[2] and submitted for qPCR duplex analysis for PEDV and PRRSV. There were three replications per disinfectant treatment. Fixed effects considered the disinfectant treatment, surface treatment, and virus type while random effect was truck cab defining it as the experimental unit for disinfectant treatment to account for the split plot design. Results of Ct data are reported as least squares means ± standard error of the mean. All statistical models were evaluated using visual assessment of studentized residuals and assumptions appeared to be reasonably met. A Tukey multiple comparison adjustment was incorporated when appropriate. Results were considered significant at $P \leq 0.05$ and marginally significant between $P > 0.05$ and $P \leq 0.10$.

Results and discussion

There was no evidence of a disinfectant × surface × virus, surface × virus, or disinfectant × virus interaction ($P > 0.10$; Table 1). There was a significant disinfectant × surface interaction ($P < 0.0001$) indicating that the disinfectant treatment efficacy differed based on surface (Table 2). Within rubber surfaces, 10% bleach had a greater Ct value compared to all other treatments ($P < 0.05$) with the exception of Intervention via hurricane fumigation application which was intermediate. In both fabric and

Figure 1: Fabric and plastic coupons were placed on the seat and dashboard respectively, within the semi-truck cab to determine the efficacy of decontamination procedures. Placement of each coupon was chosen prior to the start of the experiment to best represent actual locations of these surfaces within semi-truck cabs.

Figure 2: Fabric and rubber coupons were placed on the seat and floor respectively, within the semi-truck cab to determine the efficacy of decontamination procedures. Placement of each coupon was chosen prior to the start of the experiment to best represent actual locations of these surfaces within semi-truck cabs

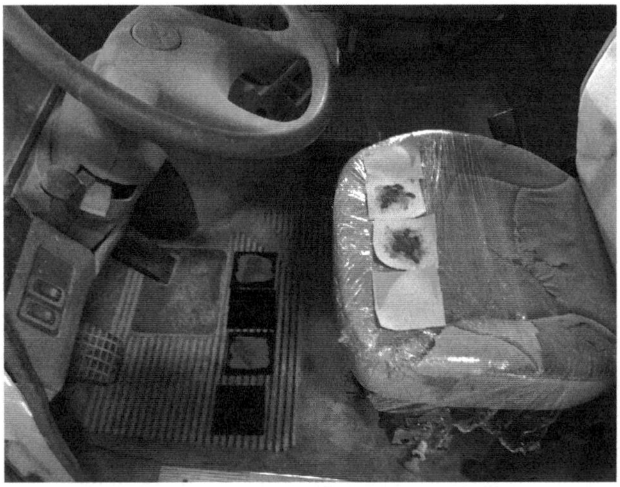

plastic surfaces, there was no evidence ($P > 0.05$) of a difference in Ct value between any of the treatments. Additionally, for the no-disinfectant treatment the Ct value was greater on fabric surfaces compared to plastic and rubber ($P < 0.05$); fabric and rubber greater than plastic in the Intervention via hurricane fumigation application treatment ($P < 0.05$); fabric and rubber greater than plastic in the 10% bleach treatment ($P < 0.05$); and fabric greater than plastic and rubber in the 10 hr of downtime and gaseous fumigation treatments ($P < 0.05$). There was a significant main effect of disinfectant treatment ($P = 0.016$; Table 3), where 10% bleach had a greater Ct value compared to the no-disinfectant, 10 hr downtime treatment, and Intervention applied using the pump sprayer ($P < 0.05$). There was a main effect of surface ($P < 0.0001$) where rubber had a greater Ct value compared to plastic ($P < 0.05$), and fabric had a greater Ct value compared to both rubber and plastic ($P < 0.05$). This could indicate that disinfectant applications can

Table 1: Effect of surface, disinfectant, and virus on the detection of viral RNA during semi-truck cab decontamination

Item	*P* =
Disinfectant × surface × virus	0.959
Surface × virus	0.926
Disinfectant × virus	0.508
Disinfectant × surface	< 0.0001
Virus	< 0.0001
Surface	< 0.0001
Disinfectant	0.016

soak into the fabric to reduce the amount of PEDV or PRRSV on the fabric or that fabric surfaces limit the amount of virus picked up by environmental swabbing. Finally, the Ct value for PRRSV was greater than PEDV ($P < 0.0001$) when averaged across all surfaces and disinfectant treatments. In summary, these findings highlight that differences in disinfectant efficacy were observed for different surfaces. This study also highlights that methods utilized to disinfect semi-truck cabs, whatever the application method or disinfectant type, should be standardized and consistently applied. Disinfectants will decrease the quantity of detectable virus of locations where applied but won't sterilize the surface altogether, which underscore that biocontainment of pathogens is critical within biosecurity programs.

References

1. Greiner LL. Evaluation of the likelihood of detection of porcine epidemic diarrhea virus or porcine delta coronavirus ribonucleic acid in area within feed mills. *J. Swine Health Prod.* 2016;24(4):198-204.

2. Elijah, C. G., Trujillo, J. D., Jones, C. K., Gaudreault, N. N., Stark, C. R., Cool, K. R., Paulk, C. B., Kwon, T., Woodworth, J. C., Morozov, I., Gebhardt, J. T., and Richt, J. A. Evaluating the Distribution of African Swine Fever Virus Within a Feed Mill Environment Following Manufacture of Inoculated Feed. *PLoS ONE*, 2021:16(8): e0256138. **doi:10.1371/journal.pone0256138**

Table 2: Effect of surface and disinfectant on the detection of viral RNA during semi-truck cab decontamination[1]

Item	Proportion PCR positive			Ct value		
	Fabric	Plastic	Rubber	Fabric	Plastic	Rubber
No disinfectant[2]	6/6	6/6	6/6	35.9[c,d,e,f,g,h]	28.6[a,b]	29.0[a,b]
Hurricane fumigation[3]						
Intervention[4]	6/6	6/6	5/6	35.8[a,b,c,d,e,f,g,h]	29.8[a,b,c,d]	35.4[a,b,c,d,e,f,g,h]
Synergize5	5/6	6/6	6/6	36.6[a,b,c,d,e,f,g,h]	31.9[a,b,c,d,e,f]	31.8[a,b,c,d,e,f]
Pump sprayer[6]						
Intervention	6/6	6/6	6/6	36.1[b,d,e,f,g,h]	29.8[a,c]	30.4[a,b,c,d,e,f]
Synergize	5/6	5/6	6/6	38.0[e,f,g,h]	32.8[a,b,c,d,e,f,g]	32.0[a,b,c,d,e,f]
10% bleach[7]	2/6	6/6	1/6	42.9[h]	29.0[a,b,c,d]	43.1[h]
10 hr downtime[8]						
No chemical	5/6	6/6	6/6	38.4[f,g,h]	28.8[a,b,c,d]	30.0[a,b,c,d]
Gaseous treatment[9]	4/6	6/6	6/6	40.6[g,h]	30.1[a,b,c,d,e]	30.9[a,b,c,d,e,f]

1 Surfaces were inoculated with 1 mL of PEDV or PRRSV, randomly placed within the truck cab, and subjected to a randomly assigned disinfectant treatment. Samples with no detectable RNA were assigned a value of 45. Disinfectant × surface, $P < 0.0001$; SEM = 1.45.

2 Surfaces were inoculated with pure virus and allowed to sit within the semi-truck cab for 15 min. These surfaces were not treated with a disinfectant application.

3 Semi-truck cabs had a hurricane fumigation system placed in the passenger sit and directed towards the driver's side. The hurricane fumigation system was filled with respective disinfectant and was allowed to run 5 min for each treatment.

4 Virox Technologies Inc., Oakville, ON.

5 Preserve International, Lexington, KY.

6 Semi-truck cabs had disinfectant applied with a conventional pump sprayer with the designated disinfectant.

7 Household bleach (10% dilution; The Clorox Company, Oakland, CA; 7.55% sodium hypochlorite).

8 Surfaces were inoculated with pure virus and allowed to sit within the semi-truck cab for 10 hr. These surfaces were not treated with a disinfectant application.

9 Semi-truck cabs had gaseous chlorine dioxide (ProKure G; ProKure Solutions, Phoenix, AZ) placed on the passenger side seat and allowed to fumigate the semi-truck cab for 10 hr.

a-h Means lacking common superscripts differ, $P < 0.05$.

Table 3: Main effects of disinfectant, surface, and virus on the detection of viral RNA during semi-truck cab decontamination[1]

Item	Proportion PCR positive	Ct Value	SEM	*P* =
Disinfectant			1.11	0.016
No disinfectant[2]	18/18	31.2[a]		
Hurricane fumigation[3]				
Intervention[4]	16/18	33.7[a,b]		
Synergize[5]	17/18	33.4[a,b]		
Pump sprayer[6]				
Intervention	18/18	32.1[a]		
Synergize	16/18	34.3[a,b]		
10% bleach[7]	9/18	38.3[b]		
10 hr downtime				
No chemical[8]	17/18	32.4[a]		
Gaseous treatment[9]	16/18	33.9[a,b]		
Surface			0.53	< 0.0001
Fabric	39/48	38.0[c]		
Plastic	47/48	30.1[a]		
Rubber	41/48	32.8[b]		
Virus			0.48	< 0.0001
PEDV	62/72	32.0[a]		
PRRSV	62/72	35.3[b]		

1 Surfaces were inoculated with 1 mL of PEDV or PRRSV, randomly placed within the truck cab, and subjected to a randomly assigned disinfectant treatment. Samples with no detectable RNA were assigned a value of 45.

2 Surfaces were inoculated with pure virus and allowed to sit within the semi-truck cab for 15 min. These surfaces were not treated with a disinfectant application.

3 Semi-truck cabs had a hurricane fumigation system placed in the passenger sit and directed towards the driver's side. The hurricane fumigation system was filled with respective disinfectant and was allowed to run 5 min for each treatment.

4 Virox Technologies Inc., Oakville, ON.

5 Preserve International, Lexington, KY.

6 Semi-truck cabs had disinfectant applied with a conventional pump sprayer with the designated disinfectant.

7 Household bleach (10% dilution; The Clorox Company, Oakland, CA; 7.55% sodium hypochlorite).

8 Surfaces were inoculated with pure virus and allowed to sit within the semi-truck cab for 10 hr. These surfaces were not treated with a disinfectant application.

9 Semi-truck cabs had gaseous chlorine dioxide (ProKure G; ProKure Solutions, Phoenix, AZ) placed on the passenger side seat and allowed to fumigate the semi-truck cab for 10 hr.

[a,b,c] Means within main effect lacking common superscripts differ, $P < 0.05$.

https://doi.org/10.54846/am2022/6

Early detection of trade-impacting swine pathogens: An epidemiological modeling study

G. Trevisan[1], DVM, MBA, PhD; P. Morris[2], MS; M. Culhane[3], BA, DVM, PhD; S. Malladi[3], BT, MS, PhD; C. Wang[1,2], PhD; R. Main[1], BS, MS, PhD; G. Silva[1], DVM, MS, PhD; J. Zimmerman[1], DVM, PhD, DACVPM
[1]Department of Veterinary Diagnostic and Production Animal Medicine;
[2]Department of Statistics, Iowa State University. College of Veterinary Medicine, University of Minnesota

After the experience of several recent pandemics, it is reasonable to conclude that North America is within reach of a FAD. If this were to occur, the effect would be major: the detection of ASFV in the US will result in an immediate 40-50% drop in live hog prices.[1] Problematically, most swine FADs produce clinical signs resembling endemic pathogens, eg, differentials for ASFV and CSFV include erysipelas, salmonellosis, bacterial septicemia, PRRSV, PRV, and others.[2-4] Thus, ASFV and CSFV will not be identified by clinical signs alone - laboratory testing will be required.

In this study, we explored early FAD detection based on collecting/testing a few samples from each of many swine farms in a large geographic region. This approach is new in that the goal is to determine the status of the region without trying to prove that each farm is negative. Thus, negative testing supports continuity of business in the region and a positive result triggers a USDA FAD response.

The USDA Animal Disease Spread Model[5] (ADSM) was used to simulate the spread of a "generic" FAD in a population of 17,522 farms (51,515,699 pigs) in a geographic region (1,615,246 km^2) composed of the states of CO, IA, KS, MN, MO, NE, OK, SD. Disease spread was modeled for various scenarios (100 iterations each) by systematically changing ADSM transmission parameters, including the location of the index case. Thereafter, the number and location of infected farms was used to model the probability of detection over time post introduction for specific sampling schemes, eg, percent farms sampled, location of farms, and timing of sampling.

As an example, a scenario with restrained transmission parameters and in which the index case was initiated in a swine-dense region resulted in 0.04% (n = 7), 0.18% (n = 32), and 0.74% (n = 130) of farms becoming infected by 21, 42, and 63 days after the introduction of the pathogen. Results from the active collection/testing model predicted probability of disease detection in the region with estimates of 32%, 70%, and 97% at the same time points.

Initial results suggest that an approach based on systematically collecting/ testing a few samples from many herds in a region could achieve early detection of a FAD, provide on-going evidence of freedom from disease outside of Control Areas, and facilitate interstate/international commerce over the course of the response and recovery period.

References

1. Carriquiry et al. 2020. CARD Working Papers. 618.

2. Sánchez-Vizcaíno et al. 2019b. Manual of Diagnostic Tests and Vaccines for Terrestrial Animals. World Organisation for Animal Health.

3. Sánchez-Vizcaíno et al. 2019a. Diseases of Swine 11th edition. John Wiley & Sons, Inc., p. 443-452.

4. Schulz et al. 2019. Viruses 11:866.

5. Harvey et al. 2007. Prev Vet Med 82:176-197.

https://doi.org/10.54846/am2022/7

Comparison of 5 ASF point of care assays against a standard OIE-based laboratory PCR using field samples

C. Goodell[1], DVM, MS, PhD, DACVPM; E. Schacht[1], PhD; V. Tran[1], DVM; H. Nguyen[1], DVM; Q. Tran[1], DVM; C. Maala[1], DVM; D. Lai[2], DVM; T. Nguyen[2], DVM; H. Le[2], DVM, MSc, PhD; N. Nguyen[2], DVM, MSc; D. Do[2], DVM, MSc, PhD; T. Nguyen[2], DVM, MSc, PhD; N. Nguyen[3], PhD; J. Zimmerman[4], DVM, PhD, DACVPM; O. Gomez-Duran[1] DVM, MS, PhD, MRCVS, Certificate in Pig Medicine, Diplomate ECPHM

[1]Boehringer Ingelheim Vetmedica GmbH, Ingelheim, Germany; [2]Animal Sciences and Veterinary Medicine, Nong Lam University, HCM City, Vietnam; [3]Research Center for Genetics and Reproductive Health, School of Medicine, Vietnam National University, HCM City, Vietnam; [4]Iowa State University, Ames Iowa

Introduction

Rapid detection of African swine fever virus (ASFV) in the field would enable immediate removal of positive animals. Commercial point-of-care (POC) ASF detection tests consist of lateral flow devices (quick tests, QT) and real-time polymerase chain reaction (PCR) assays. Known limitations of POC tests include lower sensitivity (the ability to detect a true positive as positive) and lower specificity (the ability to detect a true negative as negative) versus laboratory-based tests. A further complication for POC PCRs is the requirement to extract genetic material from the sample prior to DNA amplification. Extraction requires precision and quality control if the test is to produce valid results. The objective of this study was to compare two commercial ASF POC QT and 3 commercial ASF POC PCRs against an OIE laboratory-based ASFV PCR using whole blood samples collected from sows exhibiting clinical signs suggestive of ASFV infection, their direct and indirect contact neighbors, and known ASFV-negative blood samples.

Materials and methods

"Suspected ASF events" (n = 46) were defined as a sow exhibiting signs suggestive of ASFV infection as determined by farm caregivers (**https://www.oie.int/en/document/african_swine_fever/**). In each suspected ASFV event, whole blood was collected in EDTA from the index sow plus 14 neighbors. The collected blood samples were chilled, transported to the laboratory and tested for ASFV DNA by an OIE real-time ASFV p72 PCR (STAND PCR) within 24-hours of arrival. Subsequently, each POC test was performed according to the manufacturer's instructions and if a PCR, using the recommended POC PCR equipment, extraction procedures (if necessary), and PCR assay. For DNA extraction, ASF PCR POC A included a field-ready extraction device; ASF PCR POC B utilized a manual syringe and cartridge purification step; ASF PCR POC C did not require extraction. POC tests were run in parallel using either fresh (≤ 5 days at 2-8°C) or frozen (one freeze-thaw cycle (-80°C)) samples. Whichever the case, all POC testing was performed using samples handled in the identical manner and all testing took place in the laboratory performing the real time PCR.

Results

The 46 ASF events amassed 673 whole blood field samples. Additionally, 50 known ASF negative whole blood samples were tested. From the field samples, STAND PCR identified 86 positives and 637 negatives. After completing STAND PCR, 426 samples were subsequently tested fresh by the POC tests, and 297 underwent one freeze-thaw cycle prior to POC testing (Table 1). POC devices tested samples of the same status, for equivalent comparison.

Conclusions

- Relative to the STAND PCR, POC tests exhibited low (< 90%) diagnostic sensitivity (Se) and/or low diagnostic specificity (Sp).
- Technically, QTs were easy to use (the PCRs were multi-step processes), however all tested commercial POC assays were not sufficient for use in ASFV control or eradication programs.

Table 1: POC Test performance compared to OIE (p72) real time PCR (STAND PCR), by sample status at POC testing (fresh or frozen)

STAND PCR Comparison Field Samples (637 neg, 86 pos)	Overall (n = 723)		Fresh (n = 426)		1 Freeze-thaw (n = 297)		Known neg (n = 50)
	Se	Sp	Se	Sp	Se	Sp	Sp
POC QT A	60%	88%	53%	88%	63%	88%	100%
POC QT B	53%	74%	59%	73%	48%	76%	100%
POC PCR A	85%	98%	88%	98%	83%	98%	100%
POC PCR B	84%	95%	81%	96%	85%	93%	100%
POC PCR C	84%	98%	88%	99%	81%	97%	98%

https://doi.org/10.54846/am2022/8

Efficacy of the "Tooth Extraction" (test-and-remove) protocol in commercial swine farms in Vietnam

C. Goodell[1], DVM, MS, PhD, DACVPM; E. Schacht[1], PhD; V. Tran[1], DVM; H. Nguyen[1], DVM; Q. Tran[1], DVM; C. Maala[1], DVM; D. Lai[2], DVM; T. Nguyen[2], DVM; H. Le[2], DVM, MSc, PhD; N. Nguyen[2], DVM, MSc; D. Do[2], DVM, MSc, PhD; T. Nguyen[2], DVM, MSc, PhD; N. Nguyen[3], PhD; J. Zimmerman[4], DVM, PhD, DACVPM; O. Gomez-Duran[1], DVM, MS, PhD, MRCVS, Certificate in Pig Medicine, Diplomate ECPHM

[1]Boehringer Ingelheim Vetmedica GmbH, Ingelheim, Germany; [2]Animal Sciences and Veterinary Medicine, Nong Lam University, HCM City, Vietnam; [3]Research Center for Genetics and Reproductive Health, School of Medicine, Vietnam National University, HCM City, Vietnam; [4]Iowa State University, Ames Iowa

Introduction

Vietnam has lost ~6 million pigs to African swine fever virus (ASFV) since the first reported outbreak in February 2019. The national swine herd has returned to 88.7% of its pre-ASFV level, but the risk of ASFV recurrence is high because of new and on-going ASFV outbreaks.[1] Because of the increased value of pork in Vietnam due to reduced supply, standard ASFV control measures have centered around a modified test-and-remove or "Tooth Extraction" protocol. A common, "Tooth Extraction" protocol for a sow farm is to remove any sow exhibiting clinical signs compatible with ASF (taking a whole blood sample for ASFV PCR testing) plus the stalled sow on each side of the index (clinical) animal.[2] The objective of this study was to test the efficacy of the "Tooth Extraction" protocol.

Materials and methods

Fifty-two suspected ASF events on Vietnamese sow farms were included in the assessment. "Suspected ASF event" was defined as a sow exhibiting OIE-established clinical signs as determined by a farm caregiver (OIE Technical Disease Card: African swine fever - OIE - World Organisation for Animal Health). For each suspected ASFV event, whole blood samples were collected from the index sow plus 14 animals in gestation stalls on each side of the index sow (Figure 1). Samples were tested for ASFV DNA by an OIE-approved real-time ASFV PCR within 24-hours of arrival to the laboratory. The proportion of positive animals was analyzed as a function of the gestation stall distance from the index sow.

Results

a. In 17 of the 52 events (33%), the index sow and 14 neighbor sows were ASFV PCR negative. Of the 35 events where ASF PCR was positive, in 19 (54 %), removal of the index sow and her direct contact neighbors still left one or more ASFV-positive sow.
b. 86 of 764 whole blood samples were ASFV PCR positive.
   35 of the 86 positives were from the index sow.
   16 of the 86 positives were from sows next to the index sow (direct contact neighbors).
   35 of the 86 positives were from sows outside of the index sow and her direct contact neighbors.
c. ASFV DNA was detected in blood from sows showing no clinical signs.

Conclusions

"Tooth Extraction" is not sufficient to eliminate ASFV from a sow farm.

References

1. Report # VM20221-0042, Foreign Agricultural Service, USDA, May11, 2021. Available at: **https://www.fas.usda.gov/data/vietnam-african-swine-fever-update**

2. Yaros, Leman 2019. Available at: **Dr. Joseph Yaros - Our experience with ASF in China - YouTube**

Figure 1: Example of blood sampling protocol around a suspected ASF clinical sow in gestation.†‡

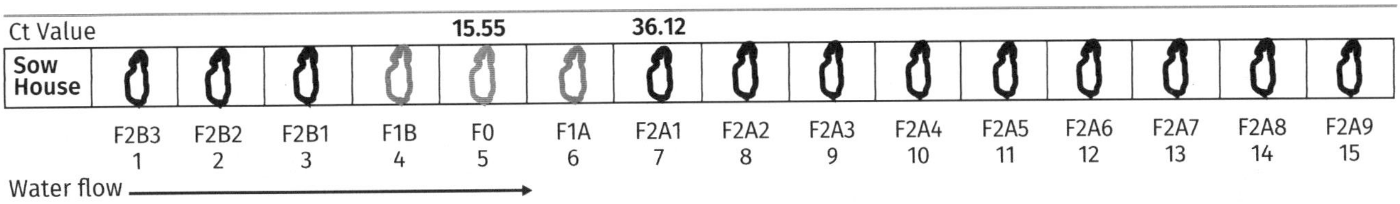

† F0 = index sow (red), F1 = direct/closest contact neighbors (blue), F2 = indirect contact neighbors (black); A = "down" row, B = "up" row

‡ The rationale for the sampling distribution was due to an assumption drinking water flowed down a common water trough

https://doi.org/10.54846/am2022/9

Practical implications of pooling family oral fluids on the probability of PRRSV detection by RT-rtPCR

Onyekachukwu H. Osemeke[1], DVM, MS; Marcelo N. Almeida[1], DVM, MS, PhD; Giovani Trevisan[1], DVM, MBA, PhD; Eduardo de Freitas Costa[2], DVM, MS, PhD; Gustavo S. Silva[1], DVM, MS, PhD; Daniel C. L. Linhares[1], DVM, MBA, PhD
[1]Veterinary Diagnostic and Production Animal Medicine, College of Veterinary Medicine, Iowa State University, Ames, Iowa; [2]Department of Epidemiology, Bioinformatics and Animal Models, Wageningen Bioveterinary Research, Lelystad, The Netherlands

Introduction

The PRRSV status of due-to-wean (DTW) pigs is an important metric for the US swine industry.[1] Piglet sera are considered the gold standard sample for ascertaining PRRSV status of DTW pigs; however, this method is relatively more laborious, requires scores of pigs to be frequently sampled in low prevalence scenarios, and can be stressful to the piglets. Family oral fluids (FOF) are a more convenient and cost-efficient alternative for determining the PRRSV status of DTW pigs, however, even FOF-based monitoring can be cost-prohibitive for large herds when prevalence is near-zero. Pooling is a commonly used option when there are cost constraints for monitoring. A concern with pooling is the possibility of obtaining false negatives when positive units within the pool are diluted beyond detectable levels. This study investigated the effect of FOF pooling on the probability of PRRSV RNA detection by RT-PCR, and a practical implication on sampling decisions.

Materials and methods

Positive FOFs were obtained from PRRSV-positive herds and negative FOFs from PRRSV-naïve herds. Six replicates of different dilution levels (undiluted, 1:3, 1:5, 1:10, and 1:20) for each of 30 FOF-positive samples were produced using PRRSV-negative FOF samples. All samples (undiluted FOF replicates, n = 180 and diluted replicates, n = 720) were tested for PRRSV RNA detection by RT-PCR. The brglm[2] function on R[3] was used to fit a probit regression model to characterize the probability of detection (POD) for each dilution level, and least-square means were obtained. Twenty-five positive FOFs were eligible for analysis and were divided into three categories based on the Ct (Cycle threshold) values of their undiluted solutions (Category A for Ct values < 34, Category B for Ct values ≥ 34 ≤ 36, and Category C for Ct values > 36). Additionally, a fixed budget scenario was created to demonstrate the implications of pooling choices, the scenario assumed 1) a 60-crate farrowing room with a $75 weekly testing budget (ie, sufficient for 3 PCRs at $25/test); 2) the litters were randomly selected; 3) the final probability to detect at least one PRRSV positive litter was the product of the probability of sampling a truly PRRSV-positive litter (calculated using a conventional sample size formula) and the probability that RT-rtPCR detects PRRSV RNA in at least one submitted sample (calculated using the principles of counting, addition of probabilities, and interpolations from our fitted regression model). Consequently, the final probability to detect at least one PRRSV-positive litter was obtained for each of the three Ct categories for each scenario and averaged; these were weighted averages with weights assigned according to the Ct distribution of all PRRSV positive FOF samples our research team obtained from the field.

Results

The probability of detecting PRRSV RNA by RT-rtPCR in pools decreased with increasing Ct of positive samples within pools and intensity of pooling. The mean probabilities of detection for 1:3, 1:5, 1:10, and 1:20 dilution levels were respectively 0.99, 0.99, 0.97 and 0.87 for Ct Category A; 0.97, 0.93, 0.91 and 0.68 for Ct Category B; and 0.79, 0.65, 0.54 and 0.26 for Ct Category C. The results for the analyses of the hypothetical scenario are described in Table 1.

Implications

Pooling FOFs is a favorable option when there is a cost constraint for monitoring. In low PRRSV prevalence scenarios, regardless of the pooling strategy chosen, the probability of detecting PRRSV increases proportionally to the number of litters sampled. For the scenario described, pooling up to 1:20 improved the probability of PRRSV detection by RT-rtPCR, as it largely increased the population sampled (ie, number of litters) while maintaining the fixed testing cost.

References

1. Holtkamp DJ, Torremorell M, Corzo CA, Linhares DCL, Almeida MN, Yeske P, Polson DD, Becton L, Snelson H, Donovan T, Pittman J, Johnson C, Vilalta C, Silva GS, Sanhueza J. Proposed modifications to porcine reproductive and respiratory syndrome herd classification. *J Swine Health Prod* 2021, 29(5), 261–270.

2. Kosmidis, I., and Firth, D. Jeffreys-prior penalty, finiteness and shrinkage in binomial-response generalized linear models. Biometrika 2021, 108, 71–82. **https://doi.org/10.1093/BIOMET/ASAA052**

3. R Core Team, R: A Language and Environment for Statistical Computing: R Foundation for Statistical Computing Vienna, Austria, 2019.

Table 1: Probability of sampling at least one PRRSV-positive litter and detecting PRRSV RNA by RT-rtPCR respectively in a 60-crate room, given assumed numbers of positive litters within the room and only 3 PCR tests on any sampling strategy chosen

	Probability of sampling ≥ 1 viremic litter **(Probability of detecting ≥ 1 FOF PCR-positive)**				
Number of litters with positive FOF	3 FOF (individual)	9 FOF (3 pools of 3)	15 FOF (3 pools of 5)	30 FOF (3 pools of 10)	60 FOF (3 pools of 20)
1 of 60	5.0 **(5.0)**	15.0 **(14.7)**	25.0 **(24.10)**	50.0 **(47.0)**	100 **(81.0)**
2 of 60	9.8 **(9.8)**	28.0 **(27.4)**	44.1 **(42.7)**	75.4 **(71.9)**	100 **(93.3)**
3 of 60	14.5 **(14.5)**	39.4 **(38.4)**	58.3 **(56.9)**	88.1 **(85.0)**	100 **(97.12)**
4 of 60	19.0 **(19.0)**	48.8 **(47.9)**	69.5 **(67.7)**	94.4 **(91.8)**	100 **(98.2)**
5 of 60	23.3 **(23.3)**	57.0 **(56.0)**	77.6 **(75.9)**	97.4 **(95.4)**	100 **(98.7)**
10 of 60	42.7 **(42.7)**	83.1 **(82.2)**	95.8 **(94.9)**	100 **(99.4)**	100 **(99.6)**
20 of 60	71.1 **(71.1)**	98.2 **(97.9)**	99.9 **(99.8)**	100 **(100)**	100 **(100)**

https://doi.org/10.54846/am2022/10

Detection of *Mycoplasma hyorhinis* in dams and piglets from birth to weaning age

Cipriano De Abreu[1], DVM; Mark Schwartz[1,2], BS; Maria Pieters[1,3,4], DVM, PhD
[1]Veterinary Population Medicine Department, College of Veterinary Medicine, University of Minnesota, St. Paul, Minnesota; [2]Schwartz Farms, Inc., Sleepy Eye, Minnesota; [3]Veterinary Diagnostic Laboratory, College of Veterinary Medicine, University of Minnesota, St. Paul, Minnesota; [4]Swine Disease Eradication Center, University of Minnesota, St. Paul, Minnesota

Introduction

Mycoplasma hyorhinis (M hyorhinis) is considered a commensal that inhabits the swine upper respiratory tract.[1] However, *M hyorhinis* can induce in some instances polyserositis and polyarthritis in nursery-age piglets through mechanisms of invasion that remain still unknown.[2] *Mycoplasma hyorhinis* can cause disease even in farms where other mycoplasmas have been eradicated.[3] Detection of *M hyorhinis* in dams and piglets prior to weaning has been reported using different sample types.[4,5] Tonsillar swabs represent a sensitive sample to detect *M hyorhinis* genetic material.[5] Nevertheless, knowledge regarding colonization and distribution of this important swine bacterium in modern pig production systems is still lacking. In the present study, we aimed to evaluate the dynamics of detection of *M hyorhinis* in dams and piglets during the lactation period at different time points using various sample types.

Materials and methods

The study was conducted in a commercial 5,000 sow farm. Twenty-four dams in a weekly farrowing group were randomly selected based on parity (1 to 4) and enrolled in the study. Four (4) samples were collected from each dam, including a vaginal swab pre-farrow, and three (3) oropharyngeal swabs: prior to farrowing, at mid lactation and at weaning. Two (2) piglets born to each dam were randomly selected, and ear tagged. Piglets were monitored and sampled during the lactation period. Six samples were collected from each piglet: nasal and tonsillar swabs both collected at birth, mid-lactation and weaning. All samples collected during the study were tested using a real-time PCR (QuantiFast Probe PCR kit; Qiagen Inc., Germantown, MD, USA), and employing custom primers and probe for *M hyorhinis* as previously described.[6] Detection of *M hyorhinis* between sample types was analyzed using a Pearson's Chi-squared test for equality of proportions, with Yates's continuity correction.

Results and discussion

Mycoplasma hyorhinis genetic material was not detected in vaginal swabs of dams. In oropharyngeal swabs 8 % of the dams (2/24) resulted positive prior to farrowing, 0% (0/24) at mid lactation and 4% (1/23) at weaning. *Mycoplasma hyorhinis* was detected in parity one dams, with no older parities being positive for the bacterium. Piglet nasal swabs were positive for 2% of samples at birth (1/48), 4.35% at mid-lactation (2/46) and 44% at weaning (20/45). Tonsillar swabs were negative in all piglets at birth, whereas amplification was detected in 2% of the samples at mid-lactation (1/46) and 53% at weaning (24/45).

Detection of *M hyorhinis* in dams using oropharyngeal swabs was unconsistent overtime. The proportion of *M hyorhinis* positive dams prior to farrowing and at weaning was similar to previous reports.[4,5] Detection of *M hyorhinis* in piglets' nasal and tonsillar swabs increased from mid lactation until weaning, consistent with previous reports.[4-6] Detection of *M hyorhinis* in nasal and tonsillar swabs did not differ statistically ($P > 0.05$). A clear pattern of dam to piglet colonization was not observed in this study. Despite efforts to assess colonization and circulation patterns of *M hyorhinis* in modern swine production systems, knowledge gaps are still present, and need to be adressed in order to characterize the epidemiology of this bacterium, to be able to offer improved prevention and disease control in the field.

References

1. Switzer, W.: 1955, Am J Vet Res. 61, 540-4.
2. Pieters et al: 2019, in *Diseases of Swine*. John Wiley & Sons, Inc., pp. 863–883.
3. Luehrs et al: Vet Microb. 203, 1-5.
4. Clavijo et al: 2017. Vet Rec. 181: 68–68.
5. Roos et al: 2019. PloS one, 14, 1.
6. Clavijo, et al: 2014. J Vet Diagnostic Investig. 26: 755–760.

https://doi.org/10.54846/am2022/11

Characterization of an experimental *Mycoplasma hyopneumoniae* aerosol infection model in pigs

Beatriz Garcia-Morante[1], DVM, PhD; Cipriano De Abreu[1], DVM; Greg Underwood[2], DVM, PhD; Jesús Horacio Lara Puente[3], DVM, MS; Maria Pieters[1,4,5], DVM, PhD
[1]Veterinary Population Medicine Department, College of Veterinary Medicine, University of Minnesota, St. Paul, Minnesota; [2]Bioproperties Pty Ltd., Melbourne, Australia; [3]Laboratorios Avi-Mex S.A., Mexico City, Mexico; [4]Veterinary Diagnostic Laboratory, College of Veterinary Medicine, University of Minnesota, St. Paul, Minnesota; [5]Swine Diseases Eradication Center, College of Veterinary Medicine, University of Minnesota, St. Paul, Minnesota

Introduction

Mycoplasma hyopneumoniae (M hyopneumoniae) is the causative agent of the enzootic pneumonia in pigs,[1] a highly prevalent respiratory disease affecting the swine industry.[2] Animal infection models provide excellent opportunities to study the disease caused by *M hyopneumoniae*, which has been experimentally reproduced by means of different inoculation methods.[3] Aerosol is the least employed method in experimental infections, although it seems to mimic natural infection conditions. The purpose of the present study was to develop and characterize an aerosol model to reproduce the infection and respiratory disease caused by *M hyopneumoniae* in gilts.

Materials and methods

Twelve 6-week-old gilts were obtained from a *M hyopneumoniae* and porcine reproductive and respiratory syndrome virus negative source and were housed at a veterinary isolation facility. Gilts were allocated to four experimental groups (1-4; Table 1) and were individually exposed to *M hyopneumoniae* via infectious aerosols in an isolator. At day 0, groups 1 and 2 received 10mL and 20mL of infectious inoculum with 10^5 CCU/mL, respectively. Gilts in groups 3 and 4 received 10mL and 20mL of 10^6 CCU/mL inoculum, respectively. Starting at 7 days post exposure (DPE) until the end of the study (28 DPE), cough was daily monitored.[4] Blood samples were obtained from all gilts at 0 and 28 DPE. Nasal (NS) and laryngeal (LS) swabs, and deep tracheal catheters (DTC) were collected weekly from each gilt at 0, 7, 14, 21 and 28 DPE. Gilts were humanely euthanized at 28 DPE. At necropsy, lung lesions were scored,[5] and a bronchial swab (BS) and broncho alveolar lavage fluid (BALF) were obtained from each gilt. Blood was used to assess seroconversion by means of an indirect ELISA test (IDEXX Laboratories, Westbrook, ME, USA). Samples from the respiratory tract were tested for *M hyopneumoniae* by qPCR.[6] The local humoral immune response against *M hyopneumoniae* (IgG and IgA) was evaluated in NS, DTC and BALF by modifying the abovementioned ELISA kit and using a commercial IgA conjugate (Bethyl Laboratories, Montgomery, TX, USA). All data was tabulated per gilt, treatment group and time point of examination and descriptive statistics were performed.

Results and discussion

Coughing was observed in groups 2, 3 and 4, with onset at 18, 12 and 10 DPE, respectively. All gilts became infected at a similar time irrespective of the experimental group, as determined by *M hyopneumoniae* detection in DTC from 7 dpe onwards. All BS collected at necropsy resulted positive for detection of *M hyopneumoniae* genetic material and gilts in all groups exhibited macroscopic lung lesions suggestive of mycoplasmal pneumonia at some extent. While seropositive gilts were only observed in group 3, IgA and IgG was detected in NS, DTC, and BALF at 14, 21 and 28 DPE in all experimental groups. IgA values were greater than the IgG values in the assayed sample types, and greater antibody levels were detected in DTC and BALF than in NS. This preliminary study describes kinetics of infection and humoral immune response following an aerosol inoculation with *M hyopneumoniae* in gilts, and further supports the utility of infectious aerosols as an appropriate model of mycoplasmal pneumonia. While only the exposure of gilts to high-dose aerosols consistently reproduced typical clinical signs and severe lung lesions, gilts from all experimental groups became infected at a similar time post-inoculation and developed similar mucosal humoral immune responses regardless the infectious dose at which they were exposed.

References

1. Mare and Switzer: 1965. Vet. Med. Small Anim. Clin. 60, 841–6.
2. Pieters and Maes: 2019. In *Diseases of Swine*. John Wiley & Sons, Inc., pp. 863–883.
3. Garcia-Morante et al: 2016. Vet. Res. 47, 54.
4. Nathues et al.: 2012. Vet. J. 193, 443–447.
5. Pointon et al.: 1999. In *Diseases of Swine*. Iowa State Press, p. 1118.
6. Vilalta et al.: 2020. Vet. Rec. 8, 1045.

Table 1: Experimental groups. Twelve 6-week-old, *M hyopneumoniae*-free gilts were randomly allocated into four different groups (1 – 4) of three gilts each and exposed via aerosol to different concentrations and volumes of inoculum containing the *M hyopneumoniae* strain 232.

Experimental group	n	Titer (CCU/mL)	Total volume (mL)
1	3	10^5	10
2	3	10^5	20
3	3	10^6	10
4	3	10^6	20

CCU= color changing units

https://doi.org/10.54846/am2022/12

Detection dynamics of *Mycoplasma hyopneumoniae* under controlled aerosol exposure for gilt acclimatization

A. M. Betlach[1], DVM, PhD; M. Pieters[2,3], DVM, PHD; P. Yeskel, DVM, MS
[1]Swine Vet Center, St. Peter, Minnesota; [2]Department of Veterinary Population Medicine, College of Veterinary Medicine, University of Minnesota, St. Paul, Minnesota; [3]Swine Disease Eradication Center, College of Veterinary Medicine, University of Minnesota, St. Paul, Minnesota

Introduction

Mycoplasma hyopneumoniae (*M hyopneumoniae*), the causative agent of enzootic pneumonia in swine,[1,2] continues to cause significant economic and health losses in the industry worldwide.[3] Gilt acclimation methods for *M hyopneumoniae* control, such as vaccination,[4] a "seeder" model,[5] or controlled exposure,[6] are employed with the intent to minimize dam-to-piglet transmission and to establish Day 0 for elimination strategies.[7] Controlled exposure of gilts via the aerosolization of *M hyopneumoniae* herd-specific lung homogenate[8] has been attempted at an increasing rate in the industry to induce infection under field conditions.[6] However, limited information regarding the validation of such administration technique to obtain a reliable and uniform exposure is available. Therefore, the objective of this study was to evaluate the spatial and temporal detection of *M hyopneumoniae* under controlled aerosol exposure conditions using various clinical and environmental samples.

Materials and methods

Spatial and temporal detection dynamics of *M hyopneumoniae* were evaluated by conducting a longitudinal prospective study. A total of five rooms (A-E) from three *M hyopneumoniae* negative gilt developing units (GDUs) were selected. The GDU sites were managed in an all-in/all-out fashion and housed a similar population size. For each room, a *M hyopneumoniae* herd-specific lung homogenate was procured based on a diagnostic criterion[8] and was aerosolized from a single location within each room using a hurricane fogger. At 0, 7 and 14-days post-exposure (dpe), environmental and air particle deposition (APD) samples were strategically collected at different locations (up to 9.8 meters) from the origin of exposure (ie, hurricane fogger). Tracheal secretions were collected from randomly selected four- to seven-week-old gilts (n = 30/room) at 0, 7, 14, and 28 dpe. Sampled gilts (n = 5/pen) were housed in pens throughout the room to assess *M. hyopneumoniae* detection dynamics. Samples were stored at -20°C and later processed individually for DNA extraction and tested for *M hyopneumoniae* using a real-time PCR. Samples with a ≤ 39 Ct value were considered *M hyopneumoniae* positive. Data was descriptively analyzed.

Results

Lung homogenate samples were *M hyopneumoniae* positive with Ct values ranging from 27.2-30.3. Prior to exposure, all samples were *M hyopneumoniae* negative by PCR. At 0 dpe, *M hyopneumoniae* was detected in environmental and ADP samples up to 5.8 meters from the hurricane fogger, in which the relative bacterial load numerically decreased as distance increased. For three rooms (A, B, and E), gilts remained *M hyopneumoniae* negative until approx. 14 dpe, in which 10% of pigs housed in neighboring pens located within 4.6 meters of the exposure origin became positive. In comparison, *M hyopneumoniae* was detected in 20 and 26% of gilts in rooms C and D, respectively at 7 dpe. At 28 dpe, 20-57% of gilts became *M hyopneumoniae* positive in rooms C-E, in which a high proportion (80-100%) of infected gilts per pen was observed in pens located within 5.8 meters from the exposure origin.

Conclusions

Overall, *M hyopneumoniae* detection dynamics appeared to be spatially associated, however, inconsistent results were observed across the five rooms. It is important to mention that *M hyopneumoniae* strain differences among the GDU sites could not be ruled out as a contributing variable. Additional research is warranted to confirm findings and to provide further insight towards the optimization of gilt acclimation procedures via aerosol exposure.

References

1. Mare CJ & Switzer WP.: 1965, Vet Med Small Anim Clin. 60, 841-846.
2. Goodwin RF et al.: 1965, Vet Rec 77, 1247-1249.
3. Pieters MG & Maes D.: 2019, Mycoplasmosis. Diseases of Swine. John Wiley & Sons, 863-883.
4. Garza-Moreno L et al.: 2018, Vet Microbiol. 219, 23-29.
5. Roos L et al.: 2016, Vet Microbiol. 184, 51-58.
6. Toohill E.: 2017, Proceedings from Allen D. Leman Conference. St. Paul, MN.
7. Pieters M & Fano E.: 2016, Vet Rec. 178(5), 122-123.
8. Robbins RC et al.: 2016, J Swine Health Prod. 27(4), 221-227.

https://doi.org/10.54846/am2022/13

Diagnostic performance of a commercial *Mycoplasma hyopneumoniae* serum antibody ELISA using processing fluids samples from three commercial swine farms

R. Magtoto[1], BS, PhD; B. Armenta-Leyva[1], DVM; P. Dizon-Magtoto[2], DVM, MS; T-Y. Cheng[1], DVM, PhD; M. J. Clavijo[1], DVM, PhD; C. Johnson[3], DVM; W. Lopez[1], DVM, PhD; D. Baum[1], DVM, PhD; J. J. Zimmerman[1], DVM PhD; L. G. Giménez-Lirola[1], BS, MS, PhD
[1]College of Veterinary Medicine, Iowa State University, Ames, Iowa; [2]College of Veterinary Medicine, Pampanga State Agricultural University, Magalang, Pampanga, Philippines; [3]Carthage Veterinary Service, Ltd, Carthage, Illinois

Introduction

Mycoplasma hyopneumoniae (MHP) serum-based antibody surveillance in the breeding herd is constrained by the labor required for collecting blood samples from individual sows. Processing fluids (PF), the tissue exudates recovered from testicles and tails collected during piglet processing, contain detectable antibody against MHP and present a potential alternative to MHP serum-based antibody surveillance in breeding herds.[1] Therefore, the purpose of this study was to evaluate the diagnostic performance of a commercial MHP serum antibody ELISA using PF samples from three commercial swine farms.

Materials and methods

Processing fluid samples were collected from three commercial farms. Based on intervention program and/or historical monitoring, one farm was classified as MHP-endemic farm (n = 246) and two farms were classified as MHP-free farms (n = 248). Processing fluid samples were tested at a 1:10 dilution using a commercial MHP ELISA designed to detect anti-P46 antibodies in serum (M hyo Ab test, IDEXX Laboratories, Inc.). Results were analyzed by receiver operating characteristic (ROC) analysis and diagnostic sensitivities and specificities were estimated over a range of ELISA sample-to-positive (S/P) cutoffs.

Results and discussion

A frequency distribution of MHP ELISA S/P responses by farm status is shown in Figure 1. At a cutoff of S/P ≥ 0.40, diagnostic sensitivity and diagnostic specificity were estimated by ROC analysis as 97.6% (95% CI: 95.5, 99.2) and 100.0 % (95% CI: 100, 100), respectively. That is, all samples (n = 248) from MHP-free herds produced S/P values of < 0.3, while 3 of 246 samples from the MHP-endemic farm had S/P values of < 0.4. Similarly, the first report by Boettcher et al[1] described an "excellent" agreement in processing fluids and sow serum. Overall, the findings of this study suggest that PF could be used as an alternative specimen for routine MHP antibody-based breeding herd surveillance.

Reference

1. Boettcher J, et al. Proceedings of the 21st IPVS Congress. International Pig Veterinary Society. 2010.

Figure 1: Frequency distribution of *Mycoplasma hyopneumoniae* (MHP) antibody ELISA* S/P responses by farm MHP status

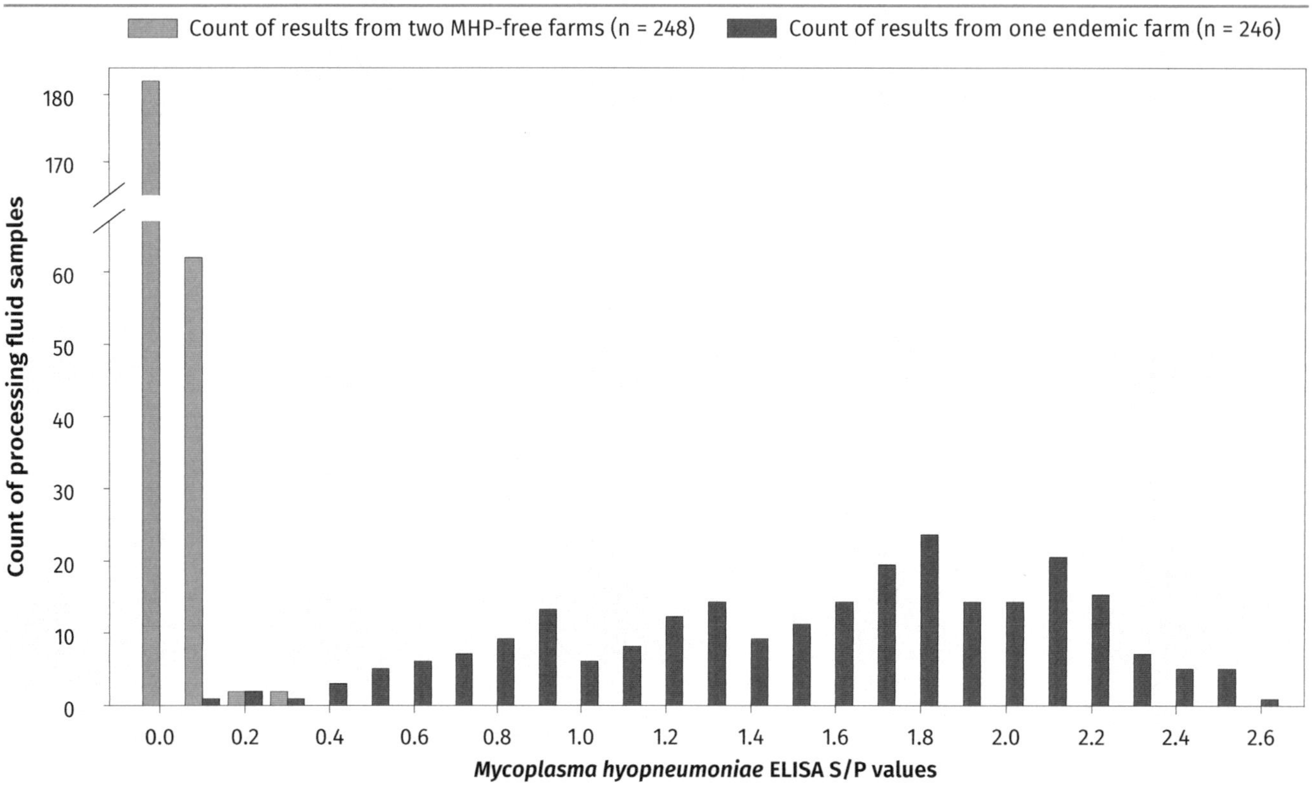

* M hyo Ab test (IDEXX Laboratories Inc., Westbrook, ME, USA)

https://doi.org/10.54846/am2022/14

Prevention and control of *Streptococcus equi* subspecies *zooepidemicus* infection in pigs

Samantha J. Hau; Alexandra Buckley; Susan L. Brockmeier
National Animal Disease Center, USDA, Ames, Iowa

Introduction

Streptococcus equi subspecies *zooepidemicus* (SEZ) is a zoonotic bacterium causing infections in a wide range of mammalian species. SEZ has historically caused outbreaks of severe disease in swine in China and Indonesia; however, it was an infrequent cause of infection in pigs in North America.[1,2] In 2019, high mortality events in swine due to SEZ were reported in the US and Canada.[3,4] Pigs developed high fever, severe lethargy, and reluctance to rise, and mortality rates of 30-50% were observed in affected animals.[3,4] Because SEZ has not been a frequent cause of disease in North American pigs, there are no approved vaccines or intervention strategies to prevent the losses caused by SEZ. To begin evaluating intervention strategies to limit the losses following SEZ introduction into a naïve group of pigs, we evaluated three different SEZ control strategies: pre-exposure to an avirulent SEZ isolate (study 1), vaccination with an autogenous vaccine containing inactivated bacteria (study 2), and the use of long-acting ceftiofur after the development of clinical signs (study 3).

Methods

Study 1

Five sows were inoculated with 10^9 CFU of a non-virulent SEZ isolate from a horse and after 21 days were challenged with a 2019 mortality event isolate. Serum antibody titers were assessed prior to inoculation with the non-virulent strain and at the time of challenge. Following challenge, pigs were monitored for the development of clinical signs and euthanized if disease became severe. At necropsy, gross pathology was assessed, and samples were collected to evaluate systemic distribution of SEZ.

Study 2

Forty-seven, six-week-old pigs were vaccinated twice with one of the following: an autogenous vaccine generated from a highly mucoid, 2019 high mortality event isolate (n = 15); an autogenous vaccine generated from a genetically similar, non-mucoid SEZ isolate from a guinea pig (n = 16); or sham vaccinated with adjuvant only (n = 16). Autogenous vaccines contained 10^9 bacteria and EmulsigenD was utilized as the adjuvant. Pigs were challenged two weeks following boost vaccination with a 2019 mortality event isolate. Serum antibody titers were measured prior to vaccination, at boost vaccination, and prior to challenge. Animals were monitored and sampled as described for Study 1.

Study 3

Thirty-two, 10-week-old pigs were challenged with a 2019 mortality event isolate. Following the development of clinical signs, 16 pigs were treated with ceftiofur crystalline free acid (Excede, Zoetis) and the other 16 were not treated. Animals were monitored and retreated as needed. Antibody titers were evaluated prior to challenge and 30 days post-challenge. Following recovery, two non-antibiotic-treated pigs (one vaccinated, one sham vaccinated from study 2) were commingled with three naïve contact pigs and, in addition, two antibiotic-treated pigs (from study 3) were commingled with three naïve contact pigs 30 days post-challenge to evaluate SEZ transmission from previously exposed, healthy pigs.

Results

Study 1

Sows exposed to the non-virulent horse SEZ isolate had a small, non-significant rise in titer to SEZ after 21 days. The prior exposure to SEZ resulted in a delayed onset of clinical signs and prolonged survival after exposure to the mortality event isolate; however, there was no difference in survival between animals pre-exposed to SEZ and animals without prior SEZ exposure.

Study 2

Both groups of pigs vaccinated with autogenous SEZ vaccines had a significant rise in mean SEZ titer compared with the sham vaccinated group. A rise in titer was observed after primary vaccination; however, no anamnestic response was noted after boost vaccination. Challenge resulted in severe disease and > 90% mortality in all groups. No statistical differences were observed in disease onset, median survival time, or mortality rates between either bacterin vaccine group and the sham vaccine group.

Study 3

All challenged animals developed clinical signs by 24 hours post-challenge. Animals in the treatment group were given a weight appropriate dose of ceftiofur. Most (15/16) treated animals returned to normal behavior and normal body temperature by 24 hours post-treatment. One pig developed neurologic signs and was euthanized due to worsening condition. At 5 days post-challenge (4 days post-treatment), surviving pigs began exhibiting clinical signs (fever, lethargy) and all pigs were retreated with ceftiofur. Retreatment resolved clinical signs; however, 10-11 days post-challenge (5-6 days post-retreatment) 7/15 pigs required a third treatment. After 11 days post-challenge, animals remained healthy and no further treatments were required. Serum antibody titers evaluated at challenge and 30 days post-challenge revealed a significant rise in titer in treated animals. Commingling of two surviving untreated pigs with naïve contacts resulted in severe disease in one contact pig and evidence of transmission in a second contact pig (rise in antibody titer). In contrast, commingling of two antibiotic-treated pigs with naïve contacts did not result in clinical disease and no evidence of transmission to contact animals (no rise in titer).

Conclusions

This study evaluated intervention strategies for SEZ disease in pigs. Study 1 and 2 evaluated two different vaccine platforms, an avirulent live vaccine and an autogenous inactivated vaccine. These vaccine strategies did not provide protection against morbidity or mortality with SEZ. Serum antibody titers at challenge in both studies may not have been sufficient to provide protection against challenge. For the avirulent live vaccine (study 1), this suggests multiple inoculations may be necessary to generate protective immunity or maintain colonization for competitive inhibition. For the inactivated autogenous vaccine (study 2), the vaccine may have contained insufficient antigen or a different adjuvant may be needed to adequately stimulate the immune system. While the vaccines were ineffective at preventing mortality with SEZ, antibiotic treatment improved clinical disease and reduced mortality following SEZ challenge (study 3). Clinical signs recurred following initial treatment and 2-3 treatments (providing 12-18 days of coverage) were necessary to control clinical disease. It is important to note that animals in the acute phase of disease do not eat or drink and initial antibiotic treatment should be provided parenterally. Finally, we found that animals surviving SEZ exposure can continue to carry SEZ at least four weeks post-exposure and can transmit to naïve contacts while appearing clinically healthy; however, antibiotic treatment of exposed animals may reduce colonization levels and help prevent transmission to naïve contact animals.

References

1. Salasia SI et al. Persistent occurrence of a single *Streptococcus equi* subsp. *zooepidemicus* clone in the pig and monkey population in Indonesia. *J Vet Sci*. 2004;5:263-265.

2. ISUVDL. Case series of high porcine mortality associated with *Streptococcus equi* ssp *zooepidemicus* in assembly yards. 2019. Available at: **https://www.swinehealth.org/wp-content/uploads/2019/10/Streptococcus-equi-ssp-zooepidemicus-as-cause-of-high-mortality-of-swine-in-assembly-yards-10-8-2019.pdf.** Accessed 23 September 2021.

3. Sitthicharoenchai P et al. Cases of high mortality in cull sows and feeder pigs associated with *Streptococcus equi* subsp. *zooepidemicus* septicemia. *J Vet Diagn Invest*. 2020;32(4):565-571.

4. Costa MO, Lage B. *Streptococcus equi* subspecies *zooepidemicus* and sudden deaths in swine, Canada. *Emerg Infect Dis*. 2020; 26(10):2522-2524.

https://doi.org/10.54846/am2022/15

Senecavirus A in the environment of sow slaughter plants

A. Buckley[1], DVM, PhD; K. Hoffman[2], PhD; K. Faaberg[1], PhD; N. Humphry[3], DVM; J. Korslund[3], DVM; K. Lager[1], DVM, PhD
[1]National Animal Disease Center, Agriculture Research Service, US Department of Agriculture, Ames, Iowa; [2]Oak Ridge Institute for Science and Education and National Animal Disease Center, US Department of Agriculture, Ames, Iowa; [3]Veterinary Services, Animal Plant Health Inspection Service, US Department of Agriculture

Introduction

Senecavirus A (SVA) infection in swine can lead to the development of vesicular lesions that are clinically indistinguishable from foot-and-mouth disease (FMD). FMD is not currently in the US; therefore, a foreign animal disease investigation (FADI) must be performed every time a vesicular lesion is observed in swine to rule out FMD. With the increase of SVA cases in the US since 2015, there has been a corresponding increase in FADIs. A large portion of these investigations have been centered at sow slaughter plants, with some plants requiring investigations weekly. These investigations are both a time and economic burden and may result in complacency in reporting vesicular lesions in swine. The objectives of this study were to understand the environmental burden of SVA in slaughter plant lairage, correlate PCR, virus isolation (VI), and swine bioassay results, and compare 2020 SVA sequences to previous SVA isolates.

Materials and methods

Four sow slaughter plants participated in the study: Plants 1-3 were experiencing a high incidence of FADIs due to reports of vesicular lesions and Plant 4 reported no FADIs. Swab samples were collected from the environment including flooring, waterers, gating, and trailers as well as swabs samples from the coronary bands of pigs that did not have vesicular lesions, on a regular basis between June and December of 2020. Swab samples were tested for SVA nucleic acid by PCR and live virus by VI on swine testicular cells over three passages. PCR and VI results were used to select samples to test via swine bioassay to correlate PCR results to VI results and infectivity in swine. Eighteen samples were used to inoculate individually housed pigs (2 pigs/sample). Blood and swab samples collected from animals were tested for SVA by PCR and a serum virus neutralization (VN) assay was performed to determine neutralizing antibody response. Finally, two isolates were chosen from plants 1-3 (n = 6 total) for next-generation sequencing. A phylogenetic analysis was performed with all other whole-genome SVA sequences available in GenBank.

Results and discussion

Plants 1-3, high FADI incidence plants, consistently had SVA PCR positive samples collected from the environment throughout the sampling period (n = 450), whereas Plant 4, no reported FADIs, was negative for SVA by PCR in all tested samples (n = 69). PCR data from Plants 1-3 was combined and more than 50% of samples collected throughout the sampling period were PCR positive. When data was separated by sample type, samples collected from the floor of lairage were the most common positive sample type and had lower C_t values compared to the other environmental samples (Figure 1A). One explanation could be SVA shedding in swine fecal material or other bodily secretions from infected sows after arriving in holding pens in the slaughter plant. This could also explain why samples at the pig level like the gating and trailer swabs were less likely to be PCR positive. Data from Plants 1-3 was also analyzed by month and showed that August had the highest percentage of positive samples and the summer months tended to have more positive samples than the fall (Figure 1C). This data supports previous data from diagnostic labs that found increased SVA cases in the summer months. As expected, samples with lower C_t values were more likely to be VI positive with 67.6% of samples with a C_t value below 30 VI positive and only 18.5% of samples with a C_t value between 30-35 that were VI positive (Figure 1B). The discrepancy between PCR positive samples and VI samples may be the result of low levels of virus, time virus has been exposed to the elements, and application of disinfectants.

Eighteen samples were used to inoculate pigs for a swine bioassay and had C_t values ranging from 24 to 33 (Table 1). Although half of the samples were VI positive, only one sample with the lowest C_t value (24.2) was able to infect piglets and produce a neutralizing antibody response. The minimum infectious dose of SVA is unknown and this study suggests that cell culture is more sensitive than swine bioassay.

Five of the six 2020 SVA sequences in this study clustered together. However, one isolate, clustered with contemporary US SVA isolates from 2017. Percent nucleotide identity of the six SVA isolates ranged from 96.5% to 99.9%, but removing the dissimilar isolate, the remaining isolates had a 99.0% nucleotide identity. Bayesian inference analysis estimates that the common ancestor of the five similar 2020 isolates was in late 2016, while the other sequence and select 2017 SVA sequences had a different common ancestor on a separate clade arising in mid-2013. The 2020 SVA isolates appear to have a limited number of non-synonymous amino acid changes from SVA isolates found during 2015 SVA US outbreaks, suggesting that exposure to previous isolate should provide cross-protection against newer strains.

Conclusion

This study demonstrated that SVA is circulating in the cull sow market chain and can be found in the environment of slaughter plants. A better understanding of the epidemiology of SVA in the marketing chain may help reduce disruptions related to FADIs and direct control measures to reduce the spread of SVA. An endemic vesicular disease in the US swine population is dangerous for the industry and could lead to complacency when it comes to FADIs and preventing FMD from entering the US swine herd.

Figure 1: PCR and VI results from Plants 1-3. **A)** Percentage of SVA positive samples by sample type, **B)** number of SVA PCR positive and VI positive samples, and **C)** percentage of SVA positive samples by month.

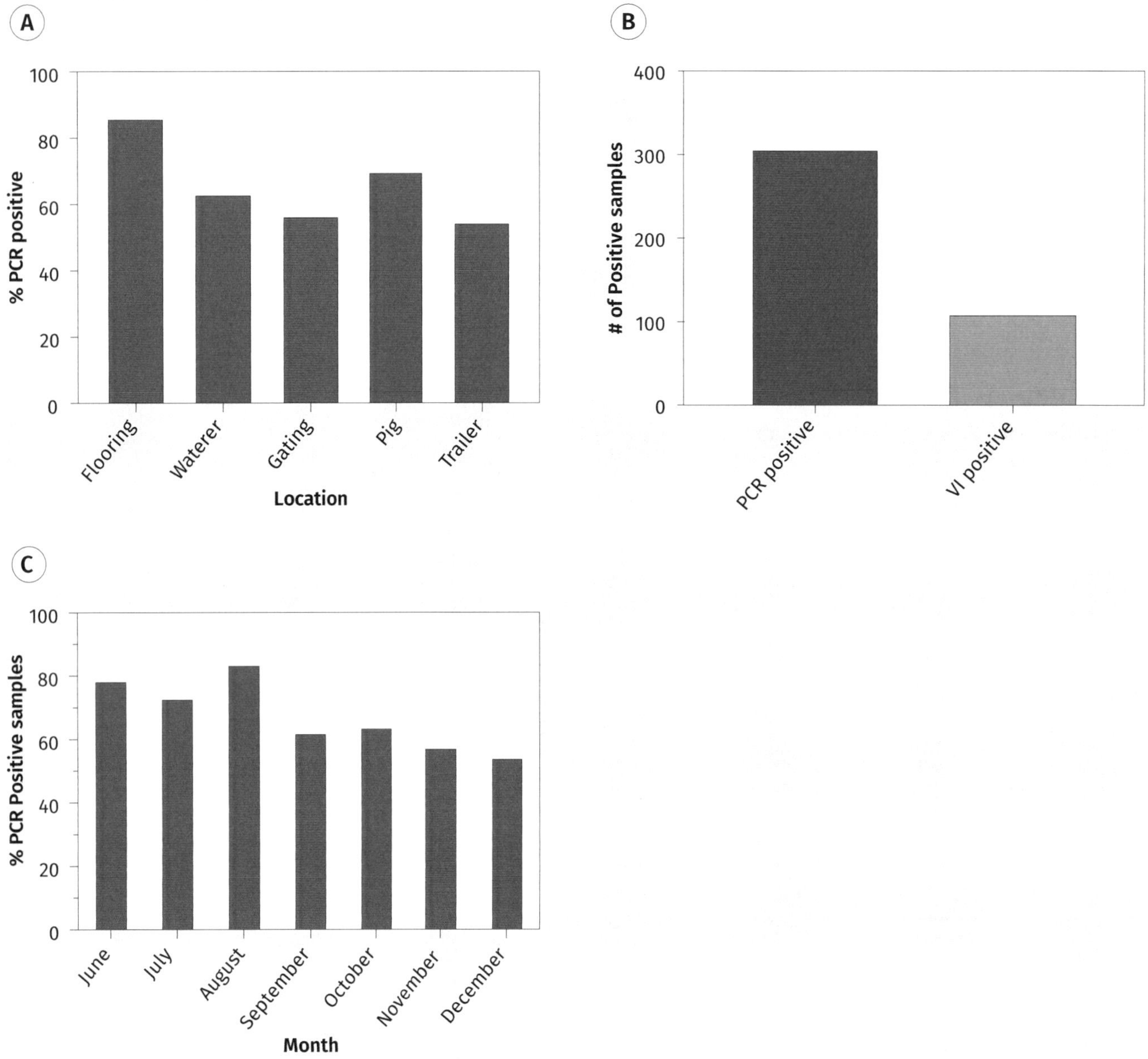

Table 1: PCR Ct value, genomic copies (GC) per mL, virus isolation (VI), and swine bioassay results from select environmental samples from slaughter plants

Slaughter plant #	Date	Sample type	qRT-PCR C_t value	GC/mL	VI Result	Bioassay
Plant 3	09/08/20	Waterer	24.18	1.14E + 06	+	2/2
Plant 1	10/29/20	Flooring	24.29	1.06E + 06	+	0/2
Plant 3	09/14/20	Flooring	25.99	3.49E + 05	+	0/2
Plant 3	07/27/20	Pig	26.72	1.84E + 05	+	0/2
Plant 1	06/24/20	Flooring	26.85	1.98E + 05	+	0/2
Plant 1	09/09/20	Flooring	26.95	1.85E + 05	+	0/2
Plant 2	08/28/20	Pig	27.23	1.54E + 05	-	0/2
Plant 3	07/27/20	Waterer	28.03	7.26E + 04	+	0/2
Plant 2	07/22/20	Gating	29.55	1.02E + 05	+	0/2
Plant 1	07/21/20	Gating	29.75	2.16E + 04	-	0/2
Plant 1	09/09/20	Trailer	30.83	4.00E + 04	-	0/2
Plant 3	08/31/20	Flooring	31.07	8.62E + 03	+	0/2
Plant 1	06/30/20	Flooring	31.13	8.11E + 03	-	0/2
Plant 1	06/24/20	Flooring	31.20	7.74E + 03	-	0/2
Plant 2	07/22/20	Waterer	32.01	4.39E + 03	-	0/2
Plant 3	08/10/20	Gating	32.08	4.33E + 03	+	0/2
Plant 3	08/10/20	Waterer	32.50	3.13E + 03	+	0/2
Plant 2	07/14/20	Flooring	33.14	2.09E + 03	-	0/2

https://doi.org/10.54846/am2022/16

First assessment of the time-to-negative processing fluids in breeding herds after a Senecavirus A outbreak

G. Preis[1]; F. Vannucci[1,2]; A. Rovira[1,2;] C. A. Corzo[1]
[1]Veterinary Population Medicine, University of Minnesota, St. Paul, Minnesota;
[2]Diagnostic Laboratory, University of Minnesota, St. Paul, Minnesota

Introduction

Senecavirus A (SVA) has been responsible for a rampant increase in foreign animal disease investigations in the United States. Besides vesicles in adult animals, neonatal mortality and diarrhea have been associated with SVA infection in sow herds in different countries. However, the pathogenesis and epidemiology of this disease are still unclear. Currently, swine practitioners have virtually no information about the presence and dynamics of SVA in processing fluids in sow farms during an outbreak, and very little is known about its transmission within the farrowing room. Additionally, the role of heat-check boars on the persistence of the virus in the breeding herd after an outbreak is also unknown. The objectives of this AASV-Foundation funded study were to: 1) estimate the time to SVA-negativity in processing fluids after an outbreak; 2) assess the role of heat-check boars in the persistence and transmission of SVA within a farm; and 3) to estimate the production losses associated with an SVA outbreak.

Materials and methods

This study was designed as an observational study, where a cohort of 10 sow farms was conveniently selected and invited to participate in the study. A total of 310 processing fluids from 10 different sow farms were collected weekly or bi-weekly and tested by SVA RT-qPCR. Processing fluids samples were tested at different weeks before and after detecting an SVA outbreak in each farm. Only one farm agreed to collect semen samples from heat-check boars. Participating companies shipped the samples to the University of Minnesota (UMN) Veterinary Diagnostic Laboratory (VDL) for testing.

Results and discussion

The follow-up time in all farms ranged from 16 to 30 weeks, with an average follow-up of 22.5 weeks. Some farms were SVA-positive by PF testing even before clinical signs of SVA were detected, with the earliest detection of the virus in PF up to three weeks before the outbreak. Sow farms had at least one positive PF for an average of 11.8 weeks (Std Dev = 5.2) after the outbreak. Most farms had a varying number of consecutive negative weeks between positive results, ranging from 1 to 10 weeks. Weekly PF SVA statuses for all tested sow farms are shown in Figure 1. Heat check boars can also be a source of SVA infection to naïve gilts and sows. In the first semen collection at week 7 after the outbreak in one farm, four and three boars (out of nine collected boars) had PCR positive and suspect samples, respectively. In a second semen collection at week 18 after the outbreak, one out of sixteen collected boars had a PCR-suspect result. This suspect boar from week 18 was euthanized and necropsied at week 22. Surprisingly, both the tonsil and testicle samples were positive for SVA after testing, with Ct values of 30.5 and 17.8, respectively. These incredibly low Ct values, especially from the testicular sample, indicate that heat-check boars might act as carriers of SVA after an outbreak in a sow farm. At the time of writing, only one farm had provided weekly production data, and pre-weaning mortality (PWM) was the only production indicator that suffered a significant change. PWM quickly increased to 18.1%, 23%, and 42.7% in the first three weeks after the outbreak, and then quickly returned to being close to the farm's average of 14.3%. In general, caution should be taken if weaning SVA negative piglets is desired as the number of consecutive negative weeks needed to consider the sow farm as stable requires more investigation. Understanding the within-herd epidemiology of this disease can lead the industry to implement aggressive programs for elimination at a system level. This project shows novel information about SVA infection dynamics, providing the first blocks for the construction of said elimination programs.

Figure 1: Weekly status to Senecavirus A (SVA) by processing fluid samples in 10 sow farms before and after SVA outbreak detection.

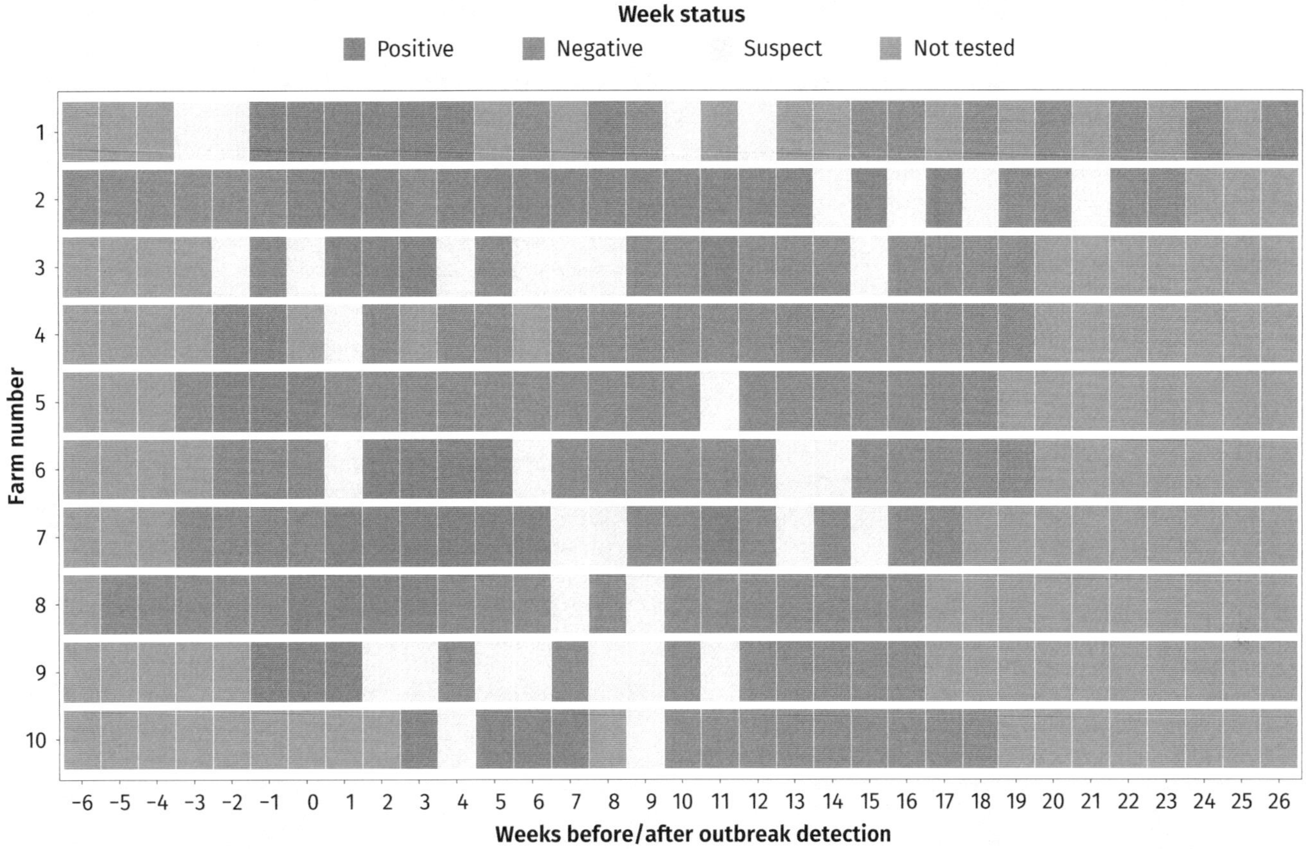

https://doi.org/10.54846/am2022/17

Microbial analysis of joints in sows euthanized for lameness

S. Haryslak; M. Pierdon, VMD
New Bolton Center, University of Pennsylvania School of Veterinary Medicine

Introduction

Lameness is a serious issue in swine and is one of the main reasons for euthanasia of sows, resulting in significant production and economic losses.[1] Although the underlying cause of lameness may vary, previous work has demonstrated a link between clinical lameness and septic purulent joint infections.[2] Through better understanding of the septic joint microbiome and prevalent pathogenic bacteria, targeted treatment and prevention measures can be further identified, and the sources of these infections could be indicated. The objective of this study was to observe microbial composition of infected joints causing lameness through aerobic and anaerobic culture methods.

Materials and methods

Lameness was identified by herd veterinarians or barn staff and sows were euthanized on farm. All 4 limbs were obtained through removal of the scapula, and disarticulation of the coxofemoral joint. Joints were grossly examined for swelling and lesions prior to disarticulation and joint spaces were observed for abnormalities in fluid consistency, articular surfaces, and underlying bone. Swabs of the articular surface and synovial membrane were aseptically collected from the affected joint and the corresponding contralateral unaffected joint to be used as the control. Samples were cultured aerobically on 5% Sheep Blood Agar, Columbia-CNA agar, and MacConkey agar and incubated at 37° C for 2 days. Samples were also cultured anaerobically on Anaerobic CNA agar, Brucella Blood agar, Laked Brucella Blood agar with Kanamycin and Vancomycin, and Bacteriodes Bile Esculin agar and incubated with an AnaeroPack™ at 37° C for 7 days. Colonies were isolated and species identified using Matrix-Assisted Laser Desorption/Ionization-Time of Flight (MALDI-TOF) mass spectrometry. Colonies were determined to be normal flora or environmental based on known microbial ecology. Colonies unidentifiable on MALDI-TOF were Gram Stained and observed microscopically. Unidentifiable colonies were analyzed by Sanger sequencing using Purelink™ Genomic DNA Mini Kit and PCR. All isolated colonies were saved on Microbank™ beads at -80° C for later analysis. Statistics were performed using Stata-IC16. Risk was measured using logistic regression including whether or not a joint was infected as the outcome and the presence or absence of each bacterial type as the fixed effects. *P*-values of less than 0.05 were treated as significant and the joint was the statistical unit.

Results

Limbs from 15 sows on 5 farms were obtained and cultures performed on 8 control and 16 affected joints. Aerobic microorganisms were cultured from 15 affected and 5 control joints. Cultures were determined to have moderate to heavy growth of mixed bacterial colonies upon initial observation. 48 aerobic colonies were identified from 24 samples, 33 of which were considered part of the sow normal flora (33/48) and 15 were considered to be of environmental origin (15/48). *Trueperella pyogenes* was isolated from 11 affected (11/16) and 2 control joints (2/8) and *Escherichia coli* isolated from 5 affected (5/16) and 3 control joints (3/8). Anaerobic microorganisms were cultured from 2 control and 7 affected joints. Cultures that had growth were determined to have heavy growth of mix bacterial colonies. 15 anaerobic colonies were identified from 24 samples, 14 of which were normal flora (14/15) and 1 environmental (1/15). *Prevotella heparinolytica* was isolated from 3 affected (3/16) and 1 control joint (1/8) and *Fuscobacterium necrophorum* was isolated from 5 affected (5/16) and 0 control joints. Affected limbs had a greater risk for normal flora contamination (94%) compared to control limbs (50%) (OR = 15.0, CI: 1.3 - 174.4), ($P < 0.05$). Affected limbs had a lower risk of infection with environmental bacteria (18%) (OR = 0.14, CI: 0.02 - 0.93), ($P < 0.05$) compared to controls (63%). There was no difference in risk for aerobes between infected (93%) and control joints (63%) (OR = 9.0, CI: 0.75 - 107.4) ($P = 0.08$).

Discussion

The lack of identification of known systemic bacterial causes of lameness, such as *Mycoplasma, Glaesserella, Erysipelothrix,* and *Streptococcus* species,[3] and the greater likelihood of normal flora identification in affected joints could indicate that opportunistic normal flora originating from the swine skin, mouth or gut microbiome, such as *T pyogenes, E coli,* and *F necrophorum* are more significant pathogenic causes of septic arthritis. The prevalence of bacteria cultured from control limbs shows a systemic component to joint infection, further indicating opportunistic normal flora may be the causative agent of most septic arthritis cases. Further investigation into identification of subclinical infection elsewhere in the body of sows euthanized for lameness, continued cataloging of the sow joint microbiome, and prospective studies to investigate joint flora prior to pathological lameness, are potential next steps for this project.

Acknowledgments

A special thanks is extended to the Pennsylvania Pork Producer's Council for their support and funding.

References

1. Nalon E et al.: 2013, Livestock Science 156 p. 10-23.
2. Engblom L et al.: 2008, Acta Veterinaria Scandinavica 50.
3. Neilsen EO et al.: 2001, Transboundary and Emerging Diseases 48 p. 475-486.

https://doi.org/10.54846/am2022/18

Comparing sample types and testing methods for endemic influenza in growing gilts

A. Sisk[1], MS; M. Billing[2], DVM; B. Ideus[3], DVM; S. Storms[4], DVM; J. Lowe[4], DVM, MS, DABVP
[1]North Carolina State University College of Veterinary Medicine, Raleigh, North Carolina; [2]Boehringer Ingelheim Animal Health USA Inc., Duluth, Georgia; [3]Hord Family Farms, Bucyrus, Ohio; [4]Applied Epidemiology Laboratory, College of Veterinary Medicine, University of Illinois, Urbana, Illinois

Statement of problem

Influenza A viruses (IAVs) circulate endemically in swine populations contributing to decreased performance and novel variants with serious zoonotic potential. Undetected subclinical infections result in inter-herd transmission and sporadic outbreaks restricting animal movements, disrupting herd continuity, and incurring economic losses.[1] Timely and accessible on-site detection methods are needed for predicting outbreaks and preventing disruptions in gilts movements. Reverse transcription loop-mediated isothermal amplification (RT-LAMP) offers potential over RT-qPCR as a pen-side diagnostic due to its simplicity and rapidity with similar sensitivity and specificity.[2] In addition, collecting IAV from nasal wipes is a proven method that is less intrusive and labor-intensive than nasal swabs.[3]

Case history

The nucleus of a 30,000-sow high-health multiplication system tested positive for IAV by RT-PCR on oral fluids during week 42 of 2020. Finisher gilts broke with the identical strain in week 19 of 2021. Prior to this outbreak, growing gilts (8 to 15 weeks of age) sporadically tested positive for IAV despite negative sow farms. Current IAV testing protocols include RT-PCR on oral fluids before movements to sow farms and when clinical signs are present. Gilts receive two doses of an autogenous IAV vaccine at five months of age.

Diagnostics and data

A total of 678 samples (390 nasal swabs, 210 nasal wipes, 78 oral fluids) were collected from clinically healthy gilts at 6 to 22 weeks of age between weeks 23 and 30 of 2021 from seven wean-finish barns. A random-number generator determined sample pens. Within a sample pen, 30 out of 60 pigs were randomly chosen and a nasal swab and nasal wipe was obtained from each. Oral fluids were collected from the entire pen. Additional ropes were placed on the dividers of spatially distributed pens. Nasal wipes were individually immersed in 3mL brain heart infusion media with added penicillin and streptomycin and frozen prior to shipment. Samples were pooled into groups of five, extracted, and tested by RT-qPCR at the Applied Livestock Epidemiology Lab at the University of Illinois Urbana Champaign. Positive pooled samples were individually tested by RT-qPCR. Individual RT-qPCR positive samples were further tested by RT-LAMP. Nucleic acid extraction, RT-qPCR, and RT-LAMP were performed using the commercial Sera-Xtracta Virus/Pathogen Kit (Cytiva, Marlborough, MA), Superscript™ III Platinum One-Step qRT-PCR Kit (Invitrogen™, Carlsbad, CA), and Warmstart® LAMP Kit (New England BioLabs Inc., Ipswich, MA), respectively. All oral fluid samples were tested by RT-PCR at the Animal Disease Diagnostic Laboratory within the Ohio Department of Agriculture.

Significant and applicable findings

By RT-qPCR, IAV was detected in 12.8% of nasal swab samples and 3.3% of nasal wipe samples. By RT-LAMP, IAV was detected in 6% of nasal swab samples but was not detected in nasal wipe samples. All oral fluids were negative. Full data comparison between sample type and assay are located in Table 1. The average RT-qPCR cycle threshold value was 28.8 for nasal swabs and 29.4 for nasal wipes, indicating low levels of virus. All positive samples came from 6 and 7 week old gilts from barns 1 and 5, identifying two sow farms as potential sources. Because the onset is unclear, it is possible that piglets infected prior to weaning or shortly after movement to wean-finish facilities may be perpetuating IAVs in this endemic herd.

Lessons learned

Frequent testing of preweaned piglets on sow farms and sampling prior to 6 weeks of age within wean-finish facilities may establish the origin and onset of the IAV challenge. Oral fluids were not adequately sensitive to detect IAV in this endemically infected population. Based on our RT-qPCR results, nasal wipes are more sensitive than oral fluids and an easy technique to implement. Because RT-LAMP was performed on individual positive RT-qPCR samples and yielded only 3 positive results, sampling endemically infected animals shedding low amounts of virus may not be a practical application of the RT-LAMP assay. While IAV-targeted RT-LAMP has potential to produce results faster than RT-qPCR, more research on the efficacy of nasal wipe collection methods and direct sample testing is needed prior to on-farm application. The advancement of RT-LAMP technology as a point-of-care pathogen detection system could prove efficacious for herd health monitoring and rewarding for the swine industry.

References

1. Bakre, AA et al. Detection of swine influenza virus in nasal specimens by reverse transcription-loop-mediated isothermal amplification (RT-LAMP). J Virol Methods. February, 2021. Accessed September 5, 2021. **https://doi.org/10.1016/j.jviromet.2020.114015**

2. Moehling, Taylor J et al. LAMP Diagnostics at the Point-of-Care: Emerging Trends and Perspectives for the Developer Community. Expert review of molecular diagnostics. January 1, 2021. Accessed September 7, 2021. **doi:10.1080/14737159.2021.1873769**

3. Nolting, J. M. et al. Nasal Wipes for Influenza A Virus Detection and Isolation from Swine. Journal of visualized experiments. December 4, 2015. Accessed September 7, 2021. **https://doi.org/10.3791/53313**

Table 1: IAV positive results by assay and sample type. Pooled nasal samples represent five individuals. Pooled oral fluids represent two pens, individual oral fluids represent one pen, each pen consists of 60 individuals.

	RT-qPCR		RT-LAMP	Total individual samples collected
	Pooled	Individual	Individual	
Nasal swab	12/78	50/60	3/50	390
Nasal wipe	2/42	7/10	0/7	210
Oral fluids	0/71	0/7	NA*	78

IAV = Influenza A virus

* oral fluids not evaluated by RT-LAMP

https://doi.org/10.54846/am2022/19

Comparison of whitewash vs. Synergize® on production metrics in farrowing rooms

Kaci Way[1], BS; Kayla Henness[2], DVM; Melissa Hensch[2], DVM; Chris Puls[3], PhD; Robert Evelsizer[3], DVM; Amy Maschhoff[2], DVM
[1]The Ohio State University, Columbus, Ohio; [2]The Maschhoffs, LLC, Carlyle, Illinois; [3]Elanco Animal Health, Greenfield, Indiana

Introduction

Many pathogens harbored in farrowing rooms can lead to enteric infections. According to the 2012 National Animal Health Monitoring System, the average preweaning mortality (PWM) rate in the US swine industry is 9.7%, and of these deaths, 10.2% are due to scours.[1] Current industry sanitation practices are variable in pathogen inactivation which has resulted in whitewashing (WW) as an additional form of sanitation. WW is a combination of hydrated lime and water that serves as a final sanitation step by covering all surfaces of the farrowing crate. Impacts of WW on production metrics have not previously been measured, although, one study suggested that WW was effective for use as a visual affirmation of crate coverage and its basicity may have been effective as a form of disinfection.[2] The objective of this study was to investigate WW vs. Synergize® Disinfectant (SY; Preserve International, Lexington, KY) in farrowing rooms on production metrics including PWM and average daily gain (ADG).

Materials and methods

Three commercial sow farms were included in this study. Two of the herds were porcine reproductive and respiratory syndrome virus (PRRSV) vaccinated and one herd was PRRSV positive. Piglets at all farms received an antibiotic at birth and piglets at one farm also received a coccidiostat treatment at processing per system standards. Treatments were applied to the room, and data were collected for 25 WW rooms and for 27 SY rooms. Rooms were cleaned of organic material, a degreaser was added, followed by a rinse step, and SY or WW was applied. SY was applied at 1:256 dilution and WW solution was created by adding one 50-lb bag of hydrated lime ($Ca[OH]_2$) to a 44-gallon barrel. The lime was slaked by filling the barrel with water, setting the mixture for 8-12 hours, and refilling the barrel with water. Scour prevalence (yes/no) was recorded by room at days 3, 10, and at weaning. Rectal swabs (RS) and environmental fecal samples (ES) were collected throughout the study as pathogen surveillance. At weaning, feces were collected directly from piglets and submitted for detection of *Cystoisospora suis* via fecal float. At one site, 110 WW and 166 SY litters were weighed at birth and weaning to calculate ADG. Total PWM was recorded for each room.

Results and discussion

Diagnostic sampling confirmed presence of Rotavirus (A, B or C), *Clostridium perfringens*, *Escherichia coli*, and *Salmonella enterica*. Refer to Table 1 for a comparison of SY and WW pathogen surveillance. *Cystoisospora suis* was detected only for SY. This suggests that WW's alkaline properties may be effective at inactivating coccidia. Compared to SY, scour prevalence was reduced 10.3 ($P = 0.02$), 3.5 ($P = 0.13$), and 4.5 ($P = 0.07$) percentage units for WW on day 3, 10, and at weaning, respectively (Table 2). Although scour prevalence was reduced with WW, there was no difference ($P > 0.05$) in PWM or ADG when grouped across all farms in the study. However, when evaluated by individual farm, there was a 2.6% reduction in PWM for the PRRSV positive farm for the WW treatment. This observation suggests there may be a greater impact of WW when used in sites that face greater disease challenges. In conclusion, this study suggests better enteric outcomes for WW compared to SY, although PWM and ADG differences were not statistically different. Recommendations drawn from this study include using WW on farms with greater scour prevalence and those that face greater epidemic disease challenges. Significant scour prevalence reduction could precedent a future study to monitor piglet performance through market. This would determine if pigs are less susceptible in contracting future enteric illness due to less compromised neonatal gut health. There are very few studies that evaluate the efficacy of WW as a form of disinfection, and the results of this study support future research in this area.

References

1. The US Department of Agriculture, Swine 2012 Part I: Baseline Reference of Swine Health and Management in the United States, 201245–48 (2015). USDA–APHIS–VS–CEAH–NAHMS.

2. Korty, R., DeDecker, A., Boyer, P., Battrell, M., Robbins, R. (2014). A cost effective alternative to depopulation for swine dysentery. 2014 AASV Annual Meeting: Our Oath in Practice, 181–183.

Table 1: Whitewashing (WW) vs. Synergize (SY) pathogen surveillance

Pathogen	SY	WW
Rotavirus A	+	+
Rotavirus B	+	-
Rotavirus C	+	+
Clostridium perfringens	+	+
Escherichia coli	+	+
Salmonella enterica	-	+
Cystoisospora suis	+	-

(+) = pathogen present; (-) = pathogen not detected

Table 2: Whitewashing (WW) vs. Synergize (SY) pooled analysis (all farms)

LSMeans for effects of cleaning treatment

	Treatment			
Item	**SY**	**WW**	**SEM**	***P*-value**
No. rooms	27	25	-	-
Scour prevalence, % litters				
Day 3 post-farrowing	37.2	26.9	8.03	**0.02**
Day 10 post-farrowing	12.7	9.2	2.69	0.13
Weaning	10.9	6.4	2.62	0.07
Pre-weaning mortality, %	11.7	12.3	0.78	0.57
Piglet growth performance				
Piglet birth weight, lbs	3.6	3.3	0.15	0.22
Piglet weaning weight, lbs	9.6	9.4	0.58	0.90
Piglet ADG, lbs/d	0.44	0.44	0.017	0.93

https://doi.org/10.54846/am2022/20

A survey to investigate the implementation methods of rotavirus ice cubes on swine farms

S. Melson[1], BS; E. Kettelkamp[2], DVM; A. M. Betlach[2], DVM, PhD
[1]University of Minnesota College of Veterinary Medicine, St. Paul, Minnesota; [2]Swine Vet Center, St. Peter, Minnesota

Introduction

The swine industry continues to investigate best practices to control pig neonatal diarrhea. Porcine rotaviruses groups A, B, and C are some of the etiologic agents of neonatal diarrhea, which is primarily observed during farrowing and early nursery.[1] Natural planned exposure of rotaviruses to gestating dams using rotavirus ice cubes is a practice implemented in swine barns aimed to help reduce the occurrence of neonatal diarrhea.[2] Rotavirus ice cubes are made by harvesting piglet intestines, blending the intestines with a diluent to make a homogenate, and later freezing. The intestinal homogenate is then thawed and administered to gestating dams.[2]

Research on the effect of different rotavirus ice cube procurement and implementation methods on swine farms have been limited. Rotavirus associated disease has been described to increase mortality by 3 to 20% and decrease weaning weight by 0.5 to 1.0 pounds.[3] As rotavirus continues to be a major cause of economic losses attributed to neonatal diarrhea, advancing the current rotavirus control method efficacy is critical.[3] Therefore, the objective of this survey was to summarize current procedures in the industry for the preparation, storage, and use of rotavirus ice cubes.

Materials and methods

A web-based anonymous survey using a targeted sampling approach was sent to nineteen swine veterinary clinics, production systems, and veterinarians, representing over two million sows during July 2021. The survey was composed of multiple-choice questions (n = 19) focused on different preparation, storage, and administration procedures for rotavirus ice cubes. Data was compiled and summarized using descriptive statistics. Relationships between different preparation, storage, and administration procedures on the respondent's overall perception of the procedures' effectiveness for neonatal diarrhea control was assessed by conducting a Chi-Square test. Results with a P-value ≤ 0.05 were considered significant.

Results

A total of 36 questionnaires were voluntarily received. Of the 36 responses, 69% and 31% did and did not use rotavirus ice cubes for neonatal diarrhea control, respectively. Of the respondents that utilized rotavirus cubes, 40%, 52%, and 8% thought the procedure was very effective, somewhat effective, or neutral, respectively. Most commonly, 86% of respondents used on farm specific rotavirus ice cubes, while 14% of respondents used a system wide stock or both.

Regarding donor selection for material preparation, 74% of respondents harvested intestinal material from colostrum-deprived neonatal pigs, whereas 16% chose 1-5 day-old scouring pigs and 10% selected 6-10 day-old scouring pigs. Procedure effectiveness for neonatal diarrhea control was not associated with age at donor selection ($P = 0.72$). Various diluents were utilized with intestinal material to create a homogenate, namely, phosphate buffered saline (PBS; 31%), milk replacer (27%), sterile or well water (30%). Procedure effectiveness for neonatal diarrhea control was associated with diluent type ($P = 0.02$). Estimated ratios of intestines to diluent for composition of intestinal homogenate were: 1) mixed < 10% intestines (11%), 2) 10% intestine to 90% diluent (31%), 3) 25% intestine to 75% diluent (11%), 4) 50% intestine to 50% diluent (9%), and 5) 75% intestine to 25% diluent (37%). Estimated sows per cube when administering the homogenate were 11-20 (6%), 21-30, (19%), 31-50 (6%), 51-60 (25%), and 60 or more (44%). Several methods for rotavirus cube storage were described: top of a fridge freezer (62%), chest freezer (27%), or a -80 Celsius freezer (11%). Moreover, different testing protocols were mentioned, in which samples that tested positive for PRRS, PED, TGE, and Delta Coronavirus were commonly discarded (33%). Figure 1 depicts timing of administration of the first and last dose of intestinal homogenate pre-parturition. The most common week to administer the first dose of homogenate was six weeks (38%), followed by the second dose at three weeks (48%) pre-parturition.

Discussion

The aim of this study was to summarize current procedures in the industry for the preparation, storage, and administration of rotavirus ice cubes for neonatal diarrhea control. Overall, there was a considerable amount of variation in the way rotavirus ice cubes are prepared and administered on swine farms. The differences amongst homogenate preparation may contribute to variable levels of rotavirus control throughout the industry. This survey was limited by the sample size, which potentially contributed to some

Figure 1: Timing pre-parturition of the first and last dose of rotavirus ice cube homogenate by percent of respondents.

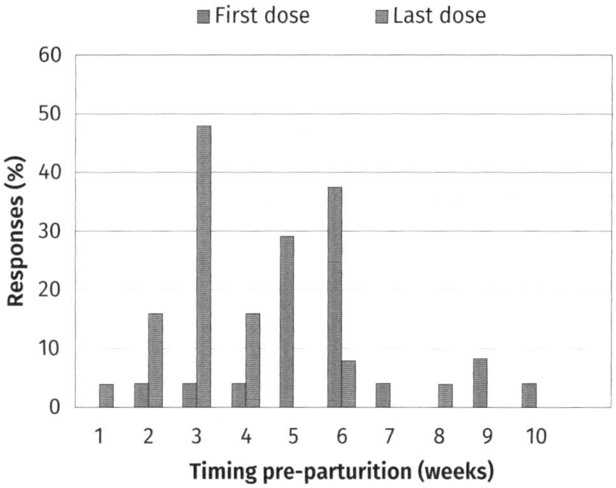

insignificant results. As this questionnaire was voluntary, there may have been inherent biases in the responses since all perceptions of effectiveness were neutral or better.

This survey highlighted the application of seven different diluents, which had a potential effect on the perceived effectiveness of the rotavirus ice cube on controlling neonatal diarrhea. The ratio of intestinal material to diluent when creating the intestinal homogenate also varied widely. Most commonly, a 60 dams per ice cube dose was used when administering the rotavirus ice cubes, leading to a relatively diluted dose given to each dam. A majority of respondents had a on farm specific stock of rotavirus ice cubes that were stored in a top of refrigerator freezer. Most also used colostrum deprived intestines as intestine donors to make a homogenate. Further research on the effect of diluent type, along with other preparation, storage, and usage methods for rotavirus ice cubes are critical to assess the efficacy of this control measure for neonatal diarrhea.

References

1. Vlasova AN, Amimo JO, Saif LJ. 2017. Porcine Rotaviruses: Epidemiology, Immune Responses and Control Strategies. Viruses 9(3):48.

2. Pittman, Jeremy; Anderson, Amanda; Boyd, Will; Madigan, Jamie; Wolfe, Zoe. 2019. Update on Rotavirus Control from the Field **https://www.aasv.org/library/swineinfo/item.php?ISU/2019/24-Pittman-Update_on_Rotavirus_Control_from_the_Field.pdf**

3. Groth D. 2014. Clinical management of rotavirus. American Association of Swine Veterinarians. Dallas, Texas pp. 561-562.

https://doi.org/10.54846/am2022/21

Evaluating the probability of PRRSV-positive oral fluids in due-to-wean piglet litters

Nathan VanKley[1], BS; Claire LeFevre[3], DVM; Onyekachukwu Osemeke[2], DVM, MSc;
John Waddell[4], DVM; Daniel Linhares[2], DVM, MBA, PhD
[1]Michigan State University, College of Veterinary Medicine, Lansing, Michigan; [2]Iowa State University, College of Veterinary Medicine, Ames, Iowa; [3]Carthage Veterinary Service, Ltd., Carthage, Illinois;
[4]Boehringer Ingelheim Animal Health USA Inc., Duluth, Georgia

Introduction

Porcine respiratory and reproductive syndrome virus (PRRSV) surveillance is a valuable strategy in understanding the health status of breed to wean herds. A study in 2012 found that the US swine industry loses over $1 billion annually due to PRRSV.[4] This significant loss to the industry is due to factors such as increased abortions, mummies, and low weaning weights, making surveillance a necessity in management of PRRSV. Implementing an effective PRRSV surveillance program is paramount for swine veterinarians to determine the status of the herd and assess the degree of viral shedding. Being able to understand the health status of the herd enables producers and veterinarians to determine the most efficient management program to optimize swine health and productivity.

The gold standard for surveillance is through serum sampling. The AASV PRRS Task Force stated in 2011 that commercial breeding herds can be classified as "positive stable" after receiving 4 negative real-time polymerase chain reaction (RT-PCR) tests over a 90-day period in due-to-wean piglet sera. For a 95% confidence, at least 30 samples of sera must be collected for each test. However, a longitudinal study was done that found PRRSV to persist in breeding herds with a PRRSV prevalence less than 10%, even though sera surveillance classified the herd as "positive stable".[5] An observational study also found 0.4% of sows part of an elimination program were PRRSV positive 15 months post-elimination.[3] Since PRRSV is known to persist at a low prevalence in breeding herds, there is a need to evolve surveillance for PRRSV in commercial herds. A modified approach to PRRSV surveillance using serum presents itself with challenges. To detect PRRSV in low prevalence herds, the number of piglet sera samples needed to maintain a 95% confidence would be nearer to the population size. Large number of samples needed may pose fiscal constraints for commercial producers, along with time and labor resources needed for such a massive surveillance program.[6] Sera collection in wean age piglets also requires a skilled individual trained to collect the sera in a low stress manner. Availability of trained sample collectors can be a limitation to gathering the sera samples. These concerns yield opportunity to explore alternative methods in PRRSV surveillance.

Processing fluids-based (PF) monitoring of PRRSV in sow farms are practical and effective. However, PF may not reflect the true status of piglets at weaning.[1] Thus, alternative surveillance methods, such as family oral fluids (FOF) and oral swabs (OS) can be implemented in due-to-wean piglets as less costly tools for surveillance of PRRSV. FOF are efficacious in PRRSV detection in farrowing rooms with moderate prevalence.[1] OS have not been used notably in surveillance of PRRSV, though a study has shown they were able to detect PRRSV antigens and antibodies at different age groups in pigs.[6] The objective of this field study was to evaluate the probability of detecting PRRSV RNA by real-time PCR in family oral fluids and oral swabs of due-to-wean litters in a low prevalence scenario.

Materials and methods

Study design

This field study was conducted on due-to-wean piglets (18-21 days of age) in a 5300-head commercial breed-to-wean farm. Litters were randomly assigned via Excel using a random number generator, in which sets of 5 to 8 litters were collected in a single room each day. Sows ranged from parity 1 to parity 8 in the allotment process. Nurse sows were not included in the allotment. Sets of FOF, OS, and piglet sera were collected from 51 total litters.

Sample collection and testing

FOF samples were collected by first tying cotton ropes to the front of each farrowing crate for interaction by sows and their respective litters, and then wringing oral fluids into individual tubes. The rope remained in the farrowing crates for at least 30 minutes to an hour. OS samples were collected using a sterile non-synthetic swab, which was rotated 360 degrees in the buccal region of the mouth for each individual piglet in a litter. Jugular serum was then collected from each piglet in a litter, and samples of OS and sera were matched with the FOF of the assigned litter, bagged together as a set, and submitted for RT-PCR testing at a veterinary diagnostic laboratory. In summary, for each study litter the following were collected: one FOF, OS and serum were obtained individually from all piglets in the litter.

Data analysis

Cohen's Kappa analysis was used to assess agreement in the PCR results between sample types. When at least one piglet tested positive by OS or serum, that litter was considered positive by that sample type. In determining agreements, values of 0 were ranked as no agreement, 0.01-0.20 none to slight, 0.21-0.40 fair, 0.41-0.60 moderate, 0.61-0.80 substantial, and 0.81-1.00 almost perfect agreement.[2] A linear regression model was used to assess the relationship between within-litter prevalence (WLP) by OS and WLP by serum. For this model, OS WLP was used as the response variable while sera WLP and "mean cut-off threshold (Ct) of positive sera samples per litter" were used as predictor variables. For the linear regression plot, the "mean Ct of positive sera samples per litter" was highlighted and categorized into three groups: Ct below 25, Ct between 25 and 30, and Ct above 30. Logistic regression models were used to assess changes in odds of PRRSV detection by each alternative sample type with changes in WLP of PRRSV by serum.

Diagnostic testing

All samples were submitted for PRRSV RNA testing using RT-PCR to an accredited VDL, using standard procedures to detect all known PRRSV-1 and PRRSV-2 strains. For differentiating samples that were positive or negative for PRRSV, the Ct cut-off value of 37 was used.

Results

Of the 51 litters that sample sets were collected from, the litter size ranged from 7 to 15 piglets per litter, with an average of 12.22 piglets per litter. 24 litters had at least 1 viremic pig and were considered truly positive. Sera was considered the gold standard in the 2 × 2 analysis of sample types. None of the alternative sample types produced a false positive result, hence a calculated specificity of 100% for each of these samples. In comparing OS to serum (Table 1), 15 true positives, 27 true negatives, 9 false negatives were observed. The sensitivity of OS samples was calculated to be 62.5%. Cohen's kappa analysis gave a value of 0.638, indicating substantial agreement between OS and serum samples. Of the 15 litters that had at least 1 pig having a PRRSV positive OS, only 6 (40%) of them had OS within litter prevalence above 10%. The mean Ct of positive OS per litter ranged from 28.2-36.9.

In comparing FOF to serum (Table 2), there were 5 true positives, 27 true negatives, and 19 false negatives. The sensitivity using FOF samples was 20.83%. Cohen's kappa analysis showed an agreement of 0.218, indicating a fair agreement between these sampling methods. OS and FOF were also compared to each other using the 2 × 2 analysis (Table 3). Cohen's kappa analysis revealed an agreement of 0.414, indicating a fair agreement between these sampling methods. When comparing pooled OS to serum, there were 5 true positives, 27 true negatives, and 19 false negatives. The sensitivity of pooled OS sampling was calculated to be 20.83%. There was a significant linear relationship between the within litter prevalence by OS and within litter prevalence by sera ($P < 0.0001$) with a coefficient of determination (R-squared) value of 73.8%. The R-squared value of 73.8% indicates that the predictor variables in the model accounted for 73.8% of the variation observed in the response variable.

The partial regression coefficient for within litter prevalence by sera in the fitted model was statistically significant ($P < 0.0001$) and had an estimated value of 0.49, indicating that, for every 1% increase in mean within litter prevalence by serum, there was a 0.49% increase in mean within litter prevalence by OS, when the mean Ct was kept constant. The partial regression coefficient for the mean Ct categories was also statistically significant ($P = 0.011$) and had an estimated value of -0.013, indicating that for every unit increase in mean Ct of positive sera per litter, there was a 0.013% decrease in mean within litter prevalence by OS, when within litter prevalence by sera was kept constant (Figure 1).

For the logistic regression models, the odds of having a positive OS in at least one piglet within a litter increased by 17% for every 1% increase in within litter prevalence by sera, while the odds of having a positive FOF from a sampled litter increases by 5% for every 1% increase in within litter prevalence by sera (Figure 2).

Table 1: Sera vs oral swabs

		Oral swabs		
		Neg	Pos	
Sera	Neg	27	0	27
	Pos	9	15	24
		36	15	

Table 2: Sera vs family oral fluids (FOF)

		FOF		
		Neg	Pos	
Sera	Neg	27	0	27
	Pos	19	5	24
		46	5	

Table 3: Oral swabs vs family oral fluids (FOF)

		FOF		
		Neg	Pos	
Oral swabs	Neg	36	0	36
	Pos	10	5	15
		46	5	

Discussion

When evaluating the ability of each sampling method to detect true positives, taking individual OS samples from every piglet within the litter yielded a 62.5% sensitivity (table 1). Cohen's kappa analysis was applied to evaluate the agreement between sampling methods over chance and showed substantial agreement between sampling with OS and sera. Though this sampling method did not have a perfect agreement with sera, this substantial agreement establishes that OS can be used to detect PRRSV RNA by RT-PCR with comfortable reliability. Evaluating FOF in comparison to sera showed much different results, FOF sampling had a sensitivity of 20.83% and a Cohen's kappa value indicating a fair agreement with sera sampling (Table 2). FOF sampling does not guarantee equal interactions from all piglets within a litter with the cotton rope provided. In some cases, only the dam may interact with the provided rope; resulting in the relatively poor sensitivity when compared to testing sera of every pig within a litter.

Pooled OS sample had a sensitivity of 20.83%, and a Cohen's kappa value that showed fair agreement with sera sampling. We expected a better sensitivity than observed for this sample type. For some litters, there was a time lag (and at least 1 freeze-thaw cycle) from when the individual OS samples were first tested to when these samples were pooled and tested. This may have accounted in part for the poor performance observed.

The regression model assessing the linear relationship between within litter prevalence by OS and two predictors, which compared within litter prevalence by sera and mean Ct of positive sera samples per litter, had an extremely small *P*-value. This indicates strong evidence of a statistically significant linear relationship between the predictor values and within litter prevalence by OS. Ct values using PCR act as good indicators of the amount of virus within a sample. The lower the Ct value, the higher the

Figure 1: Oral swab comparison to within-litter prevalence by sera

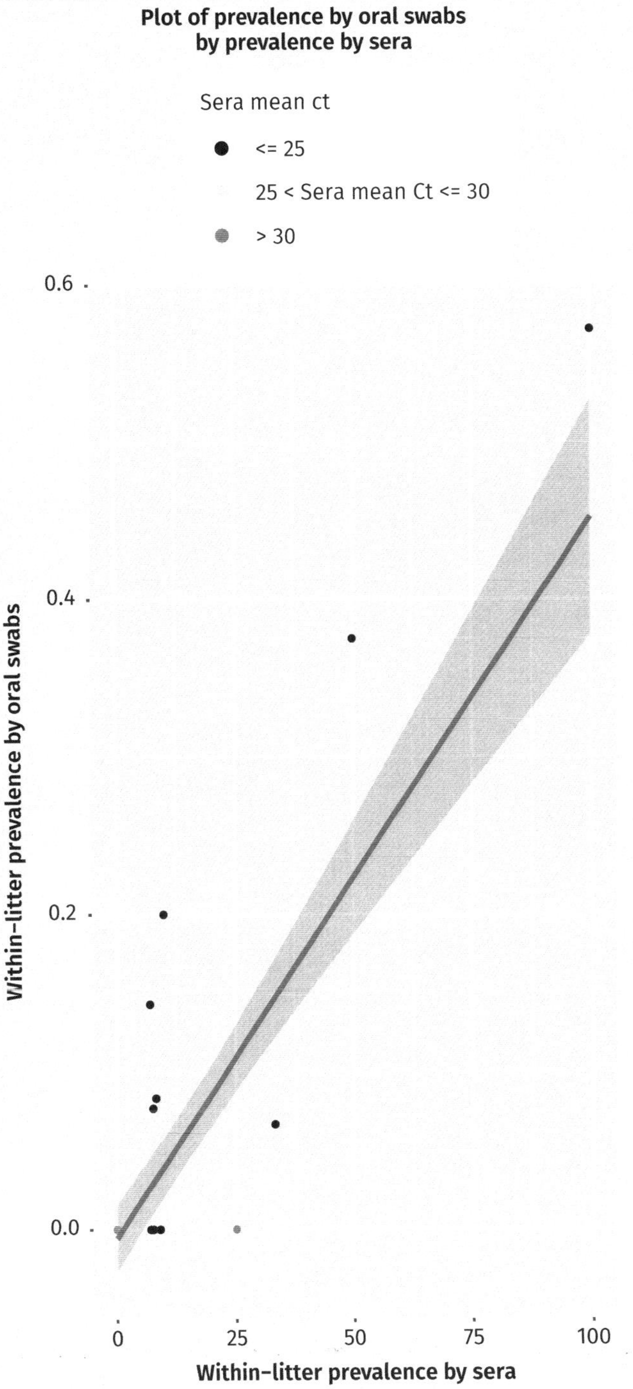

number of viral RNA within the given sample. As expected, we observed from the model estimates and the linear regression plots that the lower the mean Ct of positive sera per litter, the higher the chances of a relatively high within litter prevalence by OS. We observed a 0% within litter prevalence by OS in a litter where the mean Ct of positive sera had a value above 30, this highlights the possibility of false negative OS results when the viral load in piglets is low.

Improvements to this study may have been made by collecting more samples over a longer time frame to increase the statistical power of the study. The low prevalence scenario presented difficulty when it came to finding litters with at least one PRRSV positive piglet in the litter to allow comparisons between the different sample types to be made. A farm with a slightly higher PRRSV prevalence may afford this opportunity in the future.

A current limitation with OS is the lack of formal standardization for collecting the OS sample in regards to depth and surface of the buccal region to that of the swab. The goal of the study was to swab the caudal-most portion of the hard palate and tongue, though not as far caudal to risk damage to the epiglottis and laryngeal tissues. Without a clearly defined process for sample collection, there is currently potential for variation in the regions sampled between individual sample collectors. Compared to serum collection, the OS sampling seemed to incur less stress during collection as the piglet did not need to be inverted or repositioned.

This study has shown that OS samples may be used as a surveillance method to detect PRRSV within due-to-wean litters. FOF were also capable of detecting PRRSV but lacked in their ability to detect within litter prevalence as well as OS samples did. Pooled OS were shown to be capable of detecting PRRSV like that of FOF, but still required individual piglet samples. OS samples were found to have a substantial agreement with serum samples, and a significant positive linear relationship in PRRSV detection within litter prevalence to that of the gold-standard. FOF also had its advantages, as many ropes could be placed across a room at the same time, allowing sampling of multiple litters at once. Where OS samples are more capable of projecting the true prevalence of PRRSV within a litter, FOF samples were able to be sample larger populations in less time, though they were not as capable of projecting the true within litter prevalence of PRRSV as shown in the study.

For producers, the use of OS may provide an alternative method of surveillance for PRRSV in due-to-wean litters. FOF could also be a useful diagnostic tool in PRRSV surveillance, though supplementing with OS could yield results more similar to that of the true within litter prevalence. For litters with low PRRSV prevalence, there needs to be a larger sample size to adequately assess PRRSV prevalence. The use of OS and FOF together could increase this sample size, and more accurately assess true within-litter PRRSV prevalence. These surveillance methods would also require fewer labor resources and decrease stress of piglets during sample collection. This study may also have the potential to be expanded to a larger scale in surveillance of PRRSV prevalence in the farrowing house of breed-to-wean commercial farms.

Conclusion

There is a statistically significant linear relationship in the probability of detecting PRRSV by real-time PCR using oral swabs to that of serum. The odds of having a positive litter by OS or by FOF increases by 17% and 5% respectively for every 1% increase in within-litter prevalence. Though there were varying degrees of false negative results using alternative sample types, the specificity was consistently 100% as there were no false positives. This finding suggests that when obtaining PCR-positive results from OS or FOF samples, there is a great chance that there is a viremic piglet in that litter. The alternative sample types here provide practical and cost-efficient alternatives to sera sampling for routine PRRSV surveillance in due-to-wean litters.

Figure 2: Project probability of each sample type. FOF = family oral fluids; OS = oral fluids.

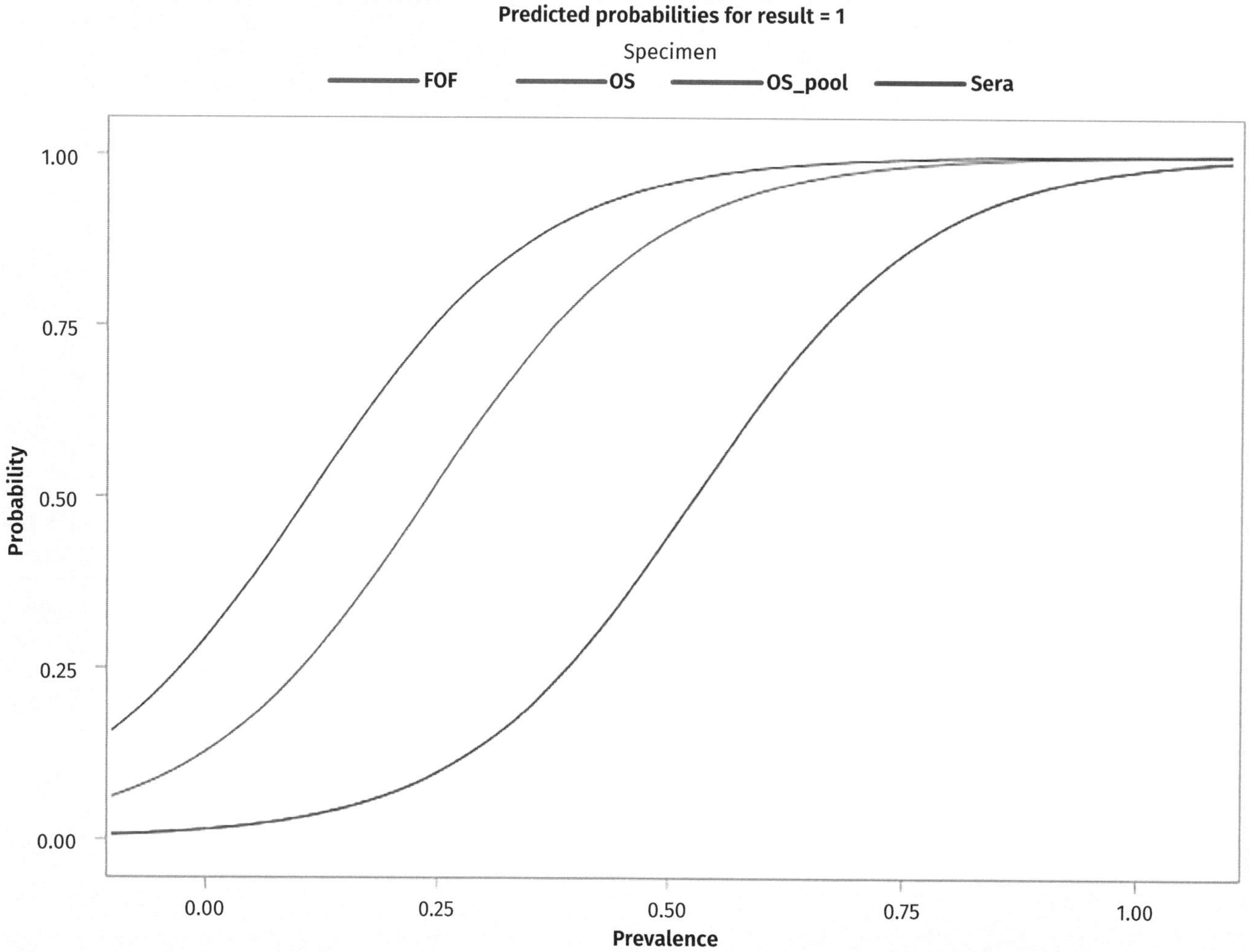

References

1. Almeida MN, et al: 2021, Finding PRRSV in sow herds: Family oral fluids vs serum samples from due-to-wean pigs. Prev Vet Med. 193:1-6.

2. Cohen, Jacob: 1960. A coefficient of agreement for nominal scales. Educational and Psychological Measurement. 20 (1):37-46.

3. Dee SA, et al: 2000, Epidemiological and diagnostic observations following the elimination of porcine reproductive and respiratory syndrome virus from a breeding herd of pigs by the test and removal protocol. Vet Record. 146:211-213.

4. Holtkamp DJ, et al: 2011, Terminology for classifying swine herds by porcine reproductive and respiratory syndrome virus status. J Swine Health Prod. 19:44-56.

5. Linhares DCL, et al: 2014, Comparison of time to PRRSV-stability and production losses between two exposure programs to control PRRSV in sow herds. Prev Vet Med. 116:111-119.

6. Prickett J, et al: 2008, Detection of porcine reproductive and respiratory syndrome virus infection in porcine oral fluid samples: a longitudinal study under experimental conditions. J Vet Diagn Invest. 20: 156-163.

https://doi.org/10.54846/am2022/22

Evaluation of water-based foaming as a mass depopulation method for swine

T. J. Williams[1]; T.-Y. Cheng[1], PhD; M. Campler[1], PhD; S. Moeller[2], PhD, MS; J. Kieffer[2], DVM; A. S. Bowman[1], DVM, PhD, MS, DACVPM; A. G. Arruda[1], DVM, PhD, MS
[1]Department of Veterinary Preventive Medicine, College of Veterinary Medicine, The Ohio State University, Columbus, Ohio; [2] Department of Animal Sciences, College of Food, Agricultural, and Environmental Sciences, The Ohio State University, Columbus, Ohio

Introduction

Swine populations are susceptible to infectious diseases, and those instances urgent responses including large-scale depopulation may be required to control and prevent farm-to-farm transmission during outbreaks. The American Veterinary Medical Association (AVMA) recommends a number of methods for swine depopulation,[1] but their efficiency for large populations is debatable. In this study, water-based foam (WBF), an AVMA-approved method in poultry, was adapted to swine. The objectives of the study were to describe "time to animal unconsciousness" and "time to cessation of animal movement" using WBF; and to provide a recommendation on the duration of WBF foaming under field conditions.

Materials and methods

Two trials (Trial 1 and 2) were conducted for a stepwise evaluation on the use of WBF in swine. Trial 1 assessed time to unconsciousness by immersing 72 feeder pigs (6 pigs for 12 replicates) in WBF for 6 time periods (2.5, 5, 7.5, 10, 12.5, and 15 minutes post-fill (MPF); 2 replicates per time period). The unconsciousness of foamed pigs was assessed via different levels of vital signs (Figure 1)[2], and the time period for which no "Level 2" were present was selected for Trial 2. In Trial 2, 135 cull sows (3 replicates of 45 sows) were loaded into an adapted rendering dump trailer (40 × 8.5 × 6 in) and immersed in WBF for the period suggested in Trial 1. Activity bio-loggers (Star-Oddi©) were implanted in 6 sows in each replicate to measure animal movements. The last movement was determined by the last external acceleration measurement beyond the third quartiles + 1.5 times the interquartile range. For both trials, pigs with any signs of consciousness were immediately euthanized by captive bolt.

Results and discussion

In Trial 1, kicking, respiratory rhythm, and convulsions were absent while regular gasping and corneal reflex were observed respectively in 3 and 12 pigs at 2.5 minutes MPF. Consciousness signs > "Level 1" were absent at ≥ 5 MPF; thus, sows were immersed for 5 MPF in Trial 2. In Trial 2, the mean durations between end of trailer WBF fill and last animal movement (min:sec) as per bio-loggers were 2:40 (SD = 1:31), 2:26 (SD = 1:11), and 1:58 (SD = 0:47) for replicates 1, 2, and 3 respectively. All sows were deemed unconscious after unloaded from the trailer.

Based on our studies, WBF could be an attractive method for emergency depopulation in swine. Given our assessment of time to unconsciousness in feeder pigs and our field validation in sows, we recommend that 7.5 minutes of submersion should be used to ensure unconsciousness and subsequent death.

References

1. AVMA. Guidelines for the Depopulation of Animals. 2019. Available at: **https://www.avma.org/sites/default/files/resources/AVMA-Guidelines-for-the-Depopulation-of-Animals.pdf.** Accessed 29 August 2021.

2. Atkinson et al.: 2012, Animal Welfare p. 487-495.

Figure 1: Swine consciousness assessment flow chart[2] (adapted from Atkinson et al., 2012)

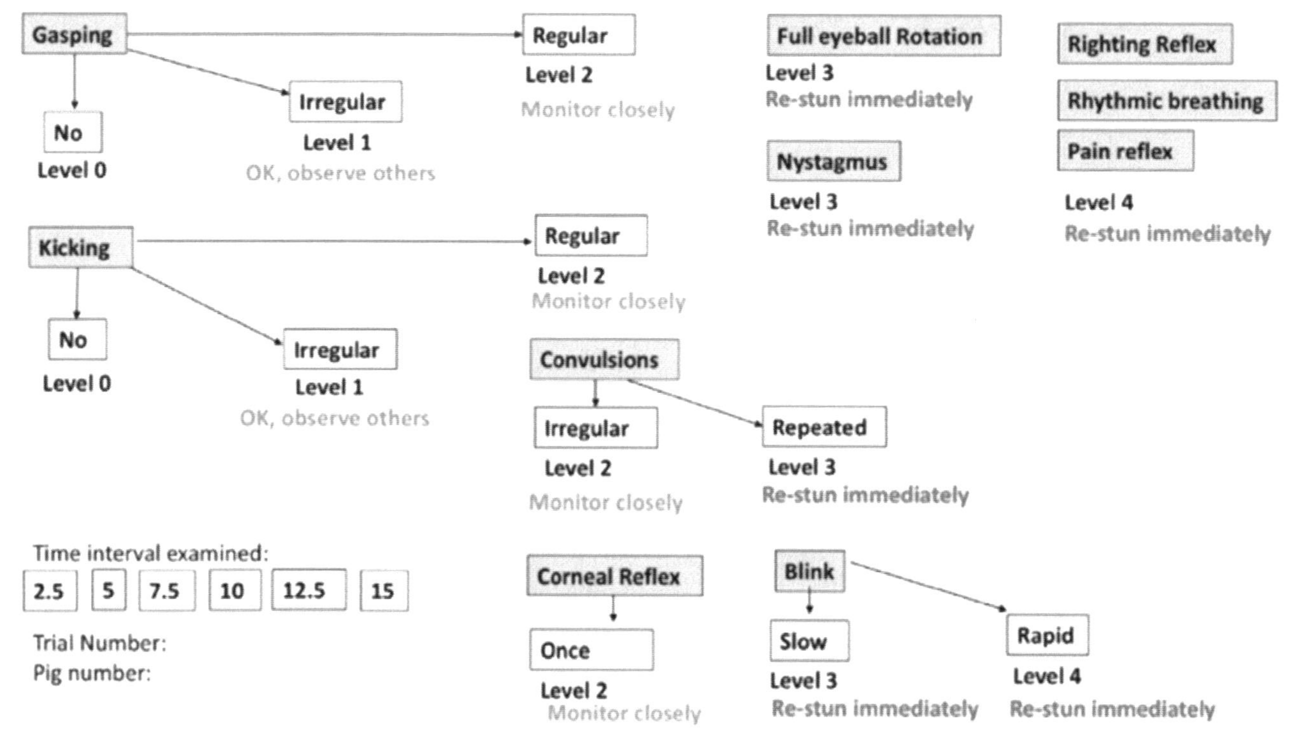

https://doi.org/10.54846/am2022/23

Assessment of market trailer contamination with porcine coronaviruses and Senecavirus A

Glorianne Vazquez[1]; Kate Dion[2], DVM; Amanda Sponheim[3], DVM
[1]Iowa State University, Ames, Iowa; [2]The HANOR Company, Webster City, Iowa;
[3]Boehringer Ingelheim Animal Health USA Inc, Duluth, Georgia

Introduction

Market trailers are commonly left unwashed due to cost, time, and truck wash availability between the harvesting facility and farms, therefore posing a risk to transmit infectious diseases between wean to market operations. This risk may cause pigs left in the barn or subsequent groups to become infected with pathogens such as porcine coronaviruses and Seneca virus A (SVA). Porcine coronaviruses, including porcine epidemic diarrhea virus (PEDV), transmissible gastroenteritis virus (TGEV), and porcine deltacoronavirus (PDCoV), are known to be highly contagious and can cause economic loss due to increased mortality and slower growth of pigs. Clinical signs of SVA are indistinguishable from multiple foreign animal diseases (FAD); the movement of pigs must be halted until confirmation that it is not a FAD. The objective of this study is to assess the frequency at which unwashed market trailers are contaminated with PEDV, TGEV, PDCoV, and SVA from the harvesting facility.

Materials and methods

Two environmental sample types were collected from a total of 100 market trailers located in Iowa during June 2021 through August 2021. Unwashed trailers were selected for sampling, after delivering their first load of pigs to a harvesting facility from a site negative for the targeted pathogens. Samples were taken at the farm before the next group of pigs was loaded onto the trailer. Four handfuls of dirty bedding were collected in a clean plastic bag from an 81.2 × 40.6-cm area of the trailer 7.6 cm from the rear door. An environmental swab of the same area was taken with a Swiffer dry cloth saturated with 20mL of Dey and Engley Neutralizing Broth (D/E Broth). Fluids extracted from the Swiffer and the dirty bedding samples were submitted for diagnostic testing. Samples were tested individually using polymerase chain reaction (PCR) for PEDV, PDCoV, TGEV, and SVA. Virus isolation was attempted on SVA-positive samples with a cycle threshold of < 35.

Results and discussion

One or more pathogens of interest were detected in 25/100 trailers sampled. Table 1: Percentage of positive samples per sample type, demonstrates the percentages of positive results for each pathogen. Two out of 10 samples submitted for virus isolation were positive. The positive results for virus isolation were obtained from Swiffer samples. The results from this study indicate that, in the summer, the frequency of trailer contamination with porcine coronaviruses is relatively low, while the frequency for contamination with SVA is higher. The increase in SVA positives during this time corresponds with the seasonal increase of cases in the United States. An increase in porcine coronavirus positive cases is expected during the winter months when conditions for survivability are more favorable. Studies during the winter months could provide a better understanding of the seasonal effects of trailer contamination, and economic assessments could be made to determine the cost of infection versus the cost of washing trailers. This study also highlights that SVA is potentially viable and, if a trailer cannot be washed, extra biosecurity steps such as "all in–all out," cleaning and disinfection, and load in/load out biosecurity precautions at both the harvesting facility and farm should be utilized to lower the risk of virus transmission.

Table 1: PCR results for trailer environmental samples

	% PEDV PCR positives	% TGEV PCR positives	% PDCoV PCR positives	% SVA PCR positives
Swiffer	3.00% (3/100)	0% (0/100)	5.00% (5/100)	18.00% (18/100)
Bedding	2.00% (2/100)	0% (0/100)	0% (0/100)	16.00% (16/100)
Positive trailers	3.00% (3/100)	0% (0/100)	5.00% (5/100)	22.0% (22/100)

https://doi.org/10.54846/am2022/24

Retrospective investigation of the relationships between sample types, fetal death outcomes and PCV3 PCR test results

Amanda Archer[1]; Brad Thacker[1,2]; Leann Presnell[1]; Channing Sebo[2]; Kelly Greiner[2]; Emily Byers[2,3]; Abby Vennekotter[3]; Allison Knox[4]; Grant Allison[4]; Nicolas Rippel[5]; Justin Cagle[2]
[1]Lincoln Memorial University, Harrogate, Tennessee; [2]Merck Animal Health, DeSoto, Kansas; [3]Prestage Farms, Inc., Clinton, North Carolina; [4]Walcott Veterinary Clinic, Walcott, Iowa; [5]Veterinary Medical Center, Williamsburg, Iowa

Introduction

Porcine circovirus type 3 was first reported as a cause of reproductive failure in 2016 and field experiences since then have supported the role of PCV3 in reproductive failure cases.[1-3] In a recent vaccine field study, PCV3 was diagnosed in fetuses from 6 out of 82 non-vaccinated gilt control litters while PCV3 was not diagnosed in 87 litters from vaccinated gilts.[4] This study (designated below as Study C) suggested that fetal death is the main manifestation of PCV3 with positive PCV3 diagnoses in both mummified and stillborn fetuses. In addition, colostrum collected at farrowing and processing fluids collected at 3-5 days of age from these 6 litters were all PCV3 PCR positive with cycle times (Ct) ranging from 21.0 to 28.6 for colostrum and 19.1 to 28.4 for processing fluids. Regarding pig sera obtained at 2-3 weeks of age and pooled by litter, half (3/6) were PCV3 PCR positive with Ct of the 3 positive samples ranging from 29.7 to 30.6.

In another field vaccine study (designated below as Study A), placental umbilical cord (PUC) blood was tested in addition to processing fluids.[5] There were no differences in the rate of PCV3 positive samples between vaccinated and control litters for either sample type in dams vaccinated in late gestation. For dams vaccinated in mid-gestation, the PCV3 positive rate of processing fluids was higher in control litters compared to vaccinated litters with no differences in PUC blood samples. In both studies A and C, correlations of testing results and the rate of fetal death on an individual litter basis were not performed.

Testing samples other than fetal tissues may be useful for PCV3 reproductive failure investigations or surveillance, especially those samples that are routinely collected for disease monitoring. Both processing fluids and serum from pigs near weaning age are used for PRRSV surveillance. In addition, colostrum has been used for assessing the status of sow herds for PCV2 while PUC blood has been mainly used for investigational purposes.[6,7] The study reported here evaluated the relationships between 1) sample type (processing fluids, colostrum, PUC blood and pig serum), 2) fetal death outcome (mummified, stillborn or both – dead born) per litter (percent within litter or whether a litter had 3 or more of each type) and 3) the PCV3 PCR result of the litter sample.

Materials and methods

PCV3 PCR test results and reproductive data from three studies were used. Studies A and C were vaccine field studies and only data from non-vaccinated control litters were included. Study B was an investigation to determine the association of PCV3 with *Streptococcus suis* disease in late farrowing through nursery age pigs. Processing fluids were obtained in all three studies. Other sample types by study were PUC blood (Study A), colostrum (Study C) and pig serum (Study C). All samples were tested by PCR for PCV3 using standard methods at the Iowa State University Veterinary Diagnostic Laboratory. Reproductive data for the sampled litters were obtained from each operation's computerized records. For statistical analysis, the Chi Square test was used to identify associations between the proportion of litters with 3 or more mummified, stillborn, or dead born (mummies plus stillbirths) fetuses (dependent variable) compared to the PCV3 PCR result (independent variable) using the laboratory's standard cut-off value (positive < 37) and a revised cut-off (positive < 32) to account for potential sample contamination during collection. In addition, the percent of mummified, stillborn, or dead born fetuses in each litter (dependent variable) was analyzed using the non-parametric Kruskal-Wallis ANOVA with the two PCR cut-off results as the independent variable. All statistical analyses were done using Statistix™ 10 (Analytic Software, Tallahassee, FL).

Results and discussion

Initial analyses found no associations or significant relationships between stillborn or dead born fetuses and PCR results, regardless of the sample type (processing fluids, PUC blood, colostrum, or pig serum). For mummified fetuses, no associations or significant relationships were found for PUC blood, colostrum, or pig serum. However, with processing fluids, significant associations and relationships between mummified fetuses and PCR results were found. For all studies combined, litters with PCV3 PCR positive processing fluid, based on either the laboratory or the revised cut-off, were more likely to have 3 or more mummified fetuses with odds rations of 7.43 and 7.45, respectively by cut-off value. For the individual studies, significant associations between PCV3 processing fluid status and 3 or more mummified fetuses were observed only with the revised cut-off in Study C with an odds ratio of 4.79. The same patterns were observed for the percentage of mummified fetuses within litter. For all three studies, the within litter percentage of mummified fetuses was 114.0% greater in litters with PCV3 PCR positive processing fluids using the laboratory cut-off and 125.0% greater using the revised cut-off. For study C, the corresponding values were 111.1% (not significant) and 96.0%, respectively.

The data from Study C was used to establish a logical PCR positive versus negative cut-off value based on the association between litters with 3 or more mummified fetuses and a progressive range of cut-off values based on the integer of the reported Ct. Cut-offs of < 28 (16-16 to 16-27) resulted in *P*-values that could not be calculated due to no cases in two cells or had *P*-values > 0.05 due to low counts in at least one cell. Starting at a cut-off of < 29 until < 32, the association between the PCR result and litters with 3 or more mummified fetuses yielded *P*-values ranging from 0.012 to 0.060 (< 29) with odds ratios ranging from 3.39 (< 29) to 5.08. Cut-offs from < 33 to < 37 yielded non-significant associations.

These data suggest that testing processing fluids on an individual litter basis may be useful for predicting the potential impact of PCV3 on reproductive failure due to fetal mummification. The failure to identify associations between PCV3 PCR litter status and stillborn or dead fetuses was most like due to the universal inability in a production setting to differentiate Type 1 stillbirths, which occur prior to the onset of parturition and could be potentially induced by PCV3, from Type 2 stillbirths, which occur during parturition and are unlikely to be due to PCV3. Furthermore, the mummified versus stillborn classification of dead fetuses with any degree of decomposition is arbitrary and variable among farm staff. It is also possible that older fetuses are able to mount an effective immune response against PCV3 and survive the infection as is observed with porcine parvovirus.[8] Accordingly, further work is needed to better understand PCV3 pathogenesis in pregnant dams and to develop methods to differentiate Type 1 versus Type 2 stillbirths at the farm level. Fetal crown-rump length to estimate gestational age along with specific parameters to describe fetal decomposition may be useful.

Further work needs to be done regarding pooling individual litter samples or using aggregate samples as is often done for PRRS surveillance for comparison with PCV3 induced reproductive failure. In addition, a revised positive/negative cut-off lower that the laboratories standard value may be indicated when evaluating PCV3 PCR results of processing fluids collected by litter. Using the data from Study C, a cut-off of < 32 is suggested.

Points of concern with this data are 1) the lower level of litters with 3 or more mummified fetuses or the within litter percentage of mummified fetuses in Studies A and B and 2) the relatively small number of litters overall and litters with 3 or more mummies in Study C.

References

1. Palinski R, Pineyro P, Shang P, et al. 2016. A novel porcine circovirus distantly related to know circoviruses is associated with porcine dermatitis and nephropathy syndrome and reproductive failure. *J Virol.* 91(1):e01879-16. **doi: 10.1128/JVI.01879-16.**

2. Phan TG, Giannitti F, Rossow S, et al. 2016. Detection of a novel circovirus in pigs with cardiac and multi-systemic inflammation. *Virol J.* 13:184. **doi: 10.1186/s12985-016-0642-z.**

3. Creel, J., Schantz, B., Armbruster, G., et al. Porcine circovirus type 3: update on field experiences with Sequivity™ vaccination for this emerging pathogen. In: Proc. 51st Annual Meeting of the American Association of Swine Veterinarians, Atlanta, Georgia, 2020, pp. 107-108.

4. Sebo, C., Knox, A., Allison, G., et al. A case study of reproductive failure associated with porcine circovirus type 3 in an endemically infected commercial sow farm. In: Proc. 52nd Annual Meeting of the American Association of Swine Veterinarians, San Francisco, California, 2021, pp. 131-134.

5. Vennekotter, A., Byers, E., Thacker, B., Cagle, J. Evaluation of PCV3/2d prevalence in placental umbilical cord blood and processing fluids and reproductive outcomes following mass vaccination with an RNA-particle platform PCV3/2d vaccine. In: Proc. 50th Annual Meeting of the American Association of Swine Veterinarians, Orlando, Florida, 2019, pp. 89-92.

6. Aljets, K., Rippel, N., Brinkman, M., Lehman, J., Thacker, B. Utilization of colostrum for evaluating the PCV2 status of sow herds. In: Proc. 23rd Congress International Pig Veterinary Society, Cancun, Mexico, 2014, vol. I, p. 241.

7. Oh, U. Kang, S., Park J. Evaluation of PCV2 breeding herd stability using placental umbilical cord serum samples and improvement after PCV2 vaccination in Korea. In: Proc. 25th Congress International Pig Veterinary Society, Chongqing, China, 2018, p. 073.

8. Leengoed, L. Vos, J., Gruys, E. et al. 1983. Porcine Parvovirus infection: Review and diagnosis in a sow herd with reproductive failure. Veterinary Quarterly 5:3, 131-141.

https://doi.org/10.54846/am2022/25

The impact of neonatal PCV3 infection on the development of *Streptococcus suis* disease in late farrowing and nursery age pigs

Leann Presnell[1], Brad Thacker[1,2], Amanda Archer[1], Lauren Wisnieski[1], Kelly Greiner[2], Nicolas Rippel[3]
[1]Lincoln Memorial University, Harrogate, Tennessee; [2]Merck Animal Health, Lenexa, Kansas;
[3]Veterinary Medical Center, Williamsburg, Iowa

Introduction

Porcine circovirus type 3 (PCV3) was first reported in the US in 2016.[1,2] The role of PCV3 in reproductive failure, mainly fetal death, has been reasonably well documented while the impact of PCV3 on growing pigs has not been well-defined.[3] In the field, PCV3 infections have been detected in growing pigs with respiratory disease, diarrhea, multi-systemic inflammation and myocarditis. Furthermore, lesions of periarteritis and arteritis in heart, kidney and mesocolon have been reported in growing pigs from field cases. PCV3 is often detected in pigs that are co-infected with a wide variety of other viral or bacterial pathogens and the virus may play a role in promoting the impact of these pathogens resulting in increased severity of disease at the herd level.[4] Of special concern is the role that vertical transmission of PCV3 from dams to their offspring may play in the development of PCV3 disease in growing pigs. The study reported here was initiated as part of a field investigation in a herd with persistent *Streptococcus (Strep) suis* disease in late farrowing and nursery age pigs. The objective of this study was to determine if vertical transmission of PCV3 from dams to piglets less than 1 week of age impacted the likelihood of getting treated for any condition, *Strep suis* incidence, and mortality.

Material and methods

The study was conducted using one farrowing group in a 1,500-sow herd located in eastern Iowa in early 2019. The herd was on a bi-weekly weaning schedule. During routine processing at 1-7 days of age (avg. 4.1), 1,424 pigs from 124 litters were individually ear tagged and testicles were collected from 3-5 male pigs per litter. The testicles were pooled by litter and then frozen and thawed for testing for PCV3 by PCR at the Iowa State University Veterinary Diagnostic Laboratory. In addition, pigs that were judged to be small within their litter were noted by the investigators at tagging, and sow ear tag and parity information was recorded. All litters were processed on one day. From processing until 40 days post weaning, the farm staffs at the farrowing site and two nursery sites that received the weaned pigs recorded the tag number, reason and date for deaths and treatments. Weaning age ranged from 17 to 23 days (avg. 20.1). Farrowing related litter size information was obtained from the farm's computerized records. All data was entered into an Excel spreadsheet.

The data set was refined to exclude the following: litters and pigs from which the reproductive records could not be retrieved (6 litters; 66 pigs), pigs that died after processing but before the first case of *Strep suis* was observed, which was 2 days prior to weaning (42 pigs), and 6 pigs that were identified with a problem/reason, including *Strep suis*, but were not treated. Accordingly, the data used for analysis was based on 118 litters and 1,314 pigs. For the PCR results, a revised cut-off of < 36 cycle time (Ct) was used rather than the laboratory's < 37 Ct cut-off to account for potential environmental contamination.

Univariate statistical analysis was done using the Chi square test with Statistix 10. *P*-values ≤ 0.05 were considered significant. Mixed logistic regression models were run in Stata version 14.2 (StataCorp; College Station, TX) for each outcome of interest: *Strep suis* case, death from *Strep suis*, treated for any reason, and death from any reason. The primary explanatory variable of interest was PCV3 litter status. Other explanatory variables included dam parity (gilt vs. sow), pig size at processing (small versus not), gender, weaning age, the number of liveborn, stillborn, and mummified fetuses in each litter, and the number of pigs in each litter that died between processing and when the first *Strep suis* cases were observed. Random intercepts for litter and litter location (room and position within room) were included in each model unless they had variance estimates that were negligible. Manual backwards variables selection was performed ($P < 0.05$) to select the final variables in the models. Only models where PCV3 litter status was significant were reported.

Results and discussion

Of the 118 litters, 16 (13.6%) were PCV3 PCR positive based on the laboratory's cut-off; 2 litters with Ct of 36.1 (not considered positive for analysis) and 14 litters (11.9%) with Ct ranging from 33.9 down to 16.8. As for parity, 8.9% of the dams were gilts and 91.1% were sows. As for pig size, 254 (19.3%) pigs were noted as small.

The overall mortality rate was 10.2% while the mortality rate due to *Strep suis* was 4.2%; 41.2% of mortality was due to *Strep suis*. The treatment success rate for *Strep suis* cases was 54.8% and for other – not *Strep suis* cases, 90.2%. Among pigs treated for *Strep suis*, the PCV3 litter status of the pigs may have impacted treatment success rates; 28.6% (2 successes out of 7 treated) if PCV3 positive and 62.5% (15 out of 24) if PCV3 negative ($P > 0.100$). For pigs treated for other reasons, there were no differences in treatment success rates between PCV3 positive (90.9%) and negative (90.0%) pigs.

Univariate analysis indicated that PCV3 litter status was associated with the proportion of pigs that died from *Strep suis*, died for any reason, or were treated for any reason. Small pig status was associated with the proportion of pigs that died from or were treated for *Strep suis* or died for any reason. The impact of parity was limited to the proportion of pigs that died from *Strep suis*, which was highest in pigs from sows. Pigs that were small and from PCV3 positive litters were significantly more likely to have *Strep suis*, die from *Strep suis*, die from any reason or be treated for any reason compared to all other pigs.

Mixed logistic regression models revealed several significant associations. After variable selection, the final multivariable for death from *Strep suis* included PCV3 litter status and litter size. Being in a PCV3 positive litter and being from a small litter (< 9) were associated with higher odds of dying from *Strep suis,* (OR [95% CI]: 2.27[1.02-5.05] and 2.72[1.03-5.65], respectively). The final multivariable model for being treated for any reason included being from a PCV3 positive litter, the number of stillborn and mummies in the litter, and having a death in the litter before the first case of *Strep suis* was observed. The odds of being treated for any reason was 2.18[1.05-4.55], times higher in pigs from litters that were PCV3 positive. Pigs from litters that had 3 or more stillborn and mummified fetuses in the litter also had higher odds 4.49 [2.18-9.26] of being treated. However, pigs from litters and that had deaths before the first *Strep suis* case was observed had lower odds of being treated. PCV3 was not significantly associated with odds of developing *Strep suis* or odds of dying from any cause.

Of interest and perhaps concern, 11 of the 14 PCV3 positive litters were housed in the back half (away from hallway door and air inlets, and towards the external exhaust fans) of one of the five farrowing rooms that housed the 118 litters. The reason for this is unclear. The sows were housed in stalls during gestation prior to entry into the farrowing facility, possibly in close proximity but their exact location in gestation could not be verified. Accordingly, the finding of clustering of PCV3 positive litters by location may need to be considered in the design of future studies and surveillance strategies.

Overall, the results of this study suggest that PCV3 infection in young pigs may increase mortality among pigs with *Strep suis* disease. However, these results should be viewed as preliminary and additional, more detailed studies are warranted.

References

1. Palinski R, Pineyro P, Shang P, et al. 2016. A novel porcine circovirus distantly related to know circoviruses is associated with porcine dermatitis and nephropathy syndrome and reproductive failure. *J Virol.* 91(1):e01879-16. **doi: 10.1128/JVI.01879-16**

2. Phan TG, Giannitti F, Rossow S, et al. 2016. Detection of a novel circovirus in pigs with cardiac and multi-systemic inflammation. *Virol J.* 13:184. **doi: 10.1186/s12985-016-0642-z**

3. Sebo, C., Knox, A., Allison, G., et al. A case study of reproductive failure associated with porcine circovirus type 3 in an endemically infected commercial sow farm. In: Proc. 52nd Annual Meeting of the American Association of Swine Veterinarians, San Francisco, California, 2021, pp. 131-134.

4. Klaumann, F., Correa-Fiz, F, Franzo, G., et al. 2018. Current knowledge of Porcine circovirus 3 (PCV-3): A Novel Virus With a Yet Unknown Impact on the Swine Industry. Front Vet Sci. 12;5:315. **doi: 10.3389/fvets.2018.00315**

https://doi.org/10.54846/am2022/26

Effect of pre-farrow injectable *Lawsonia intracellularis* vaccination of sows on maternal antibodies

H. Everett[1], BS; E. Mahan-Riggs[2], DVM; B. O'Brien[3], VMD, MS; J. Cagle[3], DVM; J. Pittman[2], DVM, MS, DABVP
[1]North Carolina State University College of Veterinary Medicine, Raleigh, North Carolina; [2]Smithfield Hog Production North Region, Waverly, Virginia; [3]Merck Animal Health, Lenexa Kansas

Introduction

Lawsonia intracellularis (LI) is an obligate intracellular bacterium that causes proliferative enteropathy (ileitis), an economically impactful disease. Infection presents in three clinical forms: a chronic form known as porcine intestinal adenomatosis, an acute form known as porcine hemorrhagic enteropathy, and a subclinical form. The latter of the three resulting in the highest economic impact.[1] One control strategy commonly employed is the vaccination of growing pigs, however control strategies for other pathogens have also utilized maternally derived immunity to limit transmission.[2] There are currently few reports in the literature evaluating the role that maternally derived antibodies (MDA) play in the protection from LI infections. Therefore, the objective of this study was to evaluate the impact of pre-farrow ileitis vaccination of sows on MDA and to characterize the decay curve in their pigs.

Materials and methods

Three weeks prior to farrowing, 55 gilts and sows were enrolled from one breeding group. They were subsequently blocked by parity (0, 1-2, 3+) then randomly allocated as either booster vaccinated with 2mL of a commercial injectable ileitis vaccine or non-boostered control. Serum samples were collected at enrollment and two weeks post vaccination. An equal number of litters were enrolled per parity group. Pig movement was not allowed in treatment litters. One pig per litter was tagged and serum samples were collected at 2-5 days of age (doa), 9-12 doa, 16-24 doa (weaning), and every 14 days until 107 doa. Oral fluid (OF) samples were collected every two weeks post weaning for surveillance of LI by polymerase chain reaction (PCR). All serum samples were analyzed using immunoperoxidase monolayer assay (IPMA) and enzyme linked immunoassay (ELISA). Data was analyzed using descriptive statistics.

Results

Gilt starting titer results were numerically higher than sows and thus data was grouped by P0 and P1+. Results of sow booster antibody response and pig MDA decay as measured by ELISA are shown in Figure 1. IPMA and ELISA results were similar in decay between treatments, but a more rapid decline was seen in the IPMA titers shown in Figure 2. The IPMA half-life was determined to be 17.9 days (SD ± 8). The ELISA half-life was 52 days (SD ± 63). OF testing detected 1 of 12 LI positive at 93 and 107 doa.

Discussion

Under the conditions of this study, vaccinated gilts produced piglets with the longest lasting maternal antibodies for LI. This suggests that a pre-farrow booster vaccination against LI along with previous titers do impact the MDA decay curve for LI with a difference noted between sows and gilts. However, MDA appears to wane before natural infection occurs at 93 doa 1, indicating that a pre-farrow vaccine is not an effective ileitis control strategy for piglets downstream. Future studies should be done to determine the success that piglet vaccination would have in the face of MDA.

References

1. Winkleman N. Subclinical ileitis: Diagnostic monitoring, R2, and economics. In: *Proceedings of the 49th AASV Annual Meeting*. American Association of Swine Veterinarians; 2018:488-493.

2. Desrosiers R. Epidemiology, diagnosis and control of swine diseases. In: *Howard Dunne Memorial Lecture*. American association of Swine Veterinarians; 2004.

Figure 1: A comparison of ELISA percent inhibition with sow vaccination status and with pig age across treatment groups.

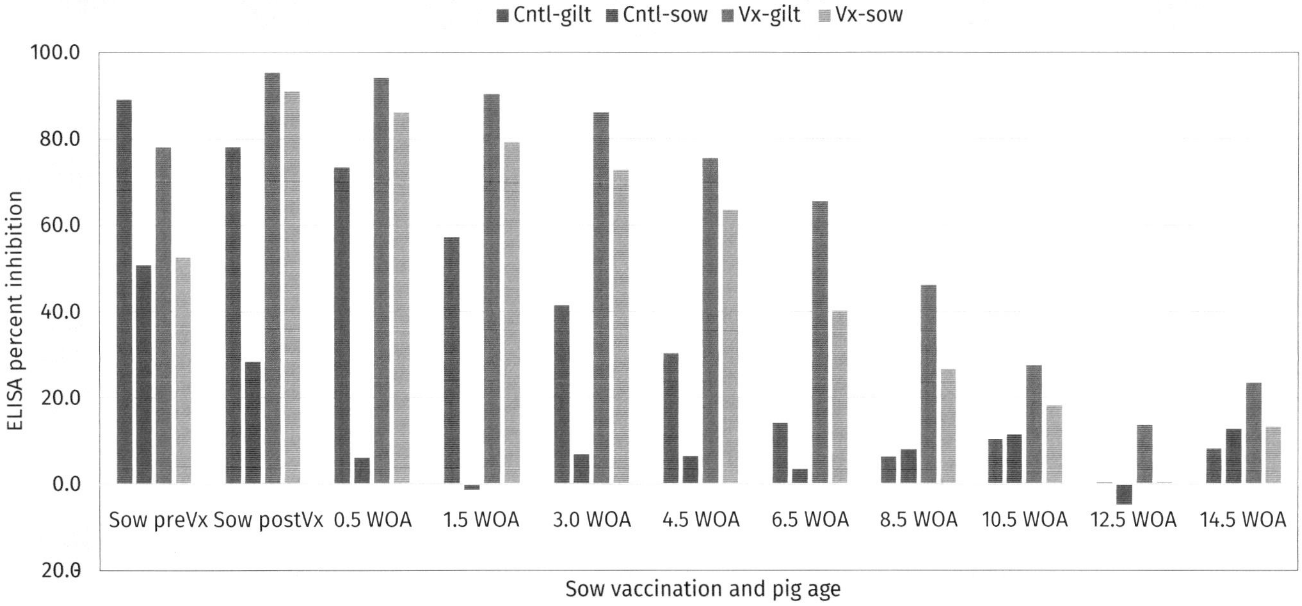

Figure 2: A comparison of IPMA titers with sow vaccination status and pig age across treatment groups.

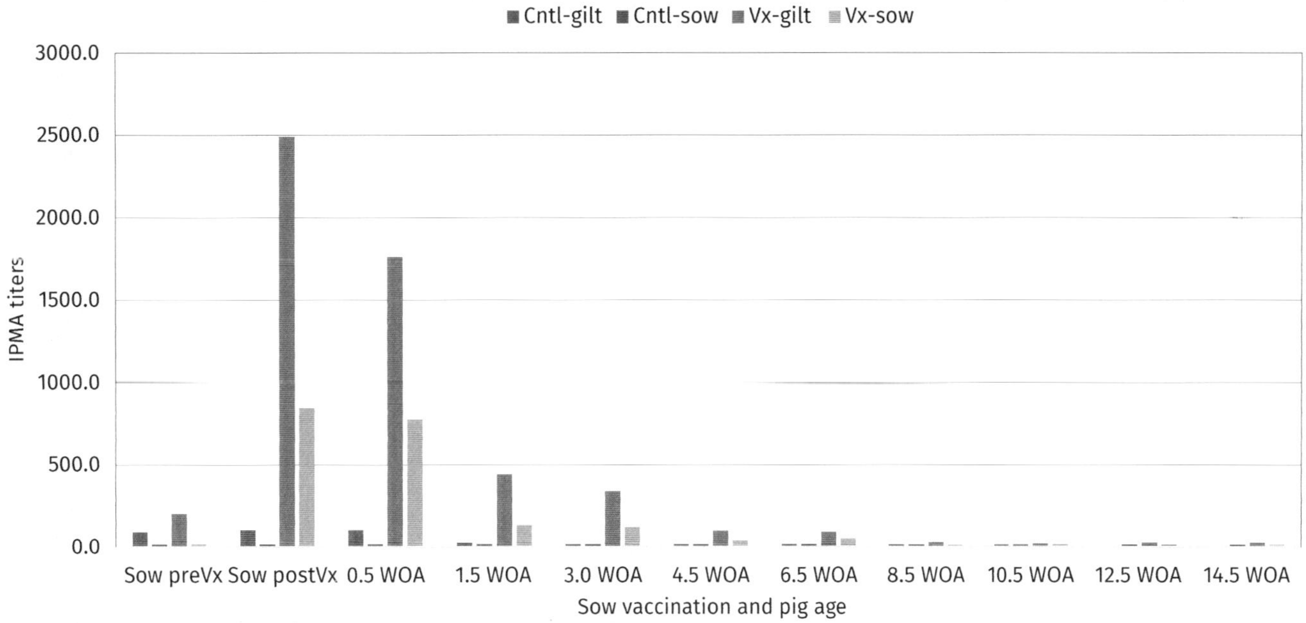

https://doi.org/10.54846/am2022/27

Evaluation of antibiotic treatment for the control of mortality and growth performance in PRRSV 1-4-4 L1C positive nursery pigs

K. Nisley[1], BS; E. Kettelkamp[2], DVM; A. M. Betlach[2], DVM, PhD
[1]Iowa State University, College of Veterinary Medicine, Ames, Iowa; [2]Swine Vet Center, St. Peter, Minnesota

Introduction

Porcine reproductive and respiratory syndrome virus (PRRSV) is one of the largest burdens to the swine industry, with losses of $361 million a year in growing-pig herds.[1] Economic losses attributed to PRRSV in growing-pig herds are primarily due to increased mortality, decreased average daily gain (ADG) and secondary bacterial respiratory infections.[1,2] During the last quarter of 2020, a highly virulent PRRSV strain was identified in swine herds in Iowa and Minnesota and later characterized as PRRSV restriction fragment length polymorphism (RFLP) 1-4-4 lineage 1C (L1C) variant strain based on open reading frame 5 (ORF5) sequence analysis.[3,4]

There are numerous documented strategies for PRRSV control, including the usage of antibiotics. To combat secondary bacterial infections, antibiotic usage in the growing phase during a PRRSV outbreak has been shown to decrease mortality and increase average daily gain.[5] However, limited research has been conducted to assess treatment options to limit the negative effects of the PRRSV 1-4-4 L1C variant. The objective of this study was to evaluate different antibiotic treatments in PRRSV 1-4-4 L1C variant positive nursery pigs for the control of mortality and growth performance.

Materials and methods

This study was conducted on a 4,000 head wean-to-finish farm with two 1,000 head rooms selected for the trial. The farm was sourced from two different sow farms (ie, source A and B), each having a similar PRRSV 1-4-4 L1C variant (100% nucleotide similarity). Pens were randomly assigned to treatment, blocked by source, weight, and location within the barn. The four treatment groups were Baytril® (enrofloxacin, Elanco Animal Health) and Dexasone® (dexamethasone, Aspen) concurrently, Draxxin® (tulathromycin, Zoetis), Excede® (ceftiofur crystalline free acid, Zoetis), and a negative control group. Treatments were administered at four weeks placed (day 0 of trial). Prior to the trial (ie, three weeks placed), three pigs from each room were selected to be sent to a diagnostic laboratory for a full necropsy. The pigs selected were chosen on a basis of being acutely ill in order to isolate active bacterial infections. At four weeks placed, pigs were individually weighed, tagged, and treated according to pen-level assigned treatment. Mortality and pulls were tracked by ear tag identification. At eight weeks placed, pigs were weighed again. At three, four, and eight weeks placed, oral fluid samples were collected from each room in a fixed-spatial manner and submitted for PRRSV polymerase chain reaction (PCR) testing. Throughout the trial, barn crew were blinded to pen-level treatment. The barn staff were permitted to treat pigs throughout the study with injectable Baytril and Dexamethasone. Data were analyzed by a generalized linear mixed model using SAS Proc Mixed Procedure (SAS 9.4, Cary, NC) with a significance of $P \leq 0.05$ and tendency for significance at $0.05 < P \leq 0.10$.

Results

Pigs in the four treatment groups had similar mean enrollment weights (Table 1, $P \geq 0.05$). At the end of the study, pigs receiving Draxxin had a tendency for higher body weights than those receiving Excede, with control and Baytril/Dex pigs being intermediate ($P = 0.09$). Mortality (%), pulls (%) and ADG showed no significant differences among treatment groups. However, analyzing by source (Table 2), pigs from source A that were treated with Draxxin had significantly higher end body weights and ADG compared to pigs treated with Excede, with pigs treated with Baytril or control being intermediate ($P \leq 0.05$).

Necropsy results showed variable results. All three source A pigs tested PCR positive for PRRSV (lowest Ct value = 17). Two of the three source B pigs tested PCR positive for PRRSV (lowest Ct value = 21). In addition, Beta-hemolytic, F18 *Escherichia coli* was identified via culture.

At three weeks placed (a week prior to antibiotic administration), 62.5% (n = 5/8) and 87.5% (n = 7/8) of oral fluids collected from source A and B, respectively, resulted PRRSV PCR positive (Ct values =27-34). At four weeks placed, 92.8% (n = 13/14) and 50% (n = 8/16) of oral fluids collected from source A and B, respectively, were PRRSV positive (Ct values = 25-34). At the end of the trial, all oral fluids were PRRSV PCR positive (Ct values = 27-35), regardless of source.

Discussion

The recently identified PRRSV 1-4-4 L1C variant has been detrimental to the swine industry thus far. While there are no available treatments that have a direct effect on the virus, secondary bacterial infections associated with PRRSV infections can be controlled utilizing antibiotics. This study sought to expand upon current research on antibiotic usage in the face of a PRRSV outbreak, specifically looking at the new variant. As it relates to this trial, the results indicated that certain antibiotics had greater efficacy in relation to the production parameters measured. Draxxin showed a tendency for higher weights in the overall trial compared to Excede, however the application of Draxxin in source A pigs showed statistically significant higher end body weights and ADG compared to Excede. Growth performance differences by source could have resulted as source A pigs had lighter body weights and presented more clinical signs compared to those from source B. Due to this, strategic selection of antibiotics based on source should be considered when developing control methods for PRRSV-challenged herds.

Throughout the trial, caretakers were permitted to spot-treat pigs, resulting in the lack of a true control (non-treated) group. Without a true control, comparisons to the control were

Table 1: Least square means and standard deviation for pig body weight (lb) and average daily gain (ADG) based on antibiotic treatment

Variable	Control	Baytril/Dex[3]	Draxxin	Excede
Day 0 weight, lbs	$25.27^{a} \pm 2.66$	$25.29^{a} \pm 2.65$	$25.62^{a} \pm 2.65$	$24.60^{a} \pm 2.65$
Day 28 end weight,[1] lbs	$59.89^{a,xy} \pm 7.91$	$59.92^{a,xy} \pm 7.90$	$61.85^{a,x} \pm 7.90$	$59.06^{a,y} \pm 7.90$
ADG, lbs/d	$1.23^{a} \pm 0.19$	$1.23^{a} \pm 0.19$	$1.29^{a} \pm 0.19$	$1.22^{a} \pm 0.19$
Mortality[2] (%)	$4.45^{a} \pm 0.02$	$4.41^{a} \pm 0.02$	$4.58^{a} \pm 0.02$	$4.61^{a} \pm 0.02$
Pulls[2] (%)	$6.97^{a} \pm 0.02$	$7.72^{a} \pm 0.02$	$7.66^{a} \pm 0.02$	$7.09^{a} \pm 0.02$

1 End weight did not consider dead or pulled pigs.

2 Mortality and pulls are expressed as a percentage of starting inventory.

3 Dex = Dexamethasone

a,b Values with different superscripts in the same row are statistically significant ($P \leq 0.05$)

x,y Values with different superscripts in the same row had a tendency for statistical significance ($0.05 < P \leq 0.10$)

Table 2: Least square means and standard deviation for pig body weight (lb) and average daily gain (ADG) for pigs sourced from source A based on antibiotic treatment

Variable	Control	Baytril/Dex[2]	Draxxin	Excede
Day 0 weight, lbs	$23.98^{a} \pm 2.91$	$23.98^{a} \pm 2.88$	$23.35^{a} \pm 2.88$	$21.92^{a} \pm 2.88$
Day 28 end weight,[1] lbs	$58.53^{ab} \pm 8.33$	$57.47^{ab} \pm 8.27$	$60.04^{a} \pm 8.27$	$54.39^{b} \pm 8.27$
ADG, lbs/d	$1.22^{ab} \pm 0.20$	$1.19^{ab} \pm 0.19$	$1.30^{a} \pm 0.19$	$1.14^{b} \pm 0.19$

1 End weight did not consider dead or pulled pigs.

2 Dex = Dexamethasone

a,b Values with different superscripts in the same row are statistically significant ($P \leq 0.05$)

confounded. Pen-level treatment was utilized because the client was interested in evaluating the effect of mass antibiotic treatment, when applied at the pen-level. Applying treatments at the pen-level decreased the number of replications, thus decreasing statistical power. This study demonstrated that certain antibiotics may be more efficacious toward controlling disease attributed to PRRSV 1-4-4 L1C variant. However, further research may be warranted in the effort to determine the best treatment options. Future studies should aim to have a negative control, as long as pig welfare is not significantly compromised, along with the administration of treatments at the pig-level. In doing so, an effect of antibiotic treatment on mortality may be identified due to the ability to capture smaller differences among treatment groups.

References

1. Holtkamp, DJ; Kliebenstein, JB; Neumann, EJ et al. Assessment of the economic impact of porcine reproductive and respiratory syndrome virus on United States pork producers. J Swine Health Prod. 2013;21(2):72-84.

2. Cho, J; Dee, S. Porcine reproductive and respiratory virus. Theriogenology. 2006;66(3):655-662.

3. Trevisan, G; Magalhaes, E; Linhares, L; Crim, B; Dubey, P; Schwartz, K; J.; Burrough, ER; Gauger, PC; Pineyro, P; Siepker, C; Main, R.G.; Linhares, D; Thurn, M; Lages, P; Corzo, C; Torrison, J; McGaughey, R; Herrman, E; Pogranichniy, R; Palinski, R; Henningson, J; Gretseth, J; Kersey, D; Clement, T; and Christopher-Hennings, J. Swine Disease Reporting System Report #36. 2021; Swine Disease Reporting System. 36.

4. Kikuti, M; Paploski I.A.D.; Pamornchainavakul, N.; Picasso-Risso, C; Rovira, A; VanderWaal, K; Corzo, C. Newly emerging PRRSV Lineage 1C variant nomenclature. Swine Health Monitoring Project.

5. Dee, S; Guzman, JE; Hanson, D; Garbes, N. A randomized controlled trial to evaluate performance of pigs raised in antibiotic-free or conventional production systems following challenge with porcine reproductive and respiratory syndrome virus. PLoS ONE. 2019;13(12): e0208430.

https://doi.org/10.54846/am2022/28

Evaluating the efficacy of exposing feeder pigs to *Mycoplasma hyopneumoniae* via fogging using pooled intra-tracheal samples

Madison Durflinger[1], BS: Brandi Burton[2], DVM; Abigail Redalen[3], DVM
[1]Iowa State University College of Veterinary Medicine, Ames, Iowa;
[2]Suidae Health & Production, Algona, Iowa; [3]Merck Animal Health, Lenexa, Kansas

Introduction

Mycoplasma hyopneumoniae (Mhp) has a huge economic impact due to the reduction in feed efficiency, average daily gain, and increased medication costs.[1] Producers will see an average reduction of $5.53 due to an Mhp infection.[2] Many are researching and experimentally inoculating pigs to measure and find better ways to control Mhp. Herd closure techniques also incorporate mass Mhp exposure as a method of elimination. Today, the gold standard for inoculating pigs with Mhp involves the use of Mhp positive lung homogenate as the inoculum source. While proven by many to be an effective inoculum, animal welfare concerns are present and associated with the process of obtaining the inoculum source. Utilizing antemortem samples as an inoculum source would also reduce production loss. The objective of this study is to evaluate the efficacy in using Mhp positive pooled tracheal samples to expose feeder pigs using a fogger.

Methods and materials

The study was performed in an 800 head feeder-to-finish site in north central Iowa with pigs from a maternal line flow. The site is divided into two 400 head rooms separated by a center office/load out area. Site was filled with maternal line gilts and barrows prior to the start of the study.

Prior to fogging, tracheal samples were collected from a Mhp positive sow farm. The tracheal samples were pooled by 5 and 0.5 mL from each pool was collected for testing. Samples were submitted to ISU VDL for Mhp PCR and PRRSV PCR. Only samples negative for PRRSV were used in this study. The pools of tracheal samples were diluted 1:1 with Friis Media (Teknova, Hollister, CA) and then stored in a - 80°C ultra-low freezer.

The goal of this study was to expose pigs to Mhp with Hurricane Ultra II (110V) foggers using pooled tracheal samples collected at the sow farm mentioned previously. The study was conducted twice, with each trial occurring in the one of the 400-head rooms. A similar protocol was used for each trial and all of our Mhp testing occurred at ISU VDL.

First trial

For the first fogging event, the average Mhp Ct value of the pooled tracheal samples collected was 21.1 with a final total volume of 27 mL. After combining the pooled tracheal samples with 4,600 mL of Friis Media, the Mhp Ct value was 27.5. The inoculum was mixed together the afternoon before the planned exposure and incubated at 37°C overnight.

The day of fogging, 4,600 mL of Mhp inoculum was divided into two Hurricane Ultra II (110V) foggers with 40 mL reserved and used to inoculate positive control pigs.

Prior to fogging, 65 pigs were randomly selected, tagged and tracheal samples were collected for Mhp testing by PCR. These samples were pooled by 5 and tested negative.

Five of the selected and tagged pigs were intra-tracheally inoculated using PCAI catheters. 8 mL of the inoculum was used and collected from the inoculum used in the fogger. The foggers were set to a medium-high flow rate and ventilation was shut off during the fogging event. The foggers were carried throughout the pens and were held at the height of the pigs during fogging. Ventilation was shut off for a half an hour after fogging. Tracheal samples were collected from the tagged pigs at 7- and 21-days post-exposure and the tracheal samples were tested for Mhp by PCR. All Mhp PCR's in this study were performed by Iowa State University Veterinary Diagnostic Laboratory (ISU VDL).

Second trial

To save time the day of fogging, 45 pigs were randomly selected, tagged and tracheal samples were collected for Mhp testing by PCR 1 week prior to the second replicate. These samples were pooled by 5 and tested negative.

The average Mhp Ct value of the pooled tracheal samples collected for the second trial was 24.1 and the final volume of the tracheal samples was 30 mL. After combining the pooled tracheal samples with 4,600 mL of treated Friis Media, the Mhp Ct value was 30.3. The inoculum was mixed together the afternoon before the planned exposure and incubated at 37°C overnight. The 4,600 mL inoculum was divided into three Hurricane Ultra II (110V) foggers and 40 mL of the solution was saved and used for the positive control pigs.

Five of the pigs served as positive controls and were intra-tracheally inoculated with 8 mL the inoculum used in the fogger.

The foggers were set to a medium flow rate and ventilation was shut off during the fogging event. The three Hurricane Foggers were distributed throughout the room and placed on top of trash cans. Throughout the exposure period, pigs were circulated around the foggers. Three people walked pens to achieve this circulation wearing proper PPE (protective eye gear, N95 face masks, coveralls, gloves, and plastic boot covers). Ventilation was turned back on immediately following exposure. Tracheal samples were collected from the tagged pigs 21 days post-exposure, and the tracheal samples were tested for Mhp by PCR. All Mhp PCR's in this study were performed by ISU VDL.

Results

For the first fogging event day 7 and 21 results show that none of the pigs fogged tested positive for Mhp. Three of the five intra-tracheally inoculated pigs tested positive for Mhp. For the second trial, 21 days after exposure to Mhp with a fogger, 95% of the tagged pigs exposed to Mhp via fogging were positive for Mhp by PCR. 3 of the 5 intra-tracheally inoculated pigs tested positive as well. This suggests that pooled tracheal samples can be a viable inoculum source when exposing pigs to Mhp using a fogger.

Discussion

When exposing groups of pigs via fogging, it is important to evaluate technique and ensuring there is an adequate distribution of fogging fluid. Placing these foggers on an elevated surface may also be beneficial in maintaining the aerosolized inoculum in the environment for a time period long enough to allow for adequate exposure to Mhp.

Utilizing tracheal samples as an inoculum source could be revolutionary when going about herd Mhp exposure. The industry could avoid euthanizing future breeding stock and reduce production and economic losses associated with the collection of lung homogenate. Tracheal samples are a cleaner Mhp inoculum source, therefore reducing the immune response associated with the foreign lung tissue found in lung homogenate. Armed with knowledge on the effectiveness of this inoculum source, this method offers a safer, cleaner, and more economic friendly alternative when the industry thinks about herd Mhp exposure. More research identifying the ideal concentration of Mhp positive intratracheal samples would be a beneficial when implementing this exposure technique in a commercial setting.

References

1. D. Maes, J. Segales, T. Meyns, M. Sibila, M. Pieters, F. Haesebrouck, Control of *Mycoplasma hyopneumoniae* infections in pigs, Veterinary Microbiology, Volume 126, Issue 4, 2008, Pages 297-309, ISSN 0378-1135.

2. Holtkamp, Derald. "Economic impact of co-infections in growing pigs." *Feedstuffs* 26 Oct. 2015: 12. *Business Insights: Essentials*. Web. 28 June 2021.

https://doi.org/10.54846/am2022/29

Fine-tuning the preparation of lung homogenate for optimal *Mycoplasma hyopneumoniae* exposure

M. McMahon[1]; B. Garcia-Morante[1], DVM, MS, PhD; A. Betlach[1,2], DVM, PhD; R. Robbins[3], DVM, PhD; P. Yeske[2], DVM, MS; M. Pieters[1], DVM, PhD
[1]Department of Veterinary Population Medicine, University of Minnesota, St. Paul, Minnesota; [2]Swine Vet Center, St. Peter, Minnesota; [3]Independent Swine Consultant, Amarillo, Texas

Problem statement and objectives

Mycoplasma hyopneumoniae causes chronic respiratory disease in swine and is known to economically impact herds worldwide.[1] To control *M hyopneumoniae* in sow herds, gilt acclimation has become a common practice in the United States. One method for gilt acclimation utilizes herd-specific lung homogenate to achieve uniform exposure of gilts prior to entry to the breeding herd, or to create a "Day 0" for disease elimination purposes.[2-4] However, the preparation and handling of lung homogenate has been minimally evaluated, which poses risks for exposure failure.[5] Therefore, the aim of this pilot study was to evaluate the capability of *M hyopneumoniae* in lung homogenate prepared in various dilution solutions and storage conditions to induce disease in gilts.

Materials and methods

Whole lungs were obtained from two *M hyopneumoniae*-positive donor pigs after humane on-farm euthanasia and were blended, separately, in the laboratory. Donor pigs were confirmed positive by real-time PCR from deep tracheal catheter (DTC) samples prior to euthanasia. For each donor lung, two preparations were created including a 70:30 tissue:Friis medium dilution and a 70:30 tissue:saline dilution. From the dilutions, four aliquots were created and stored in different conditions: 1) Fresh at 25°C, 2) 1 week frozen at -20°C, 3) 1 month frozen at -20°C and 4) 6 months frozen at -20°C. The room temperature lung homogenate was used to inoculate the first challenge group the same day of preparation.

There were four challenge groups based on the length of storage, with eight 5-week old gilts each, resulting in 32 challenged gilts total. For each challenge group, two gilts were randomly assigned a treatment of either lung homogenate prepared with saline or medium from donor pig one or two resulting in four gilts for both saline and medium. Prior to inoculation, the lung homogenate was further diluted with Friis medium or saline at a 1:10 ratio and was filtered. Gilts were then intra-tracheally inoculated with 10 mL of lung homogenate.

For all challenge groups, DTC and sera were collected at the start of the trial and four weeks post-inoculation, followed by humane euthanasia. At necropsy, lung lesions were evaluated,[6] and bronchial swabs (BS) and tissue samples were collected. For this study, a viable lung homogenate was defined as capable of causing *M hyopneumoniae* infection in a single gilt. Infection was determined if a *M hyopneumoniae* positive PCR result was obtained from a DTC or BS sample. Results from ELISA assays and observation of lung lesions were used as accessory for testing.

Results

Deep tracheal catheter and serum ELISA results were negative for all gilts prior to inoculation, confirming the *M hyopneumoniae* negative health status at the start of the trial. Four weeks post-inoculation results, obtained from serology, pathology, DTC, and BS samples are shown in Table 1. Overall, 31 of the 32 gilts were detected infected with *M hyopneumoniae* after exposure. One gilt inoculated with lung homogenate frozen for six months resulted negative to *M hyopneumoniae* in DTC, BS and serum ELISA. Histopathological lesions compatible with *M hyopneumoniae* infection were observed all 32 gilts.[7]

Conclusions

The use of lung homogenate for gilt acclimation to *M hyopneumoniae* is a popular method utilized in the field today. Results from this pilot study indicated that either saline or Friis medium were adequate for dilution of lung homogenate, as that lung homogenate successfully infected gilts with *M hyopneumoniae*, even after being stored for six months at -20°C. However, the capability of the lung homogenate to induce disease appeared to be slightly lower after six months of storage, compared to homogenate frozen for less than or equal to one month, as evidenced in the number of infected gilts and the observed lung lesions. It is important to note that being a pilot study, including a small sample size and a single source of lung homogenate, more data would be needed in order to make conclusions that can be widely applied in the swine industry.

References

1. Goodwin RFW, Pomeroy AP, Whittlestone P. Production of enzootic pneumonia in pigs with a mycoplasma. Vet Rec. 1965;77:1247-1249.

2. Fano, E., Payne, B., 2015. *Mycoplasma hyopneumoniae* gilt acclimation and sow herd stability: essentials to the systematic control approach. In: AASV Annual Meeting. Orlando, Florida. pp. 175–178.

3. Centeno, N., Chévez, J., Fano, E., 2016. Mexican swine industry on *Mycoplasma hyopneumoniae* gilts acclimatation. In: Proceedings of the 24th IPVS Congress. Dublin, Ireland, 31, 2013.

4. Garza-Moreno, L., Segalés, J., Pieters, M., Romagosa, A., & Sibila, M. (2018). Acclimation strategies in gilts to control *Mycoplasma hyopneumoniae* infection. Veterinary microbiology, 219, 23-29.

5. Robbins RC, Betlach AM, Mondragon-Evans MR, et al. Development of a herd-specific lung homogenate for exposure to *Mycoplasma hyopneumoniae* under field conditions. J Swine Health Prod. 2019;27 (4):221–227.

Table 1: ELISA, PCR and lung lesion scoring for gilts in all experimental groups 28 days post-inoculation

Parameter	Dilution solution	Experimental group			
		Fresh	Frozen 1 week	Frozen 1 month	Frozen 6 months
Seropositive (%)	Medium	75	100	75	25
	Saline	50	100	75	0
DTC Ct (Mean ± SD)	Medium	18.68 (0.62)	18.43 (1.21)	19.21 (2.30)	20.86 (0.75)
	Saline	19.77 (3.98)	20.57 (2.56)	18.52 (0.73)	24.52 (9.28)
BS Ct (Mean ± SD)	Medium	21.03 (0.69)	21.44 (1.96)	19 (1.23)	19.09 (1.46)
	Saline	19.66 (0.70)	23.01 (2.81)	20.08 (0.41)	23.25 (10.97)
Lung Lesion (%) (Mean ± SD)	Medium	47 (16.57)	22 (10.23)	24.75 (14.89)	24.75 (9.95)
	Saline	28.50 (15.02)	25 (7.62)	17.75 (8.54)	13.25 (13.45)

SD=standard deviation. DTC= deep tracheal catheter. BS=Bronchial swab. Ct=Cycle threshold.

6. Pointon, A.M., Davies, P., and Bahnson, P.B. 1999. Disease Surveillance at Slaughter. In: Diseases of Swine. 8th Edition. B. Straw, S. D'Allaire, W. Mengeline, and D. Taylor (editors). Iowa State Press. p. 1118.

7. Lauren K. Woolley, Shayne Fell, Jocelyn R. Gonsalves, et al. (2012) Evaluation of clinical, histological and immunological changes and qPCR detection of *Mycoplasma hyopneumoniae* in tissues during the early stages of mycoplasmal pneumonia in pigs after experimental challenge with two field isolates. Veterinary Microbiology, 161, 1–2, pp. 186-195.

https://doi.org/10.54846/am2022/30

Evaluation of a backpack sprayer inoculation method compared to intratracheal inoculation for the exposure of *Mycoplasma hyopneumoniae*

K. Graham[1]; N. Lauterbach[2], DVM; R. Evelsizer[3], DVM; C. Ruston[2], DVM, MS; A. Szczotka[2], DVM; R. Haupts[3], BS; C. Puls[3], PhD
[1]Iowa State University, Ames, Iowa; [2]Smithfield Foods Midwest Region, Algona, Iowa; [3]Elanco Animal Health, Greenfield, Indiana

Introduction

Mycoplasma hyopneumoniae (MHP), the etiological agent causing enzootic pneumonia, is a costly component of the porcine respiratory disease complex.[1] The control and elimination of MHP has been achieved through intratracheal inoculation and aerosolization via a fogger.[2] While the use of a fogger is less technically challenging, it poses a risk to the animals and people performing exposure, due to ventilation being substantially reduced during the process. A potential alternative to inoculating animals with MHP is individual aerosolization via a backpack sprayer. No previous literature has shown the use of a backpack sprayer as a means to inoculate animals with MHP. Thus, the objective of this study was to test the efficacy of a backpack sprayer as a means to inoculate animals with MHP in comparison to the gold standard of intratracheal inoculation.

Materials and methods

At a commercial finishing barn in northwest Iowa, 120 16-week-old pigs were enrolled into two treatment groups, intratracheal inoculated (IT) or backpack sprayer inoculated (IA). Eight treatment pens were randomly allocated as IT or IA with one non-treatment pen in between each treatment pen. Two of the non-treatment pens were assigned as neighboring pens (NP). Figure 1 displays the layout of assigned pens. On day-1 post-inoculation (DPI), 15 pigs from the IT and IA pens and 10 pigs from the NP pens were ear-tagged and deep tracheal swabbed (DTS) to test for MHP via Polymerase Chain Reaction (PCR) at Iowa State University Veterinary Diagnostic Lab. MHP homogenate was made per industry standards.[3] On 0 DPI, animals were then inoculated via IT or IA. For the IT group, 8 mL of inoculum was administered to each animal using a syringe and PCAI catheter. For the IA group, a battery-operated backpack sprayer was used. The MHP inoculum was mixed with a blue re-load pack within the tank of the backpack sprayer. Prior to being exposed, individual animals in the IA group were moved from their respective pens to an area surrounded by solid paneling. The nozzle of the sprayer was placed in each nostril for 1.5 seconds to ensure each animal was being inoculated with approximately 8 mL of MHP inoculum.

Results and discussion

Results of DTS are displayed in Table 1. Animals that were treated or euthanized were removed from the study. While the IA method did not result in the same percent positive as the IT method, both methods showed an increase in MHP PCR positives as the trial progressed. The difference in MHP PCR percent positives between the IT and IA groups may be attributed to a higher confidence that there was no uncontrolled wastage with the IT group. Both neighboring pens had a percentage of animals that went positive by day 28 of the trial. The neighboring pen in between two IT pens had a greater percent positive by day 28. Further work would need to be done to determine if animals in neighboring pens are testing positive due to direct contact or aerosolization droplets in the environment.

Overall, the IA method of MHP exposure may serve as an alternative method to inoculate animals with MHP. Additional work is necessary to determine the most effective IA dose to achieve a similar DPI percent positive to the IT method as well as compare the IA method to a fogger.

References

1. Pieters, M.G. et al. Chapter 56: Mycoplasmosis in: Disease of Swine. Ed:Zimmerman, J.J. Karriker, L., Ramirez, A., Schwartz, K., Stevenson, G., Zhnag, J. 11th ed., John & Wiley Sons, Inc., 201, pp 863.

2. Maes, D. et al. Control of Mhp infections in pigs. Veterinary Microbiology. 2008;126:297-309.

3. Robbins, R.C. et. al. Development of a herd-specific lung homogenate for exposure of *Mycoplasma hyopneumoniae* under field conditions. J Swine Health Prod 2019;27(4):221-227.

Figure 1: Diagram depicting the barn layout and which pens were assigned to respective treatment groups; blue = backpack sprayer inoculated, yellow = neighboring pen, red = intratracheal inoculated.

Table 1: Percentage pigs positive for MHP PCR based on intratracheal swabs

	Days post-inoculation (DPI)				
Treatment	**-1**	**7**	**14**	**21**	**28**
Intratracheal (IT)	0% (60/60)	28% (17/60)	28% (17/60)	76% (45/59)	95% (56/59)
Backpack sprayer (IA)	0% (60/60)	12% (7/60)	13% (8/60)	21% (12/56)	61% (34/56)
Neighboring pen IT	0% (10/10)	0% (10/10)	10% (1/10)	80% (8/10)	90% (9/10)
Neighboring pen IA	0% (10/10)	0% (10/10)	0% (0/10)	0% (0/10)	40% (4/10)

https://doi.org/10.54846/am2022/31

Porcine epidemic diarrhea virus survivability and infectivity in hog earthen manure storages

D. Banks[1]; M. Battrell[2], DVM
[1]College of Veterinary Medicine, North Carolina State University, Raleigh, North Carolina;
[2]Smithfield Foods, Kenansville, North Carolina

Introduction

When porcine epidemic diarrhea virus (PEDV) was first introduced into the United States in 2013, it rapidly spread through the naïve swine population. Although the incidence of PEDV outbreaks has steadily declined since 2013,[1] they do occur suggesting that there is a reservoir of virus that leads to these sporadic breaks. A grow-finish population may serve as the PEDV reservoir, particularly in a large continuous flow site where naïve animals are routinely added to the population. Yet, routine biosecurity practices make every effort to limit contact between growing pig sites and sow farms.

Another possibility we explored is the wastewater stored in earthen manure storages (EMS) and recycled to flush hog waste from barn pits. The EMS are located outside the perimeter of the farm, where biosecurity practices tend to be less stringent. Tun and associates revealed that PEDV could survive up to 9 months in the infected EMS after the initial outbreak in farms in the province of Manitoba during the winter.[2] They also reported that viral load in the top layer of the EMS was low and mostly non-infective and speculated the diminished infectivity to be due to UV and sunlight. This study was conducted to better understand the potential duration of PEDV infectivity of EMS during the hot, sunny days of a North Carolina summer.

Materials and methods

EMS samples were collected biweekly from seven sow farms starting the 21st week in 2021 and continuing through week 29. All sow farms associated with these EMS began showing clinical signs of PED and had infection confirmed by polymerase chain reaction (PCR) between weeks 9 and 16 of 2021. Intentional sow exposure was done to build herd immunity. Once sows had time to build sufficient immunity from intentional exposure to protect suckling pigs (2 weeks on Parity 2+ farms and 3 weeks on all gilt farms), farrowing houses were emptied of pigs, cleaned, disinfected, allowed to dry, and whitewashed. EMS sample collection did not begin until farms were believed to be making PEDV negative pigs.

Samples were collected from the EMS halfway between the farrowing house discharge line and the recycle line, approximately 12 inches below the surface using a 9-foot-long pole with a 200mL collection cup. Samples were then transferred to two 50mL polypropylene conical tubes, stored in a cooler on ice, and transferred to the Smithfield Veterinary Diagnostic lab. One sample was sent to the Iowa State University Veterinary Diagnostic Laboratory (ISU-VDL) for detection of PEDV nucleic acid via PCR, while the other was stored at -80 degrees Fahrenheit at the Smithfield Veterinary Diagnostic lab. All EMS samples were PEDV positive with cycle thresholds (ct) ranging between 26.6 to 29.4 week 21 and 31.0 to 35.5 week 29 (Table 1).

Seven 3-day old pigs were removed from a PEDV naïve sow farm for a bioassay. Each pig was housed separately in a plastic tote for the duration of the trial and administered Similac mixed with 10 cc of EMS sample from the PEDV infected farm that was paired with that pig. Pigs were fed at the start of the trial and again at hours 12, 18, and 24. Once EMS samples were depleted, Similac mixed with distilled water was offered twice daily. Fecal swabs were collected from each pig at 0, 12, 36 and 48 hours following the initial exposure.

Results and discussion

Pigs administered EMS samples from week 29 were humanely euthanized at hour 120 and samples were sent to ISU-VDL for PEDV testing using PCR. All pigs fed EMS samples collected week 29 remained PEDV negative. Seven additional pigs were obtained, and the process was repeated using samples collected from week 21. PEDV was detected in the fecal swab collected from one pig as early at 12 hours post exposure (EMS ct of 26.6). A second pig tested positive at 36 hours post exposure (EMS ct of 27.5). The remaining EMS samples failed to cause fecal shedding or clinical signs of PEDV within 48 hours post exposure (Table 2).

When farms are believed to have recovered from PEDV outbreaks and are no longer exhibiting clinical signs of infection, viable virus may remain in the EMS even in the hot sundrenched environment of an eastern North Carolina summer. Producers should treat the EMS as PEDV positive for several months following a PEDV outbreak and use extreme caution to ensure that pigs do not have access to this water either through a clogged pit, splash up while flushing, EMS water transfer, or through shared equipment.

References

1. "Porcine Epidemic Diarrhea Virus (PEDv)." AASV, 2 Apr. 2018, **www.aasv.org/Resources/PEDv/PEDvWhatsNew.php.** Accessed 10 June 2021.

2. Tun HM, Cai Z and Khafipour E (2016) Monitoring Survivability and Infectivity of Porcine Epidemic Diarrhea Virus (PEDV) in the Infected On-Farm Earthen Manure Storages (EMS). Front. Microbiol.7:265. Available at: **www.frontiersin.org/articles/10.3389/fmicb.2016.00265/full.** Accessed 15 May 2021.

Table 1: EMS samples

Farm	Week 21 (PEDV) Ct value	Week 23 (PEDV) Ct value	Week 25 (PEDV) Ct value	Week 27 (PEDV) Ct value	Week 29 (PEDV) Ct value
1	29.4	30.8	29.9	29.4	31.0
2	27.5	31.5	33.5	35.3	35.5
3	26.6	28.5	31.8	34.1	34.1
4	27.1	28.6	28.6	31.3	31.8
5	29.4	30.6	32.1	33.2	34.4
6	29.2	30.9	32.3	32.6	34.2
7	28.0	29.1	30.2	31.3	31.3

Table 2: Fecal swab results

Pig/Farm	Week 29 Fecal Swabs Results (PEDV)	Week 21 Fecal Swabs Results (PEDV)
1	Negative	Negative
2	Negative	Positive at 36 hours
3	Negative	Positive at 12 hours
4	Negative	Negative
5	Negative	Negative
6	Negative	Negative
7	Negative	Negative

https://doi.org/10.54846/am2022/32

A whole inactivated vaccine for IAV-S was shown to provide protection in the presence and absence of maternal antibodies

J. Griffel, BSc; T. Wolff, DVM, MPH; P.M. Winter, PhD, M.Res, BSc
Phibro Animal Health Corporation, Teaneck New Jersey

Introduction

Swine influenza A (IAV-S) is one of the top 3 respiratory diseases affecting pigs. IAV-S can affect greater than 70% of growing pigs per year, causing significant economic losses within the Pork Industry. While Influenza viruses are common with high rates of infection, they display low rates of mortality. Symptoms include coughing or barking, sneezing, high fever, difficulty breathing, runny nose and going off feed. However, the biggest obstacle to mass piglet vaccination is the induction of enhanced disease in the presence of maternal antibodies. This problem was further compounded by the wide window of maternal antibody waning within any given herd, making it difficult to vaccinate animals at low risk from developing serious complications. With this in mind, it was deemed that a critical path for the control of IAV-S would be to develop a vaccine and a vaccination schedule, which would be capable of inducing memory immunity in the face of maternal antibodies.

Materials and methods

The defining study utilized 250 pigs from seropositive sows, which had received a typical pre-farrow program. These 250 pigs were randomly assigned to 6-test groups, ensuring at least one piglet from each litter was assigned to each test group. These test groups looked at differing antigen levels and routes of immunization. The antibodies for both pigs and sows were screened and monitored throughout the study, this data showed an equitable distribution of piglets between test groups. The antibody levels measured were to the nucleoprotein using a blocking ELISA. The primary measurables in this study were magnitude of viremia post-challenge, consolidated pneumonic scores, antibody titers (Homologous Hemagglutinin Inhibition test) and magnitude of shedding when compared to the control group.

Results and discussion

Based on the current program of investigation and vaccine development, our aggregated study results from multiple IAV-S challenge studies have allowed Phibro Animal Health Corporation (PAHC) and MVP Adjuvants® to formulate a safe and effective whole inactivated IAV-S vaccine in both the presence and absence of maternal antibodies. This is demonstrated when the vaccine with lower antigen inclusion is administered intramuscularly to piglets between 1-4 days of age (Day 0) and boosted at 21-days; memory immunity is induced, and protection is displayed by challenge at 42 days in the absence or presence of maternal antibodies (HAI > 40). In addition, consolidated pneumonic scores were reduced, when compared to the control groups (ANOVA *P*-value 1.240×10^{-5}), so were viral loads within the lungs, which were measured by live titer (ANOVA *P*-value 7.372×10^{-6}) and PCR (ANOVA *P*-value 1.072×10^{-11}). These reduced viral loads can be extrapolated to suggest there is a significant reduction in viral shedding. While all test groups showed improved viral loads when compared to the control group a significant disparity was observed in consolidated pneumonic scores between groups and in comparison, with the control groups.

Summary

In summary; higher vaccine antigen inclusion rates, induced greater consolidated pneumonic scores when compared to control groups when administered intramuscular/intramuscular. In addition when the vaccine was administered intranasal/intramuscular (IN / IM) all groups were found to present elevated levels of consolidated pneumonic scores. With the presented program data, a clear vaccine formulation and schedule was defined, capable of overcoming maternal antibody interference and enhancement.

https://doi.org/10.54846/am2022/33

Pursuing American Board of Veterinary Practitioners (ABVP) certification in Swine Health Management

Jeffrey A. Husa, DVM, Diplomate ABVP
Swine Health Management representative to the Council of Regents

Introduction

Are you dedicated to the health and well-being of your patients and clients? Are you a life-long learner? Do you enjoy an intellectual challenge? If you answered "yes" to any of these, then ABVP credentials are right for you! Swine Health Management Diplomates of the American Board of Veterinary Practitioners report a desire for "personal challenge" and "increased knowledge" as their motivation to credential. Our Diplomates found that the journey "made them better practitioners." No matter what sparked your interest in this topic, we encourage you to take the ABVP plunge. I personally found the process to be rigorous and rewarding. Becoming a Diplomate has been far more impactful to me, my clients, and their animals than I ever expected.

Applicant requirements*

To be eligible, a veterinarian must:

A. Be a graduate of an AVMA accredited / approved college of veterinary medicine, hold a certificate from an Educational Commission for Foreign Veterinary Graduates (ECFVG), or be legally qualified to practice in the US or other country.
B. Demonstrate unquestionable moral character and ethical professional behavior.
C. Have a background in swine medicine and production in one of the following ways:

a. At least 5 years of clinical practice experience prior to first credential submission, and at least 6 years in practice prior to sitting the Certification Examination. The first year need not be in swine practice, however the applicant must have worked in a predominantly swine-oriented practice during the 5 years prior to examination.

OR:

b. Complete an ABVP-approved Residency or MS degree program, plus at least one year of swine practice.

OR:

c. Complete a swine-related PhD. The applicant must provide evidence of swine industry consulting.

The path to credentials begins with our Entrance Exam

Board certification in Swine Health Management begins by successful completion of an Entrance Examination. This 100-question, multiple-choice test is offered each Spring at the AASV Annual Meeting. It assesses your basic knowledge of key topics such as reproduction, disease, pathology, economics, epidemiology and more. This test also helps to set the stage for your success throughout the remainder of the credentialing process. To sit the exam you must submit an application and fee, due January 15th of the year in which you intend to test. Additional details about the examination, schedule, topics covered, and more can be found at **https://abvp.com**. The Applicant Handbook is available at this address, and is an excellent guide throughout the entrance and credentialing processes. Applicants who fail may retake the exam in succeeding years through the same process (application plus fee).

The certification process

After you pass the Entrance Exam, you will select an ABVP mentor, subject to the approval of the SHM Regent or their delegate. You will then create a plan of study in collaboration with your mentor, and you will complete the requirements for your credentials application. You will write two publications: either two case reports written in the ABVP format (Applicant Handbook, p22-27), or one case report and one peer-reviewed article accepted for publication. You will also begin studying for the Certification Examination during this time. Many Diplomates have found these studies to go hand-in-hand with daily clinical experiences. The credentialing process will push, stretch, and energize you!

Then, by September 1st of the year preceding your intended credentials submission, you will complete an on-line application at **https://abvp.com**. This alerts us of your impending submission. Your mentor and other Diplomates will be available to you throughout the journey. In particular, we recommend review of your materials prior to submission to better ensure compliance with ABVP guidelines. You will submit your credentials not later than 11:59PM CST on January 15th. Then, it's time to study for the next 14 months! The Certification Exam is administered during our AASV Annual Meeting. It includes written and oral evaluations with short answer and problem-solving components. A suggested credentialing timeline is shown in Table 1.

Maintenance of Certification*

Your ABVP credentials will then remain valid for 10 years. During the 9th or 10th year after you initially credential, you will recertify by fulfilling the following:

- Provide proof of veterinary licensure (or proof of employment which does not require license to practice)
- 250 hours of CE across ten years. 165 hours must be directly applicable to swine

* Note: For current credentialing and maintenance requirements, see the ABVP website for handbooks and other guidance.

Table 1: Timeline for successful board certification in Swine Health Management through the American Board of Veterinary Practitioners

	Jan	Feb/Mar	Apr	May	Jun	Jul	Aug	Sep	Oct	Nov	Dec
Year 1	Submit Entry Exam application by 11:59 PM CST, Jan 15	Pass Entry Exam at AASV Annual Meeting	*work work work work work work work work work!*								
Year 2	*work work work work work work work work work work work work work work!*										
Year 3	*work work work work work work work!*							**Submit credentials application by 11:59 PM CST, Sep 1	*work! work!*		
Year 4	Submit credentials by 11:59 PM CST, Jan 15	*study! study! study! study! study! study!*						Submit Certification Exam application by 11:59 PM CST, Sep 1	*study! study!*		
Year 5	*study! study!*	***Sit for Certification Exam at AASV Annual Meeting									

** We recommend that you submit your credentials application within 3 years after passing the Entry Exam. You may, however, submit your application any time after passing the Entry Exam.

***You must pass all 3 sections of the Certification Exam within 3 years of credentials application acceptance. You may take each section up to 3 times if necessary.

- Accumulation of 250 points across ten years. Points can be earned through a variety of options including:
  - Volunteer service on ABVP committees or as Regent
  - Attending the annual ABVP Symposium
  - Publication in an approved peer-reviewed journal
  - Lectures to veterinarians, vet students, veterinary technicians, vet tech students
  - Attending CE beyond the baseline 250-hour requirement
  - Mentoring a successful ABVP applicant
  - Achieving credentials through another specialty board
  - Passing the Certification Exam during year 8, 9, or 10
- Review and pass quizzes for 4 of 6 journal articles posted annually.
  - "Journal Club" articles are selected and quizzes are created by SHM Diplomates. This requirement promotes active engagement in current swine medical literature and builds a community of like-minded learners.

Complete details about this process can be found in the Maintenance of Certification Handbook, also available at the ABVP website under Forms and Documents on the Members-Only page. Credentials are maintained throughout the Diplomate's career through this rolling 10-year process. The Council of Regents urges all Diplomates to update their MOC records each time points are earned to best ensure continued success.

Conclusion

So why should you consider certification in Swine Health Management through the American Board of Veterinary Practitioners? It will stretch you! You will be recognized among your peers as an expert in this field. The journey will help you to hone new skills and enhance old ones. You will be asked to demonstrate not only your knowledge, but also your ability to solve critical problems...problems which may make all the difference for you, your producers, their pigs, and even consumers worldwide. Come join us!

References

1. Website: **https://abvp.com**

2. Applicant Handbook: **https://abvp.com** –> Create account –> Member Login –> Forms and Documents -> Applicant

3. Maintenance of Certification (MOC) **Handbook:https://abvp.com** –> Member Login –> Forms and Documents -> Maintenance of Certification (MOC)

https://doi.org/10.54846/am2022/34

Heritability of sow uterine prolapse in a commercial maternal line

J. R. Dunkelberger[1,2], PhD; T. Stevens[1], MSc; E. F. Knol[1], PhD
[1]Topigs Norsvin Research Center B.V., Beuningen, The Netherlands; [2]Topigs Norsvin USA, Burnsville, Minnesota

Introduction

Sow mortality has drastically increased for swine breeding herds in the US in recent years. The top three reasons for sow mortality are reported as: 1.) unknown/sudden death; 2.) feet/leg structure; and 3.) uterine prolapse.[1] Increased incidence of uterine prolapse is of particular concern, given the economic and animal welfare implications of this reproductive issue. Nearly every case of uterine prolapse results in euthasia, which is emotionally devastating for farm employees, disrupts pig flow, and results in substantial economic losses for the producer. Suspected causes of uterine prolapse include, but are not limited to: water intake/quality, farrowing crate size/structure, mycotoxins, and vitamin or mineral deficiencies.[2] These potential factors, and others, were investigated as part of the Pelvic Organ Prolapse Project, conducted by the Iowa Pork Industry Center. Analysis of potential risk factors was conducted using data collected from over 100 commercial sow farms located across the US. The conclusion from this study was that uterine prolapse is a multifactorial issue and, of the factors evaluated, appeared to be most influenced by feeding strategies during late gestation, body condition score, water treatment system usage, and perineal score.[1] Data were not available to investigate the potential role of host genetics on susceptibility to uterine prolapse, thus providing motivation for this study. The objective of this study was to estimate the heritability of uterine prolapse in a purebred, commercial maternal line.

Materials and methods

Data

Records used for this study were collected on 16,434 purebred Topigs Norsvin females from a single commercial production facility (Farm A) and 4,096 purebred animals of the same, purebred genetic line at Farm B. Phenotypes for uterine prolapse were recorded as removal due to uterine prolapse (1) or removal due to another reason (0). The incidence of removal due to uterine prolapse was 7.8% and 3.5% for Farms A and B, respectively.

Heritability estimation

Heritability of uterine prolapse was analyzed as a binary or continuous trait using a full or reduced dataset. The full dataset consisted of all 16,434 records collected from Farm A and the reduced dataset consisted of 986 cases (i.e. prolapsed animals) and 986 matched controls, selected from Farms A and B. Case/control pairs were defined as records collected from females of the same farm, parity at removal, and inseminated within the same month of the same year.

Heritability was estimated using animal models with or without fixed effects fitted in the model, when using the full or reduced dataset, respectively. Fixed effects for analyses conducted using the full dataset included "parity at removal" and "herd-year-season of insemination". Fixed effects were not fitted for analyses conducted using the reduced dataset, since the sampling approach of the case/control design pre-adjusted for these effects. Genetic relationships between individuals were accounted for by fitting an A matrix, constructed based on three generations of pedigree information.

Validation study

A validation exercise was performed using data collected from Farm A as the reference population and Farm B as the validation population. To conduct this analysis, "Reliable" sires of animals at Farm B were identified ($n = 150$), defined as sires with 10 or more daughters within the validation dataset. A breeding value for uterine prolapse was estimated for each Reliable sire using an animal model and genomic relationship matrix (GRM), to account for relatedness between individuals. The GRM was constructed using 25K single nucleotide polymorphism (SNP) genotypes collected on each Reliable sire and individual within the training population. Offspring performance was then calculated for each Reliable sire as the average incidence of uterine prolapse (corrected for parity and herd-year-season) among his progeny and regressed on sire estimated breeding value (BV).

Results and discussion

Results from this study show that uterine prolapse is heritable, whether analyzed as a binary or continuous trait, using all available data, or a case/control approach. Heritability was estimated at 0.13 (0.02) and 0.31 (0.03), when analyzed as a continuous variable using the full vs. reduced dataset, respectively. When analyzed as a binary variable, heritability estimates were similar, whether using all available data (0.22[0.02]) or the reduced dataset (0.24[0.03]). Therefore, dataset structure had a greater impact on heritability estimation for uterine prolapse when analyzed as a continuous vs. binary trait. The standard error of heritability estimation remained consistent across all four scenarios, indicating that a case/control dataset containing only 12% of the data included in the full dataset did not decrease the accuracy of heritability estimation.

The validation exercise was performed by regressing corrected offspring performance on sire BV for prolapse, resulting in a slope parameter of 0.40 and corresponding 95% CI of [0.27 - 0.54]. This result indicates that, for every 1% increase in sire BV for prolapse, corrected offspring performance increased by 0.40%. This shows that a sire's BV for prolapse, estimated using individuals of a purebred, training population, is predictive of the incidence of uterine prolapse among his progeny in an independent, validation population. Further, the 95% confidence interval for the slope parameter [0.27 - 0.54] contains the expected value (0.5) of a parent/offspring relationship.

In conclusion, this is the first study to show that uterine prolapse in sows is heritable and, therefore, can be addressed via genetic selection. Heritability of uterine prolapse was estimated using different models and with a full or reduced (ie, case/control) data set. Results of these analyses show that heritability of uterine prolapse ranged from 13 to 31%, indicating that, at the least, 13% of the total phenotypic variation in the incidence of uterine prolapse is due to genetics and the remaining 87% to non-genetic factors. Therefore, genetic approaches can (and should) be used to address the genetic component of uterine prolapse. However, identifying and mitigating environmental triggers remains the greatest opportunity for reducing the incidence of uterine prolapse within commercial sow herds.

References

1. Ross J. Identification of putative factors contributing to pelvic organ prolapse in sows. Improving Pig Survivability. 2019.

2. Suprakorn C et al. Prolapse Incidence in Swine Breeding Herds Is a Cause for Concern. *Open Journal of Veterinary Medicine*. 2017;85-97.

https://doi.org/10.54846/am2022/35

Skin recovery and care with activated plant phenolic solutions

C. B. Schmidt[1], DVM; Robert Tims[1]; Andrew Bents[2], DVM; Ryan Cain[4], General Manager; Deb Murray[5], DVM
[1]Precision Health Technologies, LLC. Technical Team; [2]Hubbard Feeds;
[3]Old Fashion Pork, LLC; [4]Veterinary Services Manager, New Fashion Pork

Introduction

Precision Health Technologies LLC (PHT) is North America's leading supplier of activated natural plant phenolic solutions for animal health and wellness. PHT phenolic products include four Grazix products, as well as several advanced phenolic formulations that are currently in development and field testing. PHT has effectively used its patented manufacturing process to produce safe and effective formulations, that prevent as well as improve an animal's ability to fight off unwanted cellular and immune challenges to the intestinal track as well as stimulating rapid healing of skin bruising, contusions, and open wounds. One of these new products, *PHT Skin Recovery*, is an advanced medicinal formulation being specifically developed to accelerate the healing of damaged skin. *PHT Skin Recovery* formulations capture the properties of Reactive Plant Immunity (RPI) using activated polyphenols to accelerate the healing of damaged skin – just as plants use the same process to rapidly heal and replace damaged cells. All PHT formulations are all-natural helping to reduce our reliance on chemicals and antibiotics in our livestock production facilities. By extending the rapid healing properties of plants, PHT is helping the industry to find effective therapies for those pursuing "antibiotic free" (ABF) or "no antibiotic ever" (NAE) production strategies.

Plant phenolic skin solutions

Why do plants heal so quickly? It's all about the healing properties of activated polyphenols – the plants natural response to cellular damage and injury. Its common knowledge that plants heal quickly after severe damage from hail, wind, and other traumatic events during their growth cycle. The building blocks of this rapid healing, plant polyphenols, are found in many plants at varying levels. PHT has studied a wide variety of plants and the phenolic response to identify those best able to support healing in animals. Specifically, PHT has discovered that animals with damaged skin as well as active skin lesions will respond quickly to activated epidermal cellular cytoplasm which in turn boosts and activates the peroxidase cellular enzyme pathway to reduce, modify and stimulate faster healing and cellular recovery. In other words, we've found the activated polyphenols that are best suited to accelerating skin healing and repair.

Dr. Deb Murray, DVM is the Managing Veterinarian for New Fashion Pork, based in Jackson, MN. She has been a leader in evaluating our newest products and I'm grateful that she is able to present some encouraging findings.

Protocol

The studies presented today are from two of five field evaluations that were conducted to test the new skin care and recovery formulation developed by PHT. Selected Veterinarians as well as production managers assisted PHT in its evaluation of *PHT Skin Recovery* for use as a topical application to treat skin lesions. Our objective was to promote granulation and size reduction in the lesion.

Evaluation protocol

Our directions for application for the product were simple and straight forward: Spray a light coating of PHT Skin Recovery on identified skin abrasions.

Label directions

- On day one and day two, apply a lite coat of PHT Skin Recovery twice a day to clean the wound and stimulate lesion healing.
- On day three plus, reduce application to once a day until the lesion is healed.

A 1 mL dose was delivered with each spray. The trial used 450 mL bottles of *PHT Skin Recovery*.

Each study location was supplied with the *PHT Skin Recovery* product. Study participants were trained to use the product as outlined above. Each participant was directed to use the solution on damaged areas such as/resulting from:

1. GDU and sow skin lesions
2. Castration
3. Tail docking
4. Other applications of skin lesions

Each participant was directed to conduct the trial with controls in place, eg, a clear "no treatment" and or a "production SOP" group for comparison purposes. Participants were further directed to take pictures of the control or non-treated lesions vs the treated lesions. All observation and pictures were to be provided to PHT.

Results

PHT Skin Recovery was observed to have very positive effects in the production units where it was tested. Observable lesions were treated with a light spray of 1 mL of skin recovery directly to the affected area twice daily. The spray applications worked rapidly to promote healing and reduce irritation in the GDU and with Sow lesions. As shown in photos in Figure 1 and Figure 2, notable positive effects are observable after only 24 hours. It should be noted that shoulder lesions may require additional daily application to stimulate the healing response.

Castration and tail dock lesion response

Evaluation one

PHT Skin Recovery was observed to a have a positive effect on post castration and tail docking wounds – reducing irritation, swelling, and accelerating healing. Castration and tail docking was performed on litters between 4 and 6 days of age. A single dose of *PHT Skin Recovery* was applied to the affected area post procedure and was compared to a control litter that received the unit's Standard Operating Procedure (SOP) for pre- and post-procedure wound care. The SOP included the spraying of iodine and lidocaine prior to castration and an injection of Iron and penicillin following castration. The photos in Figure 3 show post procedure healing with *PHT Skin Recovery* and the Photos in Figure 4 show post procedure healing using the units SOP. The litters treated with PHT Skin Recovery showed significantly better healing, including granular sealing, reduced irritation, less swelling and improved pig comfort overall.

Evaluation two

Evaluation two was designed to compare three castration procedures as described below in Figure 5, Figure 6, and Figure 7.

Additional applications

Application to a torn and irritated vaginal canal after farrowing: (24 hr. response)

The subject sow was given a 1ml spray application of *PHT Skin Recovery* when the torn vaginal canal was observed by staff (Figure 8). After 24 hours the treated area showed marked healing, reduced swelling, and less inflammation and irritation (Figure 9).

Overview and discussion

It is well established that natural based plant extracts have been successfully applied to human as well as animal health applications. PHT developmental work with a variety of plant extracts have shown significant positive effects in healing skin abrasions and lesions in livestock. While product based on plant extracts can make no specific therapeutic claims, various plant extracts and activated plant phenols selected by PHT have exhibited observable medicinal healing to damaged tissues. This early clinical research shows the positive effects of PHT patented processes to advance skin health and wellness applications. Field evaluations consistently showed accelerated healing as applied. The actual results shown above demonstrate the cellular benefits of *PHT Skin Recovery* on the farm and in live production units.

PHT R&D will be working to identify additional applications for *PHT Skin Recovery*. As part of that effort, we are seeking additional collaborators to help us study the effects of Reactive Plant Immunity on the following skin disorders and conditions:

- Ear necrosis
- Tail biting
- Mammary lesions
- "Greasy pig"
- Shoulder wounds/abrasions
- Facial abrasions in Piglets
- Fly and insect irritation
- Sow prolapse (prevention)

Reported benefits

Through field and barn trials, Skin Recovery clearly accelerates healing of skin wounds in livestock. This new product is:

- Safe, effective, and field verified
- All natural
- Easy to use and apply
- Accelerates healthy skin recovery
- Reduces skin inflammation
- No chemicals or dyes

Figure 1A-B: Lesion healing over 24 hours

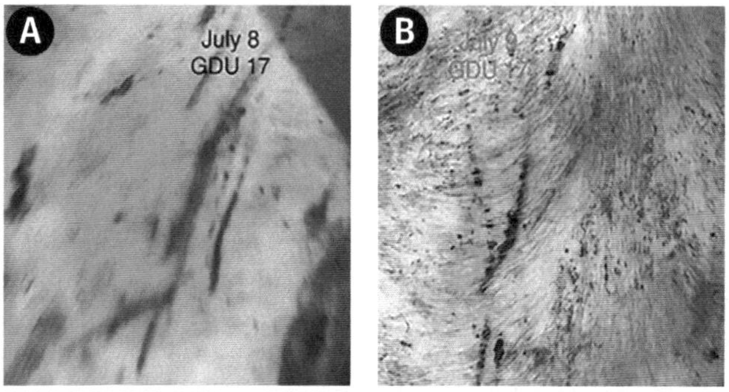

Gilt Development unit (GDU) Lesions

1 mL of spray applied twice in 24 hours

Observed results in 24 hours:

The wounds treated with PHT Skin Recovery showed significant and rapid improvement over 24 hours. There was significant granulation and wound closure. There was also a clear reduction in lesion size and irritated tissue.

Figure 2A-B: Lesion healing over 24 hours

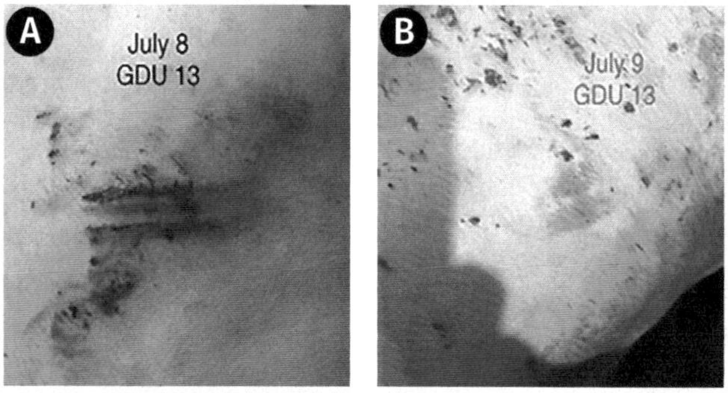

Gilt Development unit (GDU) Lesions

1 mL of spray applied twice in 24 hours

Observed results in 24 hours:

The wounds treated with PHT Skin Recovery showed significant and rapid improvement over 24 hours. There was significant granulation and wound closure. There was also a clear reduction in lesion size and irritated tissue.

Figure 3A-B: Post Procedure healing over 24 hours using PHT Skin Recovery. 24h post-castration and tail dock results are recorded with the photos below.

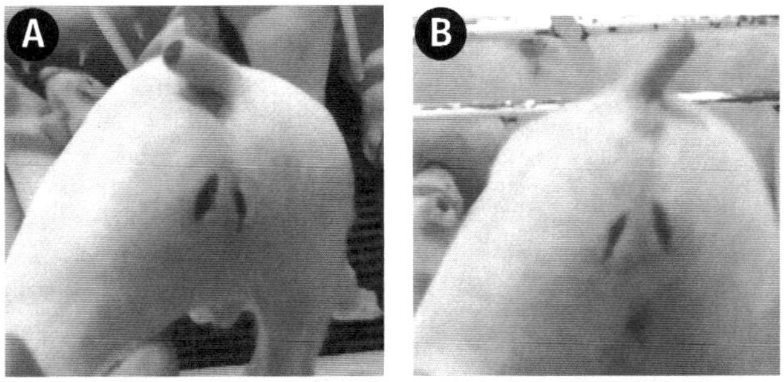

1 mL of PHT Skin Recovery applied after castration and skin docking.

Observed results in 24 hours

Piglets treated with PHT Skin Recovery showed significant granulation and wound closure. Their tails showed rapid healing and granular sealing. The wound areas showed reduced swelling and irritation leading to improved animal comfort

Figure 4A-B: Post procedure healing over 24 hours using Production Unit SOP

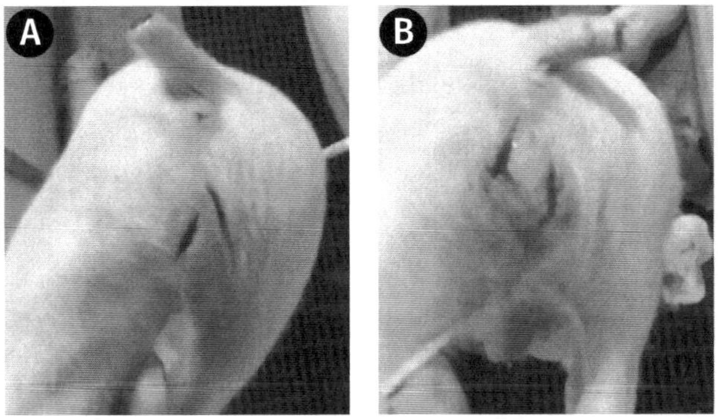

Iodine and Lidocaine applied (spray) prior to procedure

Iron and Antibiotic injected post procedure

Observed results in 24 hours

Piglets treated with the unit's SOP, there was increased swelling and observable irritation. Piglets show that tails remain redder and swollen post-docking. There was more topical discomfort in the pigs when compared to the pigs who received PHT Skin Recovery

Figure 5: Post procedure healing over 48 hours using Lidocaine applied pre-castration

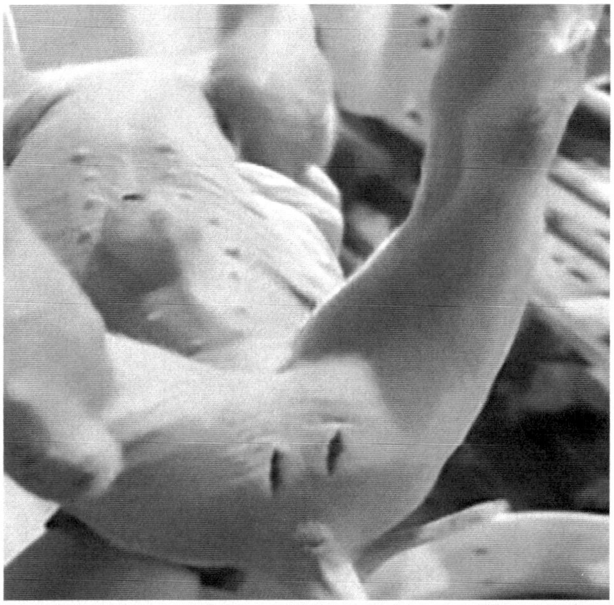

Observed results over 48 hours:

The active lesions treated with Lidocaine only, were still slightly swollen and were healing slower than the PHT Skin Recovery piglets.

Figure 6: Post procedure healing over 48 hours using PHT Skin Recovery pre-castration

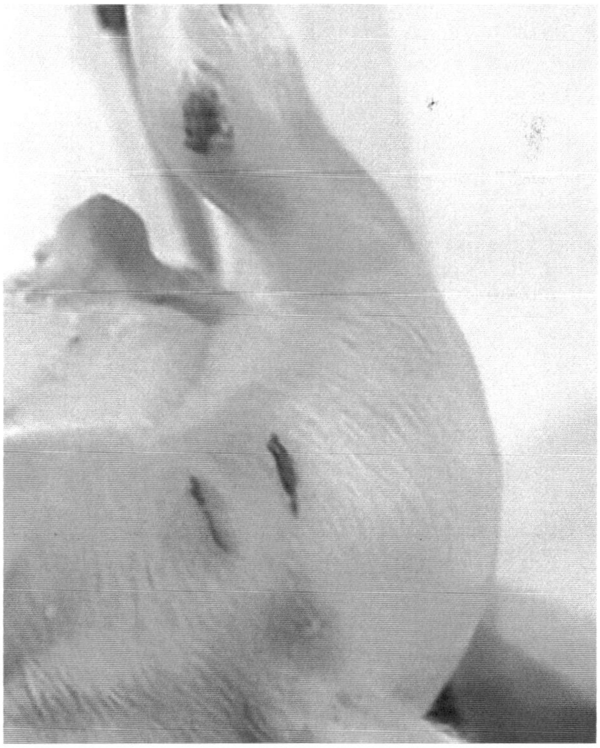

Observed results over 48 hours:

The lesions treated with PHT Skin Recovery prior to procedure showed granulation, wound closure, and improved healing with less swelling.

Figure 7: Post procedure healing over 48 hours using PHT Skin Recovery pre & post castration

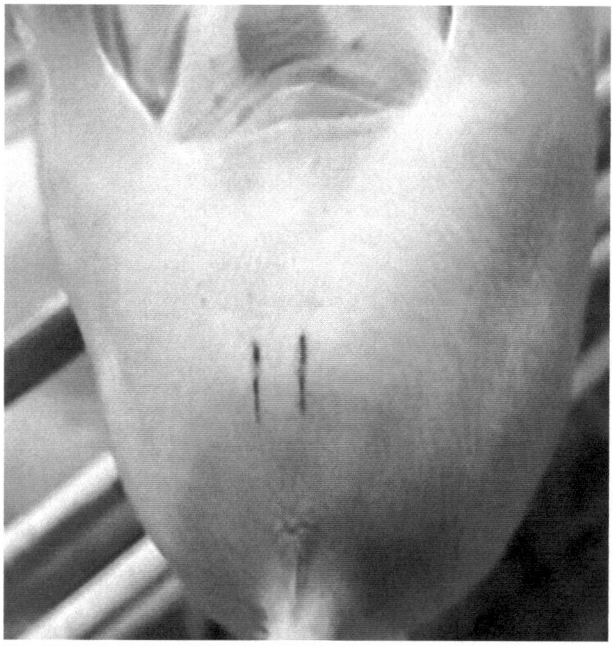

Observed results over 48 hours:

The lesions treated with PHT Skin Recovery prior to, and post-procedure showed significantly improved granulation, wound closure, and improved healing with less swelling.

The animal was given one application of Skin care when observed by staff.

Figure 8: Torn vaginal canal pre-treatment with PHT Skin Recovery

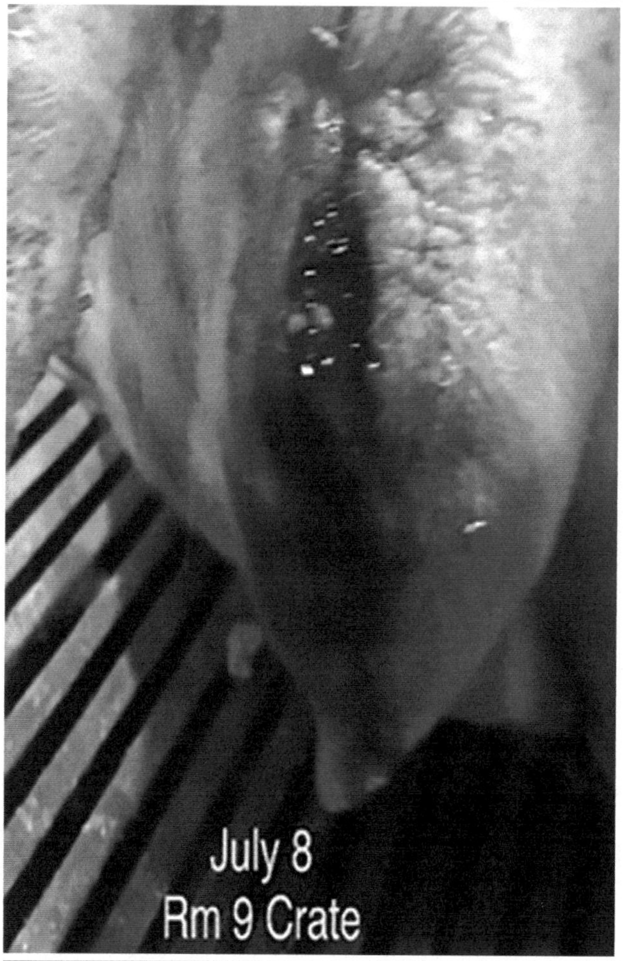

The 24-hour picture shows the response with Skin care.

Figure 9: Torn vaginal canal 24 hours post treatment with PHT Skin Recovery

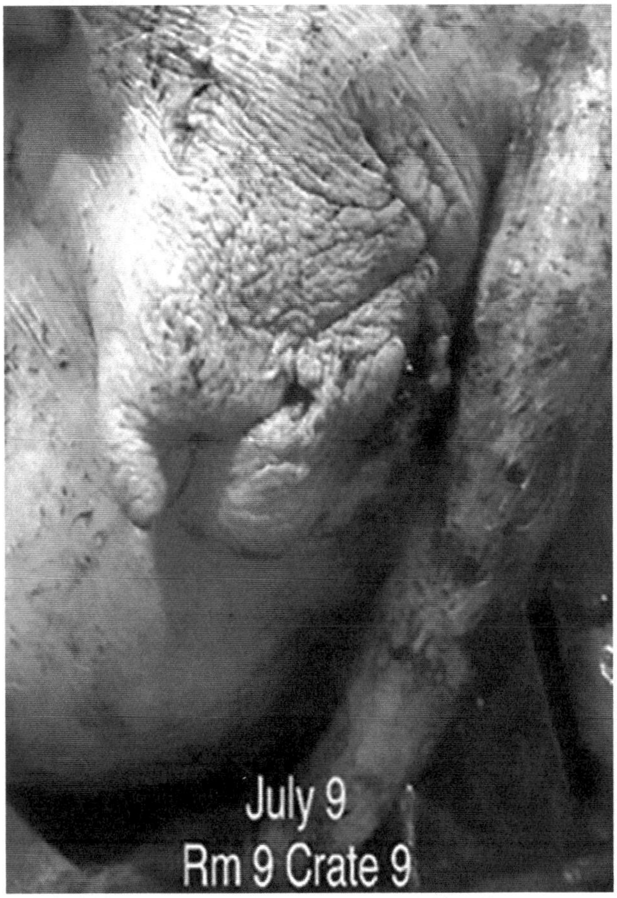

https://doi.org/10.54846/am2022/36

Safety evaluation of transitioning a previously modified-live PRRS vaccinated-positive stable sow herd to Prevacent® PRRS and effects on downstream nursery performance

M. Ackerman[1], DVM; D. Lape[1], DVM; J. Stevens[2]; J. Risser[3], DVM; C. L. Puls[3], PhD
[1]PorkVet Solutions, New Palestine, Indiana; [2]Ag Production Enterprises, Greensburg, Indiana;
[3]Elanco Animal Health, Greenfield, Indiana

Introduction

Porcine reproductive and respiratory syndrome virus (PRRSV) is a significant disease in swine and continues to contribute significant cost to the swine industry each year. PRRSV is a RNA virus that has been shown to mutate and recombine, potentially causing signs of clinical disease, and can reside in tissues of animals for up to 250 days after exposure.[1] Consequently, many swine producers may be hesitant to transition use of modified-live PRRS vaccines on sow farms due to potential shed and spread, recombination between circulating PRRSV strains, or other negative PRRSV effects associated with production. Prevacent® PRRS (Elanco Animal Health) is a USDA licensed, lineage 1, modified-live PRRS vaccine (MLV) that has shown limited shed and spread and relatively short viremia in breeding and growing swine herds.[2,3] As such, Prevacent may be an effective tool for producers to use on their herds with minimal production disruptions. This study evaluated the field safety and potential production disruption when transitioning a previous quarterly MLV vaccinated sow herd to Prevacent PRRS and the subsequent effects on downstream nursery production. Furthermore, varying vaccination programs were evaluated in the nursery pig population to evaluate effects on production parameters and nursery mortality.

Materials and methods

Part I. A 6,000 head, category II-vx, sow herd was selected for this study. The sow herd had previously received quarterly whole-herd vaccinations with a competitor vaccine for approximately two years prior to this study. For the transition of the sow herd, two whole-herd vaccinations (1 mL) with Prevacent PRRS were given to all gilts/sows 4 weeks apart. Suckling piglets received 1 mL Prevacent PRRS 1 to 3 days prior to weaning beginning approximately 1 week after the first whole-herd vaccination. Weekly piglet processing fluids were collected separately for gilt and sow litters. In addition, weekly family oral fluids were collected separately for gilt and sow litters within 3 days prior to weaning. Sow production metrics (eg, conception rate, farrowing rate, incidence of stillborn and mummified piglets) were recorded on a weekly basis. Data were compared for the 6 weeks prior to the first whole-herd Prevacent PRRS vaccination, the 4-week transition period between the first and second whole-herd vaccination, and the 6 weeks after the second whole-herd vaccination. For nursery, oral fluids were collected within 1 week of placement and again 5 to 6 weeks later. Closeout nursery exit weight, average daily gain (ADG), average daily feed intake (ADFI), feed conversion ratio (FCR), and mortality were compared.

Part II. Three weaned-pig vaccination programs were sequentially evaluated in the downstream nursery pig population to evaluate effects on production parameters and mortality: 1) 1 mL Prevacent PRRS administered 1 to 3 days prior to weaning (15 weeks), 2) 1 mL Prevacent PRRS administered at 10 days of age (21 weeks), or 3) 0.5 mL Prevacent PRRS administered at 10 days of age followed by 0.5 mL Prevacent PRRS administered 1 to 3 days prior to weaning (21 weeks). Oral fluid collection continued, through the various weaned-pig vaccination programs, at 5 to 6 weeks in the nursery. Both the average cycle threshold (Ct) value and the frequency to achieve a Ct-value of greater than 37 was evaluated. Closeout nursery exit weight, ADG, average daily feed intake (ADFI), FCR, and mortality were compared.

Results and discussion

Part I. Changes in sow herd and nursery production metrics resulting from transitioning the sow herd to Prevacent PRRS are presented in Table 1. Changes in sow herd and nursery diagnostic measures are presented in Figure 1. Sow herd production parameters remained within expected and historical farm ranges throughout the evaluation. Conception rate, farrowing rate, and the incidence of stillborns and mummified piglets remained within 1 standard deviation of the previous period and remained largely unaffected (Table 1). Piglet processing fluids from sow litters remained negative throughout the evaluation period (Figure 1). Processing fluids from gilt litters were intermittently positive occurring within 2 weeks of the whole-herd vaccination, albeit relatively high cycle threshold (Ct) values suggesting relatively low viremia. Pigs transitioned to Prevacent PRRS one week after the initial sow herd vaccination with Prevacent. Oral fluids collected from these nursery animals were PRRS positive but transitioned to negative after sows received the second whole-herd vaccination with Prevacent PRRS. Consistent lateral introduction of RFLPs 1-3-4 and 1-176-4 (Lineage 1) PRRSV occurred throughout the evaluation period (Figure 1). Nursery pig performance generally improved with Prevacent PRRS compared to the previous MLV vaccine period (Table 1). Nursery exit weight increased 3.1 lb while nursery ADG improved 8.5% and FCR improved 1.3% for Prevacent PRRS compared to the prior vaccine program. However, nursery mortality increased 1.5% after the transition to Prevacent (Table 1). While relatively similar results have been shown with other vaccines, this is the first study to evaluate transitioning a sow herd to Prevacent PRRS MLV vaccine.[4,5]

Part II. The effects of varying nursery pig vaccination programs on nursery production parameters are presented in Table 2. Generally speaking, pigs vaccinated with Prevacent PRRS at 10 days of age had the poorest nursery performance. This was evidenced

Table 1: Part I: Effects of transitioning sow herd to Prevacent PRRS on production metrics

Sow herd production metrics	Previous MLV	Transition	Prevacent PRRS
No. production weeks	6	4	6
Conception rate	92.9 ± 1.1%	92.0 ± 0.8%	91.7 ± 1.1%
Farrowing rate	89.0 ± 1.9%	89.7 ± 1.8%	88.4 ± 2.2%
Stillborn and mummified piglets	6.4 ± 0.6%	6.7 ± 0.7%	6.7 ± 0.5%
Nursery production metrics	**Previous MLV**	**Prevacent PRRS**	**Difference**
Nursery entry weight, lb	13.9	13.7	-0.2 lb
Nursery exit weight, lb	50.7	53.8	+3.1 lb
Overall ADG, lb/d	0.82	0.89	+8.5%
Overall FCR	1.50	1.48	-1.3%
Days on feed	46.1	46.4	+0.3 days
Mortality, %	3.2	4.7	+1.5%

Figure 1: Part I: Changes in sow herd and nursery diagnostic measures after transition to Prevacent PRRS.

Processing fluids	Previous MLV						Transition				Prevacent					
Gilt litters	>37	>37	>37	>37	33.4	>37	>37	35.7	>37	>37	>37	>37	31.6	>37	>37	>37
Sow litters	>37	>37	>37	>37	>37	>37	>37	>37	>37	>37	>37	>37	>37	>37	>37	>37
Family oral fluids	Previous MLV						Transition				Prevacent					
Gilt litters		>37	>37	>37	>37	>37	>37	>37	35.1	>37	>37	>37	>37	>37	>37	>37
Sow litters		>37	>37	>37	>37	>37	>37	32.4	34.9	34.0	>37	36.6	>37	>37	35.6	>37
Nursery oral fluids	Previous MLV							Prevacent								
1 week of placement	>37	>37	>37	36.2	33.3	>37	31.7	34.7	31.8	33.8	34.3	>37	>37	>37	>37	>37
Exit (5-6 weeks later)	28.7	27.7	28.3	30.2	28.6	31.2	28.9	31.1	36.5	32.5	30.0	32.4	>37	29.8	35.6	30.2

by the lowest numerical weight at the end of nursery and lowest overall ADG and ADFI. Of note, however, is that changes in farm management procedures (eg, creep feeding, tulathromycin injection at weaning, etc) approximately 6 to 8 weeks into this 21-week period markedly improved performance. Performance parameters (body weight, ADG, ADFI, FCR) were similar between the 1 to 3 days prior to weaning and the two-dose programs, although nursery mortality was lowest for the group vaccinated at 10 days of age and again 1 to 3 days prior to weaning (Table 2). Others have demonstrated that an early wild-type PRRS introduction leads to a higher mortality rate, when compared to mid or late wild-type introduction; additionally, the production losses can be mitigated by a two-dose PRRS vaccination approach.[6] The average Ct-value was 2 to 4 cycles higher on Prevacent PRRS vaccinated pigs versus the previous MLV, as shown in Figure 2. This represents an approximate 10-fold decrease in viral quantity from the previous MLV vaccinated pigs to the Prevacent PRRS vaccinated pigs. The pigs vaccinated 10 days of age and 1 to 3 days prior to weaning resulted in the highest frequency of nursery groups closed with PRRS PCR Ct-value greater than 37.

Conclusions

Prevacent PRRS was found to be a safe MLV PRRS vaccine in sow herds and can be successfully used in stable herds without disruptions in sow herd or downstream production metrics. Furthermore, vaccinating pigs at 10 days of age and/or 1 to 3 days prior to weaning provided relatively similar performance in the nursery.

References

1. J. Risser, M. Ackerman, R. Evelsizer, S. Wu, B. Kwon, J. M. Hammer. 2021. Porcine reproductive and respiratory syndrome virus genetic variability a management and diagnostic dilemma. Virology Journal. 18:206-217. **https://doi.org/10.1186/s12985-021-01675-0**

2. Elanco. Data on File.

3. C. L. Puls, G. Labarque, A. Cosgrove, J. Herbert, J. M. Hammer, and R. Evelsizer. 2020. Efficacy and safety of a modified live porcine reproductive and respiratory syndrome virus vaccine in breeding pigs (Prevacent™ PRRS). Pg. 30 2020 Allen D. Leman Swine Research Abstracts.

Table 2: Part II: Effects of vaccination program on nursery pig performance

Item	Vaccination program[1] Pre-wean	10 DOA	10 DOA + pre-wean	SEM	*P*-value
No. groups	15	21	21	-	-
Days on feed	46.0	45.1	47.2	0.62	**0.09**
Body weight, lb					
Start of nursery	13.6[c]	14.2[b]	14.5[a]	0.10	**< 0.001**
End of nursery[2]	56.4	55.0	56.8	0.82	0.22
Overall ADG, lb/d[2]	0.94[a]	0.87[b]	0.92[ab]	0.018	**0.05**
Overall ADFI, lb/d[2]	1.37[ab]	1.31[b]	1.43[a]	0.023	**0.01**
Overall FCR[2]	1.46	1.50	1.56	0.030	0.23
Mortality, %[2]	4.3[a]	4.5[a]	1.8[b]	0.42	**0.003**

1 Vaccination program: Pre-wean = pigs vaccinated with 1 mL Prevacent PRRS 1 to 3 days pre-wean; 10 DOA = pigs vaccinated with 1 mL Prevacent PRRS 10 days of age; 10 DOA + pre-wean = pigs vaccinated with 0.5 mL 10 days of age and with 0.5 mL 1 to 3 days pre-wean

2 Data were corrected for differences in days on feed and start of nursery weight

Figure 2: Part II: Changes in PRRS PCR average Ct-value of late nursery oral fluid samples.

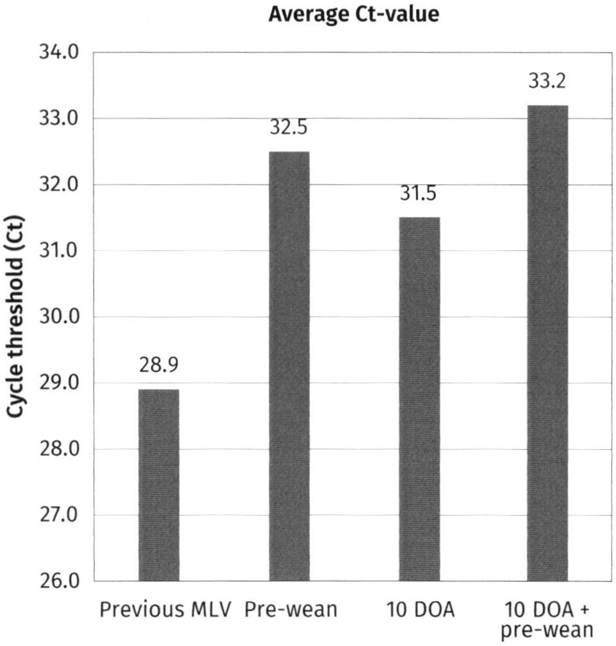

4. G. Trevisan, C. Johnson, N. Benjamin, L. Bradner, D. C. L. Linhares. 2021. Description of changes of key performance indicators and PRRSV shedding over time in a naïve breeding herd following a PRRS MLV exposure. Transboundary and Emerging Diseases. 1-6. **https://doi.org/10.1111/tbed.14327**

5. C. A. A. Moura, C. Johnson, S. R. Baker, D. J. Holtkamp, C. Wang, D. C. L. Linhares. 2019. Assessment of immediate production impact following attenuated PRRS type 2 virus vaccination in swine breeding herds. Porcine Health Management. 5:13-19. **https://doi.org/10.1186/s40813-019-0120-2**

6. C. A. A. Moura, R. Philips, G. S. Silva, A. Ramirez, P. C. Gauger, D. J. Holtkamp, D. C. L. Linhares. 2021. Association of wild-type PRRSV detection patterns with mortality of MLV-vaccinated growing pig groups. Preventative Veterinary Medicine. 189:105270. **https://doi.org/10.1016/j.prevetmed.2021.105270**

https://doi.org/10.54846/am2022/37

Performance and diagnostic evaluation of a lineage 1 (1-8-4) PRRSV vaccine in naïve sows with subsequent field exposure

T. Petznick[1]; A. Witt[2]; C. L. Yoder[3]; R. Evelsizer[3]; J. Risser[3]; D. Linhares[2]; J. M. Hammer[3]
[1]Arkcare, Omaha Nebraska; [2]Iowa State University, Ames, Iowa; [3]Elanco, Greenfield, Indiana

Introduction

Porcine reproductive and respiratory syndrome virus (PRRSV) is an RNA virus that causes reproductive failure in gestating sow and systemic disease in the growing herds. It is transmitted vertically, causing weak, small, weaned piglets with respiratory disease.[1] Lateral spread also occurs from area spread, transportation of infected stock, and biosecurity lapses.[2] The sow performance impact of PRRSV is significant with the literature reporting 1.7 pigs lost / sow /year in infected herds.[3] The estimated total cost of productivity losses due to PRRSV in the US in the national breeding and growing-pig herd was estimated to be $664 million annually.[4] As managing PRRSV is a difficult and serious matter for veterinarians and producers, difficult and complicated decisions need to be made. These decisions encompass financial, animal welfare, best care, and are typically emotional to caretakers. Although mitigation of a clinical outbreak is difficult to predict, some producers will vaccinate their herds to mitigate the high production loss associated with PRRSV. In this case study, five sow farms were vaccinated to mitigate PRRSV losses. One herd (herd A) was vaccinated when the PRRSV infection was detected. The remaining four PRRSV naïve herds (herds B-E) were vaccinated about one month later, but before natural exposure to wild-type virus. Unfortunately, three of the four naive herds were subsequently PRRSV infected. The four PRRSV-infected herds were included in a time to baseline production analysis to measure PRRSV-attributed production losses in vaccinated herds. Diagnostic surveillance was initiated to evaluate the impact of vaccination on piglet processing fluid status as well as monitor for wild-type PRRSV infection. A partial analysis is presented to evaluate time to PRRSVstable status in infected herds and the uninfected vaccinated herd. Estimated piglet loss during each farm's PRRSV infection are also presented to help determine the vaccination mitigation effect in infected sow herds.

Materials and methods

Study herds

A previously PRRSV-free geographical area was having incursions of wild-type PRRSV. A producer with five PRRSV-naïve (AASV category IV[5]) sow farms in this area had a farm, designated farm A, become PRRSV infected with a wild-type strain. This herd was vaccinated with Prevacent PRRS in the face of the PRRSV outbreak. To mitigate PRRSV losses, the other four PRRSV naïve herds were vaccinated with Prevacent PRRS within approximately four weeks of the initial outbreak in farm A.

PRRSV outbreak and vaccination history

Farm A tested PCR positive for PRRSV on calendar week 45 of 2020 after observing clinical signs consistent with PRRSV. Open reading frame (ORF)-5 sequencing revealed that the wild-type strain as a Lineage 1 of RFLP 1-8-4. Subsequently, farm A was closed to replacement gilts and the sows on farms A were vaccinated with a full dose of Prevacent PRRS on week 45 of 2020. Farm B is located one mile from farm A therefore farm B was also vaccinated week 45 of 2020. Farm B tested PCR positive for PRRSV on week 47 of 2020 and was subsequently closed to replacement gilts. On week 49 of 2020, farm A was given a second whole herd vaccination of Prevacent PRRS. Also, week 49 of 2020, sows on farm C and D were given an initial full dose of Prevacent PRRS. On week 52 of 2020, the sows on farm B were given a second full dose whole herd vaccination. On week 01 of 2021 sows on farm C were revaccinated at 70 days of gestation to farrowing with a full dose of Prevacent PRRS. Farm D was given a full dose a whole herd vaccination on week 52 of 2020. At this time Farm D sows had two full doses of Prevacent PRRS. On week 12 of 2021, farm C tested PCR positive for PRRSV. Because of a third trimester vaccination program 80% of farm C's sows had received two full doses of Prevacent PRRS when diagnosed with PRRSV. When diagnosed as PRRSV infected week 19 of 2021, 75% of farm E's sows had two full doses of Prevacent PRRS because of the third trimester vaccination program.

Diagnostics

The Iowa State University Veterinary Diagnostic Laboratory (ISU-VDL) performed all diagnostics according to standard procedures. Processing fluids (PF) were collected on all litters and pooled for screening PCR and Prevacent PRRS preferential PCR. The screening PRRSV PCR and Prevacent PRRS vaccine-like preferential PCR were performed on processing these fluids as directed by the herd veterinarian. Serum from ready to wean pigs (RTW) was collected for PCR testing when "near negative" PF were detected, as determined by the herd veterinarian. In addition, samples from aborted sows and Prevacent PRRS vaccinated replacement gilts at 28 weeks of age were collected to aid in determination of PRRSV status. The herd PRRSV status as determined utilizing the published criteria and supporting evidence for the published PRRSV classification scheme.[5,6] Time to Stability (TTS) will be defined as obtaining low prevalence of virus circulation in suckling pigs during a period of 13 weeks with a combination of RTW sera and PF testing.[6]

Productivity impact

To establish baseline productivity, farrowing parameter data was collected from June 2018 to October 2020. Farrowing data was collected by farm through August of 2021 from infected farms to determine time to baseline production (TTBP) using statistical process control (SPC).[7] More specifically, TTBP was defined as the number of weeks it took for each farm to recover the same number of weaned pigs that the farm had prior to the outbreak (ie, farm-specific baseline). Pig losses were also calculated from this data and reported as total pigs not weaned following the outbreak, until TTBP was reached. To standardize, the estimated piglets lost was divided by TTBP to generate the weekly pigs lost till TTBP statistic.

Results and discussion

A naïve system was faced with a PRRSV outbreak in farm A which required immediate action and subsequent efforts to protect animal welfare in the remaining 4 farms and the financial position of the producer.

Table 1 provides a summary of TTBP, estimated pigs lost, estimated weekly pigs lost until TTBP, and time to stability (TTS). Farm A's TTBP was 27 weeks, estimated total piglet loss being 53,686, and estimated pigs lost per week till TTBP was 1,977; Farm B's TTBP was 23 weeks with a 35,888 total piglet loss, and estimated pigs lost per week till TTBP was 1,329; Farm C's TTBP was 13 weeks with 11,634 total piglet loss, and estimated pigs lost per week till TTBP was 431; Farm E's TTBP was 22 weeks and lost 20,405 total piglet, and estimated pigs lost per week till TTBP was 756. Farm D never became wild-type PRRSV infected as determined by monitoring and the lack of clinical signs in sows and pigs. At the time of this writing, Farm C is at 33 weeks post infection (PI), and farm E at 27 weeks PI. They have not reached TTS to date. Farm A's TTS is presumed to be 32 weeks PI with 10/13 pooled RTW negative tests at week 22. The presumed TTS for farm B is week 42 PI with the 5/12 pooled RTW sera negative tests at week 21 PI. The accuracy of the presumed TTS may be extended due to testing lapses and further testing. Further data is being collected to establish more accurate TTS, but this preliminary TTS is encouraging.

The diagnostic results are detailed in Table 2. Because of pooling the PF collections may represent several weeks of production, therefore the statements about these results may have some variability. Farm A begins testing PF negative for WT PRRSV week 32 post infection (PI)and negative to Prevacent PRRS week 25 PI. Farm B begins testing PF negative for WT PRRSV week 44 PI and negative for Prevacent PRRS week 10 PI. Farm E begins testing PF negative for WT PRRSV week 29 PI and negative for Prevacent week 7 PI. Farm C is still PF WT positive at weeks 25 and 31 PI, but Prevacent PRRS negative as of week 16 PI.

Farrowing production parameters are presented in Table 3. The comparison of Farms D, C and E pre-vaccination as PRRSV Naïve (category IV herds[5,6]) versus post-vaccination with Prevacent PRRS (category II-vx) indicate little to no impact on the farrowing performance parameter.

Producers must balance the high production loss associated with PRRSV infection with immunization with PRRSV vaccination and the potential to mitigate a clinical outbreak. In this case study, the impact of wild-type PRRSV infections was evaluated on four sow farms vaccinated with Prevacent PRRS. Vaccination of naive sows is not without risk. This experience indicates Prevacent PRRS had little to no impact on farrowing performance or weaned pig production as farms D, C, and E transitioned from naïve breeding herds to a category II-vx herds. All three farms demonstrated comparable means between farrowing performance parameters pre- and post-vaccination. Furthermore, sow farms that were PRRSV infected following 2 doses of Prevacent PRRS to had shorter time to baseline production and suffered fewer lost piglets. A recent report indicates MLV vaccination of PRRSV naïve herds can potentially affect farrowing parameters leading to a decrease in pigs weaned per litter for a six-week period and increase repeat services creating lower farrowing rate.[8] However, based on findings represented in Table 3, this field study does not support that statement. The incongruitous findings need further study but are encouraging as to the low transmissibility of Prevacent PRRS. As with all MLV vaccines and wild-type PRRSV infection a persistent lymphoid period post vaccination needs to be considered in decision making about PRRSV status and transmission risk.

Finally, Farm D which was not infected with wild-type PRRSV but Prevacent PRRS vaccinated was evaluated through processing fluids and RTW sera. Table 4 summarizes these findings. Processing fluids became negative week 27 post vaccination and have remained negative through week 42 post vaccination. At 12 weeks post vaccination all RTW samples were negative, and testing was discontinued. Again, these results demonstrate the low transmissibility of Prevacent PRRS but lymphoid persistence needs to be further evaluated.

References

1. Karniychuk UU, Nauwynck HJ. Pathogenesis and prevention of placental and transplacental porcine reproductive and respiratory syndrome virus infection. Vet Res. 2013 Oct 7;44(1):95.

2. Pileri, E. and E. Mateu (2016). "Review on the transmission porcine reproductive and respiratory syndrome virus between pigs and farms and impact on vaccination." Veterinary Research 47(1): 108.

3. Nieuwenhuyse, N., et al. (2012). "Economic analysis of outbreaks of porcine reproductive and respiratory syndrome virus in nine sow herds." Vet Rec 170(9): 225.

4. Holtkamp DJ, Kliebenstein JB, Neumann EJ, et al. Assessment of the economic impact of porcine reproductive and respiratory syndrome virus on United States pork producers. J Swine Health Prod. 2013;21(2):72-84.

5. Holtkamp DJ, Polson DD, Torremorell M, et al. Terminology for classifying swine herds by porcine reproductive and respiratory syndrome virus status. J Swine Health Prod 2011;19(1):44-56.

6. Holtkamp, D., et al., Proposed modifications to porcine reproductive and respiratory syndrome virus herd classification. Journal of Swine Health and Production, 2021. 29(5): p. 261-270.

7. Linhares DC, Cano JP, Torremorell M, Morrison RB. Comparison of time to PRRSv-stability and production losses between two exposure programs to control PRRSv in sow herds. Prev Vet Med. 2014 Sep 1;116(1-2):111-9.

8. Trevisan, G., et al. (2021). "Description of changes of key performance indicators and PRRSV shedding over time in a naïve breeding herd following a PRRS MLV exposure." Transbound Emerg Dis: 1-6.

Table 1: Time to baseline production (TTBP), estimated pigs lost, estimated weekly pigs lost during TTBP, and time to stability (TTS) as of 9/27/2021

Farm*	Wk infected	Wks post vac. Infected#	TTBP	Est pigs lost	Est weekly pigs lost during TTBP	TTS (wks PI)
D	not infected	N/A	N/A	N/A	N/A	N/A
A	2045	0 wks	27 wks	53,386	1,977	32 weeks
B	2047	2 wks	23 wks	35,888	1,329	42 weeks
E	2107	9 wks	22 wks	20,405	756	too early to determine
C	2113	16 wks	13 wks	11,634	431	too early to determine

* Ordered by weeks post vaccination till diagnosed PRRSV infected; PI - post infection

Table 2: Post PRRS exposure diagnostic results by sample type as of September 27, 2021 (week 40 of 2021)

Farm	Processing fluids (PF) (yrwk)#	Weaned pigs (RTW) (yrwk)#
D	N/A*	N/A*
A		wk 2115 10/13 pools neg (wk 22 PI)
	wk 2125 6/6 pools neg (wk 32 PI)	N/C**
	wk 2127 5/6 pools neg (wk 34 PI, Ct 33.3, no prev preferential)	
	wk 2135 5/5 pools neg (wk 42 PI)	wk 2135 12/12 pools neg (wk 42 PI)
	Wk 2136 6/6 pools neg (wk 43 PI)	N/C**
	wk 2137 3/4 pools neg (wk 44 PI)	wk 2137 12/12 pools neg (wk 44 PI)
B	wk 2115 9/11 pools neg (wk 21 PI)	wk 2115 5/12 pools neg (wk 21 PI)
	wk 2136 1/5 pools neg (wk 42 PI)	wk 2136 12/12 pools neg (wk 42 PI)
	wk 2137 9/12 pools neg (wk 43 PI)	N/C**
	wk 2138 7/7 pools neg (wk 44 PI)	N/C**
	wk 2139 5/5 pools neg (wk 45 PI)	wk 2139 11/11 pools neg (wk 45 PI)
	wk 2140 7/7pools neg (wk 46 PI)	wk 2140 12/12 pools neg (wk 46 PI)
	wk 2133 4/4 pools neg (wk 26 PI)	N/C**
E	wk 2134 2/3 pools neg (wk 27 PI)	N/C**
	wk 2135 1/3 pools neg (wk 28 PI)	N/C**
	wk 2136 3/3 pools neg (wk 29 PI)	N/C**
	wk 2137 1/1 pools neg (wk 30 PI)	wk 2137 10/11 pools neg (30 wk PI)
	wk 2138 1/1 pools neg (wk 31 PI)	N/C**
C		wk 2113 0/2 pools neg (wk 0 PI)
	wk 2138 0/2 pools neg (wk 25 PI)	N/C**
	wk 2144 0/4 pools neg (wk 31 PI)	N/C**

#yrwk – year week (ie: 2125 is the 25th week of year 2021); * N/A - Not Applicable; **N/C - No Collection

Table 3: Pre- and post- Prevacent PRRS vaccination on farrowing parameter - MEAN (SD) in naïve (Category IV) sow herds

Farm	D		C		E	
Period Evaluated*	210 days		98 days		56 days	
	pre**	post**	pre**	post**	pre**	post**
Total born	17.05 (3.68)	16.84 (3.50)	16.13 (3.85)	16.81 (3.62)	16.88 (4.00)	16.85 (4.13)
Live born	15.42 (3.32)	15.23 (3.33)	14.88 (3.61)	15.33 (3.42)	15.38 (3.69)	15.23 (3.82)
Stillborn	1.16 (1.55)	1.00(1.57)	0.89 (1.45)	0.93 (1.54)	0.94 (1.55)	1.04 (1.67)
Mummies	0.47 (0.87)	0.6 (1.10)	0.36 (0.75)	0.54 (0.95)	0.56 (1.18)	0.57 (1.12)
Dead (calc)	1.43 (1.61)	1.34 (1.50)	1.2 (1.58)	1.73 (1.99)	1.75 (1.79)	2.21 (2.31)
Total weaned	11.93 (2.94)	12.05 (2.53)	12.19 (2.51)	11.97 (2.66)	11.01 (3.66)	10.46 (3.97)

* Periods are post- Prevacent vaccination and pre- PRRSV infection as applicable.

** Comparisons are similar time periods prevaccination as post vaccination. A one-week period pre- PRRSV infection was dropped to account for possible PRRSV infection prior to diagnosis.

Table 4: PRRSV surveillance in a Prevacent PRRS-vaccinated Naïve herd (II-vx) not exposed to WT PRRS as of September 27, 2021 (week 40 0f 2021)

Farm	Processing fluids (PF) (yrwk)#	Weaned pigs (RTW) (yrwk)#
D		to wk 2109 (wk 11 post vac.) piglets were Prev. positive
		wk 2110 to wk 2111 (wk 12 to wk 13 post vac.) ALL piglets were Prev. negative. Piglet testing was discontinued.
	wk 2121 3/3 pools pos for Prev. (wk 23 post vac)	
	wk 2122 1/5 pools pos for Prev. (wk 24 post vac)	
	wk 2123 1/6 pools pos for Prev. (wk 25 post vac)	
	wk 2124 1/1 pools pos for Prev. (wk 26 post vac)	
	wk 2125 0/2 pools pos for Prev. (wk 27 post vac)	
	wk 2126 0/2 pools pos for Prev. (wk 28 post vac)	
	wk 2127 to 2140 ALL PF were negative to Prev. (wk 28 to wk 42 post vac)	

#yrwk –year week (ie, wk 2109 is the 9th week of year 2021) ; WT - wild type

https://doi.org/10.54846/am2022/38

Review of detergent and disinfectant chemical exposure risks for swine veterinarians and farm production personnel – impact of chemical choice

Jose A. Ramirez, MS, PhD
Chief Science Advisor, Clipper Distributing Co.

Introduction

Swine veterinarians and their staff are exposed to a variety of occupational health and safety risks. These risks include physical injuries, allergic or irritant reactions, infections and chemical exposure.[1] Many of the chemical exposure incidents relate to detergents and disinfectants. Detergent and disinfectant chemicals in use today can be carcinogenic, teratogenic, corrosive, flammable, irritants to the skin, eyes or respiratory tract, or allergenic. In addition, compliance with use recommendations of personal protective equipment (PPE) in animal health settings has been reported to be lower than 40%.[2]

In this paper, we review the more common chemical exposure risk situations associated with the daily use of detergents and disinfectants commonly in use today in US swine farms. The risk-impact in selecting specific detergent or disinfectant technologies and how to mitigate these risks is explored. Options with favorable occupational safety profiles that can minimize safety risks and exposure are discussed.

Cleaning and disinfection chemicals in swine production

Swine production operations require a strong commitment to environmental cleaning and disinfection in order to reduce losses due to infectious diseases and maximize fed conversion ratios. Due to the nature of the operation, different types of chemical products are used in the farm or transportation station depending on the cleaning task objective. For gross soils of an organic nature alkaline detergents are typically employed. Alkali detergents typically combine high pH with one or more detergents and chelating agents. The pH value is typically above pH 10, which allows for the saponification of fats (ie, the conversion of fatty acids to soap), the neutralization of acidity from certain soils, and ensuring a natural negative charge on both substrate and soil particles (hence creating an electrical repulsion contributing to soil release).

For inorganic deposits and scale, especially as it concerns the dulling of steel surfaces, acid washes are the norm. Acid conditions (typically pH below 2) solubilize hardness deposits, and furthermore, many chelating compounds (compounds that bind to dissolved metal ions) used to remove hardness or rust deposits are acid in nature. Acid cleaning products may also include detergents and other ancillary ingredients for the purposes of reducing solution surface tension, foaming and emulsification of soils.

A wide variety of disinfectants are used in the swine sector, and many of them are rotated in cycles in order to avoid "resistance" build-up or prevent selection of pervasive microorganisms. Disinfectants based on quaternary ammonium compounds, glutaraldehyde, phenolic compounds, hypochlorite and more recently hydrogen peroxide are amongst the most common.

Perhaps the most common disinfectant compounds in use today in the swine sector worldwide are based on a mixture of quaternary ammonium compounds and glutaraldehyde. Quaternary ammonium compounds (QAC) are a workhorse chemical in many applications requiring sanitized or disinfected surfaces including food processing, health care and veterinary care. QACs are cationic surfactant molecules, first synthesized in the early 1900s. They are recognized to have good biocidal activity against gram positive bacteria and enveloped viruses, with evidence of lower activity against non-enveloped viruses and gram negative bacteria.[3] For this reason, they are typically formulated with other antimicrobial compounds. Glutaraldehyde is a powerful biocidal agent developed in 1963 as a substitute to the highly toxic formaldehyde in medical sterilization.[4] Glutaraldehyde exhibits broad spectrum biocidal activity, particularly in alkaline conditions. However, Glutaraldehyde is unstable at higher pH, undergoing polymerization with the resulting loss of biocidal activity. In most cases, commercial solutions are provided as a slightly acidic solution, or as two-part solutions that are "activated" by addition of alkali at point of use. Formulated products based on QACs and Glutaraldehyde are quite common in swine farms and for transport trailer disinfection. Trade names for these products are Synergize™, Virocid™ and Ag Forte™ (this last one, QAC-only).

Chlorine-based disinfectants are also common, with different powdered chlorine precursors like dichloroisocyanurate, powdered bleach powders like calcium hypochlorite (eg, Traffic C.O.P.™), liquid bleach like sodium hypochlorite (eg, Clorox™ solution), or in-situ generated hypochlorites based on the reaction of peroxymonosulfate and sodium chloride (e.g. Virkon S™). Hypochlorites have a broad spectrum of activity and because it is widely available, their use in swine operations is quite common.

Another class of disinfectants in use today in certain applications in the farm are mixtures of phenolic compounds. This class encompasses a wide variety of chemical compounds, but the ones most commonly used in swine production contain aryl phenols such as ortho-phenyl phenol (OPP) and chlorinated benzyl phenols. Trade name product examples include BioPhene™ and Tek-Trol®.

In some instances, strong bases are used for the purpose of disinfection, typically Sodium Hydroxide (Caustic Soda or Lye), Sodium Carbonate (Soda Ash) or Calcium Oxide (Quicklime). Aqueous solutions are typically reconstituted from powder or flakes, or the flaked material is applied directly to surfaces, such as floors.

Key regulatory aspects related to risk and safety of chemicals

In the United States, the Occupational Health and Safety Administration (OSHA) sets and enforces the standards necessary to ensure workplace safety and health. Whether a particular operation needs to comply with OSHA standards depends on the number of employees, but as a practical matter, it is generally recognized that "...a strong commitment to the health and safety of your workforce can pay dividends to your operation..." because "... people who work in operations where health and safety is a priority will often be more productive and loyal to the operation." [5] Additionally, for many of the top pork producing states, there may be state OSHA regulations and enforcement to account for. Furthermore, insurance companies typically would like to see an operation that promotes a health and safety culture and is safe for employees.[5]

The key OSHA directive of relevance here is the Hazard Communication Standard (HCS), which requires that information about the identities and hazards of the chemicals must be available and understandable to workers. Chemical manufacturers bear the responsibility of evaluating the hazards and prepare labels and safety data sheets (SDS) that convey the necessary hazard information to users. Employers using such chemicals must make the labels and safety data sheets available to their workers, as well as train them on how to handle the chemicals appropriately and how to protect themselves while doing so. In March 26, 2012, OSHA revised the HCS to amongst other things adopt the 3rd edition of the United Nations' Globally Harmonized System of Classification and Labeling of Chemicals (GHS). This implies a set of harmonized criteria for classifying chemicals according to their risk and hazards, including a specific, 16-section format for presenting that information on the SDS.

Analysis of chemical exposure risks

The easiest tool to understand the hazards and risks associated with use and exposure to a particular chemical is the SDS; which as mentioned above, has to be made available by the supplier by law. Supplementary information may also be found on the product label. All 16 sections of the SDS contain very useful information to deduce hazard and risk in using a product, but sections 2 (Hazards Identification) and section 4 (First aid measures, specifically, section 4.2, symptoms and effects) deserve particular attention. A summary of language extracted from Sections 2 and Section 4.2 of the SDS from some of the more common chemistries used in swine farms is included in Figure 1. It is seen how all products contain hazard statements pertaining to corrosion or burns in skin and eye. This hazard is not only relevant to operators breaching compliance with PPE requirements, but also to animals that could come in contact with a wet, freshly disinfected surface. Most disinfectants also have high oral toxicity – while this may not pose a serious threat to people, it is a relevant risk for animals exposed to wet, freshly disinfected surfaces. But asides from the well-known and accepted risks above, there are other risks associated with these disinfectants that have not resulted in regulatory classification due to the ongoing back-and-forth between toxicology researchers and the chemical manufacturing industry. Some of these are explored below.

Other factors not considered on the Safety Data Sheet

A study on 194 pig farmers in the Netherlands[6] found clear evidence of increased atopic sensitization (as measured by serum Immunoglobin, IgE) to common allergens among farmers using disinfectants containing QACs. In addition, atopic sensitization in combination with QAC use or with exposure to high endotoxin levels increase the risk of developing symptoms consistent with asthma. Another study[7] demonstrated a clear correlation between bronchial responsiveness in pig farmers to exposure to wood shavings, dry/pelleted feed and air exchanger design, but interestingly also to the use of QACs. Key factors influencing the effect were total duration of exposure to the QAC disinfectant and the pressure at which disinfectant is applied.[7,8] Numerous other studies document the relationship between QACs and occupational asthma in janitorial occupations.[9-12]

More recently, studies with mice that have been exposed to a QAC disinfectant have shown decreased fertility due to targeting both male and female reproductive systems.[13,14] Hrubec et al. found ambient QAC exposure was teratogenic to rodents and called for further evaluation of their safety to humans.[15] QAC manufacturers were quick to publish their own health and safety study, basically reviewing the existing health and safety literature on which current safety guidance is based[16], and mostly refuting these findings on the grounds that the studies are purely academic, do not conform to GLP (the industry-defined code of good laboratory practices), or have not been reviewed by regulatory agencies.

Glutaraldehyde as also been extensively associated in the literature with different chronic effects, including irritant contact dermatitis, chronic dermatitis, systemic sensitization and occupational asthma.[17] It is noteworthy that most of the occupational data with Glutaraldehyde has been with health care workers. In healthcare, use patterns consist of cold soaking or filling a disinfector reservoir, and not the higher risk exposure scenario of open spraying that is practiced in swine farms. The literature on the adverse health effects of OPP (an active ingredient in the most common phenolic disinfectants used in swine) is so extensive, that it is has been classified as a type 2B carcinogen by USEPA and the International Agency for Research on Cancer (IARC).[18,19]

A safer alternative

In the past three years, Intervention™, with a novel hydrogen-peroxide disinfectant technology originally developed as a safer alternative for human and small animal hospitals has been made available for disinfection in swine production and other farm applications. It has been previously discussed in the context of practical application in barns and transportation,[20] and more recently after recommendation by USDA APHIS for use vs. AFSV.[21]

A review of the SDS for the concentrated solution reveals that its worst health and safety hazard is "...mild skin irritation and eye irritation." There are no hazard pictograms, given these are only required at higher hazard levels, ie when irritation is more severe or when corrosive to skin and/or eyes. Furthermore, the in-use dilution is not classified as a chemical hazard, meaning that personal protective equipment is not required. In practical terms, when the system is used with the concentrate container connected to an automated chemical injection and foaming system, any non-compliance with existing farm protocol for personal protective equipment will most likely not have any serious health

Figure 1: Summary of SDS hazards for common disinfectant chemistries.

The meaning of pictograms is as follows:

 Substances with chronic hazards such as carcinogens, respiratory, reproductive or organ toxicity

 Materials causing skin corrosion/burns or eye damage on contact

 Flammable materials

 Toxic to aquatic life

Chemical	Hazard/Classification	Pictogram	Health effects on exposure
Accelerated Hydrogen Peroxide (eg Intervention™ concentrate)	May be harmful if swallowed. Causes mild skin irritation. Causes eye irritation.	No pictograms required	Skin : Mild skin irritant Eye : Causes moderate eye irritation.
QAC (eg AgForte™, Quatcide™)	Harmful if swallowed Harmful in contact with skin Causes severe skin burns and eye damage Very toxic to aquatic life		Causes eye damage and moderate skin irritation. May be fatal if inhaled. Harmful if swallowed.
Peroxymonosulfate + sodium chloride (eg Virkon S™)	Causes skin irritation Causes serious eye damage Harmful to aquatic life with long lasting effects		Eye: causes irritation with symptoms of reddening, tearing, stinging, swelling. Skin: causes irritation with symptoms of reddening, itching, and swelling.
Alkali substances – Na/Ca hydroxide, Na bicarbonate, Calcium Oxide (Lye, Soda Ash, Quicklime)	Causes skin irritation. Causes serious eye damage. May cause respiratory irritation	 	Skin: Potentially hazardous. Causes severe irritation of mucous membranes and wet skin. Contact dermatitis. Eyes: Extremely hazardous in eye contact (corrosive/irritant). Possible lesions and blindness if untreated for prolonged period. Inhalation: Nose, oral cavity and throat irritation, coughing and sneezing, and inflammation of breathing passages, ulceration and perforation of nasal septum, bronchitis, possible pneumonia. Ingestion: Intense burning and edema of digestive tract, abundant salivation, difficulties in swallowing and breathing, vomiting blood, drop in blood pressure and possible perforation of esophagus or stomach irritation and pain, vomiting, diarrhea, drop in blood pressure
Phenolic mixtures (eg Tek -trol™, BioPhene™)	Causes severe skin burns and eye damage May cause damage to kidneys and liver through prolonged or repeated exposure if swallowed. May cause drowsiness or dizziness Flammable liquid and vapor.	  	Eyes: Causes severe irritation experienced as discomfort or pain, excess blinking and tear production, with redness and swelling of the conjunctiva. Skin: Brief contact may cause slight irritation. Prolonged contact may cause more severe irritation with pain, local redness and swelling and possible tissue destruction. INHALATION: High vapor concentrations may be irritating to respiratory tract. INGESTION: Harmful or fatal if swallowed. Corrosive. Can cause severe burns and complete tissue perforation of mucous membranes, mouth, throat and stomach.
QAC/Glutaraldehyde (eg Synergize™, Virocid™)	Flammable liquid and vapour. Harmful if swallowed, in contact with skin or if inhaled. Causes severe skin burns and eye damage. May cause an allergic skin reaction. May cause allergy or asthma symptoms or breathing difficulties if inhaled. Very toxic to aquatic life.	   	Inhalation: Cough. May cause allergy or asthma symptoms or breathing difficulties if nhaled. Skin contact: erythema. Causes burns. Eye contact: Conjunctival redness. Impairment of vision. Causes tears. Ingestion: Causes burns. Gastrointestinal complaints. Cough. Causes cramps.

and safety consequences to the user. Likewise, risk of injury to animals is eliminated given animals can safely be exposed to wet, freshly disinfected surfaces and premises.

Conclusion

Swine industry practitioners and operators have had limited choice concerning safe, lower health risk chemicals for disinfection. Heightened awareness of occupational safety issues in farms and increased interest from consumers on chemical exposure of animals during the production process will pressure producers and veterinarians to employ safer, less hazardous disinfectant chemistries. The availability of one such option in Intervention™ provides swine producers and veterinarians with a safer alternative that was not at their disposal in the past.

References

1. Hafer, L., Langley, R.L., Morrow, W. M., and J.J. Tulis (1996) Occupational hazards reported by swine veterinarians in the United States, Swine Health & Production, May/June 1996, 128-141.

2. Hill, D. J., Langley, R. L. and W. M. Morrow (1998) Occupational Injuries and Illnesses Reported by Zoo Veterinarians in the United States, J. of Zoo & Wildlife Med., 29(4), 371-385.

3. Merianos, J.J. and G. McDonnell (2020) Surface-active agents, in Block's Disinfection, Sterilization, and Preservation, G. McDonnell (Ed.), 6th Ed., Wolters Kluwer Health.

4. Kelly, S. A., Gorman, S. P. and F. Gilmore (2020) Aldehydes, in Block's Disinfection, Sterilization, and Preservation, G. McDonnell (Ed.), 6th Ed., Wolters Kluwer Health.

5. Schutske, J. and M. Schermann (2011) What you need to know about OSHA, Fact Sheet PIG 16-01-11, Pork Information Gateway.

6. Preller, L., Doekes, G., Heederik, D., Vermeulen, R., Vogelzang, P.F.J. and J.S.M. Boleij (1996) Disinfectant use as a risk factor for atopic sensitization and symptons consistent with asthma: an epidemiological study., Eur. Respir. J., 9, 1407-1413.

7. Vogelzang, P.F.J., van der Gulden, J.W.J., Preller, L., Tielen, M.J.M., van Schayck, C.P. and H. Folgering (1997) Bronchial hyperresponsiveness and exposure in pig farmers, Int. Arch. Occup. Environ. Health, 70, 327-333.

8. Preller, L., Heederik, D., Boleij, JSM, Vogelzang, PFJ and MJM Tielen (1995) Lung function and chronic respiratory symptoms of pig farmers: focus on exposure to endotoxins and ammonia and use of disinfectants., Occup. Environ. Med., 52, 654-660.

9. Burge, P.S. and M.N. Richardson (1994) Occupational asthma due to indirect exposure to lauryl dimethyl benzyl ammonium chloride used in a floor cleaner, Thorax, 49, 842-843.

10. Bernstein, J.A., Stauder, T., Bersntein, D.I. and I.L. Bernstein (1994) A combined respiratory and cutaneous hypersensitivity syndrome induced by work exposure to quaternary amines., J. Allergy Clin. Immunol., 2, 257-259.

11. Purojit, A.. Kopferschmidt-Kubler, M.C., Moreau, C., Popin, E., Blaumeiser, M. and G. Pauli (2000) Quaternary ammonium and occupational asthma, Int. Arch. Occup. Environ. Health, 73, 423-427.

12. Rosenman, K.D., Reilly, M.J., Schill, D.P., Valiante, D., Flattery, J., Harrison, R., Reinisch, F., Pechter, E., Davis, L., Tumpowsky, C.M., Filios, M. (2003) Cleaning products and work-related asthma, J. Occup. Environ. Med., 45(5), 556-563.

13. Melin, V. E., Potineni, H., Hunt, P., Griswold, J., Siems, B., Werre, S.R. and T.C. Hrubec (2014) Exposure to common quaternary ammonium disinfectants decreases fertility in mice, Reprod. Toxcol., 50, 163-170.

14. Melin, V. E., Melin, T. E., Dessify, B. J., Nguyen, C. T., Shea, C.S. and T. C. Hrubec (2016) Quaternary ammonium disinfectants cause subfertility in mice targeting both male and female reproductive processes, Reprod. Toxicol., 59, 159-166.

15. Hrubec, T. C., Melin, V.E., Shea, S. C., Ferguson, E.E., GArofola, C., Repine, C. M., Chapman, T. W., Patel, H.R., RAzvi, R. M., Sugrue, J. E., Potineni, H., Magnin-Bissel, G. and P. A. Hunt (2017) Ambient and dosed exposure to quaternary ammonkium disinfectants causes neural tube defects in rodents, Birth Defects Res., 109(14), 1166-1178.

16. DeLeo, P.C., Huynh, C., Pattanayek, M., Schmid, K.C. and N. Pecachek (2020) Assessment of ecological hazards and environmental fate of disinfectant quaternary ammonium compounds, Ecotoxicol. Environ. Saf., Dec 15; 206:111116. **doi:10.1016/j.ecoenv.2020.111116**

17. Smith, D.R. and RS Wang (2006) Glutaraldehyde exposure and its occupational impact in the health care environment, Environ. Health Prev. Med., 11, 3-10.

18. International Agency for Research on Cancer (IARC, 1999). IARC Monographs on the Evaluation of Carcinogenic Risks to Humans. Volume 73 as reported on the IARC website at **http://193.51.164.11/past&future/OCT98.html**. Some Chemicals that Cause Tumours of the Kidney or Urinary Bladder in Rodents, and Some Other Substances. IARC, Lyon, France.

19. US Environmental Protection Agency (U.S. EPA, 1994). Memorandum: Carcinogenicity Peer Review of Orthophenylphenol (OPP) and Sodium Orthophenylphenol (SOPP). Office of Prevention, Pesticides and Toxic Substances. August 24, 1994.

20. Ramirez, J.A. (2020) Theory and practice of cleaning and disinfection in swine biosecurity, Proceedings of the 51st Annual Meeting of the AASV, Atlanta, March 7th-10th, 2020, 180-183.

21. Ramirez, J. A. (2021) A new weapon for decontamination of surfaces and premises from African swine fever virus: accelerated hydrogen peroxide., Proceedings of the 52nd Annual Meeting of the AASV, Feb. 27-March 2, 2021, 179-183.

https://doi.org/10.54846/am2022/39

PK/PD relationships of Tylovet® (tylosin) oral administration for the treatment of *Clostridium perfringens* in pigs

Ulrich Klein[1]; Wouter Depondt[1]; Thomas Fangman[2]; David Nolan[?]
[1]Huvepharma® NV, Belgium; [2]Huvepharma® Inc. Peachtree City, Georgia

Introduction

Clostridium perfringens (CP) is a widespread pathogenic bacterium associated with fatal necro-haemorrhagic enteritis seen in 1-5 day-old piglets up to 3-4 weeks old. In fattening pigs *C perfringens* causes sudden death and painful gut inflammation.[1,2]

CP infections can occur in peracute-to-acute and subacute-to-chronic forms. The disease can spread rapidly in a herd and lead to significant economic losses in pig farms.

In cases of disease outbreaks fast therapy is needed based on impending massive damage of the intestinal mucosa within a few hours after infection onset.[3]

The purpose of this paper was to put Tylovet® pharmacokinetic data into relationship to published data on tylosin pharmacodynamics. Understanding of the pharmacokinetics (PK) and pharmacodynamics (PD) of Tylovet® is key to predicting its clinical efficacy according to the guidelines of antibiotic prudent use policy. There is no Tylosin PK/PD data available to prove its clinical efficacy in cases of Necrotic Enteritis infection treatment in pigs.

Materials and methods

Pharmacokinetics: Two pharmacokinetic (PK) studies were conducted based on oral administration of Tylovet® via feed (PK study No.1) and via water (PK study No.2).

PK study No. 1: Pharmacokinetic behavior of tylosin phosphate (Pharmasin®/Tylovet® Premix) was investigated after feed medication in eleven mixed breed pigs (20-25kg bw) at a dosage of 10mg/kg bw. Medicated feed was offered ad libitum for 5 consecutive days.

PK study No. 2: Pharmacokinetics of tylosin tartrate (Pharmasin®/Tylovet® WSG) in the gut was investigated after water medication in eleven mixed breed pigs (20-25kg bw) at a dosage of 10mg/kg bw. Fresh medicated drinking water was offered ad libitum daily prepared for 5 consecutive days.

On day 5 in both studies treated pigs were euthanized (2 animals at 5 different time points) and blood / organ and tissue samples from respiratory and gastrointestinal tract were collected. PK parameters (AUC_{0-12} µg.h/mL (g); T_{max} (h); C_{max} (g) were determined based on the tylosin concentrations measured by validated HPLC method. The AUC per hour (AUC µg.h/g/hour) was calculated in both studies to determine the tylosin colon and ileum concentrations at registered enteric infection treatment dose (5mg/kg =100ppm) and at 10mg/kg (200ppm) dose.

Pharmacodynamics: Tylosin pharmacodynamic data (MIC`s) were generated based on antimicrobial susceptibility testing of *Clostridium perfringens* toxinotype A strains originating from Italy and Brazil. Tylosin MIC`s were determined by agar dilution method as described in CLSI documents.[5] 51 CP Type A strains were used in the Italian study and tylosin beside other antibiotics was tested in a range of compound dilutions (0.06 - 512 µg/mL). In the Brazil MIC study 50 CP Type A strains from several swine producing regions of Brazil were used. Tylosin and a range of other antimicrobials were tested at different dilutions (0.25 - >512 µg/mL). The MIC_{50}, MIC_{90} and MIC ranges were determined. CLSI clinical breakpoints were not available for MIC result interpretation.

Results and discussion

Pharmacokinetics

Results generated in the Tylovet® infeed PK study are summarized in Tables 1[6] and 2. C_{max} tylosin concentrations in the colon (89.4 µg/g) and ileum (31.4 µg/g) were much higher versus the blood, bronchial mucosa & secretion and lung concentration. This is based on the relatively poor absorption of tylosin (22.5% oral bioavailability) in the intestine.[9]

The colon contents concentration calculated based on AUC per 1 hour indicate a tylosin concentration at treatment dose of 38.2µg/g and at double treatment dose of 76.3µg/g. The ileum contents concentrations were slightly lower at tylosin concentration of 14.2µg/g (treatment dose) and 28.4µg/g (2x treatment dose). The average ileum/colon concentration ratio was 37%.

The results of the Tylovet® water medication PK study are summarized in Tables 3[7] and 4. Tylosin C_{max} concentrations in the colon (76.7µg/g) and ileum (27.2) were much higher versus the blood, bronchial mucosa & secretion and lung C_{max} concentrations. The average ileum/colon concentration ratio was 35%.

The colon contents concentration calculated based on AUC per hour indicate a tylosin concentration at lower treatment dose of 34.0µg/mL and at higher treatment dose of 68.1µg/mL. The ileum contents concentrations were slightly lower at tylosin concentration of 12.0µg/g (treatment dose) and 24.1µg/g (double treatment dose).

Minor differences were determined for the tylosin colon/ileum concentrations achieved based on in feed and water medication. The tylosin concentrations at enteric treatment dosage in the colon at feed medication (38.2/14.2µg/g) were slightly higher vs. the concentration after water medication (34.0/12.0µg/g)

Tylovet® registered dose for enteric disease treatment via feed is 5mg/kg bw and 5-10mg/kg bw for water medication.

Table 1: Tylosin oral pharmacokinetics – feed medication PK study[6]

	Plasma (µg/mL)	Colon contents (µg/g)	Ileum contents (µg/g)	Bronchial mucosa & secretion (µg/mL)	Lung (µg/g)
AUC 0-12h	5.34	915.90	340.80	10.87	9.69
Cmax (g)	0.53	89.4	31.4	1.09	0.95
Tmax (h)	8	4	12	4	4

Table 2: Pharmacokinetics of tylosin in the colon and ileum calculated from mean concentration from each time of kill following ad libitum in feed application

Parameter unit	Colon contents (µg.h/g)	Colon contents (average conc. µg.h/g/hour)	Colon contents (average conc. µg.h/g/hour): tylosin dosage 5mg/kg bw=100ppm	Colon contents (average conc. µg.h/g/hour): tylosin dosage 10mg/kg bw=200ppm
AUC 0-12h µg.h/g	915.90	E 76.3	E 38.2	E 76.3
Parameter unit	Ileum contents (µg.h/g)	Ileum contents (average conc. µg.h/g/hour)	Ileum contents (average conc. µg.h/g/hour): tylosin dosage 5mg/kg bw=100ppm	Ileum contents (average conc. µg.h/g/hour): tylosin dosage 10mg/kg bw=200ppm
AUC 0-12h µg.h/g	340.80	E 28.4	E 14.2	E 28.4

Table 3: Tylosin oral pharmacokinetics – water medication PK study[7]

	Plasma (µg/ml)	Colon contents (µg/g)	Ileum contents (µg/g)	Bronchial mucosa & secretion (µg/mL)	Lung (µg/g)
AUC 0-12h	5.19	816.85	288.55	11.28	9.75
Cmax (g)	0.65	76.7	27.2	1.05	0.93
Tmax (h)	2	4	4	2	2

Table 4: Pharmacokinetics of tylosin in the colon and ileum calculated from mean concentration from each time of kill following ad libitum in water application

Parameter unit	Colon contents (µg.h/g)	Colon contents (average conc. µg.h/g/hour)	Colon contents (average conc. µg.h/g/hour): tylosin dosage 5mg/kg bw=100ppm	Colon contents (average conc. µg.h/g/hour): tylosin dosage 10mg/kg bw=200ppm
AUC 0-12h µg.h/g	816.85	E 68.1	E 34.0	E 68.1
Parameter unit	Ileum contents (µg.h/g)	Ileum contents (average conc. µg.h/g/hour)	Ileum contents (average conc. µg.h/g/hour): tylosin dosage 5mg/kg bw=100ppm	Ileum contents (average conc. µg.h/g/hour): tylosin dosage 10mg/kg bw=200ppm
AUC 0-12h µg.h/g	288.55	E 24.1	E 12.0	E 24.1

Pharmacodynamics

In the Italian study tylosin MIC_{90} values of 64 and MIC range of 0.12-128 were determined for the CP type A strains tested (Table 5).[4,8] In the Brazil study the tylosin MIC_{90} value was 16 and the MIC range 0.25-32 (Table 5).[4,8] The tylosin MIC results cannot be interpreted based on missing CLSI interpretation criteria for *C perfringens*.

Tylosin PK/PD relationships

The tylosin PK/PD relationships for the treatment of *C perfringens* infections following oral administration were studied based on the determined MIC values for *C perfringens* isolates originating from Italy and Brazil and the tylosin colon content concentration following oral administration at the registered dose levels of 5mg tylosin phosphate/kg bw (feed medication, A) and 5-10mg tylosin tartrate/kg bw (water medication, B). (Figure 1).

The tylosin colon concentrations after feed and water medication at therapeutic doses are high and exceed the MIC values of all Brazil *C perfringens* strains. In the case of the Italian *C perfringens* strains the tylosin gut concentration exceed the tylosin MICs of 85% of the tested Italian *C perfringens* strains. An excellent therapeutic and metaphylactic effect of Tylovet® in cases of oral *C perfringens* infection treatment can be expected based on the available PK/PD data.

References

1. Posthaus, H. *et al.* (2020). *Clostridium perfringens* type C necrotic enteritis in pigs: diagnosis, pathogenesis and prevention. J.Vet. Diag. Inv. Vol.32 (2), 203-212.

2. Ngamwongsatit, B. *et al.* (2016). Multidrug resistance in *Clostridium perfringens* isolated from diarrheal neonatal piglets in Thailand. Anaerobe 38, 88-93.

3. Koehler, B. (1998). Bekämpfung der Nekrotisierenden Enteritis der Saugferkel (*C perfringens* Type C Enterotoxaemie). Praktischer Tierarzt, 79, 124-137.

4. Agnoletti, F. (2021). Drug susceptibility of *Clostridium perfringens* field strains isolated from swine. Istituto Zooprofilattico Sperimentale delle Venezie, final report.

5. CLSI (2009/2018). Documents M11-07/M11-A8 – Methods for Antimicrobial Susceptibility Testing of Anaerobic Bacteria.

6. Karanikolova, M. et al. (2010). PK and PK/PD of Pharmasin®/Tylovet® 250mg/g Premix (tylosin phosphate) following multiple oral administration in pigs. Proceedings 21St IPVS Congress, Vancouver, Canada, P.684, p.990.

7. Vesselova, S. *et al.* (2010). Pharmacokinetics of Pharmasin®/Tylovet® 100% WSG (tylosin tartrate) following single oral administration in pigs. Proceedings 21St IPVS Congress, Vancouver, Canada, P.687, p.993.

8. Salvarani, F.M. et al. (2012). Antimicrobial susceptibility of *Clostridium perfringens* isolated from piglets with and without diarrhoea in Brazil. Brazilian Journal of Microbiology, 1030-1033.

9. EMEA (1997). EMEA/MRL/205/97. Committee for Veterinary Medicinal Products – Tylosin – Summary Report 3.

Table 5: In vitro susceptibility (µg/mL) of in Italy and Brazil generated *C perfringens* Type A strains to tylosin (MIC ranges, MIC_{50}, MIC_{90})

Publication/report	Region	N strains	MIC range	MIC_{50}	MIC_{90}
Agnoletti, F. 2021[4]	Italy	51	0.125-128	0.25	64
Salvarani, F.M. et al. 2012[8]	Brazil	50	0.25-32	0.5	16

Figure 1: Tylosin colon content concentration after feed medication (5mg/kg bw) and water medication (5-10mg/kg bw) at registered dosages in relation to the tylosin MIC susceptibility patterns of Italian and Brazil *C perfringens* strains.

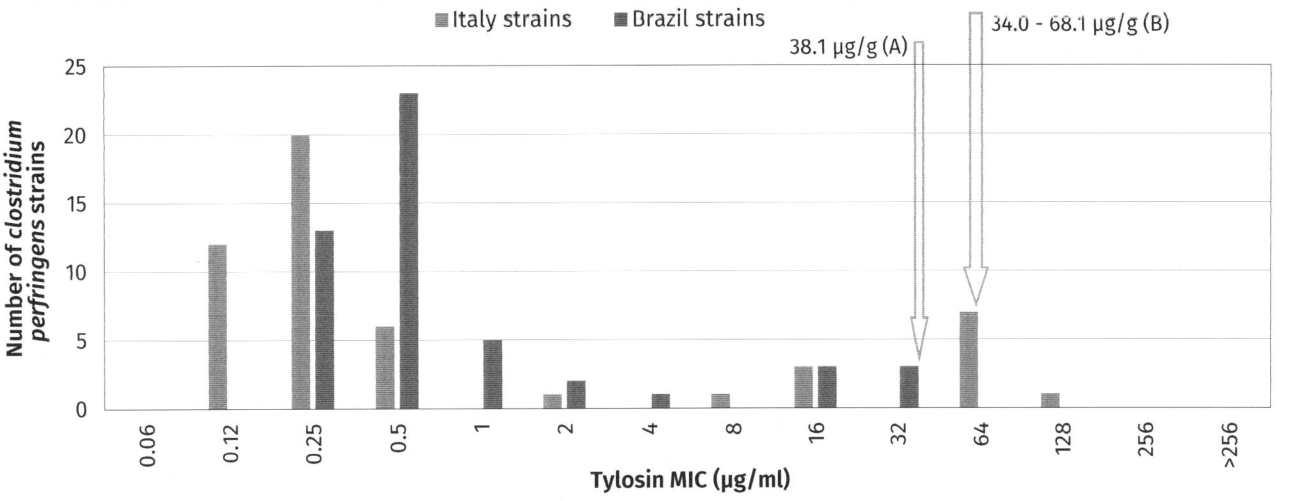

https://doi.org/10.54846/am2022/40

Production of maternal PRRSV antibody titers, against a homologous and a heterologous virus, measured in piglets from sows vaccinated with two autogenous vaccines

D. Nolan, DVM; T. Fangman, DVM, MS, DABVP
Huvepharma Inc, Peachtree City, Georgia

Introduction

PRRS control strategies continue to be difficult and elusive to veterinarians and producers battling the most economically significant disease the US pork industry has faced in at least the past 50+ years. The use of killed PRRS vaccines has been variable with less than consistent results over time; however, when the Huvepharma® autogenous killed PRRS vaccine adjuvanted with ENABL® has been used by swine veterinarians in the upper Midwest they have often (anecdotally) reported a reduced time to PCR negative weaned pigs.[1] Evaluating a reduced time to negative for PRRS is difficult to test on a given farm since the creating of a control group isn't practical or desirable. This study is designed to measure maternal antibodies passed to the piglets from vaccinated sows as a measure for the theory supporting a reduced time to negative status. Two autogenous products were used.

Materials and methods

An independent investigator[2] selected a sow farm that had previously been exposed to wild type PRRS virus strains but was currently weaning PCR negative piglets. 4 treatment groups and a control group of 10 sows each were randomly selected (Table 1).

Sows were vaccinated with either a Huvepharma multivalent autogenous vaccine at 3 (Huv 3) or 3&6 (Huv 3&6) weeks pre-farrow, or a Competitor multivalent autogenous vaccine at 3 (Com 3) or 3&6 (Com 3&6) weeks pre-farrow. The control group of sows was not vaccinated for PRRS pre-farrow. Both Huv and Com products were produced using the same PRRS isolates. After farrowing was complete, 5 (five) 14-day-old pigs were randomly selected for blood collection from each litter enrolled in the study. Serum was sent to SDSU for the measurement of neutralizing antibodies using fluorescent focus neutralization (FFN). The FFN tests were developed by SDSU to test specifically for neutralization against a 1-8-4 and a 1-4-4 strain of the PRRS virus. The 1-8-4 used to develop the FFN test was homologous with the strain used to create both vaccines. The 1-4-4 used to develop the FFN test was 96.2% similar to the 1-4-4 strain in the vaccine, measured in ORF 5 region.

Results

The 14-day old piglets from sows vaccinated in the Huv 3 and Huv 3&6 treatment groups had antibody titers that were statistically significant over Ctrl and both Com groups (Figure 1). The Huv groups demonstrated statistically significant higher serum neutralization titers (Table 2) measured by FFN in piglets ($P = .008$ and ($P = .001$) when compared to Ctrl and Com vaccinated groups. Huv 3&6 demonstrated an improvement in FFN piglet titers compared to Huv 3 ($P = .11$).

Conclusion and discussion

The use of the Huvepharma autogenous PRRS vaccine with the ENABL adjuvant pre-farrow in sows stimulated maternal antibodies to both homologous and heterologous strains of PRRS that were passed to their offspring as measured by FFN serum titers at 14 days of age. A two dose Huvepharma pre-farrowing vaccine program appears to be trending towards a significant improvement over 1 dose pre-farrow. Not all killed vaccines are equal as the Com product was unable to generate a measurable antibody response to either the homologous or heterologous PRRS virus that was passed to their offspring. A killed autogenous vaccine is a tool that may be used to assist in overall immunity, herd viral stabilization, or clean-up of a herd previously exposed to PRRS.

References

1. Wagner M et al. Field experience with an adjuvanted autogenous hard-to-grow PRRSV vaccine, *AASV* 2019; 162-163.

2. This study was designed and conducted by a third-party with no initial knowledge or funding from Huvepharma.

® ENABL is a registered trademark of Huvepharma, Inc.

® Huvepharma is a registered trademark of Huvepharma EOOD

Table 1: Treatment groups

a	No PRRS vaccine during gestation (control)	Ctrl
b	Competitor autogenous at 3 weeks pre-farrow	Com 3
c	Competitor autogenous at 3- & 6-weeks pre-farrow	Com 3&6
d	Huvepharma autogenous w/ENABL at 3 weeks pre-farrow	Huv 3
e	Huvepharma autogenous w/ENABL at 3- & 6-weeks pre-farrow	Huv 3&6

Figure 1: Fluorescent focus neutralization (FFN) titers for homologous and heterologous PRRS strains, measured in 14-day old piglets while still nursing on treatment sows.

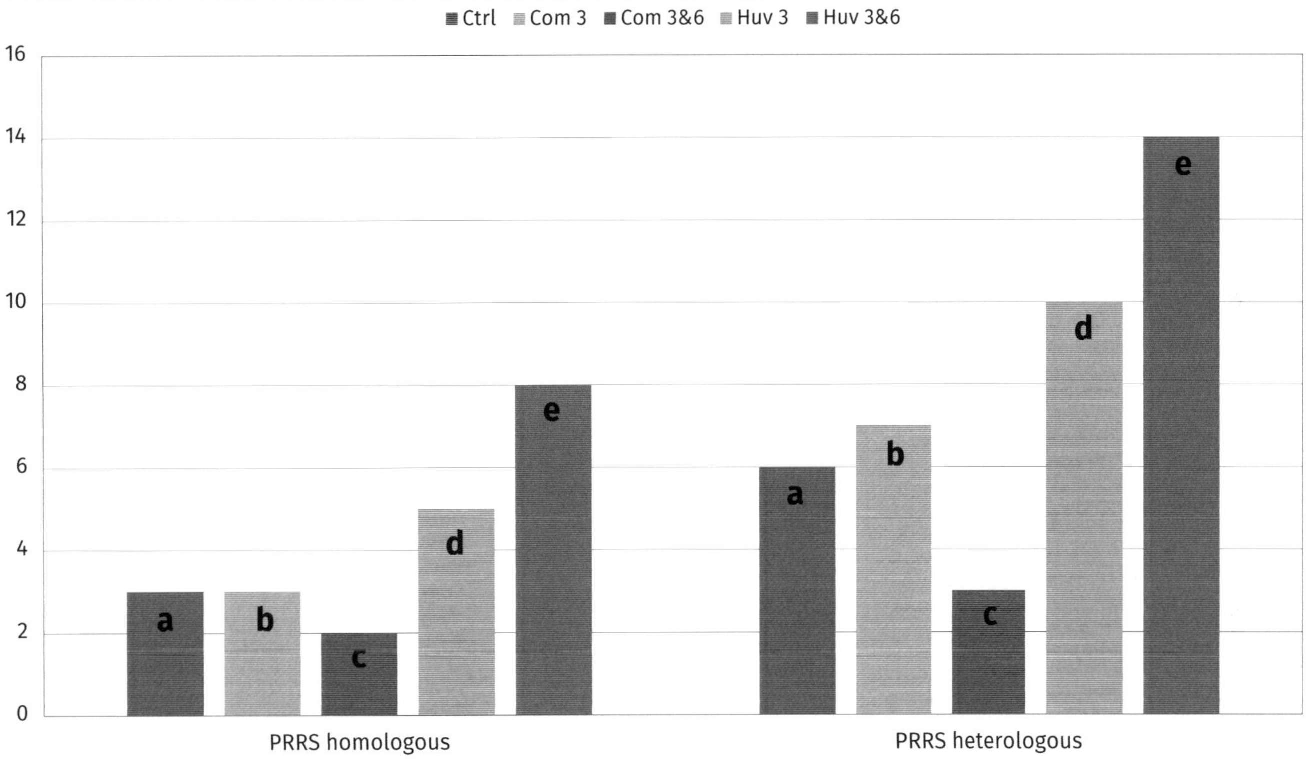

Table 2: Analysis of results

Treatment	*P*-value
(a vs d) Ctrl vs one dose HUV 3	.001
(a vs e) Ctrl vs two dose HUV 3&6	.008
(d vs e) Huv 3 vs Huv 3&6	.11

https://doi.org/10.54846/am2022/41

Trends of mycotoxin contamination in 2021 United States corn

L. Zheng[1], PhD; P. Gott[1], PhD; S. Ramirez[2], PhD; U. Hofstetter[3]; G. R. Murugesan[2], DVM, PhD
[1]DSM Nutritional Products, Parsippany, New Jersey; [2]DSM Nutritional Products, Kaiseraugst, Switzerland;
[3]DSM Nutritional Products, Getzersdorf, Austria

Introduction

Mycotoxins are naturally occurring harmful secondary fungal metabolites, which can be detrimental to animal health even at low levels. The classic signs such as reduced feed intake and vomiting used as indicators of exposure often underestimate other costs of mycotoxicosis. Mycotoxins can increase susceptibility to enteric[1] and respiratory[2] diseases by impairing gastrointestinal environment,[3] and immune dysfunction[4] which can contribute to vaccine failure[5] and increased severity of diseases.[6] Additionally, mycotoxins impact reproductive organs,[7,8] reduce semen quality,[9,10] and impact piglet health through transmission of mycotoxins via colostrum and milk.[11,12] Previous mycotoxin surveys indicate increasing trends of prevalence and co-occurrence of toxins produced by *Fusarium* spp. in US corn.[13] Historically, the *Fusarium* spp. produced mycotoxins (deoxynivalenol, fumonisins, and zearalenone) have been the main mycotoxins detected in the US corn-based feed ingredients. The data herein present preliminary information regarding mycotoxin contamination in samples of the 2021 corn crop in comparison to the previous year trends.

Materials and methods

Forty-four new crop corn samples were submitted as of October 2021. Samples were analyzed for six major mycotoxin groups: aflatoxins (Afla), type A trichothecenes (A-Trich), type B trichothecenes (B-Trich), fumonisins (FUM), zearalenone (ZEN), and ochratoxin A (OTA) utilizing liquid chromatography and tandem mass spectrometry (LC-MS/MS). Data were analyzed using GLIMMIX procedure of SAS with harvest year as fixed effect and sample as the experimental unit.

Results and discussion

To date, FUM is the most prevalent mycotoxin group (80% positive in 2021 vs. 65% in 2020) (Table 1) with average contamination level 5,883 ppb, which is significantly ($P < 0.05$) higher than the 2019 and 2020 harvested corn (3,348 ppb and 2,783 ppb, respectively) (Table 2). Fumonisin is highly impacted by geographic location, so regional evaluation of risk is paramount. The percentage of B-Trich and ZEN contaminated samples are numerically decreased in 2021 compared to 2020 (B-Trich: 59% vs. 50%; ZEN: 20% vs. 11%), whereas Alfa contaminated corn has increased numerically (5% vs. 23%). Contamination level of B-Trich has been vary ($P < 0.05$) over the survey period; however, the level in 2021 corn is similar to 2019 and 2020. Contamination levels for Afla, A-Trich, ZEN, and OTA have remained consistent over the years, and thus should be part of an annual mitigation plan. The 2021 corn risk profile is likely to change drastically as the sample pool expands. Weather-related events of the 2021 growing season including hot weather, drought, and storm events resulted in crop stress and damage, ultimately leading to grain quality and mycotoxin contamination concerns.

The preliminary results from the 2021 corn harvest suggest a continued risk due to mycotoxins produced by *Fusarium* fungal species, emphasizing the importance of risk mitigation strategies that are effective for multiple mycotoxins. Expanding mitigation strategies beyond adsorption such as mycotoxin deactivation with biotransformation and additional support of immune and liver function is essential.

References

1. Oswald, I.P., et al. 2003. Mycotoxin Fumonisin B1 increases intestinal colonization by pathogenic Escherichia coli in pigs. *Applied and Environmental Microbiology*. 2003. 69: 5870-5874.

2. Pósa, R., et al. Use of computed tomography and histopathologic review for lung lesions produced by the interaction between Mycoplasma hyopneumoniae and fumonisin mycotoxins in pigs. *Veterinary Pathology*. 2013. 50: 971-979.

3. Pinton et al. Toxicity of Deoxynivalinol and its acetylated derivatives on the intestine: differential effects on morphology, barrier function, tight junction proteins, and mitogen-activated protein kinases. *Toxilogical Sciences*. 2012. 130: 180-190.

4. Pierron et al. Impact of mycotoxin on immune response and consequences for pig health. *Animal Nutrition*. 2016. 2: 63-68.

5. Savard, C. et al. Deoxynivalenol (DON) naturally contaminated feed impairs the immune response induced by porcine reproductive and respiratory syndrome virus (PRRSV) live attenuated vaccine. *Vaccine*. 2015. 33: 3881-3886.

6. Moreno Ramos, C. et al. Experimental Trial of the Effect of Fumonisin B1 and the PRRS Virus in Swine. *Journal of Animal and Veterinary Advances*. 2010. 9: 1301-1310.

7. Cheng, Y.H. et al. Toxicity of different Fusarium mycotoxins on growth performance, immune responses and efficacy of a mycotoxin degrading enzyme in pigs. *Animal Research*. 2006. 55: 579-590.

8. Acda, Sonia P., et al. Effects of mycotoxin-deactivating agent on the growth performance of pigs fed ochratoxin and zearalenone-contaminated diets. *Philippine Journal of Veterinary Medicine*. 2009. 45:1.

9. Gbore, F. A., and G. N. Egbunike. Testicular and epididymal sperm reserves and sperm production of pubertal boars fed dietary fumonisin B1. *Animal Reproduction Science*. 2008. 105: 392-397.

10. Savard, Christian, et al. Prevention of deoxynivalenol-and zearalenone-associated oxidative stress does not restore MA-10 Leydig cell functions." *Toxicology*. 2016. (341-343):17-27.

11. Hennig-Pauka, I., et al. Current challenges in the diagnosis of zearalenone toxicosis as illustrated by a field case of hyperestrogenism in suckling piglets. *Porcine health management*. 2018. 4: 18.

12. Benthem, de Grave, et al. Transmission of Zearalenone, Deoxynivalenol, and Their Derivatives from Sows to Piglets during Lactation. *Toxins*. 2021. 13:37.

13. Zheng, L., et al. Trends of mycotoxin contamination and mitigation strategies in the US swine industry. AASV 2021, pp. 121.

Table 1: Prevalence of mycotoxin contaminated corn samples

Year	Number of samples	Mycotoxins in corn, %					
		Afla	A-Trich	B-Trich	FUM	ZEN	OTA
2016	288	7	1	79	81	40	0
2017	374	4	1	70	52	24	0
2018	447	10	3	76	79	45	1
2019	329	3	2	81	78	46	0
2020	302	5	4	59	65	20	1
2021	44	23	0	50	80	11	0

Afla: aflatoxins; A-Trich: type A trichothecenes; B-Trich: type B trichothecenes; FUM: fumonisins; ZEN: zearalenone; and OTA: ochratoxin A.

Table 2: Contamination level of mycotoxins in corn by year

Year	Mycotoxins in corn, ppb			
	A-Trich	B-Trich	FUM	ZEN
2016	314	1245[a]	3893[b]	485
2017	104	571[c]	2404[b]	266
2018	1366	1162[ab]	3305[b]	310
2019	330	883[abc]	3348[b]	415
2020	167	1128[ab]	2783[b]	353
2021	0	431[abc]	5883[a]	91
Pooled SEM	754.8	178.9	352.3	210.7
P-value	0.400	< 0.001	< 0.001	0.627

[a-c] Means with differing letters indicate significant difference ($P < 0.05$). A-Trich: type A trichothecenes; B-Trich: type B trichothecenes; FUM: fumonisins; and ZEN: zearalenone. Aflatoxins and ochratoxin A levels are not presented (no differences were observed).

https://doi.org/10.54846/am2022/42

The successful use of pure benzoic acid (VevoVitall) in swine feeds, alone or in combination with nature-identical flavorings (VVC), for diet acidification, improved gastrointestinal functionality, and sustainable pig performance

S. Hough, DVM; J. Bergstrom, PhD
DSM Nutritional Products

Introduction

The addition of acidifiers to certain swine diets, such as starter feeds, has been practiced for many years. The acidification of animal feed is a preventative control to reduce the diet pH and minimize the risk of contamination with bacteria that may cause foodborne illness in humans or animals.[1] In 2014, pure benzoic acid (VevoVitall) was approved for use in the US at up to 0.5% for acidifying swine feeds. Research and application have also shown there are associated benefits for improved nutrient digestibility and performance using swine feeds acidified with 0.5% VevoVitall. More recently, it has been shown that this source of pure benzoic acid can be used in combination with a nature-identical flavorings product (VVC) and retain its ability to control pathogens in feed. When this combination is used, concentrations of benzoic acid below 0.5% in the complete feed have demonstrated an ability to sustain similar benefits for pig production and performance. The benefits of utilizing pure benzoic acid, as well as benzoic acid combined with nature-identical flavorings will be shared here.

Background

Diet acidification has been proven useful for diets fed immediately post-weaning.[2] This is primarily because natural acidification of the stomach from HCl secretion is less developed in young, suckling pigs. Suckling pigs have an immature digestive tract and maintain functional gastric acidity via the microbial fermentation of milk lactose to lactic acids.[3] When pigs are abruptly weaned from milk to solid diets, there is a rather sudden dysbiosis with transient changes to gastrointestinal morphology that impair nutrient utilization and reduce energy metabolism.[4] When a normal, beneficial gastric pH is not maintained post-weaning, there may be increased susceptibility to enteric infections.[5] During such a stressful period requiring adaptation, the dietary addition of some organic acids has helped to sustain digestive health, reduce the lag in growth from abrupt weaning, and improve growth performance.[6]

Benzoic acid

Benzoic acid and benzoates are among the earliest and most common organic acids to be used as food or feed preservatives Generally, benzoic acid and benzoates have the broadest spectrum of activity against spoilage bacteria, fungi, and yeasts.[7] Since the year 2000, there is increasing research and evidence to demonstrate that benzoic acid is well-suited for acidifying swine feeds and it was approved for use at up to 0.5% for swine diets in the United States in 2014. Pure benzoic acid is a white, solid crystalline flake product. In this form, benzoic acid is stable when properly contained and kept dry, is non-corrosive, mildly odorous, and does not require a coating or carrier to accommodate handling when compared to using other concentrated organic acids. Adding pure benzoic acid to feeds is known to reduce the pH and buffering capacity of swine diets in a linear fashion, whereas the addition of buffered acids (or salts of acids) may not reduce the diet pH and can increase the buffering capacity of the diet.[8,9]

Recent research

Recent trials performed in North America have shown consistent improvements in post-weaning pig performance when 0.5% benzoic acid was used for feed acidification. Bradley et al. reported that ADG and ADFI were increased for pigs fed diets with 0.5% benzoic acid from day 0 to 21 post-weaning, even when diets contained high levels of zinc oxide, and F/G was also improved with 0.5% benzoic acid when there was not a high level of zinc oxide in the diet.[10] In another post-weaning study, Bergstrom et al. also found that diets with 0.5% benzoic acid improved ADG and ending BW during days 0 to 13 and 13 to 20 post-weaning. Feed efficiency was also improved by feeding 0.5% benzoic acid from day 0 to 13, and there was a tendency for pigs fed diets containing any of the acids used in the study to have greater ADFI from day 13 to 20.[11]

Research and interest for using benzoic acid to acidify nursery swine feeds has been occurring since before its introduction to North America, with an increasing number of experiments demonstrating its benefits for pigs in Europe, Asia, or Latin America since the year 2000. In an early experiment conducted in Denmark, pigs given a diet with benzoic acid (2% in the first 2 weeks and 1% in the last 4 weeks) were treated significantly less for diarrhea and had a higher production value (based on weight gain, feed intake, and feed efficiency) when compared to pigs fed the control diet or a diet containing a combination of formic acid and lactic acid.[12] There is increasing research showing benefits from feeding benzoic acid during grow-finish phases as well. Benzoic acid has been demonstrated to improve ADG, ending BW, and either ADFI, F/G or both.[13-16] These effects seem to be more consistent than those obtained with other acidifiers in the peer-reviewed research. When using a combination of benzoic acid and nature-identical flavorings in the grower phase, there was a trend for lower removals and an improvement in ADG and F/G for the pigs that remained versus the control diet.[17]

Most of the sow research with benzoic acid has been completed abroad. In a field study completed in the Philippines in 2014, there was an improvement in lactation ADFI, piglet survivability and mean litter weaning weights when sows were fed 0.5% VevoVitall two weeks prior to parturition and throughout lactation. In a 2016 Canadian study, sows receiving 0.5% VevoVitall in the lactation diet had less body weight loss at the end of farrowing compared to control sows. Because benzoic acid is metabolized differently than other organic acids, it can result in reduced urinary pH since it is converted to hippuric acid in the liver.[18] Hippuric acid that is excreted through the kidneys can potentially reduce the ammonia levels in the manure pit by reducing the pH.

There is increasing interest in using acidifiers and various oils for controlling viruses in feed, beginning with the 2013 entry of porcine epidemic diarrhea virus (PEDV) from China and the more recent threat of African swine fever (ASF). Two levels of a combination of VevoVitall and nature-identical flavorings (0.5% or 0.3%) were applied to feed before exposure to known quantities of PEDV, porcine reproductive and respiratory syndrome virus (PRRSV) and Seneca Valley virus (SVA) via an ice block dispensed into the feeds. Although some of the feed samples were found to be positive for all three agents using PCR testing, the pigs did not display clinical signs nor was virus recovered from rectal swabs, serum or tonsillar tissue. ADG was also statistically improved compared to pigs fed the control.[19] The higher inclusion rate from this study was similarly applied to swine feed that was inoculated with ASF in a BSL-3 laboratory at South China Agricultural University in Guangzhou, China. Hemadsorption tests on day 1 post inoculation yielded negative on day 1 with the addition of VevoVitall and nature-identical flavorings, and there was a higher qPCR when compared to the inoculated control.[20]

Discussion

The benefits in productivity from using benzoic acid in swine diets result from its acidification effects and the maintenance of multiple digestive mechanisms as demonstrated through research. Other acids, either used alone or in combination with other acids, have not consistently demonstrated an ability to maintain optimal conditions for digestion throughout the entirety of the gastrointestinal tract. The ability for a portion of ingested benzoic acid to reach the distal small intestine is likely responsible for differences found in the normal microbiota populations measured in the stomach, duodenum, jejunum, ileum, or caecum contents when pigs fed benzoic acid were compared to those fed a control.[21,22] Consequently, when measured, pigs fed benzoic acid often had less *E coli* bacteria in these segments of the gastrointestinal tract.[23]

Feeding benzoic acid has also been found to support normal morphology and development within the gastrointestinal tract of nursery-aged pigs. Several studies have reported that feeding benzoic acid to pigs resulted in an improved maturity of the epithelium in the duodenum, jejunum, and/or ileum; as indicated by measurements of villus height, crypt depth, or villus height-to-crypt depth ratio.[16,21,22]

Conclusions

Organic acids and diet acidification are among the most viable additives that are being considered to help maintain or improve the nutrition for pigs, both now and into the future. The results from acidifying pig diets with benzoic acid are more consistent when compared to other acids, especially when the initial buffering capacity and other ingredients are considered in diet formulation. Research with pure benzoic acid has shown that it helps to maintain performance with reduced dietary pH and healthy digestive functioning throughout most of the gastrointestinal tract. The versatility of benzoic acid in swine feeds has been demonstrated, and it's been effective in diets with added phytogenic compounds or flavorings and enzymes.

References

1. Hazard Analysis and Risk-Based Preventive Controls for Human Food: Draft Guidance for Industry, Chapter 4, US FDA 2014.

2. Tung, C.M. and J.E. Pettigrew. 2006. A Critical Review of Acidifiers. National Pork Board Animal Science Research Report, NPB #05-169.

3. Cranwell, P.D., D.E. Noakes, and K.J. Hill. 1976. Gastric secretion and fermentation in suckling pig. *British Journal of Nutrition*, 36:71-86.

4. Li, Y., Y. Guo, Z. Wen, X. Jiang, X. Ma, and X. Han. 2018. Weaning Stress Perturbs Gut Microbiome and Its Metabolic Profile in Piglets. *Scientific Reports*, 8:18068.

5. Heo, J.M., F.O. Opapeju, J.R. Pluske, J.C. Kim, D.J. Hampson, and C.M. Nyachoti. 2013. Gastrointestinal health and function in weaned pigs: a review of feeding strategies to control post-weaning diarrhoea without using in-feed antimicrobial compounds. *Journal of Animal Physiology and Animal Nutrition*, 97(2):207-237.

6. Pettigrew, J.E. 2006. Reduced use of antibiotic growth promoters in diets fed to weanling pigs: dietary tools, part 1. *Animal Biotechnology*, 17:207-215.

7. Wedzicha, B.L. 2003. Preservatives/Food Uses. In B. Caballero, *Encyclopedia of Food Sciences and Nutrition* (2nd ed), pp 4776-4780. Cambridge, MA: Academic Press.

8. Kluge, H., J. Broz, and K. Eder. 2006. Effect of benzoic acid on growth performance, nutrient digestibility, nitrogen balance, gastrointestinal microflora and parameters of microbial metabolism in piglets. *Journal of Animal Physiology and Animal Nutrition*, 90:316-324.

9. Partanen, K., H. Siljander-Rasi, and K. Suomi. 2002. Dietary preferences of weaned piglets offered diets containing organic acids. *Agricultural and Food Science in Finland*, 11:107-119.

10. Bradley, C.L., J. Bergstrom, J. Nemechek, J. Hahn, K. Bottoms, T.C. Tsai, and C.V. Maxwell. 2019. The effects of adding a combination of benzoic acid, a coated sodium butyrate, and phosphoric acid to nursery diets, either with or without high levels of zinc or copper, on post-weaning pig performance. Abstract and presentation at the "Zero Zinc Summit 2019" in Copenhagen, Denmark.

11. Bergstrom, J, J. Landero, M. Young, and J.D. Hahn. 2020. The effects of pure benzoic acid, an acid blend product, or their combination in nursery diets on post-weaning growth performance. *ASAS-ADSA Annual Meeting*, abstract 182.

12. Maribo, H., L.E. Olsen, B.B. Jensen, and N. Miquel. 2000. Products for weaners: benzoic acid or the combination of lactic acid and formic acid. The National Committee for Pig Production, Danish Bacon and Meat Council, Report no. 490.

13. Chen, J.L., P. Zheng, C. Zhang, B. Yu, J. He, J. Yu, J.Q. Luo, X.B. Mao, Z.Q. Huang, and D.W. Chen. 2016. Benzoic acid beneficially affects growth performance of weaned pigs which was associated with changes in gut bacterial populations, morphology indices and growth factor gene expression. *Journal of Animal Physiology and Animal Nutrition,* 101(6):1137-1146.

14. Gräber, T., H. Kluge, F. Hirche, J. Brož, and G.I. Stangl. 2012. Effects of dietary benzoic acid and sodium-benzoate on performance, nitrogen and mineral balance and hippuric acid excretion of piglets. *Archives of Animal Nutrition,* 66(3):227-236.

15. Kim, S.W., J.O. Vaughn, and D.A. Monson. 2008. Effects of dietary supplementation of benzoic, formic, and lactic acids on growth and health of pigs. *Journal of Animal Science,* 86(3):66.

16. Silveira, H., L. Amaral, C. Garbossa, L. Rodrigues, C. Silva, and V. Cantarelli. 2018. Benzoic acid in nursery diets increases the performance from weaning to finishing by reducing diarrhoea and improving the intestinal morphology of piglets inoculated with *Escherichia coli* K88+. *Journal of Animal Physiology and Animal Nutrition,* 00:1-11.

17. Thayer, M. T.; Langemeier, A.; Scotten, S.; Nelssen, J. L.; Ou, Z.; and Bergstrom, J. R. (2017) "The Effects of Feeding Benzoic Acid and Essential Oils on Growth Performance, Survivability and Fecal Characteristics of Grower Pigs," Kansas Agricultural Experiment Station Research Reports: Vol. 3: Issue 7.

18. Kluge, H., J. Broz and K. Eder. Effects of dietary benzoic acid on urinary pH and nutrient digestibility in lactating sows. 2010. *Livestock Science.* 134 (119-121).

19. Dee, S., M. Niederwerder, R. Elder, D. Hanson, A. Singrey, R. Cochrane, G. Spronk and E. Nelson. 2021. *Transboundary and emerging diseases.* Vol. 68 (2), pp 833-845.

20. Zhai, H., J. Chihai, M. C. Walsh, J. Bergstrom, S. Potot and H. Wang. 2021. The additional of natural identical flavorings accelerated the viricidal effect of pure benzoic acid against African swine fever contamination of complete feed. *Animals.*11, 1124.

21. Chen, J.L., P. Zheng, C. Zhang, B. Yu, J. He, J. Yu, J.Q. Luo, X.B. Mao, Z.Q. Huang, and D.W. Chen. 2016. Benzoic acid beneficially affects growth performance of weaned pigs which was associated with changes in gut bacterial populations, morphology indices and growth factor gene expression. *Journal of Animal Physiology and Animal Nutrition,* 101(6):1137-1146.

22. Diao, H., P. Zheng, B. Yu, J. He, X.B. Mao, J. Yu, and D.W. Chen. 2014. Effects of dietary supplementation with benzoic acid on intestinal morphological structure and microflora in weaned pigs. *Livestock Science,* 167:249-256.

23. Knarreborg, A., N. Miquel, T. Granli, and B.B. Jensen. 2002. Establishment and application of an in vitro methodology to study the effects of organic acids on coliform and lactic acid bacteria in the proximal part of the gastrointestinal tract of piglets. *Animal Feed Science and Technology,* 99:131-140.

https://doi.org/10.54846/am2022/43

The relationship of ruptures between purebred and crossbred progeny from the same Duroc sires

John P. Sonderman; Tom Rathje; Brady M. McNeil; Diaz, I.D.P.S; Amanda J. Cross; Muhammed Walugembe; Emily D. Mauch
DNA Swine Genetics

Introduction

A low proportion of congenital defect pigs is of importance both from an animal welfare and economic point of view. In this sense, the observation and the diagnosis are very important points to correctly evaluate these defects. A hernia is defined as a protrusion of part of the intestines or any other internal organ through an opening of the body wall. Hernias may be observed in newborn pigs or the defect can appear later in life. The mode of inheritance of hernias is not well known and the diagnosis is very complex. Thus, the accuracy of registration of the presence of defects in the litters has a decisive role for the ability to select against these defects. Linked with that, is also the question to what degree these defects have a hereditary background or an environmental importance. Since, livestock production is the result of the combined action of genetic and environmental factors, it is therefore essential to evaluate the isolated effects and combination of factors that impact ruptures in genetic breeding programs. This study concerns umbilical and scrotal hernias in a herd with purebred Duroc (DD) and crossbred Duroc × Landrace ×Yorkshire (DLY) animals and the relationship of ruptures in both herds. The challenge is to assess the extent of the interaction and if it is large enough to be of economic and biological importance. The main objective of this study was to investigate the relationship of ruptures between purebred and crossbred progeny and to determine its effect on the performance and heritability's of purebred Duroc and commercial pigs.

Table 1: Number of sires within each herd of Purebred (DD) and Crossbred (DLY) animals (diagonal) and in both herds (triangular superior) for scrotal and umbilical hernia

	DD	DLY
DD	1029	255
DLY		287

Materials and methods

Data of recorded scrotal and umbilical hernias from progeny of purebred Duroc sires born between 2016 to 2021 was analyzed. Two types of herds were evaluated: purebred Duroc progeny (DD) and crossbred Duroc progeny (DLY). In a DLY herd, commercial progeny are the result of crossing Duroc sires with a Yorkshire x Landrace sow. Purebred progeny are 100% Duroc. To evaluate the relationship between both herds, Duroc sires had to have progeny with records in both herds. The number of sires present in each herd and the common number of sires in the two herds are shown in Table 1. To account for the dam/effect on commercial progeny, a pedigree with dam information and genotypes were added in the analysis, with a total of 718,745 animals when all available relationships were considered (up to 11 generations). The total genotypes used were for scrotal hernia 6,754 and 6,239 for DD and DLY respectively. For umbilical hernia a total of 6,758 and 6,441 genotypes were used for DD and DLY environments, respectively. Only animals that had records of ruptures were evaluated. A total of 352,606 (DD) and 202,184 (DLY) animals for DD and DLY were analyzed, respectively. For scrotal hernia, a total of 762 incidences were analyzed in the DD herd, while in DLY a total of 2,676 incidences were considered. For umbilical hernia, a total of 8,175 and 702 incidences were evaluated for DD and DLY herd, respectively (Table 2). A set of graphics was used to evaluate differences of environment on incidences (Figure 1 to 6) to orientate the subsequent statistical and genetic analysis. The statistical model evaluated the fixed effects of breed, herd, season, parity number, and total number of born. A sex effect was added to the umbilical hernia model. For the genetic analysis, the effect of the animal

Table 2: Total number of animals, number of incidences (inc), total number of litters and average of litters with defects for scrotal and umbilical hernia for herds with purebred (DD) and crossbred (DLY) animals

	Scrotal hernia			Umbilical hernia		
	Total number of animals	Number of inc (%)	Average of litters with inc		Number of inc (%)	Average of litters with inc
DD	352,606	762 (0.21%)	728	DD	8175 (2.31%)	6739
DLY	202,184	2676 (1.32%)	2002	DLY	702 (0.34%)	641
Total	554,790	-	-	Total		-

and the litter was added as a genetic effect and the residual as non-genetic effect. It was assumed that the same trait becomes a different trait when genetic correlation is below 0.80 between environments.[1]

Results and discussion

Figure 1 to 6 and Table 2 show that there were wide variations in the total number of incidences and proportion of litters, among the environments. The difference of the amount of data found for umbilical hernia is due to recording differences and castration timing between a purebred herd and a commercial herd. These differences may reflect the influence of factors such as genetic variation among herds and the effect of heterosis, which gets pronounced in different environments. This variation results in different estimates of the (co)variance components and, consequently, differences in the genetic parameter estimates.[2] The average of incidences in the litters (Table 2) of scrotal and umbilical hernia may reflect the effect of the sow on both traits for DD and DLY animals. According to Sevillano et al[3] litters with any defect present and number of defect piglets in litters tend to be higher in sows in their first parity, the authors consider that if sows are not physiologically prepared to produce many piglets, they cannot provide a favorable environment for the development (both before and after birth), thus producing piglets with weaker organs that could lead to defects. Another factor related to the sow physiology is the litter size. We know that in most of the cases, litter size increases with increased parity number (up to the fifth parity) and as result, the probability to find litters with defect piglets also increase, however this also depends on the environment provided to the development of the piglets, related to the sow itself and her environment. Thus, this also could be related if the dam has an ability to continuously produce pigs with strong organs at each parity. According to Table 2, there is variation of records between DD and DLY progeny, and this might reflect different attitudes on recording defects. Defects in the umbilical region have a complex background, therefore any defect or protrusion at the umbilical region is often referred to as an umbilical hernia. The actual content of the umbilical sack is not easily determined on live pigs and umbilical hernias might be mixed up with umbilical abscesses. Figure 1 and 2 implicates that there are different genetic backgrounds for the defects in different environments for the same group of sires. The finding that genetic variance differed among DLY and DD may indicate that genotypes react differently to different environments, a fact implying the existence of genotype by environmental interaction. Another explanation is regarding the continuous years of selection made in the environment with DD animals which leads to a reduction of the genetic variation of the trait in that population. Additionally, we need to consider that DLY progeny have the effect of the breed of the dam that, although was considered in the genetic analysis, should not be underestimated on a commercial herd. These differences in genetic variances among herds may also reflect the fact that although there is a connection between the groups of sires, there may be differences between environments due to management of animals and defect recording. The across-environment estimates for genetic correlations (Table 3) between DD and DLY animals were low and positive (0.31 and 0.35) for scrotal and umbilical hernia, respectively. These estimates indicate that the relationship between the environments is biologically important. These results show that the animals selected will not perform the same depending on the considered herd, and the sow effect must be considered on commercial pigs. Further investigation of genotypic and environmental factors influencing scrotal and umbilical hernias need to occur mainly considering the effect of the sow and the effect of ruptures on purebred and crossbred animals.

Figure 1: The proportion of sires with at least one progeny with a scrotal hernia in their crossbred (DLY) and purebred (DD) offspring.

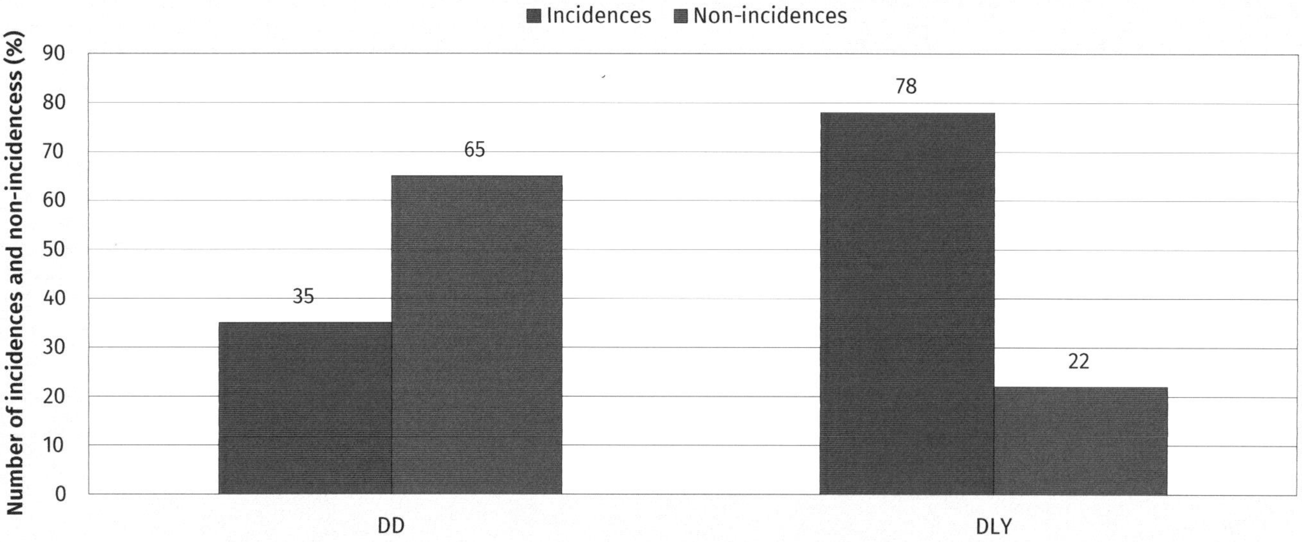

Figure 2: The proportion of sires with at least one progeny with an umbilical hernia in their crossbred (DLY) and purebred (DD) offspring.

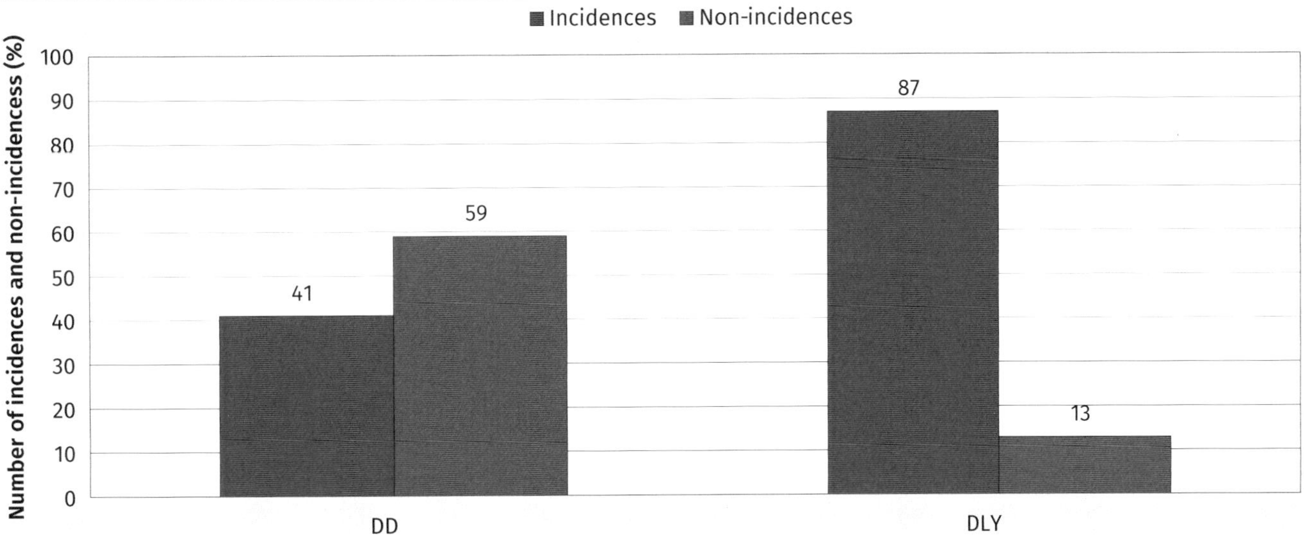

Figure 3: Scrotal hernia incidences for each stage of growth for herds with purebred (DD) and crossbred (DLY) animals.

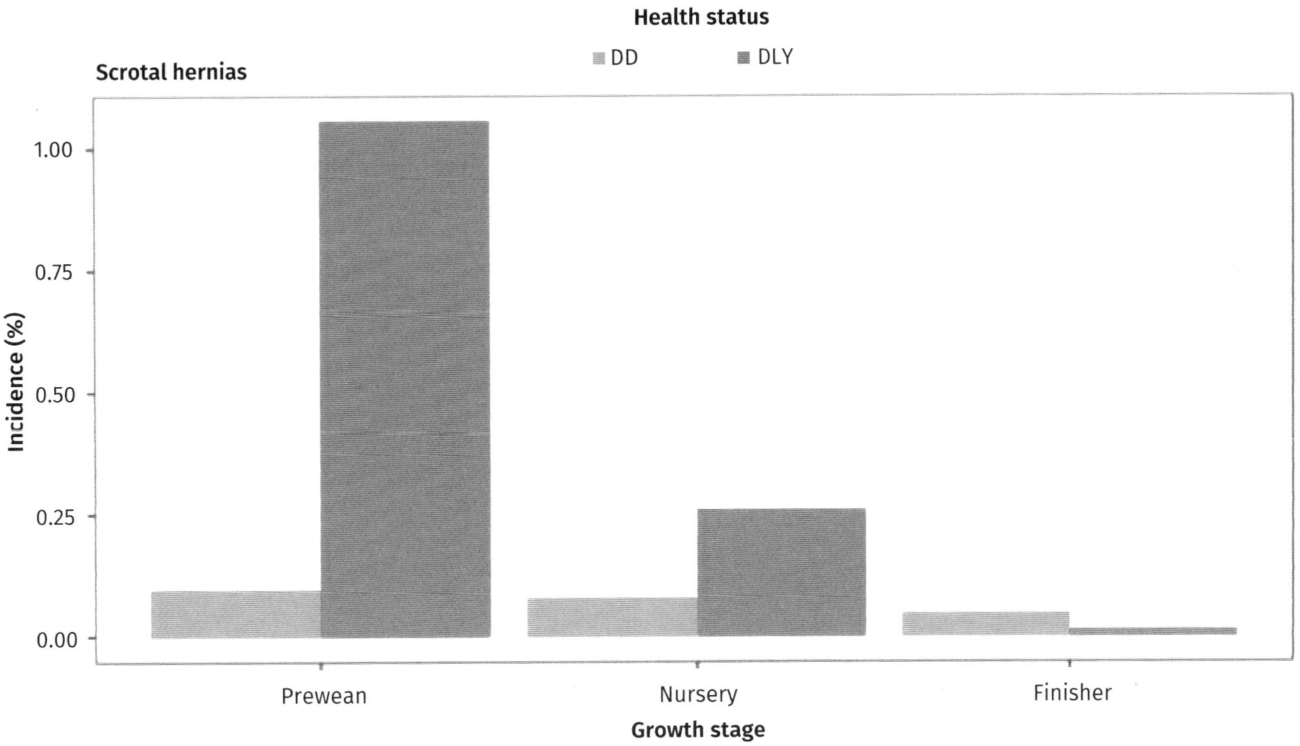

Figure 4: Scrotal hernia incidences by month of sires in their purebred (DD) and crossbred (DLY) offspring.

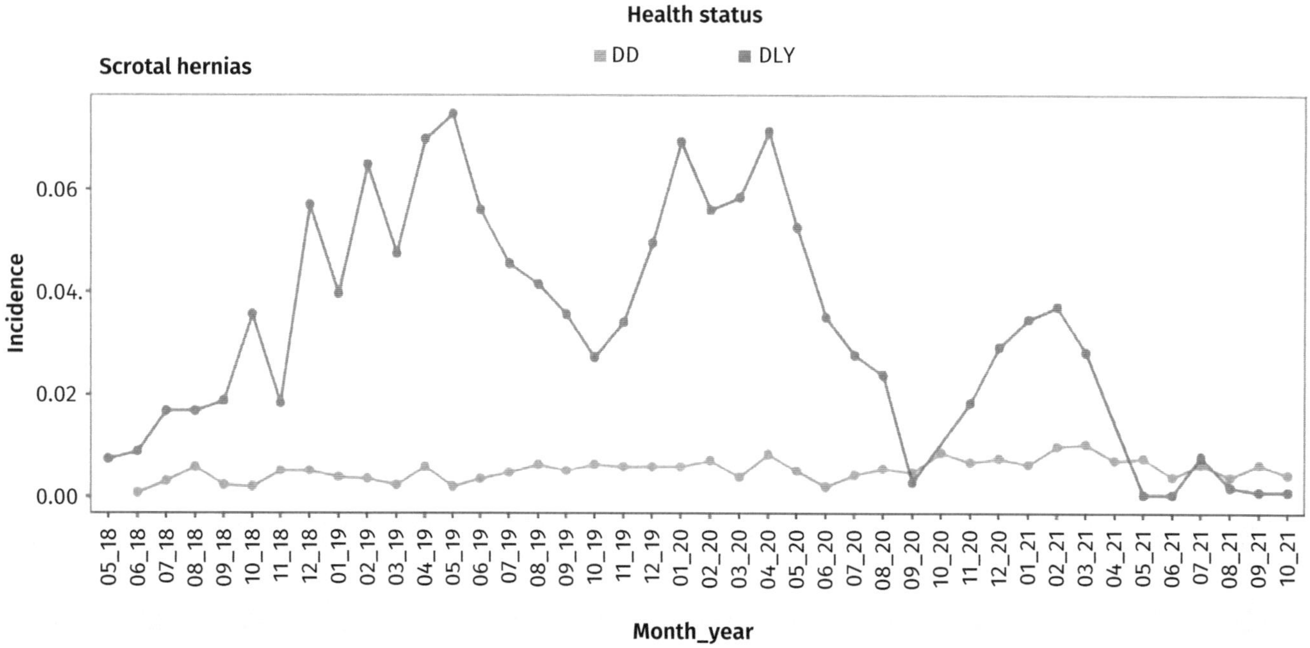

Figure 5: Umbilical hernia incidences of sires in their purebred (DD) and crossbred (DLY) offspring.

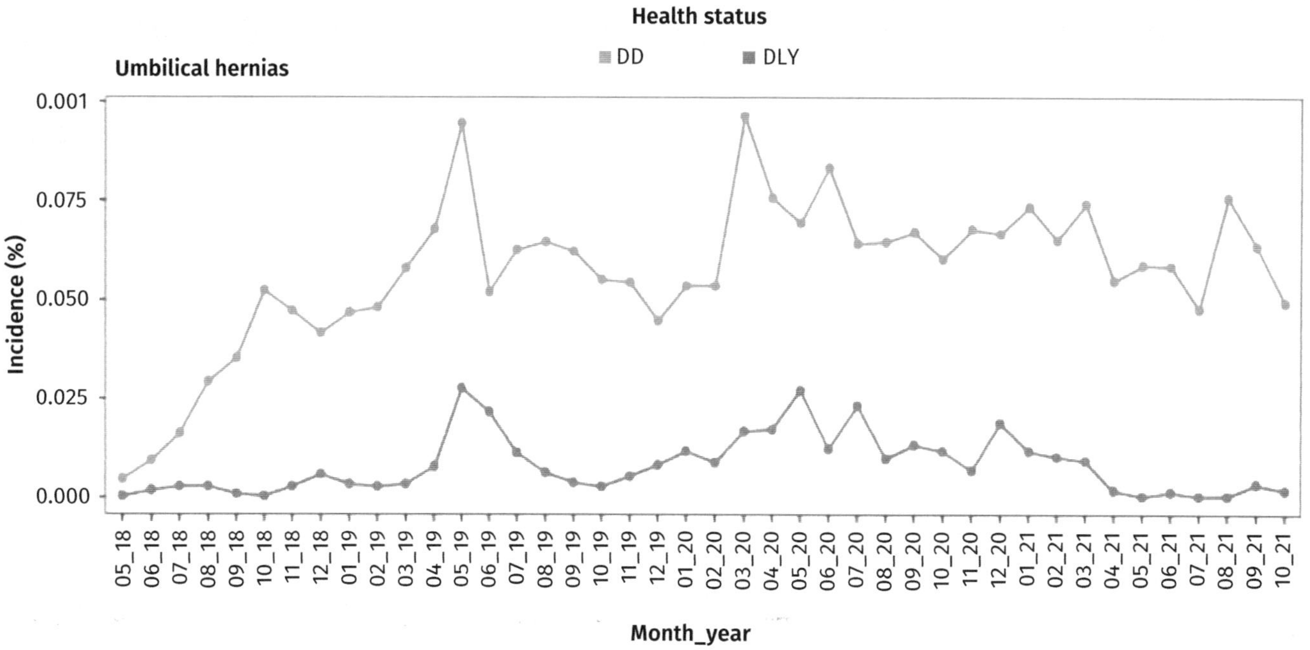

Figure 6: Umbilical hernia incidences for each stage of growth for herds with purebred (DD) and crossbred (DLY) animals.

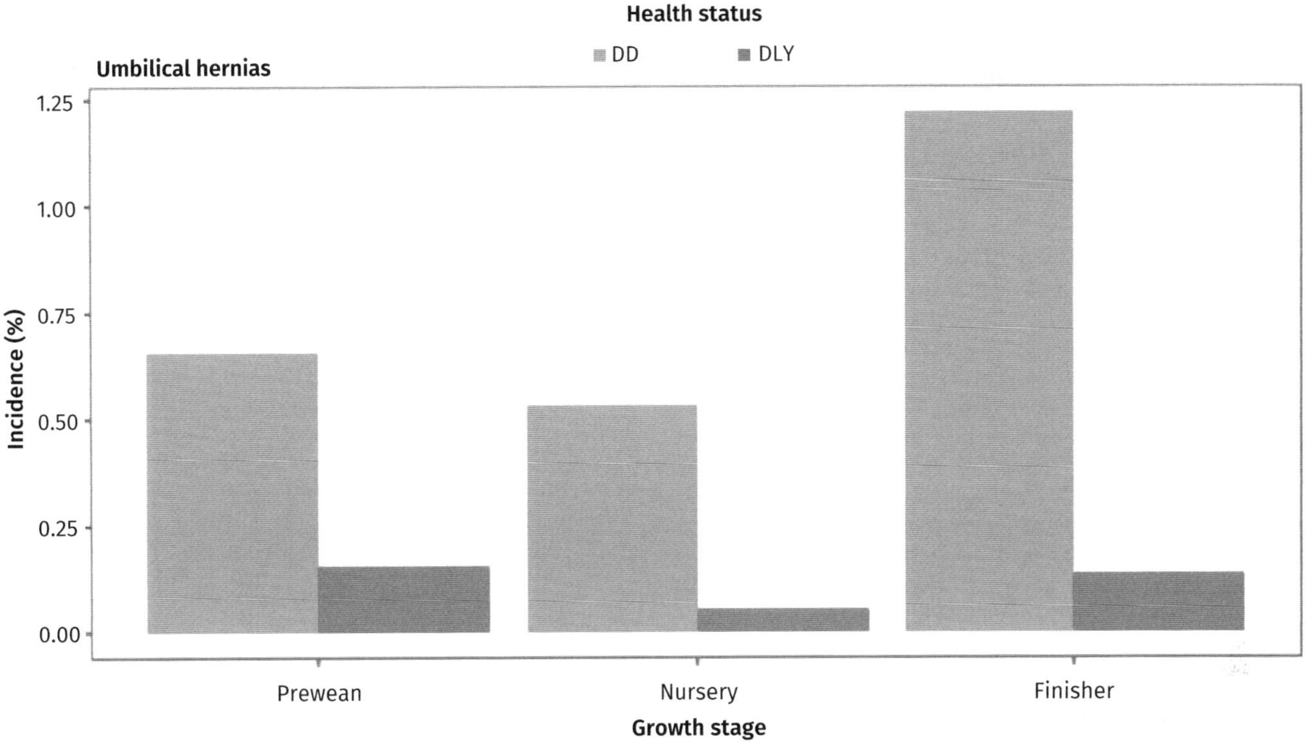

References

1. Robertson, A. The sampling variance of the genetic correlation coefficient. Biometrics, v.15, n.3, p.469-485, 1959.

2. Carvalheiro, R., Costilla, R., Neves, H.H.R. *et al.* Unraveling genetic sensitivity of beef cattle to environmental variation under tropical conditions. *Genet Sel Evol* 51, 29 (2019). **https://doi.org/10.1186/s12711-019-0470-x**

3. Sevillano, C.A., Lopes, M.S., Harlizius, B. *et al.* Genome-wide association study using deregressed breeding values for cryptorchidism and scrotal/inguinal hernia in two pig lines. *Genet Sel Evol* 47, 18 (2015). **https://doi.org/10.1186/s12711-015-0096-6**

Table 3: Estimates of genetic correlations (diagonal) between both DD and DLY animals for scrotal and umbilical hernia

	SC_DLY	UMB_DLY
SC_DD	0.31	-
UMB_DD	-	0.35

SC – Scrotal hernia; UMB – Umbilical hernia; DD, DLY– Purebred progenies and Crossbred progenies, respectively

https://doi.org/10.54846/am2022/44

TransRite Sow Ultra™, an innovative product to decrease farrowing stress

A. Bents, DVM; S. Galloway, PhD
Hubbard Feeds, Mankato, Minnesota

Background

Our industry is faced with a critical need to find ways to improve survivability of piglets in the farrowing house.[1] While direct, intensive care of the sow and newborn piglets is the best method for improving the farrowing outcome, this is not always possible. We must continue to evaluate options that will aid the sow in the farrowing process when direct help may not be readily available. Maternal fatigue and reduced piglet viability due to hypoxia are significant factors that contribute to reduced piglet survivability during the farrowing process.[2]

TransRite Sow Ultra is an innovate product that was developed to help address the challenge of farrowing house management. In the realm of nutrition, we have made strides in understanding the needs of the periparturient sow. However, we still face the challenge of the sow being able to fully utilize the nutrients that have been provided for her leading up to farrowing. In this sense, we evaluated strategies employed in the dairy industry that were developed to help the cow to better utilize endogenous calcium by feeding DCAD (dietary cation anion difference) diets to induce a mild compensatory metabolic acidosis.

Through a series of trials, we developed an encapsulated Calcium Chloride product that can be fed as a top-dress to sows prior to farrowing, that is palatable for the sow and will not reduce feed intake if fed correctly.[3] Through repeated urinalysis of sows consuming the product, we were able to demonstrate that two to three days of feeding TransRite Sow Ultra was indeed able to induce a mild, compensatory metabolic acidosis. Finally, through ongoing on-farm trials, we have been able to demonstrate that providing this metabolic alteration can indeed reduce stillborn pigs, and in some cases, shorten the duration of farrowing as well.

The advantage of feeding TransRite Sow Ultra as a top-dress prior to farrowing is that it is a relatively easy process, is non-intrusive to the sow, and avoids the need for additional injections to the sow prior to or during the farrowing process.[4] The easy-to-use pellets come in a plastic bag that can be introduced biosecurely into a farrowing facility, is easy to store with a long shelf life at room temperature, and the value of reduced stillborn piglets and lower farrowing stress far outweighs the cost to feed this product to sows.

TransRite Sow Ultra is a specially formulated top-dress designed to provide the proper cation-anion balance in sows due to farrow. Adjusting metabolic pH prior to farrowing through the use of TransRite Sow Ultra has been shown to reduce farrowing stress and duration, lower stillborn rates, and improve post-partum recovery time. As such, TransRite Sow Ultra has been demonstrated to be a valuable tool to implement in the farrowing house to aid in improving day one piglet care and reducing labor needs during the farrowing process.

Material and methods

This project was conducted at a commercial 340-sow farrow to finish operation in southwest MN. The sow herd was comprised of Compart genetics, and currently utilize twice per day feedings starting at lockup into farrowing. The farrowing room is heated with a ceiling radiator system to ensure consistent temperatures throughout the room. Three consecutive batches were evaluated from September through October, and the 160 sows enrolled in the project were randomly distributed between control and treated sows. Treated sows were started on TransRite Sow Ultra (TRU) on the Wednesday prior to farrowing, which ranged between 1- and 4-days pre-farrow, and was top-dressed on the feed prior to feed drop to the sows. Once each sow farrowed, TRU was no longer fed to that individual sow. Normal farrowing protocol for the farm was followed for both treatment groups, and all sows not yet farrowed by Sunday morning were induced. Induction effects were not evaluated in this trial.

Results and discussion

Results can be seen in Table 1. Stillborns were reduced by an average of 0.35 per litter, or approximately 2.28% as a percentage of total born. Anecdotally, the manager commented on reduced farrowing duration and less oxytocin usage during the farrowing process as well, but these data were not collected during this study. Further work to strengthen and support these results is underway, but the preliminary results demonstrate a potential for another great tool to utilize on-farm to improve farrowing performance. We continue to evaluate the benefit of feeding TransRite Sow Ultra to sows prior to farrowing, and have multiple trials that demonstrate that this is an effective tool that can be easily implemented into a farm's farrowing protocol.

References

1. Probst-Miller, S. "Day one critical care management: Critical tasks to maximize full value pigs." AASV Lecture, Hilton Orlando Buena Vista Palace, Orlando, FL, March 2015.

2. Farmer, C. (Ed). (2020). *The Suckling and Weaned Piglet*. Wageningen Academic Publishers.

3. Röcker, B. and Coenen, M. 2004. "The Use of Calcium Chloride in diets of pregnant sows as an acidifier". Proceedings of the 18th IPVS Congress, Hamburg, Germany.

4. Hintz, A.S. and Billing, A.F. "Hypocalcemia and calcium gluconate use in sows at parturition". Poster presentation at AASV, Manchester Grand Hyatt, San Diego, CA. March 2013.

Table 1: Impact on farrowing performance when feeding TRU

Control

80 sows	Parity	Gest Length	LB	SB	Mum	TB	SB as % TB
Group 1	2.75	116.1	10.4	0.88	0.67	11.92	6.47%
Group 2	2.73	114.6	11.9	1.19	0.22	13.26	9.11%
Group 3	2.77	115.8	12.9	1.22	0.74	14.85	7.88%
Average	**2.75**	**115.5**	**11.7**	**1.09**	**0.54**	**13.34**	**7.82%**

TRU

80 sows	Parity	Gest Length	LB	SB	Mum	TB	SB as % TB
Group 1	2.5	115.6	10.1	0.50	0.29	10.88	4.66%
Group 2	2.64	115.2	12.1	0.50	0.31	12.41	3.60%
Group 3	2.8	115.9	11.9	1.22	0.48	13.63	8.36%
Average	**2.6**	**115.6**	**11.4**	**0.74**	**0.36**	**12.30**	**5.54%**

https://doi.org/10.54846/am2022/45

Precision ag "lessons learned" when implementing LeeO™

Tom Petznick[1], DVM; Joel Stave[2], BA; Tyler Holck[3], DVM, MS, MBA
[1]Ark Care, Omaha, Nebraska; [2]Prairie Systems, Spencer, Iowa; [3]Feed His People LLC, Gilbert, Iowa

Introduction

Precision pork is a term used in our US swine industry and yet most, if not all, of our information platforms are based upon group data and assessing group means. Individual weights are rarely collected in the farrowing house let alone in the grow-finish stage. Individual animal data has been used in Europe to allow for better decision making on gilt selection and sow culling, piglet viability criteria, weaned pig feeding, market pig selection to maximize income, and traceable attributes for finished pork products.[1] LeeO™ is an individual animal tracking system designed for paperless data collection in "real time", with initial use in the US focused on field research and some commercial applications.[2-4] Utilizing a radio frequency identification (RFID), LeeO™ scans tags and records individual animal weights quickly using a proprietary LeeO™ O Reader and blue tooth technology. This paper will describe "lessons learned" during the implementation of LeeO™ in a start-up commercial sow farm intended for internal multiplication and the journey toward precision pork.

Implementation

In March 2021, a start-up 3000 internal multiplication sow farm in Nebraska began populating with breeding age gilts and utilizing the LeeO™ system. The initial implementation involved placing a LeeO™ RFID sow tag in the ear of each of the incoming gilts as well as applying RFID identification to each of the individual stalls in breeding/gestation and farrowing. When the sows farrowed each live piglet was also given a LeeO™ UHF ear tag and registered into the LeeO™ system including its gender and birth weight using a LeeO™ piglet scale. Mortality and cross-fostering (adoptions in LeeO™) were then recorded using the LeeO™ O Reader and mobile app.

Lessons learned and future considerations by role

In November of 2021, a review of the LeeO™ implementation was conducted by a series of interviews with the various roles involved by the attending herd veterinarian. The interviews included open-ended questions regarding the successes and failures in the implementation of a new technology. The following is a summary of the key responsibilities and "lessons learned" for each of the key roles.

Owner: The critical role of the owner is to establish a vision and purpose for implementing new technology, providing adequate resources to successfully implement, and clarifying key measures of success for all involved. As with most farms and production systems, everyone is very busy, and labor can be a limiting factor. Ensuring that the vision and purpose are well understood especially as modifications are made is critical to all activities downstream at the farm.

Project lead: The owner assigned an internal project lead who was not the farm manager to serve as a liaison with the LeeO™ team. The project lead was not able to observe LeeO™ in another farm to "see it, try it, teach it" which can be detrimental to the confidence and understanding of the system as incongruencies and issues arise – this should be mandatory in future implementations. By visiting a farm with LeeO implemented, the project lead can have some preparation and previous knowledge of the potential hurdles expected in implementation and have a support system available to clear those hurdles later on down the line. In addition, it is important for the Project Lead to have the ability to put full resources of responsibilities toward the project. Oftentimes, those that we ask to "champion" the process have other system commitments. Ensuring that the Project Lead can (and does) commit full resources to the implementation will root out problems early as they arise.

Farm manager: The farm manager plays a multi-functional role during the implementation of new technology and may have the most difficult position in that they have a business to run. The farm manager obviously needs to be attuned and engaged with the implementation of the new technology but cannot realistically commit their full resources to that process. The farm manager rather needs to perform as the liaison between the project lead/ownership, the data management and the farm personnel carrying out the procedures directly. They are the true "middleman" in the process and will need the most support in order to maintain normal farm activity while the implementation of the new project progresses. They act as the sounding board in the entire process to indicate if the process is progressing appropriately or not.

Data management: Any time you introduce a new data collection system, there will be discrepancies between the new and old systems. It is critical that any "corrections" get investigated to understand if there is "operator error" at the farm level with the new technology, or if the previous system gathered or transcribed information that resulted in either grouping the data differently (definitions of weekly groups) or transcription errors. It is critical that the data management role understands and "buys in" to "why are we doing this" and there is timely troubleshooting when discrepancies in the data occur.

Farm personnel: While the ultimate "buy in" first must happen at ownership and upper management levels, formal meeting and presentation to the farm personnel that will be actively carrying out tasks needs to also take place to create an attitude of "owning it". Establishing that ownership in the process will favor correct implementation as well as early identification of hurdles to the proper task implementation. The farm personnel are the "end users" of the technology and getting "buy in" and transparency of any mistakes that may occur is a key component of implementation success.

Conclusions

One of the key learnings was the importance of clear purpose or the "why are we doing this" communicated routinely from the owner and project lead. When implementing new technology, we typically emphasize the importance of training to the "how", but we should never underestimate the importance of the "why". Table 1 summarizes the importance of communicating and understanding "the why", the scope of training needs, and the key measures that each role identified as outcomes of the new technology. It was also clear during this review that there needs to be a project champion both internally and externally during the critical training and implementation periods to ensure issues are addressed in a timely and effective manner. Implementing new technologies such as LeeO™ requires a commitment from leadership in vision and resources as well as timely response to issues as they arise on the farm and in the office. The continued success of new technologies requires clear outcome measures to provide continued motivation and curiosity to improve.

References

1. ter Linde P et al. The value of LeeO, an individual animal identification system adding value to the pork supply chain the Netherlands. AASV, 2020; 337.
2. Yeske P et al. Utilizing LeeO individual animal identification system in commercial research. AASV, 2020; 139.
3. Burton B et al. LeeO individual animal identification system implementation in a commercial sow farm. AASV 2021; 118-120.
4. Lerner A et al. LeeO individual animal data system to track and analyze birth to harvest information. AASV 2021; 352-353.

Table 1: Key roles and responsibilities when implementing technology

	Purpose, "The Why"	Training needs	Key measures
Owner	Develop & communicate	Scope of functionality, visit with other user leadership	Overall farm KPI's
Project lead	Communicate & liaise with provider	Full functionality, critical to visit farm using technology	Timely implementation, minimal downtime with technology
Farm manager	Communicate & oversee personnel	Detailed use of system, frequent external support	Capture accurate data, create on-farm reports, and able to trouble-shoot
Farm personnel	Understand & implement	Specific use functions, frequent external support	Capture accurate data, visualize daily activities, trouble-shoot
Data manager	Understand & implement	Full functionality, critical to know other data experiences	Create accurate systems reports, share data between systems

https://doi.org/10.54846/am2022/46

Distribution of injected fat-soluble vitamins in plasma and tissues of nursery pigs

Y. D. Jang[1], PhD; D. Hammer[2], PAS, BS; R. L. Stuart[2], PhD
[1]University of Wisconsin - River Falls, Wisconsin; [2]StuartProducts, Inc., Bedford, Texas

Introduction

Vitamin A, D, and E play critical roles in health, immunity, and reproduction in swine as fat-soluble vitamins. Vitamin A functions for vision and reproduction, vitamin D is important for Ca/P homeostasis, bone integrity and immunity, and vitamin E is a strong antioxidant that helps body protection against oxidative stress and has positive effects on reproductive performance and immunity in pigs. Fat-soluble vitamins are always supplemented in the swine diets due to modern husbandry practices in confinement housing with little exposure to sunlight and grass from which pigs could obtain fat-soluble vitamins, and suckling piglets obtain these vitamins via sow colostrum and milk. However, piglets are born with low levels of the fat-soluble vitamins, and colostrum and milk are low in these vitamins even though colostrum contains more fat-soluble vitamins than milk. In Jang et al,[1] plasma vitamin D and E concentrations in suckling pigs were declined or constant after a slight increase by colostrum intake within the first 5 days after birth (Figure 1 and 2). This means that piglets may not receive a sufficient amount of fat-soluble vitamins through the sow milk during the suckling period because there is no further increase after the colostrum intake. In addition, even though nursery pigs consumed the diets containing 5 times greater amount of fat-soluble vitamins (11,000 IU of vitamin A, 1,100 IU of vitamin D_3, and 77 IU of vitamin E per kg diet) than the NRC[2] requirement estimates, plasma α-tocopherol concentrations were reduced after weaning, and plasma 25-OHD_3 concentrations were constant after a slight increase in the first week postweaning (Figure 1 and 2).[3] Also, vitamin D level is low in pigs raised in confinement housing than outdoor housing as Arnold et al,[4] reported with a survey that the pigs in an age of 2-4 weeks had about 8-13 ng/mL in confined housing whereas the pigs raised in outdoor housing had about 60 ng/mL. This indicates that dietary fat-soluble vitamin requirements in NRC[2] may not be able to meet actual needs for pigs to ensure their health and immunity and that there is still a room for improving fat-soluble vitamin status for pigs. Since these fat-soluble vitamins could be destroyed readily if diets contain pro-oxidants such as inorganic trace minerals and unsaturated fatty acids, swine industry needs to consider vitamin stability when they formulate the diets and other administration routes rather than dietary supplementation could be considered to enhance fat-soluble vitamin status in pigs. Therefore, the purpose of this paper is to review experiments that evaluated the effect of fat-soluble vitamin injection, mainly vitamin D and E on plasma vitamin status and its distribution in tissues of nursery pigs.

Improving fat-soluble vitamin status in pigs

How can we improve fat-soluble vitamin status in pigs? Pigs can get fat-soluble vitamins through several routes such as oral administration, injection, milk, diet, drinking water, etc. Several previous studies demonstrated that fat-soluble vitamin supplementation regardless of the routes significantly increased plasma levels of these vitamins in pigs. When the vitamin D

Figure 1: Plasma α-tocopherol concentrations of suckling and nursery pigs without additional vitamin E administration (Jang et al[1,3]).

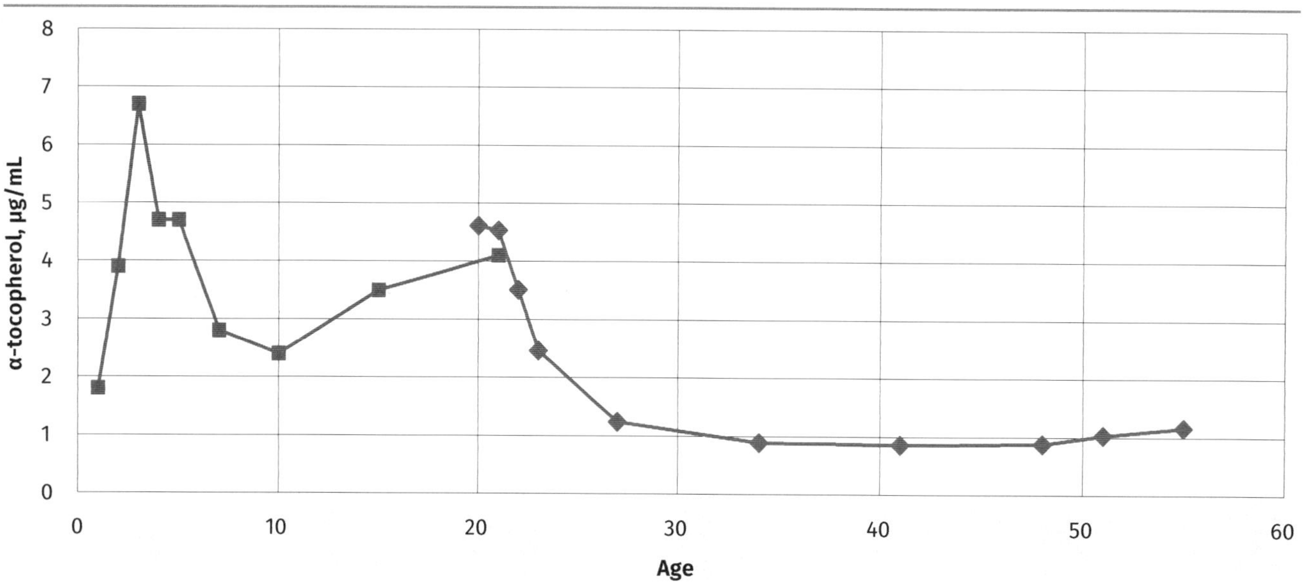

and E were supplemented to pigs via intramuscular injection (IM) injection (100,000 IU of vitamin D_3 and 1,000 IU of vitamin E as d-α-tocopherol using VITAL E®-NEWBORN) or drinking water (5,493 IU of micellized vitamin D_3 and 92 IU of micellized d-α-tocopherol per L of drinking water using EMCELLE® E-D_3 Liquid), it significantly increased plasma α-tocopherol and 25-OHD_3 concentrations and when they are supplemented via both injection and drinking water, there was an additional increase in plasma concentrations (Figure 3 and 4).[3] However, Jang et al[1] reported that intramuscular injection (VITAL E®-NEWBORN) had 1.6-1.8 times greater relative bioavailability to oral administration (EMCELLE® NEWBORN EAD; micellized vitamin D_3 and d-α-tocopherol) for vitamin D and E in newborn pigs when 40,000 IU of vitamin D_3 and 400 IU of d-α-tocopherol were administered since digestion and absorption process are not necessary with the injection, and the IM. injection route had slower plasma disappearance rate than the oral route. This is an important finding in this research that once fat-soluble vitamins are administered, the pig's body metabolized it so that these vitamins disappeared from the blood. This means that these vitamins are distributed to and stored in the tissues of pigs since fat-soluble vitamins are not readily excreted once they are absorbed. Each vitamin has their own metabolism when they are administered in the body so that the disappearance rate from the blood could vary depending on the type of fat-soluble vitamins in which it was the slowest with vitamin D, faster with vitamin E than vitamin D but the fastest in vitamin A for retinyl palmitate.

Figure 2: Plasma 25-OHD_3 concentrations of suckling and nursery pigs without additional vitamin D_3 administration (Jang et al[1,3]).

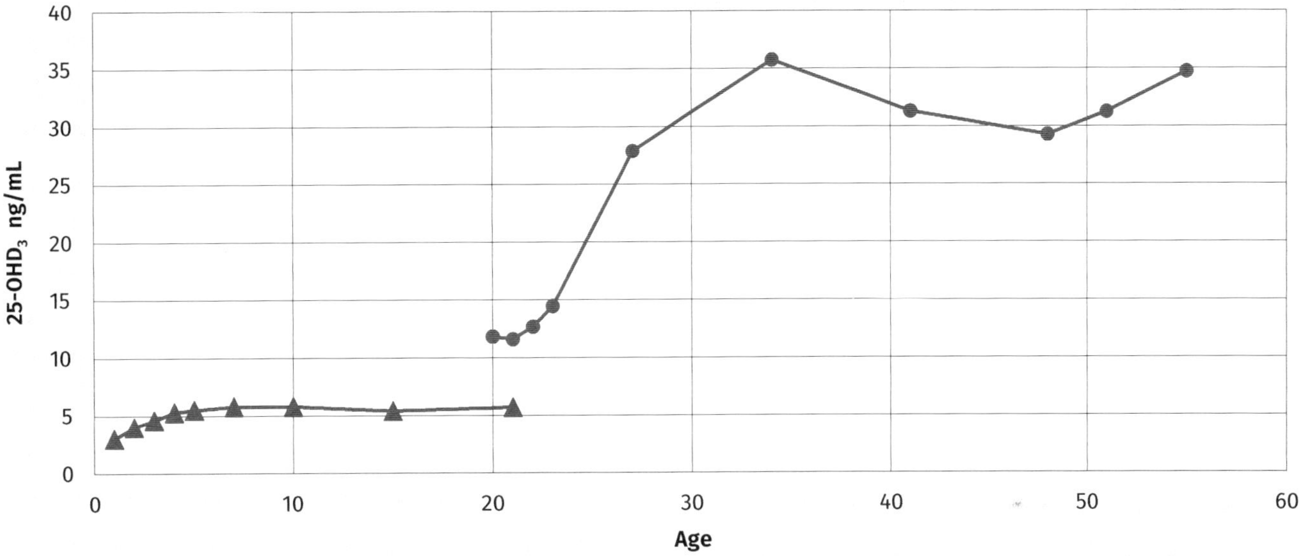

Figure 3: Effect of vitamin E injection to nursery pigs on plasma α-tocopherol concentrations (1,000 IU of vitamin E as d-α-tocopherol at weaning on d 20 of age).

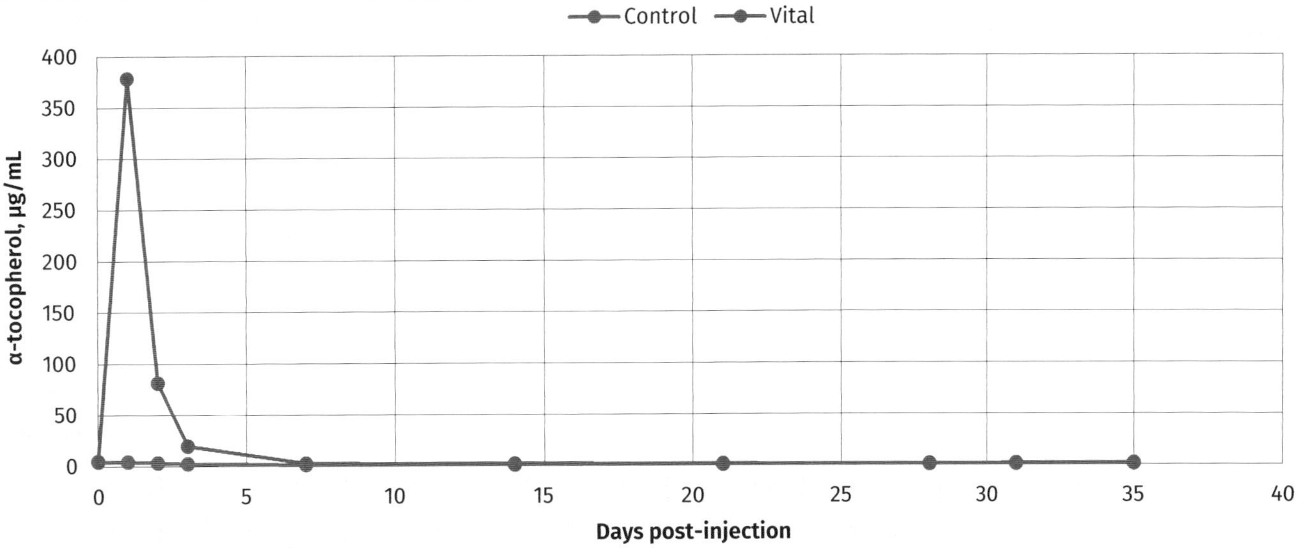

Injected fat-soluble vitamins are distributed in the body of pigs

Once fat-soluble vitamins are injected intramuscularly, it is released into blood and transported to the liver. Then, liver converts fat-soluble vitamins into storage, circulating, and active forms, stores and releases them to any other tissues to store. When 900 IU of d-α-tocopherol (VITAL E®-A+D) was injected intramuscularly to nursery pigs, plasma α-tocopherol concentrations were significantly greater than the control pigs up to 2 weeks but the level decreased from the peak at d 3 post-injection so that the difference between treatments disappeared at d 14 post-injection.[5] Then, there was a statistically significant increase in liver, muscle, and heart α-tocopherol concentrations that followed the disappearance of α-tocopherol from the blood (Figure 5). This is evidence that injected fat-soluble vitamins is traveled to liver, metabolized to be certain forms, and then stored in tissues where it is needed. Therefore, a corresponding increase in tissue α-tocopherol concentrations could explain the rapid decline in its plasma concentrations. A similar result was observed in Hidiroglou et al,[6] that reported that α-tocopherol concentrations in plasma, spleen, liver, kidney, heart, lung, and hip muscle at d 28 of age increased significantly after IM injection of d-α-tocopherol (VITAL E®-500) to suckling piglets at d 3 (500 IU), d 7 and 14 (1,000 IU) of age. These results indicate that all tissues actively accumulate vitamin E and vitamin E is important for all body tissues and organs to protect them from the oxidative stress. Interestingly, tissue α-tocopherol concentrations were reduced as time goes by from d 7 to 14 post-injection, which means that body stores injected fat-soluble vitamins and then metabolizes them for their functions resulting in the reduction in the tissue concentrations. Therefore, these vitamins may need multiple administration to maintain their high levels in the body. Regarding brain α-tocopherol concentrations, there was no difference between the injected and control pigs.[5] Lauridsen et al,[7] reported that when labeled vitamin E (RRR-α- or all-rac-α-tocopheryl acetate) is orally administered to sows, their progeny could receive α-tocopherol from sow milk and deposited it in order of liver>lung>heart>kidney>brain on an organ basis with a preference for RRR-α-tocopherol. In the other study, the suckling rats from dams fed vitamin E fortified diets had greater liver and plasma α-tocopherol concentrations than control rats except for brain that had no difference in α-tocopherol concentrations.[8] This result indicates that brain may control α-tocopherol accumulation and it may take a longer time to accumulate vitamin E into brain than the other tissues. It should be noted that synthetic vitamin E is a racemic mixture of eight stereoisomers (all-rac-α-tocopherol) and only RRR-α-tocopherol in the mixture is a natural form that has higher relative bioavailability than the synthetic vitamin E (all-rac-α-tocopherol). Laruidsen et al,[7] suggested that the relative bioavailability of the natural compared with the synthetic form of vitamin E is 2:1 instead of 1.36:1 in that study. Therefore, the form of vitamin E should be considered when administered to pigs to maximize its efficiency in enhancing vitamin E status.

There is a positive correlation between dietary vitamin D_3 level and tissue and serum 25-OHD_3 concentrations.[9] When pigs were fed diets with various levels of vitamin D_3 (5 to 50 μg/kg), serum 25-OHD_3 concentrations linearly increased with increasing dietary vitamin D_3 levels. Then, serum 25-OHD_3 concentrations were fully correlated with tissue 25-OHD_3 concentrations including liver, adipose tissue, white and red muscle tissues where tissue levels increased with increasing serum 25-OHD_3 concentrations. The concentrations of 25-OHD_3 in tissues were 3-6 times greater in liver and adipose tissues than white and red muscle tissues indicating that liver and adipose tissues are major organs for vitamin D storage. Mawer et al,[10] in human study with radioactively labeled vitamin D_3 injection (1,750 to 5,920 IU), reported that vitamin D injection increased serum 25-OHD_3 concentrations followed by a rapid decrease in serum level by physicochemical mechanisms such as distribution to adipose tissue and binding to tissue protein so that the potential toxicity of excess vitamin is limited. Then, this disappearance of a large portion of injected vitamin D from the blood means that liver took them up temporarily and produced vitamin D metabolites for redistribution to all tissues as radioactively labelled vitamin D and various metabolites were detected in all tissues. Therefore, although serum

Figure 4: Effect of vitamin D_3 injection to nursery pigs on plasma 25-OHD_3 concentrations (100,000 IU of vitamin D_3 at weaning on d 20 of age).

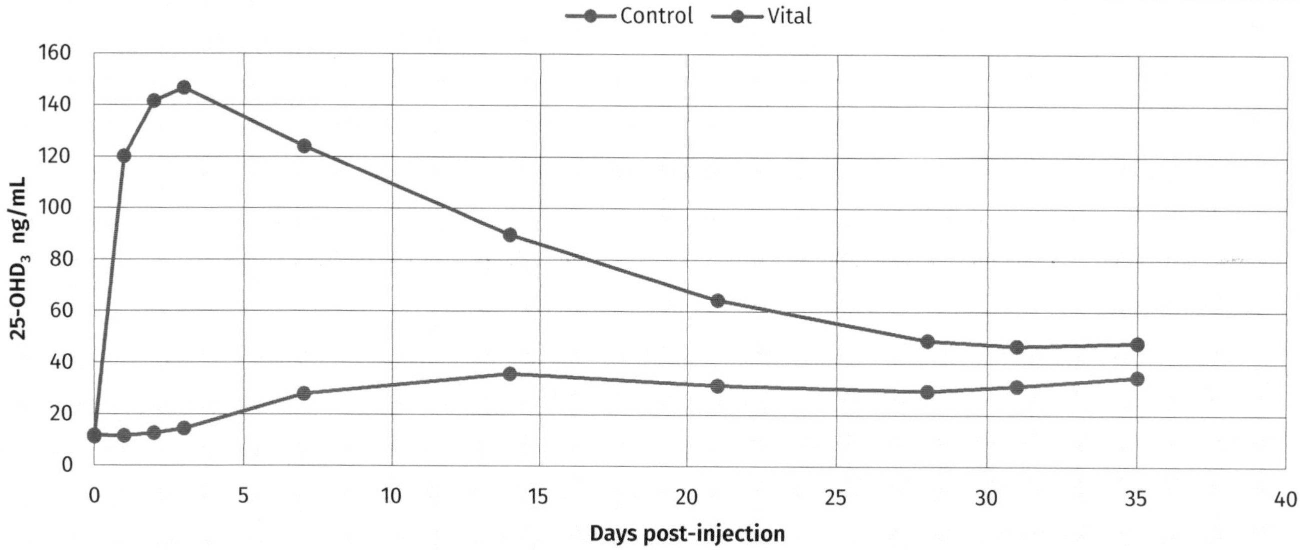

Figure 5: Tissue α-tocopherol concentrations nursery pigs after vitamin E injection

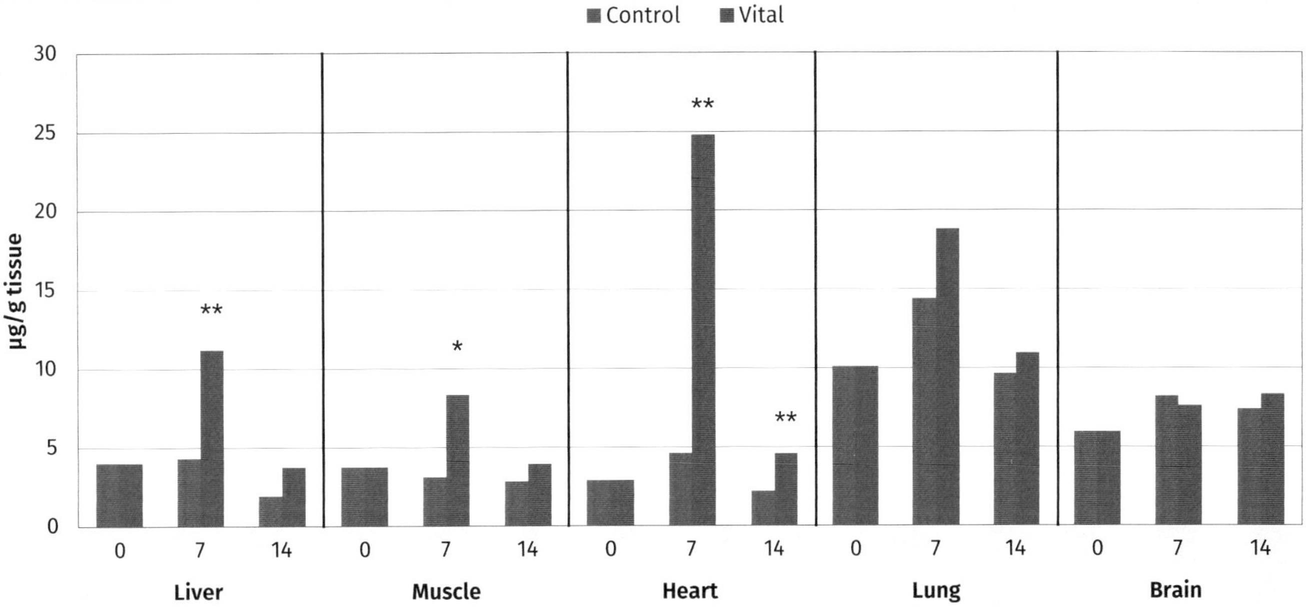

* $P < .05$ ** $P < .10$

$25\text{-}OHD_3$ is a good indicator for vitamin D status, the analysis of tissue vitamin D metabolites should follow to evaluate the corresponding distribution of injected vitamin D to pigs.

Regarding to vitamin A, in the calf study, retinyl palmitate injection at 150,000 and 300,000 IU to calves at birth and one month of age improved weaning weight and marbling score at harvest, which suggest that vitamin A may play role in muscle development and intramuscular fat deposition.[11] Jang et al,[5] reported that 300,000 IU of retinyl palmitate (VITAL E®-A+D) injection to nursery pigs significantly increased plasma retinol concentrations followed by a statistically significant increase in liver retinol and retinyl palmitate concentrations but not in the other tissues where vitamin A content was under the detection limit. Therefore, injected retinyl palmitate is mainly deposited in the liver for further metabolism as a primary organ of vitamin A storage.[5]

Importance of maintaining optimum vitamin D and E status

It has been suggested that low vitamin D and E status is associated with several disease infection in pigs. Nonnecke et al,[12] reported an inverse association between vitamin D and E status and serum amyloid A in infected calves suggesting that the infection-induced acute phase response contributed to the reduced vitamin status of these animals. Also, low blood vitamin D status was identified in periweaning failure to thrive syndrome (PFTS) in pigs.[13] Therefore, it is important to maintain optimum status of vitamin D and E to protect pigs from oxidative stress and disease infection for healthy swine herds. Then, vitamin D and E injection could be an effective way to enhance their status in pigs as it has been reported that administration of injectable A, D, and E combination decreased oxidative stress in pigs at 60 days of age.[14]

Vitamin E is known to protect pig's body from free radical damage as a strong antioxidant and plays a role in immunocompetence as it increases immune protection and resistance to bacterial infections. Hidiroglou et al,[6] reported that plasma α-tocopherol concentrations and antibody titers to Keyhole limpet hemocyanin of piglets injected with vitamin E were significantly higher than the control piglets. It has been also demonstrated that vitamin E has an influence on enhancing cell-mediated and humoral immune responses of pigs.

It has been known that the majority of immune cells such as T-cells and antigen presenting cells expresses vitamin D receptor, which apparently indicate that vitamin D plays an important role in immune response as it is involved in T-cell activation and modification of the phenotype of antigen presenting cells.[15] Yang et al,[16] reported that mega-dose of dietary $25\text{-}OHD_3$ (155.5 µg/kg equivalent to 6,220 IU vitamin D_3/kg) to weaning pigs before a porcine epidemic diarrhea challenge resulted in improvements in intestinal integrity, down-regulation of mRNA abundance of inflammatory cytokines and alleviation of diarrhea severity even though the growth was not restored back to the control pigs. Petersen et al,[17] also reported that there was an increase in the level of antibodies for PRRS virus when PRRS virus was challenged to weanling pigs receiving oral vitamin D_3 (40,000 IU) at birth and weaning indicating that vitamin D administration may stimulate a faster and more robust immune response for weaning pigs.

Conclusions

Vitamin D and E administration can enhance plasma and tissue vitamin levels and the injection is more effective to increase plasma concentrations than oral administration. Optimum level of vitamin D and E in the blood of pigs has not been established yet but it is clear that low fat-soluble vitamin status is related to certain disease condition. Increased plasma vitamin D and E concentrations by the administration is declined as time goes by and tissue concentrations of these vitamins are reduced after an increase by corresponding distribution of injected vitamins. Therefore, pigs may need multiple injections and/or continuous supplementation of fat-soluble vitamins to ensure that the pigs have sufficient fat-soluble vitamins available in the body for their best health and immunity.

References

1. Jang YD et al.: 2015, J Anim Sci -93:5273.
2. National Research Council-Nutrient requirements of swine: 2012, National Academy Press.
3. Jang YD et al.: 2018, Asian-Australas J Anim Sci -31:278.
4. Arnold J et al.: 2015, JSHAP -23:28.
5. Jang YD et al.: 2020, Asian-Australas J Anim Sci -33:1985.
6. Hidiroglou M et al.: 1995, Reprod Nutr -35:443.
7. Lauridsen C et al.: 2002, J Nutr -132:1258.
8. Amusquivar E et al.: 2000, J Nutr -130:2855.
9. Burild A et al.: 2016, J Nutr Sci -5:e3.
10. Mawer EB et al.: 1972, Clin Sci -43:413.
11. Harris CL et al.: 2018, J Anim Sci Biotechnol -9:55
12. Nonnecke BJ et al.: 2014, J Dairy Sci -97:5566.
13. Huang et al.: 2011, JSHAP -19:340.
14. Lima et al.: 2012, Preq Vet Bras -32:727.
15. Prietl et al.: 2013, Nutrients -5:2502.
16. Yang et al.: 2019, Animals 9:627.
17. Petersen ATS et al.: 2013, AASV p.355.

https://doi.org/10.54846/am2022/48

A new prescription platform to manage emerging and diverse pathogens

Ashley Petersen, MS; Amber Schueler; Dean Dau, DVM; Alan Young, PhD
Medgene Labs, LLC Brookings, South Dakota

Introduction

The USDA Center for Veterinary Biologics (CVB), with the most recent in a long history of regulations designed to support innovation in veterinary biologics, has established guidance documents (VSM 800.213 and 800.214) allowing the development and licensure of Prescription Platform (RxP) vaccines. The regulatory path of platform products is designed to facilitate rapid responses to evolving pathogens with limited risk of the loss of vaccine safety or efficacy.

A Prescription Platform vaccine is a platform-based, standardized vaccine that is nonviable and non-replicating (killed). By CVB definition, a production platform is a manufacturing process that relies on a single vector or expression system (VES "backbone") and a standard process for inserting a gene or genes of interest (GOI) into the system to generate different recombinant Master Seeds or Sequences, which are then used to produce product following a standardized method of manufacturing.[1] The GOI is selected to be one that codes for a protein that is known to convey immunity against the agent in question, so that the final product is expected to produce disease coverage in vaccinated animals.

In order to provide RxP vaccines, a biologics manufacturer must first develop an Initial Product, which is a fully licensed product that establishes factors including a standardized manufacturing method, the "backbone" used for the product, and other parameters such as minimum age, maximum antigen level, dose size, route of administration, and adjuvant type. This Initial Product must also demonstrate both safety and effectiveness. Once the Initial Product is completed, the same licensed procedures are followed to manufacture RxP vaccines using other GOIs. These RxP vaccines can then be prescribed by attending veterinarians for use in at-risk animals in the field.

Materials and methods

The Initial Product license was developed with an influenza A virus-swine (IAV-S) H3 subunit vaccine derived from a recombinant baculovirus vector with the H3 gene inserted. Two vaccines were tested, one with ARMOR™ adjuvant, a proprietary Medgene Labs adjuvant, and Emulsigen®. A standardized manufacturing method was initiated to produce the antigenic protein combined with an adjuvant to achieve the final product. A USDA requirement for fully licensed products to confirm enough antigenic content is included to ensure efficacy, a potency assay was developed and subsequently validated and approved by CVB for the Platform vaccine (Initial Product) manufacturing process. Efficacy was accomplished via homologous challenge in pigs. The primary variable was lung lesions. Safety was established through pig vaccination at three separate geographical locations.

To further validate the ability of Medgene's prescription platform vaccines as a valuable tool in vaccination programs, a study was conducted by a commercial swine unit and designed to evaluate a RxP vaccine formulated by Medgene Labs, containing an inactivated PCV2d antigenic protein produced by a baculovirus vector and adjuvanted with ARMOR™. Twenty-five hundred (2,500) pigs were randomly divided into control and vaccinate groups, vaccinated once at 3 weeks of age, and intranasally challenged with PCV2d approximately 4.5 weeks post vaccination. Following challenge, pigs were observed and data collected by swine unit personnel.

In addition to work with IAV-S and PCV2d, Medgene has obtained other GOIs for RxP vaccines through diagnostic submissions by veterinarians. Using next generation diagnostic methods at diagnostic laboratories, viruses detected by either isolation or PCR were sequenced for target GOIs. With permission from the attending veterinarian, the sequences are provided to Medgene Labs, a licensed RxP manufacturer, for RxP vaccine development.

Results

The lung lesions in both the ARMOR™ and Emulsigen® vaccinated groups of pigs in the IAV trial were significantly reduced ($P < 0.0016$ and 0.0021, respectively) compared to the unvaccinated pigs (Figure 1). Both vaccine formulations were well tolerated by the vaccinated animals. As a result of successful efficacy and safety studies, Medgene Labs has recently been granted a license for a Platform vaccine (Initial Product) and subsequently were granted a license for a RxP vaccine for swine in the ARMOR™ adjuvant. A license for the Emulsigen® adjuvanted Initial Product and RxP vaccine is pending.

The results of the PCV2d challenge study showed a significant reduction in mortality. Mortality in the vaccinated group was 7.16% compared to 18.24% in the unvaccinated group. Morbidity was determined as the number of pigs requiring treatment for illness. The vaccinated group had a morbidity rate of 42.4% while the unvaccinated group had a morbidity rate of 100%. Average daily gain was significantly ($P < 0.0001$) improved in the vaccinated group, 1.81 lbs./day, compared to the unvaccinated group, 1.70 lbs./day.

Further work to produce GOIs of interest to veterinarians seeking RxP vaccines, the following have been established as Master Seed constructs for use in RxP vaccines developed by Medgene Labs: influenza A virus-swine (H1, H3); rotavirus A, B and C; porcine circovirus 2 (a, b, d); sapovirus; porcine epidemic diarrhea virus; transmissible gastroenteritis virus; porcine parainfluenza virus; porcine delta coronavirus; porcine astrovirus; atypical porcine pestivirus; and porcine parvovirus.[2,3]

Discussion

Commercial vaccines can take years to develop, and may not be available when needed for new, emerging diseases especially if the emerging virus cannot be grown in vitro. Prescription vaccines overcome this obstacle by using only the GOI from the emerging pathogen which can be derived from molecular

Figure 1: Percent lung lesions of vaccinated versus unvaccinated pigs after direct IAV challenge.

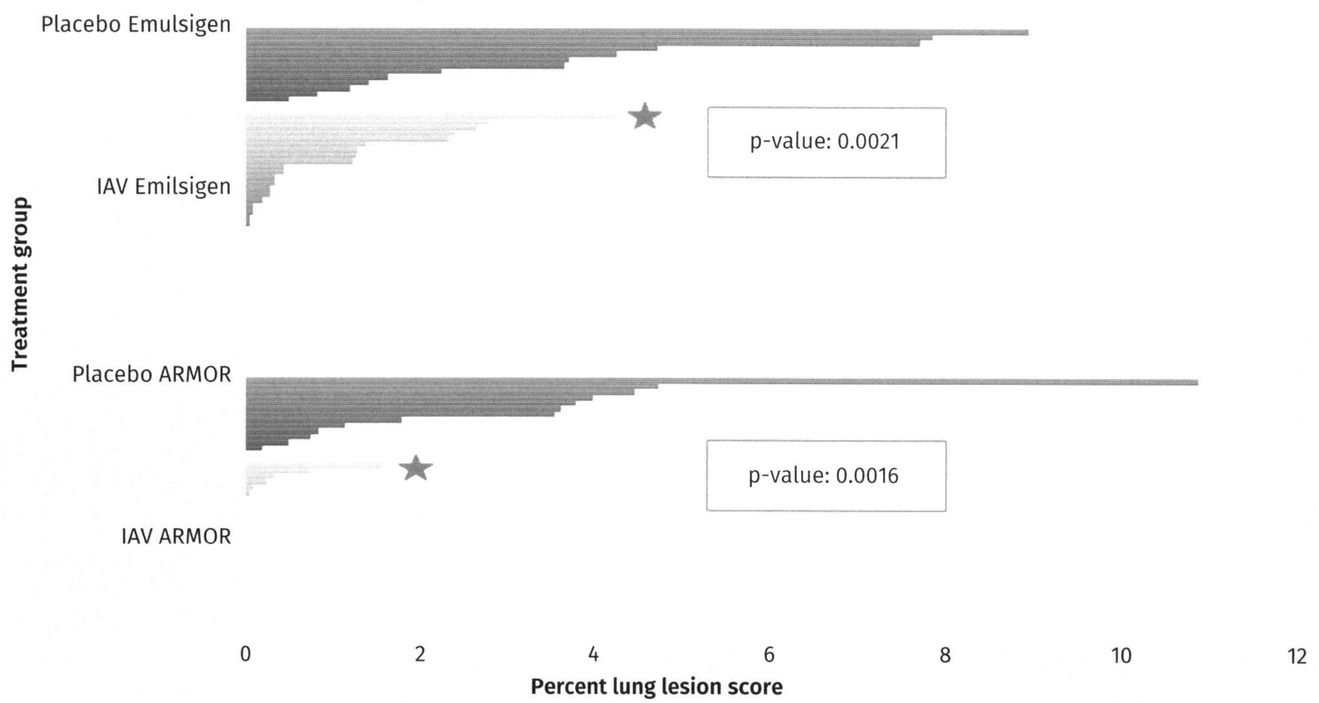

sequencing. In addition, commercial vaccines that have been available for many years can contain outdated antigens that do not reflect the strains currently causing issues in the field. Prescription Platform vaccines, on the other hand, can use GOIs from these new strains to develop current vaccines through the same manufacturing process as the original commercially licensed platform vaccine without the long wait times. In most cases, development of a new RxP vaccine can be done in approximately twelve weeks from start to finish.

Medgene Labs has recently been granted a license for a Platform vaccine (Initial Product) and subsequently was granted a license for a RxP vaccine in swine. The Medgene platform is based on a baculovirus expression system and the prescription platform license allows insertion of GOIs for other relevant viruses. Each RxP vaccine is manufactured on a custom basis using the USDA-approved manufacturing process. Medgene's conserved production process begins from inception and all inputs are certified in the creation of new vaccine targets. New targets are field tested and based on the best available scientific information. In addition, Medgene's process is completely free of ingredients of animal origin, eliminating any potential for contamination with extraneous agents such as ASF and others. Prescription platform vaccines are a tool that can be used by licensed veterinarians in herds where they have a valid VCPR and are especially useful for highly diverse pathogens such as IAV or PCV2d when commercial vaccines do not have the most current strains and for emerging pathogens such as rotavirus C (RVC) that cannot be grown in vitro for vaccine production.

References

1. Erdman, MM, Clough NE, Hauer PJ. Review of Updated Regulations and Product License Categories for Veterinary Vaccines in the United States. *JAVMA* 2020;257(11): pp 1142-1147

2. Van Reeth, K, Vincent, AL. (2019). Influenza Viruses. In Diseases of Swine 11th edition (eds J.J. Zimmerman, L.A. Karriker, A. Ramirez, K.J. Schwartz, G.W. Stevenson and J. Zhang). **https://doi.org/10.1002/9781119350927.ch36**

3. Neumann, EJ, Ramirez, A, Schwartz, KJ, editors; Swine Disease Manual, Fifth Edition. pp. 78-80. Copyright 2020, American Association of Swine Veterinarians.

https://doi.org/10.54846/am2022/49

Evaluation of pigs vaccinated with PRRSGard® or Prevacent® PRRS under a natural challenge model

B. Burton[1], DVM; J. Okones[2], DVM; C. Smith[2], DVM; R. Kanefsky[3]
[1]Suidae Health & Production, Algona, Iowa; [2]Pharmgate Animal Health, Wilmington, North Carolina;
[3]Tufts Cummings School of Veterinary Medicine, Grafton, Massachusetts

Introduction

Porcine reproductive and respiratory syndrome (PRRS) continues to be an extremely costly and endemic challenge throughout the swine producing regions of the world. In the US alone, annual economic losses have been estimated as high as $1billion.[1] Vaccination with modified-live PRRSV (MLV) vaccines has been shown to reduce clinical signs, viremia and lung lesions, which improve health and performance in growing pigs. The swine industry has several different commercial PRRSV MLV products available today. All have shown partial efficacy, this variance in efficacy can be due to the difference in virulence of PRRS field isolates and timing between exposure to the virus. The objective of this study was to evaluate the efficacy of two newer Lineage 1 PRRSV MLV vaccines, PRRSGard® or Prevacent® on a SW MN commercial site, with expected natural challenge.

Materials and methods

Pigs from a 2500 head sow farm located in northwest Iowa were used in this study. This sow farm has LeeO® implemented, so all pigs had sow ID, sow's parity, piglet sex, birth weight and wean weight recorded for each pig individually. This sow farm had previously broke with PRRS and was in the middle of herd closure to eliminate the virus.

At weaning, pigs were placed in a 2400 head wean-to-finish site with two 1200 head barns located in southern Minnesota, a very pig dense area. 625 pigs were weaned twice a week to fill the site (4 weans total). At each weaning, pigs were randomly and evenly split between the two barns, so each barn had equal amounts of pigs that were the exact same age. Parity and weaning weight were similar between each group.

The day after the second and fourth wean group, pigs were vaccinated for PRRS. One barn was vaccinated with a full dose of PRRSGard® and the other with a full dose of Prevacent PRRS®. Coveralls and boots were changed between each barn during vaccination and every day going forward after that. Serum was also collected from 60 pigs (30 pigs per barn) prior to vaccination. Serum was sent to Iowa State University Veterinary Diagnostic Lab (ISU-VDL) for PRRSV X3 ELISA and PRRS PCR to establish the status of the pigs prior to vaccination.

Oral fluids were collected 4 weeks after vaccination and every 2 weeks until a wild-type PRRS was diagnosed. If there were any health challenges throughout finishing, diagnostics were taken then as well. All samples were sent to ISU-VDL for PRRS PCR and ORF5 sequencing if results were positive.

Individual mortalities and daily treatments by barn were tracked from weaning to marketing. Individual weights were also taken on all pigs at approximately 140 days on feed (DOF). All cull pigs were identified with a reason for culling, and all full-value pigs are scanned prior to their market load. Late finishing and marketing data is still pending due to the timing of the study.

Results and discussion

Initial status of the pigs

Serum from 60 pigs revealed 67% of the pigs being positive on PRRS X3 ELISAs with an average S/P ratio of 0.612. The same serum samples were pooled by 5 and tested for PRRS by PCR and was 100% negative.

Mid-nursery PRRS status

4 weeks post-vaccination, oral fluids were taken and 3 of 4 oral fluids were positive with average Ct 35. ORF5 PCR was negative, so sequencing was unavailable. 2 weeks later (6 weeks post-vaccination), oral fluids were taken again and at this time, they were 100% positive with an average Ct 31 in both barns. ORF5 sequence came back as 1-7-4 wild type and was a 100% match to the same strain that was isolated at the sow farm during the initial PRRS break.

Morbidity and mortality through nursery phase

Daily number of head individually treated with injectable antibiotics was tracked for each barn. Figure 1 demonstrates the trends in daily shots given for the first 8 weeks on feed or the "nursery phase" by barn/treatment group. There were minor differences during this timeframe between the barn vaccinated with PRRSGard® versus the barn vaccinated with Prevacent®.

Mortality by barn was also recorded through the nursery phase. Figure 2 shows the percent mortality by week on feed between the two groups. Oral fluid results would explain why there were spikes in mortalities around week 4 and 6 as those were the first time PRRS was detected and then detected at high enough levels to sequence. Overall, the mortality after 8 weeks on feed for the PRRSGard® group was 3.69% whereas the mortality for the Prevacent® group was 7.25%.

Daily treatments, mortalities, and individual pre-market (140 DOF) weights will be recorded through marketing to evaluate the full wean-market performance effects between these two groups.

Summary

This study uses a natural challenge model utilizing pigs from a sow source with a low prevalence of PRRS positive pigs at weaning. A natural challenge model can be a better representation of what is seen out in the field when it comes to PRRS exposure post-weaning laterally or similar to this case as low prevalence vertically. The initial results up to the the nursery phase would suggest although pigs did not seem more affected or sick in one group versus the other by the number of individual treatments given on a day-to-day basis, the PRRSGard® pigs were more likely to survive the health challenge compared to the pigs vaccinated with Prevacent®. Further data collected in this study will continue to build the story on the performance differences, if any, between pigs vaccinated with PRRSGard® versus Prevacent® when challenged with PRRS.

Reference

1. Holtkamp DJ. et al. Assessment of the economic impact of porcine reproductive and respiratory syndrome virus on United States pork producers. J Swine health and Prod. 2013. 21:72.

Figure 1: Individual daily treatments by barn.

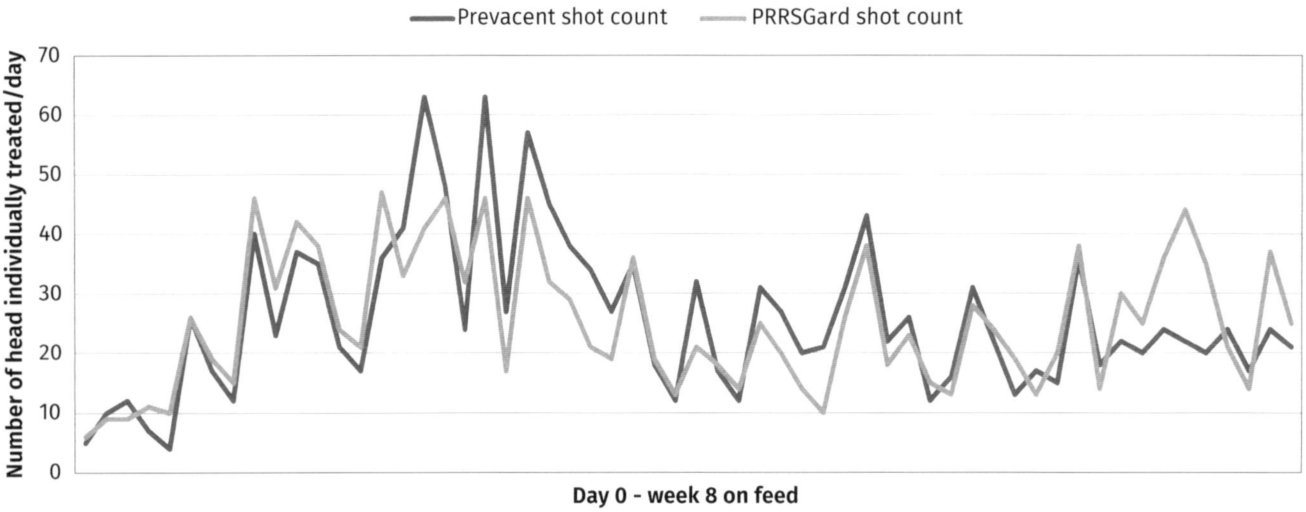

Figure 2: Percent mortality by week on feed.

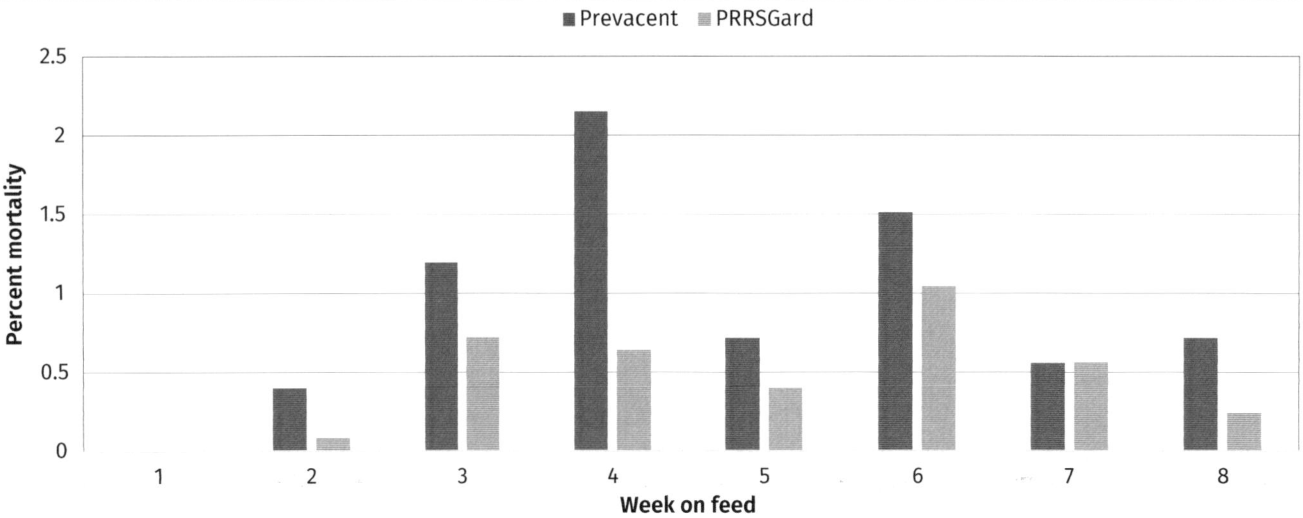

https://doi.org/10.54846/am2022/50

How GlobalVetLink assists swine veterinarians with health and movement documentation management

Aaron Lower, DVM
Carthage Veterinary Services, Ltd.

Introduction

GlobalVetLink is the industry leader in digital animal health and movement documentation creation and management. The GlobalVetLink platform streamlines animal health compliance documentation making it easy, accurate and accessible to veterinarians. This oral presentation will provide an overview of the solutions that GlobalVetLink provides to swine veterinarians and how they can be used in varying practice settings to improve processes and streamline communication (Figure 1).

Electronic Certificates of Veterinary Inspection simplify animal movement

GlobalVetLink digital Certificates of Veterinary Inspection (CVIs), also known as health certificates, help veterinarians keep their clients moving with easy-to-use software for complete, accurate certificates.[1] GVL CVIs are approved for use in all 50 US states and US territories and are a secure alternative to paper certificates. Through the GlobalVetLink platform, you can auto-fill client and animal information, and built-in state movement requirements ensure you are creating accurate certificates. Immediate submission to state officials saves you time and you can easily share completed certificates with clients and transporters online via MyVetLink.

The greatest benefit to veterinarians using GlobalVetLink CVIs is the built-in regulatory compliance. GVL CVIs are automatically checked against the regulations based on the state of origin and state of destination entered on the certificate, ensuring compliance with state entry requirements. GlobalVetLink works with state veterinarians to maintain the regulations to the highest standards, taking the worry of compliance off the shoulders of the veterinarian. The immediate submission of completed CVIs to state animal health officials also saves time and streamlines the CVI process for veterinarians and clinic staff. Electronic tracking of animal movements that pulls information from an accurate site database will be critical information for state animal health officials to make decisions regarding animal movements in the event of a foreign animal disease outbreak.

Streamlined veterinary feed directive management

Paper and word processing methods of writing VFDs and scripts involve a large amount of time, paperwork, and resources to ensure clear communication, accuracy of drug label requirements, completeness of information and compliance with regulations. The industry standard VFD software is provided by GlobalVetLink, which has been available since 2005.

With the GVL platform, veterinarians, producers, and feed distributors gain access to their VFDs in a secure, CFR 21 Part 11 compliant database, thus fulfilling the two-year storage requirements in Guidance for Industry 120.[2] Feed distributors receive copies of VFDs from veterinarians through GVL and have the option to upload digital copies of paper, emailed or faxed VFDs to their account. GlobalVetLink's VFD system also allows distributors to communicate directly with veterinarians and request VFDs for customers, and it notifies all parties as VFDs near expiration for easier renewal.

The VFD solution is another example of how GlobalVetLink is assisting veterinarians with regulatory compliance and easing the stress that comes with the variety of rules and regulations that must be followed. Electronic VFDs through GlobalVetLink streamline the process, from completing the VFD to sharing with the producer and distributor. The renewal notifications and renewal process assist veterinarians in ensuring that their clients have the medicated feed products that they need when they need them.

Simple solution for complex diagnostic analysis

LabLink HIMS is a real-time diagnostic solution that enhances herd health management for disease monitoring, analysis, and pathogen response. With LabLink HIMS, veterinarians can rapidly collect and aggregate diagnostic results from multiple labs. You can also organize data for efficient project management and track pathogen prevalence over time. LabLink HIMS allows veterinarians to increase herd health management efficiency and enable more effective treatment plans. As a veterinarian, easy access to diagnostic data is important for making herd health decisions.

The use of sequencing to compare viruses and bacteria has exponentially grown as a tool that veterinarians are using. The integration of LabLink HIMS and Disease BioPortal from the University of California, Davis allows veterinarians and livestock production personnel to manage sequencing results and improve the quality and speed of health-based decision making. LabLink HIMS automatically transmits case data into the Disease BioPortal, which offers more dynamic tools for interpretation, visualization, and multi-level analysis, such as phylogenetic analysis of sequences and spatio-temporal-genomic mapping. As a veterinarian, this integration is essential for deeper analysis of the data that is collected and managed through GlobalVetLink.

Efficient autogenous vaccine management

HerdPoint is a custom-made vaccine management tool. HerdPoint gives veterinarians greater insight into their ongoing isolate management and improves the efficiency of the nonadjacent-herd approval process. HerdPoint is now integrated into GlobalVetLink's online platform, making it more user friendly for veterinarians

Figure 1: GlobalVetLink improves processes and streamlines communication.

to view and manage their clients' isolates.[3] Many veterinarians already use GVL to manage animal health and movement documents. Now they can also manage their clients' custom-made vaccine isolates with Newport Laboratories in the same platform.

HerdPoint provides veterinarians and veterinary technicians with easy access to isolates and products, as well as notifications to assist with management of isolates and requests. Veterinarians and veterinary technicians can also review and understand isolate and product information at a glance. This greater transparency of data provides the opportunity for easy decision making. The GlobalVetLink platform allows veterinarians and veterinary technicians to easily view the status of isolates and request extensions and releases. The built-in sorting features and email notifications provide easier management of isolates and non-adjacent herd requests, and the simplified non-adjacent herd request process allows for quick processing of applications.

Conclusion

In conclusion, GlobalVetLink provides many solutions to simplify regulatory compliance and streamline daily processes for veterinarians and clinic staff. Electronic CVIs and VFDs are easy to complete through GlobalVetLink and built-in regulations takes the worry of regulatory compliance off the shoulders of the veterinarian. LabLink HIMS provides an easy solution for complex data analysis and the GlobalVetLink platform also simplifies autogenous vaccine management. With the constant changes in animal health and movement regulations, the GlobalVetLink platform supports veterinarians and practice staff to make these processes as easy and accurate as possible.

GlobalVetLink

GlobalVetLink is the leader in disease tracking and compliance documentation for animal health. The GVL Compliance Assistant SaaS platform collects, communicates, and stores animal health record information and complete compliance certificate records. Veterinarians for all species have access to automated creation and electronic submission of critical compliance documents, as well as having a real-time connection to an extended network of testing labs for submission and tracking of results. The GVL Compliance Assistant is used by over 10,000 production, equine, and companion veterinary clinics to accurately create, manage, and submit digital health certificates required by the USDA, state animal health offices, and international animal health agencies. GlobalVetLink uses artificial intelligence and blockchain technology to authenticate, and securely store and communicate animal health records. For more information on GlobalVetLink, visit **globalvetlink.com**.

References

1. Rademacher: 2020, AASV Industrial Partners Poster and Abstract. Compliance and efficiency in swine health documentation.

2. US Food and Drug Administration. CVM GFI #120 Veterinary Feed Directive Regulation Questions and Answers. March 2019. Available at: **www.fda.gov/regulatory-information/search-fda-guidance-documents/cvm-gfi-120-veterinary-feed-directive-regulation-questions-and-answers**. Accessed 22 October 2021.

3. Yeske: 2021, AASV Industrial Partners Presentation and Abstract. Easier Autogenous Vaccine Management through GVL.

https://doi.org/10.54846/am2022/51

Efficacy of an adjuvanted porcine parvovirus (PPV) vaccine in a reproductive heterologous PPV challenge model

R. Smiley, MS; T. Kaiser, DVM, MS; J. Li, PhD; R. Philips, DVM
Boehringer Ingelheim Animal Health USA Inc., Duluth, Georgia

Introduction

Porcine parvovirus (PPV) is a significant swine pathogen that can cause reproductive failure in dams. Infection of pregnant sows with PPV can cause reproductive failure including fetal death and resorption, presentation of mummified fetuses at farrowing, and irregular return to estrus. Conventional killed vaccines have been available for many decades,[1] but until now attention to this important swine disease with new vaccine technologies has been lacking. The objective of this study was to evaluate the efficacy of Reprocyc ParvoFLEX®, a new adjuvanted PPV subunit vaccine (baculovirus vector, inactivated) in a reproductive challenge model with a heterologous PPV.[2]

Materials and methods

Forty (40) 5- to 6-month-old pre-breeding gilts (with no prior history of reproductive disease or vaccination against PPV) and were confirmed PPV seronegative (HI serology) and viremia negative (PCR), were ranked by weight and blocked into three groups: Reprocyc ParvoFLEX® (Group VC, N = 24), product-matched placebo (Group NVC, N = 12), and a non-vaccinated, non-challenged control group (Group NTX, N = 4). Treatments were randomly assigned within each block, and gilts were commingled so that treatment was balanced across four pens. On Study Day 0 (D0), VC and NVC treatments were applied intramuscularly (IM) as a 2-mL dose. On D21, the VC Group and NVC Group received a 2-mL booster dose of the PPV vaccine or placebo, respectively. Gilts' estrous cycles were synchronized, and they were artificially inseminated between D37-D50. Gilts in both treatment groups were evaluated for pregnancy on D74, confirming pregnancy in 22 Group VC gilts, 10 Group NVC gilts, and 4 Group NTX gilts. On D81, at approximately 40 days of gestation (dG), Group NTX gilts were sacrificed, and all fetuses tested negative for PPV infection by polymerase chain reaction (PCR). Gilts in the remaining groups were challenged with 6.77 $\log_{10}TCID_{50}$/6-mL dose of PPV1 given IM and intranasally (IN). Gilts were necropsied on D128 and D129 (approximately 90 dG). At necropsy, fetal evaluation included recording the position of each fetus in the uterus, the fetal condition, size, and weight. Thoracic wash samples were collected from each fetus and tested for the presence of PPV by PCR. Serum samples from gilts at D-3, D74, D80, D88, D95, D102, and D127 were tested by PCR for viremia. Case definitions for a gilt considered PPV positive were: a) if one or more fetuses in a litter was not normal condition and b) if any fetus was found to be positive by PCR on a thoracic wash sample. Statistical analysis utilized Prevented Fraction (PF) analysis on binary variables with the GLIMMIX procedure of SAS utilized with a binomial distribution. The study design is summarized in Table 1.

Results

Clinical condition of fetuses

A primary outcome parameter was fetal condition, with efficacy confirmed by the clinical presentation of fetuses. A gilt was considered affected if one or more fetuses in a litter was not normal condition (mummified/necrotic or stillborn). The frequency of affected gilts in Group VC was significantly lower than affected gilts in Group NVC; 5/22 affected vs 9/10 affected, respectively. The mean fetus size (crown-rump measurement) was 19.3 cm in vaccinated litters compared to 14.4 cm in non-vaccinated litters. The mean weight of fetuses in vaccinated litters was 543.8 grams compared to 245.9 grams in non-vaccinated litters. Overall, 96% of fetuses from vaccinated gilts were normal compared to 38% of fetuses from non-vaccinated gilts.

Table 2 summarizes the gilt status based on the clinical condition of fetuses.

Table 3 summarizes the fetus condition results.

PPV PCR result of fetal thoracic fluid

Another primary outcome parameter was the evidence of PPV in fetuses as determined by PCR of fetal thoracic fluid. A gilt was considered affected if one or more fetuses in a litter was PCR positive for PPV. The frequency of affected gilts in Group VC was significantly less than affected gilts in Group NVC (10/22 vs 10/10). The mean affected fetuses per litter was 1.41 in VC litters compared to 11.30 in NVC litters.

Table 4 summarizes the gilt status based on PCR result of fetal thoracic fluid samples.

Gilt PPV viremia

All gilts were negative for PPV viremia prior to challenge on D74 and D80. Post-challenge (D81), all vaccinated gilts (VC) remained negative at all days sampled, while all ten of the non-vaccinated gilts (NVC) were viremic on D88. Samplings on D95, D102, and D127 were negative for PPV viremia for all treatment groups.

Table 5 summarizes the gilt PPV viremia (PCR) results.

Discussion and conclusions

This vaccination-challenge study demonstrated that two doses of Reprocyc ParvoFLEX® given three weeks apart are efficacious in mitigating reproductive failure caused by virulent PPV when administered to gilts prior to breeding. The efficacy is confirmed by

Table 1: Study design

Treatment	#V	1st vaccination (D0)	2nd vaccination (D21)	Estrus detection/ breeding	Pregnancy evaluation	#C	Challenge (~40dG)	Necropsy (~90dG)
NVC (Placebo)	12	Placebo 2 mL IM	Placebo 2 mL IM	D37-D50	D74	10	D81 2 mL IM & 4 mL IN PPV1	D128-129 Fetal evaluation & fetal thoracic fluid sampling
VC* (Reprocyc ParvoFLEX®)	24	PPV vaccine 2 mL IM	PPV vaccine 2 mL IM			22		
NTX (none)	4	not applicable				not applicable		D81

* Reprocyc ParvoFLEX® (adjuvanted PPV, killed baculovirus vector vaccine)

NTX = Non-treated/ Non-challenged control

#V = Number of animals vaccinated

#C = Number of animals challenged

Table 2: Gilt status based on clinical condition of fetuses

Group*	Frequency of affected gilts**	Mean affected fetuses	Percent of affected gilts**
NVC (Placebo)	9/10	9.1	90%
VC (Reprocyc ParvoFLEX®)	5/22	0.55	23%
		Prevented fraction analysis, VC vs. NVC (lower and upper 95% bound)	**0.74 (0.38, 0.89)**
		***P*-value**	**< 0.05**

* Vaccinated Challenged (VC) and Non-Vaccinated Challenged (NVC)

** A gilt with at least one abnormal fetus is considered affected.

Table 3: Summary of fetus condition results

	Group*	
Clinical parameter	**NVC (Placebo)**	**VC (Reprocyc ParvoFLEX®)**
No. of gilts	10	22
Total number of fetuses	146	297
Mean litter size	14.6	13.5
Mean fetus size (cm)	14.4	19.3
Mean fetus weight (grams)	245.9	543.8
Number necrotic fetuses (%)	9 (6%)	2 (1%)
Number mummified fetuses (%)	82 (56%)	10 (3%)
Number normal fetuses (%)	**55 (38%)**	**285 (96%)**

* Vaccinated Challenged (VC) and Non-Vaccinated Challenged (NVC)

the significant differences in the clinical condition of fetuses between vaccinated and non-vaccinated/control gilts following PPV challenge. Vaccinated gilts had 96% normal fetuses compared to 38% normal fetuses from non-vaccinated gilts. This represented a significant reduction in the number of vaccinated gilts affected by challenge compared to non-vaccinated gilts affected by challenge. Efficacy is also supported by the significant reduction in the frequency of PPV detection from thoracic fluid of fetuses from vaccinated compared to non-vaccinated gilts following challenge. The mean number of affected (PPV PCR positive) fetuses per litter from vaccinated gilts was 1.41 compared to 11.30 from non-vaccinated gilts.

Reprocyc ParvoFLEX® is a new vaccine that offers protection against PPV in reproductive vaccination protocols. Together with the recently introduced FLEX ParvoPRRS® vaccine, the industry has flexible options for protection against both PRRSV and PPV in reproductive vaccination protocols.

References

1. Mengeling WL et al: 1979, Am J Vet Res. Vol 40, No 2. p 204-207.
2. Data on File. Boehringer Ingelheim Animal Health USA Inc. study #2013057.

Table 4: Gilt status based on PCR result of fetal thoracic fluid sample

Group*	Frequency of affected gilts**	Mean affected fetuses	Percent of affected gilts**
NVC (Placebo)	10/10	11.30	100%
VC (Reprocyc ParvoFLEX®)	10/22	1.41	45%
		Prevented fraction analysis, VC vs. NVC (lower and upper 95% bound)	**0.55 (0.29, 0.70)**
		***P*-value**	**< 0.05**

* Vaccinated Challenged (VC) and Non-Vaccinated Challenged (NVC)
** A gilt with at least one fetus with positive PCR is considered affected

Table 5: Percent PPV-positive (PCR) gilts by group and sampling day

	Day of study/days of gestation (dG)					
	D74	**D80**	**D88**	**D95**	**D102**	**D127**
Treatment*	**dG 32**	**dG 38**	**dG46**	**dG 53**	**dG 60**	**dG 85**
NVC (Placebo)	0/10	0/10	**10/10**	0/10	0/10	0/10
VC (Reprocyc ParvoFLEX®)	0/22	0/22	0/22	0/22	0/22	0/22

* Vaccinated Challenged (VC) and Non-Vaccinated Challenged (NVC)

https://doi.org/10.54846/am2022/52

Efficacy of Enterisol® Salmonella T/C in protecting against monophasic *Salmonella* I 4,[5],12:i:-

Fernando Leite[1], DVM, MS, PhD; Paulo Arruda[2], DVM, MS, PhD; Dianna Jordan[1], DVM, PhD; Shawn Bearson[3], PhD
[1]Boehringer Ingelheim Animal Health USA Inc., Duluth, Georgia; [2]VRI-AMVC, Ames, Iowa; [3]ARS-USDA, Ames, Iowa

Introduction

The monophasic *Salmonella* I 4,[5],12:i:- serovar has emerged as the predominant *Salmonella* serovar isolated from clinical cases in the US swine population.[1,2] Infection can cause clinical disease in pigs as well as pose a public health threat as a human foodborne pathogen.[2,3] There are no commercial vaccines against this monophasic *Salmonella* serovar and there is lack of data on the efficacy of US commercially available vaccines in protecting against this serovar.[2] This study had the objective of evaluating a commercial vaccine, Enterisol Salmonella T/C®, in conferring protection against monophasic *S.* I 4,[5],12:i:- as measured by clinical signs, enteric lesions and average daily weight gain.

Materials and methods

A randomized, blinded, controlled study was conducted evaluating pigs in three different treatment groups. Each treatment group consisted of 20 pigs, blocked for the effects of litter, sex and weight. The treatment groups were: 1) non-vaccinated challenged (NVC); 2) non-vaccinated non-challenged (NVNC); 3) vaccinated with Enterisol Salmonella T/C® and challenged (EVC). Animals were challenged four weeks post vaccination at approximately 7 weeks of age with a dose of 2×10^9 *S.* I 4,[5],12:i:- strain SX 240. This is a multidrug resistant strain of *S.* I 4,[5],12:i:- that was associated with a human foodborne outbreak from pork in the US and that also causes clinical disease in pigs.[3,4] Animals were monitored for clinical signs, macroscopic enteric lesions and evaluated for weight gain for a period of 14 days following challenge.

Results

NVC animals had a significant increase in the incidence of diarrhea (59/254), compared to NVNC pigs (0/280) and EVC pigs (12/280) ($P < 0.05$) (Figure 1). This represented a significant increase of both the duration of diarrhea as well as the number of pigs with diarrhea (Table 1). One pig in the NVC group died following challenge and had intestinal lesions compatible with enterocolitis due to salmonellosis. At 14 days post challenge, four pigs in the NVC group had intestinal lesions while no pigs had lesions in the NVNC and EVC groups ($P < 0.05$, Table 1). Average daily weight gain was significantly decreased in the NVC group (1.197 lbs) compared to NVNC group (1.693 lbs) and EVC group (1.536 lbs) ($P < 0.05$, Figure 2). The difference in ADG equated to vaccinates being 4.7 lbs heavier than non-vaccinates over the challenge period.

Discussion and conclusions

Enterisol Salmonella T/C® is the only commercially available vaccine in the US that contains a live Typhimurium serovar vaccine strain, a serovar of the same serogroup as *S.* I 4,[5],12:i:-.[5] The results suggest that the Enterisol Salmonella T/C® vaccine conferred protection to *S.* I 4,[5],12:i:- under the conditions of this study, as measured by clinical signs, intestinal lesions and improved average daily weight gain.

References

1. Naberhaus SA, et. al. 2019. J Vet Diagn Invest. 31(6):818-827.
2. SHIC. 2021. *Salmonella* 1,4,[5],12:i:-. p1-19.
3. Shippy DC, et al. 2018. Foodborne Pathog Dis. 15(5):253-261.
4. Bearson BL, et al. 2019. Microbiol Resour Announc. 3;8(40):e00791-19.
5. Leite F.: 2021. AASV p161-163.

Figure 1: Number of daily observations of diarrhea following challenge. NVNC = non-vaccinated non-challenged; NVC = non-vaccinated challenged; EVC = vaccinated with Enterisol Salmonella T/C® and challenged.

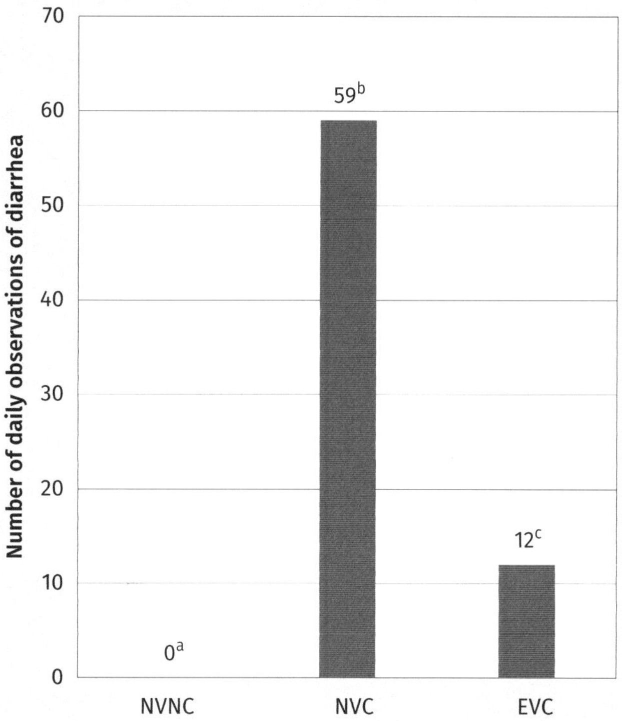

a,b,c Different superscript letters indicate statistical significance ($P < 0.05$).

Table 1: Post challenge clinical observations and gross lesions measured at necropsy

	NVNC	NVC	EVC
Number of pigs with diarrhea	0% (0/20)[a]	84% (16/19)[b]*	30% (6/20)[c]
Average duration of diarrhea (days)	0[a]	3.58[b]	1[c]
Number of pigs with intestinal gross lesions	0% (0/20)	22% (4/18)*	0% (0/20)

[a,b,c] Different superscript letters within a row indicate statistical significance ($P < 0.05$). NVNC= non-vaccinated non-challenged; NVC = non-vaccinated challenged; EVC = vaccinated and challenged. *One pig died prior to challenge (not evaluated for diarrhea) and one pig died prior to study end point (only lesions measured in study end point reported).

Figure 2: Post challenge average daily weight gain. NVNC = non-vaccinated non-challenged; NVC = non-vaccinated challenged; EVC = vaccinated with Enterisol Salmonella T/C® and challenged.

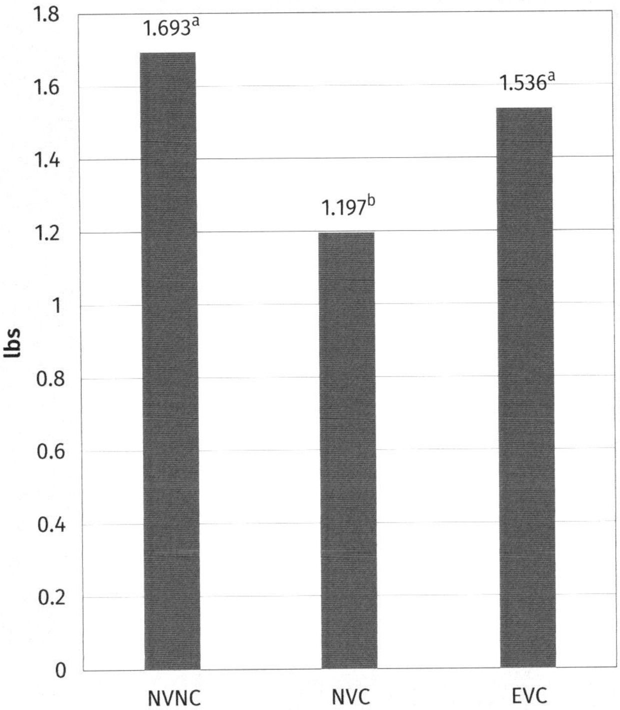

[a,b] Different superscript letters indicate statistical significance ($P < 0.05$).

https://doi.org/10.54846/am2022/53

Exploring the efficacy of a PCV2d-based vaccine under current severe PCVAD conditions

Eduardo Fano[1], DVM, MS, PhD; Pablo E. Piñeyro[2], DVM, MS, PhD; Kent J. Schwartz[2], DVM, MS; Steven J. P. Tousignant[1], DVM, PhD
[1]Boehringer Ingelheim Animal Health, Inc. USA, Duluth, Georgia;
[2]Veterinary Diagnostic and Production Animal Medicine, Iowa State University, Iowa

Introduction

Porcine circovirus Type 2 (PCV2) is the primary causative agent of porcine circovirus–associated disease (PCVAD), one of the most significant diseases affecting swine worldwide. Typical lesions include lymphoid depletion and histiocytic inflammation in multiple organ systems. Clinical disease and lesion severity are often exacerbated by co-infection with a variety of endemic pathogens. Since the introduction of PCV2 vaccines into the US market in 2006, the clinical and economic benefit of vaccination has been clearly demonstrated. Practitioners and producers have special interest in better vaccination protocols conferring protection against prevalent PCV2 strains. DNA viruses, such PCV2, have a lower mutation rates than RNA viruses. However, relative to other DNA viruses, PCV2 has the highest mutation rate, with shifts in subtype prevalence that the swine industry has been grappling with in recent years. Currently, three major subtypes including PCV2a, PCV2b and PCV2d are the most clinically relevant. First identified in 2012, PCV2d is now the predominant subtype circulating in the US, accounting for more than 90% of PCV2 infections. Scientific and clinical evidences indicate that PCV2a-based vaccines provide cross-protection against PCV2d. However, the discussion remains as to whether a homologous PCV2d vaccine could provide a more robust and targeted protection against today's most prevalent subtype, when there is severe coinfection pressure contributing to PCVAD expression. Therefore, the objective of the study was to measure the clinical efficacy of a PCV2d and PCV2a based vaccines in a PCV2d/contemporary PRRSV coinfection model performed under field conditions.

Materials and methods

A PRRSV and Mhp negative, PCV2 low prevalence sow farm was selected to source study pigs. PRRS status was based on historical PRRSV PCR negative processing fluids and PCV2 status was based on < 20% PCV2 PCR positive litters measured by PUCS (placental umbilical cords sample) over the course of 3 months. Mhp status was based on historical PCR and serology results. Pig source was IAV-S positive, and pigs were expected to carry endemic secondary bacteria. The experimental animals were housed in a wean-to-finish commercial/research swine facility. Pigs were housed in a barn appropriate for breed and age and were kept under similar conditions of climate, ventilation, temperature, humidity, and light. A total of 1410 pigs were blocked by weight and randomly assigned to one of three treatments: T1 = 470 No PCV2 vaccine; T2 = 470 PCV2a-based vaccine; T3 = 470 PCV2d-based vaccine. Wean-to-finish barns were filled in three consecutive days with weaned pigs of approximately 21 days of age. Pigs were individually weighed and identified with unique ear tags (trial day 0). On trial day 1, pigs were allocated to treatment groups and administered a single dose of their respective treatments. On trial day 28, pigs were inoculated with 1 mL IM and 1 mL IN of PCV2d (5 $\log_{10}$/2mL dose) as well as 2 mL IM of 1-7-4 PRRSV (4.0 TCID50/mL). Production data included mortality, percentage of treated pigs and body weights (day 0, 28, 56 and 140), as well as average daily gain from all periods.

On day 56 (28 days post challenge), a randomly selected subset of 20 pigs from each treatment group was euthanized. At necropsy, pulmonary gross lesions were scored. Multiple sections of lung and lymph nodes (mediastinal, tracheobronchial, inguinal lymph), were collected for PCR detection and preserved in buffered formalin, for histological and immunohistochemistry (IHC) evaluation. Lymphoid changes and amount of antigen detected by IHC was also evaluated on sections of tonsil. Viremia was evaluated weekly by serum PCR.

The overall lymphoid lesion (depletion and histiocytic infiltration) and IHC scores were analyzed as an average of the individual tissue score. This, along with other continuous outcome variables were analyzed using ANOVA where treatment and pen were used as explanatory variables. Treatment Estimated Marginal Means were calculated for each group and compared using Tukey's HSD. Mortality and the percentage of treated pigs was analyzed by testing multiple pair-wise proportions using Fishers Exact test, and a False Discovery Rate correction. Statistical significance was set with an alpha value of 0.05. Statistical analysis was conducted in RStudio version 1.4.1103.

Results and discussion

Experimental coinfection by inoculation with contemporary PCV2d/PRRSV isolates resulted in high mortality and severe PCVAD (Table 4). The roles of PRRSV, endemic IAV, and endemic, potentially pathogenic bacteria greatly influenced the outcome of PCV2d infection, thereby reproducing field conditions that represents high infection pressure scenarios. The effect of the PCV2 vaccines measured by the overall lymphoid lesion score and the detection of the virus is shown Table 1. Viremia in the post challenge period of the subset pigs is shown in Table 2. The average daily gain (ADG) from day 0 to day 140 is shown in Table 3. Mortality and percentage of treated pigs is shown in Table 4. The severity of lesions, amount of virus detected in tissue and serum, as well as production performance were significantly more severe in non-vaccinate group than the two vaccinate groups. The results of this study support previous information regarding cross-protection of a PCV2a-based vaccine against PCV2d challenge. In addition, this study demonstrates clinical benefit of a cross protection in "severe PCVAD/PRDC clinical scenario". This study also demonstrated that under these conditions, the PCV2d-based vaccine group (T3) had significantly improved the severity of lesions and amount of virus detected compared to the PCV2a-based vaccine (T2), as well on the percentage of treated animals in the acute phase of disease. Under the severe challenge conditions of this trial, that resulted in extreme variation within groups, there were no significant differences on ADG and overall mortality between both vaccinated groups.

In conclusion, under the severe PCVAD field conditions of this study (co-infections with other viruses and endemic secondary bacteria), the PCV2d-based vaccine demonstrated a reduction in lymphoid lesion severity and viral replication.

Table 1: Overall lymphoid lesion score and virus identification. Subset of 20 pigs.

Parameter	T1 NonVx	T2	T3	se	*P*-value
Overall lymphoid lesion score (average)	1.63[a]	0.77[b]	0.09[c]	0.18	< 0.0001
PCR lymphoid tissue (Ct)	12.1[a]	17.8[b]	26.1[c]	1.35	< 0.0001
PCR serum (Ct)	26.3[a]	32.0[b]	36.9[c]	1.37	< 0.0001

* Means with different superscripts indicate difference at $P \leq 0.05$

Table 2: Viremia measured by Ct value. Subset of 20 pigs.

PCV2 PCR Ct**	T1 NonVx	T2	T3	se	*P*-value
7 days after challenge	27.2[a]	33.7[b]	36.7[c]	0.79	< 0.0001
14 days after challenge	18.9[a]	29.3[b]	35.9[c]	0.93	< 0.0001
21 days after challenge	22.5[a]	31.9[b]	36.6[c]	1.35	< 0.0001
28 days after challenge	26.3[a]	32.0[b]	36.9[c]	1.37	< 0.0001

* Means with different superscripts indicate difference at $P \leq 0.05$

** Ct cutoff negative, 37

Table 3: Weight and average daily gain (ADG)

Weights	T1 NonVx	T2	T3	se	*P*-value
0	14.4	14.4	14.1	0.14	0.0879
28	37.8	36.7	36.9	0.42	0.1631
56	68.8[b]	72.9[a]	74.3[a]	1.28	0.0014
140	218.0[b]	232.0[a]	234.0[a]	2.73	< 0.0001
ADG					
0-28	0.84	0.81	0.82	0.01	0.1368
0-56	0.95[a]	1.02[b]	1.06[b]	0.02	0.0001
28-56	1.04[a]	1.21[b]	1.26[b]	0.03	< 0.0001
0-140	1.50[a]	1.59[b]	1.62[b]	0.02	< 0.0001
28-140	1.66[a]	1.79[b]	1.82[b]	0.02	< 0.0001
56-140	1.89[a]	2.01[b]	2.02[b]	0.03	0.0007

* Means with different superscripts indicate difference at $P \leq 0.05$

Table 4: Mortality and percentage of treated pigs

% Treatments	T1 NonVx	T2	T3
Day 28 -56	87.0[a]	44.2[b]	36.6[c]
Day 56 -140	6.4	5.7	3.3
% Mortality			
Day 28 - 56	61.7[a]	9.7[b]	8.2[b]
Day 56 - 140	1.3	2.9	3.1

* Means with different superscripts indicate difference at $P \leq 0.05$

https://doi.org/10.54846/am2022/54

"Boil it down": Effectively manage alerts, warnings, and alarms from multiple barns simultaneously

Tom Stein, DVM, MS, PhD
Maximus Systems & Maximum Ag Technologies, Saint-Bruno, Quebec, Canada and Jenison, Michigan

As pork producers look to the future, they have important decisions to make now about how to get ready for emerging smart technologies. Which barn control platform to adopt will be one of the critical success factors for future-proofing your production system. Producers will want to have a single system that can connect to or integrate with these new technologies. It will be smart barn control systems – "brains in the barn" – that become the single management and control platform for integrating all these new in-the-barn technologies.

Maximus controllers in nursery and finishing barns continuously generate hourly and daily environmental, feed, and water data. This includes inside barn high and low temperatures, humidity, and water usage.

The highest priority is to know what is happening with the environment, including water consumption, against what you programmed for the room or building. Managers should be able to see quickly out-of-process events and be able to drill-down into the details to find out what's wrong and intervene immediately.

The head of production for a large (> 100K sows) production system recently told us that he finds the consolidated, aggregated, roll-up reports from the Maximus Enterprise Software to be "extremely helpful. I run the report weekly and use it to make phone calls and manage at a high level. It's real-time, not after the fact."

Figure 1 shows the weekly Site Overview report where 3 rooms on 2 different sites all show warnings (red) or alerts (yellow) for high. But good news on the inside temperature daily range, it's within the range set by this production system (green). The data all come automatically from Maximus controls, with multiple data points recorded each day. It means you can run this report for a day's worth of time instead of weekly, for more intense system management.

It's possible to enter deads into the control but this company doesn't do it. They enter starting inventory and group start date so you can see Days on Feed. You can also enter sales or movements in/out. By doing this and entering deads each day, a system could have an up-to-date daily inventory by using nothing more than the control as a data entry device.

Figure 2 shows the same report for the same production system with a snapshot of some other rooms. In this example, several rooms on a site are running well, no issues with the environment. This report allows a production system leader to do a quick scan of which sites/buildings/rooms are working well and which are not. They can use their supervisor team more efficiently and direct resources to the problem sites.

Figure 1: Site Overview Report showing 3 site-building-rooms with highlighted humidity too high for each day during the week. Phone calls will follow to the grower and supervisor. Green - within range; Yellow - warning; Red - too high.

Maximus Site Overview

Large Production System

Interval: Last 7 Day(s) Include current

Sites:

Day	Site	Building	Room	Min. Inside Temp	Max. Inside Temp (°C)	Inside Temp Set Point (°C)	Inside Temp Range	Avg. Outside Temp (°C)	Max. Outside Temp (°C)	Humidity (%)	Heater Run Time	Deaths	Head Count	Group Start	Age
10/26/2021	Ash	Ash	Room #29	18.4	19.2	16.2	0.8	12.1	12.6	86.17			1,288	09/21/2021	92
10/27/2021	Ash	Ash	Room #29	18.4	19.3	16.2	0.9	12.3	13.0	86.93			1,288	09/21/2021	93
10/28/2021	Ash	Ash	Room #29	17.9	19.0	16.2	1.1	10.3	10.6	85.57			1,288	09/21/2021	94
10/29/2021	Ash	Ash	Room #29	18.3	19.2	16.2	0.9	10.5	11.8	84.67			1,288	09/21/2021	95
10/30/2021	Ash	Ash	Room #29	18.0	19.0	16.2	1.0	8.8	9.0	82.87			1,288	09/21/2021	96
10/31/2021	Ash	Ash	Room #29	18.2	18.8	16.2	0.6	6.7	7.5	84.35			1,288	09/21/2021	97
11/01/2021	Ash	Ash	Room #29	18.2	18.5	16.2	0.3	5.3	5.4	86.29			1,288	09/21/2021	98
10/26/2021	Bern	Bern	Room 25	11.5	14.1	15.0	2.6	5.3	5.9	91.54	3d 2m				
10/27/2021	Bern	Bern	Room 25	13.5	14.0	15.0	0.5	5.1	6.7	86.89	2d 19h 29m				
10/28/2021	Bern	Bern	Room 25	11.0	13.3	15.0	2.3	2.1	3.7	81.11	2d 21h 42m		600	06/29/2021	
10/29/2021	Bern	Bern	Room 25	12.8	14.3	15.0	1.5	3.6	4.7	80.67	2d 22h 15m		600	06/29/2021	
10/30/2021	Bern	Bern	Room 25	13.4	13.8	15.0	0.4	1.3	1.9	76.02	2d 22h 57m		600	06/29/2021	
10/31/2021	Bern	Bern	Room 25	11.5	12.4	15.0	1.0	-1.9	0.2	74.00	3d 2m		600	06/29/2021	
11/01/2021	Bern	Bern	Room 25	11.6	11.7	15.0	0.0	-2.4	-2.3	74.00	21h		600	06/29/2021	
10/26/2021	Bern	Bern	Room 26	18.2	18.8	16.0	0.6	5.3	5.9	77.37	2d 1m				
10/27/2021	Bern	Bern	Room 26	17.3	18.7	16.0	1.4	5.1	6.7	77.24	1d 11h 34m				
10/28/2021	Bern	Bern	Room 26	13.2	16.1	19.8	2.9	2.1	3.7	82.63	1d 11h 2m		1,288	07/04/2021	
10/29/2021	Bern	Bern	Room 26	12.7	13.4	21.7	0.7	3.6	4.7	83.80	3d 2m		1,288	07/04/2021	
10/30/2021	Bern	Bern	Room 26	11.3	11.7	21.7	0.4	1.3	1.9	83.26	3d 2m		1,288	07/04/2021	
10/31/2021	Bern	Bern	Room 26	7.9	10.0	18.1	2.1	-1.9	0.2	92.91	3d 2m		1,288	07/04/2021	
11/01/2021	Bern	Bern	Room 26	7.7	7.8	15.0	0.1	-2.4	-2.3	98.00	21h		1,288	07/04/2021	

Generated: 11/01/2021 07:01 AM

Figure 2: Site Overview Report showing a site with no issues for the week.

Maximus Site Overview

Large Production System

Interval: Last 7 Day(s) Include current

Sites:

Day	Site	Building	Room	Min. Inside Temp	Max. Inside Temp (°C)	Inside Temp Set Point (°C)	Inside Temp Range	Avg. Outside Temp (°C)	Max. Outside Temp (°C)	Humidity (%)	Heater Run Time	Deaths	Head Count	Group Start	Age
10/26/2021	Hanson	Hanson 1	Room #2	24.7	25.4	26.3	0.7	5.8	6.6	70.70	2d 14h 24m		600	10/18/2021	9
10/27/2021	Hanson	Hanson 1	Room #2	24.4	25.1	26.1	0.7	5.0	6.6	70.24	2d 15h 16m		600	10/18/2021	10
10/28/2021	Hanson	Hanson 1	Room #2	24.2	24.9	25.9	0.7	1.9	5.0	69.89	2d 15h 2m		600	10/18/2021	11
10/29/2021	Hanson	Hanson 1	Room #2	24.1	25.0	25.6	0.9	7.0	9.2	72.96	1d 13h 53m		600	10/18/2021	12
10/30/2021	Hanson	Hanson 1	Room #2	23.9	24.6	25.4	0.6	-0.7	2.4	71.26	2d 3h 56m		600	10/18/2021	13
10/31/2021	Hanson	Hanson 1	Room #2	23.8	24.1	25.1	0.3	-3.6	-1.7	70.39	2d 18h 53m		600	10/18/2021	14
11/01/2021	Hanson	Hanson 1	Room #2	23.7	23.8	25.0	0.1	-6.8	-6.1	70.00	1d		600	10/18/2021	15
10/26/2021	Hanson	Hanson 1	Room #3	25.4	26.1	27.0	0.7	5.8	6.6	71.50	2d 8h 47m		750	10/21/2021	6
10/27/2021	Hanson	Hanson 1	Room #3	25.2	26.0	26.8	0.7	5.0	6.6	70.28	2d 11h 34m		750	10/21/2021	7
10/28/2021	Hanson	Hanson 1	Room #3	25.1	25.7	26.6	0.6	1.9	5.0	68.87	2d 9h 8m		750	10/21/2021	8
10/29/2021	Hanson	Hanson 1	Room #3	24.8	25.7	26.3	0.9	7.0	9.2	72.83	1d 18h 19m		750	10/21/2021	9
10/30/2021	Hanson	Hanson 1	Room #3	24.6	25.2	26.1	0.6	-0.7	2.4	71.70	2d 12h 26m		750	10/21/2021	10
10/31/2021	Hanson	Hanson 1	Room #3	24.4	24.7	25.9	0.3	-3.6	-1.7	70.67	2d 20h 22m		750	10/21/2021	11
11/01/2021	Hanson	Hanson 1	Room #3	24.2	24.3	25.8	0.1	-6.8	-6.1	68.13	1d 1m		750	10/21/2021	12
10/26/2021	Hanson	Hanson 1	Room #5	25.4	26.1	27.0	0.7	5.8	6.6	75.89	2d 13h 12m		750	10/25/2021	2
10/27/2021	Hanson	Hanson 1	Room #5	25.5	25.9	27.0	0.4	5.0	6.6	72.85	2d 18h 35m		750	10/25/2021	3
10/28/2021	Hanson	Hanson 1	Room #5	25.4	25.8	27.0	0.4	1.9	5.0	72.13	2d 20h 25m		750	10/25/2021	4
10/29/2021	Hanson	Hanson 1	Room #5	25.5	26.0	27.0	0.6	7.0	9.2	72.33	2d 13h 7m		750	10/25/2021	5
10/30/2021	Hanson	Hanson 1	Room #5	25.0	25.6	27.0	0.6	-0.7	2.4	72.00	2d 23h 35m		750	10/25/2021	6
10/31/2021	Hanson	Hanson 1	Room #5	24.8	25.2	26.8	0.3	-3.6	-1.7	69.65	3d 3m		750	10/25/2021	7
11/01/2021	Hanson	Hanson 1	Room #5	24.8	24.8	26.7	0.1	-6.8	-6.1	69.00	1d		750	10/25/2021	8

Generated: 11/01/2021 07:01 AM

https://doi.org/10.54846/am2022/55

Effect of tulathromycin on growth and mortality in PRRS-positive weaned pigs

J. Hocker[1], DVM, MS; J. Ellingson[1], DVM, MS; R. Swalla[2], DVM; D. Amodie[2], MS; E. Guzman[2], MS
[1]AMVC, Audubon, Iowa; Zoetis Inc., [2]Parsippany, New Jersey

Introduction

A study was conducted under field conditions to determine if PRRS positive weaned pigs administered a single injection of DRAXXIN® 25 (tulathromycin 25mg/mL) upon entering wean to finish barns would show improved average daily gain (ADG) and mortality after three weeks in the nursery and at the end of the finishing period. DRAXXIN 25 is indicated for the treatment swine respiratory disease (SRD) associated with *Actinobacillus pleuropneumoniae, Pasteurella multocida, Bordetella bronchiseptica, Haemophilus parasuis*, and *Mycoplasma hyopneumoniae*; and for the control of SRD associated with *Actinobacillus pleuropneumoniae, Pasteurella multocida*, and *Mycoplasma hyopneumoniae* in groups of pigs where SRD has been diagnosed. Previous work showed improved ADG and mortality in DRAXXIN treated pigs through the nursery phase and improved ADG from wean to finish in pigs with a "low" level of PRRS involvement.[1]

Materials and methods

Study design

The study was conducted in a large midwestern production system and the pigs received their normal feed and water medications. A 2,400 head, slatted flooring, wean to finish barn with tunnel ventilation was utilized for this study. Weaned pigs from a 5,000-sow farm in Colorado with a history of a recent PRRS break were placed by the contract grower. The next day (Day 0) pigs were weighed and tagged and returned to their pens. On Day 1, a treatment of either 0.60cc of DRAXXIN 25 or 0.60cc of saline was administered IM and pigs were returned to their pens. Both treatments were equally represented within each pen. Pigs were weighed after 19 days in the nursery and just prior to the first marketing cut on day 130 (Table 1).

Statistical analysis

This study was conducted as a randomized complete block design with animal as the experimental unit. Animals were blocked within pens in multiple barns.

Weight and ADG were analyzed by a generalized linear mixed model (GLMM) approach. Using the SAS Proc Mixed Procedure (SAS 9.4, Cary NC), weights and ADG were analyzed with a model that considered the fixed effects of treatment and the random barn, pen(barn), block(pen*barn) b and the residual error.

Table 1: Treatment groups

Group	No. of pigs	Product administered	Weigh days
T1	980	DRAXXIN	0, 19, 130
T2	980	Saline	0, 19, 130

Treatment was assessed at the 5% level of significance. Treatment Least Squares Means (LSMeans) were calculated for all groups. Comparison of LSMeans was completed by the two-sided Student's t-test at the 5% level of significance.

Mortality, retreatments and culls were defined as binary (yes = 1, no = 0) and analyzed by a generalized linear mixed model approach. Using the SAS Proc Glimmix procedure responses were analyzed and LSMeans compared as described above. This analysis utilized a binomial error and logit link.

Results and discussion

Nursery and finisher weight gain

Pigs in the study treated with DRAXXIN 25 at weaning weighed 25.84 pounds after 19 days in the barns versus 25.25 pounds in the saline treated control pigs, for a difference of 0.59 pounds ($P = 0.0001$) (Table 2). ADG for DRAXXIN treated pigs after 19 days in the barn was 0.64 pound per day versus 0.61 for saline treated control pigs, for a difference of 0.03 pounds per day ($P = 0.0001$). Pigs in the study treated with DRAXXIN 25 at weaning weighed 234.83 pounds after 130 days in the barns versus 231.34 pounds in the saline treated control pigs, for a difference of 3.49 pounds ($P = 0.0029$). ADG for DRAXXIN 25 treated pigs after 130 days in the barn was 1.69 pounds per day versus 1.66 pounds per day for saline treated control pigs of 0.03 pounds per day ($P = 0.0101$). In other words, in this study the 0.03 pounds per day improvement in ADG that was established in the first 19 days post-treatment with DRAXXIN 25 continued through day 130 post-placement.

Mortality

Pigs treated with DRAXXIN 25 at placement had 0.41% mortality after 19 days in the barns versus 0.93% mortality for the pigs that received saline ($P = 0.1744$) (Table 3). Pigs treated with DRAXXIN 25 at placement had 3.45% mortality after 130 days in the barns versus 5.06% mortality for the pigs that received saline ($P = 0.0791$). Mortality was low in this study, but mortality was trending toward significance with a *P*-value of 0.0791 by 130 days placed in the barns.

Nursery treatments and finisher culls

The number of pigs treated in the first 19 days was recorded by the contract grower. 4.56% of the pigs treated with DRAXXIN 25 at weaning required an additional antibiotic injection in the first 19 days versus 4.03% of pigs treated with saline ($P = 0.5787$). In addition, an attempt was made to determine the number of culls that would need to go to a secondary market at closeout. In this case, pigs that weighed less than 180 pounds at the first cut were designated as potential cull pigs. Using 180 pounds as the cutoff, 3.65% of the pigs treated with DRAXXIN 25 at weaning would have been culls versus 4.55% of the pigs treated with saline ($P = 0.3177$).

Table 2: Weights and ADG

	DRAXXIN 25	Saline	*P*-value
Initial weight	13.71	13.72	0.5279
Day 19 weight, lbs	25.84	25.25	**0.0001**
Day 130 weight, lbs	234.83	231.34	**0.0029**
ADG, D0-19	0.64	0.61	**0.0001**
ADG, D0-130	1.69	1.66	**0.0101**

Table 3: Mortality

	DRAXXIN 25	Saline	*P*-value
Mortality, D0-19, %	0.41	0.93	0.1744
Mortality, D0-130, %	3.45	5.06	**0.0791**

Reference

1. King, D. *et al.* Effects of Draxxin at weaning in pigs on control of swine respiratory disease including "low" level of PRRS involvement and subsequent performance of pigs. *Proc AASV 2013.*

https://doi.org/10.54846/am2022/56

A bivalent PCV2 vaccine (PCV2a and PCV2b) offers biologically superior protection compared to monovalent PCV2 vaccines

Meggan Bandrick[1], DVM, PhD; Monica Balasch[2], DVM, PhD; Andrea Heinz[1]; Lucas Taylor[1], MS; Vickie King[1], PhD; Dennis Foss[1], DVM, PhD
[1]Zoetis, Global Biologics, Veterinary Medical Research and Development, Michigan;
[2]Zoetis, Global Biologics, Veterinary Medical Research and Development, Olot, Spain

Introduction

Porcine circovirus type 2 (PCV2) is a nonenveloped, single-stranded DNA virus with an evolutionary rate more like RNA viruses. There are 8 proposed PCV2 genotypes labeled PCV2a-h of which PCV2a, b, and d are clinically relevant and most often associated with clinical Systemic Disease (PCV2-SD).[1,2] PCV2 virus evolution has also been shaped by PCV2 vaccination and immune pressure to support evolution of newer genotypes.[3] PCV2a vaccines have successfully limited prevalence of PCV2a viruses, but likely supported increased diversity of epitopes within PCV2a.[4,5] Molecular evidence has documented the lack of complete cross-protection of PCV2 vaccines,[6] most likely explained by the genetic diversity among PCV2 virus capsid epitopes. A summary of challenge outcomes in homologous and heterologous vaccine-challenge studies provides evidence that PCV2a vaccines induce cross-protection to divergent genotypes to some extent and that homologous vaccines are better at reducing viremia.[7] PCV2 vaccine protection to a variety of field virus genotypes can be attributed in part to the shared immune epitopes between vaccine and field viruses. It is noteworthy, however, that minor differences within ORF2 (capsid) epitopes can result in differences in immune recognition.[8-10]

With the understanding that PCV2 is an economically important virus in global swine production and the observations that genetic diversification is ongoing, there is a growing need to improve existing vaccines. The hypothesis that a genotype-matched vaccine and challenge would provide better efficacy than a heterologous vaccine-challenge match was first investigated. Finding that cross-protection was incomplete and keeping in mind the inherent diversity within PCV2, it was hypothesized that a vaccine comprised of a combination of two PCV2 genotypes, PCV2a and PCV2b, would provide better protection than a monovalent vaccine in pigs challenged with either PCV2a or PCV2b.

Materials and methods

Three studies were conducted (A, B, and C); animals were housed and managed in strict accordance with the Guide for the Care and Use of Laboratory Animals by the National Institutes of Health at all times. Animal use was approved by the Zoetis IACUC. The investigator and individuals who performed observations, assays, and/or were involved in data collection were masked. Pigs, 3-4 weeks of age at Day 0, were sourced from commercial farms with no recent history of vaccination against or clinical problems with PCV2. Pigs were seronegative to PCV2 and negative for PCV2 viremia by qPCR[11] on Day 0. Vaccines in all studies, administered on Day 0, were formulated with the same final adjuvant (MetaStim®) concentration. All pigs were challenged IM and intra-nasally on Day 21-22 with PCV2a or PCV2b. Virulence of PCV2a and PCV2b challenge material were previously described.[12,13] At multiple time points, blood and fecal samples were collected for PCV2 antigen or antibody detection. On Day 42 all animals were euthanized and necropsied. Tissues were evaluated for histological changes typical of PCV2 (lymphoid depletion [LD] and histiocytic replacement [HR]) as well as PCV2 colonization (via IHC).

In study A, pigs were randomly assigned to pens and one of four treatment groups, designated T (treatment) 01-04 per a randomized complete block design (blocking factor was litter) using a random number generator in SAS®. Two groups received a chimeric PCV1-PCV2a (cPCV2a) vaccine (T02 and T04), and two groups received saline (T01 and T03); n = 10 pigs/group. Treatments were administered IM once at 23-30 days of age. Vaccine efficacy was determined against PCV2a (T01-T02) or PCV2b (T03-T04) challenge.

In studies B and C, pigs were randomly assigned to pens and one of three treatment groups, T01-03 (n = 24/group), per a randomized complete block design (blocking factor was litter) using a random number generator in SAS®. Vaccine serials included cPCV2a (Study B) or cPCV2b (Study C) and a bivalent PCV2 vaccine containing both cPCV2a and cPCV2b (Studies B and C). T01 animals received saline control; T02 received the monovalent vaccine; T03 received the bivalent vaccine. Vaccine efficacy was determined against virulent PCV2a challenge in Study B and PCV2b in Study C.

Efficacy outcome criteria

Post-challenge qualitative and quantitative viremia and PCV2 virus fecal shedding were the key outcome variables. Viremia and fecal shedding frequency distributions were calculated for each treatment and time point data were collected post-challenge. It was determined if an animal was ever viremic or ever shed PCV2 in its feces post challenge. PCV2 antibody values were transformed with an appropriate log transformation prior to analyses for all three studies (A, B, C). For study A, values were summarized with descriptive statistics. For studies B and C transformed values were analyzed with a general linear repeated measures mixed model. Post-challenge LD, HR, and PCV2 colonization of lymphoid tissues were additional variables. It was determined if an animal was abnormal for either LD or HR. Frequency distributions of normal/abnormal results were calculated for each treatment for LD and HR separately as well as either LD and/or HR. An animal was considered positive for PCV2 colonization if any tissue was positive via IHC.

T cell epitope content comparison (EpiCC) analysis

The EpiCC analysis was conducted as previously described.[14] Vaccine and challenge virus sequences were evaluated for T cell epitope content [sum of putative class I and class II T cell epitopes restricted by Swine Leukocyte Antigen (SLA)]. Putative epitopes of each vaccine to itself and against the PCV2a and PCV2b challenge strains were compared to calculate a relatedness score based on T cell epitopes shared between pairs of sequences (shared EpiCC score). The EpiCC score may be used to describe sharing of T cell epitope content between the vaccine and the challenge strains in order to estimate a vaccine's potential to confer cross-reactive cell-mediated immunity to a particular challenge strain.

Results

Study A evaluated efficacy of a cPCV2a monovalent vaccine in front of challenge with a homologous (PCV2a) and a heterologous (PCV2b) challenge strain. This was a pilot study with 10 pigs per group; statistical comparisons could not be made; summaries were made based on if an animal was "ever positive" for the efficacy outcomes. Challenge with virulent PCV2a or PCV2b induced similar viral levels at peak viremia and peak viral shedding as well as a similar range of abnormal histopathology in control animals (T01 and T03, respectively). The cPCV2a vaccine reduced the peak values and percent of animals ever positive for viremia, fecal shedding, lymphoid lesions, and lymphoid colonization following challenge with either PCV2a or PCV2b compared to controls. Pigs in the homologous treatment (T02) had biologically better efficacy outcomes compared to pigs in the heterologous treatment (T04) (Table 1).

Since biologically differential efficacy was observed, EpiCC analysis was performed to quantify the relatedness of the cPCV2a vaccine fraction to the PCV2a and PCV2b challenge. In the homologous situation the vaccine offered 100% epitope coverage. On the other hand, in the heterologous situation, the cPCV2a vaccine offered 72.05% coverage to the PCV2b challenge.

Studies B and C evaluated the potential impact to efficacy of a bivalent cPCV2a-cPCV2b vaccine compared to a challenge-type matched monovalent. On Days 29 to 42 for Study B and Day 42 for Study C, the vaccinated groups (T02 and T03) were positive for PCV2 antibodies based on SERELISA and had significantly higher PCV2 antibody levels compared to T01 saline controls. In both studies, vaccinated groups had a significantly lower ($P < 0.0001$) percentage of pigs ever viremic post-challenge compared to controls. Following PCV2a challenge, the cPCV2a-cPCV2b vaccine offered a biologically relevant decrease in the frequency of viremia (28.3%) compared to the monovalent cPCV2a vaccine. Following PCV2b challenge, the cPCV2a-cPCV2b vaccine offered a biologically relevant decrease in the frequency of PCV2 viremia (21.9%), compared to the monovalent cPCV2b vaccine. All vaccinated groups had a significantly lower ($P \leq 0.03$) percentage of animals that ever-shed PCV2 in feces post-challenge compared to T01 saline controls for Study B. The percent of animals ever viremic in Study C was not significantly different from the saline control group. In both studies vaccinated groups (T02 and T03) had significantly less HR and/or LD compared to T01 saline controls. Further, vaccinated groups (T02 and T03) had significantly ($P \leq 0.0004$ and $P \leq 0.05$, respectively) lower percent of animals positive for PCV2 colonization compared to T01 controls. In both studies B and C, the animals vaccinated with the bivalent vaccine were biologically better protected from PCV2 infection and disease than animals vaccinated with a monovalent cPCV2a or cPCV2b vaccine (Table 2).

Conclusion

PCV2a vaccines have greatly reduced the prevalence and severity of PCV2 viral infections. In fact, PCV2 vaccines represent a great success within the swine industry. Despite vaccine use and a dramatic reduction in clinical signs, PCV2 viremia continues around the world with some herds still reporting clinical problems with PCV2. Part of the reason PCV2 is still circulating is that PCV2 vaccines do not provide sterilizing immunity[10,15] and traditional single genotype PCV2 vaccines may not offer enough full coverage to evolving field viruses.[6,15] New viruses including recombinants are continually being sequenced with changes focused primarily on the capsid (ORF2) genetic sequence.[10,15] As minor changes in the capsid sequence accumulate, the shared epitopes among PCV2a-based vaccines and field strains wane. A growing dissimilarity in epitopes shared among evolving field strains and constant vaccine strains may be contributing to decreased efficacy of PCV2a vaccines.[1,8-10,15] Having high similarity in epitopes between vaccine and field PCV2 viruses is essential for broad vaccine protection in the field.

Study A was an exploratory study conducted with few animals per group, obviating the use of statistical analysis. However, biologically relevant differences were observed. The cPCV2a vaccine reduced viremia, fecal shedding, lymphoid lesions, and lymphoid colonization in pigs following either a PCV2a or a PCV2b challenge compared to controls, as expected. Importantly, the cPCV2a vaccine offered biologically relevant enhanced protection against PCV2a challenge compared to PCV2b challenge for the majority of outcomes. Per EpiCC analysis, the cPCV2a vaccine offered 36.46% greater coverage to PCV2a challenge than to PCV2b challenge [(|(heterologous-homologous)| / heterologous * 100%)]. This increased breadth of coverage is consistent with the improvement in protective efficacy observed in this study. Taken together with the clinical results of Study A, EpiCC results may be practically informative of clinical vaccine-induced protection.

Studies B and C were conducted to understand if clinical efficacy to PCV2a and PCV2b could be improved by vaccinating animals with cPCV2a and cPCV2b. In both studies, animals vaccinated with the cPCV2a-cPCV2b bivalent vaccine were protected from PCV2 disease. The benefit of the bivalent vaccine was greater than just the shared epitopes among the vaccine and field strains as the baseline percent coverage offered by the cPCV2a-cPCV2b vaccine is 100% to each strain. In this way, the cPCV2a-cPCV2b bivalent vaccine potentially offers greater efficacy than each vaccine fraction individually. The directional enhancement in efficacy may be due to the increased availability of epitopes (both conserved and unique to PCV2a and PCV2b) to interact with the immune system, resulting in a broader and more robust repertoire of PCV-specific T cells. The availability and diversity of epitopes in the bivalent vaccine may translate to boarder clinical protection including to recombinant field viruses that contain bits of PCV2a or PCV2b or PCV2d because of the epitope overlap in the vaccine and field viruses. Studies with higher power or use of more divergent challenge strains is needed to further define and elucidates the immune bases for the enhanced clinical protection. This work supports the need, and high probability of success, for updating current monovalent PCV2 vaccines to include PCV2a and PCV2b.

References

1. Franzo G and Segales J. Porcine circovirus 2 (PCV-2) genotype update and proposal of a new genotyping methodology. 2018. PLoS ONE 13 (12): e0208585. **https://doi.org/10.1371/journal.pone.0208585**

2. Opriessnig T, Xiao C-T, Gerber PF, Halbur PG (2013b) Emergence of a novel mutant PCV2b variant associated with clinical PCVAD in two vaccinated pig farms in the U.S. concurrently infected with PPV2. Vet Microbiol 163:177-183.

Table 1: Summary of Study A results - homologous vs heterologous protection to a PCV2a vaccine

Protection endpoint	Treatment group (T)			
	T01 – PCV2a challenge control	T02 – cPCV2a vaccine	T03 – PCV2b challenge control*	T04 – cPCV2a vaccine*
	PCV2a challenge (Homologous protection)		PCV2b challenge (Heterologous protection)	
	Ever positive Positive # / total # (%)			
Viremia	9/10 (90%)	5/10 (50%)	9/9 (100%)	7/10 (70%)
Fecal shedding	9/10 (90%)	3/10 (30%)	9/9 (100%)	3/9 (33.3%)
Lymphoid depletion (LD)	6/10 (60%)	0/10 (0%)	6/9 (66.7%)	4/9 (44.4%)
Histiocytic replacement (HR)	2/10 (20%)	0/10 (0%)	6/9 (66.7%)	2/9 (22.2%)
Lymphoid colonization (IHC)	7/10 (70%)	4/10 (40%)	7/9 (77.8%)	3/9 (33.3%)

* One animal from Treatment Group 3 (T03) and one animal from T04 were removed for health reasons.

Table 2: Study B and Study C –summary of improved efficacy outcomes imparted by the bivalent vaccine compared to monovalent vaccines for Studies B and C

Protection endpoint	% Improvement of cPCV2a-cPCV2b bivalent vs: cPCV2a monovalent[2]	cPCV2b monovalent[2]
Viremia	28.3%	21.9%
Fecal shedding	29.9%	4.5%
Lymphoid depletion (LD)[1]	68.5%	25.1%
Histiocytic replacement (HR)[1]	NA	100%
LD or HR[1]	68.5%	24.5%
Lymphoid colonization (IHC)[1]	53.3%	37.5%
Serology[1*]	17.7%	32.7%

1 Presented as back-transformed least square means.

* The ELISA used to determine PCV2 specific antibodies (serology) was an S/N (sample to negative ratio) ELISA where seropositive < 0.5. Serology data noted are from a blood sample collected at necropsy.

2 The percent improvement of the bivalent vaccine over the monovalent vaccine has been calculated for each outcome criteria ([monovalent-bivalent]/monovalent).

3. Franzo G, Tucciarone CM, Cecchinato M, Drigo M (2016b) Porcine circovirus type 2 (PCV2) evolution before and after the vaccination introduction: A large scale epidemiological study. Nature Sci Rep 6:39458. **DOI: 10.1038/srep39458**

4. Franzo G, Cortey M, Segalés J, Hughes J, Drigo M (2016) Phylodynamic analysis of porcine circovirus type 2 reveals global waves of emerging genotypes and the circulation of recombinant forms. Mol Phylogenet Evol 100:269-280.

5. Reiner G, Hofmeister R, Willems H (2015) Genetic variability of porcine circovirus 2 (PCV2) field isolates from vaccinated and non-vaccinated pig herds in Germany. Vet Microbiol 180:41-48.

6. Haach V, Gava D, Cantão ME, Schaefer D (2020) Evaluation of two multiplex RT-PCR assays for detection and subtype differentiation of Brazilian swine influenza viruses. Braz J Microbiol 51(3):1447-1451.

7. Karuppannan AK, Opriessnig T (2017) Porcine circovirus type 2 (PCV2) vaccines in the context of current molecular epidemiology. Viruses 9:99. **doi:10.3390/v9050099**

8. Constans M, Ssemadaali M, Kolyvushko O, Ramamoorthy S (2015) Antigenic Determinants of Possible Vaccine Escape by Porcine Circovirus Subtype 2b Viruses. Bioinform Biol Insights 9:1-12.

9. Kurtz S, Grau-Roma L, Cortey M, Fort M, Rodríguez F, Sibila M, Segalés J (2014) Pigs naturally exposed to porcine circovirus type 2 (PCV2) generate antibody responses capable to neutralise PCV2 isolates of different genotypes and geographic origins. Vet Res 45:443-443.

10. Saha D, Huang L, Bussalleu E, Lefebvre DJ, Fort M, Van Doorsselaere J, Nauwynck HJ (2012a) Antigenic subtyping and epitopes' competition analysis of porcine circovirus type 2 using monoclonal antibodies. Vet Microbiol 157:13-22.

11. Opriessnig T, Yu S, Gallup JM, Evans RB, Fenaux M, Pallares F, Thacker EL, Brockus CS, Ackerman MR, Thomas P, Meng XJ, Halbur PG (2003) Effect of vaccination with selective bacterins on conventional pigs infected with type 2 porcine circovirus. Vet Pathol 40:521-529.

12. Opriessnig T, Prickett JR, Madson DM, Shen H-G, Juhan NM, Pogranichniy RM, Meng XJ, Halbur PG (2010) Porcine circovirus type 2 (PCV2)-infection and re-inoculation with homologous or heterologous strains: virological, serological, pathological and clinical effects in growing pigs. Vet Res 41:31.

13. Beach NM, Ramamoorthy S, Opriessnig T, Wu SQ, Meng X-J (2011) Novel Chimeric Porcine Circovirus (PCV) with the Capsid Gene of the Emerging PCV2b Subtype Cloned in the Genomic Backbone of the Nonpathogenic PCV1 is Attenuated in vivo and Induces Protective and Cross-protective Immunity Against PCV2b and PCV2a Subtype in Pigs. Vaccine 29:221-232.

14. Bandrick M, Gutierrez AH, Desai P, Rincon G, Martin WD, Terry FE, De Groot AS, Foss DL (2020) T cell epitope content comparison (EpiCC) analysis demonstrates a bivalent PCV2 vaccine has greater T cell epitope overlap with field strains than monovalent PCV2 vaccines. Vet Immunol Immunopathol 223. **https://doi.org/10.1016/j.vetimm.2020.110034**

15. Gava D, Serrao VHB, Fernandes LT, Cantao,ME, Ciacci-Zanella JR, Mores N, Schaefer R (2018) Structure analysis of capsid protein of Porcine circovirus type 2 from pigs with systemic disease. Braz J Microbiol 49:351-357.

https://doi.org/10.54846/am2022/57

Reproductive impact of deviations in semen dose sperm quantity and quality

B. Belstra[1], PhD; R. Ausejo[2], DVM
[1]MB Swine Reproduction, Verona, WI; [2]Magapor, Ejea de los Caballeros, Zaragoza, Spain

Introduction

In general, less than 10% of the variation in sow reproductive performance tends to be related to the boar and particular semen batch used. While individual sow and sow farm management factors typically create most of the variation in reproductive outcome, the impact of the boar and insemination doses cannot be ignored. This is especially true when the sperm quantity of the doses produced is insufficient, some aspect of their sperm quality has been compromised, or both. In these situations, the boar and the fertility of his sperm can limit and induce greater variation in reproductive results.

Doses that do not meet the targeted minimum standards might affect the first, second, or all on the inseminations given to an individual female depending on how widespread and long lasting the nonconformance is and this creates another source of variation in sow reproductive outcome. Females with normal fertility can be prematurely culled for poor reproductive performance when their connection to substandard semen dose quality or boar fertility is realized too late or missed entirely. We tend to recognize and focus on the lost pig production but not the potential added loss of fertile females from the herd.

Some currently achieved levels and future goals for the consistency and quality of insemination doses are outlined below (Table 1). Greater accuracy and precision in achieving minimum semen dose quality standards will likely be necessary as the swine breeding industry accepts the rewards and risks of additional reduction of the number of sperm per insemination dose, the number of semen deliveries per wean event (ie, increased semen age), and reduced use of pooled semen in favor of single-sire doses.

Sperm quantity

There is no consistent threshold at which the number of sperm per insemination doses begins to limit total born and eventually farrowing rate due to the myriad of interacting boar, sow and management factors involved. However, there is accumulating evidence that 2 billion sperm per dose is sufficient and 1 billion sperm per dose tends to be too few to sustain top reproductive performance in most systems. Adoption of intrauterine or post-cervical artificial insemination (PCAI), as opposed to traditional cervical AI (CAI), has fostered the transition to lower insemination dose volumes (eg, 75 vs 45 mL) and fewer total sperm per dose (eg, 3 vs 2 billion). There is typically much more sperm concentration variation among different semen batches than within the doses of an individual batch, provided care is taken to prevent sperm settling during processing and packaging, and the dose volume is reasonably close to the intended target. More accurate initial sperm concentration estimation and reduction of semen processing errors are likely where more improvement can be made in reduced variation of sperm per dose from target. It should be noted that some studs adjust and inflate the total sperm per dose for each batch to compensate for the number of non-motile and(or) abnormal morphology sperm and this creates added variation in batch-to-batch sperm concentration by design. The intention is to achieve a consistent "viable" sperm per dose target. The alternate approach is to set and maintain the same total sperm per dose target for each batch.

Sperm quality

Survival of sperm and maintenance of their fertilizing function as they age during in vitro storage and in utero prior to ovulation seems to be a significant component of a given semen batch and given boar's fertility. Initial estimates of the level of sperm motility and normal morphology are used by boar studs to reject the worst quality ejaculates. While minimum targets vary slightly stud-to-stud, many only accept and process ejaculates that possess both ≥ 75% motility and ≥ 75% gross normal morphology (ie, ≤ 25% of sperm with a retained proximal, distal cytoplasmic droplet, or abnormal tail angle). Unfortunately, the level of initial motility and normal morphology of the accepted ejaculates above these minimums has relatively weak positive correlation with the level of sow reproductive performance achieved. The decrease in motility over storage time or the level at advanced storage ages (ie, 7 to 10 days) seems to be a better indicator of fertility potential than the initial level at collection. More detailed morphology of the sperm head made on fixed and stained samples at higher magnification can also add fertility prediction power. Finally, substantial decreases in a boar's level of motility or normal morphology from previous collections can be more predictive of sub-fertility than their absolute levels.

In some situations, large quantities of ejaculates just 5 to 10% below the motility and(or) normal morphology minimums are trashed, genetic lag is created, and more boars are required to

Table 1: An example of extended insemination dose quantity and quality targets and coefficient of variation (CV) or conformance achieved based on weekly quality control samples

	Target	Achieved	New goal
Sperm concentration	40 million/mL	CV = 10 to 15%	CV ≤5%
Gross normal morphology	≥ 75%	90% ≥ 75%	95% ≥ 75%
Aerobic bacterial growth	0 cfu/mL	95% ≤ 10 cfu/mL	95% ≤ 1 cfu/mL

find usable quality semen to fill orders. No sows are inseminated with these rejected ejaculates to confirm if they are in fact subfertile. It may be time to redefine the unacceptable level of specific types of the most common morphological sperm abnormalities for current swine AI conditions and current genotypes. The high level of distal cytoplasmic droplets that plague Duroc boar ejaculates is a prime example.

Semen contamination

All semen collections contain some level of bacteria and semen extender is an excellent culture media for both bacteria and sperm. Sperm viability, semen shelf-life and sow reproductive performance can all be compromised by the level and specific type of bacterial contamination, so collection and laboratory hygiene are critical to avoiding problems. Given time and the moderate temperature of boar semen storage bacteria can multiply exponentially and some species induce sperm agglutination and loss of motility and can cause inseminated females to exhibit a vaginal discharge soon after AI. It is important to note that there are also bacteria that can negatively affected sow reproductive performance with no obvious effects on sperm viability. Granted, not all bacteria cause problems and some may even help suppress them.

Items that have direct contact with semen can also add chemicals with toxic effects. Sperm toxicity has been well documented when latex rubber or nitrile rubber gloves are used for collection instead of safer vinyl gloves. Depending on the amount of chemical residue on the gloves and the amount of contact with raw semen during the collection process, immediate sperm death or a delayed effect on sperm function can occur resulting in reduced sow reproductive performance. The ultimate safeguard is to ensure the collection glove hand catches the boar's penis with tip exposed so ejaculated semen flows directly into the collection vessel and avoids glove contact altogether. Finally, plastic semen packages can introduce chemicals to the extended semen that have no apparent negative effects on sperm yet reduce their ability to fertilize oocytes or support embryonic development. Boar sperm act as a living biosensor of chemical contamination and require protection during the entire AI processes. Magapor has a robust reproductive quality control program in place to test and select raw materials for production, as well as the finished products prior to release for sale.

Quality control

Regular random sampling of finished insemination doses is required to measure how consistently sperm quantity, sperm quality and low bacterial growth targets are being achieved. Internal testing programs can be useful to provide continuous quality control (QC) measures and opportunity to adjust and improve process. However, objective third-party testing is also necessary to add integrity to QC data and internal lab results. Having non-boar stud personnel sample finished doses or intercepting samples in route to sow farm delivery is another way to add QC data integrity. Failure to share QC data on a regular basis encourages end users to pursue their own third-party QC results and such testing is often isolated to periods when they fail to reach reproductive targets.

Part of the challenge is to transition insemination dose QC testing and reproductive management from a reactive to a proactive process. There is roughly a 4-week lag to pregnancy check data and a 4-month lag to litter size data. In addition, individual batches of pooled semen may only be traceable to the sow farm level and not to the reproductive performance of the individual females inseminated with the doses. Strategies to achieve more proactive management include frequent sampling, rapid test turnaround, as well as connection and centralization of all the data sources to a visualization dashboard. Earlier identification of non-random deviations from insemination dose standard targets and increased understanding of the root causes allows for more proactive management of processes and their critical control points. Improvement of written protocols and training certification programs are also key to error prevention. In response to this need, organizations with the specific objective to standardize production of insemination doses across cooperating boar studs have been established in Europe. More accurate monitoring of the impact of insemination doses on sow reproductive performance is needed to maintain and improve pig production efficiency as we accept the rewards and risks of working with reduced numbers of sperm per herd female.

https://doi.org/10.54846/am2022/58

The next level of needle detectability: Shifting the paradigm

G. Hastings; B. Galbreath; N. Wagner, DVM; T. Leigh; P. Shipp; J. Lye; R. Bachtel; D. Jacobs
Neogen Corporation, Slingshot PDG

Introduction

Vaccines and proper vaccine administration are an integral part of modern swine production. In the US, assuming a high herd health status, it has been estimated that roughly 120 million market hog injections occur annually.[1] Twenty years ago, Neogen introduced Ideal D3® detectable needles into the swine industry to help prevent broken needles from entering the food chain. Since then, the use of detectable needles has become a staple of The National Pork Board's PQA Plus® program.[2] Ideal D3® needles are a thick-walled cannula to help prevent breakage. More importantly, the cannula is comprised of a highly ferritic metallurgy, that was designed to be detectable in the processing plant at line speed through magnetic resonance detection. At the farm level, if a broken needle occurs in a pig, standard operating procedures call for that pig to be tagged, marked, and/or euthanized. However, there are occasions where the speed of the vaccinator, combined with the scarcity of labor, complicated by the movement of unrestrained pigs in a pen can result in the unintended consequence of a broken needle cannula found in the carcass of a pig either in the processing plant or even worse, on the dinner plate of a consumer. Therefore, Neogen reimagined the Ideal D3® needle to add an extra layer of visual detection in conjunction with a feature for the practical removal at the farm level, all while maintaining its proven reliability of detection in processing plants.

Materials and methods

Finite Element Analysis (F.E.A) simulation was conducted on standard 304SS cannula as well as Neogen 2205SS cannula to evaluate the strength in response to applied force at various points along the length of the needle to determine baseline data for the most probable break point. This was replicated on 14G, 16G, 18G, and 20G needles that varied in length from ½ inch to 1.5 inches. Each needle was analyzed under the same loading and boundary or support conditions to maintain uniformity within the study. All force and pressure placements utilized the tip of the needle as the reference point to establish consistency throughout the analysis. SolidWorks and SolidWorks Simulation[3] were the tools utilized to complete this analysis (Figure 1). A highly visible extraction collar was designed to be added to the cannula at a specific distance beyond the maximum stress point as determined by F.E.A simulation. This would allow for the needle cannula to break at the weakest point while the secured extraction collar would remain intact on the cannula (Figure 2). The polypropylene extraction collar was designed as a highly visible practical retrieval feature as well as an observable early warning indicator to alert vaccination personnel to a weakened or compromised needle.

Results and discussion

F.E.A. simulation confirms that the weakest point of the needle remains at the cannula hub juncture. The repetitive stress and pressure which a needle encounters during usage has the ability to compromise its strength and result in either bending or breakage at this identified location on the cannula. The addition of the highly visible extraction collar at a specified distance beyond the maximum stress point did not shift the break point and allows for an observable early warning indicator to alert personnel to compromised cannula integrity thereby providing the vaccination personnel the opportunity to change the needle prior to failure. If needle breakage does occur, the extraction collar is designed to stay intact on the separated component that is in contact with the animal from disappearing into the subcutaneous or deeper tissue. Further, the highly visible extraction collar flares and provides a practical means for personnel to simply retrieve, by hand, the separated component of the needle cannula from the pig without any additional equipment required (Figure 3). Therefore, confirming removal eliminates the need to tag out or euthanize the animal. In the design of the needle, it was necessary to add length to the cannula with the addition of the extraction collar to maintain the designated injection length. The highly ferritic 2205SS cannula is still up to 100% detectable in processing plants at line speed with magnetic resonance systems for an added layer of food security. This patent pending needle by Neogen is referred to as the D3X™ needle (Figure 4).

References

1. The National Pork Board 2018 Pork Industry Needle Forum Report
2. The National Pork Board PQA Plus® Education Handbook (2019 release)
3. SolidWorks and SolidWorks Simulation by Dassault Systemes

Figure 1: F.E.A simulation on standard needle.

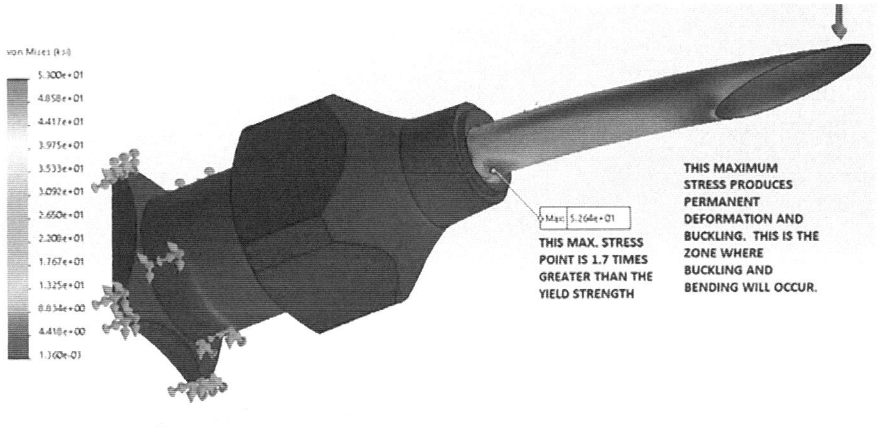

Figure 2: F.E.A. simulation on needle with extraction collar.

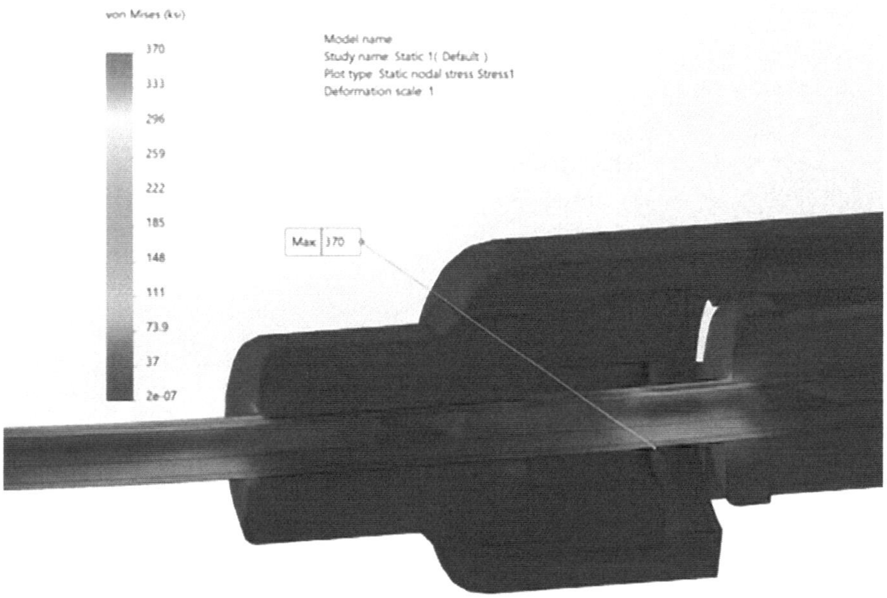

Figure 3: Highly visible extraction collar in pig.

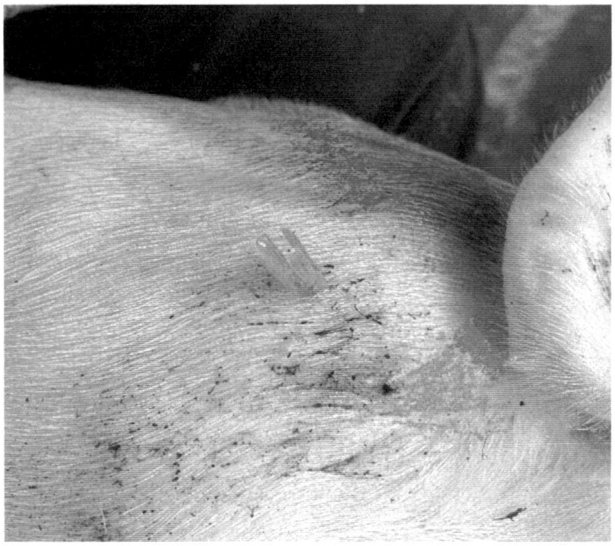

Figure 4: D3X needle assembly with logo.

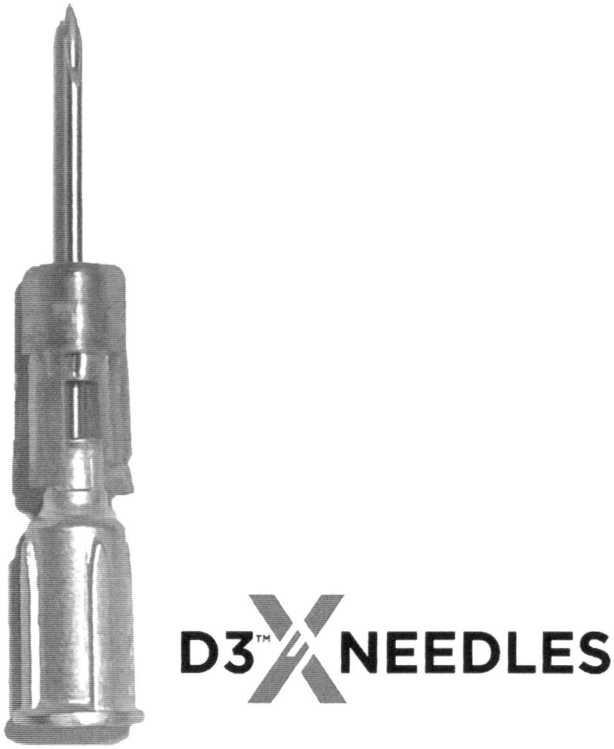

https://doi.org/10.54846/am2022/59

Efficacy of Vigilex® in diets contaminated with porcine epidemic diarrhea virus (PEDV) via a modified ice-block challenge model

R. Strobel[1], DVM; P. Yeske[1], DVM, MS; R. Saltzman[2], DVM; P. Arruda[2], DVM, MS, PhD; J. Zhang[3], MD, MS, PhD; S. Ebarb[4], MS; M. Ritter[4], PhD
[1]Swine Vet Center, Saint Peter, Minnesota; [2]Veterinary Resources Inc., Ames, Iowa;
[3]Iowa State University College of Veterinary Medicine, Ames, Iowa; [4]Cargill Animal Nutrition & Health, Lewisburg, Ohio

Introduction

Vigilex is a proprietary oil blend developed by Cargill Animal Nutrition and Health to reduce the risk of PEDV in swine feeds. Previous research has demonstrated that adding 0.4% Vigilex to complete feed challenged with PEDV at 3×10^5 $TCID_{50}$ reduced genomic copies of PEDV in feed by 1.5 logs, and when given as an oral gavage to pigs in a subsequent swine bioassay, was successful in preventing clinical signs and PEDV infection. However, a seeder pig model study showed that Vigilex did not prevent PEDV infection from non-feed sources, but Vigilex may improve growth performance traits during a PEDV challenge.[1] Therefore, Vigilex needs to be part of a comprehensive biosecurity program that covers all PEDV risk factors (feed ingredients, feed mills, feed delivery, people, pigs, and environmental factors).

Recently, there has been a lot of debate within the swine industry on how to evaluate the efficacy of swine feed mitigants. The two most common approaches are: 1) classic disease challenge model with virus administered by oral gavage[1]; and 2) ice-block challenge model where an ice-block of frozen pathogen(s) are placed in the feed bin and allowed to melt and slowly trickle to the pigs in the barn.[2] Both models have pros and cons. For example, the oral gavage challenge method ensures a consistent virus load to each pig, but the large single dose of virus in feed does not likely represent what happens in the field. Meanwhile, the slow trickle of virus challenge coming through the melting ice-block challenge may be more representative of what happens in the field. However, the ice-block model may lead to inconsistent challenge levels within study and across studies as level of challenge depends on how the ice-block melts and distributes virus through the feed. Therefore, the objectives of this study were to: 1) develop a modified PEDV ice-block challenge model that provides a more consistent virus level at the feeder and pen level; and 2) determine effectiveness of Vigilex to reduce the incidence and severity of IHC PEDV scores in a modified PEDV ice-block challenge model.

Materials and methods

Experimental design

Forty nursery pigs were utilized in a randomized complete block design with two feed mitigant treatments (control vs. 0.4% Vigilex) to determine the effectiveness of Vigilex in diets contaminated with PEDV by using a modified ice-block challenge model for 7 days.

Housing

This study utilized two rooms in a BSL-2 facility (Veterinary Resources Inc., Ames, IA). Each room had a dedicated air filtration system where air coming in and out of the room was HEPA-filtered. Within each room, there were 5 raised deck tubs (1.22 m × 1.22 m) with solid walls, and each tub had plastic slatted flooring, a nipple water, and a feeder.

Pigs

Forty crossbred males (n = 20) and females (n = 20) were utilized in this study. Upon arrival, pigs were ear tagged for individual identification. Study inclusion criteria consisted of the following: in good visual health, between 42-52 days of age, PCR negative for PEDV by fecal swab upon arrival, and serum negative for PEDV antibodies by ELISA upon arrival.

Treatment allotment

Four days prior to the modified ice-block PEDV challenge (Day -4), dietary treatments were assigned to rooms. Pigs (n = 40) were then individually weighed and randomly assigned to treatments on the basis of live weight and gender. This approach resulted in 5 mixed gender pens (2 males and 2 females) per treatment.

Dietary treatments

Two 340 kg batches of late nursery complete feed were produced by Provimi North America (Lewisburg, OH): 1) control; and 2) 0.4% Vigilex. Diets were formulated to meet or exceed the pig's requirements.[3] The commercially available Vigilex product contains a Micro-tracer to confirm presence of the additive. Control and Vigilex complete feed samples (~200 g) were sent to Micro-Tracers (San Francisco, CA) for analysis. Certificate of Micro-tracer Analysis confirmed that Vigilex was present in the Vigilex treatment diet, but not in the control treatment diet. Treatment diets were fed ad libitum from Day -4 (allotment) to Day 7 (necropsy).

Modified PEDV ice-block challenge model

PEDV challenge inoculum was prepared by Dr. Zhang at Iowa State University. The virus stock was diluted by 6.7-fold to create a challenge inoculum with a titer of 5×10^4 $TCID_{50}$/mL and a CT value of 15.8. Thirty 10 mL tubes of PEDV challenge material were then prepared and stored at -80°C until use on Day 0.

On Day 0, pigs and feeders were weighed. Feeders were emptied and refilled with 13 kg of treatment feed, and a 10 mL PEDV ice-block was placed in the top middle feed peak within each feeder (Figure 1A). At the same time, an additional 13 kg of treatment feed was weighed out and placed into a plastic storage bin assigned to each pen, and a 10 mL PEDV ice-block was placed on top of the feed and allowed to melt at barn temperature (Figure 1B). The initial 13 kg of feed in the feeder took 4 days to be consumed (Figure 1C). It was observed on Day 4 that the melted ice-block created feed balls within the storage bins (Figure 1D), and thus,

Figure 1: Images from the modified PEDV ice-block challenge model.

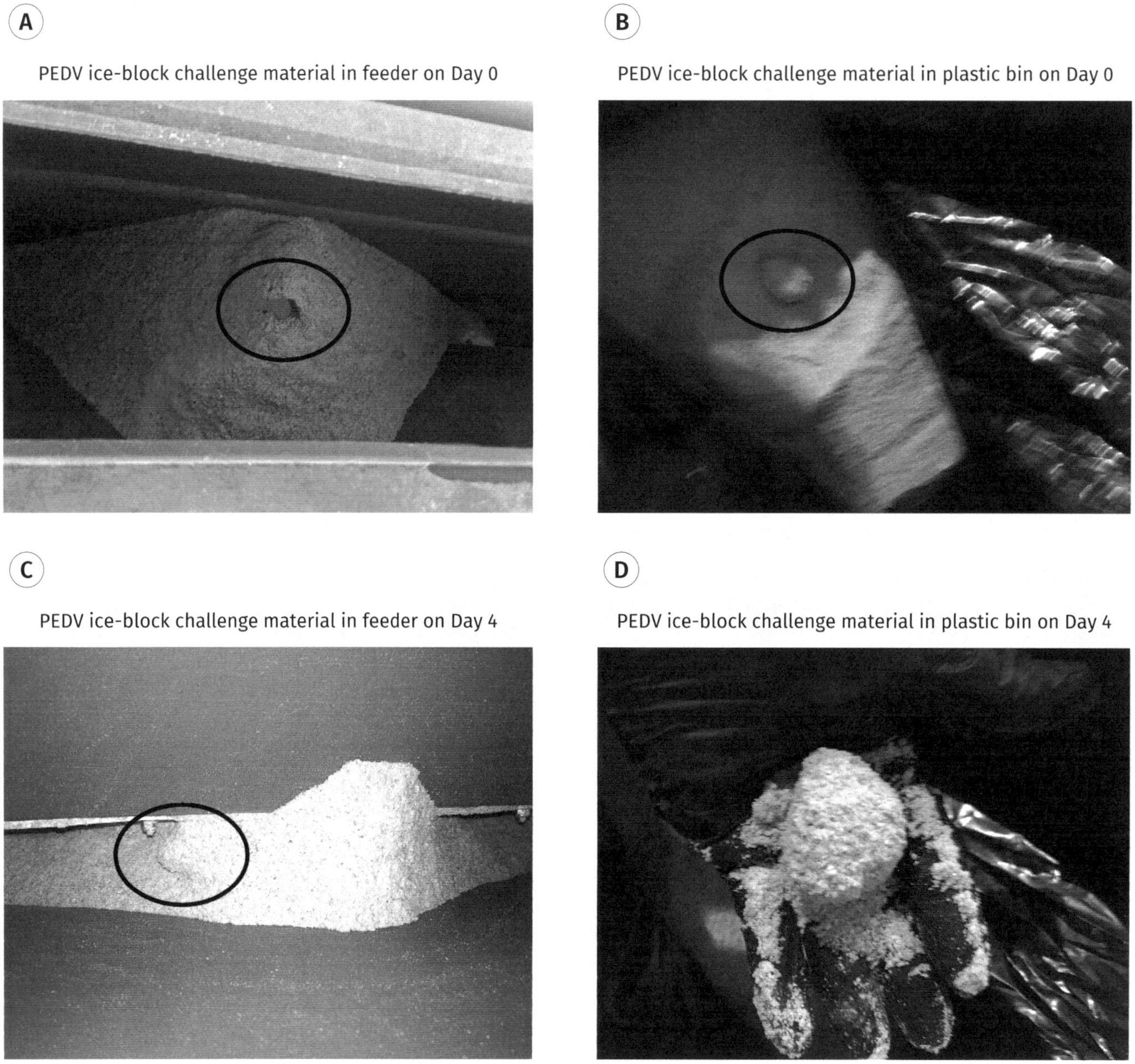

storage bin contents were thoroughly mixed before being added to feeders on Day 4. This approach matches the concentration of the viral feed challenge used in the ice-block challenge model reported by Dee et al.[2]

During the 7-day PEDV challenge period a number of performance and health measurements were collected. Growth performance traits (ADG, ADFI, gain:feed, and mortality) were collected by pen for Day 0-7. Clinical signs of PEDV diarrhea were scored on a scale of 0 to 3 (0 = firm; 3 = watery) by individual pig daily over Days 0-7. PEDV RNA was measured by qPCR in oral fluids by pen, in feed by pen, and in feces by pig on Days 0, 4, and 7 by the Iowa State University Veterinary Diagnostic Laboratory (ISU VDL).

Pigs were humanely euthanized following study site specific procedures on Day 7. During necropsy, the small intestine, cecum and spiral colon were macroscopically evaluated for content (normal, semi-liquid, pure liquid) and gross pathology (normal, gas distended, thin walled, other). Three formalin fixed sections of ileum were collected and submitted for H&E and PEDV immunohistochemistry (IHC) evaluation. PEDV severity was assessed by IHC scoring on a scale of 0 to 4 (0 = no staining; 4 = 50-100% positive enterocyte staining).

Statistical analysis

Data were analyzed with R Statistical Software (version 3.6.1) using the lme4, lmerTest, and forestinventory packages. The model included the fixed effect of treatment and random effect of block. Pen was the experimental unit for performance measurements, PEDV in oral fluids, and PEDV in feed. Pig was the experimental unit for diarrhea scores, PEDV fecal shedding, necropsy findings, and PEDV IHC staining. Statistical significance was declared for *P* values ≤ 0.05, while trends were declared at *P* ≤ 0.10.

Results and discussion

Growth performance traits

There was no morbidity or mortality observed in the current work. Dietary treatment did not affect live weight, ADG, ADFI, or gain to feed (Table 1). This was not unexpected given the small number of observations per treatment (n = 5), and the short feeding duration (n = 7 days).

PEDV diarrhea scores

Dietary treatment did not affect PEDV diarrhea scores (Table 2). In the current work, PEDV diarrhea was not detected until day 4, and there was high variability in individual pig scores from Day 4 to 7 (range: 0 to 3). This may be a reflection of how long it took for the PEDV ice-blocks within the feeders to reach the pigs.

Genomic copies of PEDV

Vigilex reduced (*P* < 0.05) PEDV RNA in feed samples (day 4 and 7) and fecal samples (day 7), but did not affect PEDV RNA in oral fluids (Table 3).

Probability of positive PEDV Results

Vigilex reduced the percentage of Day 4 PEDV fecal positives (30% vs. 60%, respectively), Day 7 PEDV fecal positives (30% vs. 100%, respectively), PEDV positive macroscopic lesions (10% vs. 50%, respectively), and IHC PEDV positives (15% vs. 60%, respectively) when compared to the control treatment (Table 4).

Modified PEDV ice-block model

The modified ice-block challenge model utilized in the current work successfully produced clinical signs of PEDV diarrhea (control: 70% of pigs had a diarrhea score ≥ 1), Day 7 PEDV fecal positives (control = 100%), Day 7 PEDV positive macroscopic lesions (control = 50%), and Day 7 IHC PEDV positives (control = 60%). The IHC PEDV scores by individual pig within pen gives additional insight into how PEDV spread within each pen (Table 5), and the solid sided 1.22 m × 1.22 m stand-alone tubs were successful in preventing cross-contamination of PEDV across pens. When analyzing the IHC PEDV scores by individual pig within pen, it is important to note the following observations: 1) 4/5 of the Vigilex pens had all 0 scores vs. 1/5 pens in the control group; and 2) multiple pens had both 0 and 4 IHC PEDV scores within the same pen (Table 5). This later point exemplifies the variation from pig to pig in the amount of active PEDV consumed from the modified PED ice-block challenge.

Conclusions

The PEDV modified ice-block challenge model was successful in inducing clinical signs, positive PEDV PCR results in feces and oral fluids, and positive PEDV IHC staining. Vigilex did not completely inactivate PEDV from feed (IHC PEDV positive pens: Control 4/5 vs. Vigilex 1/5); however, it did significantly reduce the risk and severity of PEDV when compared to controls. Furthermore, these data highlight the importance of using PEDV PCR, PEDV intestinal macroscopic lesion scores, and IHC PEDV staining to determine the efficacy of feed mitigants.

References

1. Saddoris-Clemons, K. and S. Ebarb. 2020. Efficacy of Vigilex™ feed ingredient in reducing PEDV contamination risk in complete feed and impact of PEDV transmission and nursery performance in pigs infected via non-feed vectors. AASV p. 103-106.

2. Dee et al.: 2021. An evaluation of additives for mitigating the risk of virus-contaminated feed using an ice-block challenge model. Transbound Emerg Dis. 68:833-845.

3. National Research Council. 2012. Nutrient requirements of swine. 11th ed. Washington, DC: Natl. Acad. Press.

Table 1: Effects of feed mitigant treatments on growth performance traits

Variable	Control	Vigilex	SEM	*P*-Value
Pens, #	5	5	---	---
Initial BW, kg	14.08	14.01	1.24	0.606
Final BW, kg	16.85	16.84	1.46	0.996
ADG, kg/d	0.395	0.404	0.075	0.934
ADFI, kg/d	0.695	0.646	0.086	0.700
Gain to feed, kg/kg	0.554	0.608	0.068	0.608
Mortality, %	0.0%	0.0%	---	---

Table 2: Effects of feed mitigant treatments on PEDV diarrhea scores*†

Variable	Control	Vigilex	SEM	*P*-Value
Pigs, #	20	20	---	---
Day 0-4 fecal score	0.00	0.00	---	---
Day 4-7 fecal score	0.74	0.59	0.136	0.645

* Pig was the experimental unit

† Diarrhea was scored daily on each pigs using a scale of 0 to 3 (0 = firm and 3 = watery)

Table 3: Effects of feed mitigant treatments on log transformed genomic copies of PEDV

Variable	Control	Vigilex	SEM	*P*-Value
Pens, #	5	5	---	---
Pigs, #	20	20	---	---
PEDV in oral fluids*				
d 4 post-challenge	5.316	4.435	0.579	0.343
d 7 post-challenge	2.789	1.594	0.564	0.208
PEDV in feed*				
d 4 post-challenge	7.058	4.177	0.577	0.024
d 7 post-challenge	3.420	0.585	0.439	0.010
PEDV in feces†				
d 4 post-challenge	3.039	1.662	0.940	0.122
d 7 post-challenge	6.085	1.869	0.863	< .0001

* Pen was the experimental unit

† Pig was the experimental unit

Table 4: Effects of feed mitigant treatments on positive PEDV results

Variable	Control	Vigilex	SEM	*P*-Value
d 4 PEDV fecal positive, %	60.0	30.0	11.0	0.055
d 7 PEDV fecal positive, %	100.0	30.0	10.0	< .0001
PEDv positive macroscopic lesions,%	50.0	10.0	11.0	0.004
IHC PEDV positive, % *,†	60.0	15.0	11.0	0.003

* IHC PEDV lesions were scored 0 to 4 (0 = no staining; 4 = 50-100% staining)

† Negative = no staining present; positive = 10-100% positive enterocytes

Table 5: IHC PEDV scores by pig within pen*

Vigilex Room

Pen 5A	Pen 5B	Pen 5C	Pen 5D	Pen 5E
0	0	0	4	0
0	0	0	4	0
0	0	0	4	0
0	0	0	0	0

IHC PEDV positive pigs: 3/20
IHC PEDV positive pens: 1/5

Control room

Pen 6A	Pen 6B	Pen 6C	Pen 6D	Pen 6E
4	4	0	0	0
4	4	0	1	0
4	0	0	4	1
4	4	0	2	4

IHC PEDV positive pigs: 12/20
IHC PEDV positive pens: 4/5

* IHC PEDV lesions were scored 0 to 4 (0 = no staining; 4 = 50-100% staining)

https://doi.org/10.54846/am2022/60

Discovery of a key biomarker for inflammation and associated performance economics in growing pigs

Brandi Burton[1], DVM; Rob Musser[2], PhD; Ran Song[2], PhD; J. Tyler Holck[3], DVM, MS, MBA
[1]Suidae Health & Production, Algona, Iowa; [2]NutriQuest, Mason City, Iowa; Feed His People, LLC, Gilbert, Iowa

Introduction

Inflammation has been shown to limit growth performance in growing pigs and further, serum vitamin inadequacies have been shown to be a response to chronic inflammation, rather than diet deficiency.[1] NutriQuest has been exploring the use of serum vitamin A at weaning as a potential performance indicating biomarker for negative effects of inflammation in commercial swine production.

Biomarker pathway and utilization

Inflammation directly impacts vitamin A metabolism in the liver and resulting transport to body tissues. The resulting effect is insufficient serum vitamin A being transported to the tissues for growth. Vitamin A is primarily stored in the liver and transported from liver to tissues via retinol binding protein (RBP; Figure 1). Previous work in rats has shown inflammation induced by lipopolysaccharide (LPS) challenge decreased RBP production in the liver, reducing vitamin A transport and resulting in insufficient serum vitamin A.[1] Additionally, NutriQuest commercial research[2] concluded that compared to their sufficient counterparts, pigs with insufficient serum vitamin A (< 0.15 ppm) at weaning had:

- Reduced pre-weaning piglet ADG
- Lower piglet weaning weights
- Lower post-weaning piglet survival

Inflammation comes from a variety of sources including mycotoxins, disease, and heat stress to name a few. To date, monitoring inflammation status in the field consistently and cost-effectively has been difficult. This paper will explore field experience suggesting serum vitamin A is performance indicating and may be used as a biomarker of systematic inflammation in commercial swine production.

Impact of PRRSV outbreak on biomarker levels

Serum from 72 pre-wean piglets was collected from a 2500 commercial sow farm in northwest Iowa on November 2, 2020 as part of a larger serum vitamin A survey across multiple farms. The samples were collected from 6 entire litters across parities and were analyzed for serum vitamin A levels at Iowa State Veterinary Diagnostic Laboratory (ISU VDL). The sow herd experienced an outbreak of PRRSV approximately three weeks later on November 25, 2020 and a second set of 43 sera was collected from 6 different litters (also across parities) from the same sow farm on January 1, 2021 and analyzed for serum vitamin A levels at ISU VDL. Pre-wean (19 days of age average) serum vitamin A levels prior to the PRRS outbreak averaged 0.30 ppm with a range from 0.13 to 0.44 ppm across all 72 samples (Figure 1). In comparison to other sow farms sampled at the same time in the survey, serum vitamin A levels were relatively high (unpublished data). Serum vitamin A levels collected from pre-wean pigs (22 days of age average) after the PRRS break were substantially lower with a mean of 0.11 ppm and a range from 0.03 to 0.20 ppm across the 43 samples (Figure 1).

Biomarker levels and performance

Each pig sampled in both groups was individually identified with a LeeO™ RFID tag and subsequently individually weighed at weaning, 12-13 weeks of age, and again at 22-23 weeks of age. Weights at all three time points trended higher based upon pre-wean vitamin A serum levels, therefore, there was a positive correlation between serum vitamin A levels at weaning and weights through finishing (Figure 2). This was more pronounced in the PRRS-positive pigs with significant correlation of week 13 weights to pre-wean serum vitamin A levels (Figure 3). When pigs had low biomarker levels, they were also more likely to die. For the pigs in the PRRS negative group, none of the pigs with serum vitamin A levels less than 0.2 ppm survived to market. In the PRRS positive pigs, mortality was 23% by 12-13 weeks of age and 38% mortality wean-to-finish if the serum vitamin A levels were below 0.15ppm. The number of pigs in this study was not enough to detect significant changes in mortality, but the results suggest serum vitamin A at weaning may be predictive of finishing mortality.

Serum biomarker levels and economics

Utilizing the serum biomarker for inflammation and the performance differences observed based upon pre-wean values provides an opportunity to assess the economic impact of inflammation. Figure 4 is an example of an economic calculator which estimates the financial impact of inflammation on a per head basis. This figure provides the economic impact of mortality and gain along with the populations of the pigs in each category. The comparison suggests that the variation in serum vitamin A levels in the example group results in $2.18 loss per pig average in the barn compared to the average group tested so far in the industry providing a $0.40/pig profit, and the top 50% of the farms with high biomarker levels resulted in $6.43 in profit per pig. This model utilizes the biomarker field data to provide estimated financial differences based on mortality and gain which allows producers to focus on the economic impact of inflammation and the value of interventions.

Discussion

There are many sources of inflammation that may impact growing pigs throughout their lifetime. This study aimed to use pigs that were from the same sow source and season of year to minimize the variation in the different sources of inflammation to focus on changes in serum vitamin A associated with a health challenge. This data indicates that serum vitamin A could be a biomarker for inflammation early in life which then predicts the

Figure 1: Serum vitamin A levels at weaning, PRRS negative and positive groups.

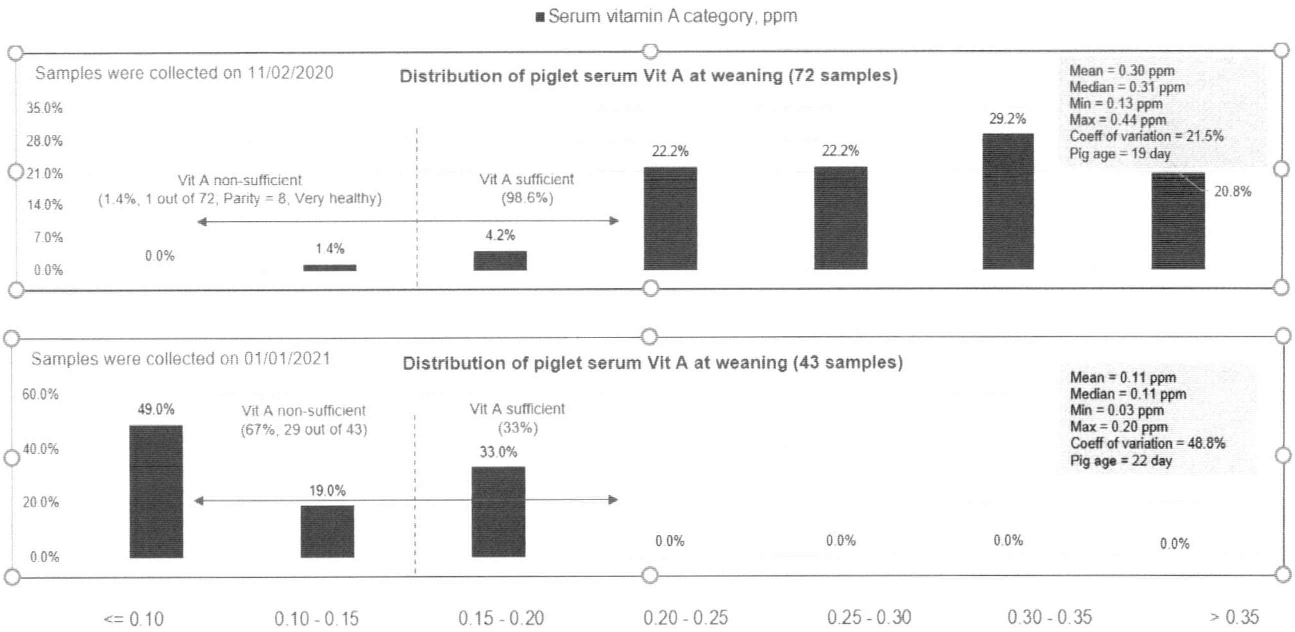

Figure 2: Relationship between vitamin A at weaning and body weights in PRRS negative pigs.

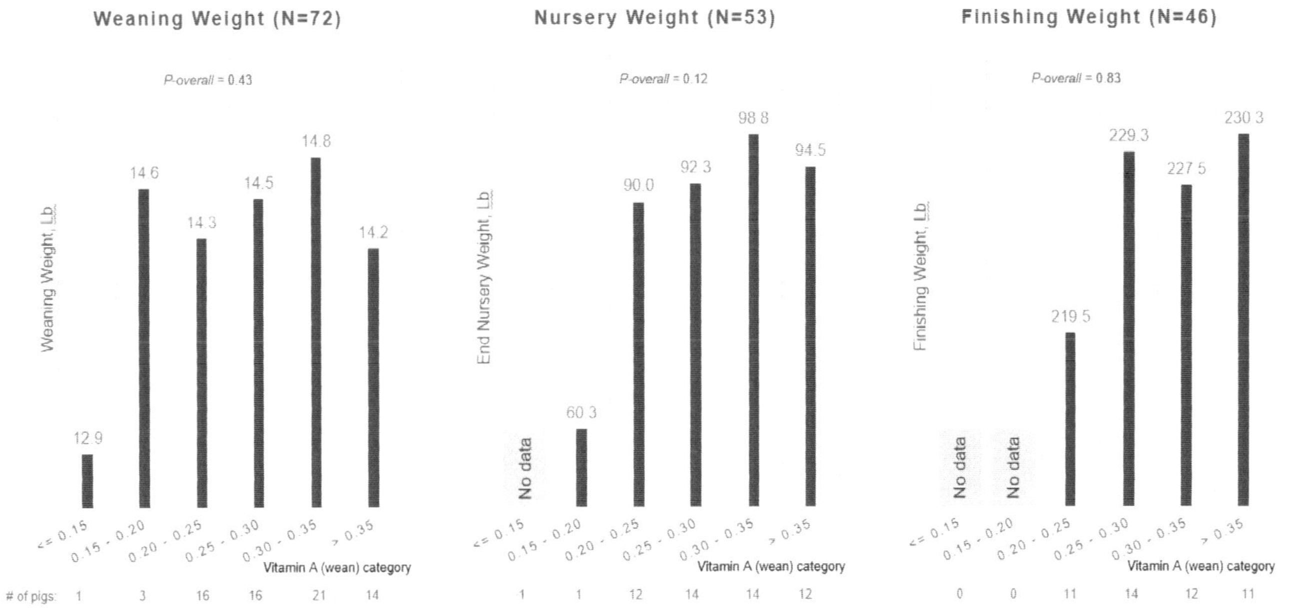

lifetime performance of that pig and related economics. Other factors and causes of early life inflammation are being investigated to completely understand the role of serum vitamin A as a biomarker and how we can utilize it as a tool to better prepare pigs in the face of inflammation. There appears to be an economic impact on pigs with low serum Vitamin A levels and therefore high inflammation, therefore identifying interventions to minimize inflammation is not only better for the pig but can also have a positive economic impact as well.

Conclusions

By utilizing serum vitamin A as a biomarker for inflammation, producers can identify opportunities for improved performance and subsequent economic gain. Intervening to remove the cause or reduce the amount of inflammation pigs are experiencing early in life can be rewarded with a higher percentage of full-value market pigs, heavier weights, and less time to market. The first step is to identify the presence of inflammation. This study and other ongoing studies support serum vitamin A as a key biomarker indicator for inflammation and can be used as a tool in the first step of this process.

References

1. Rubin, L.P., A.C. Ross, C.B. Stephensen, T.Bohn and S.A. Tanumihardjo. 2017. Metabolic effects of inflammation on vitamin A and carotenoids in humans and animal models. Adv. Nutr. 8:197-212.

2. Song R et al. Using sow technologies to drive sow and subsequent pig performance associated with stress induced inflammation. In press, AASV 2022.

Figure 3: Relationship between vitamin A at weaning and body weights in PRRS positive pigs.

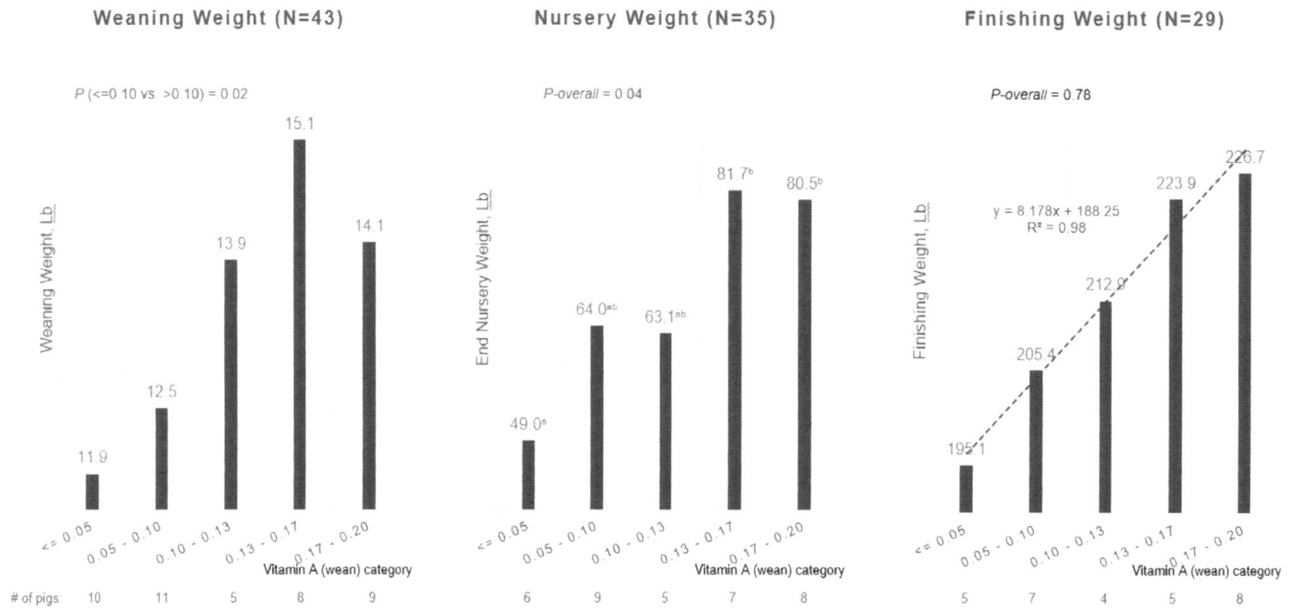

Figure 4: Economic calculator for inflammation.

Serum Vit A Level, ppm	Weight	Net Revenue	Feed cost	Discount rate	Mortality Rate	Mortality Cost	Other Costs	Net Return	Example Farm	Avg Farm	Top 50% Farms
< .10	200.20	$ 127.01	$ (60.00)	$ (15.02)	30%	$ (55.80)	$ (58.00)	$ (61.81)	5.3%	5.6%	1.9%
.1-.15	212.90	$ 135.07	$ (65.08)	$ (4.79)	10%	$ (19.66)	$ (58.00)	$ (12.46)	29.3%	24.4%	13.1%
.15-.2	225.30	$ 142.94	$ (70.03)	$ -	4%	$ (8.28)	$ (58.00)	$ 6.63	57.3%	34.2%	31.8%
.2-.25	233.5	$ 148.14	$ (73.31)	$ -	3%	$ (6.41)	$ (58.00)	$ 10.42	5.3%	18.0%	23.3%
.25-.3	243.30	$ 154.36	$ (77.22)	$ -	2%	$ (4.44)	$ (58.00)	$ 14.69	2.7%	8.8%	13.1%
.3-.35	241.50	$ 153.21	$ (76.50)	$ -	1%	$ (2.20)	$ (58.00)	$ 16.50	0.0%	5.3%	9.5%
.35-.4	244.30	$ 154.99	$ (77.62)	$ -	1%	$ (2.23)	$ (58.00)	$ 17.14	0.0%	3.6%	7.2%
									99.9%	100.0%	100.0%
Expected Net Return per pig weaned									$ (2.18)	$ 0.40	$ 6.43

https://doi.org/10.54846/am2022/61

Effect of a core antigen bacterin with an immunostimulant on piglet health and performance outcomes when challenged with enteric and respiratory pathogens

C. Cull, DVM, PhD, MBA
Midwest Veterinary Services, Inc

Introduction

Gram-negative bacteria have thousands of serotypes making it difficult to make effective commercial or autogenous vaccines that protect against a broad spectrum of gram-negative diseases. Thus, there is a paramount need for a single unique bacterin providing protection against virtually all gram-negative pathogens in swine.

Historically dairy and beef cattle are protected from numerous gram-negative diseases with ENDOVAC-Beef® and ENDOVAC-Dairy®. Therefore, viable possibilities exist to immunologically protect swine from a broad spectrum of gram-negative pathogens. ENDOVAC-Porci®, a single core antigen common to all gram-negatives combined with an immunostimulant, was tested to determine its ability to deliver such broad-spectrum protection.

Study objective

This study determines if ENDOVAC-Porci, a core antigen vaccine with an immunostimulant; provides piglets broad-spectrum protection against the enteric and respiratory effects of gram-negative bacteria.

Materials and methods

Study location

Midwest Veterinary Services, Inc in Oakland, NE. Site Manager; Kelly Lechtenberg, DVM, PhD. Investigator; Charley Cull, DVM, PhD, MBA.

Study design

This study was a single-site, randomized, prospective, blinded, comparative placebo-controlled design. Individual "Animal" was the experimental unit during the nursey phase, and "Pen" was the experimental unit through the end of the study.

Statistical analysis performed by the "Center of Outcomes Research and Epidemiology" at Kansas State University.

Animals

90 healthy piglets weighing > 2.6 pounds were selected for inclusion in the study. Animals were commercially sourced from a PRRSV stable herd.

Vaccine

ENDOVAC-Porci: (Endovac Animal Health, Columbia, MO, USA). Piglets were vaccinated IM with 1mL of either the vaccine or saline placebo; at 1 day of age (study day 0) and boostered at weaning; 21 days of age (study day 21). The following vaccines were also given at study day 21: MLV PRRS, PCV2 and *Mycoplasma hyopneumoniae*.

Challenge

All piglets were challenged 2 days following booster vaccination (study day 23). The challenge was used to mimic diseases commonly encountered when piglets enter a nursery. The gram-negative challenge organisms included: *Escherichia coli* K88 orally (5 mL of 1×10^{10} CFU/mL, MVS strain) and *Pasteurella multocida* intranasally (2 mL of 1×10^{9} CFU/mL, MVS strain).

Results and discussion

Body weights

Piglets vaccinated with ENDOVAC-Porci were 7.9% heavier at weaning age – a 1.05-pound advantage over the control group. The vaccinated piglets had a 13.6% advantage at day 42 – a 3.3-pound advantage over the control group (Table 1).

Morbidity

Daily observations for pen clinical and fecal scores occurred just prior to challenge and for 12 days following challenge. The ENDOVAC-Porci vaccinated piglets had an average "clinical score" of 0.29 versus controls of 1.19 - a 75.6% improvement over the control group; (Table 2 and Figure 1). The vaccinated piglets had an average "fecal score" of 0.96 versus controls at 1.95 - a 50.8% improvement; (Table 2 and Figure 2).

Mortality

Piglets vaccinated with ENDOVAC-Porci had zero pre-weaning mortality, while the control group experienced 11.1% pre-weaning mortality. Over the entire 42-day study vaccinated piglets experienced 61.9% less mortality (Table 3).

Summary

The ENDOVAC-Porci vaccinated piglets outperformed the control piglets in every metric. The vaccinated piglets had a 1.05 lb. higher weaning weight and a 3.3 lb. higher final (42 day) weight. The vaccinated piglets were healthier with 75.6% better clinical scores and 50.8% better fecal scores. The vaccinated piglets had a significant mortality reduction of 61.9% over the control group.

Conclusion

ENDOVAC-Porci piglet vaccination significantly improved the performance and health, while substantially reducing the mortality when compared to control piglets. This decisive advantage would provide a compelling return on investment for most swine production systems. This study indicates that broad spectrum protection from multiple gram-negative pathogens can be achieved with this product.

Table 1: Body weights

Day 0 by treatment group

Treatment	Piglets	Mean kg	Mean lbs
Saline	45	1.61	3.54
ENDOVAC-Porci	44	1.69	3.72
Effect of treatment ($P = 0.17$)			

Day 21 by treatment group

Treatment	Piglets	Mean kg	Mean lbs
Saline	40	6.04	13.29
ENDOVAC-Porci	44	6.52	14.34
Effect of treatment ($P < 0.01$)			

Day 42 by treatment group

Treatment	Piglets	Mean kg	Mean lbs
Saline	24	11.00	24.20
ENDOVAC-Porci	36	12.50	27.50
Effect of treatment ($P < 0.01$)			

Table 2: Morbidity

Day 22-35

Clinical Scores: 0 normal, 1 mild, 2 moderate, 3 severe

Fecal Scores: 0 normal, 1 soft, 2 loose, 3 watery

Treatment	Saline	ENDOVAC-Porci	*P*-value
Clinical	1.19	0.29	0.05
Fecal	1.95	0.96	0.05

Figure 1: Clinical Scores: 0 Normal, 1 Mild, 2 Moderate, 3 Severe.

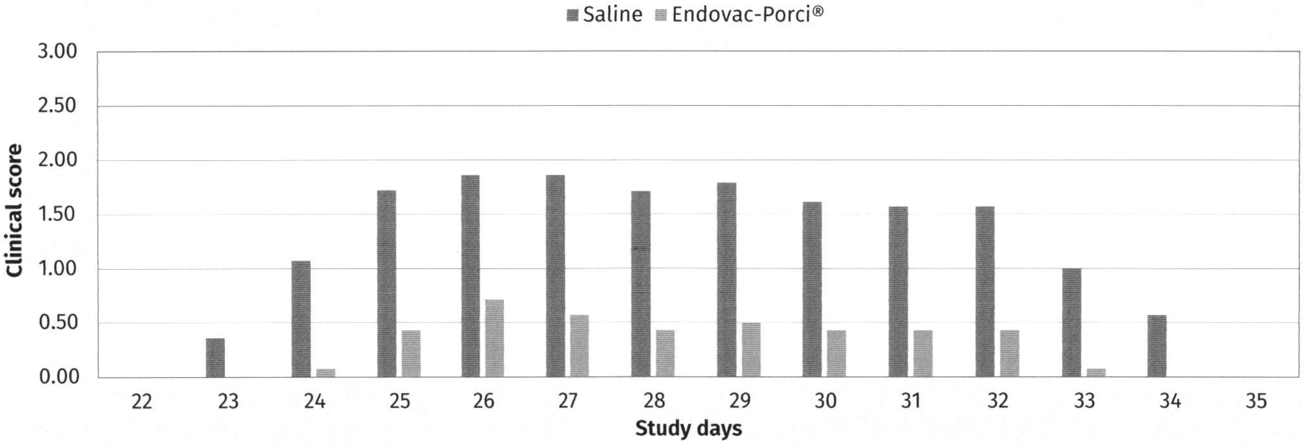

Table 3: Mortality

Day 0-21 Farrowing phase by treatment group

Treatment	Piglets	Mortalities	%
Saline	45	5	11.1%
ENDOVAC-Porci	44	0	0.0%
Effect of treatment (*P* = 0.06)			

Day 21-42 Nursery phase by treatment group

Treatment	Piglets	Mortalities	%
Saline	40	16	40.0%
ENDOVAC-Porci	44	8	18.2%
Effect of treatment (*P* = 0.03)			

Day 0-42 Overall by treatment group

Treatment	Piglets	Mortalities	%
Saline	45	21	46.7%
ENDOVAC-Porci	44	8	18.2%
Effect of treatment (*P* = 0.01)			

Figure 2: Fecal Scores: 0 Normal, 1 Soft, 2 Loose, 3 Watery.

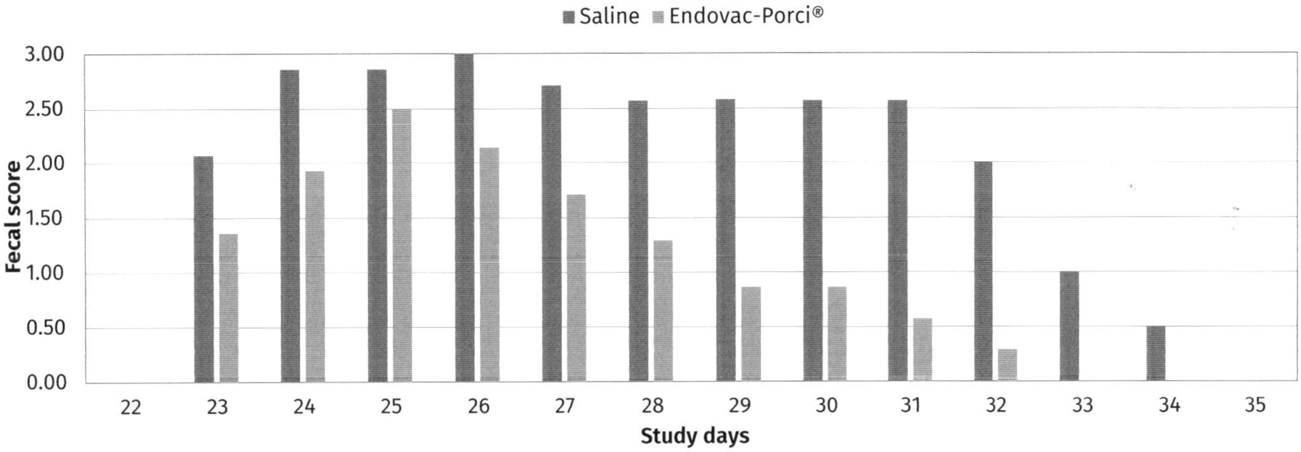

https://doi.org/10.54846/am2022/62

Detecting variation in mycotoxins within single loads of feed ingredients

A. Woodward, PhD; K. McCormick, MS
United Animal Health, Sheridan, Indiana

Introduction

The presence of mycotoxins produced by mold and present in many grain products used in livestock feed pose risks to animal health and performance by negatively impacting production and leading to reduced profitability.[1,2] Mycotoxins are one of the few naturally occurring chemical hazards that are monitored by the United States Food and Drug Administration, with guidelines provided for many swine production scenarios (Table 1).[3] Even low levels of mycotoxin contamination can cause depressed feed intake, leading to reduced gain in nursery and grow/finish pigs, or poor reproductive efficiency in sows.[4] These effects extend beyond production metrics, as mycotoxin presence can also impact immune response, increasing susceptibility to infectious disease and reducing vaccine efficacy.[5] Understanding the risks and monitoring levels of mycotoxins in feed ingredients are the first steps in any mycotoxin management program.

Mycotoxin management strategies often include routine sampling and lab analysis of toxin levels in feed ingredients to provide a snapshot of risk level prior to mixing complete diets. However, mycotoxin sampling strategies differ across feed manufacturing facilities and production systems, often with only one sample taken from a truckload of feed ingredient at delivery. While composite samples are key, if representative samples are not collected throughout the unloading process, results may provide a false sense of security if low mycotoxins are detected. Therefore, we aimed to understand the analyzed variation of mycotoxin levels within a single truckload of delivered ingredients, as well as to understand how variable mycotoxin exposure impacts nursery pig performance.

Materials and methods

Common ingredients found in swine diets at two different production locations were sectioned off and sampled multiple times to determine the variation of mycotoxin levels within a single ingredient load. The first farm received one semi-load of "low" mycotoxin corn, classified as corn with average mycotoxin results below cautionary thresholds, which was placed in a large storage silo; samples were collected as the corn was ground and moved to a small storage bin. Seven samples were collected from 12,000 lbs. of corn at two separate time points 3 weeks apart (March and April, 2021) for a total of 14 samples from 24,000 lbs. of corn. Additionally, eight soybean meal (SBM) samples representing 35,000 lbs. of SBM were collected as the semi-load was emptied into a storage bin (March 2021). Finally, DDGS were collected as they were bagged for feed mixing, with 14 samples accounting for 10,000 lbs. of ingredient (March 2021).

The second farm collected eight samples of whole, low mycotoxin corn representing 25,000 lbs. of ingredient as the corn was being off-loaded into a storage bin (March 2021). High mycotoxin corn representing 10,000 lbs. of ingredient was sampled both before and after grinding, for a total of 22 samples collected (11 whole and 11 ground samples; March 2021).

All samples were ground to a homogenous particle size, and a 20g sample was weighed into either 100 mL of 70% ethanol or 100 mL of deionized water. Mycotoxins were extracted into solution for 90 minutes while samples were stirred at 400 rpm on a magnetic stir plate. Solution containing mycotoxins was collected and analyzed for aflatoxin (AFL), deoxynivalenol (DON), fumonisin (FUM), or zearalenone (ZEA) using ELISA kits (Romer Labs) according to manufacturer's instructions.

In a separate trial, nursery pigs (n = 448; initial BW = 13.6 ± 2.1 lbs.) weaned at 21 ± 1 day were fed either constant low-level, or variable medium- or high-level, mycotoxin diets for 28 days to determine effects of different exposures of mycotoxins on growth performance. Pigs receiving medium or high mycotoxin diets were subjected to two separate 7-day exposure periods, with low mycotoxin diets provided for 7 days between exposure periods. Weekly body weights were captured to calculate growth in response to mycotoxin exposure.

Table 1: Cautionary levels of mycotoxins in feed ingredients used for swine

Mycotoxin	Feed ingredient	US Food and Drug Administration guideline
Aflatoxin	Cottonseed meal	300 ppb, regardless of age or breeding stats
	Corn and peanut products	200 ppb for swine of 100 lbs. or greater
		100 ppb for breeding swine
		20 ppb for immature swine
Deoxynivalenol	Grains and grain by-products to not exceed 20% of diet	5 ppm for swine (1 ppm for complete feed)
Fumonisin	Corn and corn-by products not to exceed 50% of diet	20 ppm for swine (10 ppm for complete feed)
Zearalenone[1]	Grains and grain-by products	0.5 ppm for swine

[1] There is no FDA guidance for ZEA; the critical level reported herein comes from the Office of Indiana State Chemist.

Results

Low mycotoxin corn

Ground corn from March 2021 at location one was considered to have low mycotoxin levels; however, all 4 major mycotoxins were represented at varying quantities throughout the 12,000 lbs. of ingredient (Table 2). Over 7 samples, AFL, DON, FUM, and ZEA varied by 2.55 ppb, 0.34 ppm, 0.61 ppm, and 0.11 ppm, respectively. Most concerning is the DON variation, as the samples ranged from 0.82 - 1.16 pm DON, indicating that some samples analyzed below standard industry guidelines for mitigation, while some samples were above the threshold.

The April 2021 ground corn samples from the same semi-load were higher in AFL and DON, but lower in FUM and ZEA, compared to the March 2021 corn (Table 2). Variation of AFL, DON, FUM, and ZEA was 1.24 ppb, 0.71 ppm, 0.50 ppm, and 0.07 ppm respectively, over 7 samples representing 12,000 lbs. of ingredient. Again, DON ranged from 0.79 - 1.50 ppm, showing some samples dissuade and others encourage the use of mycotoxin interventions.

Overall, from one semi-load of corn (14 samples from March and April 2021), analyzed range of AFL was 1.70 - 7.67 ppb, DON was 0.79 - 1.50 ppm, FUM was 0.08 - 0.69 ppm, and ZEA was 0.00 - 0.13 ppm.

At location two, which sampled whole corn in March 2021, AFL, DON, and ZEA showed more variability, and FUM less variability, than ground corn samples collected at location one in the same time frame (Table 2). Over 8 samples, AFL ranged from 2.06 - 6.28 ppb, DON ranged from 0.45 - 1.61 ppm, FUM ranged from 0.00 - 0.10 ppm, and ZEA ranged from 0.01 - 0.18 ppm.

High mycotoxin corn

Whole corn having known high levels of mycotoxins showed greater variability of AFL, DON, FUM, and ZEA compared to ground corn samples from the same load of ingredient (Table 2). Across 11 samples for each, AFL in whole corn varied by 1.82 ppb compared 1.05 ppb variation for ground corn. Deoxynivalenol ranged from 7.13 - 17.46 ppm for whole corn but was only 6.41 - 11.11 ppm in range for ground corn. Over 11 samples of each, FUM showed the most variation in both whole and ground corn, with a difference of 18.88 and 7.64 ppm, respectively, between low and high analytes. Zearalenone had a range of 0.14 - 3.28 ppm for whole corn compared to a range of 0.91 - 2.94 for ground corn across 11 samples of each.

Distillers Dried Grains (DDGS)

Only one location collected DDGS; however, less variation was noted for AFL and ZEA from 14 samples compared to corn on the same farm at the same time (Table 2). Over 14 samples, AFL ranged from 9.10 - 10.90 ppb, with an average concentration of 9.86 ppb. Deoxynivalenol ranged from 3.20 - 7.14 ppm, with an average of 5.96 ppm DON over 14 samples. Fumonisin averaged 1.14 ppm and ranged from 0.89 - 1.31 ppm over 14 samples. Zearalenone was most consistent, with an average of 0.22 ppm and a range of 0.20 - 0.23 ppm over 14 samples.

Soybean meal

Soybean meal showed some of the lowest mycotoxin variation across feed ingredients sampled (Table 2). At location one, SBM had AFL from 3.76 - 5.78 ppb; DON from 0.62 - 0.78 ppm; FUM from 0.00 - 0.10 ppm; and ZEA from 0.06 - 0.07 ppm across 8 samples taken during unloading.

Mycotoxin variation affects pig performance

When complete feed with variable concentration of DON was fed to nursery pigs, there was a negative impact on growth performance. Variation of 1.25 or 2.50 ppm added DON resulted in a 7.74 or 22.62% loss, respectively, in ADG during the first insult ($P < 0.05$) and a 16.95 or 22.65% loss, respectively, in ADG during the second insult ($P < 0.05$). Even with removal of the 1.25 or 2.50 ppm added DON, growth lost during the times of high mycotoxin resulted in 1.89 or 3.78 lb. decreased BW, respectively. Consequently, optimal performance was not restored during times when low, "clean" diets were fed after the high mycotoxin dietary exposure in nursery pigs.

Conclusion

Ultimately, variation of mycotoxin concentrations within feed ingredients exists within a single truckload, even when ingredients are considered "safe" or "clean". Higher average mycotoxin concentrations are correlated with a greater range of variation within a single load of ingredient. Less mycotoxin variation was seen in DDGS and soybean meal, but these ingredients still have the potential for high mycotoxin pockets to exist. Taken together, these variations can result in "hot spots" of mycotoxins in complete feed that, when consumed by the animal, result in negative performance impacts that cannot be completely overcome. Ensuring a representative sample or samples are collected from individual loads of ingredients will provide greater understanding of how mycotoxins, low or high, could be affecting production performance of swine.

References

1. Bryden WL. Mycotoxin contamination of the feed supply chain: Implications for animal productivity and feed security. *Animal Feed Sci and Tech.* 2012;173:134.

2. Shull LR and Cheeke PR. Effects of synthetic and natural toxicants on livestock. *J Anim Sci.* 1983;57:330.

3. U.S. Food & Drug Administration. **https://www.fda.gov/animal-veterinary/biological-chemical-and-physical-contaminants-animal-food/chemical-hazards**

4. Diekman MA et al. Mycotoxins and Swine Performance. *Pork Industry Handbook.* 1992;E2420.

5. Pierron A et al. Impact of mycotoxin in immune response and consequences for pig health. *Anim Nutr.* 2016;2:63.

Table 2: Variation of mycotoxins within a single truckload of ingredient[1]

Sample	AFL, ppb	DON, ppm	FUM, ppm	ZEA, ppm
Low mycotoxin ground corn, Location 1, time 1				
Average	2.82	0.94	0.30	0.05
Range (Difference)	1.70 – 4.25 (2.55)	0.82 – 1.16 (0.34)	0.08 – 0.69 (0.61)	0.02 – 0.13 (0.11)
Low mycotoxin ground corn, Location 1, time 2				
Average	6.91	1.02	0.20	0.03
Range (Difference)	6.43 – 7.67 (1.24)	0.79 – 1.50 (0.71)	0.08 – 0.58 (0.50)	0.00 – 0.07 (0.07)
Low mycotoxin whole corn, Location 2				
Average	2.87	0.78	0.02	0.05
Range (Difference)	2.06 – 6.28 (4.22)	0.45 – 1.61 (1.16)	0.00 – 0.10 (0.10)	0.01 – 0.18 (0.17)
High mycotoxin whole corn, Location 2				
Average	1.24	9.29	10.72	1.47
Range (Difference)	0.09 – 1.91 (1.82)	7.13 – 17.46 (10.33)	5.53 – 24.41 (18.88)	0.14 – 3.28 (3.14)
High mycotoxin ground corn, Location 2				
Average	1.41	8.32	10.76	1.67
Range (Difference)	0.88 – 1.93 (1.05)	6.41 – 11.11 (4.70)	8.29 – 15.93 (7.64)	0.91 – 2.94 (2.03)
Distillers Dried Grains (DDGS)				
Average	9.86	5.96	1.14	0.22
Range (Difference)	9.10 – 10.90 (1.80)	3.20 – 7.14 (3.94)	0.89 – 1.31 (0.42)	0.20 – 0.23 (0.03)
Soybean meal				
Average	4.70	0.70	0.02	0.07
Range (Difference)	3.76 – 5.78 (2.02)	0.62 – 0.78 (0.16)	0.00 – 0.10 (0.10)	0.06 – 0.07 (0.01)

[1] AFL = aflatoxin; DON = deoxynivalenol; FUM = fumonisin; ZEA = zearalenone.

https://doi.org/10.54846/am2022/63

Using ProVent ECL® to help solve an enteric health challenge within a sow unit

B. Mason, DVM
Country View Family Farms, Middletown, Pennsylvania

Introduction

Enteric disease and resulting scours presented in nursing piglets within a sow unit can be challenging to manage due to the multiple causative agents that may be contributing to the clinical outcome. Management and environmental factors can combine with bacterial and viral infections making it difficult to know if a specific intervention is making progress on the situation. In addition to the challenge of discovering and implementing solutions that address the root causes, it is difficult to fully understand the impact that these sow unit issues may be having on downstream metrics, like nursery mortality. We were faced with this challenge in a multiplication unit and utilized PathKinex™, a comprehensive diagnostic surveillance tool (United Animal Health, Sheridan, IN) to understand the diversity and quantitative nature of pathogens contributing to the clinical picture. These diagnostics were used to evaluate possible strategies but were also used to measure pathogen burden after the use of a direct fed microbial product, ProVent ECL® (ECL) (United Animal Health, Sheridan, IN). To quantify the impacts of ECL on specific production metrics, wean weight and antibiotic usage were measured intensively during ECL usage at the sow unit. Closeout nursery mortality was also determined in a single sow flow approach to determine the impact of ECL on the offspring when ECL was only fed to sows.

Materials and methods

Diagnostics and the ECL evaluation were conducted at a single multiplication sow unit totaling 5000 sows that weaned 2400 head per week into two, single sourced nursery sites. The sow unit had a history of rotavirus type C (RVC), *Strep suis*, and *E coli* with piglet scours and mortality starting to increase approximately 24-hrs post farrowing. An autogenous bacterin for *Strep suis* was developed and implemented prefarrow and at processing to piglets (3-5d of age). This bacterin was utilized prior to the use of other technologies of interest and continued during the course of this field evaluation.

To quantitatively understand the diversity of pathogens at the sow unit, a PathKinex™ diagnostic strategy was utilized to understand the unique pathogen profiles and virulence gene quantities in sows and piglets expressing clinical symptoms within the first 16 days post farrowing. Rectal swabs were collected from 8 scouring (sick) litters and 8 non-scouring (healthy) litters that were matched to the same piglet age for comparison. Within each litter, the sow and 2 of her piglets were rectal swabbed to collect genomic DNA and total RNA, and then tested with a panel of qPCR assays for genus or species-specific marker or virulence genes from *E coli*. Gene data were statistically analyzed to assess the differences in presence and quantity between sick and healthy pigs. This diagnostic strategy was executed twice. Once before the feeding of ECL, and the second time after ECL had been fed to sows for approximately 3 months. ProVent ECL® was fed at a dosage rate ranging from 1 to 2 pounds/ton in gestation and lactation rations for 34 weeks total. During that time, the number of litters treated with antibiotics for scours were recorded, as well as the average weight for each weaning group. Piglets were weaned into two 9600 head nurseries where nursery mortality was recorded for the cumulative 7-week nursery growing period for each room of approximately 2400 pigs.

Results

Diagnostic results prior to feeding of ECL to sows detected high levels of *E coli* adhesin genes significantly correlated ($P < 0.05$) to sick litters scouring the first 3 days post farrowing (Figure 1). Specifically, K88 was found to be in extremely high quantities in the rectal swabs from nursing piglets in these sick litters, but these log levels were also elevated in the sows nursing the scouring piglets. Rectal swabs collected from litters after ECL was fed to sows for 3 months displayed fewer and reduced levels of *E coli* adhesin genes (Figure 2) with no significant correlation of adhesin genes with sick litters.

The percentage of litters found to be scouring and treated with antibiotics was reduced approximately 25 percentage points when comparing pre ECL scour rates with scour rates after feeding (ECL) for 8 months (Figure 3). Weaning weights increased approximately 0.5 pounds (11.0 vs 11.5 lbs/pig average) during the same time period (Figure 4).

System nursery mortality closeout data prior to and during the use of ECL was analyzed using Statistical Process Control methodology (BPI Consulting, LLC; Katy, TX). After the third closeout with ECL influenced offspring, SPC analysis revealed that a system shift had occurred and a new, lower average for mortality was generated and maintained for the following nursery closeouts going forward. Nursery mortality was determined to move from an average of 8.56% in the previous nursery closeouts to 7.61% after this system shift occurred (Figure 5).

Conclusion

While the enteric issues didn't completely resolve in this field evaluation, the use of PathKinex™ and its quantification of toxin and adhesin genes allowed for the identification of technologies that could improve the health status of the sow herd. Reducing the *E coli* adhesin gene pressure (K88 specifically) resulted in reduced antibiotic usage in the farrowing crates, improved weaning weights, and reduced downstream mortality rates in the nursery. Looking at the impact of targeted technologies like ProVent ECL® to reduce bacterial pathogens should be evaluated not only in the specific phase of production where the technology is being implemented, but one should also look downstream to understand the impact on a system level basis.

Figure 1: *E coli* adhesin gene quantities detected in rectal swabs collected from sows and suckling piglets at 1-3 days and 6-11 days post farrow. Before implementing ProVent ECL® in sow diets.

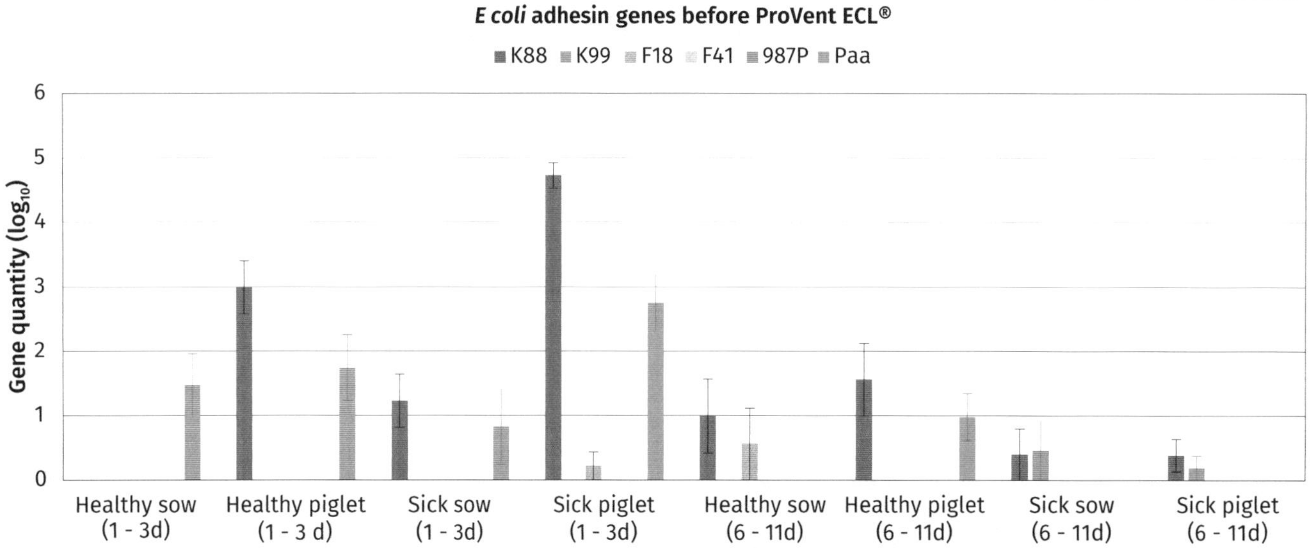

Figure 2: *E coli* adhesin gene quantities detected in rectal swabs collected from sows and suckling piglets at 1-6 days and 8-16 days post farrow. After implementing ProVent ECL® in sow diets for 3 months.

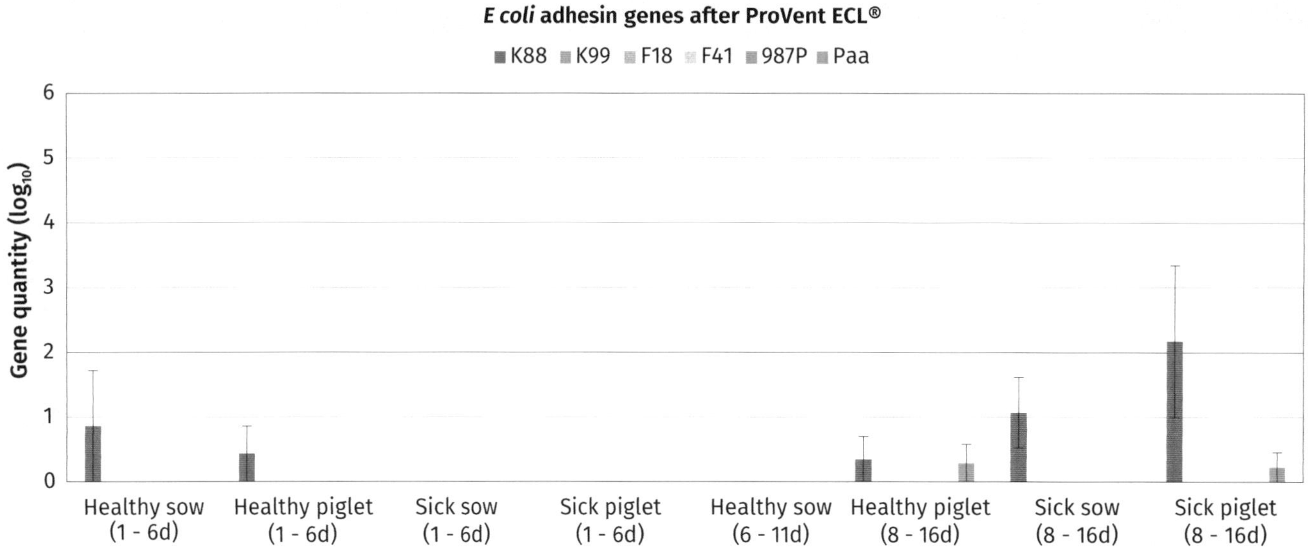

Figure 3: Percentage of litters treated with antibiotics for clinical scours in farrowing, prior to and after sows were fed ProVent ECL®.

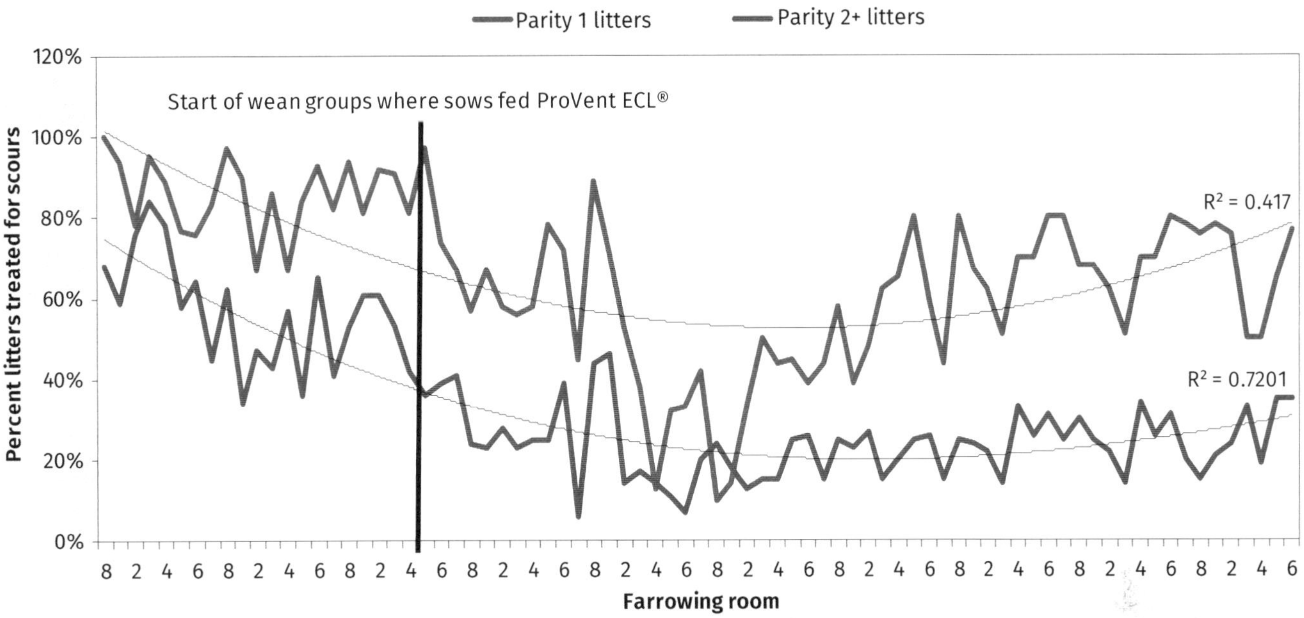

Figure 4: Average wean weights (lbs.) of pigs prior to and after ProVent ECL® was fed to sows.

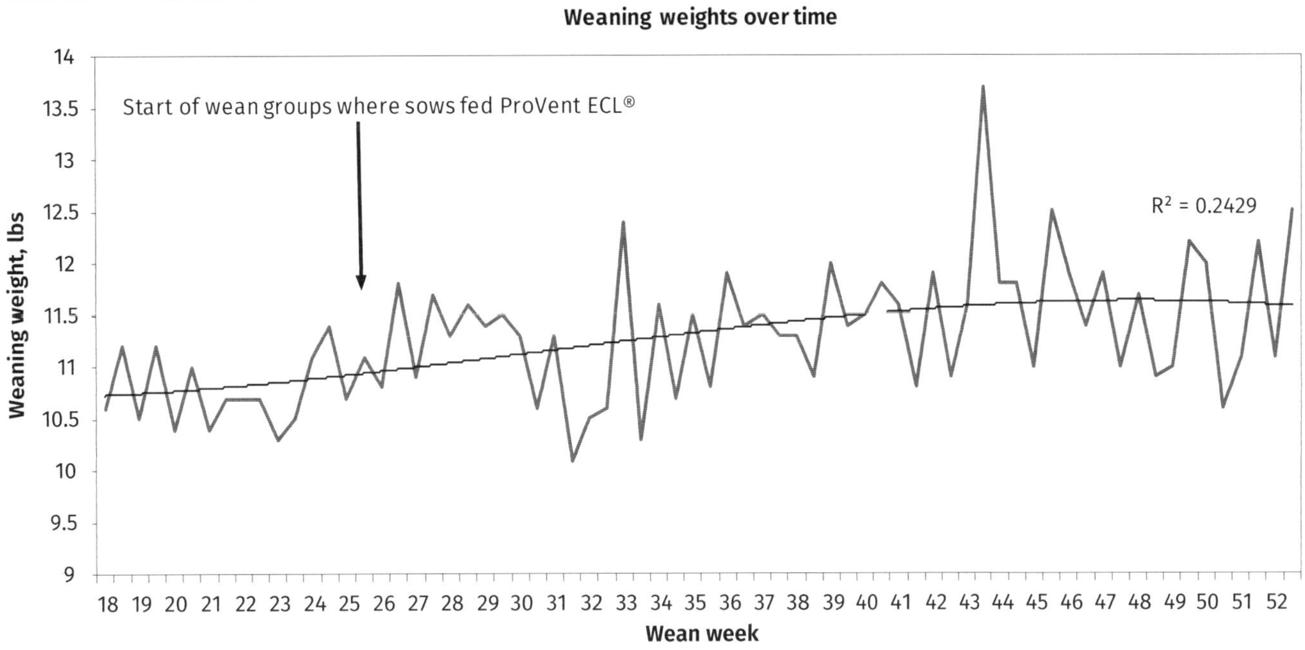

Figure 5: Statistical Process Control charting of nursery closeout mortality prior to and after ProVent ECL® was fed to sows.

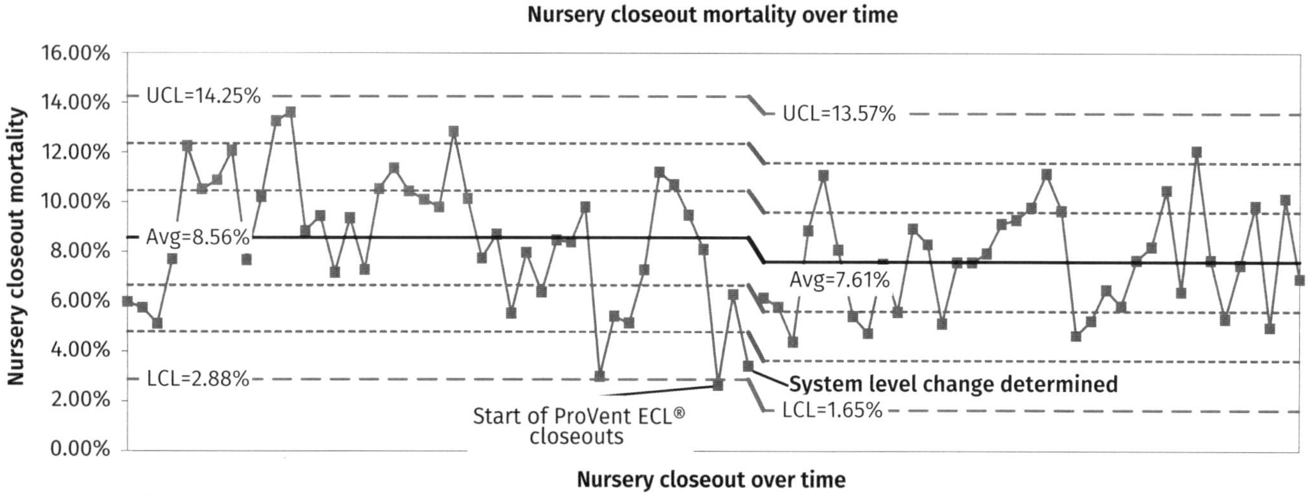

https://doi.org/10.54846/am2022/64

Fighting enteropathogenic *E coli* with algal polysaccharides

M. Gallissot[1], MSc; M. Garcia Suarez[1], DVM; M. Berri[2], MSc; P. Nyvall-Collen[1], PhD; M. Porter[1], BSc; D. Marzin[1], MSc
[1]Olmix Group, Brehan, France; [2]Infectious diseases and Public Health Unit, INRAE, University of Tours, Nouzilly, France

Introduction

Young piglets raised in commercial conditions are highly sensitive to enteropathogenic *Escherichia coli* infections (colibacillosis), which can colonize the piglets' intestinal epithelium and trigger diarrhea with important economic losses. Thus, supporting the integrity of the intestinal mucosa and the immune system is essential to ensure gut health in piglets.[1] Recent research has shown, *in vitro*, that a red algal extract from *Solieria chordalis* upregulates the gene expression of tight junction and mucin proteins[2] which are essential for the proper functioning of the intestinal epithelial barrier. Other research has shown that a green algal extract from *Ulva sp*, has the capacity to modulate the immune response as demonstrated *in vitro.*[3,4] The present work evaluated, *in vitro,* the capacity of the red algal extract to reinforce the gut barrier in the presence of an enteropathogenic *E coli* strain. Also tested, *in vivo,* was the ability of this red algal extract, together with the green algal extract from *Ulva* sp., to improve the health status and performance of suckling piglets when supplemented to sows on a commercial farm.

In vitro assays

Materials and methods

Two *in vitro* assays were conducted, using differentiated and polarized Intestinal Porcine Epithelial Cells (IPEC-1) isolated from the intestine of newborn piglets. In the first assay, IPEC-1 cells were cultured at 37°C (pig body temperature) in the cell-culture medium alone (negative control), the cell-culture medium inoculated with *E coli* K88 1305 (positive control) or the cell-culture medium incubated with *E coli* K88 1305 and the red algal extract ($MSP^{®}_{BARRIER}$). The bacteria were inoculated at a concentration of 10 bacteria per IPEC-1 cell. The Trans-Epithelial Electrical Resistance (TEER) of the cell monolayer was measured at regular intervals. Electrodes were placed on both sides of the IPEC-1 cell monolayer, and voltage was applied to them. The current between both electrodes was used to calculate the electrical resistance which qualifies the integrity of the cell-line barrier. In the second assay, IPEC-1 cells were pre-treated in 24-well plates with cell-culture medium only (control) or $MSP^{®}_{BARRIER}$ at three different concentrations (tests). After 24 hours, the cell-culture medium was removed, and cells were inoculated with *E coli* K88 968 at a concentration of 10 bacteria per IPEC-1 cell and 5% carbon dioxide and kept at 37°C for 1 hour. Cells were then washed twice with phosphate-buffered saline (PBS) and bacteria were counted in both the initial inoculate and the washed cells on agar plates. The percentage of adhesion of bacteria to the cells was obtained by dividing the number of bacteria that adhered to the cells by the total number of bacteria brought into contact with the cells (inoculate).

Results

Results of the first assay showed that the TEER was stable in the negative control (non-infected) in the first 10 hours of monitoring, while it rapidly decreased (within 4 hours) when the cells were infected with *E coli* K88 1305 (Figure 1). When the cells were in contact with the red algal extract in addition to the *E coli* infection, the TEER was maintained at a higher level than in the positive control in the first 10 hours post-infection (Figure 1). This indicates that $MSP^{°}_{BARRIER}$ preserves the epithelium integrity and barrier function in the first 10 hours post-infection. Moreover, in the second assay, the red algal extract reduced the percentage of adhesion of *E coli* K88 968 to IPEC-1 cells at the three concentrations tested compared to the control (Figure 2), thus highlighting the capacity of $MSP^{°}_{BARRIER}$ to inhibit bacterial adhesion to intestinal cells.

In vivo trial

Materials and methods

The *in vivo* trial took place in a 5,500-sow commercial unit located in Indiana. The farm was PRRS and PEDV stable but had been experiencing *E coli* problems for several months, with minimal success from the antibiotic treatments. A mix of Olmix's $MSP^{®}_{IMMUNITY}$ (green algae extract) and $MSP^{®}_{BARRIER}$ (red algae extract) has been incorporated to all sows' feeds at the rate of 1 lb/T in gestation and lactation feeds from May 10 to August 8, and at the rate of 1 lb/T in gestation feed and 2 lbs/T in lactation feed from August 9 to September 11. The evolution of the performance of the sows and piglets was monitored weekly from March 28 to September 11 with MetaFarms software. The performance monitored from March 28 to May 8 served as control.

Results

Over the 4 months of the trial, the farrowing performance was improved. The farrowing rate was 0.3 point higher in the last month (high inclusion rate of the algae-based supplement) than in the control period (Table 1). Moreover, the average number of total born and live born piglets was increased by 0.6 in the last month of trial compared to the control period (Table 1). The weaning performance was also improved along the trial. The prewean mortality decreased gradually, to the benefit of the average number of piglets weaned (Figure 3). On average, 2.2 more piglets were weaned per sow farrowed in the last month of the trial compared with the control period (Table 1).

Discussion

The studies proved the capacity of the red macroalgae extract to support *in vitro* the integrity of intestinal porcine cell lines exposed to an enteropathogenic *E coli* strain. When supplemented to sows for several months in commercial conditions, the in-feed combination of this extract with a green macroalgae extract known to activate the immune function[3,4] and to enhance the transfer of immunoglobulins from sows to their offspring[5] seemed to improve the performance of sows and suckling piglets. The trial being based on a comparison with historical data, a confirmation of the results is required, by repeating the trial in other commercial units. This in-feed algae-based supplement appears to be a promising tool to prevent and mitigate enteropathogenic *E coli* problems on pig farms.

References

1. Turner Jerrold R., 2009. Intestinal mucosal barrier function in health and disease. Nat Rev Immunol, 9, 799-809.

2. Gallissot M., Miel L., Rousseaux C., Nyvall-Collen P., 2017. Proc. Conference "5th International IHSIG Symposium on Poultry Gut Health", Bangkok, Thailand.

3. Berri M., Slugocki C., Olivier M., Helloin E., Jacques I., Salmon H., Demais H., Le Goff M., Nyvall Collén P., 2016. Marine-sulfated polysaccharides extract of Ulva armoricana green algae exhibits an antimicrobial activity and stimulates cytokine expression by intestinal epithelial cells. J Appl Phycol, 28 :1; 3-12.

4. Berri M., Olivier M., Holbert S., Dupont J., Demais H., Le Goff M., Nyvall Collén P., 2017. Ulvan from Ulva armoricana (Chlorophyta) activates the PI3K/Akt signaling pathway via TLR4 to induce intestinal cytokine production. Algal Research, 28, 39-47.

5. Bussy F., Le Goff M., Salmon H., Delaval J., Berri M., Nyvall-Collen P., 2019. Immunomodulating effect of a seaweed extract from *Ulva armoricana* in pig: Specific IgG and total IgA in colostrum, milk, and blood. Vet. Anim. Sci., 7, 100051.

Figure 1: TEER of IPEC-1 cells incubated with *E coli* K88 1305 and/or MSP®BARRIER (kΩ/cm²).

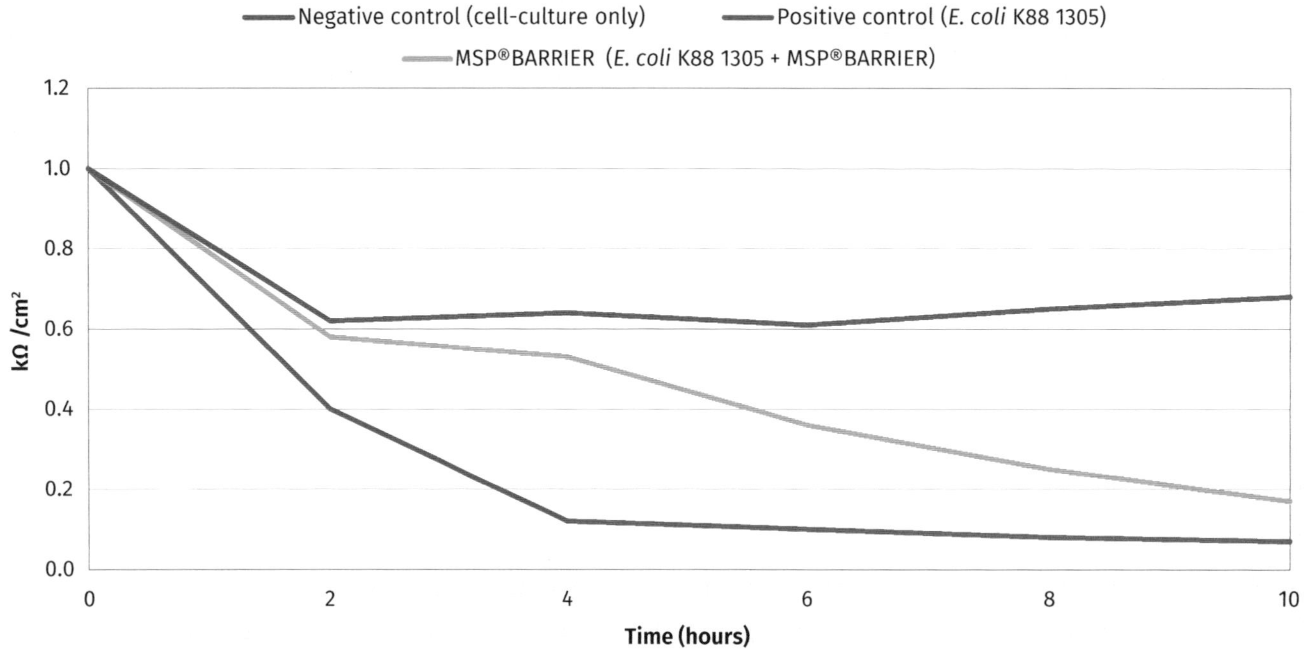

Figure 2: Adhesion of *E coli* K88 968 to IPEC-1 cells incubated in cell-culture alone or with different levels of MSP®BARRIER (%).

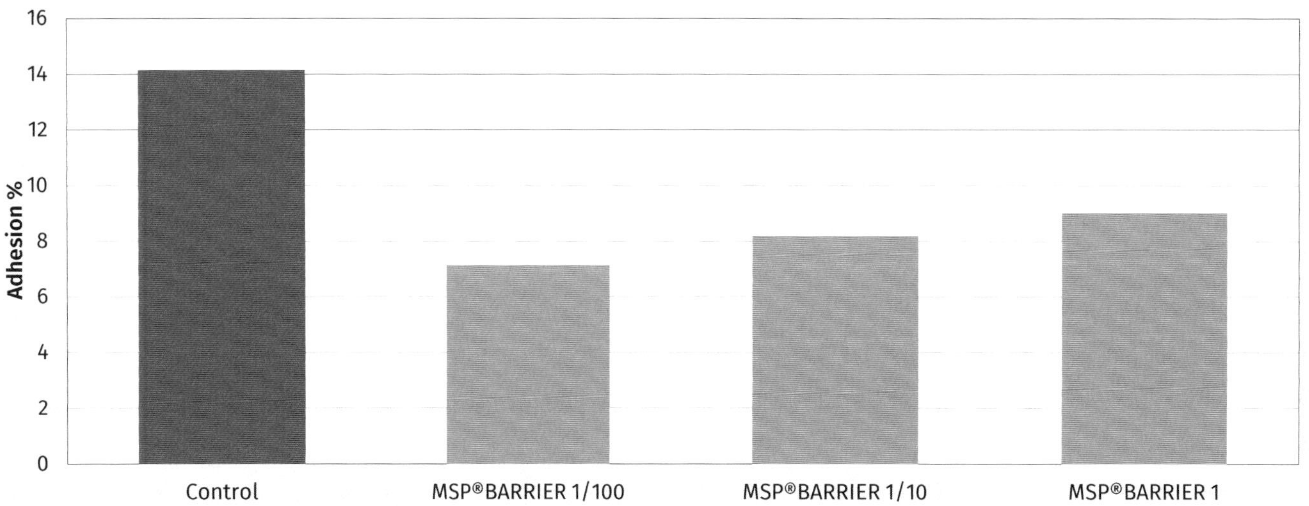

Table 1: Average farrowing and weaning performance over the three periods of the field trial

Parameters	**Control** Mar-28 to May-8	**MSP® mix 1lb/T G+L** May-8 to Aug-7	**MSP® mix 1lb/T G, 2lb/T L** Aug-7 to Sept-11	**Difference** Last period vs Control period
Farrowing rate (%)	84.8	84.5	85.1	+0.3
Average total born/sow (n)	14.8	15.1	15.4	+0.6
Average live born/sow (n)	13.6	13.9	14.2	+0.6
Pre-wean mortality (%)	17.8	13.0	5.2	-12.6
Average weaned piglets/sow farrowed (n)	11.2	12.1	13.4	+2.2

G = gestation; L = lactation

Figure 3: Weaning performance one month prior and for 4 months of MSP® mix supplementation of the sow's feed.

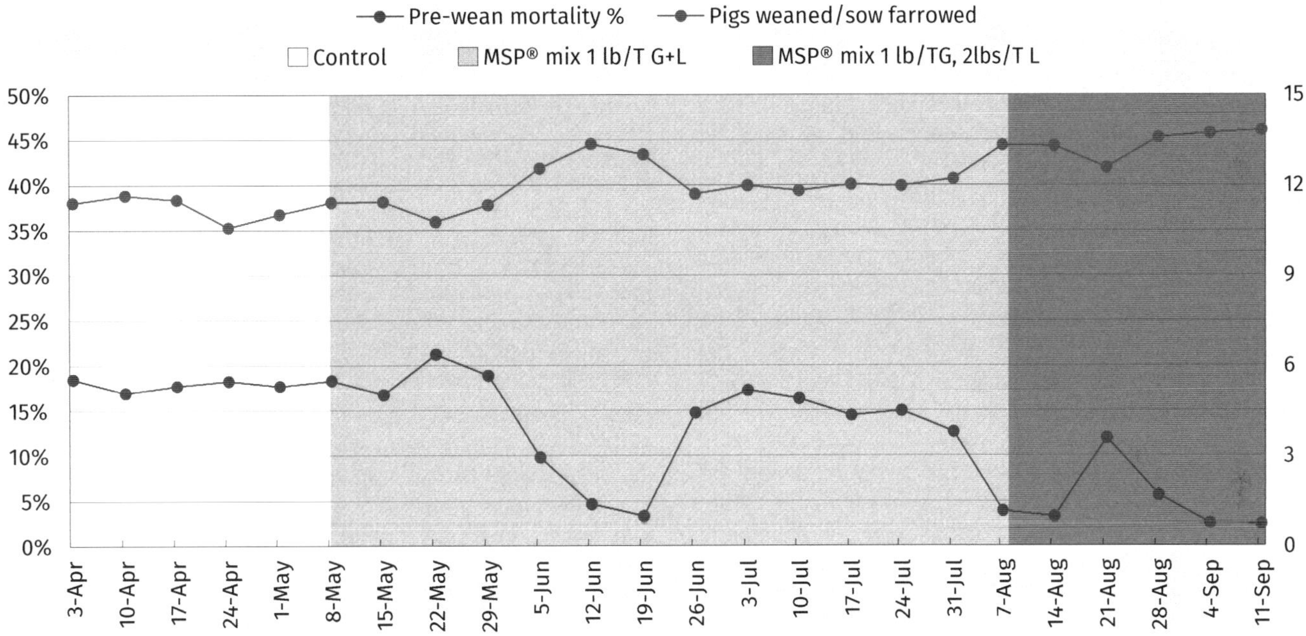

https://doi.org/10.54846/am2022/65

Validation of AndroVision® eFlow – a novel, reusable counting chamber for CASA concentration measurement of boar sperm

R. Grossfeld, PhD; Dr. Andrea Reiser; Dominika Becherer; Monika Esch, DVM; C Simmet, DVM
Minitube GmbH, Tiefenbach, Germany

Introduction

The determination of sperm concentration is an important part of the quality control of raw boar ejaculates during sperm production in every artificial insemination (AI) center. The correct evaluation of boar sperm concentration in raw semen as well as diluted semen in the doses is vital for highest efficacy in AI. A uniform insemination dose becomes more important as AI centers are aiming at reducing the number of sperm per AI dose.[1] The accuracy and precision of the sperm concentration measurement of a boar ejaculate has a direct influence on the number of producible sperm doses from a particular boar ejaculate. Available methods for the determination of sperm concentration in raw sperm include hemocytometers, spectrophotometers, flowcytometers and CASA systems with disposable counting chambers.[2]

A new method for sperm concentration analysis with a CASA system is the AndroVision® eFlow system (Minitube GmbH, Tiefenbach, Germany): a reusable counting chamber combined with a fluid management system (AndroVision® eFlow system, US Patent 10 768 087) that enables the exact determination of sperm concentration during motility assessment (Figure 1).

The chamber is filled and flushed between samples by a fluid management system (FMS) and is automatically closed during analysis to enable measurement in a closed system with defined volume. The high degree of automation decreases the influence of the user during the process of sperm concentration measurement, which could have a significant effect on the measurement result.[3]

In the presented study, the accuracy and usability of the AndroVision® eFlow system in determining sperm concentration of raw semen was compared to the NucleoCounter® SP-100™ (ChemoMetec A/S, Allerod, Denmark) as a reference method.

Material and methods

To validate the precision and correctness of sperm concentration measurement in boar ejaculates with the AndroVision® eFlow system, the concentration of raw semen (n = 185 boar ejaculates) was determined in split-samples by the AndroVision® eFlow system and the NucleoCounter SP-100™ and the concordance of both methods was analyzed.

Upon receiving the samples for sperm concentration in the semen laboratory, aliquots for the NucleoCounter™ and AndroVision® eFlow were taken directly from the same raw ejaculate (split-samples) and analyzed immediately. For the NucleoCounter™, 50µl raw sperm were pipetted into a sample cup and the sample was diluted 1 + 100 (v/v) with 5 mL reagent S100 (ChemoMetec A/S, Allerod, Denmark). Immediately thereafter, a sample of the diluted sample was aspirated into the SP1 Cassette (Chemometec) and analyzed in the NucleoCounter™. All NucleoCounter™ measurements were performed as double measurements. The mean of each double measurement was used as the reference value for sperm concentration of each boar ejaculate.

For the measurement of the sperm concentration with AndroVision® eFlow, 7.2 mL of BTS boar sperm extender without antibiotics (Minitube GmbH, Tiefenbach, Germany) was used to pre-fill a sample container. Thereafter 600 µl raw semen were added to the pre-filled sample container (dilution rate: 1 + 12 [v/v]). After that, the sample container was closed with a rubber plug and carefully turned overhead minimum 5 times, to homogenize the diluted semen. The sample container was then put into the fluid-management system (FMS) and the sperm concentration was analyzed with the AndroVision® CASA. The FMS of AndroVision® eFlow automatically transferred the pre-diluted sperm sample into the eFlow counting chamber. The measurement area of the chamber consists of two parts, one with a height of 15 µm and a second part with a height of 30 µm. Sperm concentration was measured in the part with 30 µm height. For determining the sperm concentration, the software AndroVision® (Version 1.2) was used. The videos of the sperm samples in the eFlow were acquired with an Axioscope A1 (Carl Zeiss Microscopy GmbH, Oberkochen, Germany) and 10× negative phase contrast objective.

For the evaluation of the method-agreement of sperm concentration measurements the Concordance Correlation Coefficient (CCC) was calculated.[4] The measurement results per method were additionally compared with a two-sample Student's *t*-test for paired samples. Values are expressed as mean ± standard deviation (SD).

Results and discussion

In total, 185 raw boar ejaculates were analyzed for their sperm concentration with both the NucleoCounter™ and AndroVision® eFlow. The mean value for each sample did not differ significantly between the two methods ($366.9 \pm 150.8 \times 10^6$ vs. $370.6 \times 10^6 \pm 135.0 \times 10^6$ Mio. sperm/mL, respectively; $P = 0.25$). The Concordance Correlation Coefficient (CCC) between NucleoCounter™ and AndroVision® eFlow was calculated with 0.955 (Figure 2), which corresponds to a nearly complete concordance of the measurement results of AndroVision® eFlow and the NucleoCounter® as a reference method.[5] This level of concordance with a reference method was higher as in a comparable test. Hoogewijs et al.[6] used Pearson's correlation coefficient to compare the method agreement between several disposable counting chambers and NucleoCounter™ in equine sperm concentration measurement and found correlations between r = 0.31 and r = 0.74 depending on the used counting chamber. Possible reasons for the higher concordance of the AndroVision® eFlow system with a reference method include a lower human influence on the measurement result due to a higher degree of automation with the FMS and using a larger sample volume in the 30 µm gap in the eFlow counting chamber.

The results of this trial demonstrate a high concordance of semen concentration determination of the AndroVision® eFlow system and the NucleoCounter as a reference system. The mean of 185 sperm concentration measurements in split-samples, only showed a minor, non-significant difference in all measurement results and no systematic deviation between the two methods could be detected (Figure 2). Based on these data, the NucleoCounter™ and AndroVision® eFlow could be used interchangeably for determining sperm concentration of raw semen samples in this trial.

Figure 1: AndroVision® eFlow system with fluid management system and the reusable counting chamber.

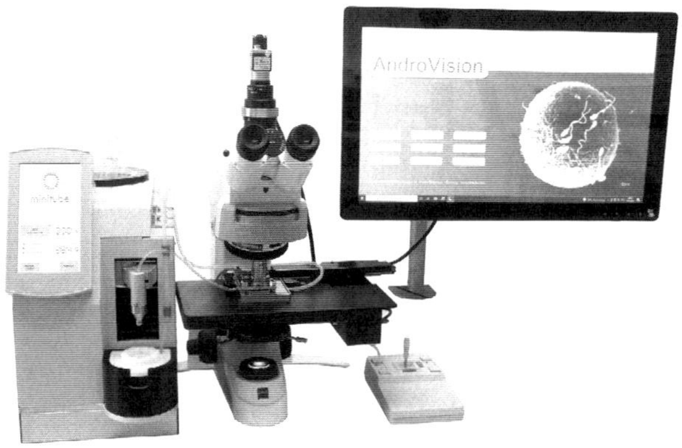

References

1. Schulze, M., Nitsche-Melkus, E., Jakop, U., Jung, M. & Waberski, D. New trends in production management in European pig AI centers. *Theriogenology* 137, 88–92 (2019).

2. Brito, L. F. C. *et al.* Andrology laboratory review: Evaluation of sperm concentration. *Theriogenology* 85, 1507–1527 (2016).

3. Ehlers, J., Behr, M., Bollwein, H., Beyerbach, M. & Waberski, D. Standardization of computer-assisted semen analysis using an e-learning application. *Theriogenology* 76, 448–454 (2011).

4. Lin, L. I. A Concordance Correlation Coefficient to evaluate reproducibility. *Biomatrics* 45, 255-268 (1989).

5. Koch, R. & Spörl, E. Statistische Verfahren zum Vergleich zweier Messmethoden und zur Kalibrierung: Konkordanz-, Korrelations- und Regressionsanalyse am Beispiel der Augeninnendruckmessung. *Klin. Monbl. Augenheilkd.* 224, 52–57 (2007).

6. Hoogewijs, M. K. *et al.* Influence of counting chamber type on CASA outcomes of equine semen analysis. *Equine Vet. J.* 44, 542–549 (2012).

Figure 2: Sperm concentration of 185 raw boar ejaculates as measured with the NucleoCounter™ as reference and AndroVision® eFlow. The Concordance Correlation Coefficient (CCC) of the method comparison was 0.955.

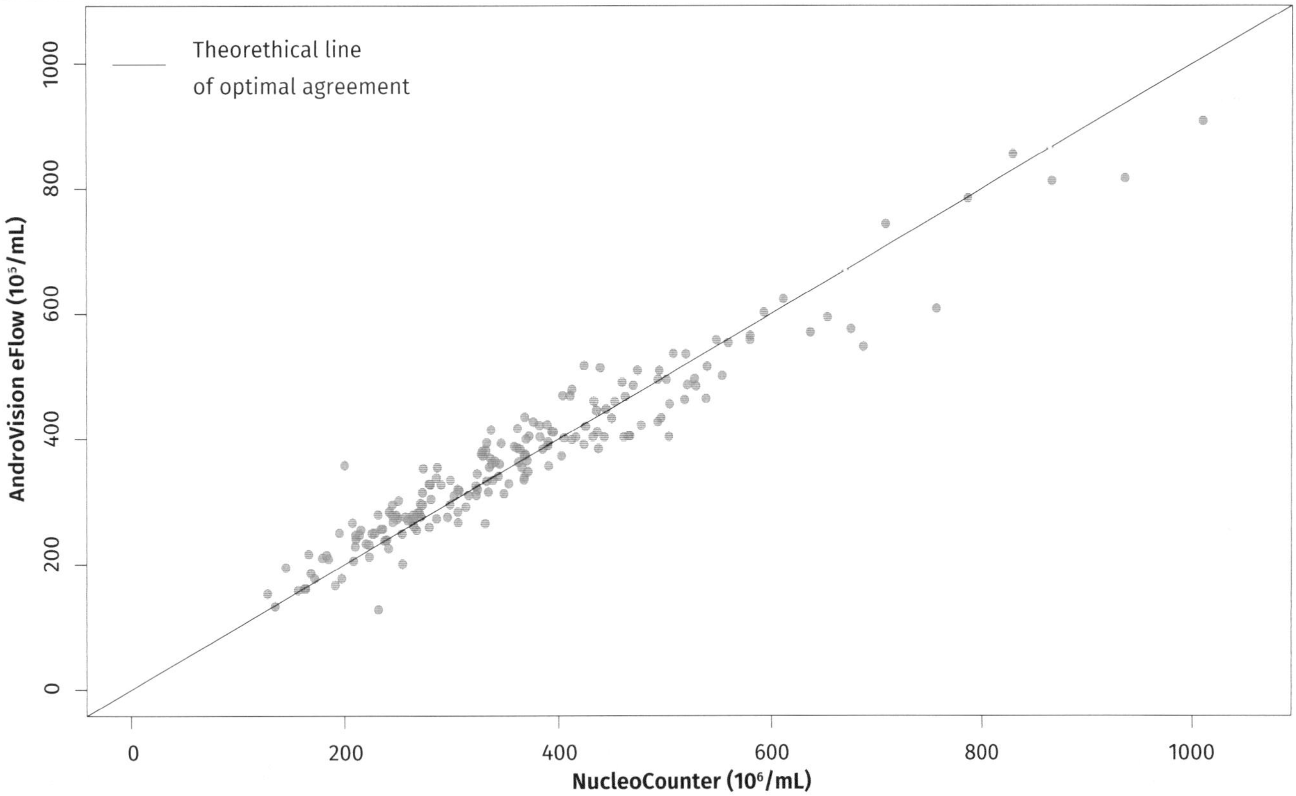

https://doi.org/10.54846/am2022/66

The impact of dietary diformates in zinc-oxide reduced piglet diets: An overview

C. Lückstädt
ADDCON GmbH, Bitterfeld-Wolfen, Germany

Introduction

During the late 1980s, it was discovered that pharmacological concentrations (1,500 to 3,000ppm) of zinc oxide (ZnO) resulted in reduced post weaning diarrhea (PWD) and increased growth in weanling pigs.[1] It had been widely used at such dosages to control *E coli* F4 infections. Several researchers have concluded however that one cannot assume a widespread antibacterial effect of ZnO as MIC values differ greatly both among and within bacterial species.[2] This has been very recently confirmed by Neath and colleagues who found minimum inhibitory concentrations of ZnO for *E coli*, but also for *S typhimurium*, *S suis* or *S aureus* to be above 31 mg × mL^{-1}, thus proving no efficacy.[3] Its impact against certain *E coli* strains for instance is understood rather as a stabilizer of the intestinal microbiota and preventing attachment of pathogenic bacteria to the intestinal villi, which can prevent many problems associated with weaning diarrhea. Unfortunately, as Zn is poorly absorbed, it becomes highly concentrated in manure. In order to minimize the risk of environmental pollution, European authorities have reduced the maximal Zn concentration authorized in pig diets. From June 2022 onwards, pharmacological doses of ZnO will be no longer authorized in the EU. Therefore, alternative strategies and/or additives to manage PWD are immediately needed. Many authors have directed their R&D effort towards the use of acidifiers – as this group of commonly used feed additives proved rather successful after the ban on antibiotic growth promoters in the EU back in 2006. Some authors even claimed, that "...Dietary acidifiers can actually become the most common and efficacious alternative solution to antibiotics, in order to improve health status and performance in pigs."[4] One of those acidifiers, potassium diformate (Formi, KDF, ADDCON), was even approved by the EU as a non-antibiotic growth promoter for use in pig diets – the first product to receive such approval.[5] It was therefore no surprise, that diformates had been tested in diets containing low levels of ZnO. The current report aims to give an overview on the usage of potassium diformate and sodium diformate (both substances produced and registered by ADDCON, Germany) in zinc-oxide reduced piglet diets.

Results and discussion

One of the first results on the usage of KDF in low zinc oxide diets was published in 2001.[6] The authors used diets with 150 ppm zinc oxide in the control, while the treatment diet contained an additional 1.2% KDF. In three successive trials in Spain, with 56 piglets per trial, the effect of potassium diformate on the performance and health status of weaned piglets (Pietrain x Dalland) was observed (Table 1). Data showed the significant performance enhancing impact of the dietary diformate, as well as a clear impact on PWD.

Another report from Spain compared a pharmacological dosage of ZnO (2500 ppm) with 1.2% dietary KDF (Table 2).[7] Both additives improved weight gain against a negative control significantly, while only the diformate had a significant impact on feed efficiency against the control. It was concluded by the authors that an improved gut health by using KDF in the critical period after weaning, was the cause of the improved production performance achieved.

A Chinese study also proved the growth promoting impact of potassium diformate in piglet diets containing only 150 ppm ZnO.[8] This time, the authors used 1.0% KDF in the diet of weaned piglets (weaning age 28 days with 5.91 kg initial weight) over a period of 35 days. Final bodyweight increased significantly ($P = 0.009$) by 9.9% from 18.17 to 19.97 kg, while the FCR was also significantly ($P = 0.022$) enhanced (1.51 vs. 1.42 for control and KDF respectively). Thus, it can be assumed, from the data above, that dietary diformate is capable of improving performance post-weaning and keeping diarrhea rates low, also in the absence of pharmacological dosages of ZnO.

Similar experience was reported by authors from the Kansas State University.[9] The researchers were running a trial to assess alternatives to replace the use of zinc oxide in nursery pig diets. The inclusion of 1.2% sodium diformate alone (Formi NDF, ADDCON) resulted in a similar feed per gain ratio and fecal dry matter as the positive control containing 3000 ppm ZnO for the first

Table 1: Daily gain and feed efficiency of weaned piglets (weaning age 24 days) fed low ZnO-diets with or without KDF

	Control	1.2% KDF
ADG – day 0 – 7 [g]	111^{a}	139^{b}
ADG – day 0 – 35 [g]	383^{a}	416^{b}
FCR – day 0 – 7	1.56	1.25
FCR – day 0 - 35	1.37^{a}	1.27^{b}
Piglets with diarrhea [n]	10^{a}	4^{b}
Diarrhea days [n]	63^{a}	22^{b}

^{a,b} Different superscripts within a row indicate evidence for difference ($P < 0.05$)

7 days, followed by 2000 ppm until day 25. Furthermore, the use of a combination of the three tested alternatives (1.2% Na-diformate, 4% coarse ground wheat bran, and 18% CP), as a replacement to ZnO, improved the appearance of feces and resulted in similar fecal DM content and fecal score to pigs fed diets containing ZnO. A similar experiment was published by the same research institute just recently.[10] This time ZnO was added till day 7 at 3000 ppm and till three weeks post-weaning at 2000 ppm. From day 0 to 21, pigs fed added ZnO had an improved ($P < 0.05$) average daily gain (ADG), average daily feed intake (ADFI), feed efficiency (FCR), and increased day 21 bodyweight (BW) compared to those fed the negative control with only 110 ppm Zn (Table 3). Adding sodium diformate (1.2%) significantly ($P < 0.05$) improved ADG, FCR, and BW. In summary, dietary addition of ZnO or sodium diformate independently improved nursery pig performance. The authors concluded that the addition of an acidifier with or without ZnO could be a useful practice to improve nursery pig growth performance.

More research was recommended by the aforementioned authors. This is currently carried out: unpublished data from the same university suggest that an agglomerate of sodium diformate and monolaurate (Formi 3G, ADDCON) fed to weaned piglets at 1% can reach similar performance parameters at day 28 post weaning to high dosages of zinc oxide and has at the same time a beneficial impact on the gut microbiome.

Finally, the latest data on a comparison of the antibacterial impact of various acidifiers, including benzoic acid, potassium diformate and sodium diformate, as well as zinc oxide (Table 4), revealed drastic differences in the minimum inhibitory concentration (MIC) in-vitro, in favor of the diformate-based acidifiers.[3] Antibacterial activity was determined using an agar plug diffusion method and quantified via broth microdilution.

In the context of the above listed data, it can be summarized that the acidifier Formi (potassium diformate), Formi NDF (sodium diformate) or Formi 3G (agglomerate of sodium diformate and monolaurate) can play a vital part in zinc oxide reduction strategies and may further help in improving the feed efficiency in post weaning piglet production – thus playing an important part in world-wide sustainable pig production approaches.

References

1. Poulsen H 1989. Proceedings of the 40th annual meeting of the European Association for Animal Production, 265-266.
2. Liedtke J and Vahjen W 2012. Veterinary Microbiology 160(1-2), 251-255.
3. Neath C, Portocarero N and Jones C 2021. Journal Applied Microbiology. **doi: 10.1111/jam.15318**
4. Papatsiros V and Billinis C 2012. *Antimicrobial Agents*, Chapter 14, 295-310.
5. Overland M 2001. Feed Mix 9 (4/5), 25-28.
6. Daza A, Ovejero I and Flox J 2001. Anaporc Cientifico 1 (1), 15-25.
7. Portocarero-Khan N, Jurado-Candaliga S, Morales J and Pineiro C 2006. Digestive Physiology in Pig. Book of Abstracts, 27.
8. Zhou Y, Wei X, Zi Z, Zou B, Xia S, Lu N, Lei H, Lu Y, Parvizi N and Xia D 2015. Livestock Science 172, 85-90.
9. Laskoski F, Tokach M and Woodworth J 2019. Kansas Agricultural Experiment Station Research Reports 5 (8/8), 13 p.
10. Hutchens W, Tokach M and Dritz S 2020. Kansas Agricultural Experiment Station Research Reports 6 (10/7), 13 p.

Table 2: Performance parameters of weaned piglets (weaning age 28 days) until 40 days of age

	Control	2500 ppm ZnO	1.2% KDF
Feed intake [g/d]	469	470	474
Weight gain [g/d]	351[a]	374[b]	386[b]
FCR	1.37[a]	1.28[ab]	1.25[b]

[a,b] Different superscripts within a row indicate evidence for difference ($P < 0.05$)

Table 3: Effects of pharmacological levels of zinc oxide and sodium diformate on growth performance of nursery pigs till day 21

	3000/2000 ppm Zinc oxide			1.2% Sodium diformate		
	None	With	*P*-level	None	With	*P*-level
ADG [g]	227	263	< 0.001	231	254	0.004
ADFI [g]	340	372	0.004	349	358	0.41
FCR	1.52	1.43	0.003	1.53	1.42	< 0.001

Table 4: MIC of various acidifiers and zinc oxide against *E coli* in mg × mL^{-1}

	ZnO	Benzoic acid	Sodium diformate	Potassium diformate
MIC *E coli*	> 31.25	15.63	3.91	1.95

https://doi.org/10.54846/am2022/67

Strain matters: Understanding how modes of action impact probiotic performance

K. Kinsley; S. Lerner
Chr Hansen, Inc., Milwaukee, Wisconsin

Probiotics are commonly used in livestock production systems to support positive economic outcomes. However, there are long held criticisms that repeatability is, at times, suspect and that benefits are sometimes difficult to measure. Whether we're evaluating the efficacy of probiotics in well-controlled research trials or less formally during on-farm observation trials, the strain of the organisms in our probiotic of choice should be considered as a critical part of the product selection process. It is well understood and accepted that there are highly important strain differences when it comes to bacterial pathogens. Therefore, it's only reasonable to assert and insist that the same is true for probiotics. The bottom line is that strain matters.

Sometimes with probiotics, the assumption is made that all are similar in their capabilities and thus are commodity products requiring little conscious selection. This is a long way from reality. Probiotic organisms and combination products come in many different shapes and sizes, or more appropriately species and strains. The reasoning behind the concept "strain matters" is simple. When describing and comparing the capabilities and characteristics of probiotic organisms, knowing their genus is simply not enough. To sufficiently elucidate the magnitude of variation in those functional features, one must look beyond genus and species to the strain of the individual, where remarkably small variations in genetics can drive profoundly meaningful changes in functions.

Going back for a moment to shapes and sizes, a more broadly based grouping of organisms should first be discussed. Initial differentiation of probiotics can begin at the genus level, where we find four major genera: *Bacillus, Lactobacillus, Clostridia,* and *Enterococcus.* These groups can be split further into spore-formers (such as *Clostridial* and *Bacillus* species) and non-spore-formers (*Lactobacillus* and *Enterococcus* species). Even this simple differentiation of spores or no spores is correctly indicative of stability under various environmental conditions – stability through pelleting, stability in storage, or stability in acidic conditions (Figure 1 and 2).

Looking specifically at *Bacillus*-based probiotics, characteristics such as rate of germination, production of enzymes, ability to interfere with activity of certain potentially pathogenic bacterial organisms, tolerance to bile salts, adaptability to culture, and, ultimately, capacity to efficiently scale up for commercial production differ not only by genus, but by genus, species, and strain.[1] Another way of stating this is that not even knowing that a *Bacillus* organism was of the species *subtilis* would be enough to know how it will perform. It's only when you get to strain, that is an individual and all of its trillions of nearly identical clonal copies, such as *Bacillus subtilis* CH201, or *Bacillus subtilis* 25841, or *Bacillus subtilis* 15549, that you have a chance of learning how the organism will work. Truly, strain matters.

Many have accepted that probiotics perform in essentially equivalent manners and do so via similar modes of action. Instead of settling for nothing more than "aiding in the maintenance of gut health," consider the mechanisms by which it can be demonstrated that *Bacillus*-based probiotics are interacting, observe the descriptions of different behaviors by strain, and choose wisely for your purposes.

Figure 1: Recovery of *Bacillus* spores from BioPlus 2B in VTM.

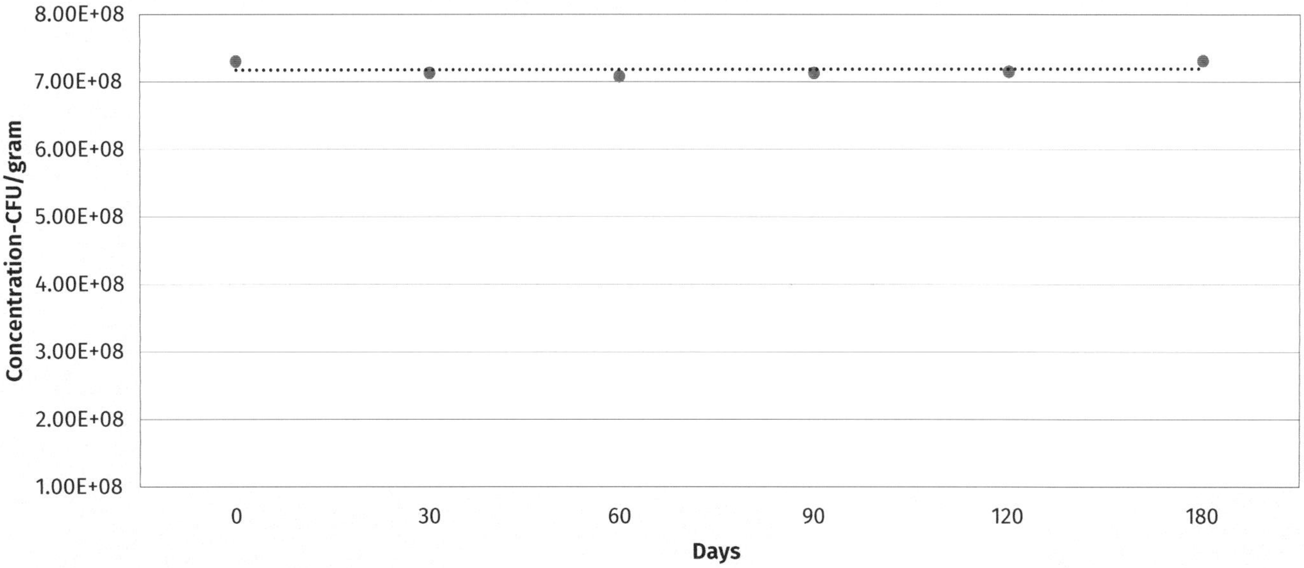

Figure 2: Spore recovery as a percentage of target, with and without MCFA.

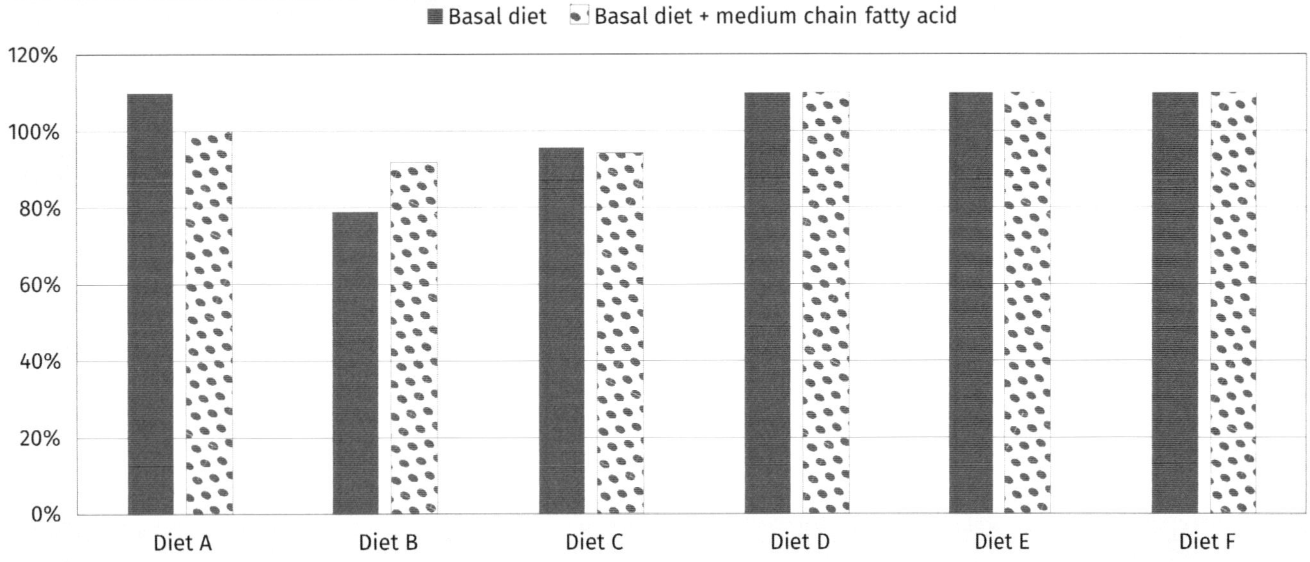

There are three primary levels at which effective probiotics interact in their environment:

1. Interactions with feedstuffs
2. Interactions with other microbes
3. Interactions with the host

Interactions with feedstuffs

Wherever *Bacillus* organisms are found, on the surface of the credenza in your living room, in the middle of the woods, or in the intestinal tract of a pig, they are interacting with the environment around them and have become quite adept at doing so. Their life's work is quite simple: find sources of nutrients that allow them to support activity and reproduction. The process of acquiring needed nutrients is benefitted by the capacity of *Bacilli* to move in their environment, to work together as a team of like *Bacilli*, and to adapt their production of enzymes to match the nature of the nutrient source that they encounter. So, thinking about *Bacilli* that may be in the GI tracts of pigs on a commercial facility, it's remarkable to see how the action of those organisms feeding themselves is also of immediate and demonstrable benefit to their hosts, because more readily absorbable nutrients, particularly from fiber sources, are being made available to those animals.

Laboratory assays can be used to demonstrate the liberation of energy or soluble proteins by *Bacilli*. To accomplish this, *Bacillus*-based probiotics are incubated with feedstuffs. Over about 48 hours, the *Bacilli* germination, become active vegetative cells, and begin digesting the complex nutrient sources in the feed. Using various chemical reagents, the amount of energy, in the form of simple sugars, and of soluble protein, based on the presence of exposed peptide-peptide bonds, can be assessed using spectrophotometry. The more intense the in vitro response, the higher the likelihood of benefit in vivo.

Various well-controlled studies show that pigs fed a basal ration plus an effective combination probiotic (*Bacillus licheniformis* and *Bacillus subtilis* – known as BioPlus® 2B) outperformed pigs fed the same basal diet without a probiotic.[2] Daily inclusion of the probiotic in rations to pigs resulted in better average daily gain (ADG) and feed-to-gain ratios (F:G) compared to those receiving only control rations. Additionally, the digestibility of certain components – crude protein, fat, and phosphorous – were greater when probiotics were provided daily. Predictably, when reduced energy diets were used, pigs tended to experience a greater magnitude of response from inclusion of probiotics compared to those receiving the lower energy diet without supplementation.

The prior evaluation was with a combination product, but what happens when looking at an individual strain? In the evaluation of two strains of a species of *Bacillus* in ileal cannulated pigs, the two strains of probiotic organisms acquired their nutrients in different locations of the GI tract.[3] One of these organisms preferred to acquire their nutrients in the small intestine and the other in the large intestine. This preference of environment, as well as unique abilities to interact with the feedstuffs further demonstrates how much strain matters.

As a means of demonstrating that these interactions are not specific to age of pig or stage of production, consider the benefits of feeding BioPlus 2B to sows prior to farrowing and throughout their lactation. Several studies looked at maintenance of weight of sows during lactation (or reduced weight loss), daily consumption of sows during lactation, and, ultimately, weights of piglets at weaning with and without the feeding of BioPlus 2B.[4,5] Feeding the probiotic for 14 days prior to farrowing and for the entire 28-day lactation benefitted daily lactation feed intake and reduced lactation weight loss compared to the control group. Additional consumption in lactation drives reduced loss of condition for sows or extra weight at the time of weaning for the piglets, which was also the case in one of the identified studies. While these evaluations were performed some time ago, recent work (Chr Hansen studies 80638 and 80707) demonstrates the repeatability of results for those same key variables: increased feed intake during lactation, reduced lactational weight loss, and the increased weaning weight of piglets.[6]

Interactions with other microbes

Inhabiting the same small space, it's a given that a probiotic delivered orally must interact at some level with other microbes in the GI tract. In the ideal scenario, the applied probiotic organisms will support the beneficial "good" bacteria in the gut and disrupt the activities of those that are potentially harmful. Both in vitro and in vivo applications can help visualize aspects of these microbe-to-microbe interactions that may be occurring. It is of benefit to understand that *Bacillus*-based probiotics are known to interact with the microbial communities of the GI tract in several ways:

1. Directly antagonizing potential pathogens,
2. Supporting a beneficial balance of organisms in the microbiome, and
3. Competitively excluding pathogenic organisms from binding to the epithelial surface of the gut.

During the process of screening potential probiotic candidates, the Chr Hansen laboratory performs evaluations of the genetic and physiologic characteristics of new strains to determine their capacity to inhibit the growth of *E coli*, *Clostridium perfringens* type A, *Salmonella* spp., *Streptococcus suis*, and *Staphylococcus aureus*[7] (Figure 3). In the case of *Clostridium perfringens* type A, a strain of *Bacillus licheniformis,* CH200, was proven to not only inhibit the proliferation of that pathogen, but to also disrupt its ability to produce alpha toxin; a phenomenon that translated to efficacy and improvement in an active production system.[8] Sows fed the *Bacillus*-based probiotic shed *C perfringens* type A organisms into the environment at only half the rate of sows in the control-fed group. Some strains are significantly better than others at outcompeting these potentially pathogenic organisms, once again speaking to the concept that strain matters.

Maintaining a beneficial balance of good versus bad bacteria in the microbiome is essential to proper functionality and optimal performance of the host. Selecting probiotic strains that positively impact that balance is the ultimate goal. A state of eubiosis is associated with proper digestion and absorption of nutrients, with normal and effective barrier functions of the gut, and with normal and effective immune functions throughout the animal. Daily feeding of effective probiotics to populations of animals increases the chances of a high number of normal, healthy individuals in those populations. The greater the number of normal, healthy individuals in any flow, then the greater the chances of economic success and the greater the sustainability of the operation.

Figure 3: Probiotic strain interfering with pathogen's zone of growth.

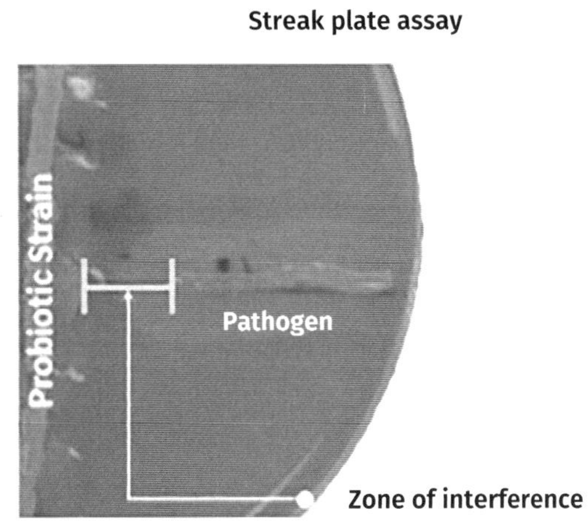

A recent finishing study provides an example of supporting the microbiome by feeding of an effective *Bacillus*-based probiotic.[9] In this evaluation, pigs fed the probiotic showed a reduction in the presence of *Enterobacteriaceae, Enterococcus, Clostridium,* and non-specific *Bacillus* species. *Bacillus subtilis* and *Bacillus licheniformis* were found at higher concentrations in pigs from the probiotic-fed group and those pigs showed high concentrations of lactic-acid-producing species compared to the pigs of the control group. It is also worth noting that the pigs in the *Bacillus*-fed group also benefited from improved survivability compared to those of the control group.

An in vitro intestinal epithelial model can be prepared by growing monolayers of mucus- or non-mucus-producing cell lines in culture wells. These cell culture systems can then be challenged with pathogenic organisms, probiotic organisms, and the combination of the two to determine the capacity of probiotic strains to exclude the binding of pathogenic organisms. The ability of two individual strains of *Bacillus* to exclude the attachment of specific pathogenic organism is shown in Figure 4. It also appears that the capacity of the cell lines to produce mucus may impact the intensity to which a probiotic strain is able to exclude a pathogen.

Practically speaking, reducing the ability of a pathogen to adhere to the intestinal epithelium can have a range of impacts in vivo: reduced clinical diarrhea scores, improved health (potentially GI and systemic), improved productivity and growth, and reduced treatment for clinical disease. An evaluation from 2016 documented the ability of an individual strain of *Bacillus subtilis* to reduce adhesion of an F5 *E coli* in vitro on several cell lines.[10] This finding was followed up with a nursery piglet challenge study that demonstrated improved ADG and reduced fecal scores when the same *Bacillus subtilis* was supplied in the feed.

Interactions with the host

Beneficial probiotic-host interactions are known to occur at the level of overall barrier functionality and integrity as well as with the host immune system. These interactions are critical in the maintenance of eubiosis, especially in scenarios where physical and infectious insults are at work. If left unchecked, a vicious downward spiral of disrupted homeostasis (dysbiosis) can draw greater proportions of a population towards the threshold of disaster, or unrealized potential. As discussed with the support of a healthy microbiome, the combination of pathogens, damaged intestinal barriers, and a flood of pro-inflammatory cytokine activity spurred on by growth of harmful bacterial species can make it nearly impossible to return to eubiotic functionality.

This breakdown of normal activity in the intestinal tract is commonly referred to by nutritionists, veterinarians, and producers as "leaky gut." In this scenario, the intestinal barrier loses its ability to maintain luminal versus systemic integrity. In other words, microbial invaders, and toxins, normally unable to cross the intestinal barrier, suddenly have access to areas of the body that had previously been off limits while this protective mechanism was working properly.

A breakdown in the integrity of the epithelial barrier can be visualized in vitro using a transepithelial electrical resistance assay, TEER for short. For the TEER assay, a monolayer of intestinal epithelium is grown on a permeable membrane in an apparatus that allows electrical resistance to be measured across the cellular layer. The resistance over time across the monolayer in the absence of microbial additions is set to a baseline of 100%. Then, resistance over time is measured as the monolayer of epithelial cells is exposed to pathogens, probiotics, and pathogens + probiotics. It is expected that the presence of probiotics should increase resistance, the presence of pathogens should decrease resistance, and the presence of the combination should increase resistance compared to that driven by the pathogen alone (Figure 5).

While electric resistance is a highly sensitive measure, it doesn't allow an accurate assessment of the size of the "leaks" created by the pathogens. To accomplish this, a fluorescent compound, fluorescein isothiocyanate (FITC)-dextran, can be added to the solution on the apical surface of the monolayer. Damaged epithelial connections permit the FITC-dextran to move across the monolayer into the solution on the basolateral side of the monolayer. Sampling that solution with spectrophotometry provides a reliable assessment of transmembrane permeability. A second method involves direct staining of one of the tight junction proteins present in the monolayers, typically, zonula occludens (ZO-1). If apical tight junctions are largely intact, then there is a strong degree of staining and a nice border structure of the cells. If these tight junction proteins are absent due to the action of the pathogens, then staining of the cell borders is much less intense and inconsistent.

Figure 4: Reduction in binding of pathogen to intestinal epithelial cells.

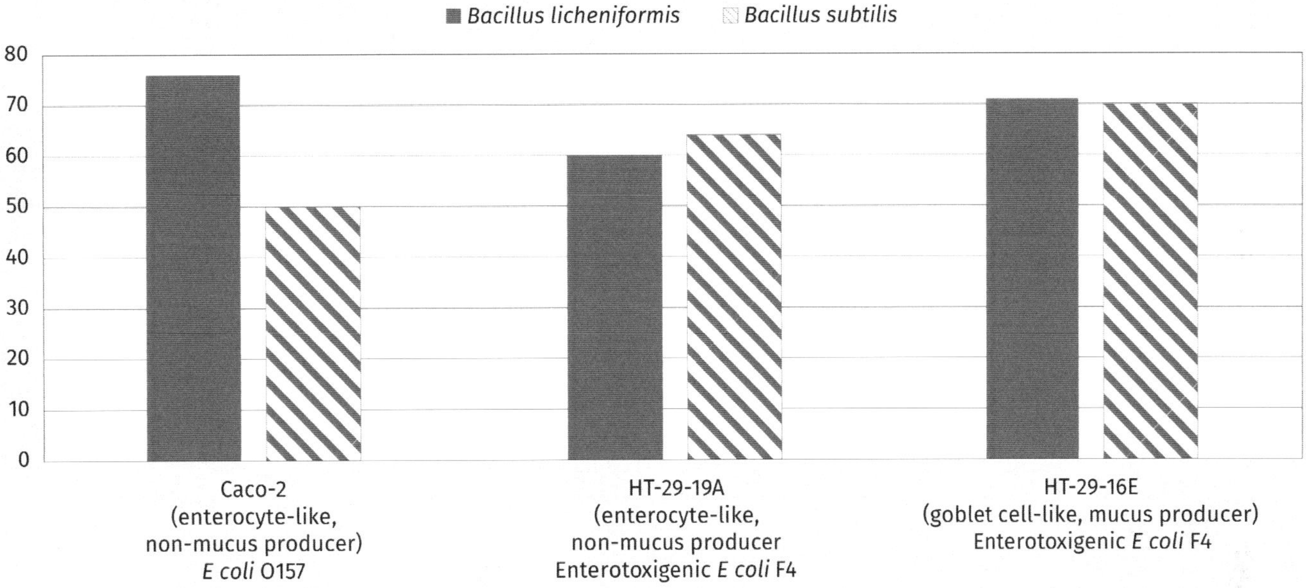

Figure 5: Transepithelial Electrical Resistance (TEER) assay

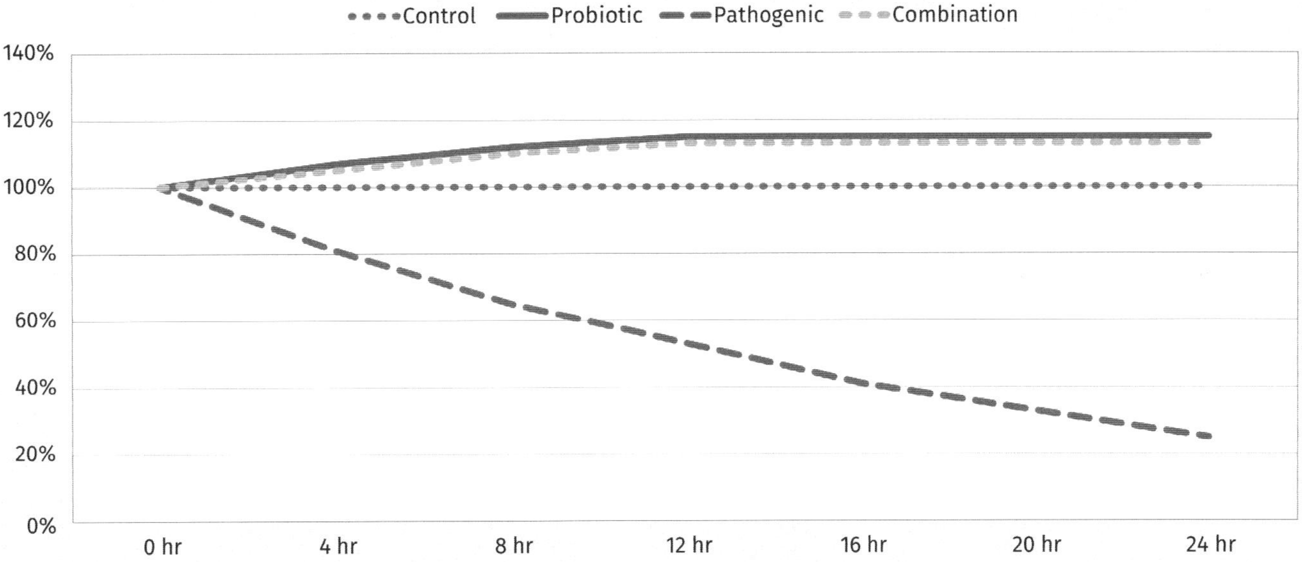

A recent *E coli*-challenge study provides some insight into the evaluation of barrier integrity.[11] In this study, pigs were allotted to one of four treatments – two levels of inclusion a *Bacillus subtilis* probiotic in feed plus challenge, an unchallenged control, and a challenged control. The three challenge groups were exposed to an F18 *E coli* at 7 days post-weaning. Piglets were sacrificed at 5- and 11- days post-challenge and samples of intestinal tissue were examined for markers of transcellular and paracellular permeability, indicative of damage to the integrity of the epithelial barrier. Pigs in the groups that received probiotics showed an upregulation of genes encoding for barrier functions (ZO-1 on day 5 and mucin 2 on day 11) compared to those in the challenged control groups. Also of note was the down regulation of genes for interleukin-1-alpha (IL-1α) and interleukin-6 (IL-6). Both are pro-inflammatory cytokines released rapidly in response to infections, tissue damage, and pathogenic invaders.

Probiotics acting to slow production and circulation of proinflammatory cytokines is not unique to infections with *E coli*. A recent Chr. Hansen study evaluated the feeding of a combination *Bacillus* product (BioPlus 2B) prior to, and during a PRRS-virus-challenge. From PRRS challenge to 21-days post-challenge, pigs fed the probiotic were heavier, had improved F:G ratios, had lower gross lung lesion scores upon necropsy (Figure 6), and appeared to respond in an immunologically beneficial manner compared to pigs in the control group.

Interferon-gamma (INF-γ), a pro-inflammatory cytokine, was measured at more moderate levels at 7 days post-challenge when BioPlus 2B was fed compared to controls and vaccine only treatments (Figure 7). Controls showed a slow PRRS-challenge specific INF-γ response that did not reach the level of other treatments over any of the observed period. Pigs in the vaccine-only group displayed the most aggressive INF-γ response in the post-challenge period. However, regardless of vaccination status, pigs fed BioPlus 2B has a moderate change in INF-γ in response to the PRRS challenge. It is interesting to note that the gross lung lesion scores were statistically reduced in pigs fed probiotics compared to those in the control group. Evidence exists from early COVID-19 information regarding cytokine storms (INF-γ and TNF-α driven) and their ability to cause severe lung damage.[12] Neutralizing the over-production of these cytokines was shown to be protective against mortality in mouse models.

Final note

It is important to be aware that measurable responses can and will vary across all species and strains of *Bacilli*, that is to say, just because we know the genus and species doesn't in any way ensure us of a predictable action or outcome – strain matters! An assumption that all *Bacillus*-based probiotics are alike, and that data garnered in vitro and/or in vivo from one is useful to describe another is a disservice to the industry, to the customer, and ultimately to the sows and pigs in our care. Furthermore, great care should be taken when considering the benefits of combining probiotic organisms. Combining organisms only works when their various modes of action are complementary and when they take no antagonistic actions against one another. In those cases, the result is greater than the sum of the parts. This necessitates careful analysis in vitro in the lab and then validation in vivo in well-controlled trials and in commercial operations. Failure to do so could result in minimally a waste of money on an ineffective product and maximally in consequential negative impacts on performance and economic returns. Bottom line, when it comes to a choice of probiotics, strain - absolutely, positively, every day, all day, in every operation, for every use class of animals – matters.

References

1. Leser, T.D., Knarrenborg, A., Worm, J. Germination and outgrowth of *Bacillus* subtilis and *Bacillus* licheniformis spores in the gastrointestinal tract of pigs. Journal of Applied Microbiology. 2008. 104: 1025-33.

2. Jorgensen, J.N., Laguna, J.S., Millan, C., Casabuena, O., and Garcia, M.I. Effects of a *Bacillus*-based probiotic and dietary energy content on the performance and nutrient digestibility of wean to finish pigs. Animal Feed Science and Technology. 2016. 221: 54-61.

Figure 6: Lung consolidation score.

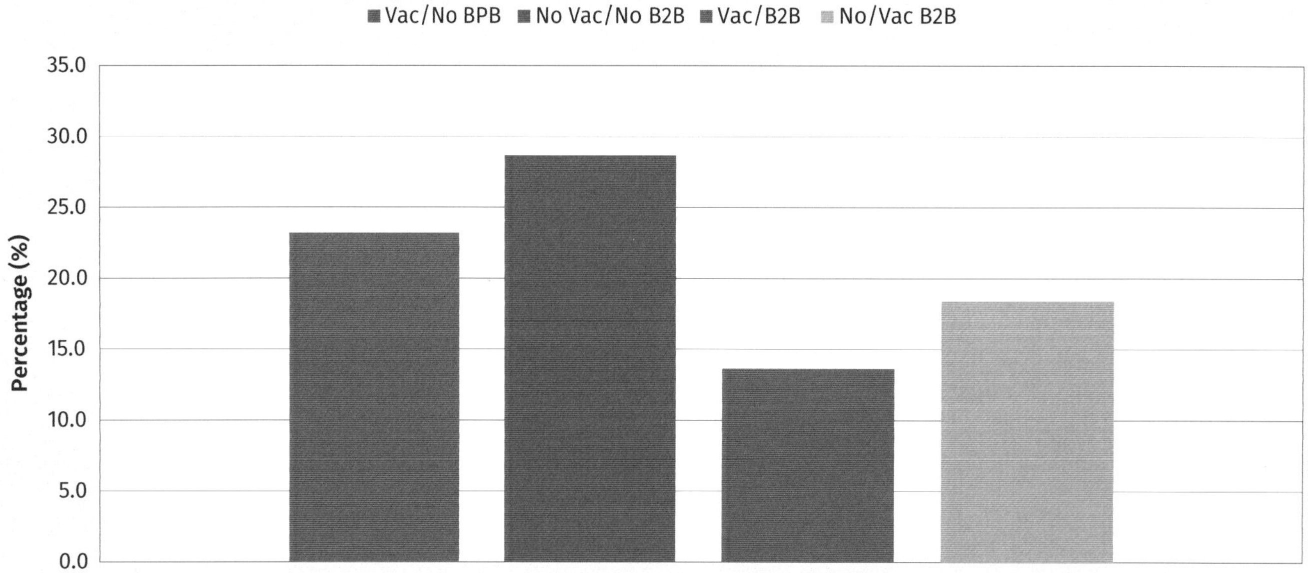

Figure 7: PRRSV-Specific INF-γ.

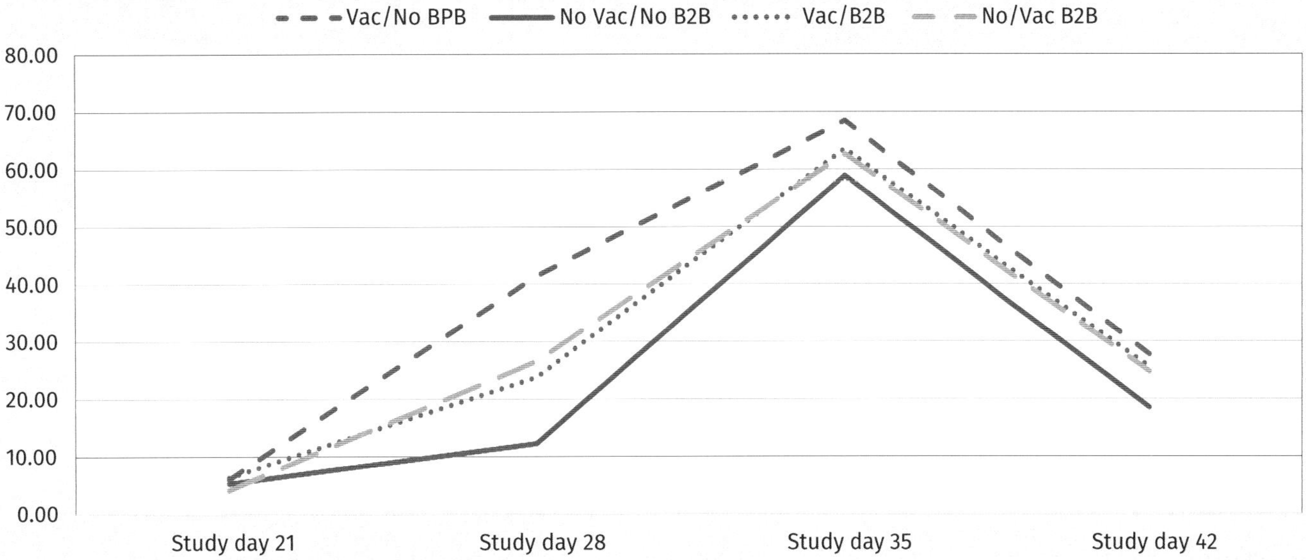

3. Blavi, L., Jorgensen, J.N., and Stein, H.H. Effects of *Bacillus* amyloliquefaciens and *Bacillus* subtilis on ileal digestibility of AA and total tract digestibility of CP and gross energy in diets fed to growing pigs. Journal of Animal Science. 2019. 97: 727-34.

4. Lilija, D., Sanita, B. Probiotic BioPlus® 2B effect on sows productivity and piglets weight. University of Agricultural Sciences and Veterinary Medicine Iasi. 2012. Pp. 266-9.

5. Alexopoulos, C. Georgoulakis, I.E., Tzivara, A., Kritas, S.K., Siochu, A., Kyriakis, S.C. Field evaluation of the efficacy of a probiotic containing *Bacillus* licheniformis and *Bacillus* subtilis spores, on the health status and performance of sows and their litters. 2004. 88: 381-92.

6. Lerner, S.P., Becker, S.R., Dee, S.A., Kitt, S.J., Bertram, M.J., Williams, T.E., Pepin, B.J., Hanson, D.B., Little, E. A., Frahm, P.F., and Elder, R. Efficacy of *Bacillus* subtilis and *Bacillus* licheniformis (BioPlus® 2B) on sow lactation performance, litter growth, nursery performance, and health of pigs challenged with PRRS virus post-wean. American Association of Swine Veterinarians Proceedings. 2021. 52: 171-3.

7. Larsen, N., Thorsen, L., Kpikpi, E.N., Stuer-Lauridsen, B., Cantor, M.D., Nielsen, C., Brockmann, E., Derks, P.M.F., Jespersen, L. Characterization of *Bacillus* spp. strains for use as probiotic additives in pig feed. Applied Microbial Biotechnology. 2013. Online.

8. Lantz, R., Knap, I., Cook, D., Minton, B. B. licheniformis for C. perfringens control in sows. International Pig Veterinary Society Poster Presentation (P.486). 2010. 21:792.

9. Rybarczyk, A., Boguslawaska-Was, E., Dlubala, A. Effects of BioPlus® YC Probiotic Supplementation on Gut Microbiota, Production Performance, Carcass and Meat Quality of Pigs. Animals. 2021. 11:1581.

10. Nielsen, B., Cernat R.C., Improvement in Piglet Performance and Gut Health with new *Bacillus* subtilis Probiotic. Allen D. Leman Swine Conference Proceedings. 2016. Abstract 18.

11. Kim K., He, Y., Xiong, X., Ehrlich, A., Li, X., Raybould, H., Atwill, E.R., Maga, E.A., Jorgensen, J., Liu, Y. Dietary supplementation of *Bacillus* subtilis influenced intestinal health of weaned pigs experimentally infected with a pathogenic E. coli. Journal of Animal Science and Biotechnology. 2019. 10:52.

12. Karki, R., Sharma, B.R., Tuladhar, S., Williams, E.P., Zalduondo, L., Samir, P., Zheng, M., Sundaram, B., Banoth, B., Malireddi, R.K.S., Schreiner, P., Neale, G., Vogel, P., Webby, R., Jonsson, C.V., Kanneganti, T.D. Synergism of TNF-a and INF-g Triggers Inflammatory Cell Death, Tissue Damage, and Mortality in SARS-CoV-2 Infection and Cytokine Shock Syndromes. Cell. 2021. 184:149-68.

https://doi.org/10.54846/am2022/68

The design, development, commercialization, and licensure of real-time PCR tests for FMDV and ASFV

Johnny Callahan, MT (ASCP), PhD; Amy Saunders, MFS; Rolf Rauh, MS; William M. Nelson, MD
Tetracore, Inc

Introduction

Tetracore, Inc. was established in 1998 after winning a multi-million dollar, multi-year government contract to develop real-time PCR tests for the detection of biothreat agents, including many agents that could be used for agroterrorism. The Tetracore Animal Health group began with a focus on foreign animal disease agents that could be detected with field deployable tests and instruments. Over the years, Tetracore has developed working relationships with key national and international laboratories. The foot-and-mouth disease virus (FMDV) and African swine fever virus (ASFV) Test Kits as USDA licensed products represent the completion of a full cycle of development efforts.

FMDV RNA test kit

Development history

In November of 2000, FMDV sequences for all seven serotypes were downloaded from GenBank and aligned with Premier Biosoft software. Primers and probes were designed and analyzed with a blast against nucleotide sequences to verify their coverage. The primers and probes were ordered, and bench development work began on November 20, 2000, using synthetic oligonucleotide sequences as a positive control. On February 8, 2001, two dried rRT-PCR assay designs for FMD viral RNA (3D and VPg3) were compared in parallel at PIADC using viral RNA extracted from FMDV serotype O (BFS) and serotype A12 viruses in the laboratory of Dr. Fred Brown, Ph.D., FRS, see figure 1.

The validation work at PIADC resulted in one publication.[1] After the publication, the test became an OIE reference method described in chapter 3.1.8 of the OIE Terrestrial Manual. The test design also became the basis for the NAHLN reference test for FMDV. In 2019, the test was converted to a wet format for use in a high-throughput, outbreak scenario. The wet FMDV RNA Test Kit was validated in studies conducted in collaboration with The Pirbright Institute, UK and the National Centre for Foreign Animal Disease (NCFAD), Canada, and was licensed by the USDA in May 2021.

Field deployment

At the height of the 2001 UK FMD outbreak, Tetracore traveled to the UK to demonstrate the feasibility of testing with a dried FMDV rRT-PCR test using a portable device at Oxford University. Attending the demonstration was David King, the scientific advisor to Tony Blair. The team included Dr. Floyd Horn, ARS Administrator; Fred Brown, Ph.D., FRS, PIADC (Figure 1); and William Nelson and Johnny Callahan from Tetracore, see Figure 2.

ASFV DNA test kit

Development history

In Q1/2000, Tetracore met with Dr. Dan Rock, of the Agricultural Research Service at the Plum Island Animal Disease Center (PIADC) to discuss the development of a real-time PCR test for detection of ASFV DNA in swine samples. The ARS team provided 20 ASFV p72 sequences from which Tetracore designed the ASFV real-time PCR primer and probe set. Tetracore returned to PIADC in Q3/2000 with a dried prototype test that underwent validation. In Q3/2004, further validation efforts were made which led to a publication by Zsak et al. in Q1/2005. The NAHLN adopted the test in 2013. In 2016, the test was modernized with a second primer/probe set in the p72 region, and this became version 2.0 of the dry assay. In 2019, the 2.0 dry version was converted to a wet assay format. Validation studies of the wet assay were conducted in Spain, Sweden, Belgium, Poland, and at PIADC. The ASFV DNA Test Kit, product code 57A5.80 was licensed by the USDA on November 1, 2021, for use on swine spleen and whole blood samples collected in EDTA.

Field use

In June 2007, in collaboration with PIADC, the test was deployed and used to establish an ASFV PCR capacity at the Laboratory of the Ministry of Agriculture (LMA) in Tbilisi, Georgia during the 2007 ASF outbreak. The test used was the ASFV DNA Test Kit version 1.0 as described in the paper.[8] During the 2007 Georgia ASFV outbreak, the PIADC assembled what they called an "Expert Team," that brought expertise and ASFV real-time PCR Test to the Laboratory of the Ministry of Agriculture (LMA) in Tbilisi, Georgia to establish laboratory capacity for the detection of the virus in swine samples (Figure 2).

In a multiple-year study, five ASF outbreak sites were identified in Uganda. This study was conducted in three stages over 3 years (2015-2017) as part of a project under the World Organization for Animal Health (OIE) in partnership with the National Veterinary Institute (SVA) in Uppsala, Sweden, an OIE Collaborating Centre for Biotechnology based Diagnosis of Infectious Diseases in Veterinary Medicine, and the National Animal Disease Diagnosis and Epidemiology Center (NADDEC) in Entebbe, Uganda. Three of the five sites had active outbreaks with clinically affected pigs. The initial laboratory diagnosis was made at the point of care, and the confirmatory diagnosis was made at NADDEC Laboratories, see Figure 3.

Dry and wet test formats

The two licensed tests are provided in a wet format, shipped on dry ice and stored at -20°C until the tests are thawed before use. The wet format is run like any other high-throughput real-time PCR assay utilizing automated extraction protocols.

Figure 1: Validation study partner/advisor, Fred Brown, PhD, FRS, 1/31/1925 – 2/20/2004.

Figure 2: The PIADC ASFV Response Team.

The rationale for using dried PCR tests (Figure 3) is that the cold chain for test delivery is broken, tests are rehydrated as needed just before use, and the tests are stable at ambient room temperature for years. This dried format is the most practical option for getting the tests through customs delays and into remote test locations.

The T-COR 8™ instrument (Figure 4), developed in-house by Tetracore engineers, is a multiplex PCR platform that currently has 5 independent dye channels. The under 10-pound platform has 8 independently programmable reaction wells and many advanced electronic features. These include innovative software that is run from a web browser, no software installation required, auto-analysis of run data, and viewing of data in real-time in a cloud-based environment that is accessible by smartphone. This instrument is ruggedized for field use and is easily transported in carry-on/backpack-type luggage for mobilization to remote locations.

Manufacturing and surge capacity

Tetracore can manufacture serials (batches or lots) of tests in quantities of up 200,000 of the USDA licensed wet assays per week on any given week but would need stockpiled reagents to maintain this production pace. The cGMP manufacturing facility is both USDA and FDA licensed.

License claims

FMDV RNA Test Kit: Cattle, sheep, and pigs; serum and epithelial samples

ASFV DNA Test Kit: Swine spleen and whole blood collected in EDTA samples

Restrictions

The sale and distribution of the tests are controlled by the USDA. The tests can only be used by USDA-authorized laboratories. Before the tests are shipped, the State Veterinarian where each authorized NAHLN lab is located must grant permission for the test to be shipped and used in their respective States.

Discussion

FMD is an ongoing food security issue where 75% of the burden of control costs is born in low and middle-income countries. FMD is estimated to circulate in 77% of the global livestock population in Africa, the Middle East, Asia, and a limited area of South America. Countries that are currently free of FMD without vaccination remain under constant threat of incursion (OIE).

Historically, outbreaks of ASFV have been reported in Africa and parts of Europe, South America, and the Caribbean. Since 2007 the disease has been reported in multiple countries across Africa, Asia, and Europe, in domestic and wild pigs. For the first time in 40 years, there are active outbreaks of ASF on the island of Hispaniola, which is of great concern to the USDA and the US pork producers.

Having USDA licensed tests that can be stockpiled in anticipation of an outbreak contingency is a valuable resource for emergency response planning if a foreign animal disease should ever be detected in the US.

Figure 3: Point of care testing with a dried, updated ASFV DNA test kit (version 2.0).

Figure 4: The T-COR 8™, a portable real-time PCR instrument

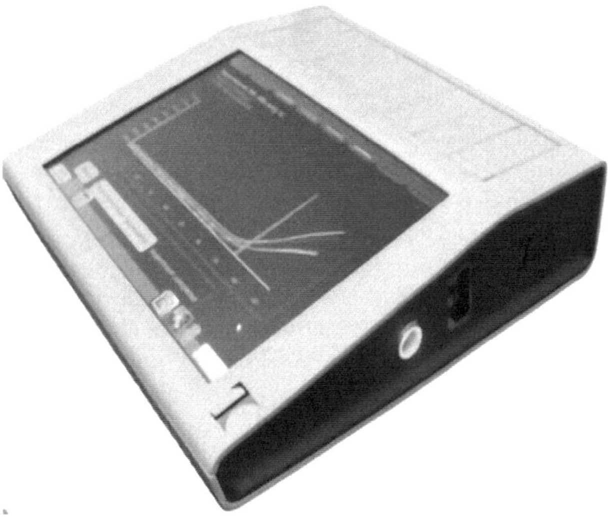

Tetracore FMD/ASF bibliography

Foot-and-mouth disease virus, 608 total citations

1. Callahan, J.D., et al., Use of a portable real-time reverse transcriptase-polymerase chain reaction assay for rapid detection of foot-and-mouth disease virus. J Am Vet Med Assoc, 2002. 220(11): p. 1636-42. 371 Citations

2. OIE Manual of Diagnostic Tests and Vaccines for Terrestrial Animals 2021. Chapter 3.1.8. Foot and mouth disease (infection with foot and mouth disease virus).

3. Boyle, D.B., T. Taylor, and M. Cardoso, Implementation in Australia of molecular diagnostic techniques for the rapid detection of foot and mouth disease virus. Aust Vet J, 2004. 82(7): p. 421-5. 22 Citations

4. King, D.P., et al., Detection of foot-and-mouth disease virus: comparative diagnostic sensitivity of two independent real-time reverse transcription-polymerase chain reaction assays. J Vet Diagn Invest, 2006. 18(1): p. 93-7. 96 Citations

5. Paixao, T.A., et al., Diagnosis of foot-and-mouth disease by real-time reverse transcription-polymerase chain reaction under field conditions in Brazil. BMC Vet Res, 2008. 4: p. 53. 45 Citations

6. Schumann, K.R., et al., Genetic characterization and molecular epidemiology of foot-and-mouth disease viruses isolated from Afghanistan in 2003-2005. Virus Genes, 2008. 36(2): p. 401-13. 46 Citations

7. Howson, E.L.A., et al., Direct detection and characterization of foot-and-mouth disease virus in East Africa using a field-ready real-time PCR platform. Transbound Emerg Dis, 2018. 65(1): p. 221-231. 28 Citations

African swine fever virus, 183 total citations

8. Zsak, L., et al., Preclinical diagnosis of African swine fever in contact-exposed swine by a real-time PCR assay. J Clin Microbiol, 2005. 43(1): p. 112-9. 171 citations

9. Liu L, et al.,. Overcoming the challenges of pen-side molecular diagnosis of African swine fever to support outbreak investigations under field conditions. Transbound Emerg Dis. 2019 Mar;66(2):908-914. 12 citations

https://doi.org/10.54846/am2022/69

Using competency-based educational technologies to develop carcass management competencies and prepare farm labor for potential depopulation catastrophes

S. P. Miller[1,2], DVM; M. Hutchinson[3]
[1]AgCreate Solutions, Inc.; [2]Pork Avenue Training Portal; [3]University of Maine

Introduction

As we raise food in the US, ensuring a competent labor force prepared to manage daily livestock mortalities and potential depopulation catastrophes, such as foreign animal disease (FAD) or that observed with COVID-19 market impacts, maximizes the ability for American producers to prosper. Feeding the world's growing population will require a doubling of the global food production by 2050. Improving farm labor competencies allows American producers to remain competitive in the global marketplace and contributes to global food security.

The use of competency-based educational technologies and employing a "See it, Do it, Teach it"[1-9] method of training is investigated in this study to determine farm labor improvements in mortality management. Further, this study focuses on the use of composting as an effective mortality management practice that can kill pathogens[10] and reduce the spread of disease such as African swine fever (ASF), provide a natural source of nutrient inputs, and contribute to environmental stewardship. This study measured baseline composting competencies to determine our level of risk and educational opportunities regarding proper methodologies.

Much is being learned during this time about what we need to be prepared for in the face of potential depopulations due to FAD, such as ASF for swine or foot-and-mouth disease (FMD) for livestock. On the ground, the industry needs skilled farm workers, machinery operators, veterinarians, and other personnel to be prepared to manage mortalities well. Education of personnel is key. Responsive methods of education thus far include written standard operating procedures, webinars, and direct communications from national and state organizations to farmers and farm staff.

It is a part of AgCreate's vision and mission to improve animal and human wellbeing on farm by creating and distributing impactful learning experiences in a way that inspires action, dependability, and trust. Education of farm labor is a key part of maximizing the ability of American agriculture to prosper.

At the end of the day, management of livestock mortality is a physical competency. While it is one thing to have information in your head, it is another to have it in your hands in a way that you are competent. There is a need for competency-based educational technology to help drive and create physical competencies for laborers on farm.

You do not teach someone how to dance by having them read a Standard Operating Procedure (SOP). Mortality management skills like composting carcasses are physical skills requiring physical competencies. Competency involves a knowledge component, a behavioral component, as well as a value component.[11]

Preliminary research shows that you create about a 25-60% level of on-farm physical competencies when you hand someone a written document of steps and ask managers to teach employees traditionally.[11,12] While national and state organizations have done a fantastic job of providing information, the technical problem is that when you are working to create competencies in farm workers, it is difficult for them to achieve a skill just by reading how to do it.

However, not all employees are able to attend hands-on group training off-farm. Biosecurity, finances, and time are possible limitations. To reach a wider audience, a learning experience that combines e-learning technologies with on-farm facilitated learning is our proposed solution to combine efforts with Professor Hutchinson's team and reach this wider on-farm labor audience. Preliminary research on a style of competency-based online learning we call, "See it, Do it, Teach it" shows movement of learners from a 25-60% competency level to a 90+% competency level[12,13] post-training. By using interactive competency-based educational technology to first see the task and create a "See it" experience, alongside creating testing that simulates the experience and assists in a mentoring, or "Do it" experience, and lastly creating learning material that facilitates a live "Teach it" experience, we have the potential to create improved measurable competencies on these skills in farm laborers.[14]

As we work to raise food in the United States, a competent work force prepared to deal with daily livestock mortalities on farm in addition to mass mortality catastrophes is crucial to our long-term success.

The first part of this study looks at baseline percent competencies of farm workers managing composting piles on swine farms.

Materials and methods

AgCreate Solutions, Inc initiated a project to create and test innovative, competency-based educational technologies to best prepare farm labor to care for daily livestock mortalities and to effectively manage carcasses during catastrophes with the use of composting. This project began with the formation of an advisory panel of mortality composting leaders, including Professor Mark Hutchinson from the University of Maine, that helped finalize baseline competencies of farm staff in charge of carcass disposal. This content was then used to create pre- and post-evaluation measures that followed these evaluation criteria: 1. Sample size calculations of at least 28 staff members; 2. Paired t-test with power = 0.8, significance level = 0.05, and Cohen effect size of 0.6; and 3. Pre- and post-competencies are evaluated by an objective and structured clinical examination that includes observation of tasks alongside workers involved in mortality management.

This enables the evaluator to pretend to be a new employee and the mortality management staff demonstrate their knowledge via teaching back.

The competency content finalized by the advisory panel was also used to create an educational training prototype that included a video script and storyboard. This training utilizes a "See it, Do it, Teach it"[1-9] learning method which incorporates the following practices:

See it

A visual animation of competencies (desired end skills) followed by a simulation of desired end skills where the learners do the task for the first time in a safe environment. Learners are informed prior to the video that they are to learn the topics well enough that they can teach it back to a verifier.

Do it

Learners continue to complete skills which are manager or mentor observed until verified. Pictorial Standard Operating Procedures supplement learning and review and are available in both English and Spanish.

Teach it

Verification of skills involves staff members teaching back what they learned to an Evaluator or Verifier. Verifiers present themselves to each employee as their student and ask them to train the Verifier as if they were a new employee learning to do the task. Verification Checklists are used to ensure all key competencies are addressed and taught back successfully according to the established protocol.

The final step of training provides the means to re-evaluate all learning and measure competency improvement post-training utilizing an additional structured clinical examination. Using the list of carcass management competencies, Dr. Probst Miller and an Instructional Technologist serve as pre-training evaluators that record competencies as yes/no (present/absent). The total score is calculated by adding the number of yes responses. While on site of each farm, Dr. Probst Miller additionally trains a farm mentor and farm verifier for sow and for wean-to-market farms who facilitate the "Do it" and "Teach it" part of the composting learning experience. The mentor and verifier are farm service people experienced in managing people and visiting the designated farm sites.

Baseline assessments were completed on five sow farms and 24 wean-to-market operations. These pre-training competency data were tallied (average percentages for all respondents shown in Table 1), staff were trained with "See it, Do it, Teach it"[1-9] learning experiences, and competencies scheduled to be remeasured.

Results

The average pre-training baseline competency for all respondents related to Composting Preparation was 37.6%. The baseline percentage for competencies related to Building the Pile was 39.6% (Table 1). The combined overall competency of farms was 38.7%. The AgCreate Solutions, Inc. team identified multiple challenges regarding farm staff base-level competencies, varying types of composting operations, and farm manager buy-in to the value of composting. These challenges demonstrate the need for targeted carcass management training and verification of competencies.

In the ongoing trial, staff will be trained with "See it, Do it, Teach it"[1-9] learning experiences, and then competencies remeasured during the "Teach it" phase.

Discussion

Preliminary research in other studies shows that it is common for baseline competencies to exist in the 25-60% range with traditional training. The second phase of this study will measure results post-training. Other research employing "See it, Do it, Teach it"[1-9] learning experiences has increased expressed competencies to the 85-100% competency level post-training. Having an educational method to ensure an 85-100% competency for a task that helps eliminate biosecurity risk when done properly is good risk management. Improving labor competencies allows American producers to remain competitive in the global marketplace and contributes to global food security.

References

1. Bear, Anne, Technology, Learning, and Individual Differences, Journal of Adult Education, 2012.

2. Dale, Edgar, Audio visual methods in teaching. Holt, Rinehart, and Winston, 1965.

3. Dale, Edgar, Cone of Experience, Education Media, 1969.

4. Gaetano, Kristian, 5 ways adults learn differently than children, Learn Kit, 2016.

5. Pork Avenue Training Portal. 2018. Available at **www.porkavenuetraining.com**

6. Probst Miller, Sarah, Impact of two training implementation plans on post-training competency expression of personnel on farm, AASV Proceedings, 2018.

7. Probst Miller, Sarah, Applying the science and physiology of how adults learn to our on-farm approach as practitioners, AASV Proceedings, 2016.

8. Probst Miller, Sarah, The impact of trained and untrained personnel on product implementation and subsequent pig health, AASV Proceedings, 2016.

9. Ryland, Bruce, Fostering Critical Reflection in Adulthood, by Jack Mezirow & Associates, San Francisco: Jossey-Bass, 1990.

10. Pepin, Brent, Williams, Todd, Polson, Dale, Gauger, Phillip, Dee, Scott. Survival of swine pathogens in compost formed from preprocessed carcasses. *Transbound Emerg Dis*. 2021; 68: 223 9– 2249. **https://doi.org/10.1111/tbed.13876**

11. O'Sullivan, N., Burce, A. 2014. Teaching and Learning in Competency-Based Education. The Fifth International Conference on e-Learning (eLearning-2014), Belgrade, Serbia, pp. 71-77.

12. Probst Miller, S., Krantz, S. 2019. Inspiring a verified culture of care post-identification of abuse on an undercover video using a baseline assessment of expressed competencies compared to post training measurement of expressed competencies.

13. Probst Miller, S. 2020. Ensuring that a company's vision and mission and core values are living on farm via pre-training, targeted training, and post-training animal care competency measurements.

14. Landry, T. 2019. Training, motivation, and a culture of safety at Maple Leaf Agri-Farms. 50th Annual Meeting of the American Association of Swine Veterinarians, 2019.

Table 1: Average pre-training baseline competency percentage for all respondents

Competency	Percentage
Composting - preparation	37.6%
Composting - creating the compost pile	39.6%
Combined overall competency	38.7%

https://doi.org/10.54846/am2022/70

Empowering a more efficient and effective workforce with the PigFlow® platform

Matthew Rooda; Ron Ketchem
SwineTech, Inc, Solon, Iowa

Introduction

Significant losses have been attributed to sow and piglet mortalities. The global swine industry is continuing to see a rise in overall unexpected sow mortality at an average of ~15%, according to Chris Rademacher, DVM, extension swine veterinarian, Iowa State University.[1] Additionally, 1.08 piglets per litter are found to be dead due to complications during the birthing process (stillborns). Once born, the first three days of the piglet's life represent the greatest danger of death. Furthermore, 70% of stillborns come from 20% of sows.[2] Approximately 70% of these piglet deaths occur during this time frame and according to the MetaFarms Production Index roughly 13.5% of all pigs born alive will not survive.[3]

Furthermore, labor turnover and access to skilled labor are plaguing the US swine industry. The National Pork Board 2016-2017 "Employee Compensation & HR Practices in Pork Production" report shows the average reported turnover rate for animal caretakers on large and small to midsize producers was 35.03% and 20.34%.[4] In many cases employee turnover will exceed 100%. This is a very costly problem for producers representing 4.05 billion US dollars in annual losses. This high turnover rate is also resulting in an annual loss of on-site knowledge, experience, and skill sets. It is leading to greater variability in processes and pig care.

Our greatest challenge is that as an industry we are unaware and unable to measure what is being completed on a daily basis, who is completing it, and whether it was completed correctly as indicated by company SOPs. A solution is needed to bring more organization, communication, and visibility to daily operations so that farms can efficiently and effectively train new team members, focus on what matters most, and improve workflow efficiencies, process compliance, and daily pig care.

Research and methods

Industry interviews were conducted with 33 of the top 40 producers in the US, and 93.9% of these producers agreed that there is a significant lack of visibility into daily workflows, process compliance, and pig care. In response to these interviews, 167 interviews were conducted with on farm personnel to understand the key stressors and issues that mattered to them. There were 103 English speaking participants and 64 Spanish speaking participants. Each participant was interviewed in their native tongue. The answers were consistent from farm to farm and fell into the following categories:

- Desire a more organized and timely onboarding experience
- Communication throughout the workplace is poor due to language barriers
- It is not clear throughout the day where people are and what needs to happen next
- Many tasks were inefficient in both preparation and execution
- The use of notepads causes information to remain invisible to the greater team
- Traditional handhelds offer no value to employees on the farm
- A strong desire to see key performance indicators tied to their individual performance

These responses indicated an opportunity to improve the quality of life of employees while also driving a more efficient workplace with improved pig care. It was deemed that in order to properly address process compliance in sow farms a solution was needed for these employees that actually makes their quality of life better.

Results and discussion

In response to these findings a group of 11 team members at SwineTech, Inc and a couple dozen industry members which included owners, veterinarians, production supervisors, farm managers, and on farm employees collaborated in the development of the PigFlow Platform. Inspired by workforce management solutions in long-term healthcare facilities, PigFlow is a specialized communication tool that enables an optimized flow of information that delivers actionable intelligence to the person most able to act in a timely enough manner to achieve the best outcome at every level of business. It also provides a record of both the communication and the resulting action in order to evaluate the effectiveness of decisions and processes. This has led to improvements in labor time to productivity, pig survivability, and the implementation and management of precision technologies. PigFlow is currently managing over 45,000 sows in the US and Canada on sow farms from 400 sows to 12,000 sows. Producers have reported a reduction in labor needs of ~20% as well as a reduction in stillborns of over 15%. It is important to note that such a comprehensive solution will see varied results from producer to producer as each producer's goals and objectives will vary. In addition to these benefits, teams are more connected than ever and have reported to be more organized and less stressed throughout the day. It is our expectation that over time we will continue to see sustained improvements in employee onboarding, development, and retention.

References

1. Wilson, Keith. "Sow Mortality Frustrates Experts." *National Hog Farmer*, 4 Feb. 2016, **https://www.nationalhogfarmer.com/mag/farming_sow_mortality_frustrates**.

2. Ketchem, Ron. "70% Of Stillborns Come from 20% of the Sows; Who Is at Risk?" *National Hog Farmer*, 5 June 2019, **https://www.nationalhogfarmer.com/reproduction/70-stillborns-come-20-sows-who-risk**.

3. "MetaFarms Production Index Analysis." MetaFarms, 2020.

4. Carroll, Heidi. "What Is the Real Cost of Employee Turnover?" *National Hog Farmer*, 17 Oct. 2017, **https://www.nationalhogfarmer.com/business/what-real-cost-employee-turnover**.

https://doi.org/10.54846/am2022/71

PCV2 management challenges in a large US production system

B. Sexton, DVM; A. Maschhoff, DVM
The Maschhoffs, LLC, Carlyle Illinois

Introduction

Porcine circovirus type 2 (PCV2) continues to be a globally impactful disease in swine populations. PCV2 infection can cause depletion and inflammation of the lymphatic system, resulting in pigs affected by a syndrome called porcine circovirus associated disease (PCVAD). Pigs with PCVAD show various signs, including decreased weight gain and increased mortality. Even though vaccines have drastically reduced the severity of clinical disease and financial impact, it continues to challenge producers in various ways. Additionally, with new reports of clinical reproductive cases associated with porcine circovirus type 3 (PCV3), practitioners have a new differential in cases where they may not have considered circovirus a factor otherwise.[1,3]

Within large commercial swine production systems, standardization of health protocols and integrated based evaluations are commonly done. Even with commercially available vaccines for PCV2 and newer vaccines for PCV3 to protect growing population and stabilize breeding herds, prioritizing the most cost-effective disease plans takes coordination, collaboration and systematic evaluations. Within the past two years, PCV2 management has been an on-going priority within The Maschhoffs, LLC (TML).

Wean to market case reports

Between 2019 and 2021, a handful of wean to market PCVAD cases materialized in several system flows, primarily as a single case or two, but also as a stretch of multiple cases within a flow. Many of these cases did not present as the traditional postweaning multisystemic wasting syndrome (PMWS), but were worked up by the herd veterinarian due to increased nursery mortality, nagging morbidity in early to mid finishing, or poorer performance compared to other flows. Upon investigation of those cases, if there was an uptick in breed to wean (BTW) farm clinical signs, it went largely unnoticed as general sow health and performance in those flows remained consistent. Early in the assessment, no protocol changes were made on replacement gilts, sows or wean pig vaccinations. Many of those cases, were treated as issues with vaccine compliance or adverse events and a "rescue" dose was applied in the wean to market sites to manage performance of those specific sites. However, the number of cases across multiple flows in a short amount of time, sparked continued investigation and more questions than answers. What if any, protocol changes needed to occur? Was the system using the correct vaccination strategy? How can the system measure PCV2 stability of the sow herd? Is there a standard protocol than can be applied based on sow farm stability to the downstream flow?

Protocol changes

The first action implemented across the system was making a standard protocol change in 2020: PCV2 vaccination of the breeding herds. Historically, TML had not routinely applied a sow herd vaccination plan, as replacement gilts received multiple doses of vaccination prior to first parity even though studies indicate dam vaccination can improve performance.[4,9] How breeding herd vaccination was applied varied across the system to evaluate protocols and flow performance in the long run. The system used a combination of commercially available PCV2 vaccines, RNA particle vaccines, and protein sub-unit vaccines. In a few herds the RNA particle and protein sub-unit vaccines, did include PCV3 along with PCV2. The timing of vaccine administration was also variable across breeding herds. In general, all farms received a onetime whole herd or a "rolling" vaccination (rolling vaccination was applied to sows/gilts either pre-farrowing as animals entered the farrowing house or at wean event). Flows identified to have more consistent PCVAD issues, were most commonly vaccinated using two whole herd shots, approximately 4 weeks apart.

The second action was altering wean pig vaccination protocols. After *Mycoplasma hyopneumoniae* eliminations, many of the system flows has switched to a full dose of commercial PCV2 vaccine at weaning which for over a year worked well with minimal clinical cases or noticed negative downstream flow performance impacts. In lieu of the uptick in wean to market cases, several flows returned to a split dose program for growing pigs. However, even with the change in sow herd and growing pig protocols, a few flows appeared to continue to have a smoldering level of circovirus associated cases which sporadically continued to pop up in wean to market with low Ct counts on real-time polymerase chain reaction (RT-PCR). Over the course of disease management evaluation, no major changes were taken on gilt vaccinations, as current protocol across the system is for gilts to receive at minimum 3 mL (most farms at 4 mL) of a commercial PCV2 vaccine prior to first service. One small alteration on a few farms was made to alter the 20 weeks of age booster into a split dose to get to the 4 mL, pre-breeding.

Measurements

Though previous reports indicate that growing pig vaccination strategies can be made independently of the sow herd status, newer diagnostic tools have become available, such as processing fluids. These new diagnostic tools may be a better indicator of how to connect piglet vaccination strategies to sow farm status, especially when used over time.[6,7] Like many diseases, but specifically PCV2, detection is not equated to clinical disease,[2] however; conversations occurred amongst the TML health team of utilizing a combination of diagnostic tools to define stability within the system. Incoming gilts, processing fluids, and wean piglets were used to help verify herd stability and thus inform potential flow vaccination protocols.

Over the course of a few months, a set of diagnostic samples were collected to assess a subset of farm "stability". Though not clearly defined what is needed to determine "stability", it has been indicated that this can be a critical factor in controlling downstream flow.[8] The diagnostic samples included in the surveillance project were: monthly replacement gilt serum and weekly processing, tested via RT-PCR, along with wean pig IFAs. The results of the monitoring indicated much better results on RT-PCR for replacement gilts than anticipated, with no pool having Ct < 20, but there were several pools across multiple farms that were positive for PCV2. Across all herds, there were variable results with processing fluids, with less than ten sets of processing fluids samples with a Ct < 25, but many herds consistently positive on this sample type. In this evaluation, it did appear that farms that had more negative tests tended to have higher Ct counts when tested positive, and vice versa with farms that had more positive tests tended to have lower Ct counts when tested positive. IFA results were similar across multiple herds with some variation week to week, but over a four-month trend indicated similar wean pig status. Between commercial and multiplication herds, there did appear to be more generalized "stable" results in multiplication, with the exception of one herd. Unrelated to stability testing, in a few herds using non-commercial vaccinations, sow titers were measured post vaccination in our collaborations with pharmaceutical companies to better understand these newer tools.

Conclusion

Across a large system, standardized health protocol approaches may be easier said than done. However, some consistently is useful in evaluating cost effectiveness of health protocol standards. As diseases changes and newer clinical presentations arise, nimble learning and critical evaluation become a useful tool in altering protocols when necessary to elicit best production performance of the growing pig. Over a two-year period, PCV2/3 discussions, protocol alterations and diagnostic surveillance has allowed TML to gain a better understanding of sow herd and flow status across the system. Additionally, there appears to be fewer PCV2 cases downstream than the previous two years. It is noted that coinfections, specifically with parvovirus or porcine reproductive and respiratory syndrome virus (PRRSV) could be a factor in all of the flows evaluated but these co-infections were not directly assessed other than the system health team having an understanding of PRRSV status throughout the evaluations.[5]

To date, the biggest protocol change has been maintenance of a breeding herd vaccination strategy and reverting to split dose growing pig vaccination. These protocols may not be needed for all herds, but the performance impact today on growing pigs allows the system to spend some "insurance" on vaccinations until there is a better understanding of the best vaccine protocol. A big lesson learned throughout system evolution of PCV2 management planning, is that sow herd stability takes time and sometimes variable actions to improve. Additionally, protocol change at one point in a production flow is no guarantee for controlling growing pig disease. Within a large production system there is the challenge of determining the best course of action from gilt to growing pig and thus continued full evaluations over time are useful in assessing the most sustainable management plans.

References

1. Byers, E. 2018. Porcine circovirus type 3: Practitioner perspective. Proc. 49th AASV Annual Meeting, 445-447

2. Madson, D. 2019. Porcine circoviruses: Are changes occurring. Proc. 50th AASV Annual Meeting, 5-8.

3. Distad, T. 2020. Circovirus-associated infertility. Proc. 51st AASV Annual Meeting, 15-18.

4. Kurmann J, Sydler T, Brugnera E, et al.: 2011, Vaccination of dams increases antibody titer and improves growth parameters in finisher pigs subclinically infected with porcine circovirus type 2. Clin Vaccine Immunol 18:1644–1649.

5. Opriessnig, T. 2009. Porcine Circovirus type 2: current understanding of pathogenesis and role of coinfections. AASV Annual Meeting, 491-498.

6. Lowe, J. 2013. The influence of fetal PCV2 infection response on lifetime performance and vaccine efficacy. AASV Annual Meeting, 39-42.

7. Johnson, C. 2021. PCV2d outbreaks and control. Proc. 52nd AASV Annual Meeting, 9-10.

8. Lehman, J. 2020. Field experiences with PCV2 unstable sow herds and downstream pig flows. Proc. 51st AASV Annual Meeting, 109-111.

9. Hobson, B. 2011. Eliminating PCV2 viremia in neonatal pigs through vaccination of the replacement gilts. Proc. AASV Annual Meeting, 311-312.

https://doi.org/10.54846/am2022/72

Porcine sapovirus: Field experiences with a potentially emerging pathogen

B. Thacker[1]; T. Petznick[2]; J. Creel[1]; M. Mogler[1]; P. Gauger[3]; K. Crawford[1]; C. Sebo[1]; J. Maurer[1]
[1]Merck Animal Health, Lenexa, Kansas; [2]Arkcare, Fairbury, Nebraska; [3]Iowa State University, Ames, Iowa

Introduction

Porcine sapovirus (PSaV) was first identified in the United States (US) in 1980.[1] Sapoviruses are RNA viruses within the family *Caliciviridae*. Sapoviruses contain two open reading frames (ORF) that code for the major capsid protein VP1 (ORF1) and a minor structural protein VP2 (ORF2). Genotyping is based on VP1 sequencing. There are 19 sapovirus genogroups of which 8 have been found in domestic swine. In the US, genogroup III is the most common and transmission occurs by oral exposure to contaminated feces.[2] The virus replicates in small intestine enterocytes resulting in lesions and clinical signs of a pasty to semi-liquid diarrhea. Clinically apparent disease appears to be limited to suckling pigs and typically occurs during mid-lactation with a slight increase in mortality and reduced weaning weights.[3] A study reported in 2010 found 44.3% of farms in Europe were infected with PSaV.[4] In the US, a 2019 survey of fecal samples obtained from diagnostic laboratory cases found that the rate of genogroup III PCR positive samples was similar for pigs exhibiting diarrhea compared to asymptomatic pigs (45.3% vs. 43.1%). However, the viral loads were significantly higher in pigs with diarrhea (average cycle time [Ct] = 26.0) compared to asymptomatic pigs (average Ct = 33.2).[5]

In 2019, customers began to express interest in PSaV RP prescription vaccines produced using Merck Animal Health's Sequivity® Technology RNA-Particle (RP) platform. Sequivity Technology utilizes an alphavirus replicon vector system to produce vaccines that contain a pathogen's gene of interest (GOI) that encodes for a protective antigen. For PSaV, the gene of interest is ORF1. The Sequivity Technology, along with supporting technical assistance and data management (Sequivity Dashboard), enable Merck Animal Health (MAH) to quickly address emerging pathogens such as PSaV and others like porcine circovirus type 3 (PCV3).[6] This paper provides an overview of field experiences as Merck Animal Health works with veterinarians to implement RP prescription vaccination programs to control PSaV in their clients' swine herds.

Field observations

Anecdotal field reports from customers reveal the following observations:

1. Diarrhea commonly starts at 7-14 days of age (DOA)
2. Diarrhea persists for approximately 2-7 days after onset in affected litters
3. Weaning weights are reduced by at least 1 pound per pig
4. Diagnostic testing detects PSaV only or in combination with other enteric pathogens
5. Mortality tends to be low, but in some herds higher, so variable overall
6. Mortality rates appear to be impacted by parity with higher rates in gilt litters
7. The disease pattern resembles rotavirus for younger practitioners and coccidiosis for older practitioners.

As for PSaV RP vaccine, the first commercial doses were distributed to customers in Q1 of 2021. To date (November, 2021), just over 20 different prescriptions have been filled and > 50% of customers have re-ordered vaccine. Most first-time orders have been for a PSaV monovalent vaccine. Overall, nearly 50% of vaccine prescriptions that include PSaV RP(s) are in combination with other enteric pathogen RPs, mainly rotavirus.

Sequencing data and RP vaccine formulation

Producing effective vaccines using the Sequivity Technology requires considerable effort to organize and analyze sequencing data. Data organization is facilitated by the Sequivity Dashboard and analyzing the sequence data is performed by MAH's Sequivity technical group. A total of 55 PSaV sequences have been obtained with 48 identified as genogroup III and 7 as other genogroups. Within genogroup III, 3 subgroups have been identified based on amino acid homology. Of the 48 genogroup III sequences, 36 fall into subgroup 1 (SV1), 8 into subgroup 2 (SV2) and 4 into subgroup 3 (SV3). Within the 3 subgroups, the average amino acid homology between sequences ranges from 97.7% to 99.4%. Between the 3 subgroups (SV1 vs. SV2, SV1 vs. SV3, SV2 vs SV3), the average homology between sequences ranges from 84.4% to 85.5%. As for RP vaccine formulation, an RNA particle construct has been made for each of the 3 subgroups. Most prescriptions include only 1 of the 3 constructs although some prescriptions request more than one construct.

Field studies

Merck Animal Health was provided the unique opportunity to participate in 4 studies in an endemic herd. The first 3 studies were surveys to characterize the disease in the herd and the fourth study was a vaccine trial.

Herd background

The herd contained approximately 4,500 sows, located in eastern Nebraska. The herd was brought to our attention by the attending herd veterinarian (co-author T. Petznick). Pigs were weaned twice a week at approximately 21 DOA. Breeding groups were designated based on the week of the year the females were served. The herd was experiencing an ongoing, persistent diarrhea in pigs starting at 7-10 DOA with a peak in severity at around 12 DOA with resolution of most clinical signs of diarrhea by 7-10 days after onset and just prior to weaning. Litters from both sows and gilts exhibited diarrhea. Repeated diagnostic laboratory submissions found PSaV in nearly all cases and in most cases, PSaV was the only pathogen detected. The farm staff attributed at least a 1 pound per pig reduction in weaning weight to PSaV.

Laboratory testing and data analysis

PCR testing for PSaV was performed at the Iowa State University Veterinary Diagnostic Laboratory. PCR results are shown as cycle times (Ct) and positive or negative. For studies 1 and 2, the positive/negative cut-off was a Ct of 37.0. For studies 3 and 4, the cutoff was 35.0. The difference in cut-off values reflects the progression of test development and validation by the laboratory.

Data was obtained by 1) observations of clinical signs, fecal sample collection and weighing of litters by the attending veterinarian, farm staff and MAH veterinarians, 2) diagnostic laboratory reports, and 3) the herd's computerized record system. Statistical analysis was performed by Kruskal-Wallis One-Way AOV (continuous variables including cycle times and percent litter mortality) and Chi Square (binary variables including positive/negative laboratory results and presence of clinical diarrhea) using Statistix™ 10, Analytic Software, Tallahassee, FL. A *P* value of < 0.05 was considered significant.

Study 1: Fecal shedding vs. diarrhea

In the initial review of diagnostic laboratory data by MAH veterinarians, the attending veterinarian provided the raw data from a survey that was performed to understand the relationship between the presence of diarrhea that was consistent with PSaV and fecal shedding of PSaV. PCR results of fecal swabs collected at approximately 12 DOA from 50 litters with diarrhea and 50 litters without diarrhea were provided. For the litters with diarrhea, 100% were PCR positive with an average Ct of 15.9. For the litters without diarrhea, 50% were PCR positive with an average Ct of 34.8. There were only 5 litters without diarrhea that had lower Ct (higher viral load) than the highest Ct (lower viral load) for litters with diarrhea.

Study 2: Fecal swab sampling time

In preparation for future studies, a survey was performed to determine if collecting fecal swabs near the time of weaning (at the same time weight data would be collected) was comparable to results obtained by sampling litters at 12 DOA (peak of disease severity). Fecal samples were collected from 30 litters at 8-12 DOA. Twenty-eight of the 30 litters were resampled 10 days later just prior to weaning. The number (percent) positive and average Ct at 12 DOA were 25/30 (83.3%) and 22.5, respectively. At re-sampling, the number (percent) positive and average Ct were 27/28 (96.4%) and 23.2, respectively. Based on this data, fecal swabs were collected 1-4 days prior to weaning in Studies 3 and 4.

Study 3: Fecal shedding vs. weaning weight

The objective of this survey was to characterize the reduction in weaning weights due to PSaV in this herd. A total of 69 litters were weighed and fecal swabs collected 1-4 days prior to weaning (average 18 DOA; range 16-22 DOA). Three swabs were collected from each litter; 1 each from small, medium and large sized pigs. The individual samples were submitted to the laboratory where the samples were then pooled by litter for PCR testing. Entire litters were weighed and the average pig weight, adjusted for pig age and dam parity, was calculated by litter.

Of the 69 litters evaluated, 57 (82.6%) had PCR positive feces and 12 (17.4%) were negative. In positive litters, the average, median and range of Ct were 20.1, 17.7 and 12.6-34.6, respectively. The average adjusted pig weight was 11.82 pounds for the positive litters and 13.29 pounds for the negative litters; a difference of 1.47 pounds ($P = 0.012$). Using lower Ct cut-offs (30 or 25) than the laboratory's standard cut-off (35) to classify positive/negative litter status resulted in similar results. These findings support the clinical assessment of farm staff that the ongoing piglet diarrhea due to PSaV in their herd was reducing weaning weights by at least 1 pound per pig. Based on this survey, weaning weight reduction due to PSaV in this herd could be estimated as the prevalence of positive litters times the pig weight difference or 82.6% times 1.47 pounds, which equals 1.21 pounds.

Study 4: Vaccine field trial

A vaccine field trial was performed to determine the impact of vaccination of dams prior to farrowing on PSaV fecal shedding and pre-weaning mortality. Originally, we had planned to evaluate individual litter weaning weights and use a study design where the dams would be randomly allocated to treatment within breeding groups. Unfortunately, personnel entrance restrictions to the farm due to Covid precluded randomization of dams to treatment and weighing litters at weaning. The RP prescription vaccine contained the SV1 construct which had 100% amino acid sequence identity to sequences previously obtained from the herd.

Study design- A total of 13 breeding groups were vaccinated (n = 7) or were not vaccinated (controls, n = 6). By order, groups 1, 2, 5, 6, 9, 10 and 11 were vaccinated and groups 3, 4, 7, 8, 12 and 13 were not vaccinated. Vaccine was administered twice by IM injection 4-5 and 1-2 weeks prior to farrowing by the farm manager.

To evaluate fecal shedding, swabs were obtained from approximately 12 litters (8 sow and 4 gilt) per breeding group 1 day prior to weaning. Three swabs were collected from each litter; 1 each from small, medium and large sized pigs. The individual samples were submitted to the laboratory where the samples were pooled by litter for PCR testing. A total of 151 litters were sampled. Thirteen litters were excluded from analysis due to errors in data recording. The number of litters with usable data by dam parity is shown in Table 1.

To evaluate pre-weaning mortality, farrowing data from the 13 breeding groups was retrieved from the farm's computerized records in Excel format. Additional parameters were calculated from the original data in order to facilitate determining which litters should be excluded from analysis. Exclusion of litters was based on the farm manager's recommendations to avoid using data from litters that were not representative of dams that mainly nursed their own pigs, were given the farm's standard range of pigs to nurse and had the farm's standard range of lactation days. Specific exclusion criteria included sows/litters 1) that weaned no pigs ("kick out" sows), 2) with no recorded wean event, 3) where the numbers of pigs that nursed, died and weaned or liveborn, transfers and nursed did not justify, 4) with total number of transfers (in or out) exceeding 9 pigs, 5) with liveborn < 6, 6) the number of pigs nursed was < 11 or > 16, 7) weaned on an "off day" (not a Monday or Thursday), 8) weaned on a day where both vaccinated and control litters were weaned, 9) weaned at < 17 or > 25 days post farrowing, and 10) gestation length < 113 days. The number of litters with usable data by dam parity is shown in Table 2. The percent pre-weaning mortality in each litter was calculated by dividing the number of pigs that died by the number of pigs the dam was given to nurse. Nearly all cross-fostering of pigs was completed by 3 days of age.

Results- Fecal shedding results are shown in Table 1. The percentage of positive litters was significantly lower in litters from vaccinated gilts, sows or both parity groups combined. As for viral load, vaccinated litters had lower viral loads (higher Ct) compared to control litters overall (gilt and sow litters combined) and with sow litters only while the difference in gilt litters was not significant ($P = 0.109$).

Pre-weaning mortality (PWM) results are shown in Table 2. In gilt litters, the average percent litter PWM was significantly lower in vaccinated litters compared to control litters with a difference of 4.07%. In sow litters, the average percent litter PWM was numerically, but not significantly, lower in vaccinated litters compared to control litters; difference was 0.80%. For both gilt and sow litters, the average percent litter PWM was significantly lower in vaccinated litters compared to control litters with a difference of 2.13%.

Conclusions- The PSaV RP vaccine reduced the rate of fecal shedding regardless of parity and reduced the level of fecal shedding in sow litters. The vaccine significantly reduced pre-weaning mortality in gilt litters but not in sow litters. The results of this study should be viewed as preliminary due in part to the COVID-related constraints on study design no (random assignment of dams to treatment) and the inability to collect litter weaning weights, which is likely the most important outcome variable to measure based on anecdotal experiences in this herd and others along with the results of Study 3.

Summary

With many emerging diseases, there is often a long period between the initial detection of the pathogen and the emergence of clinically apparent and important clinical disease. With PSaV, the virus was first detected in the US in 1980 while recognition as a potentially significant pathogen has occurred in the past few years. With some diseases, the pathogen becomes a greater clinical problem resulting in the continual need for prevention and control strategies including vaccination. In other cases, significant clinical disease occurs for period of time but then subsides for unknown reasons. In the case of PSaV, the present situation suggests that the virus should be considered a potentially significant cause of diarrhea in baby pigs resulting mainly in reduced weaning weights and possibly increased pre-weaning mortality. However, more research is needed to determine the prevalence of PSaV among herds within geographical regions and the impact of the disease in PSaV positive herds. The methods used in Study 3, swabbing and weighing litters, may provide a useful approach for determining the impact of PSaV in an individual herd as well as monitoring a herd's status over time. Time will tell if PSaV becomes a persistent problem that warrants routine vaccination and other control methods.

Acknowledgments

The authors acknowledge the generous support and assistance of the farm's staff and owners with special thanks to the farm manager.

References

1. Nagai, M. Wang, Q, Oka, T., Saif, L. 2020. Porcine sapovirus: Pathogenesis, epidemiology, genetic diversity, and diagnosis. Virus Res. **doi: 10.1016/j.virusres.2020.198025**

2. Wang, L. Marthaler, D., Fredrickson, R., et al. 2020. Genetically divergent porcine sapovirus identified in pigs, United States. *Transbound Emerg Dis.* 67:18-28.

3. Anderson, A., Burton, B., Sebo, C. Determination or porcine sapovirus prevalence and clinical correlation in suckling pigs. In: Proc. 52nd Annual Meeting of the American Association of Swine Veterinarians, San Francisco, California, 2021, p. 245.

Table 1: Study 4 - Porcine sapovirus fecal shedding results in litters from vaccinated and control dams

		Pos/Neg results			Ct results		
Group	**Parity**	**No. pos.**	**Pct. Ppos.**	***P*-value**	**Average**	***P*-value**	**Avg of pos.**
Vaccinated	Gilt	10/30	33.3	0.030	28.9	0.109	18.1
Control	Gilt	14/22	63.6		24.3		18.2
Vaccinated	Sow	10/48	20.8	0.009	32.5	0.016	23.8
Control	Sow	18/38	47.4		29.0		22.1
Vaccinated	Both	20/78	25.6	0.009	31.1	0.005	20.9
Control	Both	32/60	53.3		27.3		20.4

Table 2: Study 4 - Pre-weaning mortality results in litters from vaccinated and control dams

Group	Parity	No. litters	Pct. mortality*	Difference	*P*-value
Vaccinated	Gilt	258	13.23	4.07%	0.001
Control	Gilt	172	17.3		
Vaccinated	Sow	379	17.24	0.80%	0.448
Control	Sow	274	18.04		
Vaccinated	Both	637	15.62	2.13%	0.009
Control	Both	446	17.75		

* Average by litter

4. Reuter, G. Zimsek-Mijovski, J., Poljsak-Prijateli, M., et al. 2010. Incidence, Diversity, and Molecular Epidemiology of Sapoviruses in Swine across Europe. *J Clin Micro.* 48(2): 363-368.

5. Shen, H. Zhang, J., Gauger, P., et al. 2021. Genetic characterization of porcine sapovirus identified from pigs during a diarrhea outbreak in Iowa, 2019. *Transbound Emerg Dis.* **doi: 10.1111/tbed.14087**

6. Sebo, C., Knox, A., Allison, G., et al. A case study of reproductive failure associated with porcine circovirus type 3 in an endemically infected commercial sow farm. In: Proc. 52nd Annual Meeting of the American Association of Swine Veterinarians, San Francisco, California, 2021, pp. 131-134.

https://doi.org/10.54846/am2022/73

Experiences with two commercial ileitis vaccines during the grow-finish period for same-sourced pigs

D. Boykin, DVM
Cactus Family Farms, Osceola, Iowa

Introduction

Lesions of porcine proliferative enteropathy (PPE) were first described in field cases during the 1930s.[1] *Lawsonia intracellularis* was identified as the etiologic agent responsible for PPE during the 1990s.[2] Swine veterinarians have continued to better understand and manage the impacts of *Lawsonia intracellularis* or "ileitis." Ileitis has been associated with clinical and subclinical disease documented by anti- and post-mortem diagnostics with related influence on growth parameters as well as grow-finish economics.[3] Increasingly the swine industry has transitioned from the use of antimicrobials to utilizing vaccines for ileitis control. Two commercial vaccines are currently available for ileitis control in the United States. A field experience with Enterisol®Ileitis and Porcilis®Ileitis will be described.

Materials and methods

Pigs sourced from two multiparous 3,700 head sow farms were given either Enterisol®Ileitis at approximately 10 weeks of age per normal production system protocol, or Porcilis®Ileitis per label at weaning. Cross-sectional diagnostic samples for each treatment were taken during the grow-finish period from approximately 8 to 24 weeks of age, involving several finishing sites throughout the production system providing a snapshot overview. Each diagnostic sampling included blood samples (10 per airspace with a minimum of 20 per site) and oral fluids (1 per airspace with a minimum of 2 per site). Samples were submitted to Iowa State University Veterinary Diagnostic Laboratory with *Lawsonia* ELISA testing on serum samples and *Lawsonia* PCR testing on oral fluid samples.

Results and discussion

Figures 1 and 2 show lab results from Enterisol®Ileitis and Porcilis®Ileitis vaccinated pigs respectively. *Lawsonia* PCR cycle time negative cutoff value of 35 or greater was utilized. As shown in Figure 1 where pigs were given Enterisol®Ileitis at approximately 10 weeks of age, field exposure to *Lawsonia* typically occurred around 16 weeks of age with subsequent seroconversion due to natural exposure during the late finishing period. In Figure 2 where pigs were given Porcilis®Ileitis at weaning *Lawsonia* seroconversion occurred after Porcilis®Ileitis administration with oral fluid samples remaining *Lawsonia* PCR negative throughout the grow-finish period.

References

1. Biester & Schwarte: Am J Pathol 1931 Mar, 7(2):175-185.6.
2. Gebhart et al. Int J Syst Bacteriol. 1993 Jul; 43(3): 533-538.
3. Winkelman. 2018 AASV proceedings; 487-493.

Figure 1: Enterisol®Ileitis given at 10 WOA

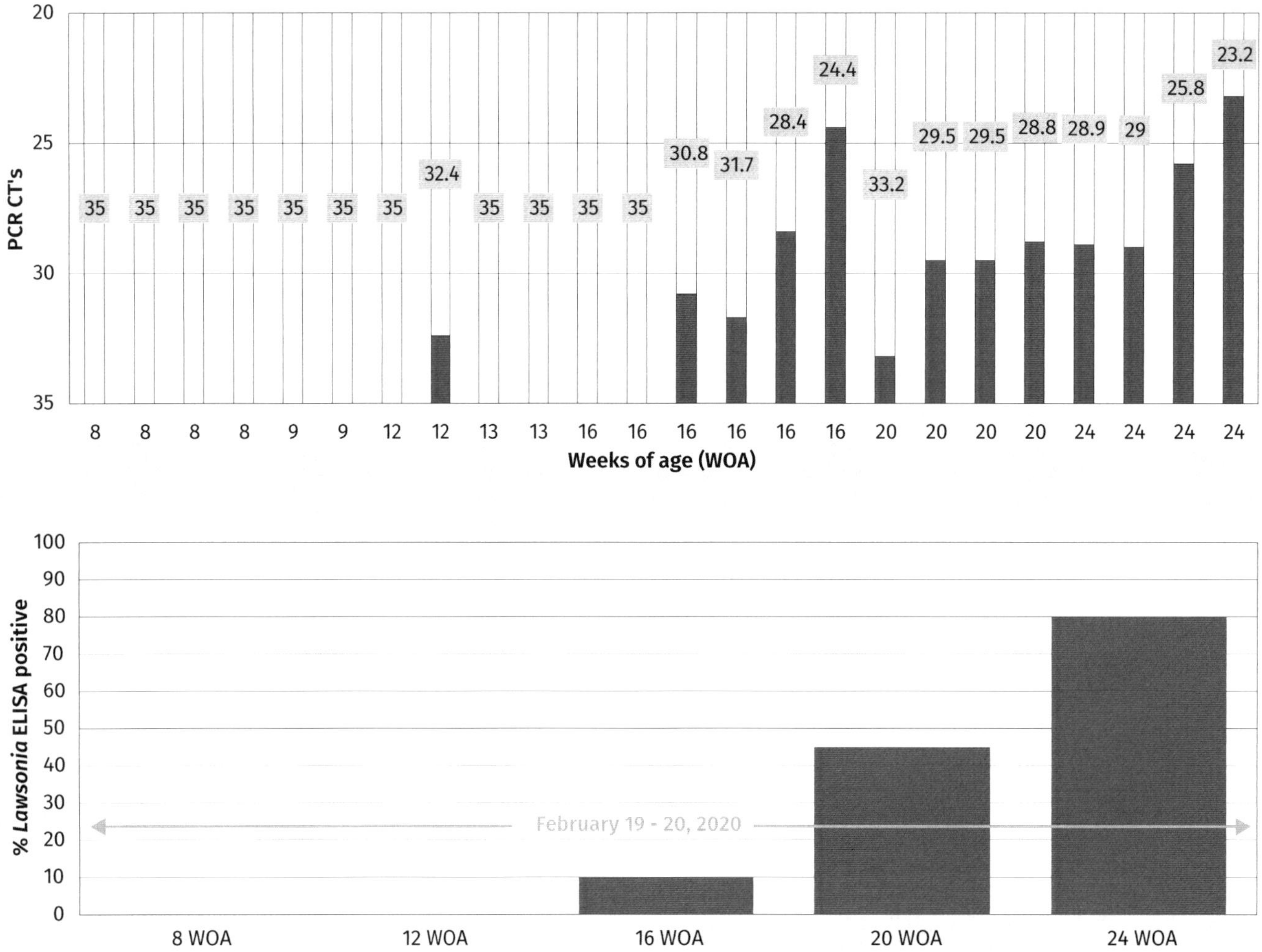

Figure 2: Porcilis®Ileitis given at weaning

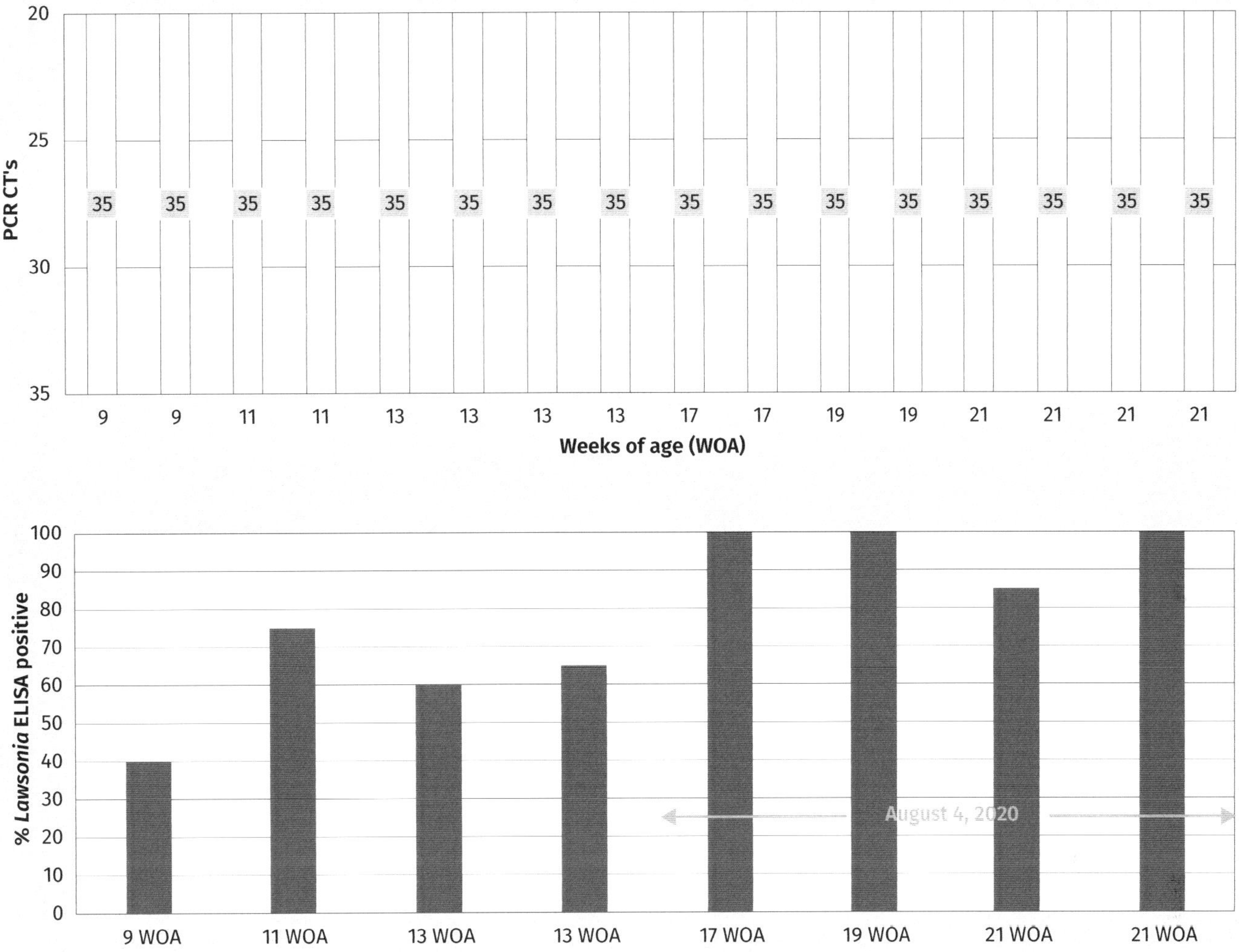

https://doi.org/10.54846/am2022/74

NextGen sequencing analysis on flies shows potential to introduce swine influenza virus in a finisher

Gene Spellman[1], BS; Grant Allison[2], DVM
[1]Central Life Sciences, Schaumburg Illinois; [2]Walcott Veterinary Clinic, Walcott, Iowa

Introduction

The common house fly, *Musca domestica*, has been shown to have the ability to transmit viral diseases of swine.[1-3] Influenza A virus of swine is a highly transmissible viral infection spread pig-to-pig via aerosols. What is less known is the risk of a viral disease such as influenza A virus starting when pigs enter an empty barn and encounter an existing fly population. Or conversely, an existing population of swine negative for influenza A virus are exposed by the introduction of influenza A positive flies. It has been shown that the legs and wings of the house fly, *Musca domestica,* display the largest microbial diversity and as such may serve as a route for the spread of infectious disease(s).[4] The potential for pigs to encounter influenza A virus (IAV) after entering an empty site will be determined using NextGen sequence analysis.

Materials and methods

The study site was a 2,400 head finisher consisting of 2 rooms in southeast Iowa that contained an abundance of houseflies. There were no clinical signs consistent with influenza A virus during the investigation. Flies were collected using seven Captivator® liquid traps (Starbar®, Central Life Sciences) that were hung throughout the room. In addition, two negative control traps were hung to allow for airborne particulate matter to enter the liquid but not flies by covering the jar with a screen (Figure 1). Flies were separated from the liquid attractant and each were submitted to the Iowa State University Veterinary Diagnostic laboratory (ISU-VDL) for Next Gen sequence (NGS) analysis. The east side was empty eight days after the west side was depopulated. The site was completely empty for two days before repopulation began, starting with the west side. Samples taken before the site was empty was from the east side. Post repopulation samples were collected from the west side of the finisher in the same fashion as before. In total, there were twenty-three days between when the samples were taken before and after the finisher was depopulated and repopulated.

Figure 1: Negative control trap.

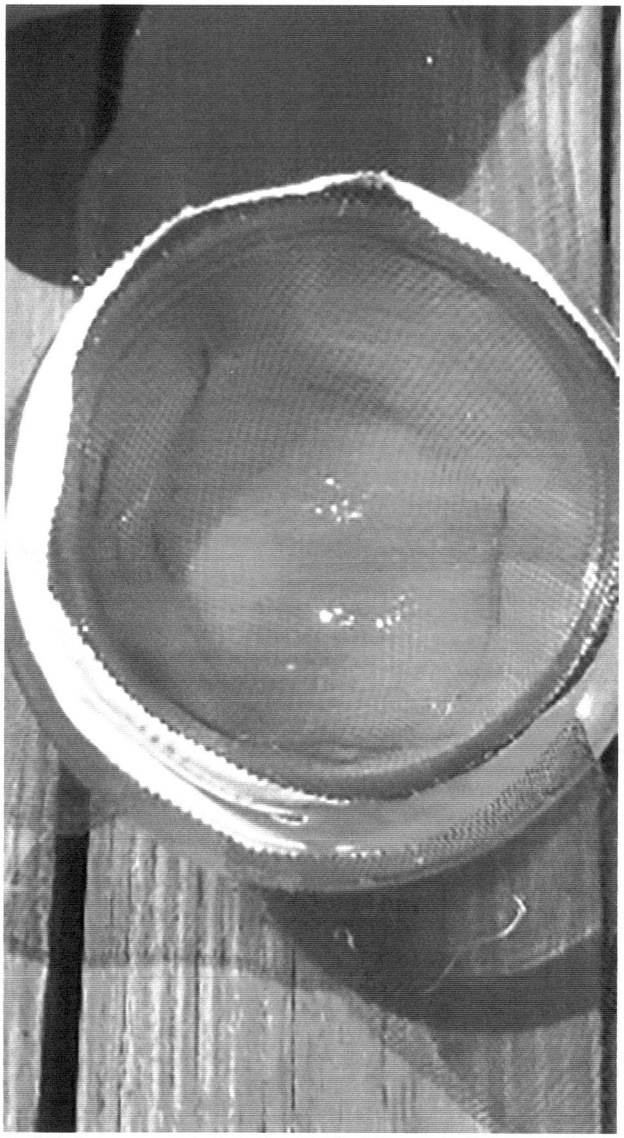

Results

With the exception of the positive fly homogenate sample taken before the finisher was empty, all sample types contained at least 1 virus. Next Gen sequence (NGS) analysis revealed the presence of four insect viruses; Wuhan mosquito virus, Wuhan fly virus 5, *Musca domestica* salivary gland hypertrophy virus and *Haemotobia irritans* Nora virus. In addition, one human virus (human papillomavirus 5) as well as two swine viruses were found; porcine bocavirus and swine influenza virus (Table 1).

Discussion

After the finisher was repopulated two insect viruses were found that had not been previously been found; *Musca domestica* salivary gland hypertrophy virus and *Haemotobia irritans* Nora virus. Whether these viruses were truly absent in the population before all swine were removed is unknown. However, it is possible that these viruses represent the introduction of a new insect population after the finisher was restocked with swine. In particular, finding *Haemotobia irritans* Nora virus in both the positive fly homogenate and positive fly liquid attractant is of concern. Unlike non-blood sucking insects such as the house fly (*Musca domestica*)

Table 1: Viruses identified by Next Gen sequence analysis before and after repopulation of a swine finisher

Condition	Samples	Viruses
East side (Before)	Negative fly attractant	Wuhan mosquito virus
		Wuhan fly virus 5
	Positive fly attractant liquid	Human papillomavirus 5
		Wuhan fly virus 5
	Positive fly homogenate	No significant viral pathogen detected
West side (After)	Negative fly attractant	Wuhan fly virus 5
	Positive fly attractant liquid	Wuhan fly virus 5
		Human papillomavirus 5
		Swine influenza virus
		Porcine bocavirus
		Haemotobia irritans Nora virus
	Positive fly homogenate	*Musca domestica* salivary gland hypertrophy virus
		Haemotobia irritans Nora virus

the Horn Fly (*Haemotobia irritans*) is an obligate blood feeder that causes considerable economic losses in livestock industries worldwide.[4] Whereas the role of *Hemotobia irritans* in the transmission of African swine fever (ASFV) has not been described, transmission of African swine fever by the stable fly (*Stomoxys calcitrans*) and horse fly (Tabanidae) have been reported.[5,6] The role of flies as a possible vector in the transmission of African swine fever cannot be overestimated as even non-blood sucking flies have been implication as potential vectors.[7] Human papillomavirus 5 is commonly found on normal skin and not listed as one of the fourteen known high-risk strains of human papillomavirus.[8] Finding porcine bocavirus was unexpected given it is a recently discovered virus that infects pigs and is classified within the Bocavirus genus (family Parvoviridae, subfamily Parvovirinae). At present, little is known about the transmission of porcine bocavirus.[9] That said, having been found only in the positive fly liquid attractant makes the case that porcine bocavirus was on the surface of the body of the fly. The house fly has the potential to fly from farm-to-farm over a distance of four miles during an expected lifespan of one to two weeks.[10] In one study it was found that influenza A virus can be transmitted for a distance of one-to one and one quarter miles.[11] The study site is within four miles from four sites housing swine and with one site being within one mile of the study site. While it is unknown whether any of these sites were positive for influenza A virus at the time of the study, the potential exists that flies were the vector for bringing swine influenza to the site recently repopulated with swine. The house fly is commonly described as a "filth fly" as they commonly eat or lay eggs on manure, garbage and carrion. The connection between house flies as a potential vector for enteric disease is easily made. In contrast, transmission of influenza A virus is via respiratory droplets expelled by coughing or sneezing. That said, house flies are commonly seen in the feeder and so come in contact with nasal secretions as the pig eats. Also, given house flies walk on surfaces common to the pigs' environment such as cup waterers it is likely that they will contact said respiratory droplets. Influenza A virus was found in the positive fly liquid attractant and as such was on the surface of the fly's body. Movement common to the fly such as walking in the feeder or near sources of water could potentially transmit Influenza A virus without ingestion of flies given the virus is on the leg or wing of the fly. Bio-exclusion is a newer concept in the discussion on what biosecurity really means. In the case of influenza A virus, the control of flies could be part of this concept; bio-exclusion or the prevention of influenza A positive flies entering a finisher housing swine negative for swine influenza virus. In the past controlling flies has been haphazard at best. Current technology offers in-feed control of the fly's lifecycle that will lower the risk of introduction of viral disease(s) such as swine influenza virus. The study shows proof of concept that flies are a potential vector for the introduction of viral disease(s) such as influenza A virus into a swine finisher.

References

1. Otake S, Dee SA et al. Transmission of porcine reproductive and respiratory syndrome virus by houseflies, (Musca domestica Linnaeus). *Vet Rec* 2003. 152: 73-76.

2. Allison G, et al. PEDV, PRRSV, and Seneca Valley virus PCR positive flies (Musca domestica). What now? *Proceedings 49th AASV 2018*. 394-396.

3. Joshi LR, Mohr KA, et al. Detection of the emerging picornavirus Senecavirus A in pigs, mice, and houseflies. *J. Clin Microbiol.* 2016. 54(6):1536-1545.

4. Junqueira, A.C.M., Ratan, A., Acerbi, E. *et al.* The microbiomes of blowflies and houseflies as bacterial transmission reservoirs. *Sci Rep* **7,** 16324 (2017). **https://doi.org/10.1038/s41598-017-16353-x**

5. Bonnet S. et. al. Putative Role of Arthropod Vectors in African Swine Fever Virus Transmission in Relation to Their Bio-Ecological Properties. Viruses. 2020 Jul; 12(7):778.

6. Olesen AS, Lohse L, Hansen MF, et al. Infection of pigs with African swine fever virus via ingestion of stable flies (Stomoxys calcitrans). Transbound Emerg Dis. 2018;65:1152–1157. **https://doi.org/10.1111/ tbed.12918**

7. Liu, J., Lu, G., Cui, Y. *et al.* An insight into the transmission role of insect vectors based on the examination of gene characteristics of African swine fever virus originated from non-blood sucking flies in pig farm environments. *BMC Vet Res* 16**,** 227 (2020). **https://doi.org/10.1186/s12917-020-02420-5**

8. Gargano J et. al. VPD Surveillance Manual.Chapter 5: Human Papillomavirus. Centers for Disease Control and Prevention.

9. Swine Health Information Center (SHIC). Center for Food Security and Public Health, College of Veterinary Medicine, Iowa State University. Porcine bocavirus. July, 2016.

10. Nazni WA, Luke H et. al. Determination of the flight range and dispersal of the house fly, Musca domestica (L.) using mark release recapture technique. *Trop Biomed* 2005 Jun;22(1):53-61.

11. Corzo C. et. al. Airborne Detection and Quantification of Swine Influenza A Virus in Air Samples Collected Inside, Outside and Downwind from Swine Barns. PLoS One. 2013; 8(8):e71444.

https://doi.org/10.54846/am2022/75

Adenosine triphosphate (ATP) bioluminescence assay for biosecurity verification

Anne Calitz[1]; David Pyburn[1], DVM; Shareef Ahmad[2]; Christoph Kunas[1]; Isabel Lechner[3]
[1]BioChek BV, Reeuwijk, The Netherlands; [2]Hygiena, Camarillo, California; [3]SAFOSO AG, Liebefeld, Switzerland

Introduction

Adenosine triphosphate (ATP) is the basic energy component molecule of all plant and animal cells and is present in all microorganisms and organic residues. Due to this, its presence on a surface can be used as an indicator of organic contamination. When ATP reacts with luciferin in the presence of luciferase it produces a release of energy in the form of light, this is called bioluminescence (also mechanism for firefly light in abdomen). The presence of ATP can be detected as a light output being emitted. Therefore, the measurement of light intensity provides a visual indicator of ATP present and of the total organic contamination from the surface tested.[16]

ATP bioluminescence is used extensively in the food industry as a tool for monitoring and maintaining cleanliness.

ATP bioluminescence is a rapid and simple method to assess surface cleanliness and microbiological contamination in food facilities, dairy industry, hospitals, animal transports and so forth.[1] ATP bioluminescence detects, among others, microbial cells, food residues, bacteria, fungi and yeast, which might persist after inadequate cleaning.[2] It is measured in Relative Light Unit (RLU) values, which is directly proportional to the amount of ATP present. Several studies have demonstrated strong correlations between RLU values and bacterial microbe counts or surfaces which were judged to be dirty.[1,3-5]

Personal hygiene, such as changing clothes and footwear as well as handwashing, prior to entering a farm, are important biosecurity measurements to prevent the entering and spreading of pathogens.[6] Several studies, especially from the pork industry, show the effectiveness of personal hygiene. Pathogens such as *Escherichia coli*[7], foot-and-mouth disease virus,[8] classical swine fever virus[9], influenza A virus[10] and porcine reproductive and respiratory syndrome virus[11] have been shown to be transmitted in directly through people from infected pigs to other pigs when no or little personal hygiene measures were implemented.

Therefore, the use of an ATP bioluminescence assay to control personal hygiene on farms would be a useful and fast method for biosecurity verification. A short literature review and pilot studies on the application of ATP bioluminescence assays on farms and the suitability to control effective hand washing was implemented. Results regarding RLU values, interpretation, and thresholds as well as possible limitations of the method are presented and discussed below.

Materials and methods

Hygiena and BioChek have undertaken field studies using the Hygiena EnSURE luminometer, MicroSnap, UltraSnap, and AquaSnap ATP bioluminescence products in the South African pork industry. The purpose of these projects was to evaluate whether the Hygiena ATP technology can be adopted at the farm level as a tool to test and validate on-farm hygiene protocols. These projects officially started on August 1, 2021 on one of the primary genetic suppliers for the South African industry. Other farms then followed this initial farm with protocols and hygiene guidelines adjusted as experience with the products and the labor force was gained. This study was undertaken to both show the potential of biosecurity verification and its effect on worker adherence to the protocols, plus to prove that the products can be effectively utilized in a farm environment. The studies focused on farm hygiene and biosecurity verification in the following areas:

- Employees and visitors pre-shower and post-shower
- Employee lunch containers brought into facility
- Water sample testing both in pens and other taps on farm
- Outside containers of semen deliveries
- Outside containers of pest control equipment and supplies delivered to the farm
- Outside containers of veterinary products delivered to the farm
- Waterers, feeders, fencing, and pen slats post cleaning and disinfection.

For each of the items or areas to be sampled detailed instructions with diagrams were presented to the employees and lead employees were trained to perform the sampling on a consistent basis.

If needed, samples were then enriched and incubated on the farm of origin using the Hygiena Snap products and the Hygiena Digital Dry block incubator. Read out of all tests was then preformed using the Hygiena EnSURE luminometer to measure light generated by ATP reacting with liquid stable luciferase/luciferin reagent and results are then reported in Relative Light Units (RLU). RLU results provide information on the level of contamination within seconds using this protocol. The higher the RLU number the more ATP present and the dirtier the surface tested.

Results

At the time of the writing of this paper early incomplete data was only available from one of the field study farms and more data will be presented in the oral presentation. Early results for the first farm can be found in Tables 1 thru 4. This preliminary data shows how the ATP bioluminescence method of biosecurity verification can be utilized to show the organic material reduction effect of cleaning and disinfection and how this pre- and post-cleaning plus disinfection data can cause labor behavioral changes for the long-term.

Discussion

Several studies have shown that the ATP bioluminescence method is useful and efficient to evaluate the cleanliness of a variety of surfaces in farm environments. For example, to monitor the hygiene standards on dairy farms by taking samples of milking equipment.[3,4,12,13] However, it is not possible to distinguish whether high RLU values originate from bacteria or other organic material, such as milk residues. Therefore, at high RLU levels, samples to measure colony forming units are still required.

Furthermore, ATP bioluminescence testing has been applied to determine the cleanliness of surfaces like floors, feeders, enrichment material, nipple drinkers and walls from farms and animal transporters.[1,5,14] These studies also show that ATP testing is a highly accurate and cost-efficient alternative to microbiological methods for rapid on-farm assessments. In addition to visual inspection, it can be used to determine the cleanliness status of farms and animal transporters and is useful to identify critical areas for subsequent washing and disinfection.

ATP bioluminescence assays are also a simple and reliable method to test handwashing compliance.[15] They are used in several areas to monitor handwashing, as for example in hospitals or food-processing companies. It needs to be considered that skin is a living organ and therefore has natural levels of ATP, even when it is clean.[16] It is thus impossible to achieve an ATP level of zero. Furthermore, due to variations of skin characteristics from person to person, it is not possible to define universal baseline ATP levels for clean hands.

Thresholds for the differentiation between "clean" and "dirty" vary depending on the surface and the environment where the sample is being taken. A collection of studies indicating threshold values and methods to define the thresholds are presented in Table 5.

One limitation of the ATP bioluminescence test is that high ATP levels are not necessarily linked to a biosecurity risk. Materials such as milk residues, organic debris, epithelial cells or body secretions can also raise ATP values.[3,4,14,17] Furthermore, the ATP bioluminescence method is not able to differ living from dead or inactivated microbes.[1,18] In addition, residues of cleaning and disinfecting agents can influence the result of an ATP test and lead to decreased ATP levels.[14,17,19] One study showed that the ability of an ATP test to detect gram-negative bacteria is lower compared to gram-positives, due to insufficient cell lysis.[17] Many pathogens that pose a risk of introduction due to inadequate personal hygiene are viruses.[8,9,10,11] ATP assays do not directly measure viruses. However, some studies suggest that ATP measurements correlate significantly with reduced viral recovery. ATP bioluminescence measurements can, therefore, be used to monitor the effectiveness of interventions taken to reduce viral contamination.[20] More work needs to be done to establish this correlation pork industry test environments.

Conclusions

Several studies have demonstrated that ATP bioluminescence is a fast and easy to use method to monitor microbial contamination and the effect of hygiene interventions for a variety of surfaces (eg, milking equipment, animal transporters, stable-surfaces, and hands). No general thresholds for the differentiation between clean and dirty exist, the differentiation needs to be defined based on the surface tested, the environment and the desired degree of cleanliness. There is no standard method for the definition of thresholds. Methods include reference values based on objects/surfaces with a low bacterial count, calculation of RLU-thresholds based on existing thresholds of microbiological assessments, or the relative drop of ATP levels after a hygiene intervention. Limitations of the method mainly concern the specificity of the assay, since high ATP-levels can also be caused by non-pathogenic or dead microbes, as well as interactions with certain disinfectant materials.

People and vehicles are major risk factors regarding the introduction of pathogens into a farm. This risk can be reduced by certain biosecurity measures, such as proper hand washing and showering or cleaning and disinfecting vehicles before entering a farm. ATP bioluminescence is a suitable tool to monitor these measurements in real time. Additionally, hand washing improves over time when ATP tests are implemented, due to the increased awareness of the on-farm personnel.[15]

Table 1: Example preliminary results reported in Relative Light Unit (RLU) values for Farm 1

Date	Pre-shower	Post-shower
02-Aug	543	914
04-Aug	746	5430
05-Aug	825	66
06-Aug	1026	8
10-Aug	252	97
11-Aug	186	78
12-Aug	908	75
19-Aug	413	33
26-Aug	2695	414

Employee 8: Raw data (RLU values from skin swabs)

Table 2: Example preliminary results reported in RLU values for Farm 1

Date	Pre-shower	Post-shower
02-Aug	161	286
03-Aug	983	1189
04-Aug	103	106
05-Aug	231	198
06-Aug	345	1378
19-Aug	5	972
26-Aug	311	10

Employee 7: Raw data (RLU values from skin swabs)

Table 3: Example preliminary results reported in RLU values for Farm 1

Date	Pre-shower	Post-shower
04-Aug	591	38
23-Aug	499	38
23-Aug	627	23

Visitor: Raw data (RLU values from skin swabs)

Table 4: Example preliminary results reported in RLU values for Farm 1

Item	Post spray disinfectant	Post spray disinfectant + disinfectant wipe down
Lunchbox 2	2933	105
Lunchbox 4	1376	239
Semen delivery exterior container	574	11
Interior semen cooler	338	84
Veterinary delivery 1	88	23
Veterinary delivery 2	417	58
Pest control 2	63	16
Pest control 3	253	11

Raw data from incoming containers (RLU values from exterior swabs)

Table 5: Methods to define adenosine triphosphate relative light units (RLU) clean/dirty thresholds

Location	Tested surfaces	Threshold	Method to define thresholds	References
Dairy farms	Milking equipment	Clean: < 150 RLU Dirty: > 300 RLU Acceptable but not good: 150 – 299 RLU	Based on recommendations by the manufacturer.	Lindell et al, 2018[3] Irie et al, 2021[4] Costa et al, 2006[21]
Dairy farms	Milking equipment	Teat cup rubbers = 152 RLU Teat dip containers = 242 RLU Milk receivers = 282 RLU Pipeline joints – 1821 RLU	Reference values for each surface were established based on mean RLU values observed in class A farms. Class A farms are farms with a very low bacterial count ($<10 \times 10^3$) in bulk tank milk.	Vilar et al, 2008[13]
Swine livestock trailer	Floors, walls, ramps, partitions, and trailer exterior surfaces	Pass (Clean): 430 RLU/100 cm^2 Critical: 431 – 850 RLU/100 cm^2 Fail: (Dirty): > 850 RLU/100 cm^2	Defined by a microbial assessment through MacConkey agar contact plates. The thresholds for the definition of cleanliness through the microbiological assessment was taken from the guidelines of the Canadian swine health board. Based on those thresholds the researchers calculated the equivalent threshold values for RLU/ 100 cm^2	Cabahug et al, 2016[1]
Food processor company	Hands	Clean: < 600 RLU Dirty: > 600 RLU Or Drop of ATP level of 80% after hand washing	Not described in the paper	Bennion et al, 2013[22]
Dentistry students	Hands	Very clean skin: < 100 RLU Clean skin: 101 – 250 RLU Somewhat clean skin: 251 – 500 RLU Dirty skin: 501 – 750 RLU Very dirty skin: 751 – 1000 RLU	Not described in the paper	Bustamante et al, 2018[2]
Hospital	Hands	Clean: < 60 RLU Dirty > 60 RLU	Not described in the paper	Hygiena[16], n.d.
Hospital	Hands	Clean: < 1000 RLU/25 cm^2 Dirty: > 1000 RLU/25 cm^2	Based on clinical trials	Marena et al, 2002[15]

References

1. Cabahug et al.: 2016, 2016 American Society of Agricultural and Biological Engineers Annual International Meeting p. 1-7.
2. Bustamante et al.: 2018, Transylvanian Review - 26:8909-8914.
3. Lindell et al.: 2018, Journal of Dairy Science - 101:2438-2447.
4. Irie et al.: 2021, The Journal of Veterinary Medical Science - 83.
5. Yi et al.: 2020, Journal of Animal Science and Technology - 62:682-691.
6. Alarcón et al.: 2021, Porcine Health Management - 7.
7. Amass et al.: 2003, JSHAP - 11:61-68.
8. Amass et al.: 2004, Veterinary Microbiology - 103:143-149.
9. Ribbens et al.: 2007, Veterinary Record - 160:687-690.
10. Allerson et al.: 2013, PloS One - 8:2-10.
11. Otake et al.: 2002, JSHAP - 10:59-65.
12. Reinemann et al.: 2000, ASAE Annual International Meeting - ASAE Paper No. 003009.
13. Vilar et al.: 2008, International Journal of Food Microbiology - 125:357-361.
14. Heinemann et al.: 2019, Individual Training for Farmers based on Results for Protein and ATP Rapid Tests and Microbiological Conventional Cultural Methods Improves Hygiene in Pig Fattening. Oxford University Press - 2:1-27.
15. Marena et al.: 2002, American Journal of Infection Control - 30:334-340.
16. Hygiena Internal Communication, no date.
17. Turner et al.: 2010, Journal of the American Association for Laboratory Science - 49:190-195.
18. Larson et al.: 2003, Food Microbiology 20:735-739.
19. Carrascosa et al.: 2012, Food Control 28:368-373.
20. Sifuentes et al.: 2017, Food and Environmental Virology - 9:14-19.
21. Costa et al.: 2006, Brazilian Journal of Microbiology - 37:345-349.
22. Bennion et al.: 2013, International Food Hygiene - 17:23-24.

https://doi.org/10.54846/am2022/76

PRRS: An outsider's viewpoint

Jeff Odle
LANXESS Corporation, Actives and Disinfectants, Pittsburgh, Pennsylvania

Introduction

For more than 40 years in North America, porcine reproductive and respiratory syndrome (PRRS) has affected the swine industry. PRRS has been the subject of numerous research studies and yet it still infects an estimated 23-41% of the USA breeding herd each year (Swine Health Monitoring Program, Dec 2015 report). Both the AASV and the NPPC have position statements around PRRS elimination. There is even a National PRRS Elimination Program framework, developed in 2016, that outlines the strategy to assist producers and the industry in achieving PRRS eradication in the US swine herd. Vaccines, air filtration, thermo-assisted drying of transportation trailers, feed management, and enhanced biosecurity protocols have helped reduce the number of outbreaks however, PRRS continues to be the number one concern to producers, even with African swine fever virus (ASFV) knocking at the door.

Animal movement is a major risk factor for disease spread from region to region, farm to farm, and even within a farm complex. During the USDA's ASFV Action Week presentations Dr. Jack Shere, Associate Administrator at USDA said, "Biosecurity is the single most important thing any producer can do on the front end to protect their herd on the back end." When asked if live haul trucks will be cleaned and disinfected before leaving processing plants he responded "one of the gaps that we have identified, and it is a serious gap, is truck washes we just don't have enough and they just don't exist." This paper will look at biosecurity practices within the swine industry, when it comes to cleaning and disinfecting as well as potential gaps that can be closed to help mitigate future outbreaks of PRRS.

Is power washing sufficient?

Swine producers in Canada understand the importance of using formulated detergents for cleaning surfaces before disinfection, whereas the use of cleaners in the USA is not a widely accepted practice. High-pressure power washing is still the primary means of soil removal in barns and transportation. How clean does a hog barn or a trailer need to be? In order to answer this question, one must first understand viral log reduction. Disinfectants in the USA are regulated and approved by the EPA. Antimicrobials are tested under a standard laboratory protocol and must demonstrate a minimum 3.75 log reduction in the pathogen of concern in order to be approved for use.

If the initial bacterial load is 1,000,000 cfu (colony forming units), then a 4 log reduction would bring the level down to 100 cfu. Ideally, one would want to get the number of cfu to less than 10,000 with cleaning which would ensure the disinfectant can kill all remaining bacteria. The same would apply for viruses like PRRS. Factors that can further affect disinfectant efficacy such as organic challenge, pH, hard water, and temperature must be taken into consideration and may require a more robust chemistry, a higher concentration of disinfectant or longer contact time. Biofilm also reacts with the disinfectant thus reducing concentration of available antimicrobial agent. It is not chemically possible to open biofilm by power washing with water alone. Cleaners penetrate soils and biofilm and loosen them from the surface where they can be easily rinsed away. This lowers the number of pathogens present and makes the job easier for the disinfectant. It is easier to kill 100 "bad germs" than it is to kill 1,000 "bad germs". Greater emphasis needs to be applied to the cleaning process, which will further assist in providing higher levels of disinfectant efficacy.

I'm pretty sure it works!

PRRS is a large enveloped virus, which in the hierarchy of pathogens should be relatively easy to kill. It is pH stable in the range 5.5 to 7.0 and can survive at cold temperature (39.4°F/4°C) for 140 hours. PRRS and PEDV are often referred to as "seasonal" diseases as the outbreaks tend to occur from October through early spring. Disinfectants that are neutral pH and are negatively impacted by cold temperature may have a harder time killing PRRS and require higher concentrations or longer contact times.

A research study[1] presented at the 2021 Allen D. Leman Swine Conference showed two disinfectant products, one a quaternary ammonium compound and the other a combination of quaternary ammonium and glutaraldehyde, were both effective at neutralizing the PRRS 144 variant with a 60-minute contact time. Both products are registered with EPA and have approval for PRRS on their labels with a 10-minute contact time. Which makes one question why the virus was still alive after 30 minutes as reported in the study? Is PRRS 144 more difficult to kill and thus requires a longer contact time or a higher concentration for these products to be effective? Is a 60-minute contact time sufficient? Will the surface being disinfected stay wet and active for 60 minutes at cold temperature? What if the application needs a fast kill such as in boot baths? Nobody is going to stand in a boot bath for 10 minutes let alone one hour. The standard test protocol used for EPA approval is conducted at room temperature with a 5% organic challenge and 400 ppm hard water, unless specified otherwise. Would a similar result be observed if these products were tested on PRRS 144 at cold temperature or with an organic presence greater than 5%? Is there a correlation between disinfectant chemistry used and PRRS outbreaks and temperature?

It is important to understand the limitations of any antimicrobial and the environmental conditions in which it is used. It may be effective under strict laboratory protocols used for regulatory approvals but what if the conditions on the farm stress those parameters. It may require the producer to switch disinfectant chemistries, which are not impacted by temperature, such as oxidative chemistry like Virkon™ S. Extensive testing with Virkon™ S has been conducted by independent laboratories proving PRRS efficacy at a 1 minute contact time, at cold temperature and low concentration.

Conclusion

Identifying the gaps in biosecurity and setting the goal of PRRS elimination is key to solving PRRS outbreaks in the USA. Making it happen takes a concerted effort by all stakeholders. Elon Musk once said "If you give yourself 30 days to clean your house, it will take you 30 days. If you give yourself 3 hours, it will take 3 hours. The same applies to your goals, ambitions, and potential." PRRS will not go away overnight. It will take more focus on cleaning and preparing surfaces for disinfection and the use of disinfectants proven to kill PRRS under all the extreme conditions encountered on the farm.

Reference

1. Dee, S. et al 2021. PRRSV 144 L1C: A Rapid Response To A National Crisis, Pipestone Research and Boehringer Ingelheim Animal Health USA Inc.

https://doi.org/10.54846/am2022/77

Trace mineral programs for performance optimization when pharmacological zinc is no longer an option

B. V. Lawrence[1], PhD; A. Hintz[1], DVM; D. Bussieres[2], Agr
[1]NOVUS International Inc, St. Charles, Missouri; [2]Groupe Ceres, Levic, Quebec Canada

Introduction

The benefit of pharmacological levels of dietary zinc (Zn), specifically zinc oxide (ZnO) during the post-weaning period on weaned pig performance and diarrhea risk have been known for more than 30 years.[1,2] Several subsequent trials were conducted in the US in the early 1990s and a review published in 1995.[3] Across nine US trials with 3,000 ppm of Zn as ZnO, a variable response was observed with a -28% to +25% impact on gain and a -8% to +11% impact on feed conversion. In 2001, a multi-experiment station study of pharmacological Zn effects in weaned pigs was published.[4] The findings indicated that for pigs weaned at 15 days, the growth rate of pigs fed 2,000 ppm Zn as ZnO growth rate was 29.5% greater and feed conversion was 7.4% lower. For pigs weaned at 21 days, a 10% greater growth rate and 4.8% lower feed conversion was observed with ZnO inclusion. A pharmacological Zn effect was observed even in pigs fed diets containing 55 ppm Carbadox although the response was lower than for pigs fed non-medicated diets. The potential modes of action of not only pharmacological Zn,[5] but other trace minerals[6] for improving digestion and gut health have been proposed. These include impacts on digestion, the immune system and inflammation, enzyme secretions, gut microflora, intestinal morphology and antioxidant capacity. However, pharmacological Zn inclusion is not without risks.[5,7] These concerns result in the elimination of pharmacological Zn as a nutritional tool in post-weaning diets in many parts of the world. Given the many modes of action of pharmacological Zn, successful ZnO replacement strategies may require a holistic approach including replacing inorganic trace minerals with highly available sources.

Materials and methods

A total of 1,152 weaned pigs were used in the study. Pigs were sorted to pens by gender and size. Average pig weight was 6.4 kg. Pens contained either all barrows or all gilts. Within gender, pens were sorted by weight into light, medium or heavy weight blocks which were subsequently used for treatment allotment. A total of 48 pens were used in the trial with 24 pigs per pen. The resulting stocking density was 3.0 ft^2/pig. Each pen had a 5-hole stainless steel feeder. Diets were fed as a crumble (2.0 kg/pig), a short pellet (5 kg/pig) and a regular pellet (remainder of the trial). All treatment groups received a diet containing chlortetracycline (400 g/ton) for the first 15.4 pounds of feed. The remainder of the feed budget for all treatment groups contained 60 ppm Salinomycin. Across phases, diets were mainly comprised of corn, wheat and soybean meal. Wheat was included in all phases at 25% of the diet. There were no distillers grains (DDGs) in phase 1 with DDGs inclusion increasing from 5% in phase 2 to 15% in phase 4. Oat hulls were included at 5% in phase 1 through 3. Lactose was provided by whey permeate with 10% lactose in phase 1 and 3% in phase 2. Dietary SID lysine varied from 1.35% in phase 1 to 1.19% in phase 4. There were three treatment groups (Table 1) and 16 pens per treatment. The treatment groups comprised different post-weaning trace mineral programs differing in source and level. In phases 1 and 2 of the ZnO Blend treatment, diets contained 200 ppm Zn as a protected Zn and 78 ppm Zn as an amino acid complex. The remaining Zn in the ZnO Blend was supplied as ZnO. In phases 3 and 4 ZnO was the sole source of Zn in the Zn Blend treatment. In the Protected Zn treatment, a protected Zn at 200 ppm was the sole source of Zn in all phases. In phases 1 and 2 of the ZnO Blend and Protected Zn treatments, copper (Cu) was supplied in equal parts by copper sulfate ($CuSO_4$) and methionine hydroxy analogue chelate of Cu (MINTREX® Cu bis-chelated trace mineral) with Cu supplied solely by copper sulfate in phases 3 and 4 for ZnO Blend and Protected Zn treatments. Manganese (Mn) was supplied solely as manganous oxide in the ZnO Blend and Protected Zn treatments. For the treatment with MINTREX®, the methionine hydroxy analogue chelate of Zn, Cu and Mn was the sole source of the respective trace minerals.

Pens were weighed weekly and feed disappearance determined. A six-point scour index (scores 0 to 5) was used. A scour index score was assigned to each pen weekly. The scour index included the following: 0 – no scour, 1 – pasty consistency with feces only in corners, 2 – pasty consistency with pens starting to get dirty, 3 – feces starting to get runny and pens dirty, 4 – watery feces with dirty pens and some pigs starting to waste and 5 – very watery feces and very dirty pens with pigs many pigs wasting. Removal weights and days were recorded.

Results and discussion

Impact of trace mineral program through the first 21 days post-weaning

Feeding MINTREX® bis-chelated trace minerals as the sole source of Zn, Cu and Mn resulted in pigs that were 350 grams per pig heavier ($P < 0.05$) at 21 days post-weaning than pigs fed the ZnO Blend diets (Table 2). The greater growth rate for pigs supplemented MINTREX® was the result of 4.2 to 4.9% higher feed intake ($P < 0.01$) compared to the Protected Zn and ZnO Blend fed pigs respectively. The higher feed intake was associated with a 4.2% and 4.5% higher growth rate ($P < 0.05$) for the pigs supplemented with MINTREX® compared with the Protected Zn and ZnO Blend pigs. There were no differences across treatments for feed conversion ($P > 0.10$).

Impact of trace mineral program on overall nursery performance

Overall, there were no differences in day 0 to 49 performance. The pigs supplemented with MINTREX® were 660 grams per pig heavier than the ZnO Blend pigs at day 49. Intake during the 49-day nursery period was approximately 2% greater for the pigs supplemented with MINTREX® compared to the other trace mineral programs.

Table 1: Trace mineral treatment program added trace mineral levels

	ZnO blend	Protected Zn	MINTREX®
Phase 1	2500 ppm Zn 125 ppm Cu 45 ppm Mn	200 ppm Zn 125 ppm Cu 45 ppm Mn	150 ppm Zn 125 ppm Cu 20 ppm Mn
Phase 2	2000 ppm Zn 125 ppm Cu 45 ppm Mn	200 ppm Zn 125 ppm Cu 45 ppm Mn	150 ppm Zn 125 ppm Cu 20 ppm Mn
Phase 3/4	200 ppm Zn 125 ppm Cu 45 ppm Mn	200 ppm Zn 125 ppm Cu 45 ppm Mn	50 ppm Zn 50 ppm Cu 20 ppm Mn

Table 2: Trace mineral treatment program impacts on nursery performance and scour index score

	ZnO blend	Protected Zn	MINTREX®	SEM	*P* =
Day 0 Weight, kg	6.44	6.44	6.43	0.05	0.98
Day 21 Weight, kg	13.89	13.93	14.24	0.09	0.02
Day 0 – 21					
Gain, g/d	353	354	369	4.18	0.02
Intake, g/d	381	384	400	4.51	0.01
Feed/Gain	1.08	1.08	1.09	0.01	0.93
Day 0 - 49					
Gain, g/d	528	535	541	4.31	0.12
Intake, g/d	752	753	767	6.08	0.15
Feed/Gain	1.42	1.41	1.42	0.006	0.22
Day 49 Weight, kg	31.44	31.83	32.10	0.22	0.12

Impact of trace mineral program on scour index score and percent of pens scouring

From a clinical standpoint, one of the primary reasons for pharmacological Zn inclusion in the diet is reducing the incidence and severity of post-weaning diarrhea and associated mortality and wasting. In the present study, mortality and removals ranged from 2.33% for the treatment with MINTREX® bis-chelated trace minerals to 2.62% for the Protected Zn treatment and were not different across all treatments ($P = 0.95$). Despite absence of differences in total removals, the ZnO Blend treatment had significantly lower scour scores over the day 0 to 21 period ($P < 0.001$), day 21 to 49 ($P < 0.01$) and over the entire trial period ($P < 0.001$). These differences are reflected in the weekly data scour scores (Table 3). The treatments with Protected Zn and MINTREX® were similar with respect to impact on scour score. Although data was not evaluated for impact of week of trial on scour score, Table 3 indicates that scour scores of the ZnO Blend were higher in weeks 4 through 7 where the pharmacological levels of Zn had been removed. For percent of pens scouring, similar results are observed as for the scour score. Inclusions of pharmacological levels of Zn in the ZnO Blend treatment resulted in a lower percentage of pens with scours (Table 4). For weeks 1 through 3, the percent of pens with scours averaged 5% ($P < 0.001$) for the ZnO Blend treatment compared to 50% to 56% for the treatments with Protected Zn and MINTREX® respectively which were not different from each other. Similar to the scour index scour, the percent of pens with scours was higher for the ZnO Blend treatment in weeks 4 through 7 averaging 25%. The percent of pens scouring in weeks 4 through 7 on the treatment with MINTREX® averaged 48% which was higher ($P < 0.01$) than for pigs fed the ZnO Blend diets. The Protected Zn treatment was intermediate to the ZnO Blend and treatment with MINTREX®.

Relationship of scour score, scour incidence and performance

In the present trial, comparison of the scour scores to the scour index score indicates that the severity of scouring was relatively mild. Although the severity was mild across treatments, feeding pharmacological Zn resulted in a detectable difference in severity compared to the other treatments. Similarly, feeding pharmacological Zn resulted in a lower percentage of pens indicated as having scouring. These scour-based measures of pharmacological Zn benefit were not achieved with either of the other trace mineral programs. However, reduced scour score and pen incidence did not result in differences in percent removals. Although the treatment with MINTREX® bis-chelated trace minerals generally had the highest scour index score and pen scour prevalence days 14 through 42, intake and growth rate were the greatest and total removals were not different from the ZnO Blend treatment.

The greatest benefit of the treatment with MINTREX® compared to both the ZnO Blend and Protected Zn treatments was during the first 21 days post-weaning when the difference in both scour score and pen scour incidence was greatest compared to the ZnO Blend treatment. Differences included feeding MINTREX® Zn at 150 ppm in the treatment with MINTREX® compared to a total of 2,500 ppm from multiple Zn source in the ZnO Blend, supplementing 125 ppm Cu solely as MINTREX® Cu in the treatment with MINTREX® compared to 125 ppm Cu as a 50/50 blend of $CuSO_4$ and MINTREX® Cu in the ZnO Blend treatment and 20 ppm of Mn as MINTREX® Mn in the treatment with MINTREX® compared 45 ppm Mn as MnO in the ZnO Blend treatment. The lower levels of Zn supplied in the treatment with MINTREX® would not be anticipated to fulfill the proposed antimicrobial role of Zn.[5] However, many other proposed roles of pharmacological Zn inclusion[5] would be dependent on Zn availability. Even with poor Zn availability from ZnO, pharmacological Zn would make an important contribution to the pig's daily available Zn supply. Minerals supplied as MINTREX® bis-chelated trace minerals have been shown to improve trace mineral retention as a percent of intake compared with minerals in the sulfate form.[8] Sulfates are considered to have greater availability than minerals supplied in the oxide form[9] therefore it would be anticipated that Zn fed as MINTREX® in this trial would have greater availability than Zn fed in the ZnO form. Similar to Zn, Cu is important for roles beyond its antimicrobial activity.[6] Feeding Cu through MINTREX® has been shown to result in greater post-weaning performance than if similar levels of Cu are fed in the sulfate form.[10] Manganous oxide is also considered to have lower availability than manganous sulfate.[9] Feeding Mn through MINTREX® has been shown to result in greater retention as a percent of intake than Mn fed in the sulfate form.[8] Oversupply of inorganic trace minerals can result in antagonisms, reducing availability of other trace and macro minerals.[8,11] Therefore the greater intake and growth rate observed for the treatment with MINTREX® in the present study during the first 3 weeks post-weaning, compared to the ZnO Blend treatment, may be an indication of better meeting the weaned pigs physiological needs for available trace minerals. The greater intake and growth rate with a similar total removal rate for the treatment with MINTREX® compared to the ZnO Blend treatment indicate that pharmacological Zn may be successfully removed under conditions of mild post-weaning scours when a strategy to supply highly available trace minerals is used.

References

1. Holm et al., Escherichia coli induced weaning diarrhea in pigs. Dietary supplementation of zinc oxide as an antibacterial method?. Dan. Veterinaertidsskr. 1988; 71:1118–1126.

2. Poulsen et al., Zinc oxide for weaned pigs. In: Proc. 40th Annual. Mtg. Eur. Assoc. Anim. Prod., Dublin, Ireland, 1989; pp 8–10.

3. Reece, D., Pharmacological levels of zinc in nursery diets – a review. 1995; Nebraska Swine Reports. 173

Table 3: Trace mineral treatment program impacts on scour index score

Scour index score	ZnO blend	Protected Zn	MINTREX®	SEM	P =
Day 7	0.05	0.30	0.15	0.04	< 0.001
Day 14	0.02	0.66	0.75	0.07	< 0.001
Day 21	0.06	0.83	1.16	0.08	< 0.001
Day 28	0.31	0.53	0.63	0.09	0.04
Day 35	0.23	0.39	0.47	0.08	0.10
Day 42	0.28	0.60	0.66	0.10	0.02
Day 49	0.16	0.20	0.14	0.05	0.71

Table 4: Trace mineral treatment program impacts on percent of pens with scouring

Scour index score	ZnO blend	Protected Zn	MINTREX®	SEM	P =
Day 7	5%	24%	19%	0.03	< 0.001
Day 14	2%	62%	63%	0.06	< 0.001
Day 21	6%	65%	85%	0.05	< 0.001
Day 28	30%	53%	60%	0.08	0.04
Day 35	23%	37%	42%	0.07	0.16
Day 42	27%	54%	66%	0.09	< 0.01
Day 49	15%	19%	14%	0.05	0.75

4. Hill et al., Effect of pharmacological concentrations of zinc oxide with or without the inclusion of an antibacterial agent on nursery pig performance. 2001; J. Anim. Sci 79:934-941.

5. Bonetti et al., Towards zero zinc oxide: Feeding strategies to manage post-weaning diarrhea in piglets. 2021. Animals. 11:642-666.

6. Broom et al., Recent advances in understanding the influence of zinc, copper, and manganese on the gastrointestinal environment of pigs and poultry. 2021. Animals. 11:1276-1288.

7. Poulsen, H., Zinc and copper as feed additives, growth factors or unwanted environmental factors. 1998 J. Anim. Feed Sci. Tech. 7:135-142.

8. Liu et al., Digestibility and retention of zinc, copper, manganese, iron, calcium, and phosphorus in pigs fed diets containing inorganic or organic minerals. 2014. J. Anim. Sci. 92:3407-3215.

9. National Research Council. Nutrient Requirements of Swine. 2012. 11th Ed.

10. Ma et al., Multitrial analysis of the effects of copper level and source on performance in nursery pigs. 2015. J. Anim. Sci. 93:606-614.

11. Zhao et al., Trace mineral nutrition in poultry and swine. 2010. Asia-Austral. J. Anim. Sci. 23:1527-1534.

https://doi.org/10.54846/am2022/78

Development of updated porcine reproductive and respiratory syndrome virus products

M. Ismail, MS; B. Qiu, PhD; R. Shah, BS; M. Swimely, MS; A Burrell, MS
Thermo Fisher Scientific, Austin, Texas

Introduction

Porcine reproductive and respiratory syndrome (PRRSV) is a viral disease that leads to abortions and weak born piglets, increased mortality in suckling and weaned piglets, and respiratory disease in weaners and finishers. PRRSV belongs to the *Arteriviridae* family within the order *Nidovirales* and contains a single-stranded, positive sense, RNA genome of approximately 15 kilobases. PRRSV is split into two genotypes, PRRSV-1 (Type 1), and PRRSV-2 (Type 2). The 2 different types only share approximately 50% sequence identity and are highly variable within each type. The current VetMAX™ NA and EU PRRSV Reagents (on-market product) contain primers/probes that enable multiplex, one-step, reverse transcription and real-time PCR (RT-PCR) targeting North American and European PRSSV RNA and internal positive RNA control Xeno™. Internal positive RNA control (IPC) is spiked into lysis solution during nucleic acid extraction to help monitoring for PCR inhibitors. Due to the high mutation rate and the emergence on novel PRRSV sequences, Thermo Fisher Scientific™ evaluated the current product bioinformatically against nearly 1600 NA PRRSV whole genome sequences publicly available from the NCBI database as well as sequences from private databases. Phylogenetic tree of PRRSV was assembled (Figure 1) using maximum likelihood method and local support values at splits were calculated using Shimodaira-Hasegawa test. 319 samples from various locations in the US were tested *in-vitro* against on-market products. Evaluation concluded that an update to the current on-market solution is required to maintain reaction efficiency when encountering novel variants of PRRSV. The aim of this project is to develop VetMAX PRRSV 3.0 Reagents and Controls (New products) which is an update to the current PRRSV reagents molecular design to achieve higher reaction efficiency for recent circulating strains of PRRSV.

Materials and methods

PCR settings (Table 1) used ROX as a passive reference dye, FAM reporter for PRRSV NA, VIC reporter for PRRSV EU, and CY5 reporter for Xeno™ (IPC). All fluorescent probes were modified with 3` non-fluorescent quencher. The newly developed products utilized FAST mode and fast thermal profile (Table 2). Testing was performed on QuantStudio™ 5 Real-Time PCR System from Thermo Fisher Scientific™. Threshold setting for analysis used control-based threshold (CBT) method of 5% for all targets. Samples were considered positive if they generated a Cq < 37. Analysis and standard curve were performed using the design and analysis (DAS) software version 2.4.3.

Limit of detection (LOD)

Limit of detection was determined by running 50 copies/reaction and 10 copies/reaction of PRRSV NA and PRRSV EU controls. Each target and concentration were tested in replicated of 88. The newly developed reagents were prepared by adding 10μl of master mix, 2μl of primer probe mix, and 8μl of sample/control to each testing well. PCR settings (Table 1) and thermal profile (Table 2) were used for this experiment.

RT-PCR efficiency

RT-PCR efficiency was tested by running a standard curve for each of the assay targets (NA PRRSV, EU PRRSV, and Xeno™ RNA control). The standard curve setup started with 8000 copies/reaction and went down to 10 copies/reaction. Each target and concentration were tested in replicates of 8. The newly developed reagents were prepared by adding 10μl of master mix, 2μl of primer probe mix, and 8μl of sample/control to each testing well.

Performance comparison between on-market product and newly developed product

1600 samples were tested *in-silico* against the new molecular design to confirm enhanced compatibility of the new design. In addition, 649 genetically diverse positive samples with variable samples types including oral fluid, processing fluid, serum, and homogenized tissue were tested *in-vitro* with on-market products and the newly developed products to compare performance. 72 negative samples were tested *in-vitro* against the newly developed products to confirm exclusivity. Moreover, 427 positive samples were tested externally with the assistance of collaborator labs. This external set of samples included 39 samples confirmed by sequencing to be L1C 1-4-4 and L1C 1-7-4 novel variants. Samples tested internally *in-vitro* were extracted using the Thermo Fisher Scientific™MagMAX™ CORE Nucleic Acid Purification Kit (A32700). The on-market product reagents and controls were prepared according to the manufacturers recommend procedure (4469533 Rev. B) and (4469535 Rev. B) respectively. The newly developed reagents were prepared by adding 10μl of master mix, 2μl of primer probe mix, and 8μl of sample or controls to each testing well. The newly developed controls don't require preparation as the control tube contains all targets, NA PRRSV, EU PRRSV, Xeno™ RNA pre-mixed. PCR settings and thermal profiles differ between the on-market and newly developed products (Table 3).

Results and discussion

qRT-PCR efficiency testing using the new product showed excellent results with efficiencies of 105.8%, 101%, and 99% for EU, NA, and Xeno™ (IPC) respectively. The efficiency for each target was calculated using the equation $E = -1+10^{(-1/slope)}$. The limit of detection was calculated based on 95% confidence level ($\alpha = 0.05$). NA and EU PRRSV controls at 50 copies/reaction and EU controls at 10 copies/reaction showed 100% detection using the newly developed products. NA PRRSV controls at 10 copies/reaction showed 98.8% detection. The overall results suggest that the LOD of the newly developed product is 10 copies/reaction for both NA and EU PRRSV targets.

Testing results demonstrated the ability of the newly developed products to generate concordant calls with the on-market products while showing enhanced performance. Specific panel testing

Figure 1: Phylogenetic tree of PRRSV based on ORF5 sequences. The phylogenetic tree used maximum likelihood method and local support values at splits were calculated using Shimodaira-Hasegawa test.

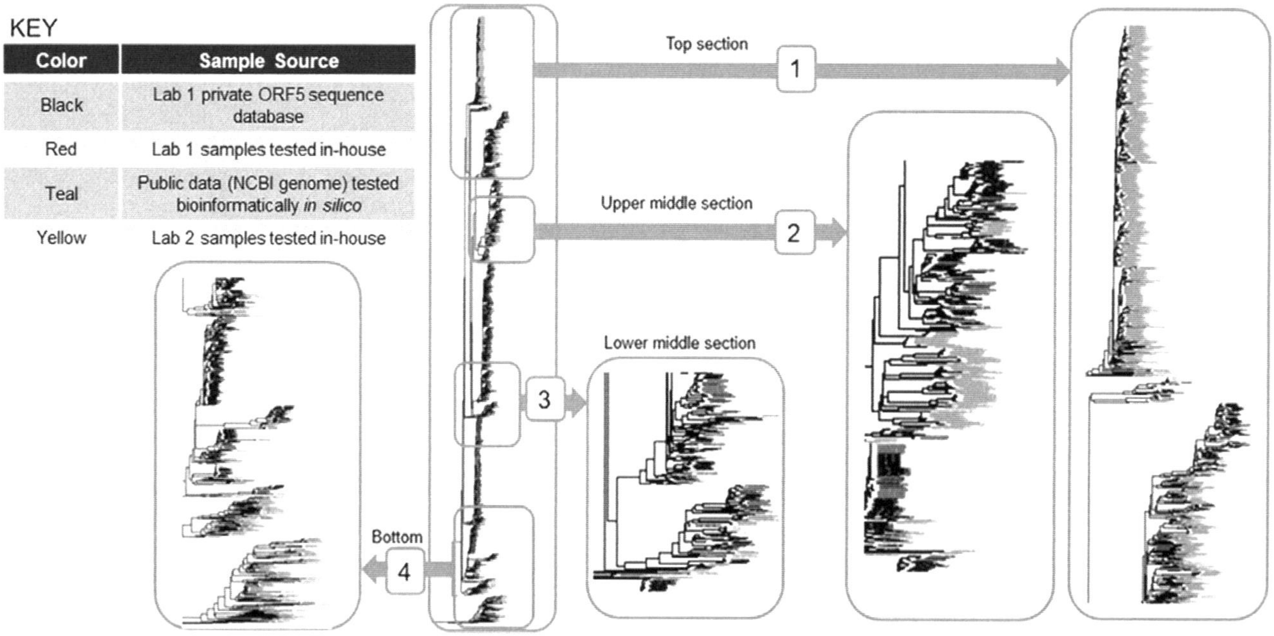

Table 1: PCR settings utilized for testing the newly developed product

PCR settings utilized		
Target	**Reporter**	**Quencher**
PRRSV NA	FAM	QSY-NFQ
PRRSV EU	VIC	
Xeno IPC	CY5	
Run mode	Fast	
Passive reference	ROX	
Volume	20µl	

Table 2: Thermal profile utilized in testing the newly developed product

Thermal profile utilized			
Stage	**Cycles**	**Temp**	**Time**
1	1	50°C	5 min
2	1	95°C	10 min
3	40	95°C	3 sec
		60°C	30 sec

of 10 samples that were reported to have lower performance using the on-market PRRSV reagents and controls show that the new molecular design is more efficient, which results in earlier Cq values (Figure 2) in most cases. Testing a large set of 330 samples from different locations in the US as well as different collection time points confirms the ability of the new updated reagents and controls to generate concordant calls with the on-market PRRSV products.

External testing generated similar findings where the new products generated concordant calls with on-market products, while being able to detect novel variants that were showing late Ct values using the on-market products. Test results of 39 positive samples confirmed by sequencing to belong to the new variants L1C 1-4-4 and 1-7-4 show the ability of the new products to generate more accurate calls with earlier Cts when encountering these variants (Figure 3). New controls do not require preparation as they come pre-mixed in one tube, and new reagents require less preparation steps than current PRRSV reagents. Overall, the new product shows excellent LOD, qRT-PCR efficiency, and enhanced performance over on-market products.

These products are for research use only (RUO) and not to be used in diagnostic procedures.

Reference

1. Kappes, M. A., & Faaberg, K. S. (2015). PRRSV structure, replication and recombination: origin of phenotype and genotype diversity. Virology, 479, 475-486.

Table 3: Thermal profiles used for different reagents

Utilized thermal profile comparisons							
On-market reagents				Newly developed reagents			
Stage	Cycles	Temp	Time	Stage	Cycles	Temp	Time
1	1	48°C	10 min	1	1	50°C	5 min
2	1	95°C	10 min	2	1	95°C	10 min
3	40	95°C	15 sec	3	40	95°C	3 sec
		60°C	1:10			60°C	30 sec
Total: 1:45:00				Total: 49:00			

Figure 2: Performance comparison between on-market products and newly developed products using specific panel of 10 variant positive samples.

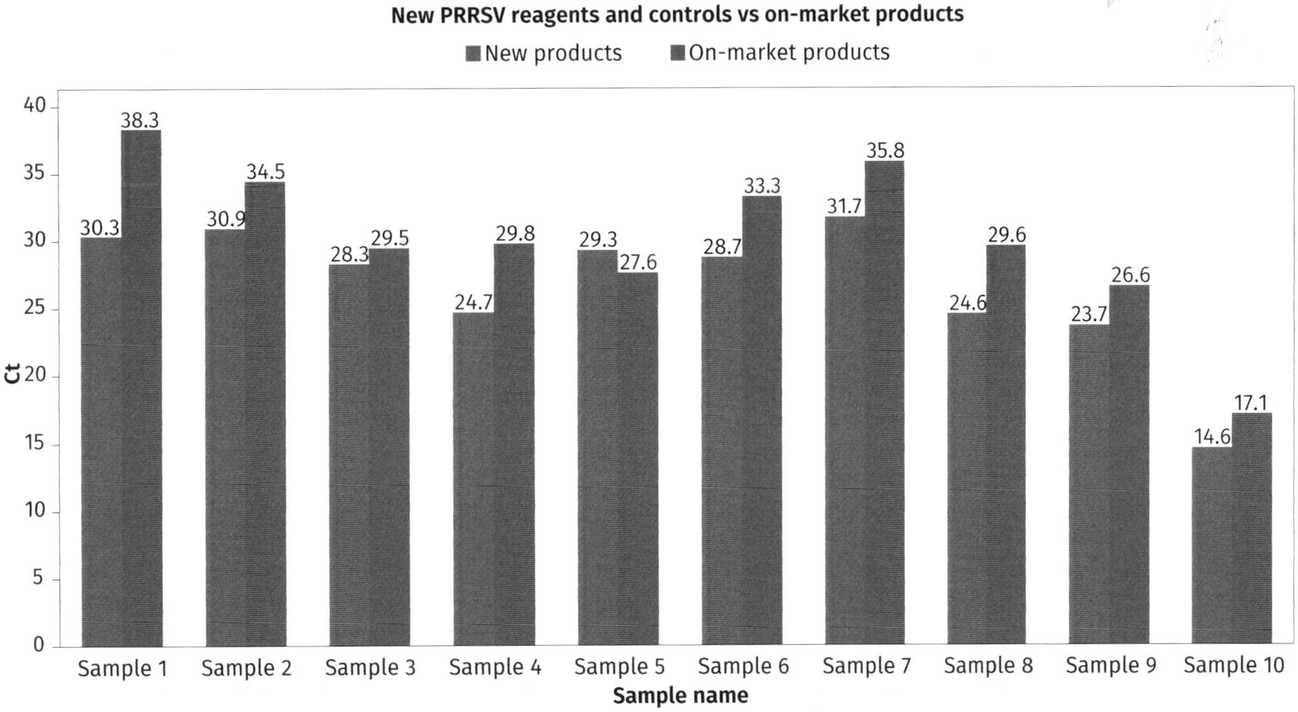

Figure 3: Performance comparison between on-market products and newly developed products using 39 positive samples confirmed by sequencing to belong to the new variants L1C 1-4-4 and 1-7-4.

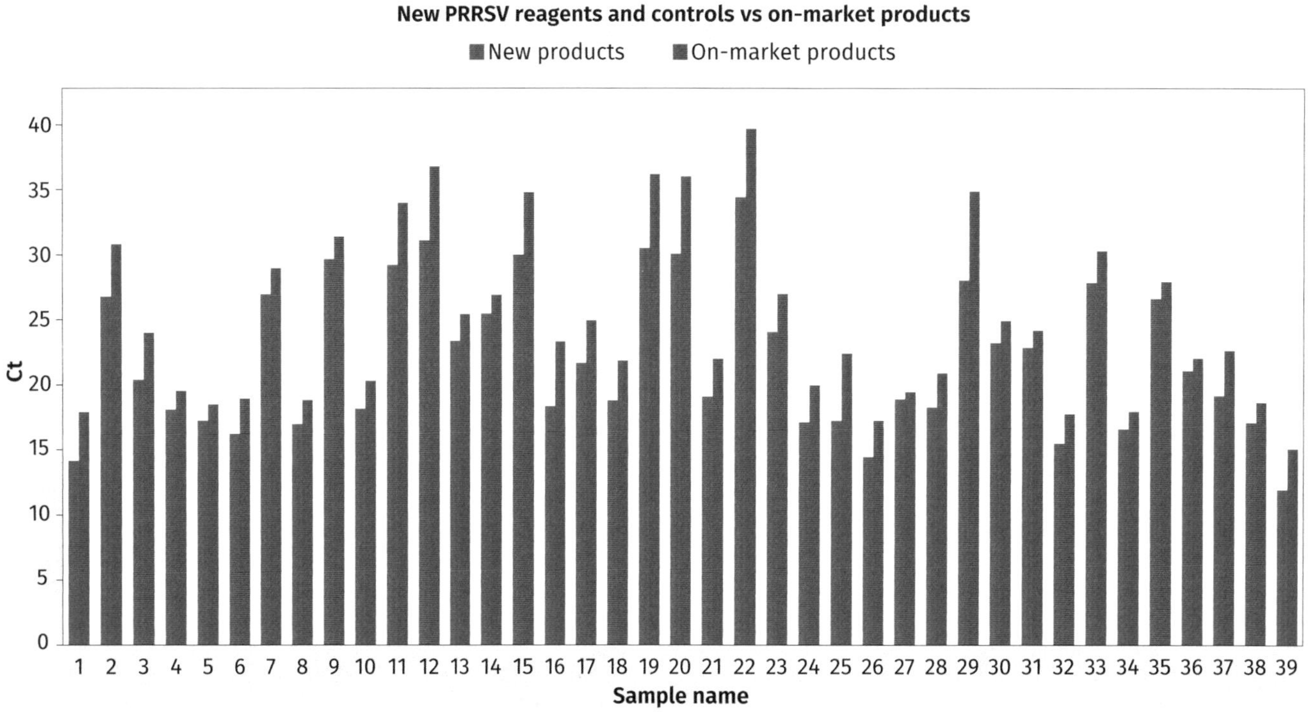

https://doi.org/10.54846/am2022/79

Monoglyceride blend reduces viability of porcine epidemic diarrhea virus in feed and prevents disease transmission to post-weaned piglets

Farrah Phillips, PhD; Dillon Mellick, MS
Kemin Industries, Inc., Des Moines, Iowa

Abstract

Outbreaks of African swine fever virus (ASFV) and porcine epidemic diarrhea virus (PEDV) have revealed the susceptibility of livestock to disease transmitted through feed. Several viruses, including PEDV, survive in feed and may introduce disease that causes significant morbidity and mortality. In 2013, PEDV, which causes severe diarrhea and vomiting, reached North America after spreading for decades across Eurasia. The global exchange of ingredients has created demand for products that prevent disease transmission from feed. Formaldehyde-based products are highly effective at inactivating enveloped viruses when applied at 3.25 kg/t. Alternative products to formaldehyde, including carboxylic acids, essential oils and medium chain fatty acids (MCFAs), have exhibited mixed efficacy against PEDV and require application rates higher than formaldehyde. Amphiphilic molecules like MCFAs disrupt the bilayer-lipid membranes that protect viral nucleic acids through the formation of micelles. Monoglycerides form micelles at lower concentrations than MCFAs, which suggests they may be more potent against enveloped viruses. The potential efficacy of monoglycerides against enveloped viruses in feed led to the development and examination of an experimental monoglyceride blend. The proprietary monoglyceride blend significantly ($P < 0.0001$) reduced PEDV viability *in vitro* after application to feed at 1.5, 2.5 and 3.5 kg/t. The monoglyceride was tested in a natural feeding behavior challenge model in piglets. The feed was contaminated with ice-blocks containing viable PEDV, and the piglets were exposed to PEDV through the feed bin for 20 days. At the end of the 20-day challenge period, all pigs were rectally swabbed and tested for PEDV by qPCR. In the untreated control group, 54.8% of the piglets tested positive for PEDV, whereas none of the MCFA-treated feed (10 kg/t inclusion) transmitted PEDV. Strikingly, the monoglyceride-treated groups (1.2, 2.5 and 3.5 kg/t) all exhibited 100% protection from PEDV. These data support the use of this proprietary monoglyceride blend in mitigation and prevention of viral disease transmission to piglets from contaminated feed.

Introduction

Lipids found in human milk, including medium chain fatty acids (MCFAs), defined as saturated fatty acids with C_6 to C_{12} hydrocarbon chains, and monoglycerides effectively inactivate enveloped viruses.[1] Several decades of research beginning in the 1800's document the efficacy of lipids against algae, bacteria, fungi, protozoa, and viruses.[2] Broad antimicrobial properties of MCFAs and related monoglycerides hold the promise of a next generation of antimicrobial agents with a low risk of pathogens developing resistance.[3] The mechanism by which MCFAs and monoglycerides inactivate viruses and kill or inhibit bacteria is through disruption of the viral envelope and bacterial cell membranes, respectively.[2] The hydrophilicity of the carboxyl group paired with the hydrophobicity of the hydrocarbon tail makes fatty acids amphiphilic. The single-chain lipid amphiphile inserts into the lipid bilayer of bacteria causing strain on their structures. This strain forces remodeling processes that generate tubules or spherical buds in the lipid bilayer.[4]

Porcine epidemic diarrhea virus (PEDV) is an enveloped single-stranded positive sense RNA virus that is part of the *Coronaviridae* family under the genus *Alphacoronavirus*.[5] In 2013, PEDV escaped its endemic locations in Europe and Asia and spread rapidly across the United States killing approximately 7 million pigs over a 12-month period.[6] PEDV causes an enteric disease that largely affects neonatal piglets resulting in severe vomiting and diarrhea, and spreads directly through fecal-oral route.[7] The virus can also be carried into barns by trailers, farm workers' hands, boots, and clothes[5] as well as animal feed.[8,9] The identification of feed as a risk factor for introduction of PEDV led to research into identifying mechanisms that inactivate enveloped viruses in feed.[8,10-14]

Several products from feed additive suppliers have been proposed as a means to inactivate viruses in feed and mitigate the risk of disease transmission from contaminated feed.[15] Formaldehyde-based products have consistently shown high efficacy against viruses in feed when applied at > 2.5 kg/t.[15-17] Proposed alternatives to formaldehyde include carboxylic acids, MCFAs, and essential oils, which have gained attention in the past few years, especially in un-pelleted feed because thermal treatments have also been shown to inactivate PEDV.[18] However, the efficacy of these alternative products is limited, and pelleting feed incurs additional costs to the producer and does not prevent recontamination during feed transport as some chemical treatments do. Recently, MCFAs were demonstrated to mitigate *Salmonella* Typhimurium[19] and PEDV in contaminated animal feed.[16] MCFAs were also shown to prevent transmission of PEDV to 100% of piglets through feed when applied at 10 kg/t.[15] However, application of MCFAs at 10 kg/t is not economical.

Although monoglycerides are known to disrupt viral envelopes,[1,4] the focus of virus mitigants in feed has largely remained on free fatty acids and the ratios of C6:C8:C10:C12 in free fatty acid blends.[16] As an alternative to free fatty acids and formaldehyde, a proprietary monoglyceride blend was tested *in vitro* and in a natural feeding behavior bioassay for efficacy against PEDV in feed. The studies compared the efficacy of the free fatty acid blend to the monoglyceride blend, which was anticipated to inactivate PEDV and prevent disease transmission at a lower inclusion rate. In the bioassay, the proprietary monoglyceride blend performed as well as a highly pure C8 and C10 MCFA blend (1%) and a formaldehyde-based commercial product (Sal CURB®, Kemin Industries, Inc., Des Moines, IA) (3.1 kg/t) when applied to feed at 1.5 kg/t, preventing 100% disease transmission from feed.

Materials and methods

Feed, feed additives and virus application for *in vitro* virus inactivation

A proprietary monoglyceride blend, a highly pure propriety blend of MCFAs, and Sal CURB® were procured from Kemin Industries, Inc. (Des Moines, IA). Soybean meal-based swine grow/finish feed was obtained from the feed mill at Southern Research and Outreach Center of the University of Minnesota (Waseca, MN) and confirmed negative for PEDV by qPCR. Feed (1 g) was weighed into round-bottom Falcon tubes (12 × 75 mm). The mitigants were diluted 1:10 in sterile distilled water, added to the feed and mixed. Each tube was inoculated with 200 µL PEDV (~$1 \times 10^{5.1}$ $TCID_{50}$/sample). The feed samples were held at room temperature for 0, 12, 24, 48 and 120 hours post inoculation. At each time point, 2 mL of eluent solution (3% beef extract in 0.05 M glycine) was added, and the sample was vortexed. The feed samples were centrifuged at 1,000 RPM for 5 minutes and the supernatant was transferred to sterile tubes and frozen at -80°C until the experiment was completed. Each time point for each treatment was performed in triplicate in two independent experiments.

In vitro virus recovery

The PEDV was propagated and titrated in Vero-81 cells using Eagle's minimum essential media (MEM) with antibiotics and fetal bovine serum. For virus titration, serial 10-fold dilutions were prepared from thawed eluates followed by inoculation onto monolayers of Vero-81 cells seeded in 96-well, flat bottom plates using 3 wells/dilution. The plates were incubated at 37°C for 90 min. Unabsorbed virus was decanted and the cells were washed once with Hanks' balanced salt solution. Fresh media (100 µL, Eagle's MEM supplemented with antibiotics, 10 µg/mL trypsin and 2% FBS) was added to each well and cells were incubated at 37°C for 5-7 days in a 5% CO_2 atmosphere. Cells were examined daily for the appearance of cytopathic effects (CPE). Virus titers were calculated as $TCID_{50}$ using the Spearman-Karber method.[20]

Bioassay design and PEDV challenge

The highly pure MCFAs and the proprietary monoglyceride blend were procured from Kemin Industries, Inc. (Des Moines, IA) and mixed into complete feed at Chandler Feed (Leota, MN). Liquid product application at the mill was conducted the day before pig placement by personnel from Kemin using a Kemin Application Solutions system. Pigs used in the study were housed in the Pipestone Applied Research biosafety level 2 facility in accordance with the institutional animal care and use guidelines approved by the investigator's ethical review board (Pipestone Applied Research). Nursery-age pigs negative for porcine reproductive and respiratory syndrome virus (PRRSV), PEDV, and *Senecavirus* A (SVA) were weighed and distributed across 6 pens in 5 rooms (10 or 11 pigs/pen; total 323). Pigs were acclimated to the environment and the feed for 5 days prior to being exposed to PEDV (pre-challenge period) and then exposed to feed containing PEDV for 20 days (post-challenge period). Feed for all treatment groups was contaminated using an ice block (454 g in weight/454 mL) containing 100 mL of PEDV Colorado-13 strain at a concentration of 10^5 $TCID_{50}$/mL.[10] The balance of the liquid for the block consisted of MEM. On day 0 and 6 of the post-challenge period, one block was dropped into each of the designated feed bins through the opening at the top of the bin.

Monitoring and sampling

Environmental samples were collected from each pen (6 per room) on trial days 6 and 15. Environmental samples included both oral fluids collected by hanging rope and pen-based feed samples collected using Swiffer dry sweeping cloths to swab inside the feeder bin. Rectal swabs were collected on day 20 from each pig for PEDV diagnosis. All samples were analyzed by qPCR at the South Dakota State University Veterinary Diagnostic Laboratory (SDSU, VDL Brookings, South Dakota), and determined to be positive or negative by the Ct (cycle time) value. Samples with a cycle number greater than 40 were determined negative for PEDV.

Results

PEDV inactivation in feed from the in vitro experiments indicated that MCFA at 10 kg/t inactivated 99.79% of the virus within 12 hours of contact (Table 1 and Figure 1). Additionally, MCFA at 10 kg/t did not further reduce virus viability in feed at 24, 48, and 120 hours of contact compared to the 12 h point (Table 1). At the 24 hour time point, Sal CURB at 3.1 kg/t inactivated approximately 2 logs (99%) of the virus (Figure 1). When used at 1.5 kg/t, the proprietary monoglyceride blend inactivated 2 logs (99%) of the virus within 24 hours. The difference in virus inactivation by 2.5 or 3.5 kg/t of monoglyceride was not significant (Table 1). At 2.5 kg/t, the proprietary monoglyceride blend inactivated 2 logs of virus within 12 hours of contact (Figure 1).

During the natural feeding behavior challenge and bioassay, the presence of PEDV RNA was detected by qPCR in feeder and oral fluid samples from all rooms on day 6 and 15 post challenge. All 6 environmental swab samples collected from the feeders and all 6 oral fluid samples in the untreated control group were positive for PEDV, demonstrating that PEDV was effectively delivered into the experimental rooms *via* the feed (Table 2). The MCFA positive control treatment effectively reduced the positivity rate in all environmental samples by 50% (Table 2). The proprietary monoglyceride blend treatment level correlated with reduced PEDV positivity in the environmental swabs on day 6 and 15 post-challenge. Specifically, on day 15 the number of positive oral fluid samples were 2, 3, and 4 out of a total of 6 for the 3.5 kg/t, 2.5 kg/t, and 1.5 kg/t monoglyceride treatments, respectively (Table 2). These data suggest the monoglyceride had a dose-dependent effect on virus stability in feed. Although a positive qPCR result from the environmental samples indicates that virus was present and does not effectively determine the viability of the virus, the data suggest that viral RNA was degraded in the feed of the MCFA- and proprietary monoglyceride blend-treated groups.

The presence of diarrhea was recorded for each pig on day 20 post PEDV challenge *via* the feed. No clinical evidence of diarrhea was noted in the MCFA control group or the 2.5 or 3.5 kg/t monoglyceride treated groups and 4 pigs showed evidence of diarrhea in the 1.5 kg/t monoglyceride group (Table 3). In contrast, diarrhea was observed in 21/62 pigs in the untreated control room. To determine if the diarrhea was related to PEDV infection, all pigs were swabbed rectally and tested for PEDV by qPCR at the end of the 20-day challenge period. In the untreated control room, 34/62 pigs tested positive for PEDV, which was a higher diagnostic incidence of PEDV than was observed by the clinical diarrhea scoring (Table 3). Conversely, rectal swabs were negative for PEDV in all the pigs from the treated groups including the lowest inclusion rate, 1.5 kg/t monoglyceride, group (Table 3). The diagnostic data indicated that the monoglyceride at the lowest tested inclusion rate of 1.5 kg/t was 100% effective in preventing transmission of PEDV over 20 days from contaminated feed.

Figure 1: Viability of PEDV over time in feed treated with mitigants.

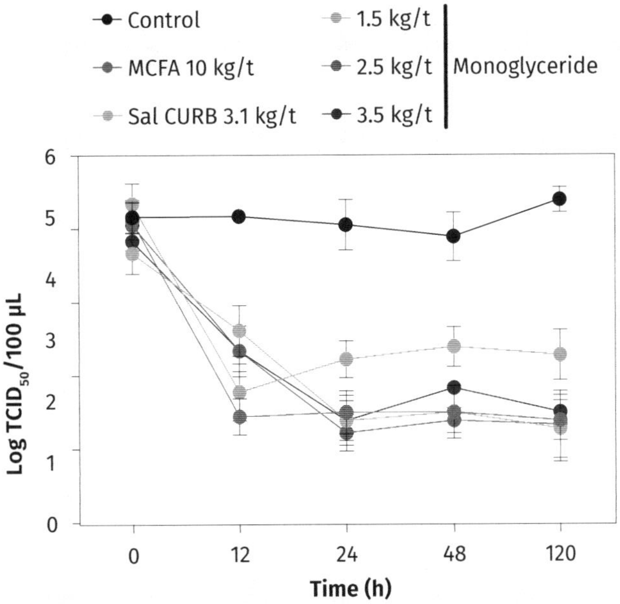

Discussion

Biosecurity is an essential element of modern pig production, which has progressively been trending toward larger farms and more pig movement across regions since the 1980s. Diseases introduced from outside the facility by people, vehicles, new animals, pests, and feed can seriously impact efficiency. One piece of the biosecurity puzzle is feed. The efficacy of MCFAs against swine viruses in feed is a promising start, yet there is still a need for a more effective, less expensive, and safe virus mitigants.

The efficacy of a proprietary blend of monoglycerides at reducing PEDV viability in feed was evaluated to find solutions to improve feed biosecurity. *In vitro* data concluded that the monoglyceride effectively reduced PEDV viability in feed from 5 logs $TCID_{50}$ to < 3 logs $TCID_{50}$ when included at 2.5 kg/t and 3.5 kg/t. The 1.5 kg/t monoglyceride treatment reduced $TCID_{50}$ of PEDV by approximately 1.5 log in the *in vitro* feed assay. In the PEDV ice-block challenge model, which delivered the virus through natural feeding behavior for 20 days, the monoglyceride blend at the lowest tested inclusion rate of 1.5 kg/t prevented 100% of disease transmission to piglets. The efficacy demonstrated in the ice-block challenge model suggests that the monoglyceride blend could be more effective than MCFAs and can be included in feed at lower doses to inactivate PEDV.

Table 1: Inactivation of PEDV by mitigants at different time points†

	Log $TCID_{50}$ /100 µL of PEDV at indicated time				
Treatment (application)	**0h**	**12h**	**24h**	**48h**	**120h**
Positive control	5.16^{ab}	5.17^{a}	5.05^{a}	4.89^{a}	5.39^{a}
MCFA (10 kg/t)	5.11^{ab}	2.45^{c}	2.50^{c}	2.50^{cd}	2.39^{c}
Sal CURB® (3.1 kg/t)	4.83^{c}	3.61^{b}	2.39^{c}	2.50^{cd}	2.28^{c}
Monoglyceride (1.5 kg/t)	5.33^{a}	3.78^{c}	3.22^{b}	3.39^{b}	3.28^{b}
Monoglyceride (2.5 kg/t)	5.05^{abc}	3.33^{b}	2.22^{c}	2.39^{d}	2.33^{c}
Monoglyceride (3.5 kg/t)	4.83^{bc}	3.33^{b}	2.38^{c}	2.83^{c}	2.50^{c}

† Values in the same column that are not connected by the same superscript letter are significantly different, $P < 0.05$. One-way ANOVA followed by Tukey-Kramer test for all pairs, n = 6 from two independent experiments. MCFA = medium chain fatty acids.

Table 2: Detection of PEDV in the environment. Environmental swabs from the pen feeder and oral fluids were analyzed for PEDV by qPCR on day 6 and 15 post challenge

	Feeders (+/total)		Oral fluids (+/total)	
Treatment (% inclusion)	**Day 6**	**Day 15**	**Day 6**	**Day 15**
Untreated control	6/6	6/6	6/6	6/6
MCFA (10 kg/t)	2/6	2/6	1/6	3/6
Monoglyceride (1.5 kg/t)	2/6	4/6	1/6	5/6
Monoglyceride (2.5 kg/t)	0/6	3/6	0/6	2/6
Monoglyceride (3.5 kg/t)	1/6	2/6	1/6	2/6

MCFA = medium chain fatty acids.

Table 3: Incidence of PEDV transmission to pigs from feed. All treatment groups were challenged via feed with an ice block containing 5 logs $TCID_{50}$ PEDV placed in the feeder bin.

Treatment (% inclusion)	†PEDV+ pigs (+/total)	‡Diarrhea incidence (+/total)
Untreated control	34/62	21/62
MCFA (10 kg/t)	0/53	0/53
Monoglyceride (1.5 kg/t)	0/64	4/64
Monoglyceride (2.5 kg/t)	0/48	0/48
Monoglyceride (3.5 kg/t)	0/57	0/57

† PEDV was diagnosed via rectal swabs collected on day 20 post-challenge from all surviving pigs and analyzed at the South Dakota State University Veterinary Diagnostic Laboratory.

‡ Diarrhea incidence was determined by facility manger on day 20 post-challenge. MCFA = medium chain fatty acids (free fatty acids).

The research approach was focused on comparing the efficacy of the proprietary monoglyceride blend with the known efficacy of the high purity MCFA blend against PEDV. Although the mode of action and efficacy against PEDV could indicate similar efficacy against other pathogens, this study did not cover other enveloped viruses and bacterial pathogens. Thus, the demonstrated efficacy of the monoglyceride is limited to one, enveloped virus pathogen, whereas the efficacy of some MCFAs has been evaluated against *Salmonella*, SVA, and PRRSV in the ice-block challenge model. Furthermore, MCFAs and glycerol monolaurate have shown some efficacy against the capsid-protected DNA African swine fever virus (ASFV) *in vitro*.[21] More efficacy information needs to be collected on the proprietary monoglyceride blend evaluated in this study expanding the list of tested pathogens. Additionally, the lowest tested monoglyceride inclusion rate of 1.5 kg/t prevented disease transmission, so the minimum effective application rate remains unknown. Additional experiments conducted using the monoglyceride at < 1.5 kg/t could reveal the point at which the material no longer provides effective protection in feed.

This discovery of the proprietary monoglyceride blend's efficacy against PEDV in feed brings into focus a new group of molecules.. The continued threat of feed-resident pathogens continues to drive research for feed biosecurity products, yet there are still few cost-effective options available. Monoglycerides further expand the possibilities for solutions to a growing problem. Intensification of pig population densities as well as personnel and facility biosecurity could be undermined by insufficient feed biosecurity practices. The use of additives and handling and transport protocols to improve feed safety are needed to ensure optimal pig performance and secure local and global food supply chains.

References

1. Thormar, H., Isaacs, C. E., Brown, H. R., Barshatzky, M. R. & Pessolano, T. Inactivation of enveloped viruses and killing of cells by fatty acids and monoglycerides. *Antimicrob. Agents Chemother.* 31, 27–31 (1987).

2. Yoon, B. K., Jackman, J. A., Valle-González, E. R. & Cho, N. J. Antibacterial free fatty acids and monoglycerides: Biological activities, experimental testing, and therapeutic applications. *Int. J. Mol. Sci.* 19, (2018).

3. Davies, J. Origins and evolution of antibiotic resistance. *Microbiologia* 12, 9–16 (1996).

4. Yoon, B. K., Jackman, J. A., Kim, M. C., Sut, T. N. & Cho, N. J. Correlating Membrane Morphological Responses with Micellar Aggregation Behavior of Capric Acid and Monocaprin. *Langmuir* 33, 2750–2759 (2017).

5. Wang, Q., Vlasova, A. N., Kenney, S. P. & Saif, L. J. Emerging and re-emerging coronaviruses in pigs. *Curr. Opin. Virol.* 34, 39–49 (2019).

6. Niederwerder, M. C. & Hesse, R. A. Swine enteric coronavirus disease: A review of 4 years with porcine epidemic diarrhoea virus and porcine deltacoronavirus in the United States and Canada. *Transbound. Emerg. Dis.* **65**, 660–675 (2018).

7. Jung, K. & Saif, L. J. Porcine epidemic diarrhea virus infection: Etiology, epidemiology, pathogenesis and immunoprophylaxis. *Vet. J.* 204, 134–143 (2015).

8. Dee, S. *et al.* An evaluation of contaminated complete feed as a vehicle for porcine epidemic diarrhea virus infection of nai¨ve pigs following consumption via natural feeding behavior: Proof of concept. *BMC Vet. Res.* 10, 176 (2014).

9. Dee, S. *et al.* Evidence of viral survival in representative volumes of feed and feed ingredients during long-distance commercial transport across the continental United States. *Transbound. Emerg. Dis.* 2021, 1–8.

10. Dee, S., Neill, C., Clement, T., Christopher-Hennings, J. & Nelson, E. An evaluation of a liquid antimicrobial (Sal CURB®) for reducing the risk of porcine epidemic diarrhea virus infection of naïve pigs during consumption of contaminated feed. *BMC Vet. Res.* 10, 1–6 (2014).

11. Trudeau, M. P. *et al.* Comparison of thermal and non-thermal processing of swine feed and the use of selected feed additives on inactivation of porcine epidemic diarrhea virus (PEDV). *PLoS One* 11, 1–14 (2016).

12. Cottingim, K. M. *et al.* Feed additives decrease survival of delta coronavirus in nursery pig diets. *Porc. Heal. Manag.* **3**, 1–7 (2017).

13. Trudeau, M. P. *et al.* Survival of porcine epidemic diarrhea virus (PEDV) in thermally treated feed ingredients and on surfaces. *Porc. Heal. Manag.* 3, 1–7 (2017).

14. Dee, S. A. *et al.* Survival of viral pathogens in animal feed ingredients under transboundary shipping models. *PLoS One* 13, 1–18 (2018).

15. Dee, S. A. *et al.* An evaluation of additives for mitigating the risk of virus-contaminated feed using an ice-block challenge model. *Transbound. Emerg. Dis.* 68, 1–13 (2020).

16. Cochrane, R. A. *et al.* Assessing the effects of medium-chain fatty acids and fat sources on PEDV infectivity. *Transl. Anim. Sci.* 4, 1051–1059 (2019).

17. Niederwerder, M. C. *et al.* Mitigating the risk of African swine fever virus in feed with anti-viral chemical additives. *Transbound. Emerg. Dis.* **68**, 477–486 (2021).

18. Cochrane, R. A. *et al.* Effect of pelleting on survival of porcine epidemic diarrhea virus–contaminated feed. *J. Anim. Sci.* 95, 1170–1178 (2017).

19. Cochrane, R. A., Huss, A. R., Aldrich, G. C., Stark, C. R. & Jones, C. K. Evaluating Chemical Mitigation of Salmonella Typhimurium ATCC 14028 in Animal Feed Ingredients. *J. Food Prot.* 79, 672–676 (2016).

20. Karber, G. 50% end-point calculation. *Arch Exp Pathol Pharmak* 162, 3 (1931).

21. Jackman, J. A., Hakobyan, A., Zakaryan, H. & Elrod, C. C. Inhibition of African swine fever virus in liquid and feed by medium-chain fatty acids and glycerol monolaurate. *J. Anim. Sci. Biotechnol. 11*, (2020).

https://doi.org/10.54846/am2022/80

Use of historical PRRS outbreak information to identify high and low risk areas in a region

K. Lenker; M. Pierdon, VMD
New Bolton Center, University of Pennsylvania School of Veterinary Medicine

Introduction

Porcine reproductive and respiratory syndrome virus (PRRSV) infects only swine, and is transmitted via all bodily secretions, including urine, feces, colostrum, milk, saliva, semen, and nasal secretions.[1] PRRSV causes losses in swine production due to respiratory disease and reproductive failure.[2] PRRSV has been shown to spread via aerosol, so spatial factors play an important part in disease transmission.[2] Regional control programs have reported on the location of outbreaks, and in some cases genetic sequences, but the information can be used to further elucidate areas of high risk and low risk as well as when times of high risk are occurring.[3] The purpose of this study was to use historical PRRS information from Pennsylvania's PRRS regional control program to investigate whether PRRS cases in sow farms clustered in space, or in time and space, and whether such analysis could illustrate high and low risk areas for PRRS outbreaks.

Materials and methods

The PRRS regional control program had a database of 93 participating sow farms that reported their PRRSV status quarterly for 10 years. From the quarterly report, Sow farms that were never PRRSV positive were considered controls, and Sow farms that were PRRS positive farms were considered positive cases. The Bernoulli case-control method was used for analysis of spatial clustering at the level of the farm. Analysis was performed using SatScan v 9.7. There were 35 case and 58 control farms included in the analysis. Each was included using the latitude and longitude. Circular scanning windows were used, and likelihood ratios and p-values were calculated by comparing the risk within and outside the scanning window. The program generates random datasets (n = 999) and compares the real data with the randomly generated data sets. If the maximum likelihood ratio was found to be in the top 5% of the clusters, then it was determined to be significant at $P = 0.05$. The null hypothesis was that case farms were spatially randomly distributed, and a cluster was identified if the null hypothesis was rejected. Both high and low levels of clustering were investigated. Relative risk was calculated by comparing the rate of cases inside the circle to those outside the circle. A relative risk greater than 1 is increased risk, and a relative risk less than 1 is decreased risk. A retrospective spatial temporal cluster analysis at the level of the quarter was also performed including only the case farms and using cylindrical scanning windows.

Results and discussion

Figure 1 represents the purely spatial analysis identified two areas of low risk and two areas of high-risk Cluster 1 had a radius of 23.45 km, an increased risk of PRRSV with a relative risk of 1.75, an expected 28.99 cases and an observed 43 cases ($P < 0.0001$). Cluster 2 had a radius of 126.10 km, a decreased risk of PRRSV with a relative risk of 0.32, an expected 24.27 cases and an observed 9 cases ($P < 0.001$). Cluster 3 had a radius of 53.88 km, decreased risk of PRRSV with a relative risk of 0.12, an expected 8.09 cases and an observed 1 case ($P < 0.01$). Cluster 4 had a radius of 29.53 km, an increased risk of PRRS with a relative risk of 1.52, an expected 29.66 cases and an observed 40 cases ($P < 0.05$). Two significant clusters were found in both space and time. One occurred from 2001-2002 with a radius of 17.44 km, 1.96 expected cases, and 12 observed cases ($P < 0.001$). The other significant cluster occurred from 2015-2016 with a radius of 4.14 km, 1.12 expected cases, and 7 observed cases ($P < 0.05$). Understanding the location of these clusters is beneficial to practitioners and farmers when analyzing PRRSV risk at sow farms. When a new sow farm is to be built, Figure 1 may be important to avoid increased risk of PRRSV introduction. Biosecurity practices in the high-risk areas could be enhanced, such as filtering facilities, to limit risk. Finally, high value animals, such as those serving as genetic replacements, could be routed to low-risk areas.

Acknowledgments

Thank you to the Pennsylvania Pork Producer's Council for providing support and funding.

References

1. Tousingnant SJP et al.: 2014 AJVR 76.1 70-76.
2. Fahrion AS et al.: 2014 Preventative veterinary medicine 114.3-4 247-258.
3. Valdes-Donoso P et al.: 2016 PloS one 11.2.

Figure 1: 314 total sow farms participating in Pennsylvania clustered based on their risk of PRRS, with high risk clusters in light red circles and low risk in light blue circles. Positive farms are indicated by smaller red circles and negative farms are indicated by smaller blue circles.

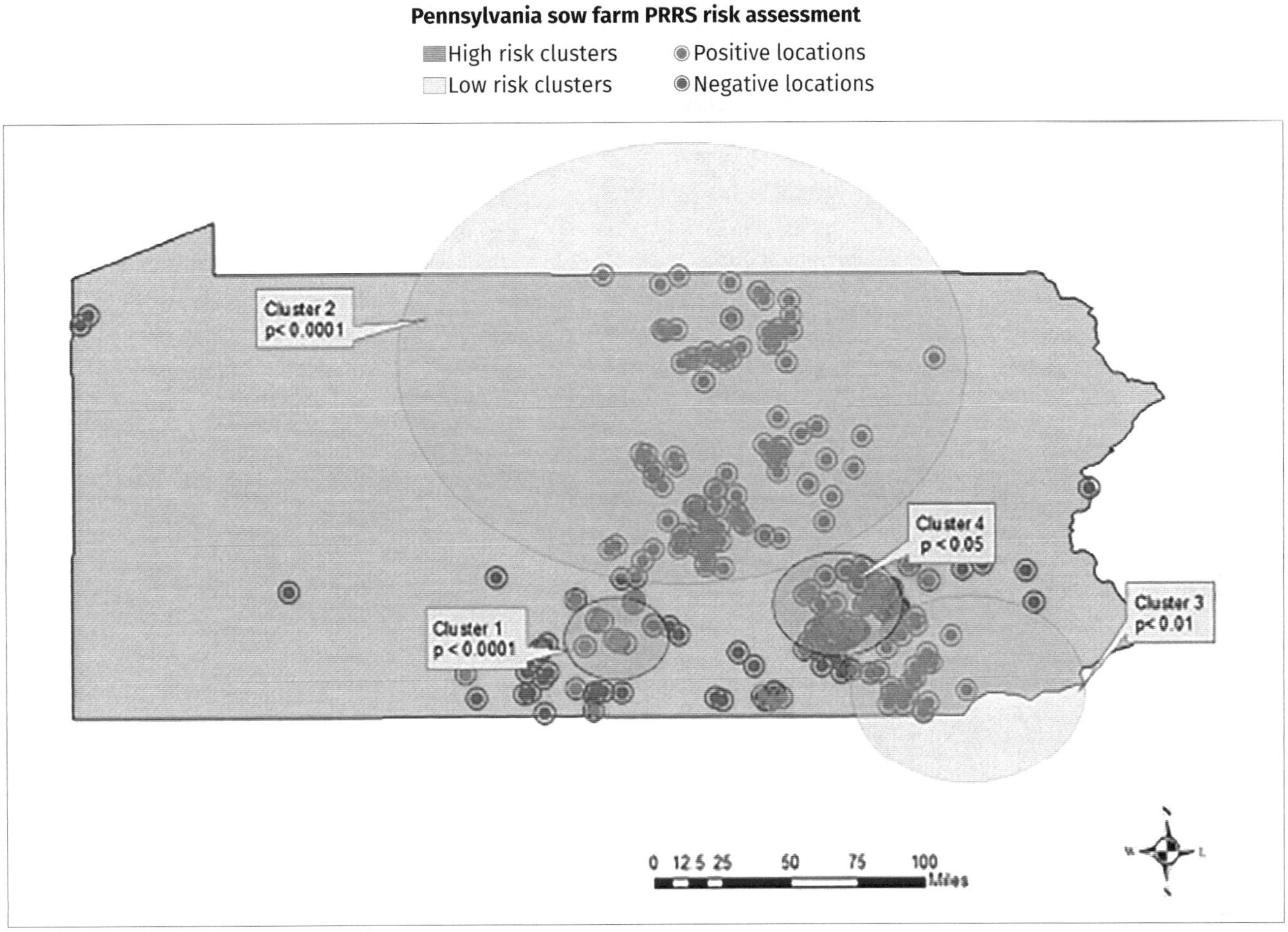

https://doi.org/10.54846/am2022/81

Field evaluation of vaccine timing regimen and genetic identity of pre-farrow SEQUIVITY® rotavirus vaccine on suckling pig performance using a challenge model

M. Brinning[1], BS; N. Lauterbach[2], DVM; C. Ruston[2], DVM, MS; J. Maurer[3], VMD; K. Crawford[3], DVM, MS
[1]Iowa State University College of Veterinary Medicine, Ames, Iowa;
[2]Smithfield Hog Production – Midwest Region, Algona, Iowa; [3]Merck Animal Health, Lenexa, Kansas

Introduction

Rotavirus remains one of the top enteric pathogens that causes piglet mortality and performance losses in the farrowing house.[1] A 2020 field study demonstrated SEQUIVITY® rotavirus vaccine given to gilts 4&1 weeks pre-farrowing (wpf) improved pig performance after a homologous rotavirus challenge.[2] It is unknown if similar benefits could be reproduced in multiparous dams using different vaccine timing regimens after a heterologous challenge. The primary objective of this study was to assess the impact of using vaccines of different homologies to a rotavirus C (RVC) challenge, while also evaluating different vaccine timing regimens.

Materials and methods

The project was conducted at an 8500-head sow farm with historical use of RVC vaccine pre-farrow and Natural Planned Exposure (NPE) in the GDU. 250 healthy dams were blocked by parity group (Gilts, P1/P2, P3+), sorted by expected due date and randomly allocated to one of 5 treatment groups: Saline at 1 wpf served as the control (T1), Vaccine A at 4&1 wpf (T2), Vaccine B 4&1 wpf (T3), Vaccine A at 1wpf (T4), Vaccine B at 1 wpf (T5). The Vaccine A and B sequences were 90.99% and 98.2% homologous to the RVC challenge, respectively; all sequences shared the same amino acid motiff at *in silico* predicted antigenic epitope regions. All injections were given as a 1 mL dose IM.

Upon farrowing, 30 litters per treatment group were enrolled in the study. Group size was determined based on the statistical power to detect a difference in final weights of 0.75 lbs between the treatment groups and control with a standard deviation (sd) of 1.25 lbs. Any dam that died, aborted, farrowed less than 10 pigs or had pigs introduced into the litter were excluded. All pigs were ear tagged, individually weighed, and challenged with a 1mL oral dose of 1:2 diluted NPE material at 2 days of age (2 DOA). Pigs were weighed again at 22 days of age (22 DOA). Prior to challenge, 11 pigs were randomly fecal swabbed from scouring litters. Data was analyzed using general and generalized linear models with litter as the experimental unit.

Results

The results are summarized in Table 1. There was not a statistical difference in weight variables among treatment groups. All 11 fecal samples taken prior to challenge were RVC PCR positive (12.5-23.6 CTs; amino acid sequence 100% homologous to the challenge virus) indicating prior exposure. Numerically, weight variables in the 4&1 wpf vaccination groups (T2 & T3) were highest. The effect of treatment group on 22 DOA Pig Weight in gilt litters was most remarkable ($P = 0.18$). The sd of 22 DOA Pig Weight in control group (2.47 lbs) was higher than anticipated and may have led to the inability to detect statistically significant differences. These results suggest the potential benefit of SEQUIVITY rotavirus vaccines– though additional studies with lower parity litters and/or more litters are needed to further evaluate performance differences and cross-protection.

References

1. Harris LD. Rotaviral enteritis in pigs. 2013 Merck Veterinary Manual.

2. Boyd W et al. Comparison of pre-farrow rotavirus vaccine and natural planned exposure on sucking pig performance. AASV 2020; 228-229.

Table 1: Weight gain data

		Weight data – all parities				
		Treatment groups				
Parameter	**Value**	**T1: control**	**T2: 4 &1 wpf Vaccine A**	**T3: 4&1 wpf Vaccine B**	**T4: 1 wpf Vaccine A**	**T5: 1 wpf Vaccine B**
No. litters	Count	30	30	29	30	30
2 DOA pig weight	Lb., Avg.	3.23	3.33	3.45	3.38	3.28
22 DOA pig weight	Lb., Avg.	11.60	12.18	12.19	11.66	11.61
Gain per pig	Lb., Avg.	8.37	8.84	8.74	8.29	8.33
Daily gain per pig	Lb., Avg.	0.419	0.442	0.437	0.414	0.417
		Weight data – gilts				
		Treatment groups				
Parameter	**Value**	**T1: control**	**T2: 4 &1 wpf Vaccine A**	**T3: 4&1 wpf Vaccine B**	**T4: 1 wpf Vaccine A**	**T5: 1 wpf Vaccine B**
No. litters	Count	6	7	7	6	5
2 DOA pig weight	Lb., Avg.	2.72	3.00	3.16	2.85	3.08
22 DOA pig weight	Lb., Avg.	10.00	11.31	11.83	9.56	11.48
Gain per pig	Lb., Avg.	7.28	8.31	8.68	6.71	8.39
Daily gain	Lb., Avg.	0.364	0.416	0.434	0.335	0.420

https://doi.org/10.54846/am2022/82

Use of digital thermography to identify scrotal and inguinal hernias in piglets at processing

A. Tatnall[1]; A. Leonardi[2], DVM; J. Lowe[3,4], DVM, MS, DABVP
[1]College of Veterinary Medicine; [2]Department of Pathobiology; [3]Department of Veterinary Clinical Medicine, University of Illinois at Urbana-Champaign; [4]Lowe Consulting Ltd.

Introduction

Pork systems are reliant on the production of healthy piglets. Inguinal and scrotal herniation is estimated to affect between 1.5% and 6.7% of male piglets.[1] At castration, piglets are inspected for hernias and treated accordingly. However, a missed herniation resulting from human error can ultimately lead to decreased weight gain and other chronic issues; affecting producer profitability.

Objective

This objective of this study was to determine if digital thermography can be used as a diagnostic tool for improved hernia recognition and treatment at the time of processing and castration.

Materials and methods

Two digital thermal cameras were used: A Fluke TI-55 and a FLIR TX-165. Fluke SmartView software version 4.2 was used for initial processing of Fluke images, while standard windows image viewing software was used for the FLIR images. Image J was used for analysis, measurement, and labeling of the processed images. Microsoft Excel was used for the data analysis and statistical calculations. 24 hours prior to imaging, piglets were visually and physically evaluated for the presence or suspicion of a hernia. Four-to-seven-day old male piglets were imaged at the time of castration, using both cameras, at a distance ranging from 15 cm to 40 cm. Twenty-eight pigs in total, with suspicion of a hernia, were restrained in dorsal recumbency for thermal imaging with the Fluke camera. The FLIR camera was used to capture traditional photographic images to show the physical appearance of the region of interest. After capturing the images, each of the piglets was castrated per the standard operating procedures of the farm, confirming the presence or absence of a hernia. Images were converted to the rainbow color gradient and the temperature ranges were manually manipulated for best contrast within the region of interest. Next, conversion to a JPEG format allowed for analysis in Image J. By defining borders of color transitions present in the inguinal and scrotal regions, the surface areas of these anatomic sites could be calculated and a value in pixels was produced. The values calculated were then evaluated as a ratio in Excel comparing the left and right sides of the scrotal region in both normal and herniated piglets. The mean and SD of the ratios for both groups were determined. The means of these ratios were compared and a t-test was used to test for significance.

Results

Thermal imaging was able to successfully identify, on average, a 53% increase in surface area in animals with hernias, showing a cooler temperature on the affected side as compared to the unaffected side. The affected area to unaffected area ratio within a single animal has a value of 1.53 ($P < 0.0001$); whereas the average ratio on an unaffected animal has a value of 1.06. Similar characteristics are seen in Figure 1 showing an unaffected animal (Figure 1A) and an affected animal (Figure 1B). The associated measurements and ratio calculation show how thermographic images can be analyzed and compared to detect herniation in piglets.

Discussion

Thermography has not been widely examined as a method to address scrotal hernias. The results of this "proof of concept" study shows a significant difference in the surface temperature, left versus right, of an affected individual; therefore, this technology has the potential to be used in hernia detection during piglet processing and castration. Further development and testing of this digital thermography will be beneficial to producers to objectively identify hernias and produce a point-of-care system on the farm.

Reference

1. Straw B, Bates R, May G. Anatomical abnormalities in a group of finishing pigs: prevalence and pig performance. J Swine Health Prod. 2009;17(1):28–31.

Figure 1: Thermographic images **A)** normal piglet with proportional scrotal areas **B)** affected piglet with hernia and enlarged scrotal area (4)

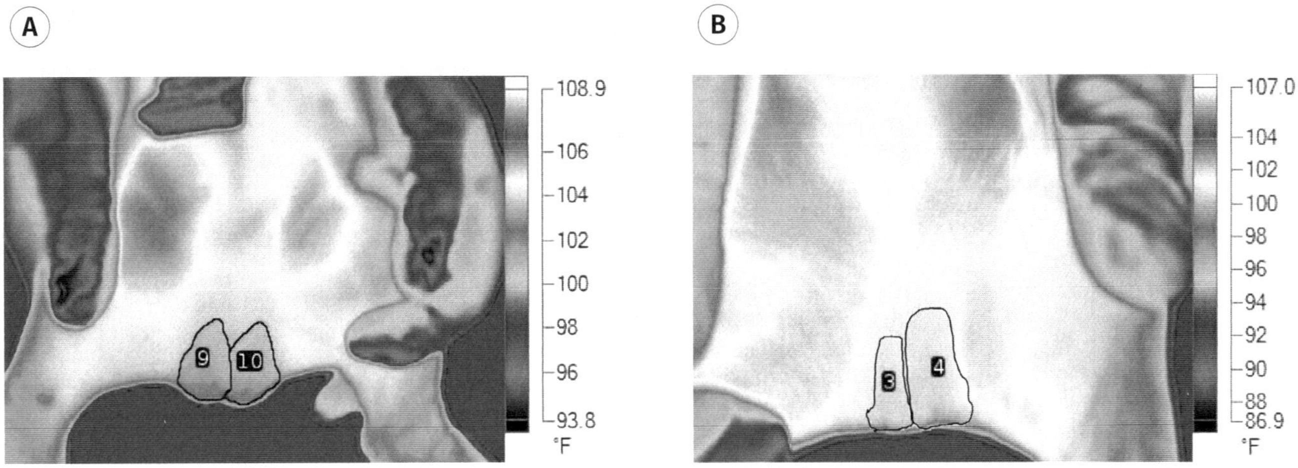

https://doi.org/10.54846/am2022/83

Assessment of non-catalytic converter gasoline engine exhaust for swine euthanasia

E. Schwarz; B. Blair, DVM; J. Lowe, DVM, MS, DABVP; S. Storms, DVM

Introduction

There is a persistent risk of African swine fever, foot-and-mouth disease, or other foreign animal disease spreading to the United States. If one of these diseases were to infect the US swine populations or pork processing plants ceased production a new form of mass euthanasia that is humane, affordable, effective, and safe will be needed. Gas euthanasia with carbon monoxide (CO) achieves hypoxia by blocking uptake of oxygen by erythrocytes, resulting in loss of consciousness without pain, hypoxemia with minimal discomfort, and rapid death if CO concentrations are above 4% to 6%.[1] Non-catalytic converter gasoline engine exhaust provides carbon monoxide levels that range from 30,000 to 100,000 ppm.[2] The AVMA Guidelines for Euthanasia of Animals: 2020 Edition[1] states the use of exhaust fumes is "associated with substantial problems such as production of other gases, inadequate production of CO, inadequate gas cooling, inability to quantify delivery rate, and maintenance of equipment." Tidemann and King demonstrated that water filtration cools engine exhaust gases leading to the humane death in mynas, starlings and sparrows.[3] The goal of this study is to effectively cool exhaust temperature, filter production of other gases, and quantify the production and distribution of CO from a non-catalytic converter gasoline engine.

Material and methods

A 4 × 8 × 2' chamber was created in Autodesk Fusion360 and Autodesk CFD to simulate how CO mixes and distributes throughout an enclosed area. A 4 × 8 × 2' wood box was built to serve as the prototype euthanasia box. This prototype included a 20' exhaust repair hose that connected to an engine that led into a 5-gal bucket that served as a water filtration system. Exhaust would diffuse through holes in a pipe entering the bucket and through holes in the bucket lid. This system is meant to filter fine particulates (diameter 2.5mm or less, PM2.5) and coarse particulates (10mm or less, PM10). The filtered gas was measured using two measuring devices: 1) MRU AMPRO 2000 gas analyzer with an upgraded CO sensor measured temperature, CO concentration (ppm), % O_2, % CO_2, NO concentration (ppm), and NO_2 concentration (ppm) 2) PCE Instruments Air Quality Meter/Particle Counter measured PM2.5 and PM10. Before connecting the exhaust to the prototype an initial reading was taken. Readings of temperature and gases were recorded every 2 minutes for 30 minutes. Highest PM2.5 and PM10 were also compared; revealing the filtration capability of the water in the 5-gal bucket. Only one test was run per day allowing the euthanasia box to reach atmospheric concentration between tests.

Results and discussion

From the CFD simulation, it was evident that CO at the given volume flow rate filled and distributed throughout the euthanasia box. The results from the prototype are observed in Table 1.

Table 1 shows CO levels reached above 4% with water filtration and without it. These high CO levels suggest that non-catalytic converter gasoline engines have the potential to ensure death of an animal humanely and rapidly. The lower temperature change showed in Table 1 suggests water filtration has the ability to cool the incoming exhaust. NO and NO_2 in the prototype decreased with water filtration as shown in Table 1; however, little is known on what ppm would cause irritation and would be acceptable. More research is needed to determine how to minimize NO and NO_2. PM2.5 in Table 1 did not show a decrease when comparing without and with water filtration, but PM10 did. The PM2.5 and PM10 results were unreliable. Variable results and fluctuations of the sensor suggest it inaccurately recorded the high PM levels. More research is needed to determine how to effectively reduce these to a reasonable amount. Non-catalytic converted gasoline engine exhaust euthanasia could provide the swine industry with a form a gas euthanasia that is safe, effective, ethical, and affordable if proper filtration occurs. More research is needed to determine a reliable filtration method that minimizes unwanted products.

References

1. AVMA Panel on Euthanasia (2020). AVMA Guidelines for the Euthanasia of Animals: 2020 Edition. **https://www.avma.org/sites/default/files/2020-02/Guidelines-on-Euthanasia-2020.pdf**

2. T.H. Greiner, Ph.D., P.E. Carbon Monoxide Poisoning: Vehicles (AEN-208). Iowa State University. Available at: **https://www.abe.iastate.edu/extension-and-outreach/carbon-monoxide-poisoning-vehicles-aen-208/**. Last accessed 10 November 2021.

3. C.R. Tidemann; D.H. King. 2009. Practicality and humaneness of euthanasia of pest birds with compressed carbon dioxide (CO_2) and carbon monoxide (CO) from petrol engine exhaust. Wildlife Research. 36. 522-527.

Table 1: Comparison of average gas concentration

	No water filtration	Water filtration
Carbon monoxide (%)	6.29	4.05
Oxygen (%)	5.81	7.45
Carbon dioxide (%)	8.55	7.6
Nitrogen oxide (ppm)	61	39.5
Nitrous oxide (ppm)	21.5	14.5
PM 2.5 (ug/m^3)	988	993.5
PM 10 (ug/m^3)	993	927.5
Temperature change (°F)	+14.75	+1.5

https://doi.org/10.54846/am2022/84

Evaluation of nursery mortality and production performance using controlled and natural *Glaesserella parasuis* exposure

K. Parker[1], BS; T. Girard[2], DVM; C.A. Moura[2], DVM, PhD
[1]Iowa State University College of Veterinary Medicine, Ames, Iowa; [2]Iowa Select Farms, Iowa Falls, Iowa

Introduction

Glaesserella parasuis is known to be a main pathogen causing polyserositis, fibrinous exudate, and systemic infection in nursery pigs. *Glaesserella parasuis* (GPS) often increases nursery mortality as a secondary disease to swine influenza and porcine reproductive and respiratory syndrome virus (PRRSV).[1] There are multiple ways to assist in controlling GPS within a herd, including commercial and autogenous vaccines, as well as using a controlled GPS exposure. In previous studies, a controlled GPS exposure program has been more effective than vaccination programs.[2] Early introduction of GPS can reduce the risk of systemic infection and improve performance. In the current study a live GPS inoculum was given with the intent of reducing nursery mortality and increasing production performance compared to those that did not receive the inoculum.

Objective

To compare levels of GPS challenge by evaluating nursey mortality, mortality with fibrin, and treatment rate between controlled GPS exposure and natural exposure.

Materials and methods

The study is a systematically randomized field trial on a wean-to-finish site including pigs from a gilt only, PRRSV endemic sow farm with a history of GPS disease in offspring post-weaning. Pigs were inoculated by litter via oral exposure at three days old with a live GPS inoculum that matched a pre-existing serotype present within the flow. The control pigs did not receive the inoculum and were identified by ear tags. Experimental and control pigs remained separate through weaning, transportation, and placement at the finisher. Pigs were placed in a single-stocked wean to finish site. The study groups were held within the same air space but did not have nose-to-nose contact. Necropsies were conducted on all mortalities to grossly detect fibrin. All treatments and mortality were recorded by pen. Four sampling events took place where lung, lymph node, and tonsil samples were taken from acutely dead pigs and submitted for histopathology and serotyping of GPS. The litters at the sow farm were the experimental units, while the individual pigs were the observational units. The proportion of mortality and mortality with fibrin between groups was evaluated by binomial regression while the treatment rate between the groups was evaluated by Poisson regression.

Results and discussion

Mortality in the experimental group was 12.44% post-weaning. Mortality in the control group was 13.00% post-weaning. Total post-weaning mortality was 12.72%. On post-mortem examination 42.93% of the control group mortalities had fibrin present while 32.95% of the experimental group mortalities had fibrin. Of the 42 pigs that samples were submitted from, 6 returned positive for GPS, all were positive for PRRSV, and one grouped sample was positive for influenza A. In addition to GPS, multiple other respiratory bacterial challenges were detected. Overall, mortality in the experimental group was less than the control group. The experimental group had significantly less treatments administered than the control group ($P = 0.0002$).

Results show improvement in mortality and treatment rate within the experimental group indicating this procedure may provide value to offspring from disease challenged farms. Implementation must be carried out with an existing serotype present within the herd to prevent further complications. Limitations of this study include minimal replication and therefore a small power.

References

1. Iowa State University. Swine Disease Index. Available at **www.vetmed.iastate.edu/vdpam/FSVD/swine/index-diseases/glasser-disease**. Accessed 7 July 2021.

2. Oliveira et al.:2004 JSHAP -123.

https://doi.org/10.54846/am2022/85

Evaluating castration techniques and effects of scrotal hernia prevalence

R. Kanefsky[1], BS; B. Burton[2], DVM; Jeff Okones[3], DVM
[1]Tufts Cummings School of Veterinary Medicine; [2]Suidae Health and Production; [3]Pharmgate Animal Health

Introduction

The most seen complications after porcine castration are abscess, hemorrhage, hematomas, and hernias.[1] It is estimated that the incidence of scrotal hernias among the swine population is around 1%, with some genetic influences.[2] Trauma to the urogenital region and genetic predisposition[2] have been shown to cause scrotal herniation, but beyond that the etiology is unknown.[4] The process of castration potentially tears a hole in the inguinal ring to allow for intestinal migration into the now empty scrotum.[1] Once a barrow has a scrotal hernia, he is more likely to have a lower weight gain prior to weaning[3] and is a zero-value pig at market. The most common method of castration used by the industry involves physical castration. There are two ways of separating the testes and spermatic cords from the body, either via physically tearing them or by cutting them with a blade. The objective of this study was to determine if there is a correlation between method of castration and prevalence of scrotal hernia.

Materials and methods

One sow farm was used for this study. Measurements were taken for 5 weeks of litters, with 168 pigs from 57 sows. There were two treatment groups per farm; one where the spermatic cord is physically pulled free when removing the testes from the incision site at castration and another where the cord is cut using a scalpel blade. Sows were randomly assigned to a treatment group as they enter the farrowing room. Information on the success of the litter was recorded until weaning. At farrowing, the sow ID, parity, farrow day on 100-day calendar, born alive, stillborn, and mummies were collected. At processing and castration, the number of barrows in the litter and number of cryptorchids in the litter were recorded. Every day until weaning, the number of deaths, cause of death (euthanized from scrotal rupture, laid on, or other), and the sex of the dead pig were recorded. At weaning, the number of pigs weaned, the number of barrows, and the number of scrotal hernias were recorded as well.

Results and discussion

Of 120 total sows in the study, the male offspring of 61 sows (51%) were castrated using the cut method and 59 sows (49%) using the pull method. Omitting for congenital defect, a resultant 182 piglets were castrated using the cut method and 135 piglets were castrated using the pull method. 2 (0.1%) piglets castrated using the cut method and 1 (0.1%) piglets castrated using the pull method were diagnosed with a scrotal hernia. Between the two methods, there was no statistically significant association between diagnosis with hernia and castration method (odds ratio = 0.67, CI: [0.11, 13.04], $P = 0.99$). While our sample size may have resulted in being underpowered to detect an association, this result can be considered preliminary evidence that method of castration has no impact on the incidence of scrotal hernias. Caretakers can castrate in whichever way is most comfortable to them that adheres to Pork Checkoff guidelines. Hernias yield pigs that are of zero value to the farmer. Determining which castration method yields less hernias can lead to less zero value pigs for the producer and better welfare for the pigs. Further research should investigate other actions taken by the caretaker and veterinarian that can prevent more scrotal hernias.

This research was supported in part by the US Department of Agriculture, Agricultural Research Service, US Meat Animal Research Center, Clay Center, NE.

References

1. St. Jean G, Anderson DE. Surgery of the swine reproductive system and urinary tract. **https://veteriankey.com/surgery-of-the-swine-reproductive-system-and-urinary-tract/**. Published September 3, 2016. Accessed June 29, 2021.

2. Vogt DW, Ellersieck MR. Heritability of susceptibility to scrotal herniation in swine. American Journal of Veterinary Research. 1990 Sep;51(9):1501-1503.

3. Searcy-Bernal R, Gardner IA, Hird DW. Effects of and factors associated with umbilical hernias in a swine herd. J Am Vet Med Assoc. 1994 May 15;204(10):1660-4. PMID: 8050950.

4. Romano GdS, Ibelli AMG, Lorenzetti WR, Weber T, Peixoto JdO, Cantão ME, Mores MAZ, Morés N, Pedrosa VB, Coutinho LL, Ledur MC. Inguinal Ring RNA Sequencing Reveals Downregulation of Muscular Genes Related to Scrotal Hernia in Pigs. *Genes*. 2020; 11(2):117. **https://doi.org/10.3390/genes11020117**

https://doi.org/10.54846/am2022/86

An investigative analysis of a biosecurity program identifying critical control points and improving the efficiency and efficacy of measuring compliance

Isaac N. Goldner[1]; BS; Brigitte Mason[2]
[1]University of Illinois College of Veterinary Medicine, Urbana, Illinois;
[2]Country View Family Farms, Hagerstown, Indiana

Introduction

Biosecurity protocols are precautions taken to minimize the risk of introducing an infectious disease into an animal population. In result, biosecurity and compliance help lessen the following effects, such as economic loss and lower production efficiency. Biosecurity program compliance is not always complete, leaving opportunities for disease to pass through mitigation and infect a herd. Biosecurity program compliance is crucial with the ever-encroaching African Swine Fever and other foreign animal diseases. This observational study will evaluate biosecurity programs within an integrated production system to identify the control points, hazards, and opportunities for intervention.

Materials and methods

A swine production system with farms in the eastern United States wanted to investigate their biosecurity program. During this study, the following biosecurity programs were chosen to be evaluated: visitor/vendor entry, supply entry, and market hauling/transportation. Written biosecurity SOPs were reviewed and organized to generate a short verbal survey focusing on the main control points. The visited sites were delivered the survey for each biosecurity program to quantify the team's knowledge of the given tasks. Later, the same sites were revisited to evaluate the execution of each process. Observation results were compared to verbal answers to determine actual understanding of the SOPs. Using these protocol evaluations, critical control points were determined, and the hazards observed were evaluated with a heat map program. The heat map program compares the difficulty (cost/labor of a hazard), frequency of occurrence, and assurance to health (current mitigation effectiveness), of which assurance to health scores were assessed through surveys completed by industry professionals and average scores were used.

Discussion and results

In conclusion, observation and hazard analysis yielded similar results with all three biosecurity programs. Similar observations within all programs identify non-compliance, absent or unnecessary protocols, and disregard for clean/dirty lines as the most frequently viewed. Main interventions strategies include increased training for all parties involved, adjusting written protocols, and increased audits with appropriate mitigation strategies to ensure increased observations of practices. Location on the heat map demonstrated that these hazards occurred frequently and had a lower difficulty to address. Market hauling/transportation was the branch of highest concern, due to the frequent hazard observation, severity of control points associated with them, and lack of biosecurity SOP protocols. Based on this project, other production systems and the industry should challenge themselves to use a similar strategy to evaluate their biosecurity programs. One non-compliance biosecurity program could mean the difference between a healthy herd of animals and an outbreak.

Acknowledgment

A huge appreciation and thanks to the veterinarians and team members at Elanco Animal Health and Country View Family Farms for their involvement and help with this project.

https://doi.org/10.54846/am2022/87

Development of national resources for the practical implementation of swine depopulation

K. Gitchell[1], BS; S. Webb[2], MS; M. Pierdon[3], VMD, DACAW; E. Strand[1], PhD, LCSW; A. Canon[2], DVM, MPH
[1]College of Veterinary Medicine, University of Tennessee, Knoxville, Tennessee;
[2]American Association of Swine Veterinarians; [3]School of Veterinary Medicine, University of Pennsylvania, Kennett Square, Pennsylvania

Introduction

During 2020, the COVID-19 pandemic in the United States forced packers to reduce slaughter capacity resulting in a market disruption which forced an estimated 2.5 to 3 million market-ready finishing pigs to remain on the farm. Producers explored alternative production practices and marketing options to manage the backlog, and some pigs had to be depopulated as a last resort. This emergency market disruption revealed key gaps in industry knowledge, resources, and preparedness regarding large-scale depopulation. The objectives of this project were to: 1) conduct interviews to gather experiential data on setup, implementation, and efficacy of swine depopulation methods; 2) develop detailed swine-specific recommendations for practical on-farm implementation of depopulation methods; and 3) develop supplemental resources to assist swine veterinarians and farmers before and after depopulation.

Materials and methods

Thirty-four swine veterinarians, animal health officials and academicians were invited to participate in an online survey and telephone interview. The telephone interview transcripts were uploaded to ATLAS.ti for qualitative analysis and coded based on common themes. Online survey and telephone interview results were used by a 9-member panel of veterinarians to develop depopulation recommendations and supplemental depopulation resources.

Results and discussion

Eighteen respondents completed the online survey, and an additional 6 partial surveys were submitted. Of the 24 respondents, there were 16 veterinarians, 5 academicians, 2 animal health officials, and 1 unknown. One respondent had < 5 years of swine industry experience, 9 had 5-10 years, 6 had 11-15 years, and 8 had > 15 years. Respondents reported their previous experience with depopulation methods (Table 1). Fifteen phone interviews were completed, and 4 common themes were identified: communications, team challenges, resources needed, and stakeholder perspectives.

The panel used these results to create depopulation recommendations addressing procedures, labor and throughput estimates, safety considerations, seasonality impacts, carcass disposal restrictions, setup configurations, and needed resources for 12 depopulation methods. A Preparing and Caring for Responders section was included to describe the human psychological and physical impacts that might result from depopulation. A depopulation decision tool was created to guide and document the decision-making process of when to depopulate and which method will work best for the operation and situation. A resiliency debrief tool was developed to aid professionals in both monitoring and taking action to address any psychological impacts of depopulation. Case studies for each method were developed to give producers successful examples of specific resources and configurations used. Recordkeeping forms and literature reference lists were also created.

From these first-hand depopulation experiences, swine veterinarians now have more resources to fill the knowledge gaps identified in 2020. These resources will assist veterinarians in decision making, planning, and implementing methods successfully. By using the debriefing resources throughout the process, teams will improve communication and address challenges. When another emergency arises and requires swine depopulation, these national education resources will be practical and beneficial for optimizing responder and animal welfare.

Table 1: Respondents previous experience with depopulation methods

Depopulation method	Respondents with experience
Gunshot	11
Penetrating captive bolt	16
Non-penetrating captive bolt	13
Electrocution	11
Ventilation shutdown plus	5
Sodium nitrite	8
Carbon dioxide	14
Carbon monoxide	6
Manual blunt force trauma	10
Injectables/euthanasia	1

Poster 09

https://doi.org/10.54846/am2022/88

The effect of zinc and copper supplementation during late gestation on subsequent farrowing house performance of commercial sows

J. Moeller[1], BS; D. Drebes[2], DVM; R. Schmitt[2], PhD; J. Flohr[2], PhD; F. Zarate[2]; R. Moreno[2]; S. Brake[2], BS

[1]The Ohio State University College of Veterinary Medicine, Columbus, Ohio; [2]Seaboard Foods, Guymon, Oklahoma

Introduction

Efficient swine production starts with reproductive performance. Recent research shows that supplementing sow diets with trace minerals, such as zinc and copper, in late gestation results in improved litter characteristics (eg, increased birth weights) and ultimately decreased pre-wean mortality.[1-4] The objective of this study was to determine if supplementing sow diets with zinc and/or copper during late gestation is an effective method to decrease pre-wean mortality in a commercial production setting.

Materials and methods

This generalized randomized block study looked at the effects of copper and/or zinc on the reproductive performance of 2,411 sows across 4 different breeding groups. The study took place on a 13,000 head, multiparous, PRRS unstable, commercial sow farm. The experimental unit of this study was the individual sow. Sows were assigned to one of four treatment groups in sequential blocks of 10. The four treatments were a control (no supplementation), zinc supplement (+ 957 ppm zinc sulfate premix), copper supplement (+ 473 ppm copper chloride premix), or both a zinc and copper supplement (+ 957 ppm zinc sulfate premix, + 473 ppm copper chloride premix). All gestating sows were fed the system's standard gestation diet and treatments were given as a ~1 oz/day top-dress, or a ~2oz/day top-dress for combined copper and zinc. Treatments began at 12 weeks of gestation and ended when they were loaded into a farrowing room, which occurred 3 days before their expected farrow date. Once in the farrowing room, sows were monitored and cared for according to Seaboard Foods' standard operating procedures with minor deviations to maintain treatment integrity (eg, cross-fostering only between same treatment groups). When a room was closed, a pig inventory census was performed, where number of pigs born alive, number of stillborn, number of mummies, and number of piglets cross-fostered were recorded for each sow. On the day before piglets were weaned, the number dead during lactation and the total number weaned were recorded per sow. After weaning, the data was analyzed using descriptive statistics.

Results and discussion

Supplementing with copper and/or zinc showed no significant effect on pre-wean mortality ($P = 0.23$, 14.36-15.35), contrary to previous studies (Table 1). One possible influencing factor is that this farm was only multiparous, while previous studies included gilts. For nutritionists, this may bring more questions to the table about the impact of copper and zinc on reproductive performance. Interestingly, even though not statistically significant, there was a tendency for sows that were supplemented with both copper and zinc to have less piglets born alive and more stillborn piglets as a percentage of total born piglets (Table 1). Further studies looking at the relationship between supplementing copper and zinc and total born breakdowns may be warranted.

References

1. Holen et al.: 2020, Transl. Anim. Sci. - 4:519-530.
2. Vallet et al.: 2014, J. Anim. Sci. – 92:2422-2432.
3. Cromwell et al.: 1993, J. Anim. Sci. – 71:2996-3002.
4. Thacker: 1991, Can. Anim. Sci. – 227-232.

Table 1: Descriptive analysis and one-way ANOVA outlining and comparing the sample populations and their respective reproductive characteristics

	Feed treatment				SEM	Probability, *P* <
	Control	Copper	Zinc	Copper & Zinc		
Sows, n	568	582	577	573	—	—
Parity, avg	3.52	3.49	3.47	3.38	0.10	0.68
Gestation length, d	115.5	115.6	115.5	115.4	0.07	0.02
Total born, n	14.90	14.81	15.12	15.05	0.16	0.55
Born alive, %	89.37	89.63	89.62	87.83	0.49	0.07
Stillborn, %	7.86	8.21	7.90	8.98	0.35	0.08
Mummies, %	2.76	2.14	2.47	3.17	0.30	0.34
Lactation length, d	17.25	17.19	17.16	17.17	0.14	0.68
Pre-wean mortality, %	15.35	14.36	15.20	14.89	0.77	0.23

Poster 10

https://doi.org/10.54846/am2022/89

Evaluating water quality and water availability parameters within a pork production system

K. Robinson[1]; L. Johnson[2]; D. Murray[2]; A. Sponheim[3]; T. Michels[3]
[1]University of Missouri, Columbia, Missouri; [2]New Fashion Pork, Jackson, Minnesota;
[3]Boehringer Ingelheim Animal Health USA Inc., Duluth, Georgia

Introduction

Though often not considered, water availability and water quality are important in maximizing pig performance and profitability. Inadequate water supply weakens physical barriers like the skin and mucociliary system, making it more susceptible to pathogen invasion. Dehydration also decreases feed intake resulting in decreased growth rates and reduced feed efficiency.[1] Conversely, excessive flow rates cause water and water medication wastage which increases manure pit volume and increases the necessary amount of water medication. Poor water quality can cause increased mineral deposits in water pipes, providing a more favorable environment for pathogens to create biofilms. This buildup eventually causes clogged water nipples and reduced water flow rates. The biofilm itself can also cause a decrease in water quality. Pathogens within the biofilms are often resistant to antimicrobials and disinfectants making it difficult to rid the pipes of theses pathogens.[2] The objective of this study was to evaluate water availability parameters and survey critical aspects of water supply that impact water availability to pigs within a pork production system's nursery and grow/finish sites.

Materials and methods

Sixteen different sites were selected for this study. Water samples were collected from the office at each site and the chlorine in the sample was immediately neutralized with sodium thiosulfate. The pH of the water was measured and recorded. Additionally, a short section of the water main pipe was removed from each site and the patent diameter of the pipe was measured. Both the water sample and the section of water main pipe were sent to The University of Minnesota Veterinary Diagnostic Lab for aerobic culture. The flow rates of all water nipples were measured and recorded in each barn. Information regarding the source of water (well/rural), the presence of inline filters, and the presence and functionality of chlorine dioxide water treatment units was recorded.

Results and discussion

The average flow rates for all water nipples within each farm ranged from 670 mL/min to 2127 mL/min. When compared to the industry standard flow rates (250-500 mL/min for nurseries; 500-1000 mL/min for wean/finish), 75% of water nipples were above the recommended range, and 6% were below the range. Water flow rates by site are recorded in Figure 1. The measured pipes ranged from no blockage to 32% blockage with an average of 8.5%. The pipe and water cultures grew a wide variety of bacteria with Citrobacter, Aeromonas, Bacillus, and Acinetobacter species being the most commonly recurring. The 75% of water nipples above the recommended range lend an opportunity for correction which could result in significant savings. Further investigation is needed to interpret the correlation between flow rates and biofilm buildup.

References

1. Menegat, Mariana B., Robert D. Goodband, Joel M. DeRouchey, Mike D. Tokach, Jason C. Woodworth, and Steve S. Dritz. 2019. Kansas State University Swine Nutrition Guide: Water in Swine Nutrition.

2. Waddilove, J. (2021, November 12). *The effect of biofilms in pig production*. The Pig Site. Retrieved November 12, 2021, from **https://www.thepigsite.com/articles/the-effect-of-biofilms-in-pig-production**

Figure 1: Variation of flow rates by site.

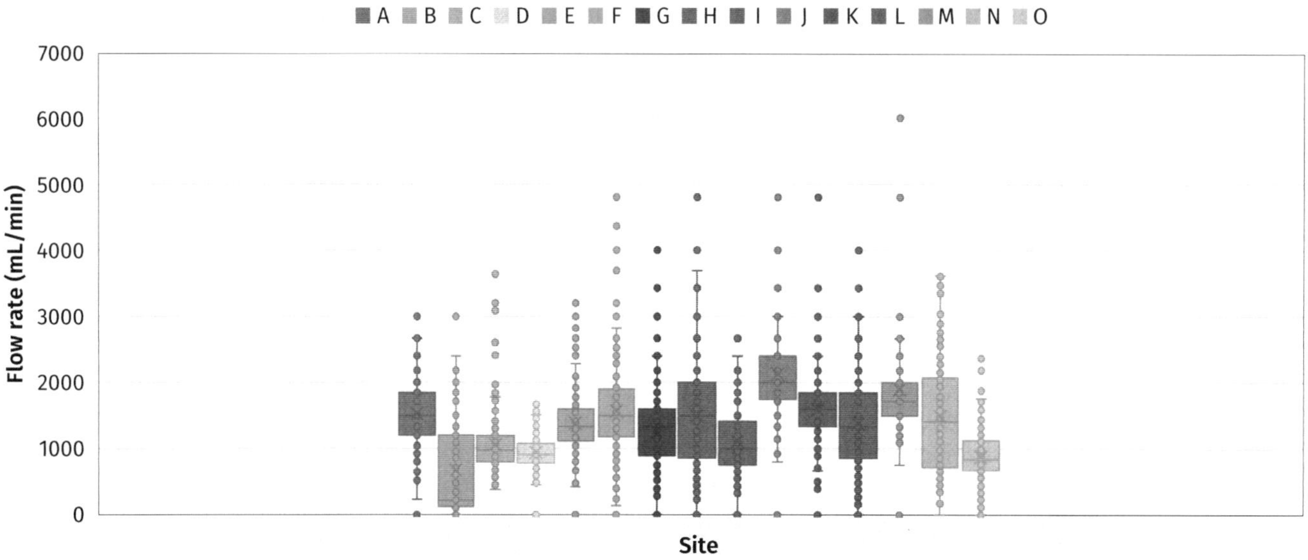

https://doi.org/10.54846/am2022/90

Differentiation of H3N2 IAV-S antigenic sites by RT-qPCR

Courtney Wangler, BS; Suzanna Storms, DVM
Applied Livestock Epidemiology Lab, University of Illinois, College of Veterinary Medicine

Introduction

IAV-S is a negative-strand RNA virus that is an endemic disease in swine production. IAV-S rapidly mutates, which can cause continuous reinfection and has zoonotic implications. PCR is used to diagnose and differentiate between IAV-S clades. To our knowledge, PCR has not been used to differentiate antigenic sites specifically. The ability to identify IAV-S based on their antigenic sites of the HA gene suggests the ability to be able to identify cross-reactive IAV-S strains. This technology would be useful in determining when new antigenic variants enter a herd and if vaccinations for one strain would provide cross-protection for a newly introduced strain.

Objective

The objective of this study is to develop PCR primers for reported antigenic sites of the hemagglutinin (HA) gene of three H3N2 strains and one H1N2 strain in order to differentiate strains based on PCR primer specificity.

Materials and methods

The HA antigenic sites of four H3N2 strains were evaluated at amino acid sites previously recognized in human IAV and swine IAV publications.[1,2] These amino acid sites were identified as the main antigenic mutation points and speculated as the reason for the loss of cross-protection between strains. HA-specific primers were designed for three IAV-S H3N2 viruses: MN/A01668936/2016, A/swine/Nebraska/A02524799/2020, and A/swine/Iowa/A02524878/2020. The H1N2 strain: A/swine/Indian a/A02478520/2019 was also included. All primers were developed following IDT and ThermoFisher primer design guidelines. The viruses were grown and passaged on MDCK cells in our lab. RNA was extracted from clarified cell culture supernatants with Cytiva Sera-Xtracta nucleic acid extraction kit using the manufacturers' instructions. RT-qPCR was completed using the Invitrogen™ SuperScript™ III Platinum™ SYBR™ Green One-Step qRT-PCR Kit and QuantStudio 3 (Applied Biosciences) according to WHO influenza PCR detection procedures for Influenza Universal Matrix gene and using IDT guidelines for designing PCR parameters. WHO Influenza Universal Matrix Primers were used to determine if cultures were growth positive for influenza. The study primers were tested to see if they were PCR positive for their specific virus, and the specificity of their annealing with non-synonymous viruses. Determination of specificity of primer binding was based on CT values, melt curve plots, and the amplification plots.

Results and discussion

The Nebraska primer was specific for the Nebraska H3N2 strain as shown as it had a CT value of 25.066 and having a proper melt curve, evident in Figure 1 Part A. The Indiana primer was only specific for the Indiana H1N2 strain as show with its CT value of 21.789 and a proper melting curve as seen in Figure 1 Part B. The Iowa primer was specific for all the H3N2 strains tested and not the H1N2.

This "proof of concept" study shows that primers can be designed based off the reported antigenic sites of a specific IAV-S strains. The primers can differentiate between clades, and some were strain specific. Currently, antigenic cross protection is measured by serum antibody hemagglutinin inhibition assays, which are time consuming, and require pigs to be challenged with each virus and produce serum antibodies. This targeted PCR technology would make it faster and easier to determine if and when new antigenically different variants of IAV-S enter a herd. PCR technology could be done on nasal swab samples, making sample much quicker and easier for practitioners and farm workers. These primers would help determine if current vaccines could provide immunity for the new strain, by annealing with the antigenic sites that are similar, or showing a negative reaction for antigenically different viruses. To prove our findings, further studies would include challenging pigs and collecting serum to confirm cross protection of antigenic sites targeted by the primers. This technology has the potential to replace time consuming lab assays, challenging pigs, and providing rapid turnaround times.

References

1. Abente, E. J., et al (2016). The molecular determinants of antibody recognition and antigenic drift in the H3 hemagglutinin of swine influenza A virus. *Journal of VIirology*, *90*(18), 8266–8280.

2. Bolton, M. J., et al (2018). Antigenic evolution of H3N2 Influenza A viruses in swine in the United States from 2012 to 2016. *Influenza and Other Respiratory Viruses*, *13*(1), 83–90.

Figure 1: Melting curve plots for Nebraska **(A)** and Indiana primer **(B)**.

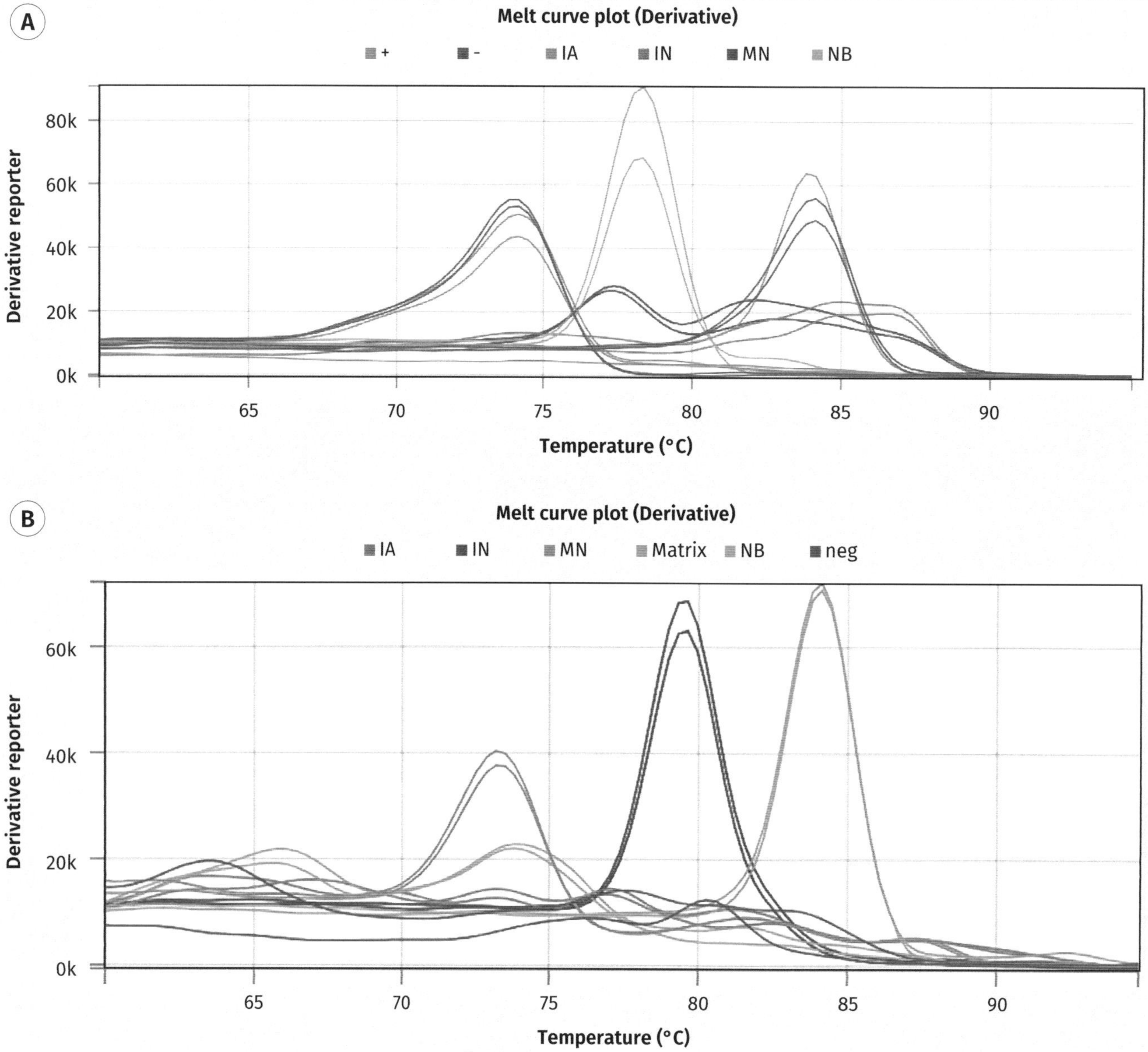

https://doi.org/10.54846/am2022/91

Nitrogen gas delivered by highly expanded foam for the depopulation of swine

Lindsay Miller[1], BS; Todd Williams[2], DVM; Brent Pepin[3], DVM, MS; Carissa Odland[2], DVM
[1]University of Minnesota College of Veterinary Medicine; [2]Pipestone Veterinary Services, Pipestone, Minnesota; [3]Cytotheryx, Inc

Introduction

An initial diagnosis of a foreign animal disease in the United States will require the depopulation of all confirmed positive and exposed swine. Depopulation is defined as the method by which a population of animals must be destroyed efficiently, with as much consideration given to animal welfare as practicable.[1] Current approved methods for the depopulation of swine are limited. Nitrogen gas (N_2) alone is currently not a commonly applied method in the field for pigs due to practicality. This study evaluates the use of a novel approach of delivering N_2 with a high expansion foaming technology as an option for the depopulation of swine. This method works by enveloping the animals in foam to rapidly deplete oxygen levels and replace with inert N_2, leading to death by anoxia. Our objectives were to verify that death with foam delivery occurs from anoxia from N_2 inhalation and not foam airway occlusion, confirm appropriate application across all stages of production, and validate alternative confinement applications during depopulation.

Materials and methods

A Nitrogen Foam Delivery System (NFDS) was applied to sows and boars (n = 31), market ready pigs (n = 135), nursery pigs (n = 184), and wean pigs (n = 325). N_2 was delivered to the animals in a high expansion foam. Three different containment methods were evaluated: a corral (steel cattle penning with plywood lining), a standard roll off dumpster, and an earthen pit (declined slope dug into the ground to a depth of 4 feet). Animals were loaded into containments, and N_2 containing foam was delivered until it reached 2.5 times the height of the animals present. Animals were monitored for cessation of movement while in the foam. As soon as it was safe to do so, animals were confirmed dead by observation of the lack of respiration and a negative corneal reflex. A subset of 10 animals per replicate of N_2 foam exposure was necropsied to evaluate airways for the presence of foam that may indicate asphyxiation.

Results and discussion

A chi-square test on the preliminary data on the frequency of observed potential asphyxiation across age groups in the roll off dumpster revealed no statistical significance ($X^2(3, n = 69) = 5.42$, $P = 0.14$). A chi-square test examining the containment type in market weight animals and frequency of potential asphyxiation revealed no statistical significance ($X^2(2, n = 48) = 0.675$, $P = 0.71$) (Table 1). Preliminary observations suggest that the NFDS primarily causes death by N_2 inhalation in the analyzed containment methods across multiple stages of production

Reference

1. American Veterinary Medical Association. Guidelines for the Depopulation of Animals: 2019 Edition. Schaumburg, Illinois: American Veterinary Medical Association; 2019. Available at: **https://www.avma.org/sites/default/files/resources/AVMA-Guidelines-for-the-Depopulation-of-Animals.pdf** Accessed 08 October 2021.

Table 1: Frequency of asphyxiation per each containment type in market weight animals

Containment type	Asphyxiation	No asphyxiation	Total
Corral	3	17	20
Earthen	2	7	9
Roll off	2	17	19
Total	7	41	48

Poster 13

https://doi.org/10.54846/am2022/92

Evaluating the ability of disinfectants to control coccidia in farrowing crates

Austin Janssen[1], BS; Marjorie Schleper[2], DVM; Steven Tousignant[3], DVM, PhD
[1]Iowa State University, College of Veterinary Medicine, Ames, Iowa; [2]Christensen Farms, Sleepy Eye, Minnesota; [3]Boehringer Ingelheim Animal Health USA Inc., Duluth, Georgia

Introduction

Coccidiosis has long been a costly issue in all pork-producing areas of the world.[1,2] Mainly caused by *Cystoisopora suis* in farrowing rooms, this condition causes diarrhea as a result of enteric damage in piglets of 7-21 days of age. Coccidiosis can be prevented with oral administration of coccidiostats to individual piglets, but this method is labor-intensive. Control of coccidiosis in farrowing rooms is rooted in sanitation. However, *Cystoisopora suis* has proven itself to be incredibly resistant against most disinfectants. The purpose of this study was to evaluate the effectiveness of four different disinfectants against coccidia in farrowing crates.

Methods and materials

Data collection took place at a sow farm with an established coccidia issue, in 5 2-row farrowing rooms. Fecal samples were collected to confirm that coccidia was present. Diagnostics were run on tissue samples from 3 piglets to confirm coccidiosis. Once empty, all crates within each room were cleaned with a degreaser and power-washed with cold water. After wash inspection, quadrants of rooms (6 crates per quadrant) randomly received 1 of 4 disinfectants: citric acid (Xtreme Bio, Midwest Ag Solutions), aqueous ozone water (AO water, BioSafe Distributors), quaternary ammonium salt (Z-71, Zoono), and copper sulfate pentahydrate (Stalosan F). Z-71 and AO water were both applied with a backpack sprayer at 1 L/80 m^2 and 0.325 L/m^2, respectively. Stalosan F was applied at 50 g/m^2 by hand, and was applied once more with sows and piglets present in the crates. Xtreme Bio was applied at a 1-to-256 oz dilution rate. Two litters per quadrant (10 litters per treatment total) of piglets were tagged and weighed the day of farrowing after litter set. Fostering was allowed within quadrants. All litters within rooms were observed from day 1 through weaning and scored on scour severity (SS, number of scour piles, 0-3, 0 being none, 1 being 1-3 piles, 2 being 4-6 piles, 3 being 7+) and fecal consistency (FC, 0-3, 0 being normal, 3 being watery). Piglets were weighed again at 16 days of age and average daily gain (ADG) was calculated. Data was analyzed using ANOVA. Final weight and ADG were adjusted for start weight.

Results and discussion

Fecal samples showed moderate coccidia presence in all rooms. Histopathology from the tissue samples of 2 of the 3 necropsied piglets showed enteric lesions and coccidian organisms within their enterocytes. 2 litters were removed from consideration for average daily gain measurements due to the sows experiencing agalactia, both being in the AO water treatment group. As shown in Table 1, no differences were shown between any of the treatments in average daily gain, fecal consistency, or scour severity. A potential limitation of this study was the absence of aisles in front of the sows, making it so someone needed to step in every crate every day for feed drop, possibly allowing for contamination. In conclusion, *Cystoisospora suis* continues to be a challenging pathogen in the swine industry and additional studies into farrowing stall hygiene or other control methods may be warranted.

References

1. Nwanta et al.: 2010, JSHAP -19:6.
2. Webster et al.: 2007, AASV- pg. 39.

Table 1: LS means, st. err., and *P*-values for performance and fecal parameters

Treatment	Xtreme Bio	Stalosan F	BioSafe AO Water	Zoono Z-71	Std. err.	*P*-value
Start weight	3.15	2.96	3.25	3.21	0.07	0.0054
Final weight (adj.)	10.5	10.8	10.9	10.8	0.19	0.3399
ADG (adj.)	0.46	0.48	0.49	0.48	0.01	0.3399
FC > 6 Days	1.0	0.8	0.8	0.9	0.09	0.1166
SS > 6 Days	0.9	0.7	0.7	0.8	0.09	0.2718

https://doi.org/10.54846/am2022/93

Caretaker movements impact on the spread of porcine coronaviruses and porcine reproductive and respiratory syndrome virus in growing pigs

Alexis Berte, BS; Kate Dion, DVM; Derald Holtkamp, DVM, MS; Daniel Linhares, DVM, MBA, PhD;
Gustavo De-Sousa-E-Silva, DVM, MS, PhD; Mafalda Pedro Mil-Homens, DVM, MS

Introduction

Biosecurity plays an important role in preventing the spread of pathogens between swine premises. Each risk event can potentially spread pathogens and cause economic loss. Caretaker movements from one swine premise to another may spread porcine reproductive and respiratory syndrome virus (PRRSV) and porcine coronaviruses. The role of caretaker movements in spreading pathogens from one premise to another is not well established in wean to market premises. The objective of this study was to investigate if the movement of caretakers from one premise to another impacted the spread of porcine epidemic diarrhea virus (PEDV), porcine deltacoronavirus (PDCoV), transmissible gastroenteritis virus (TGEV), and PRRSV.

Materials and methods

Eight oral fluids (OF) were collected every two weeks from equidistant locations (ie, fixed spatial sampling) from 63 nursery groups from May 2020-June 2021. To be eligible, the groups originated from sow farms that were Category II-vx or IV for PRRSV and negative for PEDV, PDCoV, and TGEV. The OF samples were pooled into groups of 4 and tested for PEDV, PDCoV, TGEV, and PRRSV using reverse transcriptase polymerase chain reaction (PCR) to determine status for each pathogen.

Timesheet Mobile (Freedom Telecare, Westborough, MA) was used to identify all caretaker entries into the study premises and movements between other premises three days prior. The pathogen status of the other premises visited prior to entering the study premises was collected utilizing production system diagnostic data. The number of caretaker entries to the study premises for each two-week period, and the visits to other premises visited before entering the study premises were evaluated as a ratio of the number of visits from positive premises over the total number of visits. For the analysis, a logistic regression model per pathogen was performed to determine the association between the number of caretaker entries and the ratio of the number of visits from positive premises over the total number of visits and pathogen status in R package (version 4.1.1).

Results and discussion

Out of the 63 groups, 6 tested positive for PRRSV, 8 tested positive for PEDV, and 12 tested positive for PDCoV. Four groups for PEDV and 3 groups for PDCoV were removed from the analysis after their initial test was positive. No groups tested positive for TGEV. The number of standard employee visits for the entire nursery period ranged from 39.1 to 141.9, with an average of 73.1 visits. Table 1 lists the average number of visits to other premises prior to entering the study premises. Caretaker visits and the ratio of previous site visits were not statistically significant ($P < 0.05$) in the logistic regression for any of the pathogens.

The results in this study indicate that there was no association between pathogen status and caretaker visits and the ratio of previous site visits. This result may imply that the risk of virus introduction was mitigated due to external biosecurity practices, or that the source of the virus was from other sources. Time periods other than 3 days for prior site visits should be evaluated to determine if a different time period would impact virus introduction.

Table 1: Average number of visits to other premises prior to entering the study premises

	PDCoV positive	PDCoV negative	PEDV positive	PEDV negative	PRRSV positive	PRRSV negative
Prior negative	226.75	207.40	261.63	222.80	188.50	214.79
Prior positive	52.58	60.81	42.88	42.73	87.50	55.21

https://doi.org/10.54846/am2022/94

Comparison of nebulization protocols for gilt exposure to *Mycoplasma hyopneumoniae*

S. Simmons[1], BS; J. Pittman[2], DVM, MS, DABVP; E. A. Jablonski[3], DVM, MS
[1]College of Veterinary Medicine, North Carolina State University, Raleigh, North Carolina;
[2]Smithfield Hog Production, North Region, Waverly, Virginia; [3]Zoetis, Parsippany, New Jersey

Introduction

Mycoplasma hyopneumoniae (MHP) colonizes the cilia of airways and causes enzootic pneumonia in swine. Replacement gilts that are unexposed and subsequently become exposed on positive sow farms are likely a major source of transmission to newborn pigs, as gilts continue to shed MHP for upwards of 214 days post exposure.[1] One challenge with MHP control is that the transmission rate is very slow in a population with an expected probability of infection of 0.6.[2] Therefore, one strategy utilized to control MHP is to expose replacement gilts at an age young enough to stop shedding prior to their first farrowing. One method utilized and demonstrated to be effective for exposure is nebulization of MHP positive lung homogenate into the animal airspace. The objective of this study was to identify the method of nebulization exposure of MHP homogenate that led to the highest rate of positive animals.

Materials and methods

Replacement gilts, from MHP positive sources, were exposed at 10 weeks of age in nursery buildings. Each building was 120 feet long and either 20 or 26 feet wide with 48 pens. On day 0, sixty pigs, from throughout the barn were identified with an ear tag and sampled via deep tracheal swab. The solution used to expose the gilts was prepared using one gallon of distilled water mixed with 48.7 grams of non-iodized salt and 120 mL of lung homogenate confirmed to contain MHP. Prior to nebulization the curtains were closed and ventilation fans were turned off. Two gallons of nebulization solution were used for each replicate and nebulizers were always run at maximum particle size (50 microns) and flow rate (gal/15 min). T01 (n = 2) used two nebulizers, placed in the middle of the barn facing opposite directions, towards the front and towards the back of the barn. The nebulization duration was 15 min. T02 (n = 4) used four nebulizers, two of which were placed 40 feet from the entrance of the barn, facing opposite directions and two were placed 80 feet from the entrance, facing opposite directions. The nebulization duration was 7.5 min. T03 (n = 4) had two nebulizers placed on a rolling utility cart each facing one side of the barn toward the pens. Each pen was exposed to the nebulized material for 30 sec (12 min per barn). Following exposure, the ventilation system remained off for 15 min, except in T03, where it remained off for 5 min. On day 28, the same sixty pigs were sampled as previously. Tracheal swabs and nebulization solutions were individually tested for MHP by PCR. SAS Proc Means (SAS 9.4, Cary NC) procedure was used for summary statistics.

Results and discussion

The nebulization solutions had an average PCR Ct value of 26.6 (range 23.4-29.1). On day 28, 9.2% (range 0.0-18.3%) of the pigs in T01, 24.8% (range 0.0-94.5%) of the pigs in T02, and 69.4% (range 4.7-93.3%) of the pigs in T03 were positive. Day 0 and 28 results are listed in Table 1, with ranges in parentheses. A limitation of this study was that gilts were not sourced from MHP negative sow farms, which may have resulted in variable colonization due to prior immune or exposure status. This study indicates that fogging as a method of exposure in MHP positive populations leads to highly variable results, and this study should be repeated in a MHP naïve population to more clearly discern differences in nebulization exposure techniques.

References

1. Pieters M, Pijoan C, Fano E, Dee S. An assessment of the duration of Mycoplasma hyopneumoniae infection in an experimentally infected population of pigs. Vet Microbiol. 2009 Mar 2;134(3-4):261-6. **doi: 10.1016/j.vetmic.2008.08.016.** Epub 2008 Aug 23. PMID: 18835112.

2. Roos LR, Fano E, Homwong N, Payne B, Pieters M. A model to investigate the optimal seeder-to-naïve ratio for successful natural Mycoplasma hyopneumoniae gilt exposure prior to entering the breeding herd. Vet Microbiol. 2016 Feb 29;184:51-8. **doi: 10.1016/j.vitmic.2016.01.008.** Epub 2016 Jan 7. PMID: 26854344

Table 1: Percentage of MHP PCR positive animals on days 0 and 28, by treatment group

Treatment	T01	T02	T03
Day 0 % positive	8.4 (0.0-16.7)	2.5 (0.0-10.0)	37.8 (6.7-56.7)
Day 28 % positive	9.2 (0.0-18.3)	24.8 (0.0-94.5)	69.4 (4.7-93.3)

https://doi.org/10.54846/am2022/95

Identifying pathways of entry of ASF virus into sows farms and potential improvement in biosecurity to prevent viral entry

Random Bolda[1]; R. Pogranichniy[1]; J. Richt[1]; J. Woodworth[2]; D. Ragland[4]; Jordan Gebhardt[1]; C. Jones[3]; Gustavo S. Silvia[5]; Derald Holtkamp[5]
[1]Department of Diagnostic Medicine/Pathobiology, Veterinary Diagnostic Laboratory, College of Veterinary Medicine, Kansas State University, Manhattan, Kansas; [2]Department of Grain Science and Industry, College of Agriculture, Kansas State University, Manhattan, Kansas; [3]Department of Animal Sciences and Industry, College of Agriculture, Kansas State University, Manhattan, Kansas; [4] Department of Clinical Science, College of Veterinary Medicine, Purdue State University, West Lafayette, Indiana; [5]Veterinary Diagnostic and Production Medicine Department, College of Veterinary Medicine, Iowa State University, Ames, Iowa

Introduction

African swine fever (ASF) is an economically significant virus that can lead to high morbidity and mortality. This disease is spread in a variety of ways to both domestic and wild swine, between individuals of a swine herd, pork products and by insect vectors. The ability of viral particles to remain infective in the environment makes it a particularly difficult disease for producers to control once introduced to their herds.[1] While ASF prevention is vital to the wellbeing of producers, customized recommendations based on their biosecurity on the farm are needed. The goal of this study was to identify biosecurity gaps to reduce outbreaks of ASF in Asia. Data was collected via a survey based on questions from the infected and not infected farms using the instrument developed for the Swine Health Information Center funded Rapid Response Program and adopted for ASF virus.[2]

Materials and methods

The survey for biosecurity on the swine farms was established to help with risk analysis of virus introduction into the farms. According to the answers provided by veterinarian or biosecurity staff on the farms, a risk analysis was analyzed as described with some modifications.[3] Seven major categories were examined on an individual production level including swine movement, pickup and deliveries, people movement, pork/food product entry, manure removal, domestic/wild animals and insects, and the air/water. Each major category was then broken down into several subcategories, and these subcategories rated on a scale of 0.00 to 1.00 with 0.00 being areas in which there is least risk and 1.00 being the most vulnerable (high risk) to the introduction of ASF virus to the sow farms.

Results and discussion

The subcategory results from completed surveys were then ranked as higher risk (> 0.50) and lower risk (≤ 0.49) and compiled into reports that highlighted areas that would most benefit from specific biosecurity actions to reduce the risk of a farm introducing ASF to its herds. Producers may then choose to invest in improving biosecurity efforts into areas that will produce the greatest results for their operation as well as reduce reservoirs for the virus to spread in their community between swine farms. The top five shared risks to the spread and introduction of ASF on the farms included entry of water, replacement animals import and testing, equipment used by on-farm employees, semen testing and handling, and sanitary transportation. More surveys should be analyzed from different sow farms to improve our survey.

References

1. Cwynar et al.:2019, Viruses p. 310.
2. Silva et al.: 2018, PVM p. -116:122.
3. SHIC website: **https://www.swinehealth.org/**

Poster 17

https://doi.org/10.54846/am2022/96

Development of antigen based diagnostic ELISA for *Mycoplasma hyopneumoniae*

Rachel Bradley[1]; Venkatramana D. Krishna[1]; Maria Pieters[1]; Jian-ping Wang[2]; Maxim C-J. Cheeran[1]
[1]Department of Veterinary Population Medicine, College of Veterinary Medicine; [2]Department of Electrical and Computer Engineering, College of Science and Engineering, University of Minnesota, St. Paul, Minnesota

Introduction

Mycoplasma hyopneumoniae is one of the most prevalent respiratory bacterial pathogens that causes enzootic pneumonia, a chronic infection that leaves pigs vulnerable to other respiratory pathogens. Timely detection of disease is critical for treatment, control and preventing spread. Current diagnostic testing such as culture and PCR are time consuming, require specialized laboratory facilities, and skilled personnel. Using monoclonal antibodies specific to *M hyopneumoniae*, an ELISA was optimized and evaluated for antigen detection using various clinical samples. The ELISA developed in this study will be used in a magnetic particle spectroscopy (MPS) device, which detects alterations in the harmonics of antibody-functionalized magnetic nanoparticles (MNP) when bound to mycoplasma antigen. MPS technology will give producers access to quick, reliable, point-of-care diagnostics that can be applied in the field.

Methods

Mice monoclonal and rabbit polyclonal antibodies specific to *M hyopneumoniae* were purified and confirmed to be specific to *M hyopneumoniae* with Western blot testing. Initial testing showed that the monoclonal antibody used as the capture antibody and the polyclonal antibody used as the detection antibody resulted in the best signal and lowest background noise. Using this sandwich design, the antigen-based ELISA was optimized. The optimized ELISA was then tested with serial dilutions of *M hyopneumoniae* to determine dynamic range and limit of detection. Next the concentration curve for the ELISA was established. Analytical specificity of the optimized ELISA was determined using strains common swine respiratory pathogens and commensals. Finally, the ELISA was evaluated using spiked clinical samples and *M hyopneumoniae* experimentally infected pig clinical samples.

Results and conclusions

The optimum concentration of monoclonal capture antibody was 1ug/mL. The optimum concentration of polyclonal detection antibody is 1 ug/mL. The optimum blocking buffer was 1% BSA. These concentrations resulted in the best signal with the least background absorbance readings. The dynamic range for the optimized assay was 226.57 ng/mL -7,250 ng/mL of *M hyopneumoniae* lysate, and the limit of detection is 226.57 ng/mL. The ELISA was confirmed to not cross-react with any of the other respiratory viral or bacterial pathogens or commensal *M flocculare*. *M hyopneumoniae* was detected in spiked nasal swabs and spiked tracheal wash samples. Finally, we tested 21 PCR negative and 21 PCR positive laryngeal samples from experimentally infected pigs ELISA. Figure 1 demonstrates the change in detection of *M hyopneumoniae* in the experimentally infected laryngeal swabs from 0 days post infection to 28 days post infection. Of the PCR negative samples 20/21 were negative and 1 was positive by ELISA. Of the PCR positive samples 9/21 were positive by ELISA for *M hyopneumoniae* in laryngeal swabs. Further testing of various spiked and experimentally infected clinical samples is still on going. After optimizing the ELISA for clinical sample testing, the monoclonal antibodies conjugated to MNPs. The MNPs will be used to detect antigens using MPS devices, which will be evaluated for their diagnostic ability to detect *M hyopneumoniae* in farms.

Figure 1: ELISA results of laryngeal swabs (21) of experimentally infected pigs with *M hyopneumoniae* 0 days post infection (dpi) compared to 28 dpi

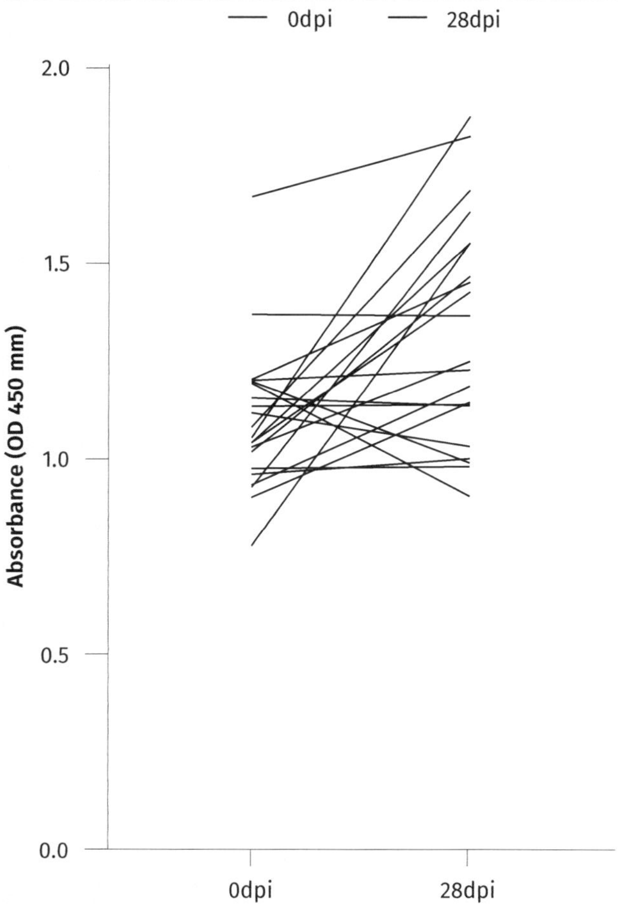

https://doi.org/10.54846/am2022/97

Evaluation of enteric disease on hemoglobin concentrations, complete blood cell counts and growth in piglets

C. Bublitz[1], BS; K. Castevens[2], DVM; B. Pecoraro[1], BS; G. W. Almond[1], PhD, MS, DVM
[1]North Carolina State University, Raleigh, North Carolina; [2]Country View Family Farms, Middletown, Pennsylvania

Introduction

Little data exists describing hemoglobin (Hb) concentrations and complete blood cell counts (CBCs) in piglets with enteric disease. The objectives of this study were to examine potential relationships between Hb concentrations, body weight, and CBCs in piglets through to weaning and to examine the effects of enteric disease on these parameters.

Materials and methods

This study was conducted on a 2000-sow, commercial farrow-to-wean farm. To compare Hb concentrations in preweaning piglets, eight litters were selected as healthy with no clinical signs of enteric disease and eight litters were chosen to represent enteric disease based on clinical signs and diarrhea found on the first day of the study for a total of 16 litters total sampled. The group with enteric disease was treated with potato starch and Drygiene®. Fifty percent of the litters with clinical enteric disease and 50% of the healthy litters had already been processed and given iron dextran (Uniferon®) injections on the first day of the study. The other 50% of litters with clinical enteric disease and 50% of the healthy litters had not yet been processed and given iron dextran injections at the start of the study. From each litter, five piglets were randomly selected to be included in the study and sampled each week. Each piglet enrolled had an ear tag placed the first week for later identification. From each sow, one of five piglets was chosen to be followed each week for a CBC in addition to the other measurements. Each blood sample for CBC analysis was taken via a jugular venipuncture and stored in EDTA tubes on ice until processing by the Clinical Pathology lab at NC State CVM. Body weights and Hb measurements, as well as selected CBCs were taken again seven days later and again eight days later, one day prior to weaning. All piglet and sow data including date of birth, sex, pig ID, sow ID, and number of piglets per liter born dead and alive were recorded. An analysis of variance (ANOVA) model was run with day and treatment taken into consideration and compared with a Tukey's test

Results and discussion

Two piglets from the healthy control group and three piglets in the enteric disease group contributed to the total mortality rate of 6.25%. Overall, the study showed that the group of healthy pigs consistently gained more weight between measurements and had lower Hb concentrations.[1] Between weeks 1 and 2, the difference in weight gain was found to be significant with a $P < 0.01$. Similarly, to other previously published papers, weight gain difference between Weeks 1 and 3 (referred to as "Overall") was also found to be significant with a $P < 0.01$.[2] Figure 1 shows lower Hb concentrations in piglets without enteric disease consistently over the three-week collection period.

CBC data showed differences in reticulocytes, platelet volume, mean corpuscular hemoglobin concentration, red blood cell distribution, red blood cells, hemoglobin, and hematocrit.[3] Multiple diagnostic tests showed rotavirus to be the confirmed diagnosis on this farm for the herd's enteric disease.

Based on this study's results, body weight seems to play a negative correlating factor in Hb concentration. Presumable dehydration of piglets with enteric disease could play a role in higher hemoglobin levels and differences in CBC values. Piglets with enteric disease consistently had higher hemoglobin concentrations than those without, leaving more questions about the physiological processes taking place that could cause sick piglets to be utilizing less hemoglobin. The variance in CBC data between healthy piglets and those with enteric disease argues that these clinical diagnostics may be more useful in swine medicine than previously thought.

References

1. Almond G et al.: 2021, AASV p. 308-312.
2. Egeli AK et al.: 1999, RVS p. 179-84.
3. Chevalier TB et al.: 2021, JSHAP p. 189-199.

Figure 1: Healthy piglets had significant differences in hemoglobin (Hb) concentrations from those piglets with enteric disease during weeks 2 and 3 of data collection with a $P < 0.05$.

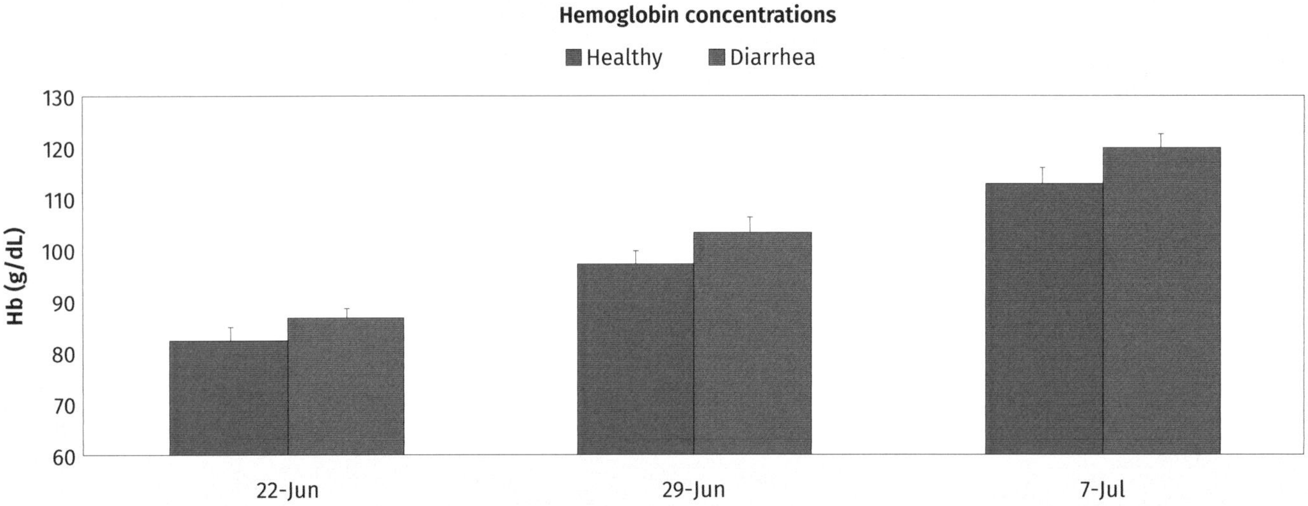

https://doi.org/10.54846/am2022/98

Use antibiotic treatments to validate *Mycoplasma hyorhinis* and *Mycoplasma hyosynoviae* titers

Lucas Buehler, BS; Taylor Engle, DVM, MS; Bryant Chapman, DVM

Introduction

Mycoplasma hyorhinis (MHR) and *M hyosynoviae* (MHS) are ubiquitous bacteria without cell walls associated with polyserositis and arthritis. MHR is part of the normal flora of the upper respiratory system in young pigs and is often diagnosed in nursery pigs.[1] MHS is diagnosed during mid-late finishing and replicates in the upper respiratory tract of pigs. Susceptible antibiotic therapies include use of tulathromycin early post-weaning or lincomycin later in the finishing phase.[2] The objective was to validate MHR and MHS ELISA tests in pigs receiving five different treatment strategies with the hypothesis that titers would be decreased in antibiotic treated pigs compared to controls.

Materials and methods

Two commercial systems with a history of lameness due to MHR and MHS in mid-late finishing were enrolled in the study. There were different treatment groups (n = 10 pigs/trt) on each farm (Table 1). The antibiotics used for treatment groups were administered at labeled dosage. Serum sampling and clinical observations were performed at 3 (ie, weaning), 6, 9, and 12 weeks and every 3 weeks until market. Serum was individually assayed via ELISA for MHR and MHS.

Results and discussion

Data were analyzed by summary statistics. At the time of abstract submission, there were few clinical cases of MHR or MHS on either farm. There was a numerical reduction in the percentage change of titer levels in T02 pigs. In Farm B, T02 had lower percent change in MHR titer levels compared to all other treatments. The timing of MHS titers is premature of normally expressed clinical signs; at this time there were few clinical cases and no differences in the patterns of MHS titers observed. Clinical signs of lameness began increasing around 12 weeks of age so we anticipate the prevalence of MHS should increase during the late finishing phase. These diseases are variable in tracking and using ELISA testing to track disease and treatments isn't necessarily a straightforward method to use. Depending on final results determination of the proper therapeutic method and timing of administration, will be deduced.

References

1. Robbins et al.: 2014 EAFS p. 261-276.

2. Scheiber et al.: 2012 Allen D. Leman Swine Conference 73-74:39.

Table 1: Treatment groups and % change of titer values from 3 weeks of age

	Grp	Treatment	Farm A								Farm B							
		Weeks of age	**3**	**6**	**9**	**12**	**15**	**18**	**21**	**24**	**3**	**6**	**9**	**12**	**15**	**18**	**21**	**24**
MHR % change	T01	Negative control	0	8	35	262	382	403	*	*	0	37	73	74	390	404	*	*
	T02	Tulathromycin IM at 3 wks	0	9	87	144	324	398	*	*	0	82	92	81	390	405	*	*
	T03	Tulathromycin IM at 3 wks + lincomycin in feed at 10 wks	0	127	585	938	823	1312	1308	1572	NT	NT	NT	NT	NT	NT	NT	NT
	T04	Tulathromycin IM at 3 wks + lincomycin in water at 10-12 wks	NT	NT	NT	NT	NT	NT	NT	NT	0	158	317	263	385	405	*	*
	T05	Lincomycin in feed at 10 wks	0	24	303	752	675	1324	1321	1591	NT	NT	NT	NT	NT	NT	NT	NT
	T06	Lincomycin in water at 10-12 wks	0	-55	77	210	331	1310	1311	1584	0	81	355	348	379	402	*	*
		Weeks of age	**3**	**6**	**9**	**12**	**15**	**18**	**21**	**24**	**3**	**6**	**9**	**12**	**15**	**18**	**21**	**24**
MHS % change	T01	Negative control	0	-51	-65	8	-59	-45	*	*	0	-75	-82	-86	-36	33	*	*
	T02	Tulathromycin IM at 3 wks	0	-69	-86	-83	-73	-62	*	*	0	-53	-69	-79	-56	70	*	*
	T03	Tulathromycin IM at 3 wks + lincomycin in feed at 10 wks	0	-62	-69	-68	-30	-36	-52	-24	NT	NT	NT	NT	NT	NT	NT	NT
	T04	Tulathromycin IM at 3 wks + lincomycin in water at 10-12 wks	NT	NT	NT	NT	NT	NT	NT	NT	0	-56	-77	-82	-63	-54	*	*
	T05	Lincomycin in feed at 10 wks	0	-63	-68	-63	-70	-5	-39	-5	NT	NT	NT	NT	NT	NT	NT	NT
	T06	Lincomycin in water at 10-12 wks	0	-41	-13	-32	-5	-12	-39	19	0	-56	-78	-80	-67	-58	*	*

NT = not tested in this farm
* = results pending

https://doi.org/10.54846/am2022/99

Comparing the diagnostic sensitivity of various samples for the detection of *Mycoplasma hyopneumoniae* in recently exposed nursery-aged piglets

M. Kellen[1], BS; B. Sexton[2], DVM; A. Maschhoff[2], DVM; J. Waddell[3], DVM
[1]Iowa State University, College of Veterinary Medicine, Ames, Iowa; [2]The Maschhoffs, LLC, Carlyle, Illinois;
[3]Boehringer-Ingelheim Vetmedica, St. Joseph, Missouri

Introduction

Mycoplasma hyopneumoniae (Mhp) contributes to enzootic pneumonia and survives in the lungs of infected animals for months. When moving replacement gilts into sow farms, the disease state of the gilt must be known prior to placement. If gilts are unknowingly Mhp positive and moved into a site, there can be profound costs to all aspects of production. Detection methods for Mhp including oral fluids (OF), tracheobronchial swabs (TBS), and serum (S) have been evaluated for their sensitivity and effectiveness, primarily in gilts that are 12 to 20 weeks old and nearing their selection weight. Few studies have looked at Mhp detection in nursery-aged pigs.

The objective of this study was to compare the diagnostic sensitivity of TBS, S, and fresh lung tissues (FLT) for detection of Mhp or antibodies in nursery-aged pigs recently exposed to Mhp.

Materials and methods

The study was conducted at a four-room nursery with 10 pens and approximately 500 pigs per room. Fourteen 3-week-old pigs in each room were identified, tagged, intratracheally inoculated with 5 mL of Mhp positive lung homogenate (1.7×10^6 genomic copies), and placed in a centrally located pen. 6 days post inoculation (DPI) TBS were collected. At 13, 26, 42, and 54 DPI, TBS, S, and FLT were collected. TBS and S were collected from inoculated and non-inoculated pigs; three inoculated pigs per room were euthanized for fresh and fixed tissue collection. Non-inoculated pigs were randomly selected from the other nine pens in each room. TBS and FLT were tested via real-time polymerase chain reaction (RT-PCR) for Mhp and S was tested via IDEXX ELISA. The non-inoculated TBS were pooled within pen; the inoculated TBS were tested individually. 76 DPI, TBS were collected from four inoculated pigs and 30 non-inoculated.

Results

Mhp was detected in 7 inoculated pig TBS, 1 inoculated pig S, and 3 inoculated pig FLT (Table 1). 13 DPI one pig was positive on FLT and TBS. One pig tested positive via TBS on 13 DPI and positive via FLT on 26 DPI. The positive FLT 42 DPI was positive via TBS.

Discussion

This study observed a low prevalence Mhp population and compared diagnostic sensitivity of TBS, S, and FLT in nursery-aged pigs. Of the positive TBS, 57% were also FLT positive. Of the positive FLT, 100% were also TBS positive. If Mhp was present in the lungs, the expectation was that it would be found via TBS. However, there was less confidence in TBS as not all of the pigs who tested positive via FLT were positive via TBS. Based on the FLT RT-PCR, Mhp was present throughout the study. Why there was a difference in shedding pattern between inoculated and non-inoculated pigs resulting in no detection of Mhp in non-inoculated pigs is unknown. Slower spread of Mhp and unique shedding patterns may have contributed to the lack of detection in the non-inoculated pigs. As Mhp continues to have economic impacts, understanding of Mhp in young pigs will be beneficial to maintaining healthy herds. As additional field studies establish the diagnostic sensitivity of various samples to Mhp, veterinarians and producers can determine monitoring strategies for Mhp in young pigs prior to movement.

Table 1: Samples Mhp positive per type submitted; tracheobronchial swabs (TBS), serum (S), fresh lung tissues (FLT)

	TBS		S		FLT
DPI	**Inoculated**	**Non-Inoculated**	**Inoculated**	**Non-Inoculated**	**Inoculated**
6	0/53	0/36	NA	NA	NA
13	5/53	0/36	1/53	0/72	1/12
26	1/40	0/36	0/40	0/72	1/12
42	1/28	0/36	0/28	0/72	1/12
54	0/16	0/36	0/16	0/72	0/12
76	0/4	0/6	NA	NA	NA

https://doi.org/10.54846/am2022/100

Identification and characterization of post-weaning *E coli* and *Salmonella* spp in a large production system

H. Lathom[1], B.S; A. Vennekotter[2], DVM; S. Copeland[2], DVM; C. Smith[3], DVM; J. Okones[3], DVM
[1]North Carolina State University College of Veterinary Medicine, Raleigh, North Carolina;
[2]Prestage Farms, Inc, Clinton, North Carolina; [3]Pharmgate Animal Health, Wilmington, North Carolina

Introduction

Post-weaning diarrhea (PWD) is a major cause of morbidity, mortality, and economic loss in pork production due to decreased pig performance and increased drug costs. Enterotoxigenic *Escherichia coli* (ETEC) with the F18 and K88 adhesion and *Salmonella* spp are two common pathogens associated with this disease process. Clinical signs of PWD include mild to marked watery diarrhea that is absent of mucus and blood, poor growth performance and septicemia.[1] To prevent PWD, vaccination using a producer manufactured live, bivalent oral product can be implemented in the nursery.[2] The producer manufactured vaccine relies on competitive exclusion by introducing avirulent F18 and K88 ETEC into the gastrointestinal microbiome so that wild-type enterotoxin producing ETEC cannot adhere. This study investigates the microbiome of nursery pigs before and after the implementation of a live, bivalent, oral, producer manufactured product.

Materials and methods

The investigation is a cross sectional survey across three nursery mega flows in a large, commercial production system in Eastern United States. The pre-vaccination sampling occurred after a three-month break in all vaccination against *E coli*. In each flow, four sites were sampled, each correlating to a different age category: 0-, 2-, 4- and 6-weeks post nursery placement. At each site, 15 pigs were necropsied to create a sample size of n = 180 for the pre-vaccination data collection and an additional 180 will be sampled for the post-vaccine data collection to make a n = 360. Complete fresh and fixed gastrointestinal sets were obtained from each pig. The samples were pooled by nursery site and age group for applicable diagnostic testing. All ETEC and salmonella isolates were genotyped and serotyped, respectfully. An enteric coronavirus PCR panel was performed when suggestive histological lesions were observed by the pathologist. Following the pre-vaccination data collection, the live, oral, bivalent (F18/K88 ETEC) vaccine manufactured by the producer was administered within 48 hours post-placement onto the nursery and when the pigs were, on average, 22-23 days old.

Results

For both *E coli* and *Salmonella* spp 10/12 sites had positive culture results. Of those 5/12 had ETEC (Table 1). All cultures obtained had the F18 pili with the most prevalent toxins being: EAST1, LT, STb, STx2, and STx2e. The most prevalent *Salmonella* serotype is 4.[5],12:1:- and muenchen. Rotavirus A, B and C was detected via PCR in all age groups. Coccidia species were obtained in the 0- and 2-week post placement groups. All the enteric coronavirus panels were negative.

Discussion

The pre-vaccination sampling results indicate a moderate to marked prevalence of both *Salmonella* spp and ETEC in this production system. The most clinically relevant *Salmonella* species being Typhimurium and 4.[5],12:i:- and F18 ETEC. The bacterial prevalence will be compared with the post-vaccination samples taken in winter 2021.

References

1. Rhouma, M., Fairbrother, J.M., Beaudry, F. et al. Post weaning diarrhea in pigs: risk factors and non-colistin-based control strategies. Acta Vet Scand 59, 31 (2017). **https://doi.org/10.1186/s13028-017-0299-**

2. Vangroenweghe, F. A. (2021). Vaccination with an E. coli F4/F18 Vaccine for the Prevention of F4-ETEC Postweaning Diarrhea Resulted in Reduced Post-weaning Mortality and Antibiotic Use. Journal of Animal Science , 99, 140. **https://doi.org/10.1093/jas/skab054.2**

Table 1: Pre-vaccination *E coli* and *Salmonella* spp results by age

Weeks placed	*E coli* growth	No. ETEC Pili and toxins	*Salmonella* spp	Serotyping	Viral PCR Protozoal	Histology
0	3/3	0/3	2/3	Typhimurium, Kuzendorf, Muenchen Infantis	Rotavirus A, B, C Coccidia	Atrophic enteritis Ulcerative colitis
2	1/3	1/3 F18, EAST1, LT, STb	3/3	Muenchen, London, 4.[5],12:i:-	Rotavirus A, B, C Coccidia	Atrophic enteritis Ulcerative colitis
4	3/3	3/3 F18, PAA, EAST1, LT, STb, STx2, STx2e	2/3	4.[5],12:1:-	Rotavirus A, B, C	Suppurative cryptitis Colitis Enteritis
6	3/3	1/3 F18, EAST1, LT, STb, STx2, STx2e	3/3	Rissen, 4.[5],12:i:- Muenchen	Rotavirus A, B, C	Purulent cryptitis Ulcerative colitis Atrophic enteritis

https://doi.org/10.54846/am2022/101

Investigation of vesicular lesions in pigs with unknown causative agents

Ethan Aljets[1], BS; Haiyan Huang[1], BS; Ganwu Li[1], PhD; Pablo Piñeyro[1], DVM, MVSc, DVSc, PhD; Kathleen Wood[2], DVM; Laura Bradner[1], MS; Chris Rademacher[1], DVM; Phillip Gauger[1], DVM, MS, PhD; Rodger Main[1], DVM, PhD; Jianqiang Zhang[1], MD, MS, PhD
[1]Veterinary Diagnostic and Production Animal Medicine, College of Veterinary Medicine, Iowa State University, Ames, Iowa; [2]Christensen Farms & Feedlots, Sleepy Eye, Minnesota

Introduction

Swine vesicular lesions may be caused by foot-and-mouth disease virus (FMDV), Senecavirus A (SVA), swine vesicular disease virus (SVDV), vesicular stomatitis virus (VSV), or vesicular exanthema of swine virus (VESV). FMDV, VESV, and SVDV are considered foreign animal disease (FAD) pathogens to the USA. FMDV is of high economic risk.[1] When FAD cases are submitted to the ISU VDL, they are always tested for FMDV with SVA testing being at client's request. Additional PCR testing is conducted at the FADDL for FMDV, SVA, SVDV, VSV, and VESV. From 2015-2021, the ISU VDL received 361 swine vesicular cases including 219 cases that tested negative for all known vesicular viral pathogens. The objective of this study was to further investigate such cases to explore potential unrecognized causative agents of swine vesicular lesions.

Materials and methods

All 228 swine vesicular cases submitted to ISU VDL from 2015 to 2020 tested FMDV-negative via PCR. Of those 228 there were 86 (37.7%) SVA-negative, 102 (44.7%) SVA-positive, and 40 (17.6%) not tested for SVA. Samples were available for 84 of these cases, which included 32 SVA-negative, 37 SVA-positive, and 15 SVA-not tested. We selected all 32 SVA-negative cases, and then selected 5 SVA-positive cases for further investigation. From January to April of 2021, ISU VDL received 133 swine vesicular lesion cases; all tested negative for FMDV and SVA via PCR, and 16 of these cases underwent further investigations. Additional investigations include bacterial culture, virus isolation (VI) using ST and PK-15 cells, metagenomics analysis via next-generation sequencing (NGS), and additional PCRs.

Results and discussion

Twelve randomly selected vesicle swabs had no significant bacterial growth on routine culture. Twenty-three randomly selected clinical vesicle swabs or fluids from the SVA-negative cases (11 samples from 2015-2020 and 12 from 2021) were tested by NGS. Numerous viruses were detected with varying numbers of reads; however, no significant sequences were assembled. In contrast, full-length genome sequences of SVA were obtained from 4 SVA-positive vesicle swabs sharing 98.2-99.3% nucleotide identity. Among the 48 SVA-negative cases, 18 were VI positive based on cytopathic effects, 30 were VI negative. NGS was conducted on all 18 virus isolates. Nearly full-length genome sequences of porcine teschoviruses (PTV) were obtained from 16 cases, including different genotypes (PTV-2, 3, 4, 7, 9, 10, and 11). Porcine sapelovirus virus (PSV) sequences were obtained from 2 cases. The 18 isolates and their clinical samples were tested for PTV, PSV, enterovirus G (EV-G), porcine parvovirus 1 (PPV1), and porcine parvovirus 2 (PPV2) via PCR. Although the virus isolates were strongly positive for PTV or PSV by PCR, their corresponding clinical samples either were negative or had high Ct values for PTV, PSV, EV-G, PPV1, and PPV2 suggesting these viruses may not be the causative agent. Experimental inoculation of pigs with these PTV and PSV isolates is needed to confirm their potential cause of the vesicular lesions. In addition, potential non-infectious factors should also be investigated.

Reference

1. Hayes D, Fabiosa J, Elobeid A, Carriquiry M, et al. Economy Wide Impacts of a Foreign Animal Disease in the United States. *Center for Agricultural and Rural Development Iowa State University.* Working Paper 11-WP 525 November 2011.

https://doi.org/10.54846/am2022/102

Longitudinal evaluation of *Mycoplasma hyopneumoniae* detection in processing fluids

Albert Canturri[1]; Carles Vilalta[1,2]; Juan Manuel Sanhueza[1,3]; Maria Pieters[1,4]
[1]Department of Veterinary Population Medicine, University of Minnesota, St. Paul, Minnesota; [2]Upnorth Analytics; [3]Departamento de Ciencias Veterinarias y Salud Publica, Facultad de Recursos Naturales, Universidad Católica de Temuco, Chile; [4]Swine Disease Eradication Center, University of Minnesota, St. Paul, Minnesota

Introduction

Mycoplasma hyopneumoniae, the primary etiologic agent of porcine enzootic pneumonia, is considered to be restricted to the respiratory system, with very occasional detection in extra-respiratory tissues reported in experimental infection studies.[1] However, recent *in vitro* studies have shown internalization of the bacterium into epithelial cells and paracellular migration.[2] In addition, the detection of *M hyopneumoniae* in processing fluids (PF) was recently described, both in a subclinically infected farm[3] and in the context of a porcine respiratory disease complex farm outbreak.[4] Although valuable, these reports included a limited number of samples collected during a reduced timeframe, in very particular farm conditions. Thus, the aim of this study was to further evaluate the PCR-detection of *M hyopneumoniae* in PF in farms with various health status and management conditions, by regularly testing for an entire year.

Materials and methods

Ten farms from three different production systems were enrolled in the study. Five farms were *M hyopneumoniae*-negative at enrollment and five were positive. Processing fluid collection was performed by farm personnel and consisted in obtaining the testicles from litters born to gilts, born to second parity sows, and born to third parity and above sows. Processing fluids were collected in individual bags on a monthly basis for an entire year, for a total of thirty PF/month/farm. Processing fluid samples were frozen at the farm until shipment to the laboratory. At arrival at the laboratory, samples were thawed at room temperature and PF were collected and aliquoted. Samples were then submitted to the Veterinary Diagnostic Laboratory at the University of Minnesota for *M hyopneumoniae* rt-PCR testing by pools of five, maintaining the same parity and month within every pool. More than 600 aggregated samples, representing near 3,000 individual litters were tested. Results were statistically analyzed by constructing a mixed effects logistic regression model. Additionally, a selection of positive pools from different farms were individually tested, and the resulting samples were submitted for *M hyopneumoniae* DNA sequencing.

Results and discussion

Based on PCR detection in PF throughout time, three farm profiles or scenarios were identified: farms that were constantly negative; farms with variable detection; and farms that were constantly positive over the entire sampling period. *Mycoplasma hyopneumoniae* farm status (negative or positive for the pathogen) was significantly associated with the detection of the bacterium in PF. The proportion of positive samples among parity groups and by month of the year was similar, hence, no association with bacterial detection in PF was observed for these two variables. The obtained six partial and one complete DNA sequences from three different farms were identical and showed 99.8-100% homology to *M hyopneumoniae* J strain. The presence of infected pigs in the farm appeared to be one key factor for *M hyopneumoniae* detection in PF. The origin of the genetic material is uncertain at this point of the investigation and further studies would be needed to determine if, environmental contamination can be ruled out.

References

1. Le Carrou J, Laurentie M, Kobisch M, et al. Persistence of *Mycoplasma hyopneumoniae* in experimentally infected pigs after marbofloxacin treatment and detection of mutations in the parC gene. Antimicrob Agents Chemother 2006;50:1959–66.

2. Raymond BBA, Turnbull L, Jenkins C, Madhkoor R, Schleicher I, Uphoff CC, Whitchurch CB, Rohde M, Djordjevic SP. *Mycoplasma hyopneumoniae* resides intracellularly within porcine epithelial cells. Sci Rep. 2018 Dec 6;8(1):17697. **doi: 10.1038/s41598-018-36054-3**

3. Vilalta C, Sanhueza JM, Murray D, Johnson L, Pieters M. Detection of *Mycoplasma hyopneumoniae* in piglet processing fluids. Vet Rec. 2019 Oct 26;185(16):510. **doi: 10.1136/vr.105475.** Epub 2019 Aug 13.

4. Vilalta C, Garcia- Morante B, Sanhueza JM, Schwartz M, Pieters, M. PCR detection of *Mycoplasma hyopneumoniae* in piglet processing fluids in the event of a clinical respiratory disease outbreak. Veterinary Record Case Reports, ISSN: 2052-6121, Vol: 8, Issue: 1, Page: e001045

https://doi.org/10.54846/am2022/103

In vivo tracheal fluid collection for *Mycoplasma hyopneumoniae* gilt exposure

Fabian O. Chamba Pardo, DVM, MS, PhD; Clint Lichty, DVM; Cathy Templeton, DVM, MBA; Gillian Greaves, BSc, MSc; Courtney Werth, BSc MSc; Guy Moser, BSc
South West Ontario Veterinary Services, Stratford, Ontario, Canada

Introduction

A farrow-to-wean producer in Ontario, Canada expanded from 600 to 1200 sows in early 2020. During that process, the herd became unstable for *Mycoplasma hyopneumoniae* (Mhp). Despite control efforts, respiratory problems were reported in nursery pigs. Euthanizing gilts to harvest lungs to prepare exposure material was not an option in this herd. A field in-vivo tracheal fluid collection method was piloted and evaluated for effectiveness in Mhp gilt exposure.

Methods

The sow herd, negative for PRRSV and IAV, introduces replacement gilts four times per year. A group of approximately 160 gilts of 3 ages (2, 3, and 4-months) are purchased quarterly from a Mhp negative source. Gilts were vaccinated with RespiSure-One® 2-weeks prior to delivery and entered an offsite gilt development barn (GDU), where they showed no clinical signs of Mhp. When gilts were moved into the sow herd, they showed mild, but ongoing, clinical signs of Mhp, despite treatment efforts. Five untreated gilts that had entered the sow farm within 14-days and exhibiting clinical signs were selected for in-vivo tracheal fluid collection. Each gilt was snared, and a flexible catheter was inserted into the trachea through a speculum where 5mL of phosphate buffer saline (PBS) was delivered. Immediately, as much fluid as possible was withdrawn from the trachea through the same catheter using a syringe.[1] The catheter was then flushed with 5mL of PBS to recover residual tracheal fluid. A tracheal fluid aliquot from each gilt was Mhp PCR tested. The total tracheal fluid collected from all 5 gilts (~35mL) was then diluted in ~450mL of PBS to create the gilt exposure material. On the same day, exposure material was used to intra-tracheally (IT) inoculate ~half of the gilts in 10 pens (70/160 gilts) in the GDU. In each pen, ~half of gilts were randomly selected and inoculated with 5mL of the tracheal fluid exposure material. Finally, 3-weeks after IT inoculation, 21 gilts (2 gilts/pen) were randomly selected for laryngeal swabbing (LS) and Mhp PCR testing.

Results

Time to collect in-vivo tracheal fluid from 5 gilts was ~1 hour. The amount of collected fluid from each gilt was ~0.5mL from 4 gilts and ~10mL for 1 gilt, without including the 5mL used to flush the catheter. Gilt tracheal fluid Mhp PCR Ct values ranged from 28 to 38 (median = 33, n = 5). The total volume collected from 5 gilts including the flushed material was ~35mL with a Mhp PCR Ct value of 33. Diluted exposure material (~35mL in ~450mL PBS) was not Mhp PCR tested. Exposed gilts in the GDU began to show mild Mhp clinical signs ~23 days post IT inoculation. Of the 21 LS samples, 16 (76%) tested Mhp PCR positive with Ct values ranging from 30 to 40 (median=36).

Conclusion

In-vivo tracheal fluid collection from gilts was feasible in a commercial sow barn where euthanizing gilts to harvest lung tissue was not possible. Mhp gilt exposure material prepared from diluted tracheal fluid was effectively used to infect gilts in a commercial GDU.

Reference

1. Pieters, M., Daniels, J. & Rovira, A. Comparison of sample types and diagnostic methods for in vivo detection of Mycoplasma hyopneumoniae during early stages of infection. Veterinary microbiology 203, 103-109, **doi:10.1016/j.vetmic.2017.02.014** (2017).

https://doi.org/10.54846/am2022/104

Mycoplasma hyopneumoniae aerosol challenge evaluation in finishing pigs

Fabian O. Chamba Pardo, DVM, MS, PhD; Gillian Greaves, BSc, MSc; Clint Lichty, DVM; Guy Moser, BSc; Courtney Werth, BSc, MSc
South West Ontario Veterinary Services, Stratford, ON, Canada

Introduction

Aerosol *Mycoplasma hyopneumoniae* (Mhp) exposure methods have been described for gilt acclimation,[1] but scarce challenge studies are available in finishing pigs. The feasibility of Mhp aerosol challenge was evaluated in finishing pigs for Mhp vaccines or antibiotic comparisons in commercial settings.

Methods

In 4 studies, at entry pigs were confirmed naïve/negative for Mhp by ELISA (serum) and PCR (laryngeal swabs, LS). The aerosol Mhp challenge method consisted of fogging Mhp positive lung homogenate (Ct range 24 to 28) for 30min using two hurricane foggers simultaneously and 1-2-hours of reduced ventilation in a 400-head (13 pen) commercial finishing site. Pigs were challenged at ~13 weeks (3-weeks after entry, study 1) and at ~10 weeks of age (immediately after entry, studies 2, 3, 4). Natural (study 1) and inoculated (study 4) PRRSV co-infection were also evaluated. Studies 2 and 3 had only Mhp challenge. In studies 3 and 4, pigs were also Mhp challenged intra-tracheally (25% pigs in study 3, 50% pigs in study 4, Ct range 28-34). Challenge was confirmed by Mhp PCR detection in LS 2-3 weeks post-challenge, lung lesions, and daily barn-level cough indices. Lung lesions were confirmed (Goodwin method)[2] at 7-weeks post-aerosol challenge/entry (3-weeks post-intra-tracheal challenge) in study 3, and 3-weeks post-challenge/entry in study 4. Lesions were not assessed in studies 1 and 2. In all studies, a daily barn-level cough index (CI) was estimated by counting: number of coughing bouts (CC), number of pigs assessed (n) and duration of assessment (t) (CI=(CC/n*t)*100). Studies 2, 3 and 4 included control pigs. Production and economic indicators were evaluated in 2 Mhp antibiotic comparisons (studies 1 and 3) and 1 Mhp vaccine comparison (study 2).

Results

Mhp was PCR detected in LS at 2-3 weeks post-challenge in study 1 (n = 24, Ct range 25 to 39), study 2 (n = 59, Ct range 29 to 40), study 3 (8 pools of 3, Ct range 35 to 40) and study 4 (5 lung tissue and 5 bronchial swabs, Ct range 21 to 31). LS were not collected in study 4. Lung lesions in study 3 (n = 7) ranged from 0.3 to 31 (median 11/55 max. score), and in study 4 (n = 5) ranged from 12 to 43 (median 29/55). In study 4, pigs tested PRRSV and Mhp PCR positive 3 weeks after challenge/entry. Study and time adjusted mean ± CI95 daily barn-level cough index was higher (0.4 ± 0.1 vs 1.3 ± 0.1, $P < .01$) and increased a week earlier (2 vs 3-weeks after challenge) with PRRSV co-infection. Removals and deaths were higher in PRRSV co-infected pigs: 3.7% (28/763) vs 8.3% (29/350), $P < .01$. Start weight (30kg), end weight (108kg), pen, block, treatment, and study adjusted mean ± CI95 feed conversion ratio (2.38 ± 0.17 vs 3.04 ± 0.47, $P = 0.04$), live weight (kg)/ton feed (586 ± 22 vs 537 ± 58, $P = 0.21$) and gross margin (CAD)/ton feed (679 ± 38 vs 592 ± 104, $P = 0.21$) were lower with PRRSV co-infection. Controls performed lower than vaccinates (study 2) and medicated groups (study 3).

Conclusion

An aerosol Mhp challenge in finishing pigs provides sufficient infection and mimics field conditions; however, lacks clinical severity similarly to recent reports.[3,4] PRRSV co-challenge and Mhp IT route increased Mhp clinical severity and production loses in finishing pigs.

References

1. Robbins, R. C., Betlach, A. M., Mondragon-Evans, M. R. & Pieters, M. Development of a herd-specific lung homogenate for exposure to *Mycoplasma hyopneumoniae* under field conditions. *J Swine Health Prod* 27, 221-227 (2019).

2. Garcia-Morante, B. et al. Assessment of *Mycoplasma hyopneumoniae*-induced pneumonia using different lung lesion scoring systems: a comparative review. *Journal of comparative pathology* 154, 125-134 (2016).

3. Garcia-Morante, B. et al. Induction of mycoplasmal pneumonia in experimentally infected pigs by means of different inoculation routes. Veterinary research 47, 54, **doi:10.1186/s13567-016-0340-2** (2016).

4. Silva, A. P. S. P. et al. Comparison of *Mycoplasma hyopneumoniae* response to infection by route of exposure. Veterinary microbiology, 109118 (2021).

https://doi.org/10.54846/am2022/105

Detection and investigation of atypical porcine pestivirus within a breed-to-finish farm and off-site nursery and finisher locations

C. Grace Elijah[1], DVM; Cassandra K. Jones[2], PhD; Jason C. Woodworth[2], PhD; Rachel Palinski[1], PhD; Chad B. Paulk[3], PhD; Tom Petznick[4], DVM; Jordan T. Gebhardt[1], DVM, PhD
[1]Department of Diagnostic Medicine/Pathobiology, Kansas State University; [2]Department of Animal Sciences and Industry, Kansas State University; [3]Department of Grain Science and Industry, Kansas State University; [4]Arkcare, Norfolk, Nebraska

Introduction

Atypical porcine pestivirus (APPV) is associated with congenital tremors (CT) and splay leg (SL) in piglets of infected dams. The major cost of this virus is increased pre-weaning mortality due to CT or SL interfering with the piglet's ability to nurse and move around the farrowing stall. A commercial farrow-to-finish farm had increased CT and SL in the farrowing rooms early 2020. Diagnostics on clinically affected pigs' samples identified APPV RNA. No histopathological diagnosis was performed based on consultation with a veterinary anatomic pathologist with experience diagnosing APPV given the PCR positive results and clinical picture. The farm had no previous clinical cases of CT or SL prior to this investigation. The two hypothesized routes were introduction of replacement gilts or incoming semen doses. Therefore, the objectives of this investigation were to determine the prevalence of clinical APPV cases at the farrow-to-finish farm, understand the route of introduction of APPV into the farrow-to-finish farm, and understand the prevalence of APPV viremia within a population of offspring from a gilt multiplication farm through an off-site nursery and finisher barn.

Materials and methods

Farrowing records from the farm were analyzed for the presence of CT or SL of single and multiparous females with affected litters. Blood samples were collected at two different times from the group of intended for replacement gilts for the farrow-to-finish and also the maternal barrows at the isolation nursery barn. Serum and oral fluids were collected from the same pigs at an off-site finisher barn to assess APPV persistence. APPV E2 sequences were obtained from a serum sample from a gilt housed at the isolation nursery intended as a replacement gilt for the farrow-to-finish farm, a semen dose utilized at the farrow-to-finish farm, and serum of a clinically affected piglet in the farrowing room on the farrow-to-finish farm.

Results and discussion

Overall, the APPV prevalence of affected litters within batch farrowing groups ranged from 0 to 31% over the course of this investigation. At the offsite nursery location, pooled serum samples (5 pigs) APPV was detected in 37.5% of pigs on arrival and 77.5% at the conclusion of the nursery period (Figure 1). When evaluating APPV prevalence in individual pigs within pens, it ranged from 20 to 40%. A subset of the outgoing pigs were tested for APPV at the finishing facility. At the start of the turn, this subset APPV prevalence was 26% while at the end of the turn, a 0% APPV prevalence was found in serum of the same animals. Oral fluids were collected from pens of all pigs present in the finisher barn. The APPV prevalence was 100% at the pen level in oral fluids sampled at the start and the end of the finisher turn (cycle threshold values ranging from 20.4-28.7). Sequencing results indicated that the virus circulating in clinically affected piglets was the most similar to an incoming semen dose (98.9% similar). In summary, introduction of APPV into a naïve herd is associated with an increase in clinical CT and SL. Furthermore, APPV RNA was detectable in serum and oral fluids of asymptomatic animals previously exposed to APPV. This research presents a rationale for quarantines of all incoming animals on a farm. Additionally, we showed semen screening for APPV is crucial to limit the potential for APPV introduction. Finally, we showed oral fluids may be a viable technique for screening incoming animals for APPV.

Figure 1: Prevalence of atypical porcine pestivirus (APPV) RNAemia in serum samples pooled by pen in an isolation nursery facility. The diagram on the left is from sampling on 8/20/2020 **(A)**, while the diagram on the right is from sampling on 9/30/2020 **(B)**. Grey indicates a suspected detectable APPV RNA, red pen indicates pooled serum with detectable APPV RNA, and white pen indicates no detectable APPV RNA.

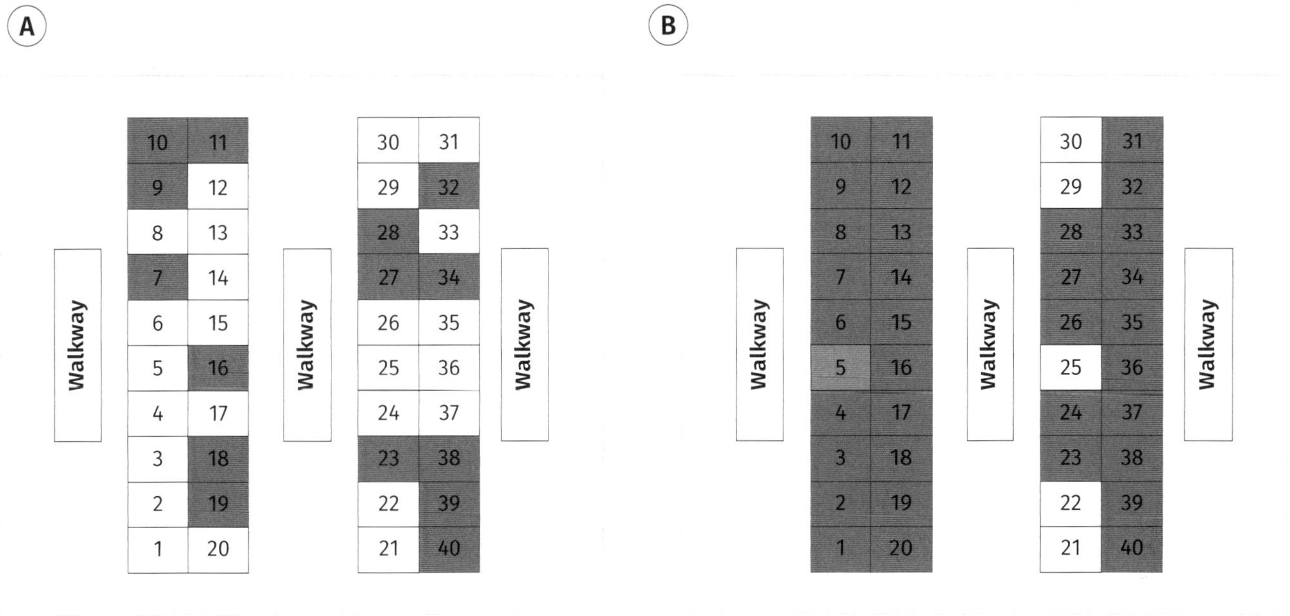

https://doi.org/10.54846/am2022/106

Evaluating the impact of organic matter and sample processing technique on RNA detection using environmental samples

C. Grace Elijah[1], DVM; Allison K. Blomme[2], BS; Olivia L. Harrison[3], MS; Jianfa Bai[1], PhD; Jason C. Woodworth[3], PhD; Cassandra K. Jones[3], PhD; Elizabeth G. Poulsen-Porter[1], MS; Chad B. Paulk[2], PhD; and Jordan T. Gebhardt[1], DVM, PhD
[1]Department of Diagnostic Medicine/Pathobiology, College of Veterinary Medicine, Kansas State University;
[2]Department of Grain Science and Industry, College of Agriculture, Kansas State University;
[3]Department of Animal Science and Industry, College of Agriculture, Kansas State University

Introduction

Environmental sampling has become a commonly accepted diagnostic sampling technique to identify breaks in biosecurity. However, environmental samples have yet to been validated for reverse transcriptase real-time PCR (RT-PCR) analysis, and there is no standardization for environmental sample processing. Therefore the objective of this project was to evaluate different types of environmental samples and if processing samples prior to RT-PCR analysis would impact results.

Materials and methods

Treatments were arranged as a 5 × 4 factorial with five types of contamination and four types of sample processing. Steel coupons, measuring 4 × 4 in were inoculated with 1 mL of pure virus (USA/Co/2013 with a titer of 1.33 × 105 $TCID_{50}$/mL), 1 mL of virus diluted into 5 mL of PBS, 1 mL of virus inoculated with 5 g of dirt, 5 g of feces, or 5 g of organic matter mixture consisting of a 1:1 ratio of dirt and feces. After surfaces were allowed to sit for 15 min, the steel coupon was environmentally swabbed with aseptic technique. Environmental samples were inverted for 5-10 s, incubated for 1hr, vortexed for 10-15 s, and then processed according to designated sample processing. Samples were either submitted as is (no processing), centrifuged for 10 minutes at 706 × g (centrifuge), syringe filtered with a 0.45 µm 25 mm syringe filter (syringe filter), or centrifuged for 10 min at 706 × g then syringe filtered through a 0.45 µm 25 mm syringe filter (centrifuge + syringe filter). After processing, samples were submitted for porcine epidemic diarrhea virus (PEDV) RT-PCR assay. Results are reported as proportion RT-PCR positive and in cycle threshold value. If there was no detectable RNA in the sample, the sample was assigned a Ct value of 45.

Results and discussion

There was a contamination type × sample processing ($P < 0.0001$; Table 1) interaction in Ct values, indicating that the contamination type and its processing method impacted the Ct value of the sample. For pure virus and virus with PBS, there was no difference in Ct values between different sample processing techniques ($P < 0.05$). For virus and dirt contamination, samples that were centrifuged had greater amounts of PEDV RNA detected compared to samples that were not processed ($P < 0.05$). For virus and fecal contamination, samples that were not processed or samples that were processed with centrifuging only had greater amounts of PEDV RNA detected compared to syringe filtered samples or centrifuged and syringe filtered samples ($P < 0.05$). For virus and organic matter contamination, samples that were centrifuged had greater amounts of PEDV RNA detected compared to all other sample processing techniques ($P < 0.05$). Main effects of contamination type ($P < 0.0001$) and sample processing ($P < 0.0001$) were also significant. In summary these results indicate that syringe filtering was the least successful at extracting PEDV RNA from samples while centrifugation performed the best with samples that either contained dirt, feces, or a mixture of both when compared to non-processed samples. These findings indicate that contamination type must be considered when submitting samples for RT-PCR.

Table 1: Proportion of PCR positive and interactive means of RT-PCR reactions based on different contamination types with PEDV and different types of sample processing[1]

Item	No processing	Centrifuge	Syringe filter	Centrifuge + syringe filter
Proportion RT-PCR positive				
Pure virus	3/3	3/3	3/3	3/3
Virus and PBS	3/3	3/3	3/3	3/3
Virus and dirt	2/3	3/3	3/3	3/3
Virus and feces	3/3	3/3	0/3	0/3
Virus and organic matter	1/3	3/3	0/3	2/3
Cycle threshold[2]				
Pure virus	24.5[a]	24.6[a]	28.9[abcd]	27.5[abc]
Virus and PBS	24.8[ab]	24.7[ab]	28.0[abc]	28.4[abc]
Virus and dirt	35.9[de]	28.2[abc]	32.0[cd]	30.8[abcd]
Virus and feces	31.8b[cd]	32.5[cd]	45.0[f]	45.0[f]
Virus and organic matter	42.4[ef]	31.3[abcd]	45.0[f]	40.9[ef]

1 Steel coupons, measuring 4 × 4 in, were inoculated with PEDV, isolate USA/Co/2013 with a titer of 1.33×10^5 $TCID_{50}$/mL. Coupons were inoculated with 1 mL of pure virus, 1 mL of virus diluted into 5 mL of PBS, 1 mL of virus inoculated with 5 g of dirt, 5 g of feces, or 5 g of organic matter mixture consisting of a 1:1 ratio of dirt and feces. After surfaces were allowed to sit for 15 min, the steel coupon was environmentally swabbed with aseptic technique. Environmental samples were inverted for 5-10 s, incubated for an hr, vortexed for 10-15 s, and then processed according to designated sample pre-processing. Samples were either submitted as is (no processing), centrifuged for 10 minutes at 706 × g (centrifuge), syringe filtered with a 0.45 µm 25 mm syringe filter (syringe filter), or centrifuged for 10 min at 706 × g then syringe filtered through a 0.45 µm 25 mm syringe filter (centrifuge + syringe filter). After processing, samples were submitted for PEDV RT-PCR assay. Results are reported as proportion RT-PCR positive and in cycle threshold value. If there was no detectable RNA in the sample, the sample was assigned a Ct value of 45.

2 Contamination type × sample processing, $P < 0.0001$; SEM = 1.41

a-f Means lacking common superscripts differ, $P < 0.05$.

https://doi.org/10.54846/am2022/107

Diagnostic performance of a porcine deltacoronavirus indirect ELISA and dynamic of the antibody response in growing pigs

Lu Yen[1]; Juan Carlos Mora-Díaz[1]; Jose Antonio Carrillo-Avila[2]; Ting-Yu Cheng[1]; Gino Castillo-Vilca[1]; Rahul K. Nelli[1]; David H. Baum[1]; Jeffrey J. Zimmerman[1]; Luis G. Giménez-Lirola[1]
[1]Veterinary Diagnostic and Production Animal Medicine, Iowa State University, Iowa;
[2]Andalusian Public Health System Biobank, Granada, Spain

Introduction

The porcine deltacoronavirus (PDCoV), genus *Deltacoronavirus* of the family *Coronaviridae*, has been consistently associated with enteric disease in nursery pigs. However, PDCoV infections are clinically and histologically indistinguishable from other enteropathogenic pathogens of pigs, including the swine coronaviruses transmissible gastroenteritis virus (TGEV) and porcine epidemic diarrhea virus (PEDV). Therefore, the differential diagnosis of porcine coronavirus-related enteric disease relies on laboratory testing. Routine diagnosis is usually carried out through clinical observation and testing by RT-qPCR testing, which target conserved regions of PDCoV M or N genes. However, applied PDCoV-related research is still scarce due to its relatively recent emergence and comparably lower pathogenicity than TGEV and highly pathogenic PEDV strains.

Material and methods

This study evaluated the performance of an indirect ELISA based on the spike (S) protein of PDCoV, including the diagnostic sensitivity, specificity, and potential cross-reactivity with other coronaviruses circulating in swine herds. Conventional growing pigs (n = 81) were acquired from a comercial farm with no history nor laboratory evidence (RT-qPCR or serology) of porcine coronavirus infections. Pigs were randomly allocated in seven groups of treatment and experimentally inoculated with PDCoV, TGEV Miller, TGEV Purdue, PEDV, porcine respiratory coronavirus (PRCV), porcine hemagglutinating encephalomyelitis virus (PHEV), or mock-inoculated with culture medium. Serum samples (n = 882) were collected at day post-inoculation (dpi) –7, 0, 3, 7, 10, 14, 17, 21, 28, 35, and 42, and tested by the PDCoV S iELISA.

Results and discussion

PDCoV-specific antibodies were only detected on the PDCoV group by dpi 10, with most of the animals seroconverting on dpi 17. The Receiver Operating Characteristic (ROC) curve analysis determined an optimal cut-off S/P value of 0.17, associated with a diagnostic sensitivity and specificity of 96.4% and 100%, respectively. Most importantly, the PDCoV S ELISA did not react with any positive serum from pigs inoculated with other porcine coronaviruses. Overall, these results support the use of S protein-derived antigens for antibody-based specific detection of PDCoV past/ongoing infection, active PDCoV surveillance, and assessment of immunity against this virus.

https://doi.org/10.54846/am2022/108

Colostrum fostering: Under practical conditions, is it useful to analyze colostrum with a Brix refractometer in the fostering protocols of hyperproliferative swine genetics?

László Gombos[2]; Botond Alpár[1,2]; Rita Horváth[1,2]
[1]University of Győr, Wittmann Antal Plant-, Animal- and Food Sciences Multidisciplinary Doctoral School, Mosonmagyaróvár; [2]Agrofeed Research and Development Division, Hungary

Introduction

Hyperproliferative swine genetics are an increasing challenge for swine producers. One of the challenges is the supply of the large number of newborn piglets with colostrum of adequate quality and quantity. The long-term impact of colostrum on animal health is unquestionable. It is generally recognized that the minimum piglet requirement is to receive 200-250g of colostrum over the first 24 hours (4000g colostrum/16 piglets in 24 hours). Unfortunately, the colostrum yield fluctuates from a minimum of 800-1000g up to 5,500–6,000g, and at least 30% of the sows do not produce enough colostrum. Research has shown that the gilt's first lactation will determine her milking capacity in her future parities, if she has the same number of functional teats. A Brix refractometer proved to be a quick and easy to use method for measuring colostrum ImmonglobulinG under field conditions.

Material and methods

In our study, colostrum quality was examined in 556 breeding animals (440 gilts+116 sows) twice with a Brix refractometer: after the first live born piglet (first) and 24-hours later(second). The value of < 20 was considered low, 21-25 adequate, and > 26 high. Postpartum Disgalactia Syndrome (PDS) and as well as suckling pig mortality, weaned piglet/sow, and fostering potential, were also recorded and analyzed individually.

Results

At farrowing, 98.9% of sows had a Brix value of > 21, of which 73.7% belonged to the high (> 26) category. Twenty-four hours after farrowing, 65% of the sows had a value of > 21, but only 15.9% had a very high level (this is notable due to colostrum fostering). In sows where the value was between 21-25 at farrowing, only 33.8% reached an acceptable value of > 21 after 24 hours. It was only 0.7% above > 26.

In sows where the > 26 value could be measured for the first time, 71.7% still achieved > 21 value. Of these, 20.1% could be measured > 26. At more than half of the sows (52.5%) were able to measure > 21 in both the first and second measurement time, and 13.5% were above 26. According to the PDS syndrome, 1 mastitis was observed at < 20 values, 30 metritis at 21-25 values and 100 pcs metritis and 5 mastitis above > 26. Taking into account the 24-hour measurements, the average piglet mortality at all three values was the same: 1 pc./farrowing. Regarding the number of the weaned piglets, approximately more than1 piglet/sow was at 21-25 (13.2) and > 26 (13.4) values compared to < 20 (12.4). We found a significant difference regarding fostering. The sows with > 26 value were able to raise 25% more piglets compared to the 21-15 or < 20 groups.

Discussion and conclusion

According to this study, 99% of the sows supplied the piglets with a good concentration, but not quality, of colostrum, based on the measured values with a Brix refractometer. More than half of the sows (52.5%) maintained this concentration for 24 hours after farrowing. Colostrum fostering is a good option for farms to raise large numbers of piglets, but without a unique examination of the colostrum, it is impossible to make an informed decision. In this farm, only 15% of the sows can be used for colostrum fostering with adequate safety.

https://doi.org/10.54846/am2022/109

Timing of euthanasia - is it an economic or animal welfare issue?

László Gombos[2]; Botond Alpár[1,2]; Rita Horváth[1,2]; László Vágó[3]
[1]University of Győr, Wittmann Antal Plant-, Animal- and Food Sciences Multidisciplinary Doctoral School, Mosonmagyaróvár; [2]Agrofeed Research and Development Division, Hungary; [3]BoldAgro Ltd., Derecske

Introduction

In practice, it is almost inevitable that there will be need for euthanasia on a swine farm. There are diseases and serious injuries that are classified as incurable, but other reasons may play a role in making the decision to euthanize. One of the biggest challenges is judging compromised, low viability piglets. In addition to animal welfare issues, economic considerations also play a role in these cases.

Farmers often complain about fatteners, that they often receive as "unsuitable for fattening" piglets. When these farmers are unhappy with fatteners, they frequently ask a veterinarian for help with animal welfare issues. In our study, we examined whether euthanasia is indeed a solution to both an animal welfare and an economic problem and, if so, when the decision should be made.

Material and methods

A total of 2,970 nursery-age, Danbred pigs in 3+1 batches were screened on a farrow to finish farm. Productivity parameters (pcs/sow/year): live borns: 39.3, weaned piglets: 33.7, sold fatteners: 32.8. Mortality: suckling piglets (28 days): 14.3%, nursery (52 days): 4.1%, fatteners (80 days): 4.5%. The collected mortality data are sorted by the cause of death. Proportion of compromised piglets that died, compared to all: suckling piglet: 28.4% (2 types: 19.4 non-viable + 9% small, viable), weaned piglets: 15.6%, fatteners: 42%. Based on these data and taking into consideration, that approx. 10-12% of underdeveloped, compromised piglets are transferred to the fattening barns, the management of the unit has the right to request an analytical audit.

According to the study, the unit managers had to choose approx. 30-30 piglets (farrowing unit: at weaning, nursery: 1 week after weaning) which will probably not be able to reach the slaughterhouse alive or in good quality. The piglets were ear-tagged (red: farrowing unit, green: weaning unit), monitored, and had their weight recorded.

Results

According to the results, the postweaning mortality was lower in piglets marked in the farrowing unit (11,1%) compared to the nursery (36,1%). The average daily weight gain (ADG) in the nursery was on average -80 g/day below the farm average in both groups. After transfer to the fattening unit, the number of the compromised pigs was < 10% with red ear tags and 20-30% with green ear tags (Table 1).

Discussion and conclusion

The study revealed that euthanasia cannot be clearly determined on the basis of the pig's current quality or state of development. The results of this study of animal health and welfare considerations can help managers of weaning units make safer, better informed decision at farms. This study also highlights the need for managers to be well prepared to make appropriate decisions about whether to euthanize or not.

Table 1: Productivity parameters in the ear tagged group

	Group no.	Ear-tagged piglets (No.)	Transferred to fattening unit	Dead piglet (No.)	Death rate (%)	Group weight (ear tagging)	Average piglet weight	Group weight (transfer to fattening unit)	Average piglet weight	ADG compromised piglets	ADG batch average	ADG difference
Red Ear-tag	M1-2	30	26	4	13.3%	141	4.7	472	18.2	258.7	383	-124.3
	M1-4	30	28	2	6.7%	106,5	3.6	568	20.3	321.8	383	-61.2
	M1-7	30	26	4	13.3%	89	3.0	475	18.3	294.3	363	-68.7
				Average:	**11.1%**		**3.7**		**18.9**	**291.6**	**376.3**	**-84.7**
Green Ear-tag	M1-1	36	22	14	38.9%	122,4	3.4	431	19.6	311.4	426	-114.6
	M1-2	42	32	10	23.8%	177,2	4.2	618	19.3	290.3	383	-92.7
	M1-4	35	26	9	25.7%	114	3.3	512	19.7	316.1	383	-66.9
	M1-7	34	15	19	55.9%	107	3.1	316	21.1	344.6	363	-18.4
				Average:	**36.1%**		**3.5**		**19.9**	**315.6**	**388.8**	**-73.2**

https://doi.org/10.54846/am2022/110

Distribution and characterization of *Streptococcus suis* strains of clinical importance within the US swine herd

Jessica G. dos Santos[1]; Kent Schwartz[1]; Rachel Derscheid[1]; Drew Magstadt[1]; Eric Burrough[1]; Philip Gauger[1]; Loni Schumacher[1]; Michael Rahe[1]; Alyona Michael[1]; Panchan Sitthicharoenchai[1]; Christopher Siepker[1]; Franco M. Ferreyra[1]; Marcelo Almeida[1]; Rodger Main[1]; Orhan Sahin[1]; Bailey L. Arruda[1,2]
[1]Iowa State University, Iowa; [2]USDA, Iowa

Streptococcus suis is a leading bacterial cause of swine mortality domestically. Yet, there remains limited vetted information concerning the distribution of *S suis* strains of importance in the US swine herd, which hinders the development of effective control and prevention measures. Herein, we aggregated and analyzed the largest collection of systemic *S suis* isolates from diagnostic case submissions using histopathology, coagglutination and whole-genome sequencing (WGS), collecting over 50 data points per case including laboratory diagnosis, age, site, state, viral status, *S suis* growth quantity and contamination level. Around 2,100 porcine cases had *S suis* isolated from the central nervous system (CNS) and histologic evaluation. Importantly, only about half of the cases had evidence of bacterial meningitis as confirmed by histopathology.

Five serotypes were the putative cause of meningitis in 64% of the animals: serotype 1 (17%), untypeable by coagglutination (14%), serotype 7 (13%), serotype 2 (11%) and serotype 1/2 (9%). The serotype distribution differed when analyzed by state with serotype 1 being the leading cause of bacterial meningitis in Missouri (36 %), Minnesota (35%), Indiana (25%), and Illinois (14%). Interestingly, untypeable strains by coagglutination were the leading cause of bacterial meningitis in both North Carolina (27%) and Iowa (16%), primarily due to multiple reactions (83%; i.e. the strain cross-reacted with different antibodies). Approximately 10% of sites had identification of two serotypes associated with bacterial meningitis and fewer than 3% of sites had three or more serotypes associated with bacterial meningitis. Sixty-five percent of the cases occurred prior to 7 weeks of age, where serotype 1 had the highest peak of cases at 3 weeks of age, serotypes 7 and 1/2 at 5 weeks of age, and serotype 2 had a later peak of 8 weeks of age.

One hundred and forty-six *S suis* isolates spanning 28 serotypes were selected for WGS based upon the isolation frequency of the serotype, whether the pig was individually examined, the growth quantities of *S suis* versus contaminating bacteria, the absence of other systemic bacterial pathogens, and PRRSV and/or IAV status. The genotype based on traditional virulence-associated genes (VAGs) was identified in 29% (*epf+/mrp+/sly+*) of the isolates, as opposed to the most recent VAG to predict virulence (*ofs+/srtF+*) with 84% of the isolates. Based on WGS analysis, almost all the isolates that were negative for *ofs* and *srtF* had an average nucleotide identity (ANI) lower than 95% when compared to the *S suis* strain P1/7 as the reference, indicating that these isolates were probably "*S suis*-like" strains. The most common sequence types (ST) were ST1 (25%), ST28 (8%) and ST977 (7%); ST1 was consisted of serotype 1 (58%), serotype 2 (25%) and serotype 14 (17%).

Based upon the data presented herein, strains associated with bacterial meningitis in US swine herds appear to be more diverse than previously thought; however, within-site variation in systemic disease-causing *S suis* strains may be limited (ie, only 10% of sites had identification of two or more serotypes associated with bacterial meningitis). Lastly, the reported variable effectiveness of *S suis* autogenous vaccines may be in part due to inappropriate strain selection (ie, almost 45% of porcine cases without any histopathologic evidence of bacterial meningitis still had *S suis* isolation from the CNS). Accurate case definitions, selection of appropriate pigs to sample, sample collection, sample management, and submission process, along with the proper interpretation of diagnostic assay results (site of isolation, bacterial growth quantity and quality, and histologic lesions) are crucial for accurate identification of the true disease-causing strain(s).

https://doi.org/10.54846/am2022/111

Diagnostic monitoring using oral fluids and sound: Complementing detection technologies

C. Goodell[1], DVM, MS, PhD, DACVPM; A. Witt[2]; J. Rudy[3]; K. Hogan[1], PhD;
C. Alonso[4], DVM, MS, PhD; D. Polson[1], DVM, MS, PhD
[1]Boehringer Ingelheim Animal Health USA Inc. Duluth, Georgia; [2]Iowa State University, College of Veterinary Medicine, Ames, Iowa; [3]Summit SmartFarms, Remington, Indiana; [4]Boehringer Ingelheim Vetmedica GmbH, Germany

Introduction

Oral fluids (OF) diagnostic monitoring programs are commonly detecting low prevalence pathogens (ie, before sufficient pigs are exhibiting clinical signs for human observation), which is particularly beneficial during control and elimination programs. Monitoring programs will increase as the industry addresses new or reemerging pathogens. Combining sensor monitoring (ie, respiratory distress indicators [RDI]) and routine diagnostic monitoring enables an earlier, quicker, and more targeted response, if the combination of these two methodologies are complementary. The objective of this study was to monitor finishing pigs using weekly OF PCR coupled with 24-hour auditory monitoring using SoundTalks™ (ST), to identify synergies between monitoring methods.

Materials and methods

August 2020, growing pigs from a single sow farm entered their finishing phase in 2, 2 barn 4 room, wean-to-finish sites housing ~2100 pigs/room. Four OF were collected weekly in the same 4 locations (under each ST sensor, systematically distributed per room), chilled and shipped overnight to the Iowa State University VDL for PRRSV, IAV-S, PEDV, PDCoV, TGEV and *M hyopneumoniae* (Mhp) PCR testing. ST color (green, yellow or red RDI indicator) was documented for each device on OF collection day, although the devices collected audio data continuously.

Results

446 OF samples were collected over a 16wk finishing period representing ~16,800 pigs, with ST RDI indicators compared on sampling day. Results of 2 rooms are shown in Figure 1. Room 5 monitors identified acute IAV at 14 weeks of age (WOA) and late onset Mhp, while Room 6 ST alerted continued RDI after IAV-S which OF PCR detected as Mhp. Finishing mortality for room 5 was ~1.2% and room 6 was ~1.8%.

Conclusions

- Routine OF diagnostics rapidly detected IAV-S, and early treatment minimized duration of cough
- Sensitivity of ST enabled early treatment for Mhp in Room 6, minimizing Mhp impact on health
- Monitoring with sensors and diagnostics lowered disease impact with early targeted interventions

Figure 1: OF diagnostic results for IAV-S and Mhp in 2 rooms, compared with ST RDI alert at sampling†.

Room 5	September 2020			October 2020				November 2020				December 2020			
Pig age (wks)	11	12	13	14	15	16	17	18	19	20	21	22	23	24	25
ST A	green	green	green	green	red	green	green	green	green	green	green	green	green	yellow	red
ST B	green	green	green	green	red	yellow	green	green	green	green	green	yellow	yellow	red	red
ST C	green	green	green	green	red	green	green	green	green	green	green	green	green	yellow	red
ST D	green	green	green	green	red	yellow	green	green	green	green	green	green	green	red	red
	IAVS PCR (Ct)														
Location A	0	0	0	37.3	31.2	0	0	0	0	0	0	0	0	0	0
Location B	0	0	0	NT	28.1	0	0	0	0	0	0	0	0	0	0
Location C	0	0	0	30.7	31.9	0	0	0	0	0	0	0	0	0	0
Location D	0	0	0	0	26.3	35.4	0	0	0	0	0	0	0	0	0
	Mhp PCR (Ct)														
Location A	0	0	0	0	0	0	0	0	0	0	0	35.6	0	34	35.3
Location B	0	0	0	0	0	0	0	0	0	0	0	34.8	0	28.5	32.5
Location C	0	0	0	0	0	0	0	0	0	0	0	0	0	0	0
Location D	0	0	0	0	0	0	0	0	0	0	0	0	0	29.4	33.5

Room 6	September 2020			October 2020				November 2020				December 2020			
Pig age (wks)	11	12	13	14	15	16	17	18	19	20	21	22	23	24	25
ST E	yellow	green	green	green	red	yellow	red	red	red	red	red	green	green	yellow	NT
ST F	green	green	green	green	yellow	green	green	green	green	yellow	yellow	yellow	red	red	NT
ST G	green	green	green	green	red	green	green	green	red	green	yellow	green	green	yellow	red
ST H	green	green	green	green	red	green	green	green	green	green	yellow	green	yellow	red	red
	IAVS PCR (Ct)														
Location E	0	0	0	29.2	29.6	0	0	0	0	0	0	0	0	0	0
Location F	0	0	0	23.9	35.1	0	0	0	0	0	0	0	0	0	0
Location G	0	0	0	29	32	37.9	0	0	0	0	0	0	0	0	0
Location H	0	0	0	29	32.4	0	0	0	0	0	0	0	0	0	0
	Mhp PCR (Ct)														
Location E	0	0	0	0	0	31.4	31.8	32.6	0	34.9	33.4	33.1	0	33.6	0
Location F	0	0	0	0	0	0	0	0	0	34.7	36.8	33	0	31.5	32.5
Location G	0	0	0	0	0	0	0	0	0	33.3	0	33.1	33.5	35.9	0
Location H	0	0	0	0	0	0	0	0	0	0	0	35	33.7	33.2	33.8

† ST RDI sensor colors, "green", "yellow", "red" indicate "no", "moderate" and "severe" distress levels respectively

https://doi.org/10.54846/am2022/112

Pathogen monitoring for early disease detection using oral fluid and air samples

C. Goodell[1], DVM, MS, PhD, DACVPM; A. Witt[2]; J. Rudy[3]; K. Hogan[1], PhD;
C. Alonso[4], DVM, MS, PhD; D. Polson[1], DVM, MS, PhD
[1]Boehringer Ingelheim Animal Health USA Inc. Duluth, Georgia; [2]Iowa State University, College of Veterinary Medicine, Ames, Iowa; [3]Summit SmartFarms, Remington, Indiana; [4]Boehringer Ingelheim Vetmedica GmbH, Germany

Introduction

Monitoring for early detection of disease benefits pigs and producers by enabling better and more timely responses. Historically caregivers identify signs of disease when walking a barn once or twice a day. Non-invasive antemortem testing like oral fluid (OF) is now commonly used for even earlier pathogen detection particularly during control and elimination programs.[1-8] Additionally, aerosols are becoming an accepted environmental monitoring sample as the industry focuses on pathogen transmission.[9-12] The objective of this study was to implement weekly OF and aerosol sampling in an 8400 head growing pig site and identify benefits these additional diagnostic observations provide to producers and veterinarians.

Materials and methods

Starting June 2020, weekly OF and aerosol samples were collected weekly in a 4 room, 2 barn wean-to-finish site housing ~2100 pigs/room. Samples were chilled and shipped overnight to the Iowa State University VDL for PCR testing to detect PRRSV, IAV-S, PEDV, PDCoV, TGEV and *M hyopneumoniae* (Mhp). Rooms 1&2 (Barn 1) were double stocked at weaning and spread into rooms 3&4 (Barn 2) after sampling week 6. At study start, 26-week-old animals were still housed in rooms 3 & 4. Sample collection and testing continued through the nursery phase of production.

Results

88 OF and 44 aerosol samples were collected over an 8wk nursery period. Detection of PEDV, PRRSV and Mhp were most notable during the first 8weeks at this site. as seen in Table 1. IAV-S, PDCoV and TGEV were not detected during this time-period. Pigs were scouring at placement into room 1 & 2 and PEDV was detected in 8 of 8 OF, and 4 of 4 air samples at that time. Although clinical signs had abated when half the pigs were moved into rooms 3 & 4 six weeks later, PEDV was also detected in 8 of 8 OF and 4 of 4 air samples in the newly placed rooms shortly after transfer. Wild type PRRSV was detected in 1 of 8 OF samples in the market pigs of rooms 3 & 4. Two weeks later, it was identified in Barn 1 OF samples (1 of 8), where it slowly spread over the next 3 weeks (3 of 8 OF and at least 1 of 4 air samples), prompting the producer to vaccinate the site against PRRSV. No other rooms were positive for PRRSV before vaccination.

Conclusions

a) PEDV was detected in OF and aerosols at weaning

b) OF and aerosols detected PRRSV contamination from pigs on the same site

c) Early detection of wild-type PRRSV by OF enabled the producer to proactively intervene and prevent disease. Resulting mortality for this phase of production in this group of pigs was < 2%

References

1. Prickett et al., JSHAP. 2008.
2. Panyasing et al., TBED. 2014
3. Panyasing et al., TBED. 2018.
4. Rotolo et al., Vet Micro. 2017.
5. Linhares et al., Prev Vet Med. 2015.
6. Trevisani et al., PloS One. 2020.
7. Trevisani et al., Porc Health Manag. 2019.
8. Lebret et al., Porc Health Manag. 2021.
9. De Carvahlo Ferreira et al., Vet Micro 2013.
10. Raynor et al., Plos One. 2021.
11. Pepin et al., TBED. 2020.
12. Otake et al., Vet Micro 2010.

Table 1: Oral fluid (OF) and aerosol diagnostic results by room and week of sampling

		Barn 1 (Rm 1 & 2)†				Barn 2 (Rm 3 & 4)††					
		PEDV		PRRSV		PEDV‡		PRRSV		Mhp	
		OF	Air	OF	Air	OF	Air	OF	Air	OF	Air
Sampling week	1	8/8	4/4	0/8	0/4	1/8	0/4	**1/8***	0/4	4/8	0/4
	2	8/8	3/4	0/8	0/4	No testing, barn marketed and empty					
	3	8/8	4/4	**1/8**	0/4						
	4	8/8	3/4	**1/8**	**1/4**						
	5	8/8	1/4	**3/8**	**2/4**						
	**6	8/8	2/4	**3/8**	**1/4**						
	7	7/8	0/4	7/8	2/4	8/8	4/4	8/8	1/8	0/8	0/4
	8	7/8	0/4	8/8	0/4	7/8	1/4	8/8	0/4	0/8	0/4

† Barn 1 contained 3wk old weaned pigs at study start (week 1)

†† Barn 2 contained 26 wk animals ready to market at study start

‡ Wk 7, rooms 1 & 2 (doublestocked) moved overstocked cohorts into now empty rooms 3 & 4

* WT PRRSV detected in wk 1 Barn 2 (market pigs) was detected (99% homology) wk 3 in Barn 1

** All pigs on site were vaccinated with Ingelvac PRRS® MLV (BI) after OF monitoring samples were collected

https://doi.org/10.54846/am2022/113

Inoculation of weaned pigs by feed, water and airborne transmission of *Salmonella enterica* serotype 4,[5],12:i:-

O. L. Harrison[1], BS; J. T. Gebhardt[2], DVM, PhD; C. B. Paulk[3], PhD; B. L. Plattner[2], DVM PhD; J. C. Woodworth[1], PhD; S. Rensing[4], PhD; C. K. Jones[1], PhD; V. Trinetta[5], PhD
[1]Department of Animal Sciences and Industry, Kansas State University; [2]Department of Diagnostic Medicine/Pathobiology, Kansas State University; [3]Department of Grain Science and Industry, Kansas State University; [4]Department of Gender, Women, and Sexuality Studies, Kansas State University; [5]Food Science Institute, Kansas State University, Manhattan, Kansas

Introduction

The number of salmonellosis cases in humans and specifically linked to pork products has increased in recent years. Swine have been identified as possible reservoir for *Salmonella*. There is concern that contamination of feed, water, or air of certain *Salmonella* serotypes may lead to food product contamination and therefore foodborne illnesses. *Salmonella enterica* 4,[5],12:i:- (*STM*), a monophasic variant of Typhimurium, is one of the most common serotypes associated with pork. While *STM* is a relatively new strain, it has become the number one cause of foodborne illness in the United States. This strain is particularly problematic because of its resistance to common antimicrobials. There has been little research on the routes of transmission for *STM* and its resulting colonization in pigs. Therefore, the objective of this study was to evaluate the transmission of *STM* when weaned pigs were provided contaminated feed, water, or aerosolized particles.

Materials and methods

Twelve weaned pigs were brought to the Large Animal Research Center (LARC) at Kansas State University. There were four treatments with three pigs per treatment: 1) control, 2) inoculated feed, 3) inoculated water and 4) inoculated aerosol. Feed and water inoculated pigs were given 10 mL of a contaminated feed slurry or a buffered peptone water solution at 1.3×10^9 CFU/g and 1.0×10^9 CFU/mL, respectively. The aerosol inoculated pigs were given 1 mL of the initial inoculum at 6.7×10^9 CFU/mL through a portable nebulizer. Fecal, rectal, nasal, and environmental samples were collected from d1-7. Environmental samples were taken from the feeder, waterers, and flooring of each pen. Pigs were euthanized on d7. Samples were collected during necropsy for bacteriology and histopathologic evaluation. Fresh tissue samples were serial diluted and plated on *Enterobacteriaceae* (EB) petrifilm. Plates showing typical EB colonies were pressed onto Xylose Lysine Deoxycholate (XLD) agar to quantify *Salmonella* typical colonies. Three random *Salmonella* positive samples per pig were confirmed as *STM* via PCR. Data were fit using a linear mixed model using the GLIMMIX procedure of SAS, v 9.4 (SAS Institute Inc., Cary, NC). Fixed effects in the statistical model included treatment, day, and interaction.

Results and discussion

Feed and water inoculated pigs tended to excrete greater quantities of *STM* ($P < 0.10$) than the aerosol inoculated or control pigs in their fecal samples (2.8, 2.9, 1.1, and1.0 CFU/g, respectively). There was a greater quantity of *STM* ($P < 0.05$) secreted in the nasal samples on day 3 than on day 5 or 6 (1.72, 0.67, and 0.72 CFU/mL, respectively); there was also a decreasing linear trend throughout the seven days of the trial ($P < 0.05$). The environment was not impacted by treatment, day, or its interaction. The aerosol inoculated pigs' mesenteric lymph nodes (ileocolic and jejunal) and the lower digestive tract (cecum and colon) had greater quantities of *STM* than the control pigs (4.0 and 4.0 CFU/g, respectively vs no detectable *STM* in the control) with the other inoculated pigs' *STM* quantities as intermediates. Tonsils of pigs inoculated through feed, but not air or water, resulted in greater ($P < 0.05$) detected *STM* (4.7, 2.3, and 1.5 CFU/g, respectively). All inoculated pigs displayed evidence of a *STM* infection in their digestive organs and lymph nodes; however, there were no signs of colony formation in the leg muscles (Table 1).

In conclusion, *STM*-contaminated feed, water, and air may cause *STM* infection in pigs and could be a potential source of STM entry into the pork supply chain.

Table 1: Detection of *STM* positive necropsy samples from each treatment[1]

	Control	Aerosol	Feed	Water
Lung	---	+++	+--	---
Ileocolic and jejunal lymph node	---	+++	+++	+++
Inguinal lymph node	---	---	+--	---
Tonsil	---	+--	+++	++-
Cecum and colon	---	+++	++-	++-
Ileum	---	++-	++-	++-
Gluteal muscle	---	---	---	---
Gracilis muscle	---	---	---	---

[1] Pigs were euthanized on d7 and tissue and organ samples were collected at necropsy. Each treatment had 3 pigs. +/- denotes the *STM* status of each pig for each tissue type.

https://doi.org/10.54846/am2022/114

Evaluation of one versus two doses of an inactivated Senecavirus A vaccine in weaned pigs

K. Hoffman[1], PhD; K. Lager[1,2], DVM, PhD; A. Buckley[1,2], DVM, PhD
[1]Oak Ridge Institute for Science and Education and National Animal Disease Center, US Department of Agriculture, Agricultural Research Service, Ames, Iowa; [2]National Animal Disease Center, US Department of Agriculture, Agricultural Research Service, Ames, Iowa

Introduction

Senecavirus A (SVA), also known as Seneca Valley virus, is a causative agent of vesicular disease in swine that is clinically indistinguishable from other vesicular diseases of swine, including foot-and-mouth disease (FMD). Due to the risks associated with the establishment of a foreign animal disease in the United States, an investigation is required when a vesicular lesion is observed to rule out FMD. These investigations can impart a substantial economic burden on local and federal resources. An efficacious vaccine could reduce the number of these investigations being performed. Previous work has shown the effectiveness of two doses of an inactivated SVA vaccine but administering one dose of vaccine could be more practical to implement in industry; therefore, the objective of this study was to evaluate the efficacy of one dose of a whole-virus inactivated vaccine for the control of SVA.

Materials and methods

Weaned pigs were randomly placed into 4 groups: Group 1 (one dose + sham challenge, n = 12), Group 2 (one dose + challenge, n = 12), Group 3 (two dose + challenge, n = 12), Group 4 (sham vax + challenge, n = 12). All pigs were vaccinated intramuscularly (IM) with 2 mL of either whole-virus inactivated SVA (2020 isolate) and adjuvant or sham. Those groups receiving two doses had three weeks between each dose. Subsequently, two weeks after each group's final dose, pigs were challenged intranasally with either a 2020 SVA isolate (10^7 $TCID_{50}$/mL) or sham. Pigs were rectal swabbed and observed for clinical signs on 0-7, 9, 11-, 14-, 18-, and 21-days post challenge (dpc). Serum was collected from animals on -35, -14, 0, 3, 5, 7, 14 and 21 dpc. Serum and swabs were tested for SVA nucleic acids by PCR. Sera was tested by virus neutralization assay and by western blot to monitor antibody responses to SVA.

Results

Results are pending as the vaccine study is currently ongoing. Previous research from this group has demonstrated the efficacy of two doses of an inactivated SVA vaccine against challenge with a 2015 SVA isolate. In those vaccinated animals, neutralization titers were observed after the first dose; however, neutralization titers were significantly higher after the second dose. Currently, the neutralizing titer that correlates with protection against SVA challenge is unknown. We hypothesize that one dose of a whole-virus inactivated SVA vaccine will protect pigs against clinical disease and reduce or eliminate viral shedding.

Conclusion

An effective single-dose inactivated SVA vaccine would be economical and easier to apply in the field. Further, an effective inactivated SVA vaccine would provide a positive impact on the welfare of swine and reduce economic costs due to foreign animal disease investigations.

https://doi.org/10.54846/am2022/115

Measuring the effectiveness of mitigants against PRRS 1-4-4 in a simulation of an emergency feed outage

M. Hood[1], DVM; D. Reicks[1], DVM; S. Kitt[2], PhD
[1]Reicks Veterinary Research and Consulting, Saint Peter, Minnesota; [2]First Choice Livestock, LLC, Columbus, Nebraska

Introduction

With the introduction of porcine reproductive and respiratory syndrome virus (PRRSV) 1-4-4, disease prevention and mitigation has become an industry priority. Historically viral infection via feed has been a concern for hog producers; perhaps the most infamous example being the introduction of porcine epidemic diarrhea virus (PEDV) into the United States in 2013 via imported feed ingredients.[2] In response to this concern, the use of feed mitigants has become a growing practice in the swine industry. Previous research has shown that pigs exposed to a viral disease challenge in their feed may have improved health and performance if their feed contains a mitigant.[1] However, the spread of PRRS in feed has up until now been viewed as a low-risk avenue of exposure. An experiment was conducted to determine the effectiveness of feed mitigants against the PRRS 1-4-4 strain in a simulated feed outage emergency – 1 hour from feed mill to feed bin.

Materials and methods

The study consisted of 9 groups of 4 individually housed pigs: 6 treatment groups each receiving a different feed mitigant, a negative control, a positive feed control, and a positive oral control. Groups were placed in the barn according to air flow to reduce the risk of aerosol transmission from the positive control groups. Blood was collected from pigs on a farm known to be infected with PRRS 1-4-4 and individually tested via PRRS PCR to confirm that the blood was suitable to use as inoculum. The blood was mixed so that it reached a concentration of 10^4 $TCID_{50}$. Mitigants were weighed on a gram scale and added to a designated amount of feed weighed with a fish scale per the label. Pigs were fed their designated diet for 7 days pre-test. On day -1 all pigs were tested for PRRS PCR and ELISA. Feed was withheld from all pigs 12 hours pre-test. On day 0 all 4 pigs per treatment group were dosed as follows: 1 mL inoculum (10^4 $TCID_{50}$) was added to ½ cup of feed according to treatment group and mixed in a disposable cup. This feed was enclosed in a plastic bag and refrigerated for one hour to simulate a feed outage emergency in winter conditions. The negative control was given basal feed mixed with 1 mL sterile saline. The positive feed control was given ½ cup of basal feed with 1 mL inoculum. The positive oral control was given a direct oral dose of 1 mL inoculum. Pigs were tested individually for PRRS via PCR at 24, 48, and 72 hours post inoculation. An additional test was performed on the 4th day – after pigs had been comingled in their treatment groups for 24 hours.

Results and conclusions

On day -1 all pigs tested negative for PRRS on both PCR and ELISA. On day 1 one pig tested positive for PRRS from the positive oral control group. 2 pigs from the positive oral control group tested positive on day 2. The same 2 pigs tested positive on day 3. On day 4, 3 pigs from the positive oral control group tested positive with cts 21.7, 21.8 and 27.5. No pigs from any other group tested positive for PRRS.

These results illustrate the challenge of infecting pigs with PRRS via their feed and align with previous assumptions that PRRS infection via feed is a low risk.

References

1. Dee SA, Niederwerder MC, Edler R, et al. An evaluation of additives for mitigating the risk of virus-contaminated feed using an ice-block challenge model. *Transbound Emerg Dis.* 2021; 68:833–845.

2. Jones C, et al. Investigating feed and feed mill risks. *50th Annual Meeting of AASV* (Orlando; March 9-12, 2019).

https://doi.org/10.54846/am2022/116

Occurrence of wild type and vaccine-like porcine reproductive and respiratory syndrome virus 2 (PRRSV-2) strains in the United States

M. Kikuti[1], DVM, MPH, PhD; C. Vilalta[1], DVM, PhD; J. Sanhueza[2], DVM, MS, PhD; I. A. D. Paploski[1], DVM, MS, PhD; K. VanderWaal[1], BS, PhD; C. A. Corzo[1], DVM, MS, PhD

[1]Department of Veterinary Population Medicine, University of Minnesota, Saint Paul, Minnesota;
[2]Departamento de Ciencias Veterinarias y Salud Pública, Universidad Católica de Temuco

Introduction

Nucleotide sequencing of the open reading frame 5 (ORF5) region of the viral genome has become a popular methodology to differentiate genotypes and strains, and to understand viral diversity across time. Additionally, PRRS is one of the main infectious diseases affecting swine, and causes an average economic loss of US$664 million per year in the US.[1] Thus, efforts in immunizing the herds have been made in the past few decades, particularly in high density swine regions. Our goal therefore was to describe PRRSV-2 genetic diversity and the frequency of vaccine-like strain detection in the course of over 10+ years of routine monitoring in the US swine industry.

Methods

Porcine reproductive and respiratory syndrome virus genotype 2 (PRRSV-2) genetic diversity in the US was assessed using a database comprising 10 years' worth of sequence data obtained from swine production systems routine monitoring and outbreak investigations. A total of 26,831 ORF5 PRRSV-2 sequences from 34 production systems were included in this analysis. Within group mean genetic distance (ie, mean proportion of nucleotide differences within ORF5) per year according to herd type was calculated for all PRRSV-2 sequences. The percent nucleotide difference between each sequence and the ORF5 sequences from four commercially available PRRSV-2 vaccines (Ingelvac PRRS MLV, Ingelvac PRRS ATP, Fostera PRRS, and Prevacent PRRS) within the same lineage over time was used to classify sequences in wild-type or vaccine-like.

Results

The mean ORF5 genetic distance fluctuated from 0.09 to 0.13, being generally smaller in years in which there was a relative higher frequency of dominant lineage. Vaccine-like sequences comprised about one fourth of sequences obtained through routine monitoring of PRRS. We found that lineage 5 sequences were mostly Ingelvac PRRS MLV-like (Figure 1). Lineage 8 sequences up to 2011 were 62.9% Ingelvac PRRS ATP-like while the remaining were wild-type viruses. From 2012 onwards, 51.9% of lineage 8 sequences were Ingelvac PRRS ATP-like, 45.0% were Fostera PRRS-like, and only 3.2% were wild-type. For lineage 1 sequences, 0.1% and 1.7% of the sequences were Prevacent PRRS-like in 2009-2018 and 2019, respectively.

Discussion

The yearly increase and decrease in mean genetic distance possibly explained by the change in frequency of the dominant lineages and sub-lineages over time.[2] The overall frequency of vaccine-like strains is in accordance with what has been previously described, in which 80% and 60% of sub-lineages 5.1 and 8.9, respectively, were found to be vaccine-related.[3] Our results suggest that repeated introductions of vaccine-like viruses through the use of modified-live vaccines might decrease within-lineage viral diversity as vaccine-like strains become more prevalent.

References

1. Holtkamp D, Kliebenstein J, Neumann E, et al. Assessment of the economic impact of porcine reproductive and respiratory syndrome virus on United States pork producers. *J Swine Heal Prod.* 2013;21(2):72-84.

2. Paploski IAD, Corzo C, Rovira A, et al. Temporal Dynamics of Co-circulating Lineages of Porcine Reproductive and Respiratory Syndrome Virus. *Front Microbiol.* 2019;10:2486. **https://www.frontiersin.org/article/10.3389/fmicb.2019.02486**

3. Brar MS, Shi M, Murtaugh MP, Leung FC-C. Evolutionary diversification of type 2 porcine reproductive and respiratory syndrome virus. *J Gen Virol.* 2015;96(Pt 7):1570-1580. **doi:10.1099/vir.0.000104**

Figure 1: Yearly percent nucleotide differences between lineage 5 orf5 sequences to Ingelvac PRRS MLV; lineage 8 sequences to either Ingelvac PRRS ATP or Fostera PRRS; and lineage 1 sequences to Prevacent PRRS.

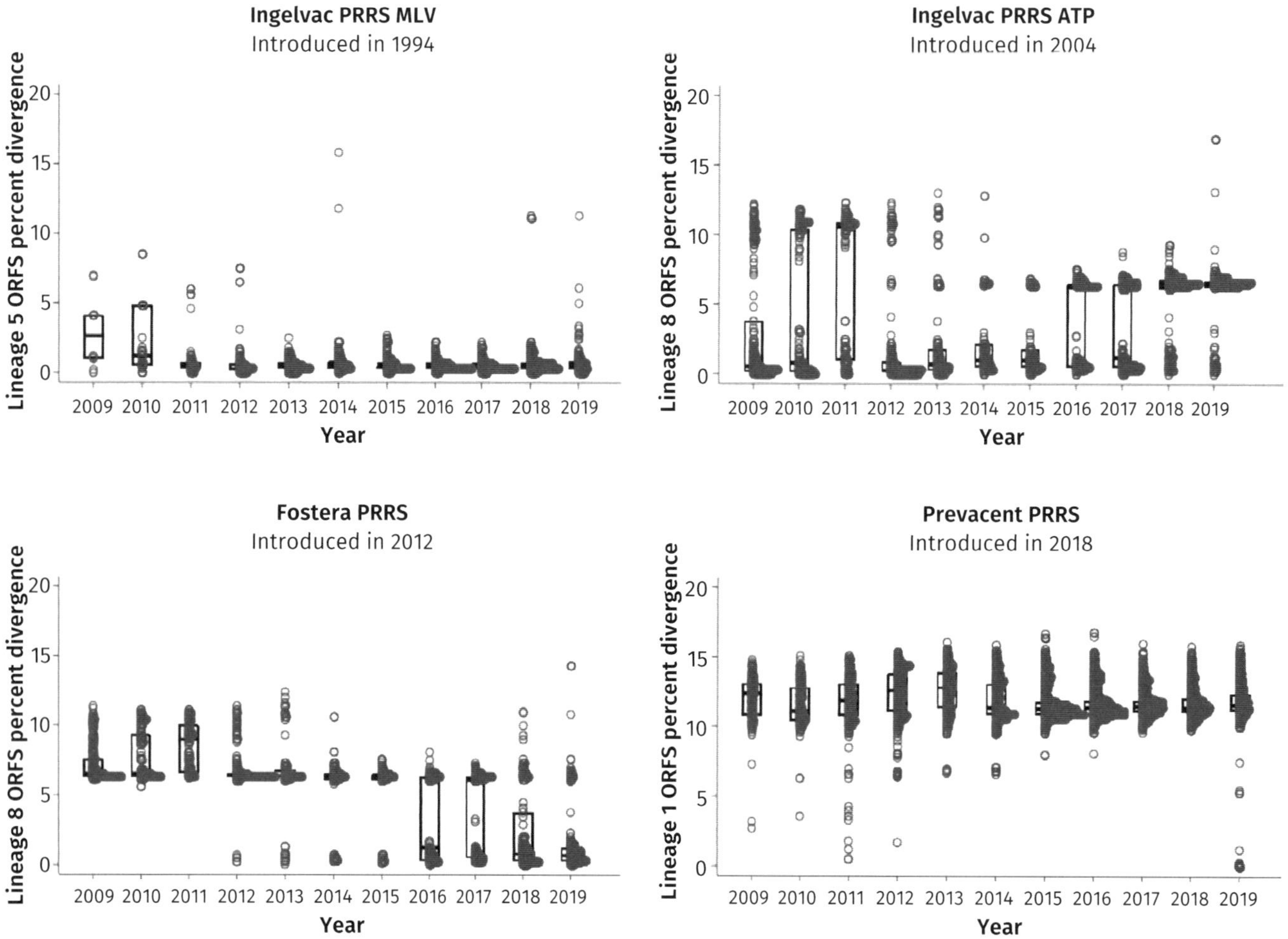

https://doi.org/10.54846/am2022/117

Shedding patterns and virus diversification of primary and secondary influenza infections in pigs infected with two distinct subtypes

C. Li[1], DVM, MS; M. Culhane[1], DVM, PhD; M. Cheeran[1], MVSc, PhD; D. Schroeder[1], PhD; L. Galina Pantoja[2], DVM, PhD; M. Jansen[2], DVM; M. Torremorell[1], DVM, PhD
[1]College of Veterinary Medicine, University of Minnesota, St. Paul, Minnesota; [2]Zoetis, Parsippany New Jersey

Introduction

Influenza is one of the most common and costly diseases in the US swine industry.[1] It happens throughout all production stages and have adverse effects on growing performance by worsening feed conversion, especially in young pigs.[2] Pigs have a wide range of susceptibility to influenza A virus (IAV) and harbor a complex genetic pool of influenza viruses, which facilitate the emergence of novel IAVs with diverse genotypes and antigenic properties in swine herds.[3] The co-circulation of multiple lineages and subtypes of IAV makes disease prevention difficult.[4] However, how co-infections with a heterosubtypic IAV impacts virus shedding and overall diversity at the individual pig level is poorly understood. In this study, we assessed the shedding patterns of IAV-infected pigs upon a secondary infection of IAV of a distinct subtype and we evaluated its impact on virus diversity.

Materials and methods

Fourteen naive pigs were inoculated with either an H1N1 or an H3N2 IAV. When all the pigs started shedding the viruses, they were evenly distributed in 7 rooms with one pig of each subtype housed together. Nasal swabs were taken daily from two to six days post contact (dpc), and bronchoalveolar lavage fluid (BALF) samples were collected during necropsy at 7 dpc. Samples were tested by matrix and HA subtyping real-time PCR. Any nasal swabs or BALF samples with Ct values under 35 were submitted for whole-genome sequencing using the Illumina Nextseq platform. Forty plaques were isolated from three BALF samples to identify virus reassortment in pigs with a confirmed heterosubtypic IAV secondary infection.

Results and discussion

During the 7-day observation period, we found that 43% of pigs (6/14) had acquired a heterosubtypic IAV secondary infection in the lungs or nasal cavities. Among these pigs, five pigs apparently cleared their primary IAV infection and at necropsy (7 dpc) they had mainly the subtype of the other pig (secondary infection). Of importance is that the simultaneous shedding of both challenge viruses was detected through the upper respiratory tract in one pig at 2 to 5 dpc, a second pig at 3 to 5 dpc, and a third pig at 2 to 3 dpc. We observed similar genetic variation patterns between the primary and secondary IAV infections as both challenge viruses exhibited excess or similar synonymous diversity than nonsynonymous diversity for all coding regions and at whole-genome level, except for M1 protein in H1N1 primary infection. In addition, 10% (4/40) of plaques isolated from 3 pigs (BALF samples) were identified as reassortants which resulted in 2 distinct genotypes. Our study demonstrated that pigs that become infected consecutively by multiple subtypes of IAVs can have extended IAV shedding patterns over time and can result in novel reassortant viruses promoting virus diversity. More research is needed to validate these results under field conditions.

References

1. USDA. Swine 2012 part II: Reference of swine health and health management in the United States, 2012. USDA-APHIS-VS-CEAH-NAHMS Fort Collins, CO; 2016.

2. Er C, Lium B, Tavornpanich S, Hofmo PO, Forberg H, Hauge AG, et al. Adverse effects of Influenza A (H1N1) pdm09 virus infection on growth performance of Norwegian pigs-a longitudinal study at a boar testing station. BMC Vet Res. 2014;10: 1–14.

3. Lewis NS, Russell CA, Langat P, Anderson TK, Berger K, Bielejec F, et al. The global antigenic diversity of swine influenza A viruses. Elife. 2016;5: e12217.

4. Nirmala J, Perez A, Culhane MR, Allerson MW, Sreevatsan S, Torremorell M. Genetic variability of influenza A virus in pigs at weaning in Midwestern United States swine farms. Transbound Emerg Dis. 2020.

Poster 39

https://doi.org/10.54846/am2022/118

Evaluation of internal biosecurity practices combined with sow vaccination to wean influenza negative piglets

Gustavo Lopez[1]; Marie Culhane[1]; Cameron Schmitt[2]; Montserrat Torremorell[1]
[1]College of Veterinary Medicine, University of Minnesota; [2]Pipestone Veterinary Services

Introduction

Influenza A virus (IAV) is widespread in swine populations worldwide.[1] Piglets of weaning age are one of the sub-populations most likely to test IAV positive in breeding herds.[2] Commonly at weaning pigs are transported to distant pig locations where pigs may become commingled with pigs from other herds creating an opportunity for co-circulation of distinct influenza viruses. As a result, new strains of IAV may emerge.

Vaccination of pregnant sows has been the main tool for controlling IAV infections in breeding herds.[3] This strategy has been effective at reducing IAV transmission and prevalence at weaning, but is not sufficient to consistently wean IAV negative pigs.[4] A recent field study that evaluated enhanced biosecurity measures resulted in a significant delay in IAV infections during lactation, but it was not enough to reduce IAV prevalence at weaning.[5] The objective of the present study was to evaluate the impact of combining both, internal biosecurity practices directed at minimizing IAV infection in piglets and sow vaccination, in IAV prevalence at weaning.

Material and methods

Six IAV positive breeding herds were selected for the study. Five herds were assigned to the treatment group, which consisted in implementing an enhanced internal biosecurity protocol combined with mass sow vaccination. The internal biosecurity protocol consisted of implementing practices of no cross fostering after the first 3 days of age, no usage of nurse sows, changing of disposable gloves between litters when handling piglets and daily disinfection of tools used in farrowing rooms. The mass sow vaccination included vaccinating all females with an autogenous herd-specific vaccine with a booster 3 weeks later. One farm served as control, in which there was no change in management or vaccination practices. Prior to the intervention, all farms were screened for IAV for 3 consecutive weeks, collecting 90 udder skin wipes from litters of weaning age.[6] Six weeks after the booster vaccination, litters at weaning were sampled for 3 more consecutive weeks to assess IAV prevalence after intervention. All collected samples were then tested individually using an IAV rRT-PCR test.

Results

Three of the five farms (60%) that were assigned to the treatment group tested IAV negative in all 3 sampling points post-intervention. One of the treatment farms had a significant decrease in IAV prevalence post intervention but IAV could still be detected in low levels in the last sampling of the study. In contrast, IAV prevalence in one of the treatment farms was not altered indicating that the treatment failed and had no effect on decreasing prevalence at weaning. As expected, the control farm tested IAV positive throughout the study.

Conclusions

The study presented here provides a protocol that combines sow vaccination and enhanced internal biosecurity practices to help wean negative pigs to IAV or pigs with low risk of influenza dissemination. This protocol can serve as a guide to pork producers that have the goal of controlling IAV infections in their breeding herds.

References

1. Baudon E, Peyre M, Peiris M, Cowling BJ. Epidemiological features of influenza circulation in swine populations: A systematic review and meta-analysis. PLoS ONE. 2017;12(6):1-25. **doi:10.1371/journal.pone.0179044**

2. Diaz A, Perez A, Sreevatsan S, Davies P, Culhane M, Torremorell M. Association between influenza A virus infection and pigs subpopulations in endemically infected breeding herds. PLoS ONE. 2015;10(6):1-11. **doi:10.1371/journal.pone.0129213**

3. Van Reeth K, Ma W. Swine influenza virus vaccines: to change or not to change-that's the question. Springer International Publishing. 2017;410(September 2012):173-200. **doi:10.1007/82**

4. Chamba FO, Matthew P, Marie WA, Morrison RB, Davies PR, Perez A. Effect of influenza A virus sow vaccination on infection in pigs at weaning : A prospective longitudinal study. 2020;(June):1-11. **doi:10.1111/tbed.13688**

5. Lopez-Moreno G, Davies P, Culhane M, et al. Evaluation of parity, personnel, and cross-fostering in influenza virus infection during the pre-weaning period. AASV. Published online 2021. **doi:10.1186/s12917-019-1805-0**

6. Garrido-Mantilla J, Alvarez J, Culhane M, Nirmala J, Cano JP, Torremorell M. Comparison of individual, group and environmental sampling strategies to conduct influenza surveillance in pigs. BMC Veterinary Research. 2019;15:61:1-10. **doi:10.1186/s12917-019-1805-0**

https://doi.org/10.54846/am2022/119

Characterizing United States boar stud biosecurity practices to assess African swine fever virus infection and transmission pathways

M. Medrano; M. Culhane; C. Alexander; M. Leonard; K. St. Charles; T. Goldsmith; A. Ssematimba; P. Bonney; S. Malladi, B. Blair, C. Cardona; C. Corzo
College of Veterinary Medicine, University of Minnesota, St. Paul, Minnesota

Introduction

Every part of the United States (US) swine industry has their own unique practices and activities, which, when combined, can create possible pathways for disease introduction or spread. To assess the overall risk of disease transmission, a systematic evaluation of these pathways is needed. Therefore, the Secure Food Systems (SFS) team at the University of Minnesota (UMN)[1] is currently conducting a proactive risk assessment (RA) that systemically evaluates the potential risk of liquid, cooled semen movements from a boar stud in a control area during an African swine fever (ASF) outbreak since information is scarce. An RA is one of the four components of the risk analysis framework as outlined by the OIE that provides a process to evaluate movement risk.[2] As part of the pathways evaluated for this RA, a thorough overview and characterization of US boar studs and their practices were required to create an outline against which pathways can be discerned, and risk via those pathways assessed.

Methods

A working group (WG) of boar stud subject matter experts (SMEs), state and federal regulatory officials, and academicians was convened to guide the RA. Ten different potential routes of African swine fever virus (ASFV) entry into a US boar stud were evaluated (Figure 1) and the WG SMEs provided information to the UMN SFS team regarding stud characteristics, existing biosecurity practices, and industry standards. The information was collected in our regularly scheduled meetings and via surveys given to WG, SMEs and a subset of boar stud managers, veterinarians, and consultants. For each pathway assessed, the current boar stud biosecurity practices were then evaluated for their ASFV introduction potential. To further decrease the likelihood of ASFV introduction, the WG then developed more targeted biosecurity practices that would be used in the event of an ASF outbreak. Finally, the likelihood ratings of ASFV introduction via the assessed pathways were then estimated based on these targeted recommendations by the WG, along with the enhanced biosecurity requirements (EBRs) for permitted movement of semen during an ASF outbreak, as set forth by the Secure Pork Supply Plan (SPS).[3]

Results and discussion

The WG discussions and survey responses allowed the UMN SFS team to characterize US boar studs and their biosecurity practices for the first time to assess the potential gaps through which ASFV introduction may occur. Assuming that targeted and enhanced biosecurity recommendations from the WG and SPS are in place during an ASF outbreak, the tentative likelihood ratings for ASFV entry, which varied from negligible, low, and low to moderate, were estimated for each pathway and qualitatively assessed. It is important to note that the likelihood ratings are only for the likelihood that a boar stud premises in a Control Area will become infected with ASFV and does not consider the pathways and risk of the subsequent ASFV spread from the boar stud, which is currently being assessed by the UMN SFS team (Figure 1). In conclusion, the US boar stud population can reduce their risk of ASFV introduction by implementing more enhanced and targeted biosecurity practices. This entry analysis can be referred to when evaluating the risk of moving risk of liquid, cooled semen movements from a boar stud in a control area during an ASF outbreak.

References

1. Regents of the University of Minnesota. Secure Food Systems. 2021. Available at: **www.securefoodsystems.umn.edu/** Accessed 12 November 2021.

2. World Organization for Animal Health (OIE). Terrestrial Animal Health Code: Chapter 2.1 Import Risk Analysis. 2021. Available at: **www.oie.int/fileadmin/Home/eng/Health_standards/tahc/current/chapitre_import_risk_analysis.pdf.** Accessed 09 November 2021.

3. Secure Pork Supply Plan. Information Manual for Enhanced Biosecurity for Pork Production: Animal Raised Indoors. 2017. Available at: **www.securepork.org/Resources/SPS_Biosecurity_Information_Manual_IndoorAnimals.pdf.** Accessed 09 November 2021.

Figure 1: Likelihood statements evaluated as part of the proactive risk assessment of the movement of liquid, cooled boar semen within, into, and outside of a control area during an ASF outbreak in the US, along with the ten pathways assessed.

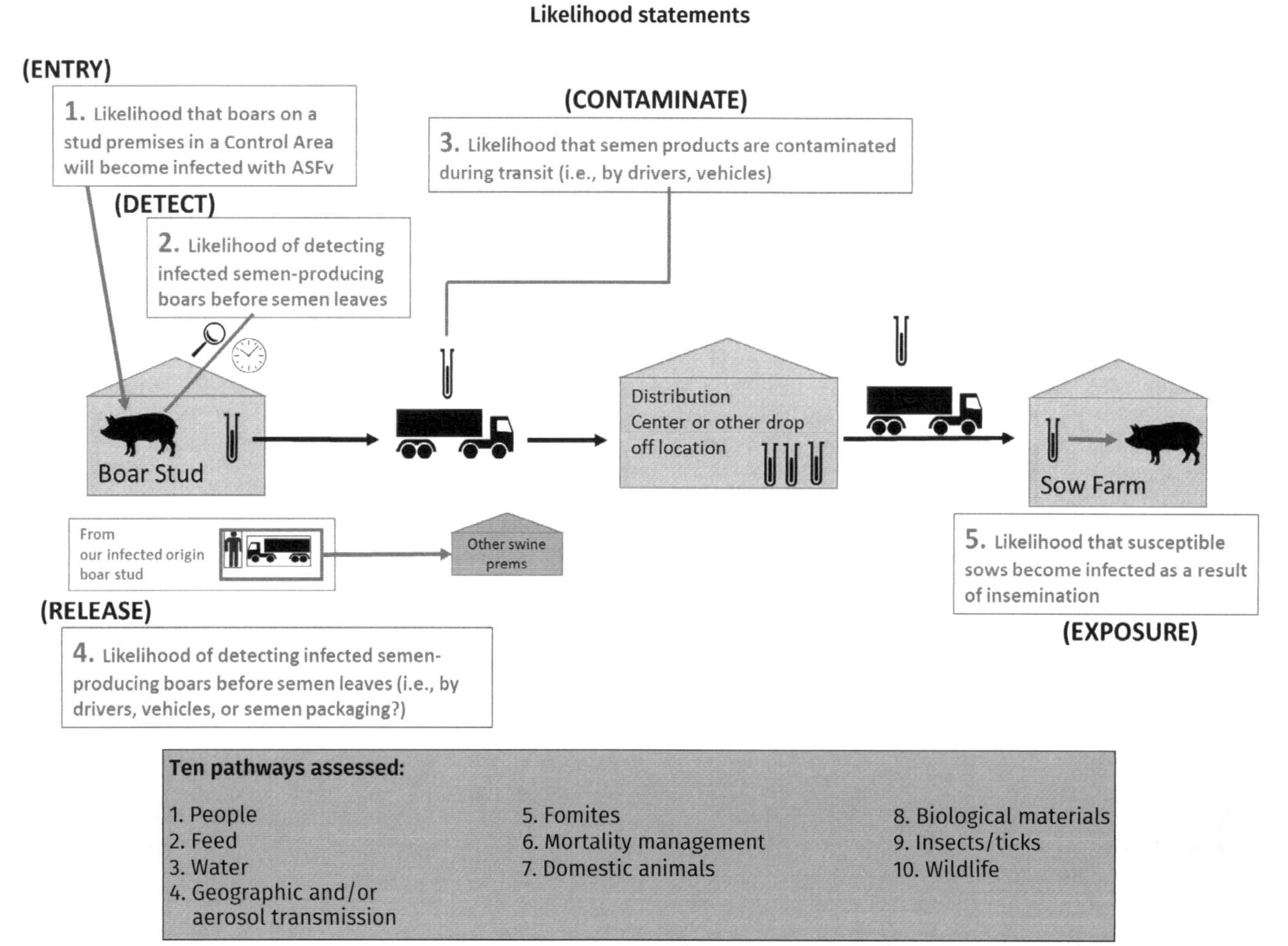

https://doi.org/10.54846/am2022/120

In situ characterization of *Mycoplasma hyorhinis* in finisher pigs with neurological signs

M. Merodio[1], DVM, MS; E. Spronk[2], DVM ; JR Lehman[3], DVM; H. Shen[1], DVM PhD; G. Li[1], DVM, PhD; R. Derscheid[1], DVM, PhD, ACVP; P. Piñeyro[1], DVM, MS, DVSc, PhD
[1]Veterinary Diagnostic and Production Animal Medicine, Iowa State University, Iowa;
[2]Swine Vet Center P.A., St. Peter, Minnesota; [3]Swine Technical Services, Merck Animal Health, Lenexa, Kansas

Introduction

Mycoplasma hyorhinis (Mhr) is a member of the class Mollicute and a saprophytic inhabitant of the respiratory tract in pigs. It can be isolated from the upper respiratory tract, tonsils, and lungs in healthy pigs. It is also causative of polyserositis and arthritis in postweaning pigs and occasionally in older animals. The bacterial pathogenic mechanisms and host predisposing factors leading to the spread of the microorganism remain unknown. Furthermore, little is known about the pathogenic potential of Mhr when it is associated with other agents capable of causing immunosuppression. Although the main clinical presentation of Mhr infection is polyserositis associated with systemic infection, there is scarce information regarding Mhr as causative of meningitis and neurological clinical signs. The objective of this study was to evaluate the role of Mhr in a clinical outbreak of neurological signs in finisher pigs.

Material and methods

A group of pigs between 18-20 weeks of age presented with an acute outbreak of lameness and shaking/trembling that progressed rapidly to lateral recumbency. Affected pigs had paresis to paralysis that was more severe in the front limbs but were centrally aware. The mortality rate associated with this clinical presentation reached 16.5%. Fresh and fixed tissue, including spinal cord, brain, liver, spleen, and limbs from two representatively affected animals, were submitted to the ISU-VDL for gross, histological, and etiological evaluation.

Results

The histological evaluation demonstrated severe lymphoplasmacytic myelitis, encephalitis, and meningitis, focally extensive, chronic-acute, with severe lymphoplasmacytic perivasculitis. In addition, there is severe lymphoplasmacytic and proliferative synovitis. Molecular evaluation ruled out the presence of porcine teschovirus, porcine sapelovirus, and porcine astrovirus 3 but confirmed the presence of *Mhr* in brain (or spinal cord) via qPCR at 39.1 Ct. Further bacteriological testing ruled out the presence of *Streptococcus suis* and *Glaesserella (Haemophilus) parasuis*. However, the bacteriological culture of sections of the brain in Friis media confirmed the presence of Mhr and ruled out the presence of *Mycoplasma hyopneumoniae* (Mhp) and *Mycoplasma hyosynoviae* (Mhs). Sequencing, phylogenetic analysis, and virulence gene detection demonstrate the difference between these three Mhr isolates from previous clinical isolates. Finally, the presence of Mhr was confirmed *in situ* by RNAscope® in sections of the spinal cord, brain, and synovial capsule (Figure 1A, B).

Conclusion

Only two studies have correlated the presence of Mhr with neurological presentation. However, this is the first report that confirms the presence of Mhr in CNS lesions by several diagnostic methods, including direct detection. The mechanism of Mhr invading the CNS is still unknown. We hypothesized that it could be either part of the systemic spread to serosal and synovial tissues or direct entry from the nasopharyngeal/middle ear mucosa to the brain. The true impact of Mhr in neurological cases affecting finisher pigs is unknown, and further studies are necessary to assess the role of Mhr on outbreaks of neurological disease. This case report shows the importance of including Mhr as a differential diagnosis for neurological signs in finisher pigs.

Figure 1: In situ characterization of *Mycoplasma hyorhinis* in finisher pigs with neurological signs. Mhr in situ detection by RNAscope®. Mhr genetic material (red staining) was detected in high concentration throughout the meningeal inflammatory exudate in the spinal cord **(A)**. The synovium also presents abundant chromogenic reactions within the inflammatory exudate, consistent with Mhr replication **(B)**.

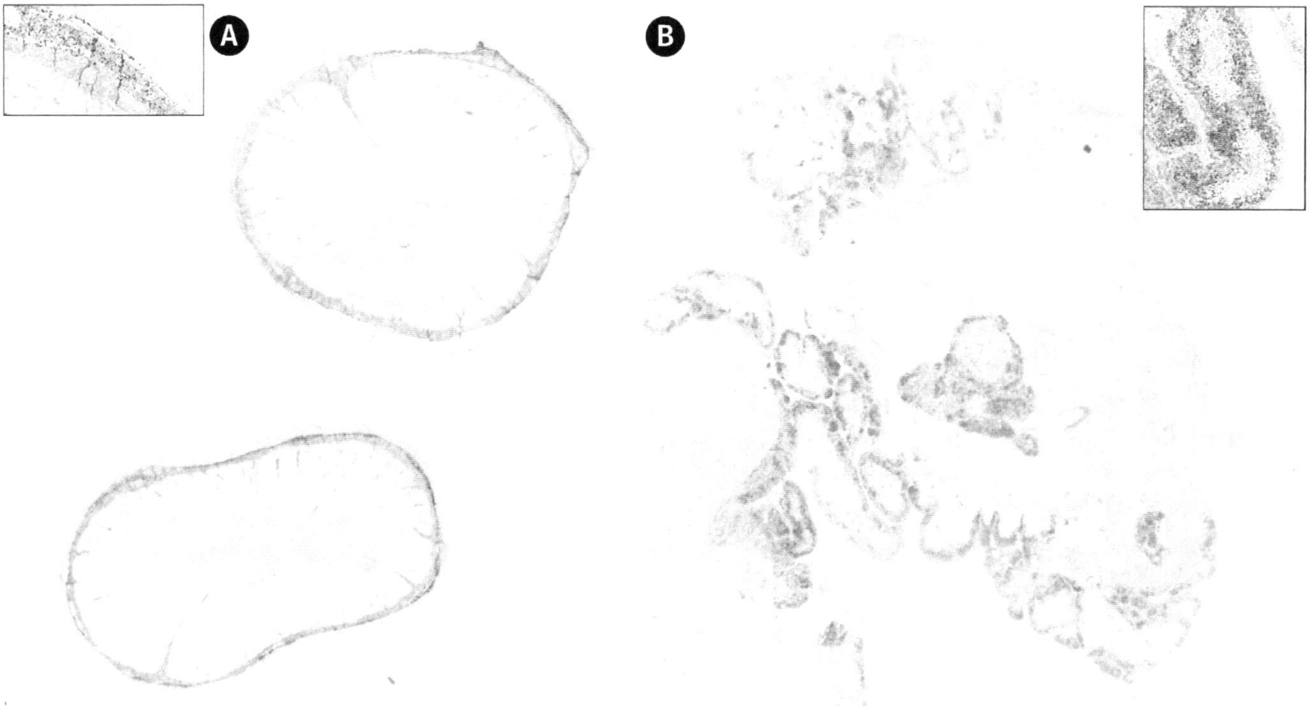

https://doi.org/10.54846/am2022/121

Retrospective analysis of *Mycoplasma hyorhinis* pulmonary and systemic infection in diagnostic cases with correlation of qPCR Ct values and detection by RNAscope®

M. Merodio, DVM, MS; J. Groeltz, BS; P. Piñeyro, DVM, MS, DVSc, PhD; R. Derscheid, DVM, PhD, ACVP
Veterinary Diagnostic and Production Animal Medicine, Iowa State University, Iowa

Introduction

Mycoplasma hyorhinis (Mhr) disease is characterized by polyserositis and arthritis in nursery pigs. It has been suggested that in the absence of other pathogens, Mhr can produce lesions in lung similar to those observed during *M hyopneumoniae* (Mhp) infections, suggesting that it could be a primary pneumonic agent. The diagnosis of Mhr disease is determined by identifying lesions consistent with polyserositis with concurrent Mhr detection from fibrin or joint fluid by qPCR or bacteriological culture. There is scarce information about the use of different diagnostic tools that correlate the role of Mhr with different lesions. Therefore, this study aims to 1) determine the coinfection rate of Mhr with different bacterial pathogens in clinical samples using retrospective diagnostic data 2) correlate *in situ* detection and quantification of Mhr by RNAscope® with the amount of Mhr nucleic acid detected by qPCR in different tissue samples in naturally infected pigs.

Material and methods

The Mhr interaction with other systemic and respiratory pathogens was assessed by a retrospective study of diagnostic cases qPCR positive for Mhr in different sample matrices submitted to ISU-VDL between January 2016 and March 2021. A subset of respiratory (n = 19) and systemic (n = 28) cases were selected for further Mhr RNAscope® *in situ* detection, quantification by digital analysis, and correlation with Mhr qPCR Ct values.

Results

Out of 7079 cases evaluated by qPCR, Mhr was detected in 53% of the total cases, with 9.8% and 62.9% of lung and fibrin, respectively. Mhr was detected as a single infection in 5.05% of respiratory cases and 13.2% of systemic cases. Mhr-bacterial coinfection observed in respiratory and systemic cases included *G parasuis*, *S suis*, *P multocida*, *A suis*, *Streptococcus* spp, *T pyogenes*, *M hyopneumoniae*, and *A pleuropneumoniae*. The proportion of Mhr-*G parasuis* and Mhr-*S suis* coinfection in systemic cases was higher than the proportion for the same coinfection in pulmonary cases. There were no significant differences in the mean Mhr Ct values in lung or fibrin. The mean Mhr Ct value in systemic cases was significantly lower in Mhr single infection than in cases with bacterial coinfection. In pulmonary cases, the mean Mhr Ct value did not show significant differences between Mhr single infection and cases with bacterial coinfection. No difference in the proportion of bronchopneumonia and interstitial pneumonia was observed between systemic and respiratory cases. In contrast, a significantly higher proportion of cases of airway-centric lesions with BALT hyperplasia was observed in respiratory cases. No correlation between the amount of Mhr detected *in situ* by RNAscope® and qPCR Ct values was observed in systemic cases, but there was a strong correlation in pulmonary cases (Figure 1).

Conclusion

The detection of Mhr without concomitant pathogens supports its role as a primary bacterial agent in systemic disease. Polymicrobial infections are common in the field, leading to the conclusion that pathogen interaction may favor Mhr infection, colonization, and replication. The high proportion of respiratory cases and the BALT hyperplasia in Mhr-positive respiratory cases supports the potential role of Mhr as a pulmonary pathogen. The use of RNAscope® as a direct detection method, in conjunction with digital analysis, showed a high correlation with qPCR detection in lung. Thus, this technique can add diagnostic value to Mhr-respiratory cases and could be an essential tool for investigating the pathogenesis of Mhr.

Figure 1: The correlation was calculated by simple linear regression. The correlation for individual compartments including bronchi (A, systemic; E, respiratory), parenchyma (B, systemic; F, respiratory), and serosa (C, systemic; G, respiratory) was calculated individually. The overall pulmonary parenchyma was calculated as an additive score of all individual compartments (D, systemic; F, respiratory). The solid lines represent the linearity of Spearman's correlation coefficient (A) $r = 0.01853$; (B) $r = 0.02031$; (C) $r = 0.02258$; (D) $r = 0.004277$; (E) $r = 0.4322$; (F) $r = 0.4024$; (G) $r = 0.1965$; (H) $r = 0.4888$. Dot lines represent 95% confidence interval. ($P < 0.05$).

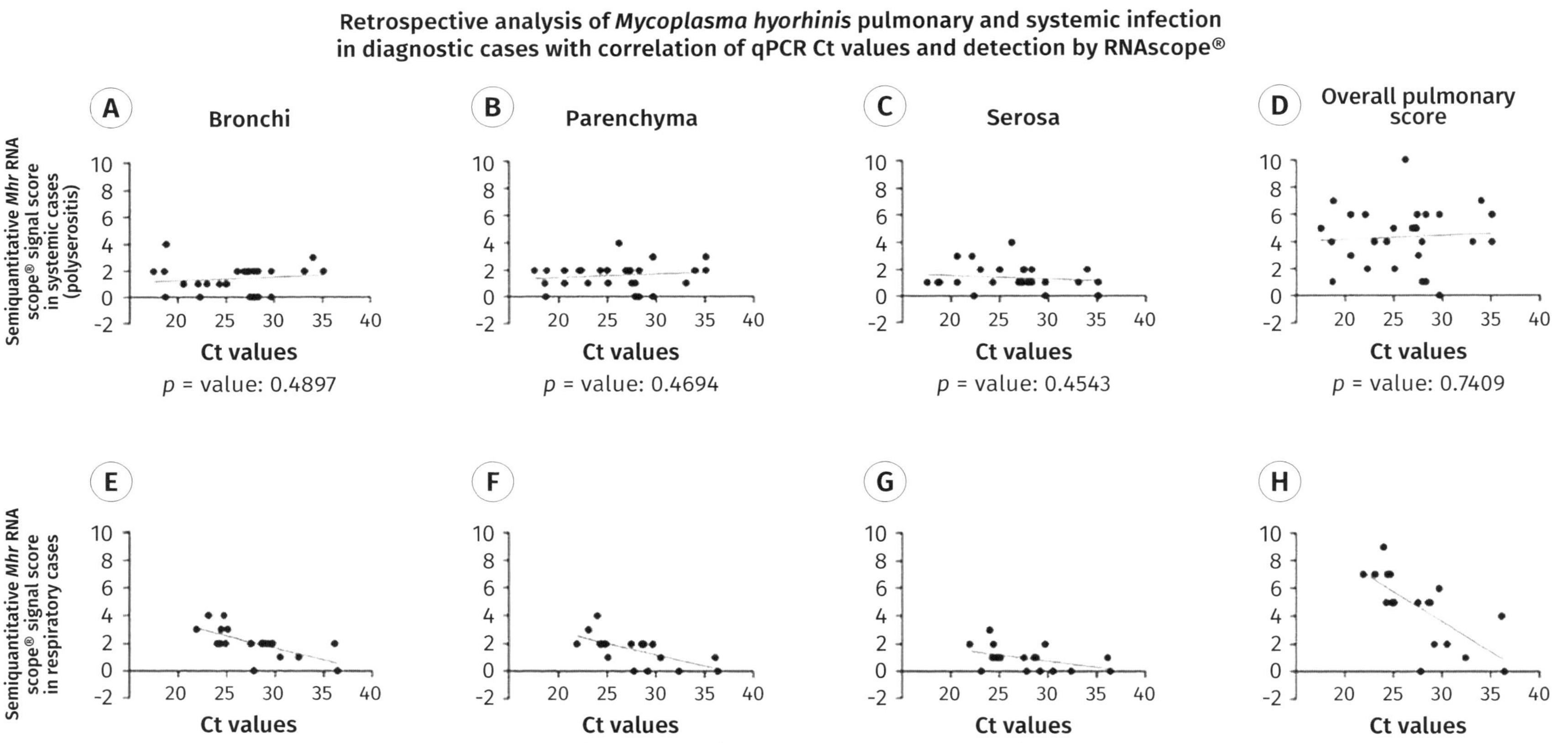

https://doi.org/10.54846/am2022/122

Detection of influenza A virus in US swine populations during 2010-2021: Implications for epidemiological investigations and control programs

D. C. A. Moraes[1], DVM, MS; G. Trevisan[1], DVM, MBA, PhD; E. S. Magalhães[1], DVM, MS; B. Crim[1], K. Schwartz[1], DVM, MS; E. R. Burrough[1], DVM, PhD, ACVP; P. C. Gauger,[1] DVM, MS, PhD; G. S. Silva[1], DVM, MS, PhD; P. Pineyro[1], DVM, MVSc, DVSc, PhD; C. Siepker[1], DVM, ACVP; R. Main[1], BS, DVM, PhD; M. Thurn[2]; P. Lages[2]; C. Corzo[2], DVM, MS, PhD; J. Torrison[2], DVM, PhD, DACVPM; R. McGaughey[3]; F. Matias-Ferreyra[3], DVM, PhD; J. Retallick[3], BS, DVM, PhD; J. Greseth[4]; D. Kersey[4]; T. Clement[4]; J. Christopher-Hennings[4], BS, DVM, MS; M. Prarat[5], DVM, MS; Yan Zhang[5], DVM, PhD; D. Summers[5], DVM; R. French[5], BS, MS, PhD; A. G. Arruda[5], DVM, MS, PhD; D. C. L. Linhares[1], DVM, MBA, PhD

[1]Veterinary Diagnostic and Production Animal Medicine, Iowa State University, Ames, Iowa; [2]Veterinary Diagnostic Laboratory, University of Minnesota, Saint Paul, Minnesota; [3]Kansas State Veterinary Diagnostic Laboratory, Kansas State University, Manhattan, Kansas; [4]Animal Disease Research and Diagnostic Laboratory, South Dakota State University, Brookings, South Dakota; [5]Ohio Animal Disease Diagnostic Laboratory, Reynoldsburg, Ohio

Introduction

Influenza A virus (IAV-S) is an important cause of swine respiratory disease worldwide. IAV-S is considered a significant contributor in the porcine respiratory disease complex along with porcine reproductive and respiratory syndrome virus and *Mycoplasma hyopneumoniae*.[1]

IAV-S subtypes are classified according to the hemagglutinin (H1 and H3) and neuraminidase (N1 and N2) genes. Changes in the antigenic characteristics within subtypes pose a significant concern to animal health from region to region.[2] An organized information hub would enable a better understanding of the distribution of the IAV-S subtypes, thus improving IAV-S control strategies and allowing strategic surveillance in the US swine industry. The purpose of this study was to investigate trends involving the detection of IAV-S subtype in samples submitted for testing at the Veterinary Diagnostic Laboratories (VDLs) participating in the Swine Disease Reporting System (SDRS).[3]

Material and methods

The IAV-S molecular testing results were obtained from diagnostic submissions from January 2010 to July 2021 at the five SDRS participant VDLs: Iowa State University, University of Minnesota, Kansas State University, and Ohio Animal Disease Research and Diagnostic Laboratory. The variables considered were geographic region (state), qPCR test and testing result, IAV subtype, age category, specimen, year, and sample receive date. Data were aggregated and organized using SAS® software (SAS v.9.4; SAS Institute), and the processed data were exported to Microsoft Power BI Pro® (Power BI) for data visualization.

Results

From the total of samples submitted for testing, 35.28% were PCR-positive for IAV-S RNA. Five specimens accounted for 96.24% of total samples submitted: oral fluid (47.98%), tissue-lung (34.0%), multiple specimens (7.57%), nasal swab (4.6%), and oropharyngeal swab (2.09%). Grow-finish age category represented the largest number of submissions (30.06%) for testing. There was a bi-seasonal pattern of IAV-S detection with increased detection during Spring (March-May) and Fall (September-November). From all samples with PCR positive results, 46.17% had IAV-S hemagglutinin or neuraminidase subtype detection (H1, H3, N1, N2). The four most frequent specimens tested by PCR for subtyping were: oral fluid (44.36%), lung (27.53%), nasal swab (16.35%), and virus isolates (7.68%). Samples identified from the grow-to-finish age category were the most frequent with subtyping requests (36.81%). The most frequent complete subtypes detected were H1N1 (31.42%), H1N2 (27.33%), and H3N2 (23.59%). Samples reporting the detection of either the hemagglutinin or neuraminidase genes included: H1 (3.53%), H3 (2.11%), N1 (1.62%), and N2 (4.09%). Finally, mixed subtype detections of the hemagglutinin and/or neuraminidase genes included: H1H3N1N2 (2.13%), H1N1N2 (1.96%), H1H3N2 (1.47%), H1H3N1 (0.23%), H3N1 (0.22%), H1H3 (0.12%), H3N1N2 (0.12%), and N1N2 (0.05%).

Discussion and conclusions

This study described the macroepidemiological aspects of IAV-S RNA detection and its distribution according to the age category, specimen, and seasonal detection trend. This work has the capability to provide real-time information on IAV-S PCR detection from 5 National Animal Health Laboratory Network (NAHLN) level 1 accredited VDLs. Additionally, this study described the diversity of IAV-S subtypes over time. In summary, data from this study suggests IAV-S seasonality trends and subtyping detection dynamics are constantly changing over time. This study outlines the importance of monitoring IAV-S and subtypes in a standardized method using SDRS, thus enabling timely surveillance and the ability to make more effective decisions regarding biosecurity measures.

References

1. Van Reeth K. and Vincent, A.L. 2012. Influenza Virus: Diseases of Swine 10th edition. Edited by: Zimmerman J.J., Karriker LA, Ramirez A, Schwartz K., Stevenson G. Wiley-Blackwell, p.557-571.

2. Lewis NS, et al. 2016. The global antigenic diversity of swine influenza A viruses. Epidemiology and Global Health. Microb and Inf Dis. (5):1-17. Available at: **http://dx.doi.org/10.7554/elife.12217**. Accessed 09 August 2021.

3. SDRS. Swine Disease Reporting System. Available at: **http://www.fieldepi.org/sdrs**. Accessed 09 August 2021.

https://doi.org/10.54846/am2022/123

Factors associated with sort loss in pork production

D. C. A. Moraes[1], DVM, MS; E. S. Magalhães[1], DVM, MS; C. A. A. Moura[2], DVM, PhD; B. Haberl[2], BS; D. C. L. Linhares[1], DVM, MBA, PhD; G. S. Silva[1], DVM, MS, PhD
[1]Veterinary Diagnostic and Production Animal Medicine, Iowa State University, Ames, Iowa;
[2]Iowa Select Farms, Iowa Falls, Iowa

Introduction

Sort loss is the monetary loss per carcass when those do not fall within the optimum established weight range.[1] The grid system changes between processing facilities and seasons, depending on the desirable carcass weight. The primary focus of the grid-based marketing system is to ensure that producers are marketing a consistent product that meets industry standards for quality, yield, and size.[2] Eleven million hogs were slaughtered in the US in 2019,[3] thus, there is a tremendous economic impact of sort loss to production systems. A systematic analysis enables a better understanding of the possible factors related to sort loss, thus improving strategies to minimize it. This study aimed to identify the factors associated with sort loss using data from a US production.

Material and methods

Data from batches of slaughtered pigs from August 2019 to April 2020 was collected. The data was organized, categorized, and analyzed using SAS® software (SAS v.9.4; SAS Institute). Briefly, the data was imported to SAS® using PROC Import, PROC SQL and DATA steps. One dataset corresponded to the information from the slaughtered animals: lot identification, carcass weight from each lot, number of animals by lot, number of days to first cut, number of loads per lot, and sort loss (in USD) from each lot. This data was matched and merged with the respective batch in a second dataset, with the following information during the grow-finish phase. The final dataset was used to perform the statistical analysis. Sort loss was the outcome and the predictors were: (a) lot identification, kill date, carcass weight standard deviation, interval days to fill the barn, lot size, interval days to cut, number of loads by lot, number of heads per load, days on feed, number of sources from the sow farm, disease status (*Mycoplasma hyopneumoniae* at weaning, PRRSV lateral, enteric disease), PRRSV at weaning (Endemic, Epidemic, Mixed and Negative), percentual mortality, group size, group type and season (spring, summer, fall, winter); (b) Linear regression with PROC REG was conducted with sort loss per head ($) as the outcome with each of these candidates to be predictors; (c) candidates were screened for significance and ranked (*P*-value); (d) Multivariable regression analysis with PROC GLM was used to identify the variables that were significant after adjusting to each other ($P < 0.05$), and Least Square Means (LS-Means) was obtained for each categoric variable in the final model. In addition, a quadratic term for average carcass weight was included in the final multivariate model based on the relationship observed in the univariate analysis.

Results

The final dataset was composed of 495 closeouts lots, with the data mentioned above summarized from the first day in the growing phase until the slaughterhouse (i.e., nursery to slaughter). The univariate analysis revealed that sort loss was significantly lower with smaller carcass weight standard deviation, average carcass weight on target, and lower grow-finish mortality. The fall season was the season with the lower sort loss, and summer the highest. The feeder to finisher group type showed the lower sort loss compared to wean-to-finish.

The final multivariable model (i.e., model adjusted for all other significant variables) explained 93.98% of the variation (adjusted R^2) of sort loss, with the final variables: standard deviation carcass weight, average carcass weight, season, percentual mortality and interaction between season and percentual mortality. More specifically, sort loss was lower when carcass weight standard deviation was lower, and when live weight was on target. For every increase in one carcass weight standard deviation, there was a sort loss increase of $0.31. There was a quadratic effect of carcass weight on sort loss, meaning that sort loss was minimum at 204 lb, and increased on lighter or heavier weight. Also, there was an interaction between the percentage of mortality and season, demonstrating higher sort loss ($2.09) when there was high mortality during summertime.

Discussion and conclusions

This study presented the factors associated with sort loss in a US production system. In summary, sort loss was minimized when the standard deviation of carcasses was smaller, carcass weight was on target (around 204 pounds), lower grow-finish mortality, and non-summer months. Additionally, it is essential to mention that potential disease effect might have been masked by marketing strategies, for instance, prolonging the days to market to add more weight. This study outlines the importance of understanding sorting loss associations, thus enabling pork producers to make more effective decisions based on these measures and reduce financial losses per lot. It is essential to address the limitation of sort loss analysis because it is not the ultimate analysis for economic performance.

References

1.USDA. Live Swine Sort Loss Reporting (**https://www.ams.usda.gov/sites/default/files/media/SwineSortLoss.pdf**).

2. Zhou & Bohrer. *Can J of Animal Science*. 99(3): 542-552.

3. USDA. Livestock Slaughter (**https://www.nass.usda.gov/Publications/Todays_Reports/reports/lstk1220.pdf**).

https://doi.org/10.54846/am2022/124

Use of an endogenous reference control in a PRRSV RT-qPCR

B. Munguía-Ramírez[1]; B. Armenta-Leyva[1]; A. Henao-Díaz[3]; F. Ye[2]; L. Giménez-Lirola[1]; J. Zimmerman[1]
[1]Veterinary Diagnostic and Production Animal Medicine, Iowa State University, Ames, Iowa;
[2]Department of Statistics, Iowa State University, Ames, Iowa; [3]Pig Improvement Company, Querétaro, México

Introduction

The performance of reverse transcriptase real-time polymerase chain reaction (RT-qPCR) in diagnosis can be affected by testing procedures, ie, extraction and/or amplification steps.[1] However, detection of nucleic acids may also be affected by sample handling (eg, sample not chilled after collection, multiple freeze-thaw cycles, etc). Ideally, a diagnostic RT-qPCR procedure would include an indicator of sample quality. Endogenous reference controls (ERC) are used for this purpose. Since they are subjected to the same conditions as the target, consistently detection of the ERC indicates both that the test procedure was conducted correctly and that the sample was of good quality.[2] ERC are commonly used in human research for gene expression analysis, but they are hardly described in veterinary diagnostics.[3] The aim of this preliminary study was to test the response of a swine ERC for a commercial PRRSV RT-qPCR using oral fluid and serum collected from pigs vaccinated against PRRSV under experimental conditions.

Materials and methods

Twelve 14-week-old pigs were vaccinated with a PRRSV modified live vaccine (Ingelvac PRRS® MLV, Boehringer Ingelheim Vetmedica, Inc.) under experimental conditions. Serum (n = 132) and oral fluids (n = 130) were collected from -7 to 42 days post vaccination (DPV) and then tested with a PRRSV RT-qPCR that detects both PRRSV RNA and pig-specific ERC (IDEXX Laboratories, Inc.). The resulting Cq values of the ERC were evaluated in serum and oral fluids within pigs through DPV's and reference intervals were calculated using R 4.1.0 (R core team, 2020).

Results

The ERC was detected in all serum and oral fluid samples (n = 262). Data for ERC Cq in oral fluids was normally distributed based on Shapiro-Wilk normality test ($P = 0.584$). Thus, the 95% reference interval using normal distribution was 23.1 to 30.1. For serum, the normality assumption was rejected ($P = 0.031$). Hence, a 95% reference interval with non-parametric percentile method was performed. Upper and lower limits obtained for serum were 23.9 to 29.4. No specific response pattern of the ERC was found in pigs over DPVs for both sample types. In oral fluids, 6 out of 130 samples were out of the reference interval values. In serum, 4 out of 132 samples were out of the reference intervals. Cq values of the ERC maintained within the reference intervals after PRRSV vaccination.

References

1. Nygard AB et al. (2007). Selection of reference genes for gene expression studies in pig tissues using SYBR green qPCR. BMC molecular biology, 8(1), 1-6.

2. Vandesompele J et al. (2009). Reference gene validation software for improved normalization. Real-time PCR: current technology and applications, 4, 47-64.

3. Cheng TY et al. (2021). Internal reference genes with the potential for normalizing quantitative PCR results for oral fluid specimens. Animal Health Research Reviews (submitted).

Poster 46

https://doi.org/10.54846/am2022/125

Changes in productivity parameters as breeding herds targeting stability changed PRRSV status over time

Onyekachukwu H. Osemeke[1], DVM, MS; Tara Donovan[2], DVM; Kate Dion[1,3], DVM; Derald J. Holtkamp[1], DVM, MS; Daniel C. L. Linhares[1], DVM, MBA, PhD

[1]Veterinary Diagnostic and Production Animal Medicine, College of Veterinary Medicine, Iowa state university, Ames, Iowa; [2]The Hanor Company, Spring Green, Wisconsin; [3]The Hanor Company, Webster City, Iowa

Introduction

The American Association of Swine Veterinarians (AASV) porcine reproductive and respiratory syndrome (PRRS) status classification for breeding herds was recently reviewed and modified.[1] This study aimed to investigate changes in key productivity parameters as breeding herds transitioned through these categories.

Materials and methods

Six breeding herds belonging to a US production system were recruited for this study. The herds were periodically vaccinated with PRRS modified-live virus (MLV) vaccine, as such could only attain three (Statuses 1A, 1B and 2vx) out of the six categories in the new classification system. Four years of matched weekly retrospective production and diagnostic data were obtained and statuses were assigned to each farm-week using the following criteria:

Status 1A (Positive unstable, high prevalence): A herd attains this status following an outbreak until there is evidence of low PRRSV prevalence. When a herd does not meet criteria for placement into any of the other two categories, it falls under status 1A.

Status 1B (Positive unstable, low prevalence): A herd attains this status when at least 75% of PCR tests for PRRSV within a 90-day period are negative.

Status 2vx (Stable with attenuated PRRSV vaccination): This is the best possible status attainable for a vaccinating herd. A herd attains this status when 100% of PCR tests for PRRSV within a 90-day period are negative.

A linear mixed regression analysis was performed with PRRS status, herd ID number and season as predictors, and each weekly productivity parameter (average born alive, average neonatal losses defined as the difference between total born and born alive, average piglets weaned per sow, and average pre-weaning mortality) modeled separately as dependent variables. The differences in least-square means of the 1A and 1B statuses compared to the 2vx status for each parameter were obtained. Whenever there was a PRRS outbreak recorded, status 1A was further divided into 2: the first 10 weeks into the outbreak and the rest of the weeks after, until the herd transitioned to 1B.

Results

It was possible to assign the new AASV PRRSV classification system to all herds based on historic outbreak, vaccinations, and diagnostic data. When compared to status 2vx; Ten weeks into status 1A, the 11th week and upwards of status 1A, and status 1B, respectively, had 1.1, 0.5 and 0.1 fewer piglets born alive per litter; 1.39, 0.36 and 0.13 more neonatal losses per litter; 1.9, 0.6 and 0.2 less pigs weaned per sow; 12.5%, 1.4% and 1.3% more preweaning mortality. (Figure 1).

Conclusion

Productivity improved as herds improved PRRS status. Comapring productivity differences between status 2vx and 11th week and upwards of status 1A, one could infer that herds could actively be shedding PRRSV even after returning to near baseline productivity following an outbreak; this underscores the importance of continued monitoring and sustaining best biosecurity practices after herds have become apparently stable clinically following an outbreak.

Reference

1. Holtkamp DJ, Torremorell M, Corzo CA, Linhares DCL, Almeida MN, Yeske P, Polson DD, Becton L, Snelson H, Donovan T, Pittman J, Johnson C, Vilalta C, Silva GS, Sanhueza J. (2021). Proposed modifications to porcine reproductive and respiratory syndrome herd classification. *J Swine Health Prod* 2021, 29(5), 261–270.

Figure 1: A graphical representation of **differences** in mean weekly productivity parameters of vaccinating herds when each of the PRRSV unstable categories (1A and 1B) is compared to the PRRSV stable category (status 2vx).

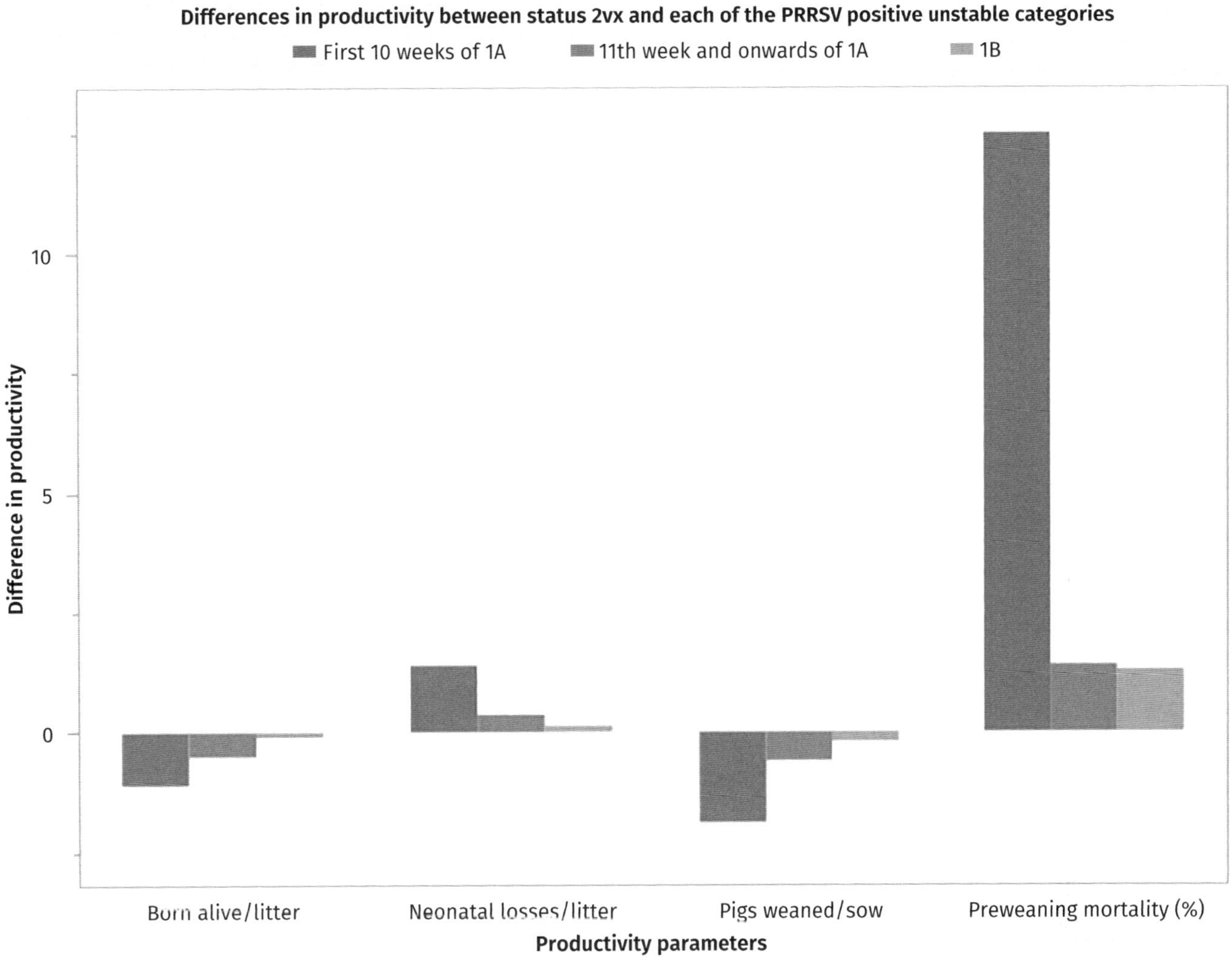

Poster 47
https://doi.org/10.54846/am2022/126

Detection and correlation of PCV2 by PCR and IHC in tissues in a PCV2d/PRRSV challenge model in the new paradigm of PCV2d infection

M. Kroeger[1], BS; E. Fano[2], DVM, PhD; K. Schwartz[1], DVM, MS; P. Piñeyro[1], DVM, MVSc, DVSc, PhD
[1]Veterinary Diagnostic and Production Animal Medicine, Iowa State University, Iowa;
[2]Boehringer-Ingelheim Animal Health Inc.

Introduction

Porcine circovirus type 2 (PCV2) is economically significant, highly prevalent, and nearly ubiquitous globally, and causes porcine circovirus-associated disease (PCVAD). Typical PCVAD lesions include lymphoid depletion and histiocytic inflammation and are often exacerbated by co-infection with endemic pathogens. Therefore, the diagnosis of PCVAD not only considers the clinical context and presence of characteristic lesions, but must be supported by relevant laboratory testing. However, the dynamics of virus replication, tissue localizations during infection, stage of infection, particular animal(s) sampled, sample type(s), diagnostic techniques applied, and vaccination status can compromise accuracy of diagnosis. The increasingly prevalence of PCV2d raises questions of current vaccine efficacy in herds reporting more PCVAD cases and/or greater magnitude of viral detection in vaccinated animals. The study objective is to evaluate the relationship of sample types, duration of infection, and results of diagnostic techniques in vaccinated and unvaccinated pigs in a PCV2d/PRRSV co-infection model under field conditions.

Material and methods

On study day 0 (D0), 100 pigs from a PRRSV/Mhp negative, PCV2 low prevalence, commercial sow farm were weaned and vaccinated with Ingelvac PRRSV MLV. Eighty of the 100 pigs also received commercial PCV2 vaccine (Vacc), leaving 20 pigs as unvaccinated controls (NoVacc). On D28 (49 days-of-age), all 100 pigs were inoculated with 1 mL IM and 1 mL IN of PCV2d (5 log_{10}/2mL dose) as well as 2 mL IM of 1-7-4 PRRSV (4.0 TCID50/mL). Blood samples were collected weekly for PCR and ELISA for both PCV2 and PRRSV. At termination (D56), tissues were collected (tonsil, lung, lymph nodes) for histopathology, IHC and PCR.

Results

The challenge resulted in 60% and 9.09% mortality in NoVacc and Vacc, respectively prior to D56. On NoVacc that died before D56, tonsil and lymph node histopathology had average lesion scores of 3 (severe inflammation and lymphoid depletion) and IHC scores of 3 (> 50% positive cells), compatible with severe PCVAD. In contrast, Vacc IHC and lesion score was 1-2 (10-50% positive cells and mild-moderate inflammation/depletion). Vacc necropsied at D56 had average lesion scores of 0-1 (none-minimal inflammation/depletion) and IHC scores of 0-1 (< 10% positive cells) compared to NoVacc that had average scores of 3 (severe) for both lesions and IHC. The average PCR Ct values on lung, lymph node, and tonsil in NoVacc were 8.1, 8.4, and 9.9, respectively. In the same tissues, Vacc had significantly higher Ct values of 25.4, 22.6, and 23.5, respectively.

PCV2 IHC score had positive correlation with PCR Ct values in lung, lymph node, and tonsil, driven by the vaccination status, due to a strong positive correlation in NoVacc and a linear correlation in Vacc (Figure 1). The Ct values in feces varied from 18.7 to 24.1 between 46 and 56 DPI in NoVacc and 25.6 and 27.6 in Vacc. Viral load in serum was consistently 2-3 logs higher in NoVacc.

Figure 1: Comparison of diagnostic correlation of antigen detection by IHC and nucleic acid detection by PCR in tonsil **(A)**, lymph node **(B)**, and lung **(C)** in non-vaccinated (left) and vaccinated (right) animals.

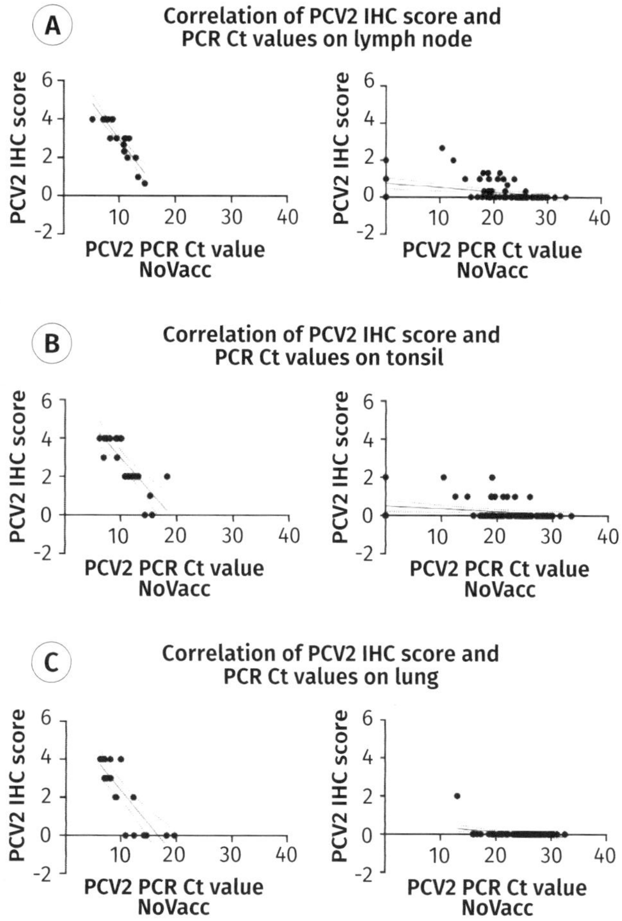

Conclusion

Experimental co-infection by inoculation with contemporary PCV2d/PRRSV isolates resulted in high mortality and severe PCVAD. Co-infection can affect PCVAD diagnosis, but NoVacc and Vacc present a different viral detection threshold in tissues and serum. The correlation of viral detection by PCR and *in situ* detection by IHC seems to be less affected by tissue type than by vaccination status. NoVacc had strong IHC positivity within an average Ct of 10, whereas Vacc had less positive staining and an average Ct of 23.3. Virus load in feces is higher in NoVacc than Vacc for several weeks post-infection but is not reliable for diagnosis of PCVAD. Death was highly correlated to viral load as determined by both IHC and PCR. Continuing production impact of PCV2 on survivors is likely less in Vacc than in NoVacc, with variation influenced by particular and present ongoing co-infections.

https://doi.org/10.54846/am2022/127

Cough associated with *Mycoplasma hyopneumoniae* DNA detection in deep tracheal swabs and pen-based oral fluids

Ana Paula S. Poeta Silva[1]; Gabriel Y. Storino[2]; Franco S. Matias Ferreyra[1]; Min Zhang[1]; Eduardo Fano[3]; Dale Polson[3]; Chong Wang[1]; Rachel J. Derscheid[1]; Jeffrey J. Zimmerman[1]; Maria J. Clavijo[1,4]; Bailey L. Arruda[1,5]
[1]Iowa State University, Iowa; [2]Universidade Estadual de São Paulo, SP, Brasil;
[3]Boehringer Ingelheim Animal Health US Inc., Georgia; [4] PIC®, Tennessee; [5]USDA, Iowa

The observation of dry, non-productive cough may suggest *Mycoplasma hyopneumoniae* (MHP), but the frequent occurrence of polymicrobial infections that may (or may not) involve MHP means that testing is required to establish a diagnosis. In this study, the value of cough as a predictor for MHP DNA detection in environmental and/or clinical specimens was evaluated in pigs of known MHP infection status.

Seven-week-old MHP-negative pigs (n = 39) were allocated to five separated rooms with one pen each: one negative control (n = 3 sham-inoculated pigs) and four other pens (1 to 4) contained 9 pigs each. Pens 1 to 4 contained 1, 3, 6, and 9 MHP-inoculated pigs (lung homogenate, 1×10^5 CFU/mL MHP 232), respectively. Pigs were monitored through 59 days post inoculation (DPI). Individual tracheal swabs (2 per week), pen-based oral fluids (daily), pen-based drinker (1 per week), and 60-minute pen-based air samples (3 per week from Pens 2 and 4, i.e., 3 and 9 MHP-inoculated pigs) were collected and tested for DNA. Cough data included counts (2 per week) by an observer blinded to pig MHP-infection status and quantitative cough counts per day using a digital system (SoundTalk®, Belgium). On DPI 59, pigs were humanely euthanized, and lung and tracheal tissue evaluated for gross and microscopic lesions.

MHP DNA was detected in individual tracheal swabs from 15 of 19 inoculated pigs on DPI 3, and from all 19 inoculated pigs on DPI 10. MHP DNA was detected in pen-based oral fluid from Pens 1, 2, and 4 on DPI 8 and pen-based drinker wipe samples in Pens 1 to 4 on 21 DPI. MHP DNA was detected in 5 or 3 of 26 total pen-based air samples collected from Pens 2 and 4. In Pens 1 - 4, coughs were first noted by the observer on DPI 10 (Rooms 2 and 4), DPI 14 (Room 3) and DPI 21 (Room 1). Thereafter, cough was continuously observed until the termination of the study in all pens, but not after DPI 45 in Pen 4. Regardless of the proportion of MHP-inoculated pigs in each pen, mixed effect logistic regression analyses showed that the presence of coughing pigs in a pen was a significant predictor for MHP DNA detection in pen-based oral fluid samples ($P < 0.001$, Odds ratio = 14.0; 95 CI% 3.7, 76.0), nearly significant in pen-based drinker wipes ($P = 0.08$; Odds ratio = 6.0; 95 CI% 0.9, 67.3), but not for MHP DNA detection in individual deep tracheal swabs ($P = 0.13$). Analysis of the pathology data revealed that tracheal cilia were shorter in MHP-infected pigs than negative controls ($P < 0.05$, linear regression) and the severity of gross lung lesions were higher in pig identified as MHP-infected after DPI 15 ($P < 0.05$, Kruskal-Wallis).

In conclusion, extensive coughing may be useful in monitoring MHP, but laboratory testing is required to confirm MHP infection. Individual tracheal swabs provided the earliest MHP DNA detection, but cough episodes were statistically associated with the detection of MHP DNA in pen-based oral fluids. Thus, in the presence of extensive cough, pen-based oral fluids may provide the most efficient means to surveil the population for MHP infection.

https://doi.org/10.54846/am2022/128

Proving competency-based learning educational strategies to create measurable carcass management competencies in farm labor

S. Probst Miller[1,2], DVM; M. Hutchinson[3]
[1]AgCreate Solutions, Inc.; [2]Pork Avenue Training Portal; [3]University of Maine

Introduction

As we raise food in the US, ensuring a competent labor force prepared to manage daily livestock mortalities and potential depopulation catastrophes, such as foreign animal disease or unexpected events like those observed with COVID-19 market impacts, maximizes the ability for American producers to prosper. The use of competency-based educational technologies and employing a "See it, Do it, Teach it"[1-9] method of training is investigated in this study to measure industry baseline competencies and determine our ability to positively impact farm labor improvements in mortality management. Further, this study focuses on the use of composting as an effective mortality management practice that reduces the spread of disease,[10] provides a natural source of nutrient inputs, and contributes to environmental stewardship.

Materials and methods

This project began with the formation of an advisory panel of mortality composting leaders that helped finalize baseline competencies of farm staff. This content was then used to create pre- and post-evaluation measures in addition to facilitating the creation of a video script and storyboard. Baseline assessments were completed on five sow farms, a gilt developer unit, and 24 wean-to-market operations. Competency data were tallied (Table 1), staff trained with "See it, Do it, Teach it"[1-9] learning experiences, and then competencies remeasured.

Results and discussion

The average pre-training baseline competency for all respondents related to Composting Preparation was 37.6%. The baseline percentage for competencies related to Building the Composting Pile was 39.6% (Table 1). The combined overall competency of farms was 38.7%. The AgCreate Solutions, Inc. team identified multiple challenges regarding farm staff base-level competencies, varying types of composting operations, and farm manager buy-in to the value of composting. These challenges demonstrate the need for targeted carcass management training and verification of competencies. In the ongoing trial, staff will be trained with "See it, Do it, Teach it"[1-9] learning experiences, and then competencies remeasured during the "Teach it" back phase.

References

1. Bear, Anne, Technology, Learning, and Individual Differences, Journal of Adult Education, 2012.

2. Dale, Edgar, Audio visual methods in teaching. Holt, Rinehart, and Winston, 1965.

3. Dale, Edgar, Cone of Experience, Education Media, 1969.

4. Gaetano, Kristian, 5 ways adults learn differently than children, Learn Kit, 2016.

5. Pepin, Brent, Williams, Todd, Polson, Dale, Gauger, Phillip, Dee, Scott. Survival of swine pathogens in compost formed from preprocessed carcasses. *Transbound Emerg Dis*. 2021; 68: 2239– 2249. **https://doi.org/10.1111/tbed.13876**

6. Pork Avenue Training Portal. 2018. Available at **www.porkavenuetraining.com**

7. Probst Miller, Sarah, Impact of two training implementation plans on post-training competency expression of personnel on farm, AASV Proceedings, 2018.

8. Probst Miller, Sarah, Applying the science and physiology of how adults learn to our on-farm approach as practitioners, AASV Proceedings, 2016.

9. Probst Miller, Sarah, The impact of trained and untrained personnel on product implementation and subsequent pig health, AASV Proceedings, 2016.

10. Ryland, Bruce, Fostering Critical Reflection in Adulthood, by Jack Mezirow & Associates, San Francisco: Jossey-Bass, 1990.

Table 1: Average pre-training baseline competency percentage for all respondents

Competency	Percentage
Composting - preparation	37.6%
Composting - creating the compost pile	39.6%
Combined overall competency	38.7%

https://doi.org/10.54846/am2022/129

Investigating differences in the source of *Serratia* and other bacteria in boar semen

Darwin L. Reicks, BS, DVM

Introduction

The presence of certain bacteria such as *Serratia* in the extended semen can cause loss of sperm viability and fertility.[1,2] In the boar stud industry, Serratia marcescens contamination is a massive concern due to its severe and unpredictable impact on downstream fertility. In some cases, farms have seen conception rates drop well below 50 percent. The antibiotics used and approved in semen extenders are not effective against Serratia. A significant effect on litter size and fertility[3] can be associated with using Serratia contaminated semen. There is frustration due to lack of information on how to prevent such occurrences. Although the barn is believed to be a primary source of bacteria, the specifics have not been thoroughly defined. A series of experiments have been conducted by the author over the last few years to help determine the effectiveness of various collection strategies to the presence of bacteria in the semen. An additional experiment was conducted using 16S sequencing techniques and presented at the 2021 AASV meeting to identify which bacteria were present in boars from four different studs in semen, prepuce fluid, feces, and environmental Swiffer. The percentage of isolates in prepuce fluid identified as *Serratia* marcescens varied from 0.03% to 2.4% in individual studs. For feces, the range was 0.05% to 1.7%. In semen, the percentage was from 0.2-0.5%, and for an environmental Swiffer, between 0.2-0.5%. Although one of the studs has never cultured Serratia from extended semen, *Serratia* was identified in 36 out of 40 samples from that stud.

Materials and methods

In this follow up experiment, funded by the AASV Foundation, three boar studs were sampled, one which has never identified *Serratia* and two that frequently identify *Serratia* in extended semen. Eight boars were sampled per stud. For the studs that frequently isolate *Serratia* in extended semen, boars were targeted who have been associated with a *Serratia* positive culture from extended semen. Raw semen (using the three-glove method to achieve a free-catch collection), prepuce fluid, feces, and an environmental Swiffer (from the boar's stall) were obtained from each boar. Sampling is taking place during the 3rd quarter of 2021. Analysis will take place as follows when all samples are collected: 16S metagenomics analysis using the Illumina MiSeq® System will be performed at South Dakota State Veterinary Diagnostic Laboratory. Results will be analyzed by qualitive statistical measures, including the presence or absence of *Serratia* marcescens in various samples, and when present, the relative quantity of *Serratia* present between sample types, between boars and the comparison of boar studs with a history of *Serratia* to those without.

Results and conclusions

The results of this experiment will be presented at the 2022 AASV meeting. This project was funded as one of the AASV Foundation research awards at the 2021 AASV meeting.

References

1. Althouse GC, Kuster CE, Clark SG, Weisiger RM. Field investigations of bacterial contaminants and their effects on extended porcine semen. Theriogenology 53 (2000) 1167-1176.

2. Ubeda JL et al. Adverse effects of members of the Enterobacteriaceae family on boar sperm quality. Theriogenology 80 (2013) 565-70.

3. Maroto Martín LO et al. Bacterial contamination of boar semen affects the litter size. Animal reproduction science 120 (2010) 95-104.

https://doi.org/10.54846/am2022/130

Oral administration of PEDV S-antigen to pregnant sows provides protection to newborn pigs

R. Sibbel[1], DVM; J. Howard[2], PhD
[1]Mazen Animal Health, ISU Research Park, Ames, Iowa; [2]Applied Biotechnology Institute, San Luis Obispo, California

Introduction

Porcine epidemic diarrhea virus (PEDV) is a coronavirus that causes severe diarrhea with a high mortality in nursing pigs. The most common target of vaccine strategies for coronaviruses is the spike protein (S) because it contains multiple epitopes for neutralizing antibodies (NABs). Because PEDV initiates its infectious cycle at the intestinal mucosal epithelial surface,[1] effective protection would optimally require vaccination which elicits an immune response at both the systemic and mucosal levels.[2] An orally administered vaccine may provide a more robust mucosal response and greatly facilitate widespread vaccination against PEDV by eliminating individual handling of the pigs for injections. Maize grain has emerged as a preferred option for oral vaccines as it provides high levels of accumulation of the recombinant proteins[3] and bioencapsulation of the protein to protect it from degradation in the digestive tract.[4,5] The objectives of this study were to (1) determine if administration of the S1 vaccine candidate to naïve sows could provide passive immunity to the newborn pigs upon a PEDV challenge and (2) better understand the immune response of sows to oral vaccination.

Materials and methods

Production of maize expressing-S1 antigen

A transgenic maize line expressing high levels of the S1 antigen (PDC18)[6] was grown to obtain grain for the animal study.

Trial

Sixteen PEDV-free pregnant sows (PIC 1050 line) were divided into four treatments: A) injection with commercial PEDV vaccine; B) maize grain without the S antigen; C) maize grain containing a low dose of the S antigen; D) maize grain containing a high dose S antigen. Piglets were challenged with PEDV and evaluated for disease symptoms. Milk and serum samples were taken at various timepoints for neutralizing activity and cytokine analyses.

Results and discussion

Oral vaccination of sows with the S-antigen protects nursing piglets against PEDV challenge

The survival rate of piglets after 11 days post challenge with PEDV is shown by sow in (Figure 1). The mean survival rates for each treatment group were: 53% in the oral low dose S antigen, 37% in the commercial injected vaccine, 7% in the oral high dose S antigen, and 10% in the control maize. The oral low dose S1 antigen had a significantly higher survival rate ($P < 0.001$) as compared with the high dose S1 antigen and with control maize treatments.

Administration of S1 antigen decreases the level of GM-CSF in sows' sera and milk

Thirteen cytokines in the sera and milk of treated animals were analyzed. There was a significant decrease in the level of a pro-inflammatory cytokine granulocyte-macrophage colony stimulating factor (GM-CSF) in the sera from animals administered oral low dose S antigen expressing maize (P-value = 0.0137) and from animals injected commercial vaccine (P -value = 0.0456) as compared to animals fed untransformed corn. Similarly, the level of GM-CSF was lower in the milk samples from sows administered oral low dose S antigen expressing maize (P -value = 0.02) and from animals injected commercial vaccine (P -value = 0.0386) as compared to animals fed untransformed corn.

Maize-produced, orally delivered S antigen administered to pregnant sows provided passive immunity to suckling piglets. Statistically higher survival rates in newborn piglets born to sows fed a low dose S antigen as compared to control animals. Higher titers of NABs in the sows' milk and lower levels of cytokine granulocyte macrophage-colony stimulating factor in sows' milk correlated with higher survival rate of piglets, suggesting that the S antigen may act as an immunogen and as an immune modulator.

References

1. Wang et al. 2019, Biosci Rep 39(3):BSR20182028.
2. Mota et al. 2006, BMC Biotechnology 6:2.
3. Hood and Howard 2009) *Molecular Genetic Approaches to Maize Improvement. Biotechnology in Agriculture and Forestry,* vol 63. Springer, Berlin, Heidelberg
4. Streatfield SJ et al. 2001, Vaccine 19(17-19):2742-9.
5. Modelska A et al. 1998, Proc. Natl. Acad. Sci 95(5):2481-5.
6. Egelkrout et al. 2020, Plant Cell Tissue Organ Cult. 11:1-8.

Figure 1: Survival rates of piglets for individual sows. Individual bars represent the percent of piglets by sow that survived the challenge with PEDV. Treatment groups: **A)** Commercial injected vaccine (positive control), **B)** Untransformed corn (negative control), **C)** Oral low dose S antigen-expressing maize, **D)** Oral high dose S antigen-expressing maize. The mean percentage for each group is also shown.

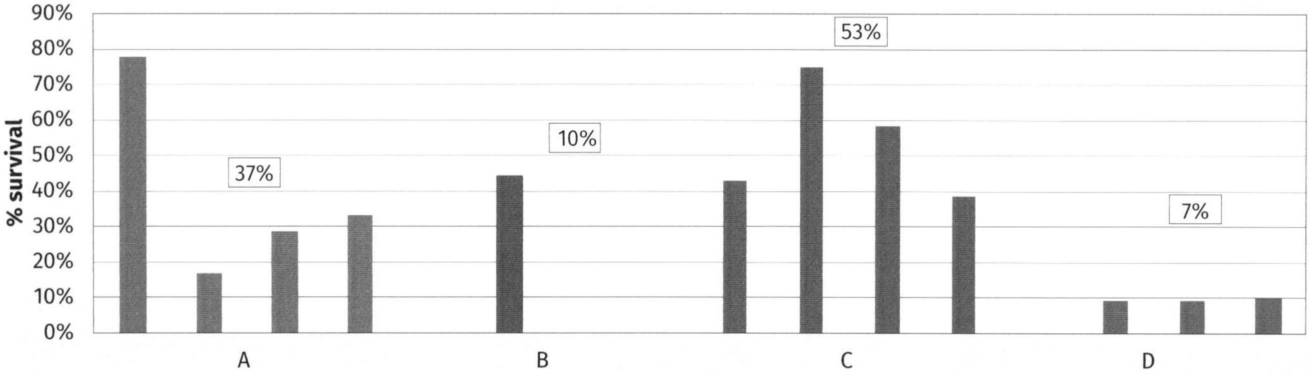

https://doi.org/10.54846/am2022/132

False positive or early detection? SPC charting improves ELISA MHP interpretation

E. Tapia[1], DVM, PhD; Marisa Rotolo[3], DVM, PhD; Andres Diaz[3]; S. Zimmerman[1], DVM; PhD; M. J. Clavijo[2,3], DVM
[1]IDEXX Laboratories, Inc., Westbrook Maine; [2]Iowa State University, Ames, Iowa;
[3]Pig Improvement Company, Hendersonville, Tennessee

Introduction

Mycoplasma hyopneumoniae (MHP) antibody detection by ELISA is routinely used to monitor MHP. However, in expected MHP-negative populations, it can be challenging to differentiate a false positive response from an early antibody response.[1] Standard Process Control (SPC) can be used to interpret test results by tracking the group mean and standard deviation (SD) of S/P values over time. SPC identifies potentially significant shifts in the distribution by delivering a signal when a value is above or below a threshold. This signal requires further investigation to identify the root cause.[2] This study investigated MHP antibody kinetics and the use of Statistical Process Control (SPC) charting as a means to improve the interpretation of results.

Materials and methods

Ten randomly selected pigs in a pen of 29 MHP-negative 9-week-old pigs were intratracheally inoculated with 10 mL (1×10^5 CCU per mL) of lung homogenate containing MHP strain 232. Uninoculated pigs (n = 19) served as contact pigs. Serum samples from all contact pigs were collected at 15, 24, 42, and 57 days post inoculation (DPI) and tested by indirect ELISA (IDEXX) as per the manufacturer's instructions. Results were reported as sample-to-positive (S/P) ratios and the results evaluated by SPC using Microsoft Excel®. Results were plotted using and *s* Shewart Chart for continuous numerical variables.[2]

Results

At 15 DPI, all contact pigs showed very homogeneous S/P values with strong negativity. The S/P mean and SD are below the LCL due to the limits were estimated using all sampling points (Figure 1). At 24 DPI the first positive contact pig is detected, however, the rest of the contact pigs maintain a similar negative S/P values. At this point is identified an increase in both S/P mean and SD. With 4 positive contact pigs at 36 DPI, there is a 5-times increase in the S/P mean value and the SD is above the control limits (Figure 1). At 42 DPI the same trend is maintained and contact pigs S/P values of negative contact pigs increase towards the test cut-off point. Now the SD is just at the border of the control limit due to the increase of positives (Figure 1). Finally, at 57 DPI, 16 contact pigs are now ELISA positive and only 1 negative and 2 suspects (Figure 1). As the S/P values of the positives are very similar, the SD is not impacted. Now a more homogeneous positive population is considered without much variation.

Conclusions

In SPC analysis, a process is considered in statistical control when the results are stable. When a special cause is present in the process (eg, seroconversion), it will trigger an alarm or warning signal as the group mean and SD overpass the control limits. According to this, a negative population should have stable S/P values. An alarm or signal will appear when seroconversion occurs. All historical and new values are taken into account in SPC and therefore the control limits are updated as more data is entered, allowing for a continuous monitoring system sensitive to the identification of alarm signals when there is a field infection. Through this analysis, the group of animals sampled is taken into consideration and the interpretation of results is at the group level and not at the individual level.

References

1. Peterson, C., Odland, C. *Mycoplasma hyopneumoniae* serological testing challenges in a negative breeding herd. 50th Annual Meeting of the American Association of Swine Veterinarians. Orlando; March 9-12, 2019.

2. Qiu, P. Introduction to Statistical Process Control. 1st ed. Chapman and Hall/CRC. 2013.

Figure 1: (A) MHP Elisa S/P mean (x-bar chart) and **(B)** MPH Elisa S/P standard deviation.

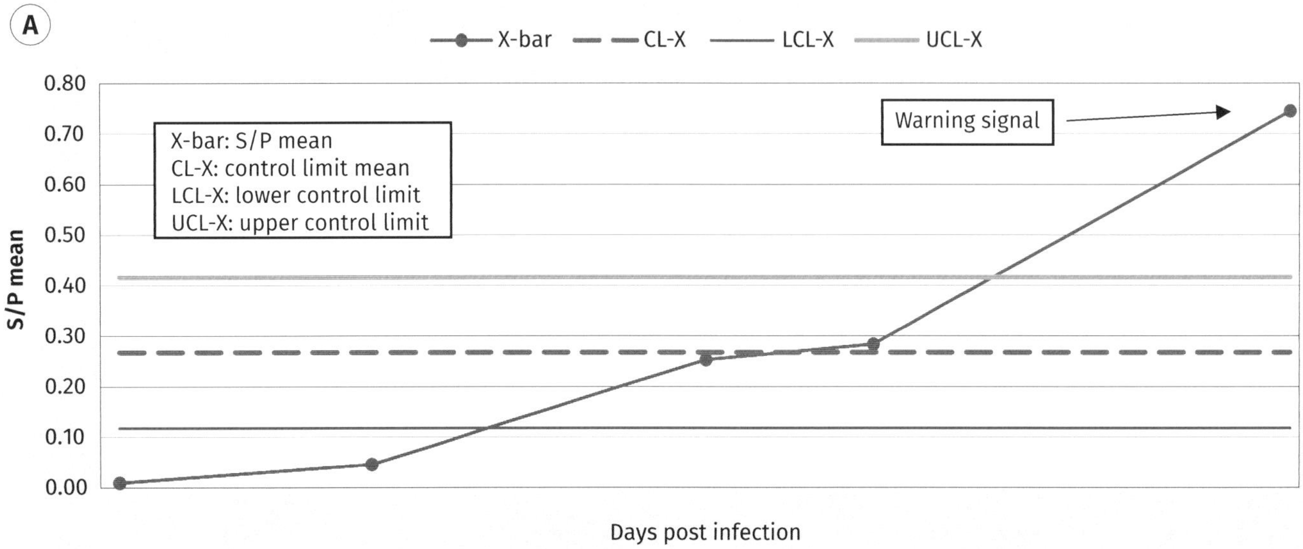

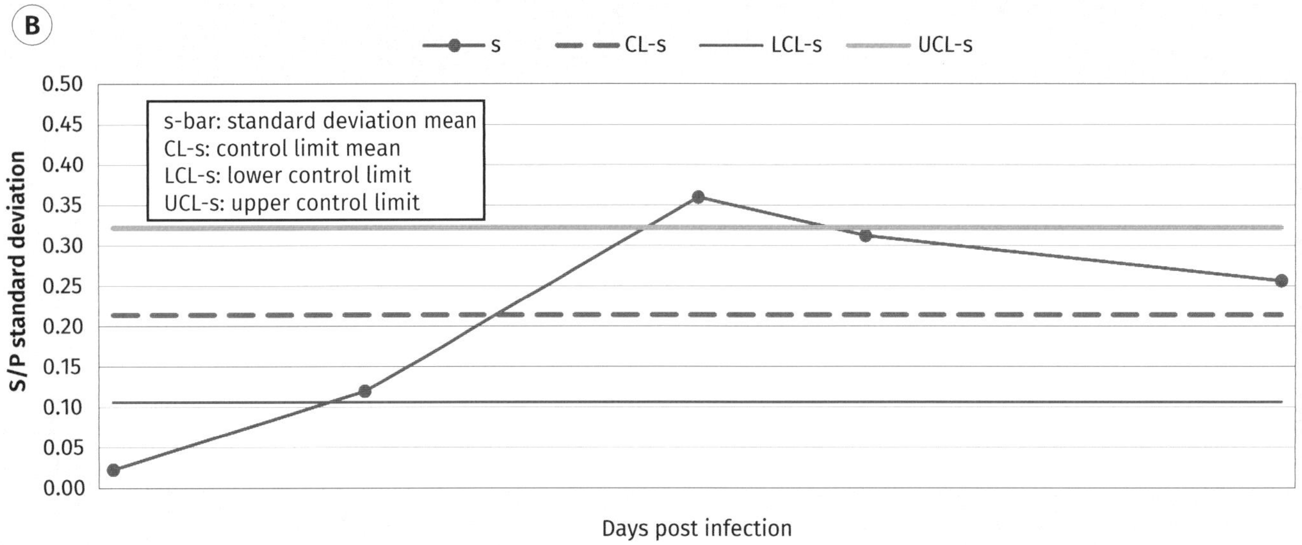

https://doi.org/10.54846/am2022/133

Improved piglet performance and reduced mortality and antimicrobial use following oral vaccination with a live non-pathogenic *E coli* F4/F18 vaccine against PWD

F. Vangroenweghe;
Elanco Animal Health - Belgium

Introduction

Post-weaning *Escherichia coli (E coli)* diarrhea (PWD), also called post-weaning enteric colibacillosis, remains a major cause of economic losses for the pig industry.[1,2] PWD typically causes mild to severe watery diarrhea between 5 and 10 days after weaning and is caused primarily by enterotoxigenic *Escherichia coli* (ETEC). The most common adhesins found on ETEC from PWD in pigs are associated with fimbriae F4 (previously called K88) and F18, while the predominant enterotoxins are heat-labile toxin (LT), heat-stable toxin a (STa), and heat-stable toxin b (STb). In addition to F4 and F18, other fimbrial adhesins, such as F5 (K99), F6 (987P), and F7 (F41), have been associated with PWD, but less frequently.[3-6] Therapy to combat PWD typically consists of antibiotic treatment. However, emergence of antimicrobial resistance in *E coli* strains and new EU regulations urge the need for alternative control measures, such as adapted feeding strategies or immunization. Recently, an oral live bivalent *E coli* F4/F18 vaccine (Coliprotec® F4/F18; Elanco) has become available on the European market, which reduces the impact of PWD provoked by F4-ETEC and F18-ETEC.

Materials and methods

Oral vaccination of suckling piglets using a live bivalent non-pathogenic *E coli* F4/F18 vaccine was performed in 10 farrow-to-finish sow farms to prevent against PWD due to F4-enterotoxigenic *E coli* (ETEC) or F18-ETEC. The vaccination strategy was compared to the standard therapeutic approach in each farm, meanwhile collecting data on Average Daily Weight Gain (ADWG), Feed Conversion Rate (FCR), mortality rate and treatment incidence with antimicrobial drugs (TI100) during the post-weaning period.

Results

Vaccine-treated groups demonstrated a significant improvement in FCR (-4.2%), mortality rate (-57.1%) and TI_{100} (-84.9%) as compared to the Control group. The ADWG only marginally and non-significantly improved (+2.3%) in the Vaccine-treated group (Table 1).

Conclusions

In conclusion, the present study demonstrated the efficacy of an oral live non-pathogenic *E coli* F4/F18 vaccine (Coliprotec® F4/F8; Elanco Animal Health) for active immunization of piglets against PWD due to F4-ETEC and F18-ETEC under field conditions. For several economically important performance parameters, such as FCR, mortality rate and TI_{100}, *E coli* vaccination performed significantly better as compared to the standard therapeutic approach. Therefore, vaccination against PWD due to F4-ETEC or F18-ETEC using an oral live non-pathogenic *E coli* F4/F18 vaccinated may be considered a good alternative to consolidate post-weaning piglet performance results while meeting the new European requirements concerning prudent use of antimicrobials in intensive pig production.

References

1. Van Beers-Schreurs et al., 1992. *Veterinary Quarterly* 14, 29-34.
2. Fairbrother et al., 2005. *Animal Health Research and Reviews* 6, 17-39.
3. Kwon et al., 2002. *Veterinary Record* 150, 35-37.
4. Frydendahl, 2002. *Veterinary Microbiology* 85, 169-182.
5. Chen et al., 2004. *Veterinary Microbiology* 103, 13-20.
6. Vu Khac et al., 2006. BMC *Veterinary Research* 2, 10-15.

Table 1: Performance results obtained in control and vaccine-treated animals from different field trials using a live, non-pathogenic, oral *E coli* F4/F18 vaccine in piglets to prevent clinical impact of post-weaning diarrhea due to enterotoxigenic *E coli* F4 or F18. Significant differences are indicated by their *P*-value. ADWG = average daily weight gain; FCR = feed conversion rate; TI_{100} = treatment incidence per 100 days in production.

Parameter	Control	Vaccine	*P*-value
ADWG (g/day)	348.14	359.29	0.1641
FCR (kg/kg)	1.72	1.66	0.0164
Mortality rate (%)	5.43	2.33	0.0083
TI_{100} (days)	17.74	2.64	0.0009

https://doi.org/10.54846/am2022/134

Evaluation of air filters in swine farms as a surveillance method to assess the spread of airborne viruses

Lan Wang; Jayaveeramuthu Nirmala; My Yang; Montserrat Torremorell
College of Veterinary Medicine, University of Minnesota, Saint Paul, Minnesota

Introduction

Porcine reproductive and respiratory syndrome virus (PRRSV) and influenza A virus (IAV) are two major respiratory pathogens that result in significant economic losses to the pig industry in the US. Airborne transmission is considered an important route of PRRSV and IAV spread, which makes the two pathogens costly and difficult to control. Air filtration of incoming air has been widely used in swine farms in the Midwestern US and has effectively reduced the incidence of airborne PRRSV infections in breeding herds.[1] Therefore, detection of viruses in used air filters from farms offers a unique largely unexplored opportunity to monitor the regional spread of PRRSV and IAV. In this study, we aimed to evaluate the use of air filters as a surveillance method to monitor the regional spread of PRRSV and IAV, and to enhance our understanding of the epidemiology and control of airborne diseases.

Methods

We selected 7 breeding herds from high pig density areas which were either PRRSV negative or stable at the beginning of the study. Twenty brand new air filters were installed in each farm, and five filters were removed each time at approximately 6, 8, 11 and 14- months post installation. Five samples were cut from each filter, with each sample consisting of six 2-by-2-inch squares. These samples were ground in liquid nitrogen and mixed with minimum essential media. Viral RNA was extracted from the supernatant of each sample and quantified by real time RT-PCR for PRRSV and IAV.[2] A filter was considered positive if at least one sample tested positive. Samples positive for PRRSV or IAV were further analyzed with whole genome sequencing.

Results

A total of 136 air filters were analyzed. These filters were installed in July 2019 at the earliest and removed successively until October 2020. Filters were cut into 680 samples for testing. Out of the 136 filters, ten (1.5%) samples corresponding to seven (5%) filters from three farms tested positive for PRRSV. During the study, PRRS outbreaks were reported in four farms, however, only one PRRSV positive filter originated from farms that had PRRS outbreaks. In contrast, sixty-five (47.8%) filters from all seven farms tested positive for IAV, with a total of 131 samples positive (19.3%). Six IAV positive samples were sequenced and one sample was successfully subtyped as an H3N2 human-like influenza virus. In addition, multiple lineages were identified for the influenza internal genes from different samples.

Conclusions

Testing of used air filters in swine farms did not result in an enhanced surveillance method for airborne PRRSV. However, used filters for influenza surveillance should be further evaluated. Overall, detection of PRRSV and IAV in the air filters showed some potential evidence of regional airborne transmission for these viruses, but additional investigations are needed to better understand the factors that contribute to airborne transmission of these viruses.

References

1. Dee, Scott, et al. Viruses 4.5 (2012): 654-662.
2. Nirmala, Jayaveeramuthu, et al. Journal of Aerosol Science 151 (2021): 105624.

https://doi.org/10.54846/am2022/135

Development and validation of a swine enteric coronavirus multiplex reverse transcription real-time PCR and investigation of virus detection frequency in US swine

Jinhui Zhu[1], BS; Gaurav Rawal[1], DVM, MS; Ethan Aljets[1], BS; Wannarat Yim-im[1], MS; Yong-Le Yang[2], PhD; Yao-Wei Huang[2], PhD; Karen Harmon[1], PhD; Phillip Gauger[1], DVM, PhD; Jianqiang Zhang[1], MD, PhD
[1]Veterinary Diagnostic and Production Animal Medicine, Iowa State University, Ames, Iowa;
[2]Department of Veterinary Medicine, Zhejiang University, Hangzhou, Zhejiang, China

Introduction

Coronaviruses (CoVs) are enveloped, single-stranded, positive-sense RNA viruses in the order *Nidovirales* and the family *Coronaviridae*. In pigs, six CoVs have been identified. Among the six CoVs, porcine epidemic diarrhea virus (PEDV), porcine deltacoronavirus (PDCoV), transmissible gastroenteritis virus (TGEV), and swine acute diarrhea syndrome coronavirus (SADS-CoV), are important enteric pathogens causing similar clinical signs in swine. SADS-CoV has only been reported in China[1] and its presence in the US remained unknown. This study aimed to develop and validate a PEDV/PDCoV/TGEV/SADS-CoV/XIPC 5-plex reverse transcription real-time PCR (RT-rtPCR) for simultaneous detection and differentiation of the four viruses (XIPC serves as an internal positive control for the PCR) and investigate the detection frequency of these viruses in US swine.

Materials and methods

Four singleplex SADS-CoV PCR assays were evaluated and one SADS-CoV PCR with the best performance was chosen for the 5-plex assay development. In addition, one PDCoV PCR, one PEDV PCR and one TGEV PCR were included for the 5-plex PCR development. The 5-plex PCR includes primers and probe for an exogenous internal positive control (XIPC) in addition to including primers and probes for PEDV, PDCoV, TGEV, and SADS-CoV (in total 10 primers and 5 probes). The analytical specificity of the 5-plex PCR was evaluated by testing 32 different swine viral and bacterial pathogens. The limits of detection of the 5-plex PCR were evaluated based on testing serial dilutions of *in vitro* transcribed RNAs. The diagnostic performance of the 5-plex PCR was compared with a commercial PEDV/PDCoV/TGEV PCR by testing 219 clinical samples. Subsequently, 1807 clinical samples were tested by the 5-plex PCR to investigate the presence of SADS-CoV in US swine and the frequency of detecting swine enteric CoVs. First, 288 clinical samples archived during 2019-2020 from diarrheic pigs negative for PEDV, PDCoV, TGEV, rotavirus A, B, C were tested. Next, 1519 samples submitted to the ISU VDL during Mar-Oct 2019, Nov-Dec 2020, and Feb-Apr 2021 were randomly selected for testing.

Results and discussion

The novel PEDV/PDCoV/TGEV/SADS-CoV/XIPC 5-plex RT-rtPCR developed and validated in this study is sensitive and specific and provides a convenient tool for simultaneous detection and differentiation of these important swine enteric coronaviruses. All the 1807 samples tested negative for SADS-CoV by the 5-plex PCR. And among samples positive for swine enteric CoVs, there were low frequency of detecting TGEV, intermediate frequency of detecting PDCoV, and high frequency of detecting PEDV. Although there is no evidence of SADS-CoV presence in the US now, the availability of the PEDV/PDCoV/TGEV/SADS-CoV/XIPC 5-plex PCR will enable us to conduct ongoing surveillance and thereby we are better prepared to respond to introduction.

Reference

1. Yang et al.: 2020. Swine enteric alphacoronavirus (swine acute diarrhea syndrome coronavirus): An update three years after its discovery. Virus Res 285, 198024.

https://doi.org/10.54846/am2022/136

Comparison of benzoic acid and potassium diformate on the performance of post-weaning piglets: A review

C. Lückstädt
ADDCON GmbH, Bitterfeld-Wolfen, Germany

Introduction

This review investigates the average impact of potassium diformate (FORMI, KDF) and benzoic acid (BA) on post weaning piglet performance.[1] Both additives are registered as zootechnical additives for pigs in the EU. FORMI however was the first organic acid salt to be approved as a non-antibiotic growth promoter in pig feed in the European Union, where it has been shown to improve growth performance and feed efficiency in pig production in numerous efficacy trials. This study analyzes all data from commercial and academic trials, where the impact of both additives on the performance parameters feed intake, weight gain and feed efficiency were compared.

Materials and methods

The final dataset contains the results of 4 documented studies, comprising 5 trials with KDF and BA-inclusion in piglets, at inclusion rates ranging from 0.5% to 1.2%. These studies were carried out between 2002 and 2014 in Denmark, Germany, Brazil and France under both commercial and institutional conditions and included more than 8762 pigs. The data were statistically analyzed and a $P < 0.05$ value was considered significant.

Results and discussion

The average level of dietary KDF from the dataset in all treated pigs was 1.0%, whereas the BA-inclusion averaged at 0.8%. Feed intake in the diformate-fed piglets was significantly increased by 5.1% ($P = 0.03$) over the BA-inclusion (Table 1). Furthermore, weight gain in the KDF-piglets was also significantly improved (7.7%; $P = 0.003$), which led to a numerically improved feed efficiency in the FORMI-piglets over BA-fed animals by 2%.

These results show that the inclusion of dietary diformate can enhance performance in piglets, even if compared against a positive control (benzoic acid), which is in full agreement with data obtained in Asia,[2] demonstrating furthermore that the additive is a valuable tool in reducing reliance on antibiotic growth promoters.

References

1. Lückstädt C and Wang H 2021. Leman Swine Conference China. Book of abstract, 407.

2. Wang Y, Long H and Kim Y 2007. Journal Animal Science 85, Suppl. 2, 103.

Table 1: Average impact (Meta-analysis of 5 trials) of potassium diformate (FORMI) and benzoic acid (BA) on piglet performance

	Feed intake [g/d]		Weight gain [g/d]		FCR	
	BA	FORMI	BA	FORMI	BA	FORMI
Mean value	567	596	365	393	1.53	1.50
SD	146	140	47	40	0.24	0.23
P-level	0.030		0.003		0.144	

https://doi.org/10.54846/am2022/137

Oral vaccination against *Lawsonia intracellularis* improves feed efficiency in experimentally challenged pigs

Emma Helm[1], MS, PhD; Fernando Leite[2], DVM, PhD; Eric Burrough[3], DVM, PhD, ACVP; Nicholas Gabler[3], PhD
[1]Virginia Tech, Blacksburg, Virginia; [2]Boehringer Ingelheim Animal Health USA Inc, Duluth, Georgia;
[3] Iowa State University, Ames, Iowa

Introduction

Porcine proliferative enteropathy or ileitis, caused by *Lawsonia intracellularis,* is an endemic disease that reduces both growth rates and feed efficiency.[1,2] Diseases that affect both these production measures have a greater economic burden, particularly when feed prices are high.[3] Thus, this study aimed to evaluate feed efficiency of pigs during *L intracellularis* challenge and investigate the impact of vaccination with an oral live vaccine in a controlled manner. Additionally, this study aimed to determine how *L intracellularis* diagnostic metrics (fecal shedding, antibody levels) correlate to production metrics.

Materials and methods

A total of 36 barrows were selected, individually confirmed negative for *L intracellularis,* and assigned to treatment groups as follows (n = 12/trt): 1) nonvaccinated, *L intracellularis* negative (NC); 2) nonvaccinated, *L intracellularis* challenged (PC); and 3) *L intracellularis* challenged, vaccinated with Enterisol® Ileitis (Boehringer Ingelheim Animal Health, Duluth, GA) via 1 mL oral drench at 1-week postweaning (VAC). On days post inoculation (dpi) 0 PC and VAC pigs were inoculated with *L intracellularis* via gastric gavage (2.7×10^8 organisms/mL). Individual feed disappearance and BW were recorded for each pig on dpi 0, 7, 14, and 19. Additionally, fecal samples and serum were collected at these timepoints to evaluate fecal shedding and serum antibody levels. At dpi 21 pigs were euthanized and scored for gross and microscopic lesions characteristic of *L intracellularis* infection. To further investigate the relationship between diagnostic parameters and performance metrics, Spearman correlation coefficients were calculated in the PC and VAC groups.

Results

For overall performance (dpi 0–19), ADG was reduced 41% in PC pigs compared with NC pigs, and was 26% greater in VAC pigs compared with PC pigs ($P < 0.001$). Similarly, ADFI was reduced 24% in PC ($P < 0.001$) pigs compared with NC pigs, and was 14% greater in VAC pigs compared with PC pigs ($P = 0.032$). Overall G:F was reduced 25% in PC pigs compared with NC ($P < 0.001$) pigs, and was 18% greater in VAC pigs compared with PC pigs ($P = 0.015$). Fecal shedding of *L intracellularis* increased over the experimental period for PC pigs ($6.4 \times \log_{10}$ genomic copies/mL at dpi 21). Vaccinated pigs did shed organism after inoculation, fecal shedding of VAC pigs peaked at dpi 14 ($6.39 \times \log_{10}$ genomic copies/mL) and at 21 dpi was $3.4 \times \log_{10}$ genomic copies/mL lower than PC pigs ($P < 0.001$). For antibody responses, PC pigs had antibody levels peaking at 61% inhibition at dpi 21. The serum antibody response of VAC pigs was milder, averaging 15–16% inhibition at dpi 14. Average lesion length was greater in PC pigs (118 cm) compared with both NC (0.0 cm, $P = 0.007$) and VAC (3.0 cm, $P = 0.045$) pigs. Lesion length was negatively correlated with ADG (-0.616, $P = 0.034$) and G:F (-0.783, $P = 0.003$) for PC pigs. Further, *L intracellularis* fecal shedding at 14 dpi was negatively correlated with overall G:F in PC pigs (-0.587, $P = 0.045$). Lesion length and fecal shedding did not correlate with overall performance in VAC pigs. Serum antibody concentrations did not correlate with ADG, ADFI, or G:F in either PC or VAC pigs.

Conclusions

Few studies have evaluated the impact of a vaccine on feed efficiency. This study shows that oral vaccination not only improves average daily weight gain, but significantly improves feed efficiency in pigs challenged with *L intracellularis,* potentially a result of lesser disease severity and earlier disease resolution. Intestinal lesion length was found to be negatively correlated to ADG and G:F, while only at the 14 dpi time point did *L intracellularis* fecal shedding levels correlate with reduced feed efficiency in challenged, non-vaccinated pigs. No diagnostic parameter appeared to correlate with growth performance or feed efficiency in vaccinated pigs.

References

1. Paradis, M et al. J Swine Health Prod. 2012; 20(3):137-141.
2. Helm, E. et al. 2018. J. Anim. Sci. 96:462–472.
3. Holtkamp, D. et al. ISU James D. McKean Swine Disease Conference. 2007. P. 91-97.

Poster 58

https://doi.org/10.54846/am2022/138

Evaluation of a *Streptococcus suis* strain that confers cross protection to serotype 1 and 1/2

E. A. Collin, PhD; E. Bumgardner, MS; D. McKeown, BS; J. Alsdurf, MS; K. Peters-Smith, BS; D. Stine, DVM, PhD
Cambridge Technologies, Diagnostics and Research Department, Worthington, Minnesota

Introduction

Streptococcus suis is a common swine pathogen that causes lameness and CNS symptoms in clinically affected pigs. Prior to the evolution and use of genomics as a method for serotyping, there were 35 recognized serotypes of *S suis*.[1] However, the use of traditional serotyping methods often yielded inconclusive results and many strains were not typeable. Over the past several years PCR, and next-generation sequencing have emerged as important methods for establishing the serotypes of isolates.[2,3] To date, no commercial vaccine is available to combat this important swine pathogen, largely owing to the fact that cross-protection amongst serotypes is difficult to achieve. Thus, inactivated autogenous vaccines have become the most widely available means for preventing infection. In the work presented here, a unique serotype 2 *S suis* strain was identified that demonstrated cross-protection when it is used as a vaccine candidate. To date, our studies have demonstrated cross-protection against serotypes 1 and 1/2.

Materials and methods

In total, forty-eight high health pigs were obtained for use in two separate studies. Pigs were approximately 3 weeks of age at study enrollment and were housed in a BSL2 research facility for the duration of each study. For each study, 8 pigs were included in each treatment group described. In the first study, vaccines included: the cross-protective strain, a quadrivalent competitor's autogenous product, and a Cambridge quadrivalent autogenous product (both containing a serotype 1 isolate). All received a serotype 1 challenge. In both studies, vaccines were delivered IM as described in Figure 1 at days 0 and 14. In the second study, the cross-protective vaccine strain was evaluated against a serotype 1/2 challenge strain. For challenge, 1mL of challenge culture was delivered via the IV/IP route (2mL total) at day 28. Serotype 1/2 and serotype 1 challenge titers were 2.05*10^9 cfu/mL and 2.6*10^9 cfu/mL respectively. Following challenge, pigs were monitored for clinical signs and scored according to the following criteria: 0 = Healthy, 1 = Signs of lameness, 2 = Significant signs of lameness, must be encouraged to rise, no CNS symptoms 3 = CNS symptoms, cannot rise, must be euthanized 4 = Dead prior to arrival of study personnel. Scores were recorded for four days. At the conclusion of the study clinical scores obtained over the course of the challenge were summed for all groups (Figure 1).

Results and discussion

The results of these studies demonstrate the cross-protective vaccine candidate confers protection against a virulent challenge with either serotype 1 or serotype 1/2. Furthermore, the protection observed in the serotype 1 challenge model was similar to that obtained from a competitor's product. Additional work is planned to assess the level of protection that this vaccine candidate can provide against the remaining serotypes that are commonly isolated in the United States.

References

1. Higgins R et al.: 1995, J. Vet. Diagn. Invest. -7:405.
2. Shun Lae Thu I et al.: 2021, Pathogens. -10:996.
3. Athey TBT et al.: 2016, BMC Microbiol. -22:162.

Figure 1: Cumulative results of clinical scoring. Treatment group names followed by serotype of challenge strain.

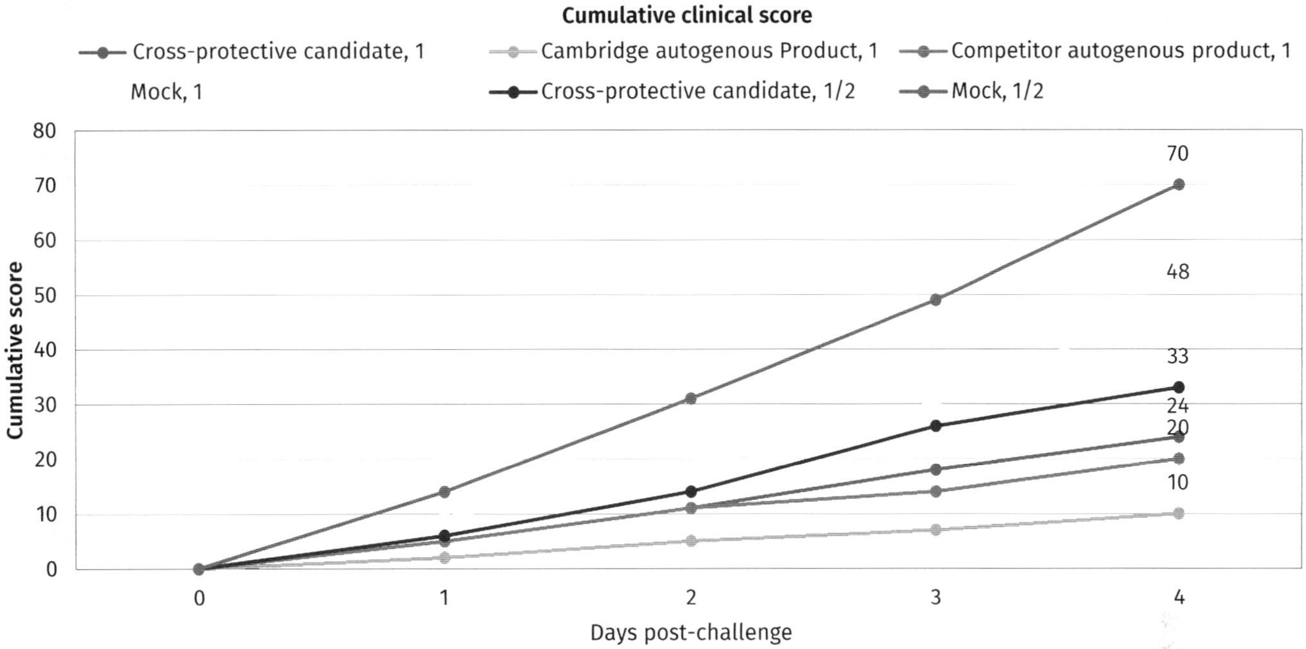

https://doi.org/10.54846/am2022/139

Bacteriological cultures show flies and gnats harbor live *Salmonella* in nursery exhibiting enteric disease

Allison Knox[1], DVM; Grant Allison[1], DVM; Gene Spellman[2], BS
[1]Walcott Veterinary Clinic, Walcott Iowa; [2]Central Life Sciences, Schaumburg, Illinois

Introduction

Non-blood sucking insects, such as the house fly, *Musca domestica*, and small gnats of the suborder Nematocera, pose a serious concern to on-farm biosecurity measures. Their small size and ability to fly allows them to easily enter and exit barns and vehicles while carrying viral, bacterial, and parasitic pathogens. Recent studies using PCR technologies have been able to identify *Musca domestica* as a vector of Senecavirus A and PRRS,[1] and bioassay studies have shown the capability of house flies carrying PEDV to transmit live virus resulting in clinical disease in pre-weaning pigs.[2] These non-blood sucking insects are also noted vectors for *Salmonella* species.[3] As *Salmonella* is pathogenic among livestock species and a zoonotic agent, the goal of this study was to evaluate the potential for flies and gnats to harbor *Salmonella enterica* when exposed to nursery pigs exhibiting clinical enteritis. Additionally, the ability of flies and gnats to transmit *Salmonella enterica* from contact with feed in nursery sites was also investigated.

Materials and methods

Nursery and wean-to-finish barns in southeast Iowa were evaluated for evidence of clinical enteritis in pigs and the presence of flies and/or gnats on site. Two 4,800 head nursery sites meeting these criteria were selected for further diagnostics. Fresh tissue and rectal swabs were collected from pigs on these sites exhibiting evidence of diarrhea. When flies or gnats were present onsite, Captivator® liquid traps (Starbar®, Central Life Sciences) were hung throughout the room to collect flies. Traps were used with the goal of reducing potential cross-contamination from the environment. Flies were allowed to enter the traps on their own volition, at which point they would become suspended in the sterile water and attractant mixture. At site 2, a negative control trap was hung in addition to other fly traps in each room. The control trap was open to the air but had a screen over the opening to prevent flies/gnats from entering the trap. Feed samples were collected from the hopper as well as the feed pan. A 4 × 4" gauze square moistened with sterile water was used to pat the superficial feed surface where fly contact was most likely to occur. Tissue, swabs, flies, gnats, trap liquid, and feed samples were submitted to the Iowa State Veterinary Diagnostic Lab (ISU VDL) for bacterial culture. Samples that were culture positive for *Salmonella* species were forwarded to the National Veterinary Services Laboratories (NSVL) for serotyping.

Results and discussion

At Site 1, *Salmonella enterica* subspecies *enterica* serovar Derby (*Salmonella* Derby) was cultured from tissues, rectal swabs, and flies. Samples collected from the fly trap liquid and feed were culture negative (Table 1). At Site 2, *Salmonella enterica* subspecies *enterica* serovar Infantis (*Salmonella* Infantis) was cultured from rectal swabs and gnats. Negative control fly traps were reported as culture negative (Table 1). Feed was not tested on this site. Conclusions made from this diagnostic summary are that flies and gnats exposed to *Salmonella enterica* have the capability of harboring and possibly transmitting these pathogens. Additionally, while *S* Derby and *S* Infantis were not isolated from the feed samples during the course of this trial, the potential for transmission of Salmonella cannot be ruled out. As there are few drugs labeled for the treatment or control of Salmonellosis and increasing frequency of resistance to these labeled drugs, new methods for prevention should be sought out. National *Salmonella* control programs across North American and Europe have been initiated by the swine industry, and including non-biting insect control measures into these programs and on-farm biosecurity protocols could positively contribute to the reduction of Salmonellosis.

References

1. Allison G, et al. PEDV, PRRSV, and Seneca Valley virus PCR positive flies (Musca domestica). What now? *Proceedings 49th AASV 2018*. 394-396.

2. Allison G, et al. PEDV positive bioassay reveals houseflies (Musca domestica) can transmit infectious PEDV to pigs. *Proceedings 50th AASV 2019*. 206-208.

3. Moon RD, et al. Flying insects as biosecurity risks and what the swine vet needs to know. *Proceedings AASV 2005*. 417-423.

Table 1: Diagnostic summary

Site	Date	Sample	Culture results	NSVL serotyping
1	10-12-2020	Feed hopper	Negative	
		Feed pan	Negative	
		Flies	***Salmonella* species Group B**	***S* Derby**
		Fly trap liquid	Negative	
	10-21-2020	**Tissue A**	***Salmonella* species Group B**	***S* Derby**
		Tissue B	***Salmonella* species Group B**	***S* Derby**
		Rectal swabs	***Salmonella* species Group B**	***S* Derby**
	10-26-2020	**Rectal swabs**	***Salmonella* species Group B**	***S* Derby**
		Feed pan	Negative	
		Flies	Negative	
		Fly trap liquid	Negative	
2	1-19-2021	**Rectal swabs**	***Salmonella* species Group C1**	***S* Infantis**
	2-1-2021	**Rectal swabs**	***Salmonella* species Group C1**	***S* Infantis**
	2-25-2021	**Rectal swabs**	***Salmonella* species Group C1**	***S* Infantis**
		Flies	Negative	
		Gnats	***Salmonella* species Group C1**	***S* Infantis**
		Fly trap liquid	Negative	
		Control fly trap	Negative	

https://doi.org/10.54846/am2022/141

An update of *Mycoplasma hyopneumoniae* prevalence at weaning age in Spain

Lorena Pérez Esteruelas[1], DVM, MSc; Miguel Claver Mateos[1], DVM; Guillermo Ramis Vidal[2], DVM, MSc, PhD
[1]Elanco Animal Health, Alcobendas, Spain;
[2]Dpto Producción Animal. Facultad de Veterinaria. Universidad de Murcia, Spain

Introduction

Mycoplasma hyopneumoniae continues to be an important primary pathogen of enzootic pneumonia and, also in porcine respiratory disease complex (PRDC). It is assumed that more than 70% of swine herds around the world are infected with *M hyopneumoniae.*[6] Piglets are born free of this microorganism and in-utero transmission has not been documented. Then, it is considered that dam-to-piglet transmission is a crucial contributor to piglet colonization and later infection development.[10] *M hyopneumoniae* prevalence at weaning is being described as a predictor of disease severity in downstream herds.[2] The aim of the present study was to evaluate the prevalence of *M hyopneumoniae* around weaning age to consider it as a risk factor for a population subsequently developing the disease.

Material and methods

The study was conducted from 2019 to 2020 in a total of 25 *M hyopneumoniae* positive Spanish pig herds. Tracheobronchial swabs were collected from 3 to 4 weeks-old piglets and tested for the presence of *M hyopneumoniae,* throughout an individual quantitative PCR using the primers[4] for the protein p97 with a thermocycler ABI 7300 (Life Technologies, USA). A total of 750 samples were taken from 30 piglets per farm. Two piglets were randomly sampled per dam (50% of multiparous and 50% of primiparous). It is important to know the status of gilts when entering a positive farm, since it is being demonstrated that around 20% of them are still negative to *M hyopneumoniae* up to first farrowing.[8]

Results

From the total of analyzed farms, 48% were positive to *M hyopneumoniae* around weaning age (12 out of 25 farms) and 10.4% yielded positive PCR results (78 out of 750 tested piglets) (Table 1). Within the positive farms, 50% showed more than 20% of infected piglets (6 out of 12 *M hyopneumoniae* positive farms) with positive results.

Table 1: Detection rate of M *hyopneumoniae* using tracheobronchial swabs in piglets around weaning age

Parameter	Herd	Animal
Number (positive/total)	12/25	78/750
Percentage	48%	10.4%

Discussion and conclusion

The present study searched for the prevalence of *M hyopneumoniae* at weaning age. Almost 50% of the farms were considered as positive. Using the same diagnostic sampling, there was another study carried out in 2011 in Spain, with 40.4% of positive farms.[1] Other studies showed a lower prevalence of *M hyopneumoniae* at weaning age.[9] Prevalence at weaning is the main risk factor for a population subsequently developing the disease.[5] Piglets who are shedding the pathogen, could be infectious to other penmates.[7] Then, it is important to know the status around weaning to stablish different strategies to reduce dam-to-piglet transmission and piglet colonization at weaning.

References

1. Bringas J et al. Prevalence of *Mycoplasma hyopneumoniae* infection at weaning age in Spanish pig herds. Bacteriology and bacterial diseases, B0-338. 2011.

2. Fano E et al. Assessing the *M. hyopneumoniae* infection pattern in a sow herd following an increase in the replacement rate. Proceeding's 19th IPVS Congress, Copenhagen, Denmark, Volume 1, 135. 2006.

3. Maes D et al. M. Mycoplasmas in swine. 2020.

4. Marois CD et al. Development of a quantitative Real-Time TaqMan PCRassay for determination of the minimal dose of *Mycoplasma hyopneumoniae* strain 116 required to inducepneumonia in SPF pigs. Journal of Applied Microbiology 108 1523–1533. 2010

5. Pijoan C. A controversial view of *Mycoplasma hyopneumoniae* epidemiology. Allen D. Leman Swine Conference, 114-117, 2005.

6. Pieters MG et al. Diseases of swine (11th ed.; pp. 863-883). Hoboken, NJ, USA: John Wiley & Sons, Inc.

7. Pieters M et al. An assessment of the duration of *Mycoplasma hyopneumoniae* infection in an experimentally infected population of pigs. *Veterinary Microbiology, 134*(3-4), 261-266. 2009.

8. Takeuti KL et al. Detection of *Mycoplasma hyopneumoniae* in naturally infected gilts over time. *Veterinary Microbiology, 203,* 215-220. 2017.

9. Vangroenweghe FA et al. *Mycoplasma hyopneumoniae* infections in peri-weaned and post-weaned pigs in Belgium and The Netherlands: Prevalence and associations with climatic conditions. *Veterinary Journal, 205*(1), 93-97. 2015.

10. Villarreal I et al. Early *Mycoplasma hyopneumoniae* infections in European suckling pigs in herds with respiratory problems: Detection rate and risk factors. *Veterinary Medicine Czech, 55,* 318-324. 2010.

https://doi.org/10.54846/am2022/143

Monitoring PRRSV infection in vaccinated pigs

K. Doolittle[1], BSc; D. Linhares[2], DVM, PhD; C. Moura[2], DVM, PhD; S. Zimmerman[1], DVM; A. Hidalgo[1], DVM, PhD
[1]IDEXX Laboratories, Inc., Westbrook, Maine; [2]Iowa State University, Ames, Iowa

Introduction

Porcine reproductive and respiratory syndrome (PRRS) continues to be one of the most economically significant diseases affecting the swine industry worldwide. In the US, its impact has been estimated at $4.67 for every pig marketed.[1] Novel strategies are needed for an optimized control of PRRSV in the herds, including disease and vaccination monitoring programs. This study reports on the use of an oral fluid ELISA to identify groups of nursery pigs at high risk of PRRSV infection and monitor vaccination.

Materials and methods

A total of 26 wean-to-finish pig farms sourcing three-week-old pigs from two different breeding herds were included in the study. The breeding herds were considered PRRS stable[2] with uncertain shedding status at the time of the study. Pigs were injected with a PRRSV vaccine (Ingelvac PRRS MLV, Boehringer Ingelheim) within one week after placement. Six pen-based oral fluid samples were collected per barn at the time of vaccination and 3 weeks after.

Samples were tested using an ELISA for detection of antibodies against PRRSV in oral fluids [PRRS Oral Fluid Ab Test, IDEXX] following the manufacturer's instructions and analyzed by PRRSV qPCR at Iowa State University Veterinary Diagnostic Laboratory. RFLP patterns were obtained from qPCR positive samples and used to differentiate between the vaccine strain and wild-type PRRSV. In addition, production parameters (feed conversion ratio [FCR], average daily weight gain [AWDG] and mortality) were recorded for each farm.

Results and discussion

At the time of the first sampling, a higher ELISA mean S/P value was shown in non-infected nurseries compared with wild-type PRRSV infected groups (Mean S/P = 4.08 and 2.93, respectively. Figure 1). In addition, non-infected farms presented a more homogenous distribution in S/P values, with a lower coefficient of variation compared to infected farms (CV% = 43 and 63, respectively). Although the oral fluid ELISA detects IgG, this finding suggests higher titers of maternally derived neutralizing antibodies in the non-infected groups having a protective effect.

Following vaccination, the mean S/P value increased to 5.16 in the non-infected groups, with a marked reduction of variation (CV% = 26). Similarly, the mean S/P value increased in the infected farms following vaccination (Mean S/P = 4.11). However, the high variability persisted (CV% = 46). This low homogeneity in S/P values 3 weeks post-vaccination is suggestive of an impaired immune response to vaccination, probably caused by the active PRRSV infection at the time of vaccination. Similarly, wean-to-finish average mortality increased to 5.02% in the farms with wild-type PRRSV infection at the time of vaccination, compared with 3.85% in the non-infected groups. No differences were recorded in FCR or ADWG between infected and non-infected groups.

Routine monitoring using an oral fluid antibody test can help to differentiate groups of pigs at high risk of PRRSV infection post-weaning (low mean S/P value and high variability) and evaluate successful PRRSV vaccination (increased mean S/P value and reduced variability three weeks post vaccination) for an optimized management of PRRSV infection. This study also suggests that monitoring PRRSV oral fluid antibodies can be a valuable tool to identify groups with wild type infection at placement, allowing a more appropriate disease control strategy.

References

1. Holtkamp et al. Assessment of the economic impact of porcine reproductive and respiratory syndrome virus on United States pork producers. J Swine Health Prod. 2013;21(2):72–84.

2. Holtkamp et al. Terminology for classifying swine herds by porcine reproductive and respiratory syndrome virus status. J Swine Health Prod. 2011; 19(1):44–56.

Figure 1: Box plots identifying differences in oral fluid PRRS ELISA S/P values in wild-type PRRSV infected (dark and light blue) and non-infected (dark and light green) groups of pigs at the time of PRRSV vaccination (4-week) and three weeks later (7-weeks).

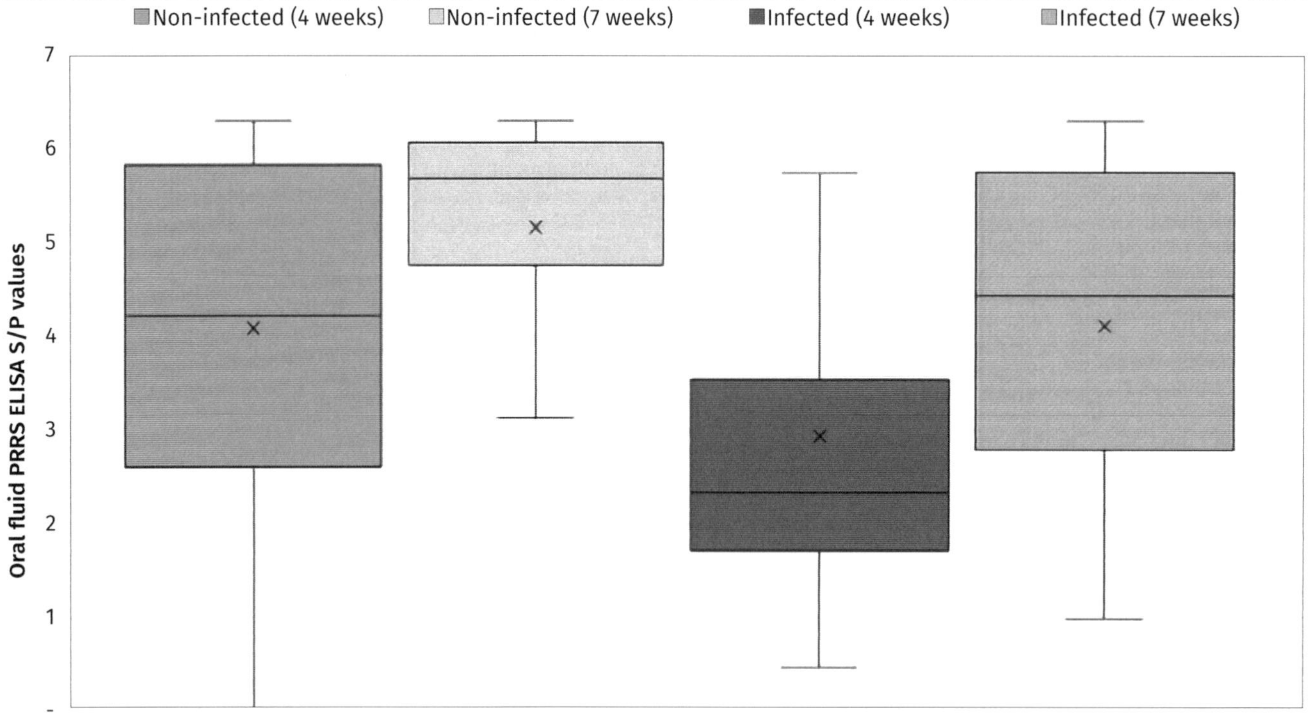

https://doi.org/10.54846/am2022/144

Evaluate the impact of beta-1,3 glucans (Aleta™) on immune response and production/health indicators in development gilts post-vaccination during the isolation period

Clayton Johnson[1]; Gustavo Silva[2]; Beau Peterson[1]; Tom Marsteller[3]; Marc White[4]; Vanessa Iseri[4]
[1]Carthage Veterinary Services, Carthage, Illinois; [2]Iowa State University, Ames, Iowa;
[3]Stellar Swine Health, Indianapolis, Indiana; [4]Kemin Industries, Inc., Des Moines, Iowa

Introduction

Aleta™ is a source of greater than 50% beta-1,3 glucans, which can be utilized to help animals minimize the impact of disease challenges by priming the immune system. The beta-1,3 glucans in Aleta are provided as bioactive granules and shown to be readily taken up by macrophages *in vitro* and enhance innate immunity *in vivo* (Kemin internal data). Aleta activated T cells, reduced gut inflammation, improved gut barrier function and boosted immune response in a swine *E coli* challenge model.[1] In addition, Aleta has been shown to positively impact growth, Feed:Gain and morbidity in porcine reproductive and respiratory syndrome virus (PRRSV)-positive weaned piglets administered in-feed antibiotics.[2] The objective of this study was to evaluate the impact of Aleta on post-vaccination and production performance in development gilts.

Materials and methods

A total of 1,252 gilts were sourced from a multiplication breed-to-wean farm, that were negative for PRRSV and *Mycoplasma hyopneumoniae*. Every four weeks, one group of (313) 6-month-old breeding gilts were received in the isolation facility and randomly assigned to 1 of 2 treatments (2 groups/treatment): negative control or negative control + Aleta feed (0.5 lb/ton). Dietary treatments started on d0 when gilts (43 gilts/group) were placed in the isolation room and vaccinated with PCV-2a, PCV-2b, PCV-2d, *Mycoplasma hyopneumoniae, Haemophilus parasuis* and *Streptococcus suis*. Dietary treatments ended on d28 post vaccination. Due to gilt selection rates at the breeding farm, not all gilt blood samples could be collected on d28. In addition, morbidity, mortality, age of first breeding, retention rate up to P2 and selection rate before breeding were collected. Statistical analyses were performed using JMP® (SAS Institute, Cary, NC) software, and significance was determined at $P \leq 0.05$.

Results

There were no differences between treatments on the selection rate, morbidity and mortality. Results for age at first breeding and retention rate up to P2 are still pending. There was a trend for Aleta to improve PCV-2d titers at d28 titers compared to control ($P = 0.12$). There was a trend for Aleta to improve *Mycoplasma hyopneumoniae* titers on d28 compared to control ($P = 0.09$). There was an effect of Aleta on *Haemophilus parasuis* titers in which Aleta improved titers at d28 compared to d0 ($P < 0.05$). There was no effect of Aleta on PCV-2a, PCV-2b and *Streptococcus suis* titers on d28 compared to control ($P > 0.05$).

Conclusions

Overall, these results provide evidence that Aleta was able to improve serum antibody titers for different vaccines in breeding gilts. Further research is warranted in understanding the effect of Aleta on vaccination titer improvement.

References

1. Kim K. et al, 2019 Algae-derived β-glucan enhanced gut health and immune responses of weaned pigs experimentally infected with a pathogenic *E coli*.

2. Marsteller, Tom. et al, 2019. Comparison of Aleta™ with and without in-feed antibiotics for growth performance, morbidity, and Full Value Pigs™ in PRRSV-positive nursery pigs. Abstract presented at Allen D. Leman Swine Conference.

https://doi.org/10.54846/am2022/145

Retrospective study: Effect of boar collection interval on semen obtained and doses produced

J. Pradieé[1], DVM, Dr; P. I. De Quadros[1], DVM; K. Calderan[1], DMV; J. C. Rocha, DVM, MS;
L. P. Bevilacqua, DMV; M. M. Oliveira[2], Animal Science, Dr
[1]Bretanha Suínos, Passo Fundo, Rio Grande do Sul, Brazil;
[2]Morgado Experimental Statistic, Passo Fundo, Rio Grande do Sul, Brazil

Introduction

Annual boar stud replacement rates near 60% are common with age and genetic index often the primary drivers. The interval between semen collections typically averages 5 to 7 days[1] and this does not maximize the sperm that could be harvested. In addition, underutilization of the boar inventory adds cost and genetic lag, especially if the highest indexing boars are not collected more frequently. Conversely, overuse of the boar inventory and reduced rest days can decrease sperm quality and increase retention of sperm cytoplasmic droplets and thus, the number of discarded ejaculates. Boar age and season are known key covariates in the effect of collection interval on semen parameters.[2] The available data set included a range of boar ages (8 to 17 months) and spanned a 3½ year period (April 2017 to March 2021) at a stud in Southern Brazil. The objective was to examine the relationship between collection interval (3, 4, 5 and 7 days) and insemination doses produced within different boar age groups.

Materials and methods

Variables examined from this database included: collection date, ejaculate volume (mL), sperm concentration of neat semen ($\times 10^6$), interval between collections (days), age at collection (months), age at first collection (months), total sperm per collection ($\times 10^9$), and total sperm collected per month ($\times 10^9$). Boar age (at collection) was split into three arbitrary classes or groups: Group 1: 8-10 months (n = 5392 collections); Group 2: 11-13 months of age (n = 4835 collections); Group 3: 14-17 months of age (n = 4163 collections). Ejaculates within each group were then filtered by collection intervals of 3, 4, 5 and 7 days. Statistical analyses were performed using the R Core Team.[3] Variables were tested for normality with the Lilliefors (Kolmogorov-Smirnov) test and non-parametric Kruskal-Wallis and Dunn tests were performed.

Results and discussion

The ejaculate volume and the sperm concentration of neat semen (Table 1) followed the same profile in all analyzed age groups. The greater the interval between collections, the greater the volume of the ejaculate and the greater the sperm concentration of the neat semen until some threshold is reached. Physiologically, since sperm production is constant, as rest days and the interval between collections increases, the epididymides accumulate sperm until there is no more space and some are lost if the population is not reduced by ejaculation. Conversely, extremely short collections (< 3 days), such as every day, could deplete the epididymal sperm reserves, as sperm are removed faster than they can be produced. Therefore, one should not expect a linear relationship across a wider collection internal range than examined. Nonetheless, the number of insemination doses produced per boar per month (Table 1) exhibited an inverse relationship with collection interval days over the range studied in all boar age groups so more sperm can be harvested. These trends held even when the percentage of doses discarded was accounted for (doses/year (7% discard) (G1: IC3 = 2100,32; IC4 = 1757,03; IC5 = 1460,98; IC7 = 1166,97) (G2: IC3 = 2556,26; IC4 = 2117,87; IC5 = 1839,71; IC7 = 1494,98) (G3: IC3 = 2986,45; IC4 = 2430,24; IC5 = 2039,06; IC7 = 1683,67).

References

1. Flowers, W.L.: 2021, Animal Reproduction Science, 106840.

2. Knecht, D., et al.: 2017, Livestock Science, p.-21.

3. R: A language and environment for statistical computing. R Foundation for Statistical Computing, Vienna, Austria. URL **https://www.R-project.org/**. Adrian Dragulescu and Cole Arendt (2020). xlsx: Read, Write, Format Excel 2007 and Excel 97/2000/XP/2003 Files. R package version 0.6.5. **https://CRAN.R-project.org/package=xlsx**. Felipe de Mendiburu (2020). Agricolae: Statistical Procedures for Agricultural Research. R package version 1.3-3. **https://CRAN.Rproject.org/package=agricolae**. R package version 0.6.5.

Table 1: Mean ± standard deviation for ejaculate volume (mL), sperm concentration of neat semen (×10⁶), total sperm per collection (×10⁹), total sperm collected per month (×10⁹) and Doses/Boar/Month by age groups (G1: 8-10 months; G2: 11-13 months and G3: 14-17 months) between collection interval (CI3,4,5 and 7) in boars

	Ejaculate volume (mL)			
	CI3	**CI4**	**CI5**	**CI7**
G1 (8-10m)	140,74 ± 54,49[a]	148,19 ± 64,67[ab]	155,39 ± 69,12[b]	167,36 ± 75,51[c]
G2 (11-13m)	167,88 ± 74,08[a]	179,90 ± 79,61[b]	186,93 ± 80,49[b]	201,65 ± 84,81[c]
G3 (14-17m)	211,70 ± 81,792[a]	225,62 ± 89,00[b]	226,36 ± 91,98[b]	245,07 ± 101,26[c]
	Sperm concentration of neat semen ($\times 10^6$)			
	CI3	**CI4**	**CI5**	**CI7**
G1 (8-10m)	249,20 ± 128,70[a]	262,81 ± 125,18[b]	259,66 ± 120,92[ab]	269,11 ± 125,26[b]
G2 (11-13m)	252,92 ± 123,21[a]	262,70 ± 120,24[ab]	267,27 ± 117,55[b]	287,57 ± 132.25[c]
G3 (14-17m)	230,04 ± 95,54[a]	235,64 ± 101,87[ab]	244,30 ± 101,77[b]	267,28 ± 114,08[c]
	Total sperm per collection ($\times 10^9$)			
	CI3	**CI4**	**CI5**	**CI7**
G1 (8-10m)	32370,46 ± 15552,69[a]	36106,28 ± 16822,34[b]	37528,13 ± 18028,77[b]	41966,37 ± 20657,67[c]
G2 (11-13m)	39397,49 ± 18014,53[a]	43521,40 ± 19337,74[b]	47256,50 ± 22704,38[c]	53762,06 ± 23768,29[d]
G3 (14-17m)	46027,78 ± 19521,33[a]	49940,39 ± 21078,33[b]	60547,75 ± 24481,79[c]	60547,75 ± 24481,79[c]
	Total sperm collected per month ($\times 10^9$)			
	CI3	**CI4**	**CI5**	**CI7**
G1 (8-10m)	323704,6 ± 155526,85[a]	270797,1 ± 126167,56[b]	225168,8 ± 108172,62[c]	179855,9 ± 88532,86[d]
G2 (11-13m)	393974,9 ± 180145,3[a]	326410,5 ± 145033,0[b]	283539,0 ± 136226,3[c]	230408,8 ± 101864,1[d]
G3 (14-17m)	460277,8 ± 195213,3[a]	374552,9 ± 158087,4[b]	314264,0 ± 143527,6[c]	259490,4 ± 104922,0[d]
	Doses/Boar/Month			
	CI3	**CI4**	**CI5**	**CI7**
G1 (8-10m)	188,20 ± 90,42[a]	157,44 + 73,35[b]	130,91 ± 62,89[c]	104,56 ± 51,47[d]
G2 (11-13m)	229,05 ± 104,73[a]	189,77 ± 84,32[b]	164,84 ± 79,20[c]	133,95 ± 59,22[d]
G3 (14-17m)	267,60 ± 113,49[a]	217,76 ± 91,91[b]	182,71 ± 83,44[c]	150,86 ± 61,00[d]

Different letters on the line differ statistically $P < 0.001$

https://doi.org/10.54846/am2022/146

Bioinformatics as a tool in prescription platform vaccine development

Ashley Petersen, MS; Alan Young, PhD; Bishwas Sharma, MS; James Young, PhD
Medgene Labs LLC, Brookings, South Dakota

Introduction

A Prescription Platform Vaccine is a platform-based, standardized vaccine that is nonviable and non-replicating (ie, "killed"). By CVB definition, a production platform is a manufacturing process that relies on a single vector or expression system ("backbone") and a standard process for inserting a gene or genes of interest (GOI) into the system to generate different recombinant Master Seeds or Sequences, which are then used to produce product following a standardized method for manufacture. The GOI (s) selected is known to convey immunity against the agent in question, so that the final product will be expected to produce disease coverage in vaccinated animals.

To select the most appropriate GOI(s) to create the vaccine Master Seed construct for a prescription vaccine, bioinformatics or a systematic evaluation of the viral sequence present in a population of animals is needed. Bioinformatic analysis is especially important for vaccine development with highly genetically diverse viruses such as influenza A-swine (IAV-S) virus or rotaviruses A, B, and C (RVA, B and C) in swine.

Materials and methods

The GOI for IAV-S is the HA gene. Gene sequences for the H3 and H1 epitopes were collected from diagnostic submissions from collaborating veterinarians and veterinary diagnostic laboratories. Sequences were also collected from the VIPr database which includes all reported IAV-S sequences in the US.

The GOIs for rotavirus include two surface proteins that are associated with development of serum neutralizing antibody titers. These sequences were also collected from diagnostic submissions as described for IAV-S.

A database was developed with these sequences and a comparative analysis was done using Geneious®.

Results

Phylogenetic trees were created to evaluate and demonstrate the diversity of IAV-S and rotaviruses. Figure 1 depicts an example of a phylogenetic tree of RVA protein 1. Master Seed constructs were subsequently developed to cover the various clades in the tree at a 95% homology level for IAV-S and a 95% homology level for RVA and C.

Discussion

Medgene's IAV-S Master Seed construct bank is extensive enough to cover approximately 98% of the known HA sequences in the US.

Although the extent of rotavirus diversity in the US pig population is not as well known as IAV-S, Medgene Labs has developed a broad Master Seed construct bank that will cover the sequences recovered from diagnostic submissions to date.

New sequences that arise will immediately be compared via Medgene Labs' automated bioinformatics system and new Master Seed constructs can be developed as needed to provide a vaccine solution.

Figure 1: Medgene's construct coverage for rotavirus A protein 1.

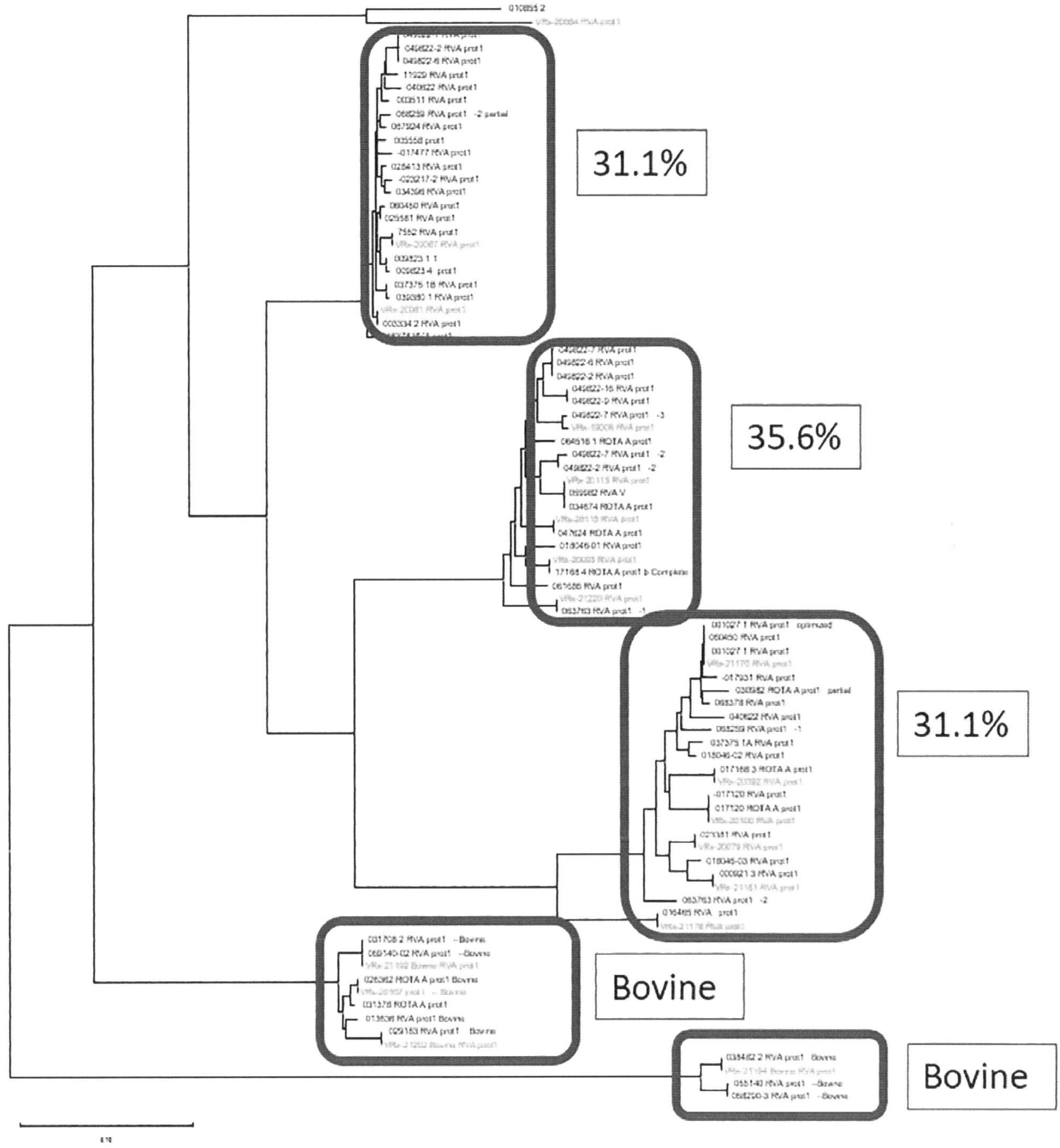

Poster 66

https://doi.org/10.54846/am2022/147

Stability of three commercial *Salmonella* vaccines in the presence of common drinking water additives

R. Fleck[1], DVM; L. Ruesch[2], MS; K. Hendrickson-Guttum[2], BS; J. Scaria[2], PhD; Z. Lau[2] BS; B. O'Brien[1], VMD MS
[1]Merck Animal Health, Lenexa, Kansas; [2]Animal Disease Research and Diagnostic Laboratory, South Dakota State University, Brookings, South Dakota

Introduction

As pork producers continue to reduce feed grade antibiotics, the application of modified-live *Salmonella* vaccines is becoming more common. While the removal of antibiotics from feed and water during vaccination is a well-known label indication, the impact of drinking water additives has been overlooked. Drinking water additives may negatively impact the vaccination process by either killing the organism or by promoting overgrowth.

Materials and methods

A pilot project was performed with three commercial Salmonella vaccines, Argus ST/SC (Merck Animal Health), Enterisol Salmonella TC (Boehringer Ingelheim Animal Health) and Nitrosal FD (Arko Labs) and one positive control organism, *Salmonella enterica* serovar *Typhimurium* Sal 54 (Microbiologics). Vaccines and control organism were reconstituted per label and plated onto *E coli*/ Coliform Count Petrifilm Plates (3M) at different dilution rates such that individual colonies could be visualized. The following drinking water additives were prepared per label or per industry standards using distilled, sterile water: 8.5% sodium hypochlorite (diluted to 4 ppm), 3% hydrogen peroxide (diluted to 30 ppm), Sky Lytes Non-Medicated Buffered Electrolytes (Skylar labs), Citric acid (Animal Science Products), Acid pack 4 way 2X (Alltech), Swine Blue Lite (TechMix), Christensen Farm Electrolyte Pack (Techmix), calcium carbonate (180 mg/liter), dried skim milk (0.92 grams per liter), and Vac-Pac Plus Stabilizer (Animal Science Products). Distilled sterile water was used as a negative control solution. Vaccines and positive controls were added to the drinking water solutions according to the dilution rate determined in the pilot project. Aliquots were taken and plated immediately after mixing (T0), at two hours post mixing (T2) and at twelve hours post mixing (T12). All solutions were maintained at 77°F during the course of the study. Plates were read at 24- and 48-hours post-plating using a Petrifilm Plate Reader Advanced (3M). Negative control solutions which did not contain live organisms were aliquoted and plated at each time point. All plates were done in triplicate. Data were analyzed by averaging the results of the three plates at the 48-hour reading for each time point. The average plate count at T2 and T12 was divided by the average plate count at T0 and multiplied by 100 to express the percentage of T0 plate count (Table 1). A number less than 100% may indicate reduced viability and a number greater than 100% may indicate sustained or increased viability

Results and discussion

All negative control solutions remained free of colonies during the study indicating sterile technique was maintained. The positive control organism and all vaccines grew out of all T0 samples indicating viability of test subjects. All plates except the skim milk, which was too numerous to count at T2 and T12, had countable colonies. The largest numerical reductions in viability at T2 were found in citric acid, hydrogen peroxide and Swine Blue Lite. $CaCO_3$, Vac-Pac plus, sterile distilled water, sodium hypochlorite 4ppm and Acid Pack 4 way performed similarly with mild reduction in plate count at T2 and loss of half viability by T12. Sky Lytes and Christenson Farm Electrolytes produced sustained viability through T12. Skim milk resulted in aggressive growth by both T2 and T12. The use of skim milk in conjunction with live *Salmonella* vaccines will cause rapid overgrowth and potential safety issues. The results of this study highlight the importance of making sure that swine consume live *Salmonella* vaccines within 2-4 hours after preparation and that prolonged times even in tap water can erode viability.

Table 1: Percentage of average T0 plate count at T2 and T12

Vaccine	Time	Citric acid	H_2O_2 30 ppm	Swine blue lite	Acid pack 4 way	Distilled sterile H_2O	$CaCO^3$ 180 ppm
A SC/ST	T2	71	75	73	88	92	98
	T12	10	19	54	52	76	47
ES TC	T2	7	56	72	93	103	82
	T12	0	14	26	54	74	21
NS FD	T2	0	39	13	48	118	58
	T12	0	4	0	72	64	6
PC	T2	28	97	50	67	107	113
	T12	0	38	4	19	92	45

Vaccine	Time	Vac Pac Plus	NaOCl 4 ppm	Sky-Lytes	CF electrolytes	Skim milk
A SC/ST	T2	96	93	102	93	101
	T12	66	45	127	98	TNTC
ES TC	T2	94	91	110	126	96
	T12	38	15	132	115	317
NS FD	T2	89	75	302	130	118
	T12	36	3	229	79	TNTC
PC	T2	111	111	134	111	104
	T12	68	48	128	102	TNTC

A SC/ST = Argus SC/ST; ES TC = Enterisol Salmonella TC; NS FD = Nitrosal FD; PC = positive control organism; TNTC = too numerous to count

https://doi.org/10.54846/am2022/148

Two-year commercial demonstration of methionine hydroxy-analogue chelated zinc, copper, and manganese (MINTREX®) on sow reproductive and production performance in a highly prolific sow farm

A. Hintz, DVM; B. Lawrence, PhD; M. Collins; C. Heiwy
Novus International Inc, St. Charles, Missouri

Introduction

Genetic companies and pork producers continue to focus on sow productivity, mainly; increased pigs weaned per sow farrowed. Sow productivity has increased, while at the same time sow mortality has risen and sow retention to later parities has diminished. To maximize sow profitability, focus should be placed on pigs weaned per sow lifetime. The optimal parity for maximum sow profitability is parity 6.[1] Trace minerals, specifically zinc (Zn), copper (Cu) and manganese (Mn) play vital roles in hundreds of physiological processes impacting growth, maintenance and function. The importance of trace mineral source on the sow's lifetime productivity has been previously demonstrated.[2,3] Furthermore, substituting 50% of the Zn, Cu and Mn in the sulphate form with a similar level of mineral provided as the methionine hydroxy-analogue chelate (MINTREX® trace minerals) has been shown to improve sow productivity and longevity in a moderately-productive herd[4] as has total replacement of all Zn, Cu and Mn with 50% of the level of trace mineral provided in the inorganic form.[2] The commercial demonstration discussed in this paper evaluated the efficacy of replacing approximately 50% of the trace mineral in sulphate form with MINTREX® on sow performance in a high-health, highly prolific sow herd.

Materials and methods

Two 2,500 sow farrow-to-wean units on the same site were used. Both units were managed as one and labor was provided by the same personnel providing a side-by-side evaluation of trace mineral source impact on sow productivity. One 2,500 sow unit remained on the current trace mineral program which provided 125 ppm Zn, 15 ppm Cu and 30 ppm Mn in the sulphate form (ITM). For sows in the other unit, 60 ppm Zn, 10 ppm Cu and 20 ppm Mn was provided as methionine hydroxy-analogue chelate (MTX) substituting for a similar amount of sulphate trace minerals with the total amount of trace mineral provided being similar between the units. Data evaluation included a 1-year historical analysis for comparative data and a 2-year evaluation period of the two trace mineral programs, concluding in mid-2020.

Results and discussion

Comparison of the impact of trace mineral programs were made first to the baseline (1-year historical production) averages and then the difference between the two treatment groups corrected for historical differences was noted. The following differences were observed for the MTX fed sows compared to the ITM fed sows when corrected for historical differences between the units: 1) a net 2.2% increase in MTX sows bred within 7-days, 2) a 1.0% net decrease in MTX percent repeats, 3) a 1.0% net increase in farrowing rate 4) a 2.0% decrease in cull rate. The MTX unit had 0.5 more pigs total born and 0.4 more pigs live born resulting in a net 0.2 more pigs weaned per litter (Table 1). The improvements in sow productivity resulted in 0.7 more pigs weaned per mated female per year and 1.4 more pigs weaned per sow lifetime in the MTX unit vs. the ITM unit. These data, while a field demonstration rather than a controlled trial, are consistent with prior field results in production units of differing health and productivity,[4] and demonstrated that, although mineral levels were consistent between sow units on the same site, mineral supplied at the methionine hydroxy-analogue chelate (MINTREX®) may result in greater sow productivity and reproductive performance in a highly prolific sow system improving sow lifetime productivity.

References

1. Gruhot et al., An economic analysis of retention in a United States breed-to-wean system. J. Swine Health Prod. 2017; 25.

2. Barea et al., Utilisation d'oligo éléments organiques dans l'alimentation des truies: effets sur la longévité et les performances des portées. Journées Recherche Porcine, 2019; 51:167.

3. Zhao et al., Chelated trace minerals support sow reproduction. 2012; Feedstuffs. June 4.

4. Lawrence et al., One-year commercial demonstration of methionine hydroxy-analogue chelated zinc, copper, and manganese (MINTREX® trace minerals) on sow productivity. AASV 52nd Annual Meeting Proceedings, 2021; 100-103.

Table 1: Sow productivity and reproductive efficiency experimental period change vs. baseline

	ITM	MTX	MTX net difference
Litter/mated female/year	+0.01	+0.02	+0.01
Pigs/mated female/year	+1.0	+1.7	+0.7
Non-productive days, d	Unavailable	Unavailable	-4 days
% bred within 7 days	+0.7	+2.9	+2.2
% repeats	-3.3	-4.3	-1.0
Wean-estrus interval, d	-0.4	-0.7	-0.3
Farrowing rate, %	-0.7	+3.0	+1.0
Avg. total born, pigs	+0.1	+0.5	+0.4
Avg. live born, pigs	+0.6	+0.7	+0.1
Prewean mortality, %	-0.3	-0.5	-0.2
Pigs weaned/litter	+0.4	+0.6	+0.2
Cull rate, %	-2.0	-4.0	-2.0
Sow mortality, %	-0.6	+1.1	+1.7
Replacement rate, %	+3.5	+3.9	+0.4

https://doi.org/10.54846/am2022/149

Using sow technologies to drive sow and subsequent pig performance associated with stress induced inflammation

Ran Song, PhD; Kim Friesen, PhD; Rob Musser, PhD
NutriQuest, Mason City, Iowa

Introduction

Managing sows for optimal productivity requires control of environmental stress during both gestation and lactation. Maintaining healthier sows has significant down-stream impact on producing more, heavier, and healthier pigs. NutriQuest Sow Platform™ (SP) is a proprietary blend of yeast, DFM and L-carnitine that supports sow immune function, improves sow performance and post-weaning pig quality. Our research has focused on evaluating the benefits of feeding SP on sow and litter performance and identifying inflammatory biomarkers that are associated with pre- and post-weaning pig performance.

Feeding Sow Platform™ improved sow reproductive performance – Meta-analysis of 10 studies

A meta-analysis was conducted utilizing data from 10 studies with a total of 2,202 sows on experiment to evaluate the effects of feeding SP on sow reproductive performance. In each study, sows (PIC 1050; Parity ≥ 2) from a single, high-health sow farm in the upper Midwest were randomly assigned to either a Control treatment or a treatment receiving SP at 1.0 lb/ton during entire gestation and lactation. Compared to Control fed sows, feeding SP improved born alive by 0.2 more pigs per litter ($P = 0.07$) and reduced PWM by 0.5% ($P = 0.04$), which resulted in 0.3 more pigs at weaning ($P = 0.009$). Individual pig birth weight was improved from 2.82 to 2.87 lb ($P = 0.008$) and weaning weight was improved from 11.40 to 11.50 lb ($P = 0.07$) in SP treatment compared to Control. These results suggest feeding SP improves sow and litter performance.

Feeding Sow Platform™ promoted healthy immune status of sows and their piglets as measured by serum pro-inflammatory cytokines and vitamin A

To further understand the immune responses of feeding SP, serum concentrations of pro-inflammatory cytokines and vitamin A were measured. Serum concentration of vitamin A is reduced under systemic inflammation due to impaired synthesis of retinol-binding protein in the liver that transport vitamin A from the liver to the circulating blood.[1] The current study reported positive correlation ($P < 0.05$) between serum vitamin A and pre-weaning growth rate in nursing pigs, suggesting that serum vitamin A could be a performance-indicating biomarker for inflammation.

Sows fed SP showed 54% and 38% lower serum concentrations of IFN-γ prior to farrowing ($P = 0.08$) and at weaning ($P = 0.31$), respectively, compared to those on Control treatment. The reduction in pro-inflammatory cytokines in sows during the stress of farrowing and lactation also resulted in a reduction in inflammation status of their piglets. Compared to the piglets born from the sows on Control treatment, piglets from the sows fed SP showed either numerically or significantly lower serum concentrations of cytokines (IFN-α, IFN-γ, IL-1β, IL-4, IL-6, IL-8, IL-10, IL-12 and TNF-α) at birth and at weaning. Additionally, feeding SP to sows led to 28% reduction on pigs with vitamin A insufficiency (< 0.15 ppm) at weaning (Figure 1). The heavier and more robust pigs from SP treatment at weaning showed different survivability post-weaning (97.4 vs. 95.6%).

Conclusions

Feeding NutriQuest Sow Platform™ improved sow reproductive performance and post-weaning performance of their offspring that were associated with reduced stress and inflammation status as measured by serum vitamin A and pro-inflammatory cytokines.

Reference

1. L. P. Rubin, et al., Metabolic effects of inflammation on vitamin A and carotenoids in humans and animal models. Adv Nutr 2017;8:197-212.

Figure 1: Feeding NutriQuest Sow Platform™ to sows reduced the percentage of piglets with serum vitamin A insufficiency at weaning.

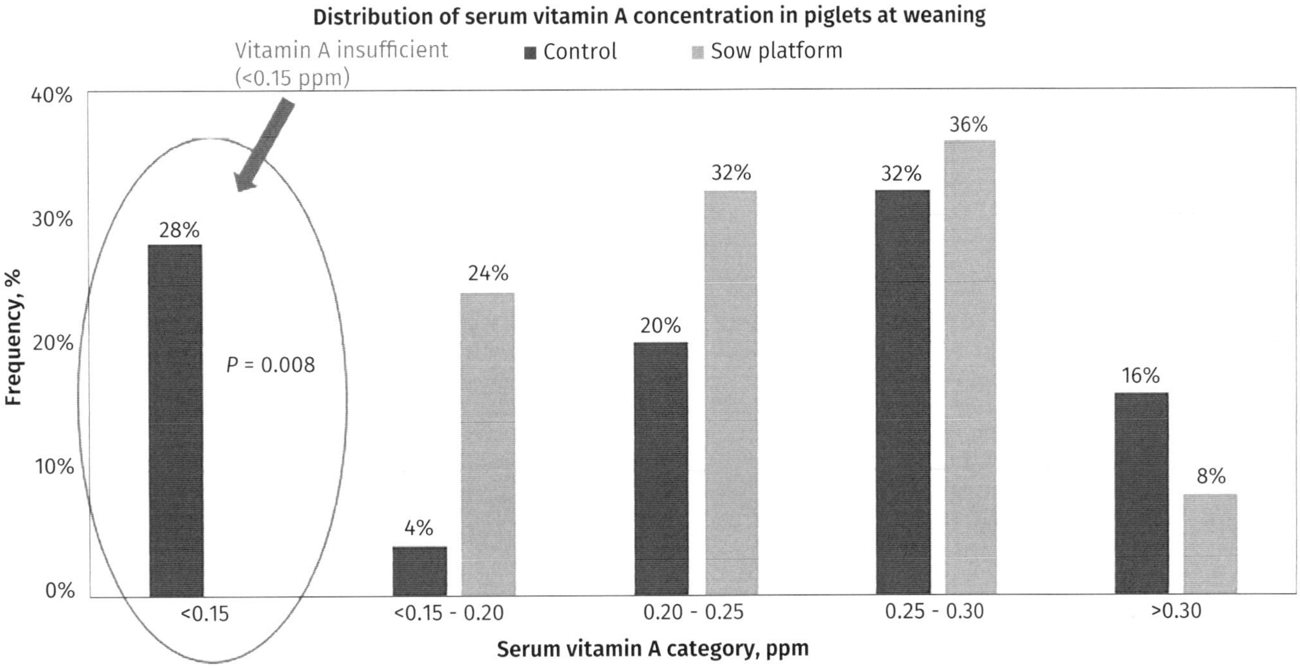

https://doi.org/10.54846/am2022/150

Improving young piglets' immunity in nursery with an algae extract

M. Gallissot, MSc; VT. Quoc, DVM; H. Le Thi, MSc; M. Le Goff, F. Bussy MSc; P. Luong, DVM
Olmix Group, Brehan, France

Introduction

The porcine circovirus type 2 (PCV2) is one of the most prevalent viruses causing clinical and subclinical infections and thus great economical losses to the pig industry.[1] To protect the piglets, vaccination is commonly performed at 3-4 weeks of age. However, in the nursery, young piglets face many stressful challenges including vaccination, environmental change and feed transitions, causing immunodeficiency and an increased risk of pathogen infections. In this context, the modulation of the innate and adaptive immunity of the piglets becomes a key parameter. Previous research has shown that a green algal extract from *Ulva sp* (MSP®$_{\text{IMMUNITY}}$), had the capacity to modulate the immune response *in vitro*[2,3] and *in vivo*.[4] The objective of the present trial, conducted on a commercial swine farm, was to evaluate the capacity of a water-supplement containing MSP®$_{\text{IMMUNITY}}$ to reinforce the adaptive immune response to the PCV2 vaccine.

Materials and methods

A total 60 piglets (Landrace/Yorkshire × Duroc) weaned at 28 days were involved in the trial and split in 2 groups: the control group and the MSP®$_{\text{IMMUNITY}}$ group. Piglets from the MSP®$_{\text{IMMUNITY}}$ group received the algae-based supplement in the drinking water for 3 days after weaning, at the rate of 1 mL/20 lbs of live weight per day; one day before and two days after the PCV2 vaccination at the rate of 0.5 mL/20 lbs of live weight; and then once a week until 78 days of age at the rate of 0.5 mL/20 lbs of live weight. The piglets were individually weighed at weaning and at the entry fattening unit (78 days). The occurrence of respiratory problems was recorded daily. Moreover, blood was sampled on 5 piglets per pen before the PCV2 vaccine and 14, 21 and 30 days after the vaccine to evaluate the vaccine uptake.

Results and discussion

The piglets' performance was similar in both groups all along the trial period. A non-significant decrease (-0.06) of the feed to gain ratio was observed in the test group compared to the control group while the average daily gain was similar in both groups (Table 1). Piglets from the MSP®$_{\text{IMMUNITY}}$ group showed fewer respiratory problems than piglets from the control group (0.88% and 2.65% respectively, $P < 0.001$) (Table 1). The antibody levels were not significantly different between the treatments but the uniformity (standard deviation) was improved by 16% three weeks after the vaccination in the MSP®$_{\text{IMMUNITY}}$ group compared to the control group (378.00 and 453.96 respectively) (Table 1). Hence, this trial proves the capacity of the algae-based water supplement to improve the PCV2 vaccine uptake by nursery piglets, and to reduce the risk of disease occurrence while maintaining performance.

References

1. Segalés J., Kekarainen T., Cortey M., 2013. The natural history of porcine circovirus type 2: from an inoffensive virus to a devastating swine disease? Vet Microbiol 165:13–20.

2. Berri M., Slugocki C., Olivier M., Helloin E., Jacques I., Salmon H., Demais H., Le Goff M., Nyvall Collén P., 2016. Marine-sulfated polysaccharides extract of Ulva armoricana green algae exhibits an antimicrobial activity and stimulates cytokine expression by intestinal epithelial cells. J Appl Phycol, 28 :1; 3-12.

3. Berri M., Olivier M., Holbert S., Dupont J., Demais H., Le Goff M., Nyvall Collén P., 2017. Ulvan from Ulva armoricana (Chlorophyta) activates the PI3K/Akt signaling pathway via TLR4 to induce intestinal cytokine production. Algal Research, 28, 39-47.

4. Bussy F., Le Goff M., Salmon H., Delaval J., Berri M., Nyvall Collén P. 2019. Immunomodulating effect of a seaweed extract from Ulva armoricana in pig: Specific IgG and total IgA in colostrum, milk, and blood. Veterinary and Animal Science 100051.

Table 1: Growth and health performance of nursery piglets

Parameters	Control (n = 30)	MSP®IMMUNITY (n = 30)	Difference	*P*-value
Performance (mean ± sd)				
Average daily gain (g/day)	1.038 ± 0.059	1.036 ± 0.042	-0.002	1.00
Feed to gain ratio	1.85 ± 0.07	1.79 ± 0.03	-0.06	0.70
Health indicators (%)				
Respiratory problems	2.65	0.88	-1.77	0.001
Antibody titers (mean ± sd)				
Before vaccination	4433 ± 680	5093 ± 696	+16	0.443
14 days after vaccination	2527 ± 429	2986 ± 399	-30	0.443
21 days after vaccination	2111 ± 454	2335 ± 378	-76	0.478
30 days after vaccination	2245 ± 393	2577 ± 402	+9	0.378

https://doi.org/10.54846/am2022/151

The power of LeeO™ individual pig data in commercial production

J. Tyler Holck[1], DVM, MS, MBA; Jack Sanders[1], BS; Joel Stave[2], BA
[1]Feed His People, LLC, Gilbert, Iowa; [2]Prairie Systems, Spencer, Iowa

Introduction

LeeO™ is an individual animal tracking system designed for paperless data collection in "real time", supporting precision pork in the commercial pork supply chain.[1] Utilizing radio frequency identification (RFID), LeeO™ users leverage the mobility of the LeeO™ ecosystem to scan tags and record individual animal weights quickly and effortlessly with a LeeO™ O Reader and bluetooth technology connecting to automated individual scales. This paper describes an individual animal birth-to-harvest dataset in a commercial swine supply chain using LeeO™.

Data set

A commercial birth-to-harvest supply chain in the Netherlands utilizes LeeO™ for traceability and precision pork data collection from the source farms to the harvest facility. Data from 312,992 individual pigs born from January 2, 2020 to March 21, 2021 and slaughtered in the prime market comprise this dataset. Individual pig data includes birth date, dam, parity of dam, dam genotype, sire, sire genotype, farm of origin, slaughter date, and carcass weight. Individual slaughter weight was calculated using 75% yield of carcass weight and an individual birth-to-harvest average daily gain (BTHADG) was calculated using birth and slaughter dates.

Summary of birth-to-harvest average daily gain (BTHADG)

The BTHADG was bell-shaped and varied dramatically in this dataset of prime market pigs, with a mean of 1.552 lbs/hd/d and a standard deviation of 0.159 (Figure 1). The harvest weight was fairly consistent with a mean of 290.5 lbs and a standard deviation of 24.0 lbs. BTHADG varied by birth month with January and February have the lowest performance (1.53 lb/d) and September through November the highest (1.57). Parity of litter for this dataset while normal for The Netherlands represented an older parity structure compared to typical distributions in the US, with 18% P1, 16% P2, 14% P3, 12.5% P4, 11% P5, 9.5% P6, 7.5% P7, and 11.5% ≥ P8. BTHADG was lowest for P1 and ≥ P11 (1.53), and highest for P3-8 (1.56). Over 100 unique dam genotypes and 65 unique sire genotypes comprised the full dataset, with ranges in mean BTHADG of 1.35 to 1.80 lbs/hd/d for dam genotypes and 1.42 to 1.72 for sire genotypes. 82 farms contributed to the dataset, with mean BTHADG ranging from 1.40 to 1.75 lbs/hd/d.

Variation in individual performance and subsequent economic impact

The individual pig BTHADG data within this large dataset emphasizes the dramatic variation within and between farms. With a target market weight of 290 lbs, the BTHADG difference from 1 standard deviation above and below the mean is 39 days to market, and 2 standard deviations is 78 days difference to achieve target harvest weight. Using a space available economic model at $0.15/hd/d suggests an opportunity to capture $1.88 per head which increases to a total of $4.07 per head in a space limited model at $0.60/lb live weight. The economic opportunity for pigs one, two, or three standard deviations below the mean is $4.45, $14.68, and $24.44 respectively (Figure 1).

Conclusions

The LeeO™ Ecosystem is very effective in gathering real-time birth-to-harvest data while providing supply chain traceability. While focusing on day-to-day animal lifecycle activities LeeO™ facilitates the collection of precision data that then allows for analysis of numerous factors impacting birth-to-harvest performance and the subsequent opportunities for economic improvement at the farm level and broader supply chain.

Reference

1. ter Linde P et al. The value of LeeO, an individual animal identification system adding value to the pork supply chain the Netherlands. AASV, 2020; 337.

Figure 1: Economic opportunity based on standard deviation from the mean (312,992 pigs)

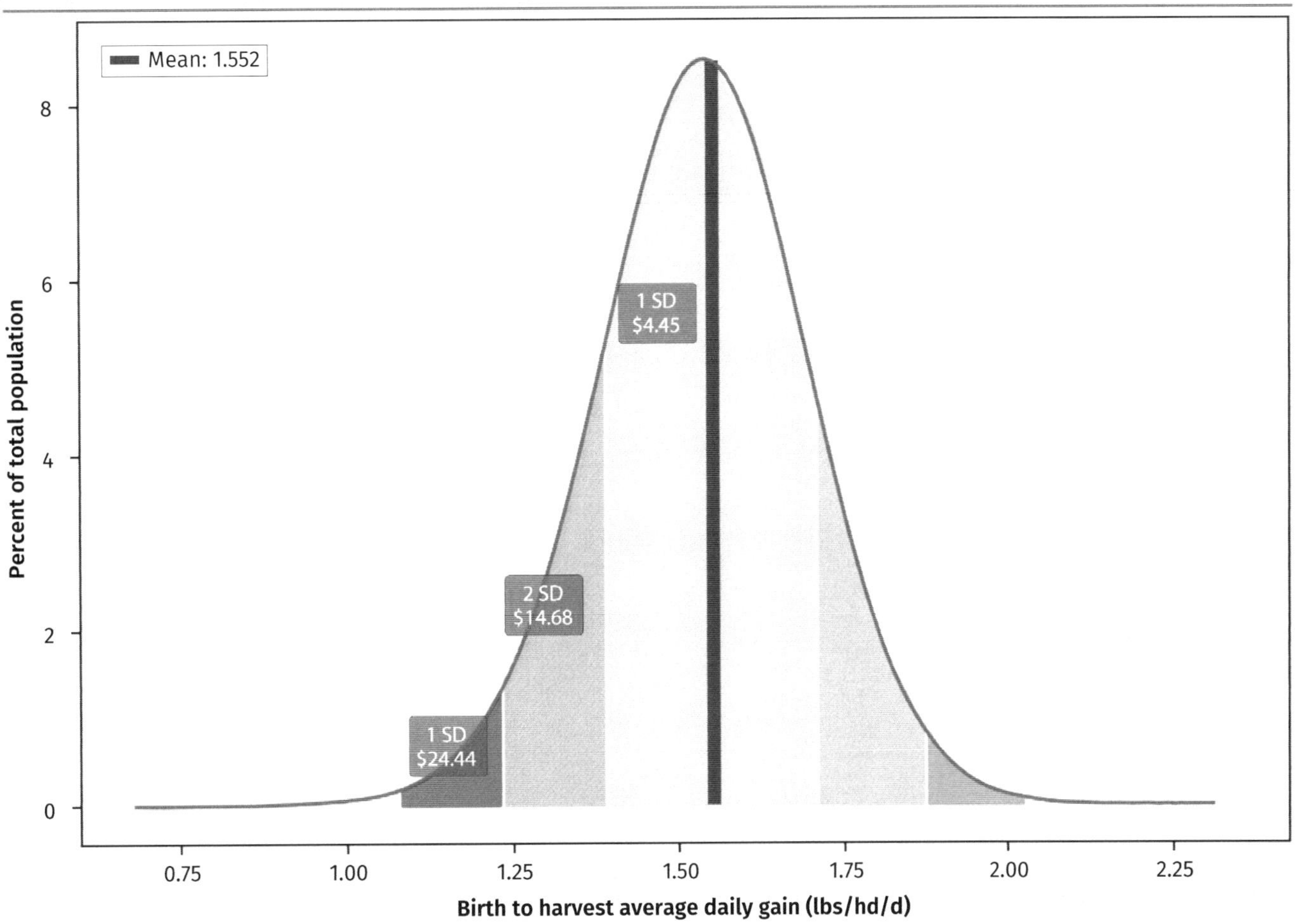

Poster 71

https://doi.org/10.54846/am2022/152

Effect of Grazix, a phytobiotics-based product, on the lactate-producing and -utilizing gut bacterial communities of nursery pigs

A. Fresno; R. Samuel; B. St-Pierre
Department of Animal Science, South Dakota State University, Brookings, South Dakota

Introduction

Weaning remains a challenging phase of production for swine operations worldwide. Post weaning diarrhea, typically associated with a high incidence of entero-toxigenic *E coli* (ETEC) proliferation, results in dehydration, growth retardation, and sudden death.[1] Plant-based extract additives, or phytobiotics, have been developed into tools available to the swine industry to mitigate these negative effects.[2] Phytobiotics have been shown to improve animal growth and health, providing benefits, such as antimicrobial, antioxidant, and anti-inflammatory activities, as well as improvement of gut function.[3] However, when considering many different types of chemical and bioactive compounds present in these plant-derived products, the consensus from reported studies is that their modes of action can be highly variable. Also, since the influence of phytobiotics can be attributed not only to a simple decrease or increase in bacterial proliferation, but also to fluctuations in microbiota composition and function,[4] further research is required to gain additional insights. Thus, the aim of this study was to evaluate the effects of Grazix, a plant-based product on the diversity and composition of fecal bacterial communities of post weaned pigs.

Materials and methods

Twenty-four weaned pigs were randomly assigned to two pens, with equal barrows and gilts distribution. Animals in the treatment pen (Gx) were provided drinking water supplemented with the phytobiotics product (Grazix) at a final dilution of 8 mL per liter of water for seven consecutive days, starting one day after their arrival at the barn. Fresh feces were collected from each pig by rectal palpation on post weaning day 1 (before treatment), then at days 4, 10, and 21. Microbial DNA was extracted from fecal samples using a repeated bead beating plus column method.[5] 16S rRNA gene sequence diversity was determined by high-throughput sequencing (MiSeq 2X300, Illumina) of PCR-generated amplicons (V1-V3). Sequences were clustered into OTUs using custom-written Perl scripts (96% cutoff). Taxonomic affiliations were determined using online tools: RDP classifier and BLAST. Alpha diversity indices and Principal Coordinate Analysis were performed using MOTHUR.[6] All statistical analyses were performed using RStudio Statistical Software (Version 1.3.959).

Results and discussion

Our results show that *Lactobacillaceae* and *Peptostreptococcaceae* are both higher in the supplemented group at D4, while *Streptococcaceae* are significantly lower in the treated group at D10 and D21. At D10, *Erysipelotrichaceae* are lower, and *Veillonellaceae* are higher in the treated samples than the control group (Table 1). Species affiliated to these families have shown a number of lactate production and utilization activities in the gut. Indeed, Of the 13 abundant OTUs that have different representation between treated and control pigs ($P < 0.05$), six are predicted to be lactate producers (affiliation to Lactobacillus or Streptococcus), and one is predicted to be a lactate utilizer, based on its high identity to *Megasphaera elsdenii*. As lactate is considered a critical microbial end product in the gut, as it can inhibit pathogens or be metabolized to propionate for utilization by host cells, these data suggest that Grazix may provide a favorable metabolic equilibrium between lactate production and utilization.

References

1. Amezcua et al.: 2002, Can. J. Vet. Res, 66, 73.
2. Sun and Kim.: 2017, Anim. Nutr, 3, 322–330.
3. Windisch et al.: 2008, 86, J. Anim. Sci, E140–E148.
4. Castillo et al.: 2006, J. Anim. Sci, 84, 2725–2734.
5. Yu and Morrison.: 2004, Biotechniques 2004, 36, 808–812.
6. Schloss et al.: 2009, Appl. Environ. Microbiol. 2009, 423, 75.

Table 1: Mean relative abundance (%) of main bacterial groups in control (Co) and treatment (Gx) samples

Taxon	D1	Co-D4	Gx D4	Co-D10	Gx D10	Co-D21	Gx D21
Firmicutes#	63.8^{a}	69.6^{ab}	66.6^{ab}	78.1^{b}	74.8^{b}	90.1^{c}	91.5^{c}
Ruminococcaceae#	22.7^{a}	22.2^{a}	22.7^{a}	16.8^{ab}	19.6^{ab}	11.2^{b}	12.3^{b}
Lachnospiraceae	13.8	18.0	17.4	17.1	20.0	16.9	20.0
Streptococcaceae#	0.5^{a}	1.2^{ab}	1.8^{b}	11.8^{c}	3.4^{b}	36.4^{d}	28.3^{e}
Erysipelotrichaceae#	4.0^{ab}	7.3^{a}	4.4^{a}	5.8^{a}	2.0^{b}	5.1^{a}	8.6^{a}
Lactobacillaceae#	6.2^{a}	0.3^{c}	1.6^{b}	8.2^{ab}	13.8^{ab}	7.0^{a}	3.6^{a}
Veillonellaceae#	0.9^{a}	0.5^{a}	0.9^{a}	0.7^{a}	4.4^{b}	1.6^{a}	4.4^{b}
Clostridiaceae[1]#	0.6^{a}	7.9^{b}	7.6^{b}	6.5^{bc}	1.2^{ac}	1.5^{c}	3.0^{abc}
Peptostreptococcaceae#	0.6^{a}	0.7^{c}	2.2^{b}	0.2^{d}	0.2^{ad}	0.^{3ad}	1.1^{abcd}
Eubacteriaceae#	2.9^{a}	0.2^{abc}	0.2^{bc}	0.1^{c}	0.5^{abc}	0.6^{a}	0.4^{ab}
Clostridiales Inc. Sedis_XIII#	1.0^{abc}	1.0^{a}	0.8^{abc}	1.0^{a}	1.3^{ab}	0.4^{bc}	0.3^{c}
Other Firmicutes$	10.6	10.4	7.1	10.1	8.5	9.1	9.6
Bacteroidetes#	22.2^{a}	24.8^{a}	30.0^{a}	18.9^{a}	20.6^{a}	8.3^{b}	7.0^{b}
Porphyromonadaceae#	2.8^{ab}	1.6^{ab}	1.2^{ad}	4.3^{b}	1.7^{ab}	0.6^{cd}	0.4^{c}
Prevotellaceae#	6.7^{a}	18.9^{bc}	26.5^{b}	11.3^{ac}	14.4^{bcd}	6.8^{ad}	6.0^{a}
Bacteroidaceae#	11.3^{a}	0.6^{bd}	0.5^{bd}	0.6^{bd}	2.0^{b}	0.2^{cd}	0.2^{c}
Other Bacteroidetes$	1.36	3.7	1.7	2.7	2.6	0.6	0.5
Proteobacteria#	8.9^{a}	1.5^{bd}	1.2^{b}	0.2^{cd}	2.1^{b}	0.2^{c}	0.4^{d}
Enterobacteriaceae#	7.1^{a}	0.3^{bc}	0.5^{b}	<0.1^{de}	< 0.1^{c}	< 0.1^{d}	< 0.1^{ce}
Other Proteobacteria$	1.8	1.2	0.7	0.2	2.0	0.2	0.3
Synergistetes#	2.8^{a}	0.1^{b}	< 0.1^{b}	0^{*c}	< 0.1^{c}	< 0.1^{c}	0^{*c}
Other Bacteria$	2.3	4.0	2.2	2.9	2.4	1.6	1.2

Note: Mean relative abundance of taxonomic groups is presented as a percentage (%) of the total number of analyzed reads per sample.

Taxa showing a statistically significant difference by the Kruskal-Wallis sum rank test ($P < 0.05$)

$ Statistical test not performed because of group heterogeneity

* no reads detected in any of the samples from this group

Poster 72

https://doi.org/10.54846/am2022/153

Investigation of pathogen and host genome variation influencing host response to *Streptococcus suis*-infection

J. R. Dunkelberger[1,2], PhD; Astrid de Greeff[3], PhD; E. van Langen – Willems[1], DVM; E. F. Knol[1], PhD; M. Vrieling[3], DVM PhD
[1]Topigs Norsvin Research Center B.V., Beuningen, The Netherlands; [2]Topigs Norsvin USA, Burnsville, Minnesota; [3]Wageningen Bioveterinary Research, Lelystad, The Netherlands

Introduction

Streptococcus suis is a gram positive bacterium that resides on the mucosal surfaces of the tonsil, nasal cavities, or gastrointestinal tract. Colonization of *S suis* may, or may not, cause disease. Therefore, clinical signs of *S suis*-infection can range from absent to severe, depending on virulence factors of the pathogen, co-infections, or various environmental triggers. When disease occurs, *S suis* is most often associated with meningitis, but arthritis, abortions, or even sudden death can result from *S suis*-infection.[1] This pathogen is of major concern to the global swine industry for two main reasons: 1.) *S suis* is zoonotic in nature and; 2.) it is extremely difficult to control using traditional disease control strategies. The latter characteristic is due, in part, to the high genetic diversity of the pathogen. Multiple serotypes of *S suis* exist, including multiple strains within a serotype, contributing to the difficulty of developing an efficacious, cross-protective vaccine.[2]

Of the 29 known *S suis* serotypes, serotype 2 has, historically, been of most concern, due to the prevalence, pathogenicity, and zoonotic nature of this serotype.[1] Other serotypes, including serotype 9, have also been increasing in prevalence (and relevance) in a number of countries, particularly within Europe and Asia.[2] In The Netherlands, serotype 9 is currently most often associated with clinical signs of disease at *S suis* positive farms.[3] To better understand the incidence, impact, and potential means of controlling this serotype, a group of researchers, veterinarians, and industry partners joined to form the Dutch *S suis* Public Private Partnership (PPP).

This project will be conducted in two phases, with the aim of better understanding the genetics of the pathogen (Phase I), and the genetics of the host (Phase II). The main objectives of Phases I and II are: 1.) to characterize the genetic population structure of *S suis* serotype 9 isolates; and 2.) to identify the prevalence of virulent serotype 9 in The Netherlands, respectively. An additional objective of Phase II (for Topigs Norsvin) is to quantify the extent of naturally occurring genetic variation in *S suis* colonization. Results addressing the latter objective will be used to determine the potential of breeding pigs for improved host response to *S suis* challenge.

Materials and methods

Phase I

Virulent isolates available for this study included isolates obtained from a previous trial ($n = 28$), isolates provided by the GD Animal Health Center ($n = 32$) or consortium partners ($n = 16$), and the reference isolate (8067). Carrier isolates ($n = 47$) were derived from samples collected at 6 unrelated "healthy" Dutch farms. "Healthy farms" were defined as farms that did not use feed medication, or practice autovaccination, and were free of meningitis, arthritis, or sudden death for a minimum of one year prior to sample collection. At these farms, tonsil swabs were collected from both sows and piglets and analyzed using a *S suis* 9-specific polymerase chain reaction (PCR) test.

As a proof-of-concept, one virulent isolate and 2 randomly-selected carrier isolates were selected to perform an experimental infection trial. One of these 3 isolates, or the reference isolate, was used to inoculate caesarean-derived colostrum-deprived piglets at 6 weeks of age and followed for the next 8 days. Results of this trial showed that piglets inoculated with a virulent isolate (including the reference isolate) had significantly greater colonization of *S suis*, more clinical signs of disease, and higher mortality rate, than pigs inoculated with a carrier isolate. These results confirmed the pathogenic and non-pathogenic nature of the virulent vs. carrier isolates used for subsequent analyses.

Next, all virulent ($n = 77$) and carrier ($n = 47$) isolates derived from diseased and healthy animals, respectively, were submitted for whole genome sequencing to: 1.) assess differences in viral genome sequence for virulent vs. carrier isolates; and 2.) to determine population structure. The latter objective was addressed by performing a Bayesian Analysis of Population Structure (BAPS), where isolates with at least 95% homology at the amino acid level were assigned to the same BAPS group.

Phase II

To investigate the extent of naturally occurring genetic variation in response to *S suis* colonization, 100 tonsil swabs will be collected per farm from Dutch multipliers containing full-program Topigs Norsvin genetics. Two tonsil swabs will be collected per individual, to facilitate genotyping of the pathogen and genotyping of the host. Genotyping of the host DNA will be performed using a custom 25K single nucleotide polymorphism genotyping panel and resulting genotypes used to construct a genomic relationship matrix, to account for relatedness between individuals. Variance components and heritability estimates for host response to *S suis*-infection will be derived using ASReml.

Results and discussion

Phase I

Results of the BAPS exercise showed that multiple, unique BAPS clusters were identified. Of the 77 virulent isolates used for this study, 73 isolates clustered in Group 2, 4 clustered in Group 4, and 1 clustered in Group 8. Carrier isolates tended to be more diverse, clustering in 9 distinct groups. Targeted analysis of the capsule locus revealed that, in general, virulent isolates clustered together, as well as with the reference strain (8067), which were genetically distinct from the carrier isolates, which clustered separately.

Further investigation of the capsule region revealed that genes *cpsA* and *cpsK* were most discriminatory between virulent vs. carrier isolates. These findings prompted a follow-up study, designed with the aim of developing a quantitative PCR (qPCR) assay capable of discriminating virulent from carrier isolates based on *cpsK* genotype. A qPCR test was successfully developed and validated using a subset of isolates evaluated during Phase I. Results of the validation exercise show that 43/44 virulent isolates and 4/34 carrier isolates were recognized as virulent. Therefore, these results suggest that the *cpsK* qPCR test has potential to be used a diagnostic tool to discriminate virulent from non-virulent *S suis* isolates identified in the field.

Phase II

Phase II was initiated in the Fall of 2021 and began with recruiting Dutch multiplication farms to participate in the study. Sample collection is expected to commence in the summer of 2022.

In conclusion, *S suis* is a major threat to the global swine industry. The Dutch *S suis* PPP was formed to better understand the genetic population structure of *S suis* type 9 isolates, including the potential of distinguishing virulent from non-virulent isolates based on genomic data. Results from Phase I show that multiple isolates of *S suis* type 9 exist, and that virulent isolates are genetically distinct from non-virulent isolates. This finding led to the development of a qPCR test capable of discriminating virulent from non-virulent *S suis* type 9 isolates. This test can be used to make more informed decisions regarding pig movement, including the ability to establish pig flows free of virulent isolates. Findings from Phase I also suggest that the severity of *S suis*-infection is due, in part, to the genetic composition of the pathogen. We hypothesize that severity of *S suis*-infection is also influenced by the genetic composition of the host. This hypothesis will be addressed during Phase II. As pressure to reduce (or even eliminate) antibiotic usage increases, alternative disease control strategies, such as eliminating pathogenic *S suis* isolates from commercial pig flows and breeding pigs for enhanced, natural robustness to *S suis* challenge, is of increasing importance for the global swine industry.

References

1. Obradovic MR et al. Review of the speculative role of co-infection in *Streptococcus suis*-associated diseases in pigs. *Veterinary Research.* 2021;52:49.

2. Segura M. *Streptococcus suis* Research: Progress and Challenges. *Pathogens.* 2020;9:707.

3. Willemse, N et al. Clonal expansion of a virulent *Streptococcus suis* serotype 9 lineage distinguishable from carriage subpopulations. *Scientific Reports.* 2019;9:15429.

https://doi.org/10.54846/am2022/154

DFM Pak®, a novel combination of *Bacillus* strains reduces environmental *E coli* bioburden, manure buildup, and odor compounds in the swine housing environment

J. Lee[1], MS; E. Galbraith[2], MS; G. Krahn[1], PhD; J. Spencer[1], PhD; S. Son[2], MBA; M. King[2], PhD
[1]United Animal Health, Sheridan, Indiana; [2]Microbial Discovery Group, Franklin, Wisconsin

Introduction

Direct-fed microbials (DFM) have been documented to influence manure characteristics when fed to pigs.[1] However, previous studies fail to reflect the modern manure composition caused by high amounts of corn fiber in contemporary diets. *Bacillus* are found in environments all over earth, exhibiting a high degree of genetic diversity between species and strains. This diversity results in a wide range of unique, functional capabilities amongst strains. The selection strategy for *Bacillus* strains in DFM Pak® (United Animal Health Inc., Sheridan, IN) involved screening for utilization of corn fiber and long chain fatty acids found in corn as a substrate for growth, resulting in reduced manure pH, lower NH_3 and H_2S production and reduced fiber.[2] Here, DFM Pak® was evaluated in a controlled, commercial scale swine research facility to evaluate its impact on manure buildup, *E coli* concentrations on slats, fiber degradation of manure, and pit gas emissions in the barn.

Materials and methods

Two trials were conducted in a slatted commercial research facility, custom designed with 48 pens, 16 individual manure pits and an air tight curtain hung in the middle of the barn to create two separate air spaces. Mixed sex grower pigs were individually RFID tagged (LeeO, Spencer, IA), blocked by entry weight, allotted to pens (10 pigs per pen) and assigned one of two dietary treatments : Treatment 1 = Control; Treatment 2 = Control + 0.05% DFM Pak® addition for a 92d feeding period in each trial. Prior to the first trial, all pens were washed with typical chemical sanitation procedures, however, pit liquid and solid contents were removed after pen washing and each pit was refilled with 12 in. of clean water. Immediately after the first trial, the second trial was conducted. Prior to the second trial, pens were washed with water only, liquids were drained from the pit, and pit sludge from the first trial remained during the second trial. *E coli* adhesin and toxin gene detections and quantifications were measured from slats at a consistent location within each pen before and after washing by swabbing inside a 5 cm^2 template with a sterile foam swab and transferring to an FTA card for stabilization of DNA prior to quantitative PCR analysis. Manure buildup was measured by hand scraping and weighing the total manure contents accumulated in a consistent 6in × 13in rectangular area in the center of each pen. Pit sludge samples were collected from each pit after liquids were drained and analyzed for viscosity and fiber concentrations. H_2S, NH_3, CO_2 and CH_4 emissions were measured using a Dräger X-am 5600 gas meter inside the barn at 60 second increments, while each pit was individually drained into a common holding tank for 35 minutes. Growth performance was measured using weekly individual pig weights.

Results and discussion

In pens where pigs were fed DFM Pak®, slats harbored numerically lower total *E coli*, and fewer *E coli* virulence genes prior to washing, when compared to Control pens in Trial 1. Similarly, in trial 2, DFM Pak® pens tended to have reduced total *E coli* ($P = 0.11$), fewer *E coli* virulence genes and lower quantities of *eaeA* ($P < 0.05$) and *EAST1* ($P = 0.06$) prior to pen washing (Figure 1A), indicating a lower *E coli* bioburden in the pen environment of pigs fed DFM Pak®. Pen slat scrapings in the second trial revealed a 30% reduction in manure buildup *($P < 0.05$)* in pens where pigs were fed DFM Pak® (Figure 1B). Sludge samples from DFM Pak pits had reduced ADF ($P < 0.05$) in Trial 1 and a numerical reduction in NDF in Trial 2. Improved sludge viscosity was observed in both trials. Reductions in NH_3 and CO_2 ($P < 0.05$), a trend for reduced H_2S, and a numerical reduction in CH_4 emissions were recorded from pits where pigs were fed DFM Pak® when compared to emissions from Control pits. Individual pig weights indicated that feeding DFM Pak® reduced the CV in final pig weight by 8.5% and 7.8% in trial 1 and trial 2 respectively. These reductions in CV were found to be from improved growth rates in the lightest pigs starting on trial (bottom 20%) and fed DFM Pak®.

Conclusion

A healthy pen environment is central to the production of high-performing pigs. DFM Pak® *Bacillus* strains were selected for capabilities in utilizing corn fiber and other residual components in modern manure. These results show promise for using DFM Pak® as a tool to create a favorable production environment by reducing slat buildup, pen bioburden and harmful gas emissions which, in turn, can benefit health and pig performance.

References

1. Davis, M.E., T. Parrot, D.C. Brown, B.Z. de Rodas, J.B. Johnson, C.V. Maxwell, and T. Rehberger. (2008). Effect of a Bacillus-based direct fed microbial feed supplement on growth performance and pen cleaning characteristics of growing-finishing pigs. J. Anim. Sci. 86:1459.

2. Lee, J.; E. Galbraith; J. Spencer; S. Son; M. King. (2020). DFM Pak®, a novel combination of *Bacillus* strains selected for corn fiber utilization and biofilm activity, shows beneficial impacts on environmental and storage characteristics of manure. AASV Annual Meeting Poster Number 76.

Figure 1: Pen cleanliness as determined by quantification of *E coli* genes **(A)** and manure buildup **(B)** on slats in pens of pigs fed diets with or without DFM Pak®.

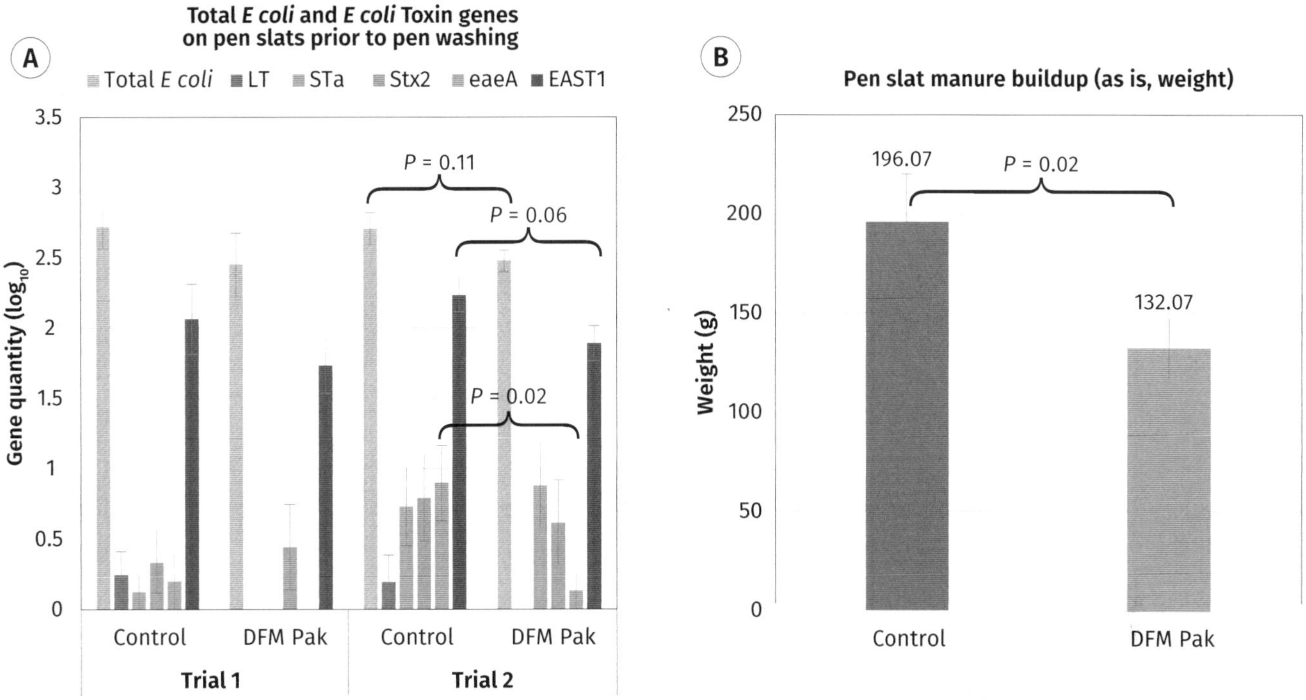

Error bars represent the standard error of the mean (SEM). Brackets indicate significance across treatment groups.

https://doi.org/10.54846/am2022/155

Retrospective analysis of antibody responses to a 2-dose PCV2 vaccination regimen initiated at 3 days or 3 weeks of age

L. Galina-Pantoja, DVM, PhD; M. A. Vasquez Hidalgo, DVM, PhD; K. Vonnahme, PhD; M. A. Mellencamp, PhD; B. Cowles, DVM, MS, MBA
Zoetis, Parsippany, New Jersey

Introduction

Vaccination of very young piglets has potential advantages. For *M hyopneumoniae* (M hyo) and porcine reproductive and respiratory syndrome virus (PRRSV) there are effective, single-dose products that can be given at processing. However, for porcine circovirus 2 (PCV2), early vaccination has proved problematic due to interference from maternally derived antibody (MDA) and immune system immaturity suggested as possible explanations. Fostera® Gold PCV MH may be administered as a single 2mL dose or as two, 1mL doses 3 weeks apart. A previously reported, multi-product comparison included pigs receiving the two-dose protocol starting at either 3 days or 3 weeks of age.[1] Although both showed efficacy, there were significant differences in serological response and viremia reduction. Additional analyses have been performed to look further at these results, particularly with respect to MDA.

Materials and methods

The full study included 880 piglets randomly allocated by litter to 8 groups at 3 days of age (D0) and experimentally challenged with M hyo on D49 and PCV2d on D58. Results from 2 groups are presented: T01 received 1mL of Fostera® Gold PCV MH on D0 and 3 weeks later, T02 received the same regimen starting at D18. All pigs had blood samples taken to measure immunofluorescence assay antibodies (IFA) before their first vaccination (D0 for T01, D17 for T02) and at the time of PCV2 challenge (D58). All were sampled for PCV2 viremia (Ct count) on D77. Seroconversion (pre-vaccination titer above the minimum staying the same or rising) and control of viremia were now assessed in relation to initial titer.

Results and discussion

As expected, given the earlier age at testing, there were more high titer piglets in T01 than T02. T01 also had a higher mean initial titer (LSM 9.36 v 8.05 after log2 transformation). The mean D58 titer, however, was lower (8.27 v 10.79, $P < 0.0001$) due to both fewer pigs seroconverting and an apparently lower response in those that did, although the longer gap between vaccination and sampling could partially explain this. As shown in Table 1, pigs with a high initial titer (2560, 5120) did not seroconvert in either group. Low titer pigs in T02 (40, 80, 160) consistently seroconverted, but in T01 some did not, including 3 which were essentially seronegative (40). Mean D77 PCV2 Ct counts were lower in T01 than T02 (34.28 v 36.13, $P = 0.0001$), and a lower proportion of pigs had no detectable PCV2 (28% v 66% at Ct37). The positive Ct counts (< 37) in both groups were typically high, at a level usually considered of no clinical significance and distributed across the range of D58 IFA titers. In relation to initial titers, both groups appeared to have more positive counts associated with the highest titer. In T01 only, this was also true for the lowest titer (40), with the 3 pigs that failed to seroconvert having Ct counts of 30, 30.9 and 32.4, which were relatively low. The study was not designed to investigate the impact of MDA and the results have limitations. Overall, however, they suggest that high MDA will inhibit seroconversion, although not necessarily efficacy in controlling PCVAD, and suggest an additional factor in at least some very young piglets. Low colostrum intake is a possible confounder but does not explain the between-group difference. Poor serological responses in 3-day old pigs, including lack of priming, have been noted for other antigens[2] and could reflect immaturity, or perhaps the presence of cytokines or corticosteroids in neonates. Starting vaccination at 3 days of age may not be optimal for PCV2 unless early protection is particularly required.

References

1. Galina L et al., 2018, IPVS p121.
2. Lessard M et al., 2002. Can J An Sci p519-526.

Table 1: Number and % of pigs seroconverting to PCV2 vaccination of 3-day (T01) or 3-week (T02) old piglets

	T01			T02		
Initial IFA titer	**No. pigs**	**No. seroconverting**	**% seroconverting**	**No. pigs**	**No. seroconverting**	**% seroconverting**
40	11	8	73	17	17	100
80	4	4	100	10	10	100
160	8	6	75	22	22	100
320	12	10	83	16	13	81
640	21	8	38	10	8	80
1280	15	7	47	6	3	50
2560	11	0	0	6	0	0
5120	19	0	0	8	0	0
Total	101	43	43	95	73	77

Seroconversion was defined as no change or an increase in IFA titer at 3 weeks post-vaccination. IFA = immunofluorescence assay antibodies.

https://doi.org/10.54846/am2022/156

Diversity, equity, and inclusion in veterinary medicine

K. Fred Gingrich II, DVM, Executive Director
American Association of Bovine Practitioners

Introduction

What is diversity, equity and inclusion and why is it important to veterinary medicine? This presentation and paper will explain not only what diversity, equity and inclusion means but also the impact of improving this space in veterinary medicine might have on rural food animal practice. We will also try to identify the barriers that may exist for under-represented groups in entering not only veterinary medicine in general, but the unique challenges of rural food animal practice.

Definitions

There are many definitions of diversity, equity, and inclusion. My definitions of these terms, as they apply to veterinary medicine are as follows:

Diversity

Diversity encompasses all the unique characteristics of an individual that represent the similarities and differences among us. Diversity can include gender, gender identity, sexual orientation, race, religion, and ethnicity. Diversity can include who we identify with and how we define ourselves.

Equity

Equity means that all persons are treated fairly, respectfully, and equally. To achieve equity between a diverse population, we must identify and remove barriers that may not be readily identified to everyone. One of the most difficult obstacles in achieving equity is to step outside of our comfort zone and identify a barrier to a group that may not exist with how we self-identify. As a white middle-aged male, I may not understand a barrier for a black new graduate female. Equity can only be achieved when these barriers and obstacles are identified and eliminated.

Inclusion

Inclusivity means that we actively invite, welcome, and accept everyone to participate in rural food animal practice. The key word in this statement is "active." We must actively invite under-represented groups to participate. We must actively pursue diversity within our organizations and profession. Passive inclusivity indicates that you welcome them to participate but you do not take action to improve participation.

Importance to veterinary medicine

The USDA identifies 221 designated shortage areas in the United States.[1] The American Association of Veterinary Medical Colleges (AAVMC) publishes an annual report of admissions to veterinary medical colleges.[2] This report demonstrates improvement in some areas of diversity in veterinary medical school applicants and enrollments. Total enrollment in colleges of veterinary medicine is 13.952 with approximately 80% of students identifying as female. This is one example of lack of diversity that is often not discussed with males being under-represented. The report also identifies that the majority of students report Caucasian/white as their ethnicity (80%). These reports indicate that veterinary medicine is one of the least diverse professions with the majority of members being white and female. However, the profession has improved ethnic or racial diversity with under-represented minorities increasing from 5% in the mid-1970's to 20% in 2020. The AAVMC also reports that approximately 12% of applicants identify as bisexual or gay/lesbian versus over 80% as heterosexual. Less than 20% of veterinary school applicants are from rural communities.

Veterinary practice owners in rural communities often report that they have difficulty with recruitment and retention of associates. One method of increasing recruitment and retention in rural food animal practice is to be more inclusive to under-represented groups. Actively engaging with under-represented groups and welcoming them to rural food animal practice not only improves the diversity of our profession but opens up the opportunity to improve on the recognized shortage and maldistribution of food animal veterinarians.

Identifying obstacles and barriers

It is often difficult to see the perspective and understand the experiences of those that are "different" than us. Have you ever felt unwelcome or uncomfortable in a certain group setting? Have you ever felt that a veterinary practice or community was not a "safe space" for you? Did you have the opportunity to extern or shadow a veterinarian in your community as a young person which developed your interest in food animal practice? Many potential colleagues simply do not have the same opportunity that many of us who are well represented in veterinary medicine had as we grew up.

A young person from an urban area may not have the opportunity to extern with a food animal veterinarian in a rural community. If we are recruiting from rural communities, we will miss the 80% of veterinary students who are not from rural communities. If our practice is not a safe space for a non-white non-heterosexual veterinarian, they will not feel welcome in your practice. Many of us have difficulty recognizing that language or action can have a negative impact on the recipient that is not intentional. It is important to recognize that your intent may have a negative impact. We must feel comfortable enough to allow feedback so we can correct the impact of our action. Examples can be inappropriate jokes, asking what language someone speaks, asking someone where they are from, or making an assumption about someone based on how they look, act or speak.

Being courageous and speaking out when these situations occur is very difficult. It is paramount for practice owners to support their associates and not reward bad behavior from clients or staff. Remaining silent is rewarding bad behavior. Speaking up by stating that a comment was inappropriate and had a negative impact on the associate is difficult but the only way to create an inclusive environment.

Associations can also create diversity by inviting colleagues from under-represented groups to actively participate in the association. Associations should evaluate if their leadership, speaker panels, and award recipients represent the diversity of the profession and the goals of diversity. Actively pursuing diversity demonstrates to all members that they belong. When members do not see colleagues within an association that do not represent them, they do not feel included.

Rural food animal practice has unique challenges, as well as opportunities, that may not be realized by other facets of the veterinary profession. Access to childcare, job opportunities for partners and spouses, historically non-diverse communities, gender bias, long work hours, and emergency duties all represent challenges that may not be realized in urban environments. There are also opportunities in rural food animal practice for under-represented groups. Many farms have a labor force that is more ethnically diverse, particularly employees from Latinx backgrounds. Improving cultural awareness and language barriers by promoting diversity in our profession can improve the relationship veterinarians have with farm staff.

Conclusion

Veterinary medicine is one of the least diverse professions and there is great opportunity to improve the diversity of our profession which will benefit the profession and rural food animal practice. Actively pursuing historically under-represented groups can improve the shortage situation by opening the door to a wider pool of applicants. Identifying the barriers that under-represented groups have for experiences in rural food animal practice and removing those obstacles can address the recruitment and retention problems in rural food animal practice. Improving our self-awareness to create an inclusive environment in our practice and accepting feedback on the impacts of our actions and words will foster a safe environment for all who want to join our practice and organization. We must continue to recognize that possibility is not the same as opportunity. It is possible for everyone to join us in rural food animal practice, however not everyone has the same opportunity. Identifying these obstacles will create more opportunities for rural food animal practitioners which will foster an environment of diversity, equity and inclusion.

References

1. USDA Veterinary Services Shortage Situations Map. Accessed November 10, 2021. **https://nifa.usda.gov/vmlrp-map**

2. AAVC 2021 Annual Report. Accessed November 10, 2021. **https://www.aavmc.org/wp-content/uploads/2021/05/2021-AAVMC-Annual-Data-Report.pdf**

https://doi.org/10.54846/am2022/157

Diversity, equity, and inclusion: Academia perspective

A. Ramirez, DVM, MPH, PhD, DACVPM
College of Veterinary Medicine, University of Arizona, Oro Valley, Arizona

Introduction

Academic institutions are known for the development and dissemination of knowledge. They also play an essential role in helping prepare all of us for our professional careers. These academic institutions involve a mixture of colleges, departments, sections, or units that encompass great diversity in subject expertise. Within each college, department, section, or unit there are faculty and staff that have a diversity of backgrounds and opinions. Besides this great diversity, these same institutions are filled with thousands of young minds that all have diversity in backgrounds and opinions and are at an age of great exploration and personal formation.[1] You combine all these different diversities and end up with what one may suggest, a great environment for implementing diversity, equity, and inclusion (DE&I) programs.

Discussion

Unfortunately, academic institutions are some of the worst offenders when it comes to DE&I issues. The entire structure and hierarchy at most academic institutions has developed over many decades primarily by those who have had great financial means seeking others just like them. Academic institutions are known for being creative when it comes to research development, but unfortunately are very traditional and "safe" when it comes to on-campus implementation of policies or approaches to education. To better understand this, it is important to understand the structure and hierarchy of these institutions. In a way it is similar to many other institutions, including the military, where individuals must move through the ranks before they can be raised to a position of leadership with great authority. This creates a structure where it is challenging to almost impossible for young minds to climb the ladder fast; which creates an environment slow for change.

The hierarchical structure at academic institutions starts with faculty having academic freedom. Academic freedom provides great leeway for faculty in some of their decision making. These faculty report to a department or unit chair who then reports to a dean of the college. The deans from all colleges then report to a provost who then reports to the president of the university. The university president then reports to a board of trustees or a board of regents. So ultimately, it is the board of regents or board of trustees that make decisions which must be implemented in the academic institution. Definitely a top-down approach to many policies. In many states, decisions made by the board of regents are implemented equally at all state institutions regardless of each institution's uniqueness creating even more challenges for these institutions.

The traditional approach to dealing with DE&I or any other issue at academic institutions is to require special training. Faculty and staff are overburdened with training which year after year continues to increase. For example, with my employment at a new institution, I have had to do over 22 different training modules so far and there are a few more to come. For each of these modules a certificate of completion is required as proof of training. There are 2 very important questions that need to be answered. 1) How effective are these trainings? and 2) Do trainings change behaviors for those who have different perspectives on the topic? This last one is especially critical when discussing DE&I.

A significant challenge is promoting DE&I is that often it is interpreted as focused on race. The true intentions and goals of DE&I applies to all aspects of diversity. The American Association of Veterinary Medical Colleges defines diversity as follows: "Diversity Definition: Increasing diversity in the health professions also requires recognition of many dimensions, including, but not limited to gender, gender identity, sexual orientation, socioeconomic status, cultural background, language, cognitive style, nationality, age, physical abilities, religious beliefs, political beliefs, and other forms of differences, both visible and invisible. In defining diversity, it is also incumbent to acknowledge the concept of intersectionality; no single dimension of diversity exists in isolation. We acknowledge that each individual is a reflection of multiple diversity dimensions."[2]

A second challenge is that often the approach to DE&I in academic institutions will swing all the way from one end to the complete other end. Although the intention is to overcome past approaches to the problem, this abrupt approach creates an environment of distrust and antagonism which is counterproductive. This is especially true in today's environment where topics of Black Lives Matter and COVID-19 masking or vaccinations can be so divisive in our nation. This divisive nature again, is counterproductive to achieving the ultimate goal of harmony and maximum productivity.

A third significant challenge is the inability to speak freely. The previously mentioned divisiveness also fuels an environment where open discussion is not happening. Any discussion that questions or even seems to question DE&I in any way, is immediately interpreted as "racist" and labels individuals as against the cause. This fear of speaking up is keeping many from speaking out and learning. This is especially true the higher you go up the chain of command. Open dialogue is not occurring. It is well established that true dialogue is what is needed to better impact changes in minds.[3]

Academic institutions are working hard to get DE&I right. Unfortunately, talk is easy, and implementation is a challenge. Some Departments, colleges, and universities are doing quite well, while many others continue to talk the talk but fail to act. The topics of DE&I are here to stay. If change is to occur, it must start at the core of the institution and work its way up. Top-down approaches will not be successful in becoming sustainable in the long run.

Disclaimer

Statements provided here are strictly the opinion of the author to generate a dialogue and are not necessarily reflecting the opinions of any academic institution the author has been affiliated with past or present.

References

1. Schultheiss DE and Blustein. J Counsel Psych p248-255.

2. American Association of Veterinary Medical Colleges. Definition of diversity. 2017. Available at: **https://www.aavmc.org/assets/Site_18/files/About_AAVMC/Definition%20of%20Diversity.pdf.** Accessed 10 October 2021.

3. Issacs WN: 1996, Ed Tech p20-30.

https://doi.org/10.54846/am2022/158

Diversity, equity, and inclusion: Student perspective

K. Hewitt, BS
College of Veterinary Medicine, Iowa State University, Ames, Iowa

Disclaimer

Thank you for taking the time to read my proceedings thoughts. I want to preface these thoughts by saying that these thoughts, perspectives and beliefs are my own, I am no expert in diversity, equity and inclusion. Each BIPOC (Black, Indigenous, People of Color) and/or LGBTQIA+ individual has their own story and own unique perspective -the best way allies (supportive to diverse individuals) can help is to listen, acknowledge and amplify diverse voices. The following headings are those that I want to emphasize during my presentation–these are common definitions that need to be understood in order to become a good ally.

Privilege

"Privilege" refers to certain social advantages, benefits, or degrees of prestige and respect that an individual has by virtue of belonging to certain social identity groups. Within American and other Western societies, these privileged social identities—of people who have historically occupied positions of dominance over others—include whites, males, heterosexuals, Christians, and the wealthy, among others.[1] I as a cis-gendered, heterosexual, Asian-American woman have plenty of privilege and I acknowledge that privilege. I have several friends in the LGBTQIA+ community and feel that it's important that I make them feel comfortable in different spaces by announcing my pronouns and calling other people out if incorrect pronouns are used–I have been called out myself for using incorrect pronouns and appreciate the correction and then moved on. As a student, I appreciate when professors and professionals add their pronouns to either name tags or introductory slides, I also appreciate when they recognize their personal perspectives may differ from others.

Microaggressions

Microaggressions are the everyday verbal, nonverbal, and environmental slights, snubs, or insults, whether intentional or unintentional, that communicate hostile, derogatory, or negative messages to target persons based solely upon their marginalized group membership.[2] I remember clearly one incident in an undergraduate class, being asked "Where are you from?" by a professor, I didn't think much about it and responded I was from Minnesota. I clearly gave the wrong answer as the professor then asked "No, where are you *really* from?" I was a little confused and taken aback. From that seemingly simple question, I heard "You are not American, you are a foreigner."[3] Stunned, I responded that I was adopted from China–looking back on the incident I realize that I didn't owe anyone that explanation. I grew up in America and see myself as American, despite my appearance. It doesn't matter where I was born and if I really thought someone needed to know, I would likely tell them.

Welcoming working environment

As a student I am always searching for future employers that see the importance of DEI and it's not just through policies and the one-off speaker, but it's through co-workers and how they are able to express themselves (freely or not). I have had many conversations with others about current events and it is disheartening to hear when others don't understand why Black Lives Matter or why Asian Americans are nervous to be out in public during COVID. It's important not only to celebrate diversity through acknowledgment of diverse holidays at work but also to be cognizant of current events and how they might affect co-workers–a simple, "I know recent events might be hard on you, how are you holding up," shows caring and understanding. Active allyship, by researching current events and effects to systemic racism and sexism are beginning steps for those in power at companies. It's not easy, nor will it be comfortable to confront privilege, without self-education diverse people won't feel comfortable in the industry. Creating a safe space for listening and understanding are just the beginning–actively seeking resources and changing any policies will make big impacts on employees.

Resources

1. Allyship: **https://guidetoallyship.com/**

2. Recognizing Microaggressions and the Messages they Send: **https://academicaffairs.ucsc.edu/events/documents/Microaggressions_Examples_Arial_2014_11_12.pdf**

3. Uncomfortable Conversations with a Black Man (Youtube) **https://www.youtube.com/watch?v=h8jUA7JBkF4**

4. Code Switch (Podcast) **https://www.npr.org/podcasts/510312/codeswitch**

References

1. García, Justin D. 2018. "Privilege (Social Inequality)." *Salem Press Encyclopedia.*

2. Diversity in the Classroom, UCLA Diversity & Faculty Development, 2014.

3. Recognizing Microaggressions and the Messages they send. Adapted from Sue, Derald Wing, Microaggressions in Everyday Life: Race, Gender and Sexual Orientation, Wiley & Sons, 2010.

https://doi.org/10.54846/am2022/159

Diversity, equity, and inclusion: Business perspective

L. M. Tokach, DVM, Dipl. ABVP
Abilene Animal Hospital, P.A. Abilene, Kansas

Introduction

Those of you that know me, know that I grew up in Barnesville, Minnesota. It was a sleepy little town of 1200 people in northwest Minnesota just outside the metropolis of Fargo, North Dakota. I had 63 kids in my 1983 high school graduating class, nearly all white, mostly poor to middle class. In those days, our diversity was limited. We had a few Hispanic children that were the children of migratory workers who had decided to stay up north year around and attend school with us. Some local churches adopted refugee families from Vietnam. We referred to them as "the boat people" because they fled Vietnam by sea. We knew nobody personally that was openly gay, bi, or trans. We saw Black athletes and actors on TV, but didn't actually see any in person except for a few African and Indian missionaries that spoke at our church. We knew a few Native Americans, mostly from surrounding reservations that mostly worked construction. That was about it: by all standards, my life was not very diverse. So, you should be thinking, "Why is she presenting in the general session of AASV on the topic of diversity, equity, and inclusion?" My answer is, the majority of us in this room are coming from a similar position of privilege and power and we need to challenge ourselves to be learners, not knowers. It is hard to admit that we have inherent biases, but we do. I am here to discuss how we try to increase self-awareness and start making better or at least more informed decisions within our practice and on farms.

That was then

Historically, agriculture has been the landing spot for immigrants coming into this country. Many immigrants that have landed in the United States or Canada left their homes for persecution reasons, war, or lack of an ability to make a living in the countries they came from. A large percentage found work in agriculture. Agriculture has been the interface for working relations between workers, land owners, commodity buyers/sellers, and supporting allied industries. It is still that way today. Many of us have probably had more exposure to more diverse populations on farms than anywhere else in our lives. Pig farms are generally a good spot to observe differences in wealth, housing, citizenship, skin color, and education. Differences in sexuality, mental health, physical ability, or gender variance were just not openly discussed back then, but that is changing.

Our own AASV Associate Director, Dr. Sue Schulteis put it best in our personal conversation when she said, "Growing up, I thought that racism, cigarette smoking, and drug use (along with domestic and child abuse, and sex trafficking) would all be things of the past by the time my generation became adults. Now I'm a grandma and dismayed to find out that all of these things are as predominant as ever – and much more complicated than I realized."

This is now

Fast forward to 2022 and whether you like it or not, the struggles with diversity are more prominent than ever. Some of it is good because we are seeing long suppressed and oppressed groups coming forth and being heard, often for the first time. We are speaking out for gender fluidity, sexuality, neurodiversity, and body positivity. However, it is disappointing and distressing that we continue to struggle with racism, income disparities, educational opportunities and political divisiveness.

After 30+ years of human resource management, I have learned the value of having diverse backgrounds within a farm or company. I have also accepted that conflict is inevitable, but prefer to focus on moving on to a better spot whenever possible using open communication as my main tool. Avoid making it us vs. them, whether that is skin color, sexuality, political party, gender, or even COVID vaccinated or not. Nelson Mandela, social rights activist said "When we dehumanize and demonize our opponents, we abandon the possibility of peacefully resolving our differences, and seek to justify violence against them." I agree, we are all in this together and it will all go better if we work together.

Let's spin the Wheel of Power & Privilege

There are many charts, tables, and diagrams available on the internet that depict the hierarchy of power and privilege. I chose the wheel as depicted by artist Sylvia Duckworth (see Figure 1).

The first step to increasing self-awareness is to study this wheel and really reflect on your own biases and then the biases of your team. The most powerful and privileged are in the center of the wheel, while those most vulnerable to our biases are on the outer margins of the wheel in each category.

After my own self- reflection, I am going to start with my self-assessed top 5 biases:

1. Education

We often assume if somebody doesn't finish high school, they won't be reliable on the job. Maybe the high school learning system wasn't ideal for them, maybe they had a child to care for, maybe they were too busy caring for younger siblings, a drug-addicted parent, or trying to figure out where their next meal was coming from.

Scenario: We might assume that training the college intern on day 1 pig care will be easy and training the new, but uneducated worker might be pointless when in reality day 1 pig care ability often has nothing to do with education level. It has to do with nurturing ability and observational skills.

2. Mental health

Having robust mental health myself, I am certainly biased to think, "Just get yourself out of bed and get going!" After assisting many employees with mental health crises, I have learned to be

Figure 1: Wheel of Power/Privilege (Permission granted by artist Sylvia Duckworth).

more patient, be a better listener, and know that some are certainly more vulnerable to the everyday stresses in life and may need your support or the support of a good mental health counselor. Why can't we view mental health more like physical health? You don't tell a diabetic person they are incompetent or unpredictable and blame them for their disorder, so we need to strive to feel the same way about variations in the brain.

Scenario: A manager asked an employee who was suffering from depression, "What do you have to be depressed about? You have a good paying job and food on the table, there is nothing to be depressed about." We need to be more empathetic about mental health issues and offer help the same as we would for a physical health issue.

3. Language

Again, my bias is "If I were living in another country that English was not the primary language, I would learn to be fluent in the native language." I am biased against Hispanic employees who have lived and worked in the US for 10 years, yet still can't master the English language. Then two years ago, I decided to try in earnest to learn to speak Spanish. The upside of this was the delight of many Hispanic farm workers I have regular contact with who take great pride in helping me learn, and great pleasure in correcting my rudimentary skills in español. I continue to support language learning in any way that I can, but I am much more patient as I struggle through verb tenses and the masculinity or femininity of objects while attempting to be conversational in Spanish.

Scenario: Admit it, we have all done it, if someone with English as a second language doesn't understand you, just speak louder! When I have gone to local cafes with a visiting veterinarian from another country, the waitress has asked me, "And what will he have?" rather than allowing them to speak for themselves. Instead of belittling their intelligence, try to value their remarkable background experiences and laugh together about language learning.

4. Wealth

I worked my way through vet school on my own. My two sisters and I each worked our way out of lower middle class to upper middle class after my parents went from poor to lower middle class. My perspective on wealth is biased to my own myopic view that everyone can improve their wealth situation if they just work hard enough. Although this was true for my sisters and me and may have been true for a lot of you in the audience today, some have much bigger obstacles to overcome and don't always get the same opportunities.

Scenario: A manager expressed his frustrations to me that all these lower-class people know how to do is file for their welfare check, they don't know how to fill out their tax forms, renew their driver's licenses, or even make a doctor's appointment, they just show up at the emergency room. Perhaps they haven't been taught? And perhaps you should check your biases at the door?

5. Body size

I am ashamed to admit that I carry this bias. I have always been a relatively fit person. I try to eat healthy, but am prone to just grabbing food from the local convenience store because I am too busy. My bias is thinking that everyone is just like me and if you are grossly overweight, it just because you make bad food and exercise choices. There are so many other factors: genetics, stress, mental health, hormonal imbalances, food availability to name a few.

Scenario: When giving logoed clothing as a gift, we have posted a sign-up sheet for people to pick the color and size of clothing they want. Imagine being the female posting 2XL for all to see. On the farm, we often assume that the large body-sized person will not be able to physically handle the work so we are hesitant to hire them, or we don't even have clothing available for them to shower into the facility. Stop the body shaming.

More biases from the wheel of power/privilege

Racism

I remember when I was in grade school and my teacher asked what color an apple could be. We all shouted, "Red, yellow, green!" She had apples in several colors and cut them all open and said "They are all the same on the inside, just like people. It doesn't matter what color they are on the outside." That simple experiment has always stuck with me. I'm sure I have an inherent bias regarding skin color, we all do. I am not claiming to be color blind; color matters, I just don't want it to be a factor for biases against someone who is not like me. In the book, *A Brief History of Everyone Who Ever Lived,* Adam Rutherford and Siddhartha Mukherjee argue that modern DNA testing shows there are no essential genetic elements for any particular "race" from a scientific point of view. Yet this is one of the biggest, loudest arguments for power and privilege bias.

Scenario: In small town Kansas, we are still not that diverse. A Black man will stand out. One employee relayed that within the short time they have lived here, her husband has had his bags searched as he left Wal-Mart and has had another couple asked to be seated elsewhere in a restaurant to apparently not be seated next to him. I have shopped that Wal-Mart for years, as has my husband and we have *never* had our bags searched. We can do better than this.

Citizenship

Sometimes non-citizen workers may feel hesitant to complain about any discrepancies in their employment because they are often here on a temporary basis and don't want to risk deportment or non-renewal of their work status.

Scenario: It is more difficult to compare apples to apples when assessing an employment package for a TN worker that includes housing, transportation and legal fees to one of a citizen employee who does not have all those additional expenses, but it needs to be fair.

Gender

Ever since I can remember, women have strived for equal pay for equal work and opportunities to be promoted at the same rate as their male counterparts. Desires to also raise children and be present for them have stirred up controversy for decades. Fortunately for me, I entered vet school at a time when the gender ratio was 50:50 and landed in a practice with a wonderful mentor, Dr. Steve Henry, who is the least biased person I know. For me, the swine clients welcomed female veterinarians with open arms. We've always known swine producers to be early adapters of change, so this was no big surprise.

Scenario: A production manager is speaking to the female herd manager and her male finishing assistant. The topic is proper ventilation. The production manager speaks in plain language to the male assistant, but continues to repeat himself in simpler terms to the female manager, basically mansplaining to her how ventilation works. Be sure women have the same opportunities to become farm managers and earn equal pay. Also don't assume women can't understand ventilation, electrical, welding or plumbing. Given the same opportunities to learn, I find women to be equally competent.

Sexuality

As I mentioned earlier, I didn't think I knew any gay, lesbian, bisexual, pansexual, or asexual people growing up because we just didn't talk about it. Although we have come a long way, we still have a lot of biases out there. We have gay and lesbian employees that are very open, others that we know about, but are very private, and I'm sure some we aren't aware of. Sexuality is not a factor in whether I hire or how I view the performance of an employee, but once they are here, I want them to feel non-threatened.

Scenario: This is a story that was shared with me. A gay farm manager was asked not to display the framed picture of him and his partner with their arms around each other because it made other people uncomfortable. Meanwhile, his female office mate has a wedding engagement picture on her desk of her and her fiancée kissing. Love is love; fair is fair.

Neurodiversity

Again, these were things that were either not previously diagnosed or at least not talked about. Now we routinely talk about ADHD (attention-deficit/hyperactivity disorder), autism, OCD (obsessive-compulsive disorder), dyslexia, and more. I am not going to pretend to be an expert in neurodivergence, but it can sometimes be a distraction from work, but it can also be an asset. The key is trying to help the person get a proper diagnosis, therapy, and medication if needed. The other key is supporting the idea that there is no "right" way or "normal" way for your brain to work and sometimes "thinking outside the box" is exactly what is needed.

Scenario: My youngest daughter is diagnosed with ADHD and generalized anxiety disorder. She regularly takes medication to help with her ability to focus and stay calm. She gets annoyed when people belittle her diagnosis as if it is not real or say things like "You don't seem crazy" or "Oh yeah, I was having a panic attack yesterday" when they were actually just nervous about something. She has experienced crippling panic attacks and they are nowhere near the same thing. Listen and learn.

Housing

How do you even function without a roof over your head? I have lived in some less than desirable rental situations, but never in my life have I actually been homeless. Certainly, this would hamper your ability to focus on a job.

Scenario: An employee got kicked out of their rental home on short notice because the landlord sold the property. The employee had to move their family quickly during the winter. The next rental property was supposed to be temporary and was less than desirable. While the family was away on the weekend, a water pipe burst and flooded parts of the house. They did not have renter's insurance, so things like the computer, TV, and food were a complete loss. Not saying there was a bias here, just illustrating that life is rough sometimes.

Ability

Obviously some parts of our world would be difficult for somebody less than able bodied to work in. I don't see working in pig farms to be very accessible for someone in a wheelchair or a blind person. However, there are increasing ways that the work force has tried to open up for those with certain disabilities.

Scenario: I once met a man who was deaf that was a breeding manager in a sow farm. He was one of the most observant gilt heat detectors I had ever seen. What he lacked in hearing, he more than made up for with his visual capabilities. In this day and age of labor shortages, we need to be open to all options.

Even more biases not included on the wheel because the bias depends on the situation

Ageism

This is commonly seen in the multigenerational work force. The younger are biased against the older because they are slower to pick up on electronic technology. The older are biased against the younger because they don't think they have the same work ethic as their generation did.

Scenario: An older employee sent me a photo of the breast pumping equipment left in the breakroom dish drying rack asking if people couldn't be more private about such things. All millennials are not lazy, just as all boomers are not technologically challenged.

Political persuasion

Here is a loaded bias depending on where you live and what your political persuasion might be. I am not interested in debating politics today, but I am interested in how we can have open on-going discussion in our daily lives without getting heated about it. I find listening to non-US citizens tell me what they think of our government and our political candidates to be a fascinating viewpoint. I also want to foster a work environment where discussion is welcome and The First Amendment is upheld, but political warfare does not ensue.

Scenario: What do you do when one employee shows up to work in a pickup with a confederate flag sticker across the back window, another has a bumper sticker that says F*%k Trump!, and a third with a shirt that says "Let's Go Brandon!"? Can we not debate issues more intelligently and more peacefully?

Religion

Many fled to this country to escape religious persecution with the desire to be able to worship however they choose, but yet here we are centuries later, still arguing about it.

Scenario: One employee is offended that another one put out an e-mail about the Xmas party because "Xmas takes Christ out of Christmas." Another farm has dueling points of view where one young lady has a travel mug with a sticker "My body, my choice" while other employees chastise her because she is Hispanic and therefore should be a "good Catholic girl" who must obviously be against abortion. Freedom of religion doesn't mean freedom to preach to everyone else about what you believe in.

Cleaning your own house

No doubt that keeping on track and avoiding the pitfalls of biases of power and privilege is a tricky labyrinth of many pathways. I am certain there is no single "right" way. I assure you that ignoring it, however, is the "wrong" way. Start making steps in the right direction.

Listen

The inherent problem of power and privilege biases is that the underprivileged do not feel heard. The first, and perhaps most important, step is to stop talking and start listening. If they are not talking, ask leading questions and truly listen to what your employees have to say. Do you ever feel like you are being discriminated against? What are the things you love the most about your job? What are the things you hate about your job? If you could change one thing about your job, what would it be?

Promote a safe space

Rather than allow inter-employee gossip to run the show, make it a rule that you shouldn't say anything about anyone that you wouldn't say to their face. Then live it. Let employees know that you intend to help them be problem solvers. Encourage them to work out their own problems, but let them know that your door is always open if they fail, feel uncomfortable/intimidated, or just need your advice.

Educate yourself

Read about DEI, attend CE on DEI, look at other organization's DEI policies, but don't copy them verbatim. You know your team's culture better than anyone, so you should customize your policies to fit your team.

Let others teach (empower the people)

If someone is doing a good job, allow them to teach. For example, if you have a receptionist that is very good at diffusing difficult clients, have them teach the team how to diffuse in a power/privilege situation. We have run some of Iowa State's DEI videos at our staff meetings if we think they apply. No sense in re-inventing the wheel. Celebrate all people. If your Latin employees prefer Latinx, then go with it. Some that I have asked, hate Latinx because they feel like it is trying to force their culture away from masculine and feminine words which they feel is part of the beauty of the culture.

Adopt a see it, stop it culture

Macro and micro aggressions will happen. Just like we do with animal welfare issues, we need to do with equity issues. If someone on your team uses a racial slur, politely correct them the first time. Issue a written warning the second time. Discuss macro and microaggressions openly. Some will disagree. For example, should you ban the use of the N word or the R word. I would certainly think so. Should you ban the use of such phrases as "to Jew down" because it is generally thought to be anti-Semitic? Probably so. Should you ban the use of the phrase "rule of thumb"? Its origins supposedly date back to the 18th century when a man was permitted to beat his wife provided, he uses a stick no thicker than his thumb. Or the phrase "master bedroom" as per its references to days of slavery. These are hotly debated and you will have to set your own standards.

Ask yourself, "What are you afraid of?" when it comes to diversity, equity, and inclusion

For me it is two things. First there is always the fear of the unknown. If I am not familiar with autism, I may be hesitant to hire a neurodiverse person because of my fear of dealing with what I might think of as aberrant behaviors even though the person could be well qualified for the job at hand. Secondly, there is the fear that I will be perceived as unfair. I try hard to be fair and be a good listener, but what happens when I am accused of being biased?

Scenario: What if I choose a white male employee to be promoted over a Hispanic female employee when time worked and work ethics are equal, but the male employee speaks English, communicates more effectively, and knows how to enter data into a computer and run farm reports? Am I biased or not? Have I spoken with the Hispanic worker about my expectations, given her opportunities to learn the skills, and listened to her opinion on the matter? Does my decision have anything to do with the wheel of power and privilege?

Recognize that being fair and inclusive is hard

Your employees should feel comfortable calling you out on your biases. Being called out is not a bad thing, it doesn't make you "an evil person". It makes you human and gives you the opportunity to grow and improve. You will make mistakes. Apologize, truly. This means not saying, "I know I did this, but..." It means truly recognizing the error, apologizing sincerely, and striving to fix the situation or at least do better next time

Book recommendations

1. *A Brief History of Everyone Who Has Ever Lived: The Human Story Retold Through Our* Genes by Adam Rutherford and Siddhartha Mukherjee
2. *The Coddling of the American Mind: How Good Intentions and Bad Ideas Are Setting Up A Generation For Failure* by Greg Lukianoff and Jonathan Haidt
3. *The Loudest Duck: Moving Beyond Diversity While Embracing Differences to Achieve Success at Work* by Laura Liswood
4. *We Can't Talk About That At Work!: How to Talk About Race, Religion, Politics, and Other Polarizing Topics* by Mary-Frances Winters

Summary

Clean your own house: promote inclusion, make learning expected, make contributing at all levels normal, make it safe to challenge the status quo. Do this by listening, educating yourself, empowering the people, and an adequate amount of self-reflection. As Muhammad Ali once said, "A man who views the world the same at 50 as he did at 20 has wasted 30 years of his life." You aren't perfect, but neither is anyone else. Start in earnest by striving to become a learner, not a knower.

https://doi.org/10.54846/am2022/160

A practitioner's perspective of PRRSV in sow herds: What has worked, what has failed and what we still have to learn?

P. Yeske, DVM, MS
Swine Vet Center St. Peter Minnesota

PRRS continues to be a challenge for the US swine industry, costing the industry from $580.68 – $663.91/sow.[1,2] With approximately 24% incidence of PRRS among the sow farms being monitored in Morrison Swine Health Monitoring Project (MSHMP) database, the number of category 1 herds is increasing as the number of category farms declines.[3] Over the last year (July 2020 to June 2021), within the MSHMP database, there was the typical Fall epidemic curve followed by a Spring epidemic curve that rivaled 3 out of the 4 previous fall PRRSv epidemics. Laboratory data shown as part of the Swine Disease Reporting System (SDRS) verified from a diagnostic perspective.[4]

Observations

Unfortunately, new PRRSV cases within sow farms continue to occur. The classical approach to limiting new virus introductions has been working to solidify herd biosecurity practices, particularly bio-exclusion to try and avoid having an outbreak. One of the best biosecurity measures has been locating sow herds in less pig dense areas. Sow herds located in increasingly pig dense areas break at a greater frequency which has been associated with the number of surrounding finishing populations.[5] Examples of this are new builds that continue to be located further north and west to avoid high density geographies.

One unnerving experience with the PRRSV 1-4-4 variant 1C was the number of sow herds in historically good areas that became infected, even in the face of common biosecurity practices that had been in place in these farms for years.

The other biosecurity practice that we've seen expand is the addition of air filtration (positive or negative pressure) to farms located in areas of high pig density. Filtration has been shown over time to successfully reduce the frequency of PRRSV introductions into herds, including herds that had a history of routine PRRSV introduction.[6-8]

Air filtration technology and the process of remodeling barns for filtration has evolved over time, with dramatic improvements in loadouts, shutter walls and supply entries just to name a few. Reducing the incidence from approximately 50% to 12% has made a difference for these herds.[7,8] The filtration system only works as well as we maintain it, so one of the most important parts of filtered farms is having an ongoing auditing process in place to ensure that everything is working as setup and that protocols are being followed.

Feed grade mitigants are a relatively new addition to our biosecurity repertoire. The introduction and transmission of PEDV showed us that feed can be a risk factor for viral movement into herds. There is still some debate on the best products to use but the formalin-based mitigants tend to be the gold standard. The acids and medium chain fatty acids have shown good control and reduction in of virus present in the feed with a mitigant. Today many sow farms at least seasonally will be fed a diet containing a mitigant to reduce risk of PRRS, PEDV, PDCoV.

Many herds continue to utilize PRRS vaccine in an attempt to reduce the impact of PRRS. Although vaccination may not stop the outbreak allows herds to return to base line production faster.[9] Vaccination continues to be a very useful tool in the wean to finish population and today even more widely adopted. It doesn't stop the outbreak but does reduce the impact and potential shedding of the virus in the area. In pig dense areas have a ring of vaccinated wean to finish barns around the sow farm has seemed to be helpful but not perfect especially with the 1-4-4 L1C variants.

As practitioners we need to continue to complete detailed systematic PRRS outbreak investigations following the introduction of a new PRRS virus to understand where the weaknesses are in in herd PPRS control programs from these outbreaks. Taking the time and effort to understand what may have gone wrong. Then how to build a better program for the future.

The elimination method that has the highest success rate continues to be depopulation and repopulation. As we analyze the impact that 1-4-4 variant 1C has had over the last 12-months, herd depopulation has looked increasing attractive. Typically, would say the losses aren't high enough to justify the cost but some systems have done this to get back on track faster. In one outbreak lost 10,000 pigs per 1000 sows before returning to baseline performance. Could have easily repaid the cost of depopulation when factoring in the cost of wean to finish.

Some systems aborted sows to try and reduce the mortality in late gestation sows and reduce viral loads in farrowing and downstream. In some cases, pigs were euthanized because healthy pigs could be purchased at a very reasonable rate due to the market and allowed better wean to finish performance in the time of higher viremic pigs; may be something to consider depending on market conditions and availability.

The load close and homogenize PRRSV elimination method does work even with the 1-4-4 1C variant. Some of the early herds have responded like we have experienced with non-1-4-4 strains of PRRSV. In some of those herds that have eliminated 1-4-4, the time to negative status has taken longer but virus elimination has been successful.

This appears to be consistent at least so far with the breaks that have been experienced. There are many variations of this procedure depending on the attributes of the farm and production system. Farms with internal gilt developers' units (GDU) that grow pigs up from the nursery have a big advantage in that they are ready to do closure almost any time just need to increase stocking numbers on youngest groups to allow for extended closure. Their biggest challenge is to find available staged animals to repopulate the GDU once the closure is complete. There are other options available to herds, just need to be creative on how to house the animals that are being exposed and negative replacements without adding more additional risk of rebreaking. Offsite breeding projects have been done since there are often more areas to house clean animals. Breeding projects can be done in finishing sites to fill the holes in breeding at the end of the closure, this has the risk that the closure is extended and then there is no place to go with these animals. Need to make sure things are progressing as expected before breeding begins or have an alternative plan for these animals just in case. Ongoing monitoring of the process has become easier and better at detecting low prevalence herds that may be at risk of rebreaking if opened too soon. The use of processing fluids allows for testing all litters and make the process much easier and if pooling samples reduces the cost of monitoring.

Due to wean pigs need to continue to be monitored to be sure that you don't miss infection occurring in the farrowing barn. McRebel and modified McRebel protocols are used to varying degrees depending on the risk that farms are willing to take and how fast they want to open the herd. Each system will have their own criteria for reopening herds. Another reference is the AASV PRRS classification system gives the currently accepted definition to categorize herds.[10] One tool that I have used is a rough guideline of what to expect during the closure, so farms don't get too worried that they are not making progress and that they have milestones to go by to evaluate their progress. Figure 1

Wean to finish health has been a definite challenge with 1-4-4 variant 1C, even as the sow herd progresses to PRRS negative status. Even though pig quality leaving the sow farms improves quickly following a break, the seroconversion in the nursery phase continues to be a challenge until pigs are negative. The finishing stage of infected flows tends to be relatively normal with most of the challenges in nursery phase. Cleaning up these sites is much more straight forward as long as site is running on all-in all-out bases. For continuous sites need to get creative with the flow to get the site empty and break the infection cycle.

Later wean to finish breaks have definitely been more common with this virus and again can have an impact on mortality and ADG. These herds again are easy to cleanup with normal cleaning and disinfection protocols. If multiple age groups on site, should depopulate the site and may have to be creative to do this.

Conclusions

It appears that PRRS 1-4-4 1C variants are acting like other PRRS viruses we have delt with in the past. Like previous strains of virus, the severity of the break depends on existing herd immunity, how quickly the virus moves through the herd and amount of time from outbreak to closure. There will be definite variations as move forward as an industry on the outcomes of closures and will all depend on each herds goals for the closure and herd cleanup plan. Herds that choose depopulation and repopulation will move faster and should have a higher likely hood of success but will also be at risk again to a re-break, so all these factors need to be considered.

References

1. Holtkamp D.J., et.al., 2013 Assessment of the economic impact of porcine reproductive and respiratory syndrome virus on United States pork producers; Journal of Swine Health and production March/ April p 72-84.

2. Miller M., 2016 Annual PRRS Costs Fall $83.3 Million: Pork p38-39.

3. Corso C., 2021 Morrison Swine Health Monitoring Project.

4.Trevisan G., 2021 Swine Disease Reporting System.

5. Angulo J., et.al., 2018 Understanding PRRSV infection dynamics in growing pigs in control and elimination programs, Allen D Leman proceedings p15-16.

6. Dee S. et.al, 2011. A four-year summary of air filtration system efficacy for preventing spread of porcine reproductive and respiratory syndrome and Mycoplasma hyopneumoniae. J Swine Health and Production 19 (5) 292-294.

7. Reicks D. 2014. Filtration ten years and 100 farms later... what have we learned. Allen D leman conference.

8. Feder J. 2020 personal communication.

9. Linhares D. et.al, 2014 Comparison of time to PRRSv stability and production losses between two exposure programs to control PRRSv in sow herds, Preventative Veterinary Medicine 116 91-2): 111-119.

10.Holtkamp D. et.al., 2021 Proposed modification to porcine reproductive and respiratory syndrome virus herd classification, Swine Health and Production p261-270.

Figure 1: PRRS timeline for herds in closure.

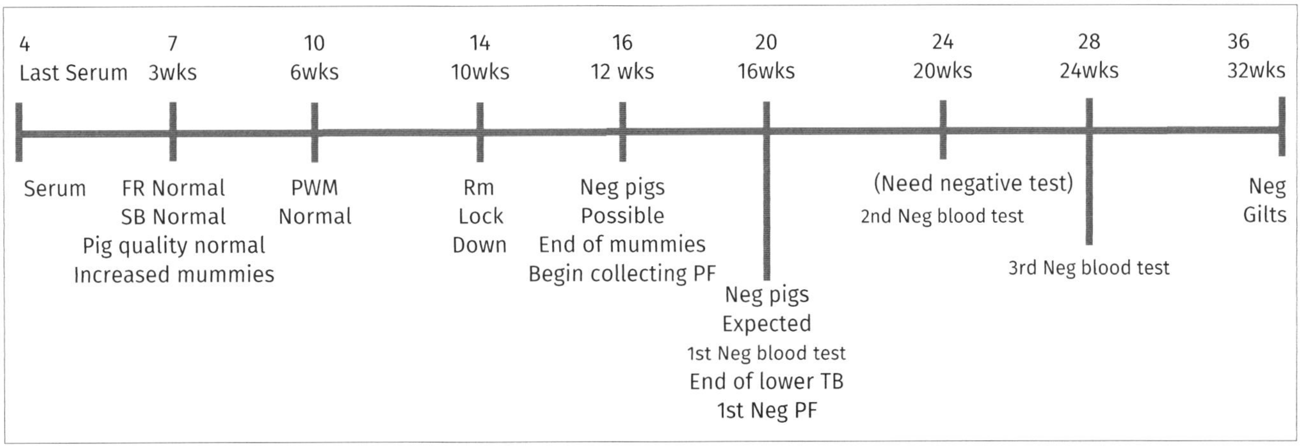

https://doi.org/10.54846/am2022/161

Mitigating the downstream effects of PRRSV and secondary infections

Dyneah M. Classen, DVM
Carthage Veterinary Service, Ltd

Introduction

Porcine reproductive and respiratory virus (PRRSV) outbreaks in wildtype negative or wildtype positive and/or PRRS MLV vaccinated sow herds results in variable effects in the post weaning phase of the downstream flow. The strain variation of a PRRSV influences the secondary infections diagnosed in the downstream flows. This paper will focus on control measures following a new wildtype PRRSV outbreak. The decision on wildtype PRRSV elimination can be made later. The secondary infections can be septicemic bacterial, opportunistic bacterial or viral pathogens. Primary bacterial and viral infections can also increase the effects of PRRSV and secondary bacterial infections.

Discussion

Diseases of Swine 11th edition offers an excellent review of the effects PRRS has on the innate, humoral, and cellular immunity. This paper is not intended to be a review of the immune system during a PRRSV outbreak.

Mitigation of primary bacterial or viral pneumoniae pathogens starts prior to the PRRSV outbreak. Primary swine bacterial respiratory pathogens include *Mycoplasma hyopneumoniae, Actinobacillus pleuropneumoniae, Bordetella bronchiseptica,* and some strains of *Pasteurella multocida*. The primary swine virus that causes pneumonia is Influenza. Eliminating or controlling these primary pathogens will lead to one less co-infection that can lower the post weaning pig's immunity making it more susceptible to those secondary infections. In the event elimination or negative replacement gilts are not available, having a robust gilt acclimation and/or vaccination program in the Gilt Develop Unit (GDU) on or off site are a crucial step in preparation for the inevitable PRRSV outbreak.

PRRS infection increases risk of septicemias and opportunistic bacterial pneumonia.[1,2] Most of the bacteria that cause pneumonia are commensals of the upper respiratory tract or tonsil. Septicemic causes of bacterial pneumonia include *Salmonella choleraesuis, Actinobacillus suis,* and *Arcanobacterium pyogenes*. Opportunistic bacterial respiratory pathogens include *Streptococcus suis, Glaesserella parasuis, Mycoplasma hyorhinis,* and most strains of *Pasteurella multocida*. In addition to bacterial disease, PRRSV also significantly enhances replication of porcine circovirus type 2 (PCV2), resulting in more severe PRRS viral pneumoniae, as well as lesions of PCV2 -AD.[1]

Bacterial and viral pneumonia should be thought of in terms of management + environment + pathogens[3] → mitigation.

Management starts prior to the anticipated wildtype PRRSV outbreak. Focus should be on prevention of wildtype PRRSV outbreaks through biosecurity measures. The Swine Health and Information Center (SHIC) has demonstrated that PRRSV are seasonal and repeatable. Commercial sow herds that are in a high risk area[4] or have a history of wildtype PRRSV outbreaks should consider PRRSV vaccination program. PRRSV MLV vaccines has shown to reduce the losses post outbreak. Killed PRRSV vaccines administered following a known viral exposure whether wildtype or MLV vaccine virus decrease shedding of the PRRSV. There are multiple PRRS MLV vaccines on the market today. History, experience, neighborhood, Area Regional Control Programs (ARC) may dictate which of those PRRS MLV vaccines are utilized in the sow herd. Administration timing of the PRRSV MLV vaccine is important. Using the sow herd's history as well as national databases such as SHIC's national PRRS incidence/prevalence report[5] support the timing of annual, biannual, or quarterly PRRS MLV mass vaccination of the show herd.

Managing the gilt acclimation protocols prior and during the PRRSV outbreak will also prove beneficial in mitigating the secondary infections. Early exposure to *Mycoplasma hyopneumoniae* increases the stability of the downstream flow. Having a robust gilt acclimation program increase sow antibodies for passive transfer and decreases the risk of sows/gilts shedding at the time of farrowing and infecting offspring. Another approach would be to eliminate *Mycoplasma hyopneumoniae* from the sow herd. The likelihood of a lateral transmission post weaning in finishing was reported at 6% versus a PRRSV break at 50%.[8]

Managing influenza infections prior to and during a PRRSV outbreak. Encouraging employees and family members to receive the human influenza vaccine. Implementing a policy with human resources following the Centers of Disease Control (CDC) guidelines to keep individuals' home when experiencing influenza or influenza like symptoms. This has the added benefit of protecting the other co-workers as well as the pigs. Utilizing commercial or autogenous influenza vaccination programs for gilts and sows. Weighing the advantages and disadvantages of each option. In the event, an influenza outbreak occurs and become endemic using an autogenous influenza vaccination program two doses prefarrowing to eliminate the influenza from the suckling pigs. Often the sows are no longer expressing clinical signs; however, the piglets are coughing post weaning. Elimination of influenza will improve the respiratory disease impact of the post weaning pig.

Managing the lactation experience for the piglet. Implementing protocols like management changes to reduce exposure to bacteria to eliminate losses (McREBEL).[6] The McREBEL process is often modified throughout the PRRSV outbreak. However, in an elimination program for PRRSV, a must, to control both the PRRSV and secondary bacterial infection as the acronym states.

Environment starts in the farrowing stall and creep area then extends into the nursery or wean to finish barns. Knowing that pigs are sick requires attention to detail in the ventilation and environment. The farrowing house needs to be set up and adjusted for the health of that sow and piglet. If ever a time to focus on sanitation a PRRSV outbreak is that opportunity. Decreasing both the PRRSV and secondary pathogens in the environment are critical. Knowing a sow farm has a PRRSV infection and communicating that to the piglet destination nursery or wean to finish barns allow that farmer to adequately clean and prepare the barn environment on their end.

Vaccination status of the pigs pre and post weaning and that of the sow herd need to be considered in mitigations of secondary infections. At the outset of an outbreak administration of PCV2 vaccines to sick piglets may not confer protective immunity. The PRRSV infection resulting in high mortality post weaning makes the financial impact of vaccinating pigs that won't go to market prohibitive. Discontinuing the PCV2 at weaning and delaying until the late finishing phase can increase the response to vaccination and decrease the cost. The implementation of this vaccine strategy requires follow up with the wean pig downstream flow. This is a change in protocol and nursery/finishers may forget the PCV2 vaccine is still needed. I don't recommend this protocol if communication between the sow farm and the downstream flow is limited or inadequate.

In personal communication the last six to twelve months have seen an increase in Hemolytic *E coli* disease post weaning. These primary infections will only worsen in the presence of a PRRSV outbreak. Utilizing the commercial *E coli* vaccines in the piglet vaccine protocol will need to continue.

The use of antimicrobials may also be warranted; injectable treatment at the time of weaning, water medications, feed medications (sows and post weaning) and individual treatments post weaning. Tulathromycin and enrofloxacin administered to the piglet at the time of weaning on the sow farm treat the secondary pathogens. Necropsy and collecting diagnostics of clinical pigs will narrow down which secondary pathogens are infecting the pigs in the downstream flow. Using tilmicosin in the lactating sows to decrease sow to piglet transmission of secondary infections. Choosing the correct antimicrobial in the nursery phase in both water and feed.[7] I often remind production managers and farmers, "don't forget the sick pens." The water used to mix the gruel feed for these pens may or may not branch off the water lines after the medicator. If the water pipes for the supplement water branch off prior to the water medicator, the most vulnerable pig will not receive the water antimicrobial included in the water medicator. Feed medications should be selected with similar consideration as water medications. Understand the secondary infections present post weaning. Pigs in the general population may not warrant pulling to the sick pen, however, may be struggling to consume a therapeutic dose of water (or feed medication) walking the pens and utilizing individual treatments.

Conclusion

"PRRS sucks." Dr. Loula.[9]

PRRSV outbreaks require interventions to mitigate the effects of secondary infections. Migration strategies need to start before the PRRSV outbreak and continue in multiple phases of production.

References

1. Zimmerman, Jeffrey J.; Karriker, Locke A.; Ramirez, Alejandro; Schwartz, Kent J.; Stevenson, Gregory W.; and Zhang, Jianqiang,(eds), "Diseases of Swine" 11th edition (2019). *Wiley-Blackwell, Ames, IA, USA.* 685-702.

2. Neumann, E.J., Ramirez, A. and Schwartz, K.J. (eds.), 2009. *Swine diseases manual. 4th edition. American Association of Swine Veterinarians, Perry IA, USA.*

3. Halbur, P.G. Top 10 rules for controlling porcine reproductive and respiratory disease complex. George A Young Swine Conference, 2000. 155-183.

4. Holtkamp D.J. 2009. Herd risk assessment: PADRAP. In: Proc. Pre-conference seminar, Allen D. Leman Swine Conference. St. Paul, Minnesota. September 2009. 43-55.

5. Swine Health Information Center. 2021. **https://www.swinehealth.org/disease-monitoring-reports/**

6. McCaw M.B. Effect of reducing crossfostering at birth on piglet mortality and performance during an acute outbreak of porcine reproductive and respiratory syndrome. *Swine Health Production.* 2000;8(1): 15-21.

7. Carroll K et. Al. Comparison of tulathromycin vs. tilmicosin at placement in porcine reproductive and respiratory syndrome virus positive pigs. AASV Poster Sessions. 2020.

8. Yeske, Paul. Assessment of the likelihood of rate of *Mycoplasma hyopneumoniae* lateral transmission. Allen D Leman Conference. St. Paul, Minnesota. September 2017. Presentation.

9. Loula, Tim. 2013. PRRS ARC and Beyond: Applying the Principles of PRRS Control, Maximizing Immunity - Minimizing Transmission, What Are We Learning? 2013 North American PRRS Symposium. Chicago, IL. December 7-8, 2013.

https://doi.org/10.54846/am2022/162

Managing PRRSV testing and immunity for gilt development

Kate Dion, DVM
The HANOR Company, Webster City, Iowa

Monitoring gilt populations for porcine reproductive and respiratory syndrome virus (PRRSV) is an important tool to prevent the introduction of PRRSV into sow herds and to understand the PRRSV immunity status of the gilt pool. The goal of a gilt PRRSV testing program may focus on antigen detection to determine freedom or exposure to PRRSV or detect intentional exposure to wild-type strains or modified live vaccine (MLV) virus. Alternatively, the testing program may determine freedom or exposure to PRRSV via antibody detection. Other goals may include determining if exposed animals have "cooled" down and are no longer shedding PRRSV or a combination of some or all the listed goals.

There is no one size fits all testing program for PRRSV. Ultimately, the monitoring program should be developed to provide a level of confidence the producer can accept at an expected accuracy and be delivered in a time frame that management decisions can still be made if needed. There are many considerations when building a program that includes but are not limited to:

- Sample type-Many sample types have been described to detect PRRSV. Some are individual animal tests, while others test populations. The technical skill needed to obtain samples may vary, so what will work best for the desired detection and the skill level of the person sampling the gilts?
- Sample frequency-Are animals tested on arrival, prior to departure, or routinely monitored throughout the growing phase? Samples repeated over time may give more confidence to determine freedom from disease given limitations in sample numbers and assay sensitivity/specificity.
- Number of samples-The number of samples required to detect one or more positives in each population size at an estimated prevalence with an expected confidence level has previously been described.[1] It is important to note that these sample numbers assume positive animals are randomly distributed in a herd, and that assumption may not be true for a population of gilts.
- Location of the samples-The location of where samples are taken can impact detection, and for population samples like oral fluids, fixed spatial sampling in multiple barns has been shown to provide better detection than random sampling and only sampling one barn on a farm.[2]
- Individual animal status vs. population status-Does individual animal status matter, or is the monitoring program's goal to understand the population's PRRSV status?
- All-in, all-out vs. continuous flow-The frequency of shipments and deliveries should be considered when determining how often to test. Does every group need to be tested in and out, or can routine monitoring provide confidence in PRRSV status? Are deliveries and shipments overlapping where you may have to accept that animals leave before animals delivered have tested negative?
- Negative vs. vaccinated vs. intentionally exposed-Are you trying to detect antigen, antibody, or both? Freedom from disease or confirmation of exposure or vaccination? Lack of shedding?
- Source history/farm history-Farms with a long history of lack of PRRSV introductions may need less intensive monitoring to have confidence in diagnostic detection, although historical freedom from PRRSV should not solely be relied on for a given group of growing gilts.
- Area risk of PRRSV introduction-Farms located in geographic areas that are pig dense or have nearby pig traffic may require a more intensive monitoring program to increase confidence to detect lateral PRRSV introduction.
- Assay selection-Assays may differ in diagnostic sensitivity and specificity, so assay selection and results must be interpreted in the context of an assay's diagnostic sensitivity and specificity to understand the potential for false-negative or false-positive results.
- Cost-What are you willing to spend?

When purchasing gilts, it should be an expectation that the source farm is tested prior to departure to understand the PRRSV status prior to leaving the farm at a detection rate the receiving operation is comfortable with.

Testing after arrival may determine if animals were infected during transportation. PRRSV viremia can be detected 6-48 hours after exposure and antibody at seven days,[3] so sampling time points should be determined in the context of when exposure occurred, but low prevalence scenarios may be hard to detect immediately after expected initial viremia or antibody response so repeated testing may be beneficial for detection.

Time points for sample collection prior to shipment must be done with enough time to receive test results and retest if unexpected results are achieved. This time lag may leave a few days where the animals are still susceptible to PRRSV introduction and untested prior to the movement that needs to be accepted.

Gilt populations that are routinely monitored may pick time points of significance related to the detection goals. After arrival, prior to departure, prior to vaccination/exposure, after vaccination/exposure, and expected time points when lack of shedding should be achieved may be likely time points to consider. All groups of gilts may be tested at these time points or on a routine basis, depending on the program's confidence required.

A plan for unexpected results is necessary for gilt populations as management decisions may need to be made quickly. Alternative assays for that sample type can be ordered on the samples already collected (remember limitations in interpretation due to assay diagnostic sensitivity and specificity.) More importantly, additional samples should be taken from the same locations, and/or a larger sample size should be obtained to either confirm or refute prior test results. ORF-5 viral sequencing or whole-genome sequencing may be utilized to determine if the strain is a wild-type virus introduction or the intended strain in exposed animals. In the case of MLV vaccinated animals, RT-PCR clamping assays may be utilized to distinguish wild-type or vaccine-like viruses.[4]

References

1. Chase C., Polson D. Sampling the Herd: When is Enough? In: *Proc of the 31st AASP Annual Meeting*. American Association of Swine Practitioners, 2000: 465-477.

2. Rotolo ML, Sun Y, Wang C, Giménez-Lirola L, Baum DH, Gauger PC, Harmon KM, Hoogland M, Main R, Zimmerman JJ. Sampling guidelines for oral fluid-based surveys of group-housed animals. *Vet Microbiol*. 2017 Sep; 209:20-29. **doi: 10.1016/j.vetmic.2017.02.004**. Epub 2017 Feb 17. PMID: 28284415.

3. Zimmerman JJ, Karriker LA, Ramirez A, Schwartz KJ, Stevenson GW, Zhang J eds. Diseases of Swine. 11th ed. Wiley-Blackwell; 2019.

4. Harmon K, Bradner L, Bieber M, Gauger P, Zhang J. Put a CLAMP on it! PCR-based strategy to selectively sequence wild-type PRRSV in vaccinated herds. *Proc of the 50th AASV Annual Meeting*. American Association of Swine Veterinarians; 2019:50-52.

https://doi.org/10.54846/am2022/163

Trading places: Valued learnings from both

Darin Madson, DVM, PhD, DACVP

Life-long learning is something we as swine veterinarians generally have in common; that's why we are here. We also have a strong work ethic, are data driven and focused on serving our clients and the industry. Don't know about you, but I sure enjoy what I do even if that means changing roles. The following are learnings that have stuck with me, both from the laboratory setting and practice; mostly because I was wrong the first time and then the lesson had greater impact.

People

1. Be respectful and kind to everyone you interact with. Too often frustrations come out to innocent bystanders on the other end of a phone. I can cite stories of practitioners calling the laboratory and being unpleasant to the person answering the phone. Do you think they deserved the rudeness and disrespect because you were having a bad day? Those experiences don't go untold. The point being, treat EVERYONE as how you would want to be treated (technician, administrative personnel and colleagues). We are more willing to help when kindness is practiced.
2. Say "thank you". It goes a long way to those who serve and are not always recognized for what they do. From the lab side, a diagnostician could have 100 cases in one day. This may be from 100 different practitioners who are thinking their case is the most important that the lab received that day. From the practice side, be gracious and show humility. We can get more accomplished on farm if you are humble and have empathy. Example, name any additional task you may wish to have done on a sow farm that is short-handed. Additionally, this shows respect.
3. Emotions are not easy to understand, but play a large role in how we communicate and are part of some decisions. From the laboratorian perspective, there is pressure to perform, but this is generally emotionless as the financial impact is hard to grasp. Alternatively, anxiousness and distress are often part of the practitioner side; eagerly awaiting results which could have huge fiscal impacts, often related to clients that are frequently friends.
4. Form relationships. Don't think you can go at it alone. This is on both sides of the coin. Our experiences teach us valuable lessons; if you don't know something reach out. Often both sides will benefit. Who is your phone-a-friend? Who is your lab go-to person? Who is your practitioner advisor? Do they know they are valued? Tell them. Relationships impact decisions. There is nothing wrong with this, though it can make certain aspects more difficult. Don't be afraid to respectfully asks questions in these scenarios.
5. Pick up the phone. Emails are a good way to record tasks or hit a larger audience, yet context and the personal interactions are lost. This is a lost art on both sides.
6. Labor and efficiencies will be needed moving forward. How do we do more with less?

Cases

1. Garbage in garbage out; as the saying goes. If you want the most accurate results, take the time. Alternatively, it is difficult to be in multiple locations at the same time. The practitioner does heavily rely on others for sample collection. The lab understands this; although we as practitioner do need to have coaching conversations on sample collection rather than only directing to collect (sample size, quantity, processing, packaging, etc).
2. Be concise. Be concise in your history and your testing. From the lab side, don't had fluff/filler, bonus points are not gained for length.
3. Know your question. Similar to the above, understand what you want to know. The diagnostician can help guide you, although you are the one with the most knowledge about the situation. To turn the question/situation around, would you want someone to order food for you at a restaurant? Work with the lab to be concise on testing; this saves money and frustrations on both sides. The laboratory wants to be a partner in the process. If the diagnostician is unaware of what is needed, inadvertent over-testing and confusion can ensue.
4. You make the diagnosis. The laboratory reports findings and test results. Use this info to make your diagnosis. Don't make the diagnostician be the practitioner too. To add clarity here, the laboratory realizes many cases are submitted by non-veterinarians and it is a difficult situation when the diagnostician knows more about what is happening in the barn than the submitting veterinarian.
5. Samples and isolates are saved for a period of time. A shotgun test approach doesn't necessarily need to happen. Tests can be added as more information is gained in during the testing process.
6. Follow-up communication with the lab is important in complex cases; both for understanding of what is occurring and the share. The laboratorian and the practitioner learn from each other. I don't think I can stress this last part enough, we all benefit.

Tests

1. Don't be afraid to question the test results. If something is not adding up, follow-up by calling. This on both fronts. Errors happen.
2. Contamination happens. More often on the practitioner side. Take the time or find a way to get samples to the lab a different way.
3. Sample quality is important; don't let samples sit in the vehicle or leave unprocessed (ie spin blood in a timely fashion).
4. Understand what the tests are looking for: DNA, antigen, protein, antibody, etc. If you don't know, phone your valued friends.
5. Don't over-interpret. Detection does not equal causation.
6. Highly sensitive tests (ie, PCR) or singular time point sampling don't always answer your question. Compounding different assays or timeframes can be useful. An example is acute and convalescent sampling. Ultimately, sampling collection timing is important.

Disease

1. Embrace the word "depends." I learned that from Dr. Kent Schwartz.
2. Pigs rarely read the book. The quicker this is realized the better. Facts are great and guide problem solving. Albeit though, most situations have some nuance or difference that adds complexity or a twist.
3. All variants of the same pathogen are not equal. The practitioner side has an advantage here as clinical difference are noted. However, this difference may not be realized as clearly on the lab side. Tissues are only able to reaction in certain ways. The full physiological effect is rarely grasped. This can be said for PRRS, IAV, *E coli*, *Strep* isolates and more. Practice listening during conversation with interacting the colleagues; useful pieces of information are often expressed.
4. Biosecurity is a culture. Non-negotiable. I was baffled this past year with PRRS transmission (particularly 1-4-4 strains). Distances seemed longer, possible explanations were less, animal movements were not always involved, and clinical impact was severe and protracted. During field necropsies, I kept wondering how severe the immune (over-) response must be; ie, the cytokine storm resulting in pulmonary edema, inflammation and unregulated secondary pathogen replication. Most animals didn't want to move. The will to over-come was not there.

Self

1. Make time for you and your family. PRRS will be there tomorrow.
2. Learn to say no, I don't have time at the moment.
3. Be honest in communications. "I don't know" is an answer. Followed by "here is what I do know" helps transition conversations.
4. Follow-through. If you said you'd do something, do it. Or at least own up to know that you forgot or don't have the time.

Economics

1. Understand the gravity of a test result. An unanticipated result could lead to several thousand dollars in transportation costs, feed or feed reclaim, animals or contractual problems.
2. Decisions need to be made fast when unanticipated findings are reported. This cost and nose in the system can last for many months. My advice for laboratories would be to respect how much money, time and resources may be involved in these sensitive times. Not saying one needs to fully grasp or understand but would promote empathy for the practitioner side.
3. Spend some money. If the lab does a good job for you or goes out of their way, buy donuts or lunch. They are worth it.
4. Spend more money on surveillance. Much cheaper in the long run.
5. Spend even more money on people. The right people in the right spot pay for themselves quickly. "Attitude trumps experience" is a something I heard recently and fully agree with.

Never stop learning, don't be afraid to ask questions and value those you work with, they have a lot of experience and wisdom.

https://doi.org/10.54846/am2022/164

Understanding the emergence of a new PRRSV variant (L1C 1-4-4) through the implementation of epidemiological tools

C. A. Corzo, DVM, MSc, PhD; M. Kikuti, DVM, MSc, PhD; I. A. D. Paploski, DVM, MSc, PhD; N. Pamornchainavakul, DVM; C. Picasso-Risso, DVM, MSc, PhD; D. Makau, DVM, MSc, PhD; A. Rovira, DVM, MSc, PhD; K. VanderWaal, PhD
Veterinary Population Medicine Department, University of Minnesota, St. Paul, Minnesota

Introduction

Since the emergence of porcine reproductive and respiratory syndrome virus (PRRSV) in the United States (US), there has been significant production losses every time a new variant emerges or is introduced into a susceptible pig population. These losses have been well characterized but in general include some degree of mortality (eg, pre-farrow, post-farrow) in every production phase, poor performance (eg, reproductive and growth efficiency) accompanied by an increase in costs on outbreak mitigation and control/ elimination expenses (eg, vaccines, antibiotics, herd-closures, pig flow segregation, depopulation). The virus has been winning the battle as it has found a way to constantly maintain itself in the US pig population by taking advantage of our daily processes within the production chain and by evolving which has led the virus to continue to adapt and rapidly disseminate throughout neighborhoods, production systems, and states. Despite all the biosecurity enhancements and control efforts implemented by producers, production teams and practitioners, we continue to witness epidemics on a yearly basis with different strains of this virus.[1]

Historically, certain pathogens have clearly taught us that evolution can occur. Besides PRRSV, swine influenza virus (SIV) and rotavirus to name a few, have exploited the overwhelmingly number of opportunities that our production systems have given these pathogens to maintain their replication levels through endless transmission chains which in turn give the virus opportunities to evolve.[2-4] These constant transmission chains, represented by endemically infected populations in continuous flow farms (eg, sow farms, nurseries, finishers) continue to be the Achilles heel of our industry when it comes to RNA viruses such as PRRSV. During the last decades, the industry dealt with the emergence of new strains of PRRSV that challenged the industry from a biosecurity, control, and elimination standpoint. However, during these unfortunate epidemic episodes there has not been a characterization of the initial spread of these strains which we believe is important to understand how and why these events occur and how we can prepare ourselves to mitigate the impact and perhaps stop dissemination.

Background of the recent PRRSV variant emergence

During the last months of 2020 an interesting event began drawing the attention of several southern Minnesota producers and practitioners. The event began with outbreaks of the disease in growing pig farms generating important respiratory clinical signs and high mortality. Interestingly, these outbreaks were occurring simultaneously throughout the same and nearby "neighborhoods". As this situation developed and the number of affected premises increased, the proactive nature of swine producers and practitioners began through active communication and data sharing through existing mechanisms such as the Morrison Swine Health Monitoring Project (MSHMP) to further understand what was occurring.[5]

Characterizing the event

Case definition and molecular data

One of the first steps needed to characterize a new infectious diseases event is to develop a case definition that will aid in deciding whether an individual unit of interest has a particular disease which helps us make sure every case is equivalent enabling us to have a consistent and comparable dataset pertaining to the outbreak. Since case definitions need to be tailored to each situation, in this case the MSHMP team decided to lead by using the first set of ORF5 sequences obtained from cases reporting the issue as a starting point to identify sequences grouping in a tight genetic cluster with 98% nucleotide identity or higher amongst each other. Sequences obtained from cases were characterized into lineages and sublineages. All sequences were classified as Lineage 1, sublineage 1C. The restriction fragment length polymorphism (RFLP) pattern for most cases was 1-4-4. However, since unrelated PRRSV strains with the same lineage/sublineage and RFLP pattern were co-circulating, we at MSHMP decided to use a phylogenetic approach for identifying cases to increase the accuracy and positive predictive value of our classification system to 81% compared to 34% or 43% if we were only to use RFLP or Lineage, respectively.

A key component of this investigation was the availability of diagnostic molecular data (eg, sequence). For this event, many ORF5 sequences were available and were the starting point to assess whether sequences obtained from these cases were similar. We took it a step further and a subset of these cases were further tested using the whole genome sequence (WGS) approach to assure the interpretation that this represented a separate genetic cluster would remain. In addition, this information was used to verify whether similarity at the ORF5 level resembled the similarity at the WGS level but most importantly, was used to further understand whether there were potential ancestors on previously existing databases and thus work towards potential emergence hypotheses.

Measuring frequency of disease

An important metric during a new infectious disease event is to quantify the number of cases over time as a means of understanding transmission. In the present case, an epidemic curve graph in which the number of sequences obtained either directly from producers/practitioners and diagnostic veterinary laboratories detected by week were summarized allowing us to see a temporal trend. These sequences were matched to a farm name and type. A farm that met the above-mentioned criteria would be classified as a case farm regardless of whether it had one or more sequences that matched the criteria. As time went by, these cases would be recorded and then allowed us to estimate the number of new cases through the cumulative incidence or the regular incidence rate. In both cases, knowing the denominator is key to either generate the cumulative number of farms that became positive to the virus or on the other hand the number of farms that became positive per unit of time. Both of these were calculated during the PRRSV outbreak. Up to September 2021, the cumulative incidence in breeding herds from Minnesota and Iowa, the two main states involved in this epidemic, was estimated at 12%.

Assessing clustering

Due to the nature of this fast-developing event, a question that was important to address was whether disease occurred more frequently in certain areas than in others and whether cases were grouped over a specific timeframe. This type of analysis can be done separately or together (eg, space-time clustering) depending on data availability.[6] This specific approach enables us to study potential interactions between location and time in relation to the onset of these outbreaks. Clustering can be assessed through different methods as there are a variety of models that can be used. However, the key ingredients for this type of analysis are the need of metadata such as date of the case and location (eg, coordinates) since these are key inputs when running these statistical models.[7,8] Altogether can lead to understanding whether there are common factors related to disease occurrence as proximity in time and space suggests some commonalities.[8]

Factors associated with occurrence

As the outbreak developed, it was necessary to investigate whether there were farm factors related to the occurrence of cases as this is important to decipher if there are potential transmission modes in which interventions can be implemented to slow spread. Therefore, the tool used to assess the frequency of risk factors in both cases and non-cases (controls) was to design a case-control study. Through a set of predefined criteria, farms were included as cases or controls. Then, a farm characteristics survey was designed to capture key farm design aspects and practices that could potentially be related to virus introduction. Once the surveys were administered and the data summarized, data analysis was conducted through logistic regression.

What did we learn from this event?

Characterization is key and risk factors

As the outbreak begins and evolves, characterizing the event as best as possible is important as metrics, visuals (eg, phylogenetic trees, maps) can be developed in a short period of time and can be shared back with industry stakeholders to make sure everyone is up to date with the latest developments. Without these basic measures it would not be possible to understand the magnitude of this event and thus some extra interventions (eg, enhanced monitoring and biosecurity) may not occur. Furthermore, by characterizing it in a specific matter with a case definition we avoided misinterpretation as the data was shared. Even though the industry thought that every 1-4-4 was the same, this event confirmed once again that we need clear and specific data to avoid these misinterpretations and confusion.

Through our case-control study we were able to confirm that there were no significant differences between the occurrence of the new variant when comparted to other co-circulating strains. Such finding suggests that there may have been no particular routes of transmission that helped this strain disseminate at a potentially faster rate.

Collaborative network

We witnessed once again that the swine industry is willing to share data, especially before emergency scenarios. Fortunately, the MSHMP was in place and active as this significantly facilitated data acquisition, sharing and further constructive discussions. Without this platform we would not have had the chance to respond quickly and provide feedback from a data summary and analysis perspective.[9] In previous PRRSV emerging events, this hadn't been done perhaps because of lack of data availability, weaker relationships, lack of communication and maybe because the event was not identified from the beginning. This reminds the industry of the importance of these large voluntary and accurate monitoring programs as they serve as a national tool just like it was seen with PEDV in 2013.[1]

Emergence of a new variant

Using current available sequence databases combined with quick collaboration with MSHMP systems to generate WGS, we were the first to conclude that this was an emergent strain and that this strain shows evidence of recombination.[5] Furthermore, we learned that WGS sequence data helped us confirm that ORF5 sequencing certainly provides good quality data as the results between these two levels of analysis were similar. The fact that information in public and private databases was available, the process of coming to this conclusion was significantly facilitated. Furthermore, current diagnostic tools allowed us to characterize the newly emerged variant to a level that evolving theories are now in the process, reminding us that this is not the first or last time that has/will happen. Our time-scale analysis clearly highlighted that the emergence variant had not been seen before and clearly represented a new clade within the evolution of PRRSV.[5]

Proximity plays a role most of the time

Proximity from a geographical and time standpoint has a significant impact on disease spread. For most these findings will not be new; however, through this investigation we were able to learn that an important number of farms located in the epicenter or in other areas where the virus spread did not get infected, regardless of whether they were filtered or not, used feed mitigants or not, were owned/managed by the same production company. This finding certainly generated more frustration as we indirectly showed that there are intrinsic factors that we still do not understand that could explain why these farms didn't break.

Not perfect

As with any other epidemiological investigation, there will always be areas of opportunity from which to learn. One of these areas is timing of the introduction as some farms clearly exhibited clinical signs but others did not, which presents a challenge to the investigative process of when the viral introduction occurred. A second area for improvement is inclusion of other cases/companies/clinics in that several were approached and invited to join the investigation, but some refused participation.

Practitioners and producers were vital during the investigation

Authors would like to acknowledge the progressive mentality of the following practitioners/producers for raising awareness and enabling us to characterize this event and for providing constant input that played a significant role during the investigation.

Mark Schwartz (Schwartz Farms), Paul Yeske (Swine Vet Center), Brad Leuwerke (Swine Vet Center), Laura Bruner (Swine Vet Center), Deborah Murray (New Fashion Pork), Brian Roggow (Fairmont Veterinary Clinic), Pete Thomas (Iowa Select Farms), Lori Feldman (Protein Sources), Matt Allerson (Holden Farms), Melissa Hensch (The Maschhoffs), Tyler Bauman (The Maschhoffs) and Brent Sexton (The Maschhoffs).

References

1. MSHMP Report.: 2021, Nov. 19.
2. Paploski IAD et al.: 2019, Front Microbiol. 10.
3. Domingo E et al. 1996, FASEB J. 10.
4. Murtaugh MP et al.: 2010. Virus. Res. 154, 18-30.
5. Kikuti M et al.: 2021, Front. Vet. Sci. – 8.
6. MMWR: 1990. 39, 1-6.
7. Stevens KB et al.: 2011, Spat Spatiotemporal Epidemiol. 2(3), 125-133.
8. Ward M: 2007, Vet Ital. 43(3), 559-570.
9. Perez A et al.: Front. Vet. Sci. - 6.

https://doi.org/10.54846/am2022/165

Breaking down PRRSV next-generation sequencing into a user-friendly format

G. Trevisan, DVM, MBA, PhD; M. Zeller, BS, MS, PhD; G. Li, MS, PhD; J. Zhang, MD, MS, PhD; P. Gauger, BS, DVM, MS, PhD; D.C.L. Linhares, DVM, MBA, PhD
Department of Veterinary Diagnostic and Production Animal Medicine, Iowa State University

Porcine reproductive and respiratory syndrome (PRRS) is a devastating swine disease caused by the PRRS virus (PRRSV).[1] PRRS was initially described as "The Mystery Swine Disease," and since its detection in the United States (US) and Europe, 2-6 PRRSV has spread globally.[7-13] In the US, the estimated losses associated with PRRS were $664 million annually,[14] and in Europe, it was estimated to range between €3 and €160 per sow.[15]

There are two distinct species of PRRSV: PRRSV1 (*Betaarterivirus suid 1*) and PRRSV2 (*Betaarterivirus suid 2*)[16,17] distributed worldwide. The PRRSV-1, colloquially known as the European type, has the Lelystad isolate[3,4] as the prototype strain. The PRRSV-2, colloquially known as the North American type, has the ATCC VR-23322 as the prototype strain. PRRSV-1 and PRRSV-2 have ~44% nucleotide differences.[1,17]

The whole PRRSV genome (whole-PRRSV) comprises ~15.1 kilobases in length[3,9,17] and is organized into at least ten open reading frames (ORFs).[7] The non-structural region of the genome is composed of ORF1a and ORF1b with subdivisions into non-structural proteins nsp1 to nsp12. The structural genome region comprises ORF2a, ORF2b, ORF3, ORF4, ORF5a, ORF5b, ORF6, and ORF7. Within the whole-PRRSV, the specific region ORF5b, also known as ORF5, has been extensively used in molecular epidemiology and evolutionary studies.[7,18-23] The ORF5 gene is composed of 603 nucleotide base pairs (bp) for PRRSV-2 and 606 bp for PRRSV-1, representing ~4% of the PRRSV genome.[1,7] Sequencing ORF5 region is generally performed using a Sanger technique,[24] and the recovered genetic sequence is a consensus from the Sanger reads. The Sanger technique generally determines the consensus virus sequence in a sample and is not an efficient tool to determine all virus sequences if more than one PRRSV strain is present in a sample.[25]

Processing fluids, a novel population-based sample type,[26,27] have been recently adopted as part of surveillance programs to detect PRRSV in newborn piglet populations and have also been validated to monitor breeding herds undergoing PRRSV elimination.[28] Next-generation sequencing (NGS) is a technique used to sequence the whole-PRRSV genome[29] and may have the potential to more thoroughly detect and characterize different PRRSV strains if present in one sample. Even though NGS seems promising, the use of NGS results by practitioners is still a dilemma.[30] Additionally, anecdotal discussions between practitioners have commonly reported the perception of more than one PRRSV strain circulating in a herd. In an effort to help veterinarians understand PRRSV genetic variability through the use of NGS results, our team has conducted a study to evaluate the feasibility of applying NGS techniques in processing fluid (PF) samples to investigate PRRSV genetic variability and characterize the presence of multiple strains in breeding herds over time.

This study enrolled breeding herds (n = 20) that faced a PRRSV outbreak and adopted measures to eliminate the virus from the herd. Serum, tissue samples, the live virus inoculation (LVI) material, or processing fluid samples were initially submitted for NGS. Each farm had a whole-PRRSV sequence detected through NGS that was used as the referent strain. NGS was performed on PF samples over time: targeting two at week (wk) 2-4 after the break, two at wk 8-12, and 2 on the last two samples with Ct values < 30 before the farm achieved low PRRSV prevalence, defined as eight consecutive weeks of PRRSV PCR-negative results in PF. A BLASTn technique compared the whole-PRRSV genome or contigs, ie, genome fragments, with the farm reference strain and commercially available PRRSV modified live virus (MLV) vaccine strains to identify nucleotide identities. The outputs were used to determine the number of PRRSV strains. Time to low PRRSV prevalence, time to baseline production, and total losses defined as the number of pigs not weaned per 1,000 sows after the break were calculated. One additional step was taken to investigate recombination events. Farm-specific strains were compared with wild-type strains and PRRSV MLV sequences available in Genbank.

Full genome referent PRRSV strains were recovered for 16 of 20 farms. A large contig (12,888 bp) was recovered and used as the reference strain in one farm. In 3 of 20 farms smaller contigs (< 10,000 bp) were recovered and used as referent strains. The mean whole-PRRSV genome nucleotide identity between herds (n = 16) was at 86.14 %. The viral load measured in cycle threshold (Ct) values from the processing fluid samples submitted for NGS were median 25.1, 25th percentile 23.1, and 75th percentile 27.5. Even though the success rate of whole-PRRSV genome recovery (~15kb) in processing fluid was 5.26% (5 out of 95 samples submitted for NGS), the contigs, recovered from different genome regions, identified at least two wild-type PRRSV strains circulating simultaneously in 18 farms and at least one vaccine-like strain in 8 of 18 farms (Figure 1). Farms with ≤ 2 PRRSV strains (n 8) reached stability 12 weeks before farms with ≥ 3 PRRSV strains (n 12) (log-rank 0.0459). Median time to the low prevalence and interquartile range (25th to 75th percentiles) for herds ≤ 2 PRRSV strains and ≥ 3 were 30.8 weeks (27.0 - 37.0) and 42.8 weeks (26.0 - 54.0), respectively. No significant differences in time to baseline production were associated with the difference in the number of strains and the presence of recombination events.

The recovery of PRRSV sequences by applying NGS in processing fluid samples provided scientific evidence supporting the presence of multiple PRRSV strains circulating simultaneously in breeding herds, as determined by having at least one contig differing at least 98% with the farm referent wild-type strain and with vaccine strains those were recovered in at least two different sequences samples, i.e., sampling days. NGS results are not restricted to the ORF5 region and provide better insights into the PRRSV genetic variability circulating in a breeding herd. Surprisingly, detecting at least two wild-type PRRSV strains in a breeding herd (18 of 20) and recombination events (11 of 17) was common. It has been previously reported that when more than one strain is circulating in a population, the more aggressive strain and with a

Figure 1: Comparison between a farm-specific wild-type PRRSV strain and genome fragments, ie, contigs, recovered from one processing fluid sample characterizing the presence of at least 3 PRRSV strains. The structure on the top represents the farm referent strain position of open read frames (ORFs) and proteins (nsp). The percentage of nucleotide identity between each contig and farm referent strain is labeled at the center of each contig. Colors represent the level of similarity with farm referent strain or with a commercially available PRRSV MLV vaccine. Contigs in green: > 99% similarity with farm referent strain. Contigs in yellow: > 99% similarity with a commercially available PRRS MLV vaccine. Contigs in red: < 99% similarity with the farm referent strain.

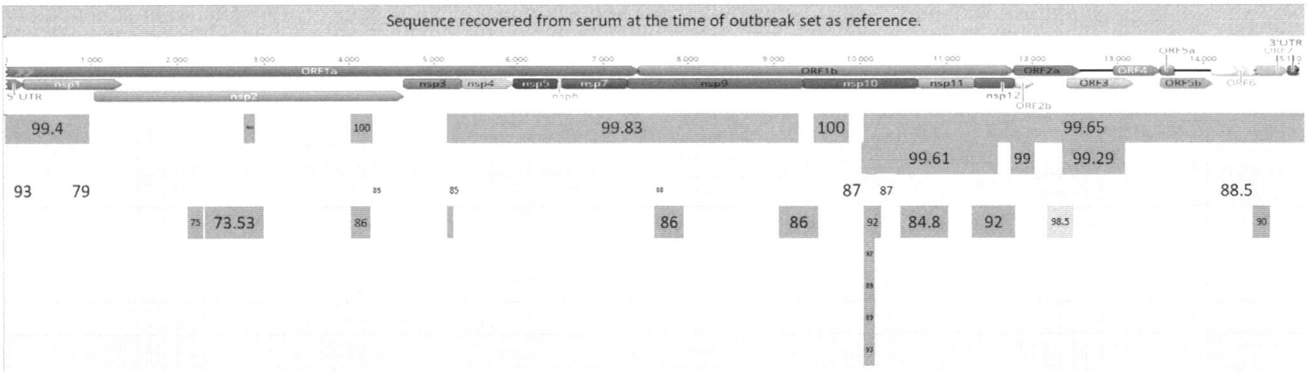

higher replication rate will suppress the less predominant virus.[7] Also, the NGS technique may only be able to obtain the consensus genome sequence but fail in recovering individual whole-PRRSV genomes when closely related strains are present in the sample, eg, two PRRSV2 strains.[29] These facts are well aligned with the results observed in this study, where genome fragments were predominantly recovered, and contigs from different genome regions were repeatedly detected over time. Recombination events seem to be common in the field. Due to the restricted number of whole-PRRSV genomes available in the GenBank database, there is potential to miss the detection of recombination events between wild-type strains. Alternatively, PRRSV MLV vaccine strains are well characterized, and identification of recombination events between wild-type and vaccine strains is more likely to occur.

This study presented a novel approach to visualize NGS outcomes that provided additional information into the PRRSV dynamics and genetic variability present in breeding herds and approaches to detect > 1 PRRSV present in a production system. Viral concentration, potential NGS inhibitors, presence of multiple PRRSV strains may have prevented the recovery of whole-PRRSV genomes in processing fluid samples. The contigs that were recovered from different regions of the genome and compared with the farm referent strain could identify if > 1 PRRSV strain was circulating in the herd. Sanger techniques are likely to be more successful sequencing an entire gene; however, these traditional sequencing methods lack the ability to consistently detect more than one sequence in a sample but may represent a consensus sequence of the gene of interest. Based on these results, the assumption that PRRSV lateral outbreaks occurring in the downstream flow could be leakage of less predominant PRRSV strains that are currently circulating in the sow farm but previously were undetected or remained uknown due to the use of traditional diagnostic methods and gene sequencing. In repetitive circumstances, the less predominant strains were detected more frequently when herds were approaching low prevalence. Adopting McREBEL procedures[31] could help minimize PRRSV transmission to other piglets and rooms, reducing potential recirculation of multiple strains on the farm and perhaps lessen the PRRSV impact in the downstream flow. A general recommendation for the most productive use of NGS is to target the recovery of a whole-PRRSV genome for each breeding herd that can be used or established as the referent strain, or strains when more than one virus is present, using samples with a higher success rate for the recovery of whole genomes through NGS, eg, serum and tissue samples that are likely to have higher concentrations of the virus in the sample. Subsequent NGS in processing fluid samples and comparing sequences, even if partial sequences are recovered, with the farm referent strain is highly valuable and can help identify additional PRRSV strains that may be circulating in a breeding herd.

References

1. Zimmerman J et al: 2019, Diseases of Swine. 11 ed. Wiley Blackwell; p 685-708.
2. Collins JE et al: 1992, J Vet Diagn Invest. 4(2):117-26.
3. Meulenberg et al: 1993, Virology. 192(1):62-72.
4. Terpstra C et al: 1991, Vet Q. 13(3):131-6.
5. Wensvoort G et al: 1991, Vet Q. 13(3):121-30.
6. Hill H, 1990, 29-31.
7. Brar MS et al: 2015, J Gen Virol. 96(Pt 7).
8. Murtaugh MP eet al: 1995, Arch Virol,140(8):1451-60.
9. Nelsen CJ et al: 1999, J Virol. 73(1):270-80.
10. Ramírez M et al: 2019, Transbound Emerg Dis. 66(3):1107-1113.
11. Ramos N et al: 2018, Transbound Emerg Dis, 65(2):352-356.
12. Rosendal T et al: 2014, BMC Vet Res. 10:83.
13. Trevisan G et al: 2019, PLoS One.14(10):e0223544.
14. Holtkamp DJ et al: 2019, JSHP p: 72-84.
15. Nieuwenhuis N et al: 2012, Vet Rec. 170(9):225.
16. Zhang J et al: 2019. Diseases of Swine. 11 ed. Wiley Blackwell; p 427-437.
17. Kuhn JH et al: 2016, Arch Virol. 161(3):755-68.
18. Lambert M et al: 2019, Infect Genet Evol. 74:103999.
19. Shi M et al: 2010, J Virol. 84(17):8700-11.
20. Paploski IAD et al: 2021, Vaccines. 9(6).

21. Paploski IAD et al: 2019, Front Microbiol. 10:2486.
22. Trevisan G et al: 2021, J Vet Diagn Invest. 10406387211027221.
23. Wang A et al: 2019, Vet Microbiol. 239:108486.
24. Sanger F et al: 1977, Proc Natl Acad Sci U S A. 74(12):5463-7.
25. Harmon K et al: 2019, AASV, p: 50-52.
26. Lopez WA et al: 2018,JSHP, 26(3):146-150.
27. Vilalta C et al: 2018, Vet Microbiol. 225:149-156.
28. Trevisan G et al: 2019, Porcine Health Manag. 5:18.
29. Zhang J et al: 2017, J Vet Diagn Invest. 29(1):41-50.
30. Risser J et al: 2021, Virol J. 18(1):206.
31. McCaw MB. 2000, JSHP 8(1):15-21.

https://doi.org/10.54846/am2022/166

Responding to animal disease outbreaks with a One Health approach

Gary Flory,
G. A. Flory Consulting, Mt. Crawford, Virginia

Introduction

Each year, our appreciation for the interconnectedness of human, animal, and environmental health increases. The One Health concept is the recognition that the health of people is closely connected to the health of animals and our shared environment. National and international organizations such as the Centers for Disease Control and Prevention (CDC), Food and Agriculture Organization of the United Nations (FAO), and the World Organization for Animal Health (OIE) continue to expand their educational and coordinating efforts to harmonize our response to zoonotic diseases and other threats to global health. However, there is still much work to be done at the field level when responding to disease outbreaks.

For those of us who manage disease outbreaks in animals, it is easy to overlook the potential human and environmental impacts of our response activities. We must remember that 60 percent of the emerging infectious diseases reported globally are zoonotic, and that 75 percent of the more than 30 new human pathogens detected in the last three decades originated in animals.[1] During the early phases of an animal disease response, we may not know exactly what is causing the symptoms we are observing, or if the disease organism is zoonotic. Zoonotic diseases are responsible for an estimated 2.5 billion cases of illness and 2.7 million deaths worldwide each year.[2] Rabies alone is responsible for approximately 59,000 human deaths annually, with an estimated 8.6 billion in economic loss.[3] Further, we cannot forget that the COVID-19 pandemic that has resulted in the death of more than 5 million individuals was caused by a zoonotic agent.

The management of animal disease outbreaks can also have significant environmental impacts. For example, improper carcass disposal can contaminate ground and surface water. Between 55 and 80 percent of an animal's body is composed of water, and when carcasses decompose, they release this liquid as leachate. For example, a 450-pound sow can release as much as 37 gallons of leachate. Leachate is a complex mixture that, if released into surface water or groundwater, can sicken humans and animals and cause long-term environmental pollution. Additionally, the viability of many disease organisms in ground and surface water is not yet understood, and the contamination of water supplies may lead to the additional spread of animal diseases.

Materials and methods

New tools are needed to address the One Health implications of animal mortality management activities. To fill this gap, research teams conducted a number of field studies looking at two new carcass disposal methods: grinding and composting and Above Ground Burial.

Grinding and composting

Composting animal carcasses transitioned from a novel idea to the main carcass disposal method in the United States during the Highly Pathogenic Avian Influenza outbreak of 2015 (Figure 1). During this outbreak, 85 percent of the nearly 50 million poultry carcasses impacted by the disease were composted.[4] Although composting proved a practical carcass disposal method for poultry, there was uncertainty about the feasibility of this method for livestock. The most significant of these concerns was the time needed to compost larger livestock carcasses. There was also concern about the lack of data demonstrating the effectiveness of composting for inactivating the African swine fever virus (ASFV).

In 2018, researchers conducted a proof-of-concept project in Virginia to evaluate the effectiveness of grinding swine carcasses with a horizontal bed grinder prior to composting. Based on the results of this project, the team conducted a series of operational scale research and demonstration projects in North Carolina and Minnesota. These projects evaluated the methodology's throughput, carbon material requirements, potential for pathogen aerosolization, and impact on the composting process.

Although composting has been shown to inactivate a variety of pathogens, studies had not been conducted to confirm its effectiveness for inactivating ASFV. To address this uncertainty, a team of US and Vietnamese researchers, with funding from the National Pork Board, conducted a research project on the campus of the Vietnam National University of Agriculture in Hanoi, Vietnam. In the study, 6 pigs (Figure 2) from local farms presenting typical ASF symptoms (high fever, skin hemorrhages, and enlarged and hemorrhagic spleen and lymph nodes) were collected and confirmed to have ASF by real-time PCR. Each of the pigs was placed in compost piles from which tissue samples were collected on days 1, 3, 7, 14, 28, 56, and 144 (Figure 3). Compost pile temperatures were monitored daily with a data logger and handheld analog thermometer. Spleen and bone marrow samples were collected for analysis by real-time PCR and cell culture.

Figure 1: Composting during the 2015 HPAI outbreak.

Figure 2: ASF infected pigs for composting project in Vietnam.

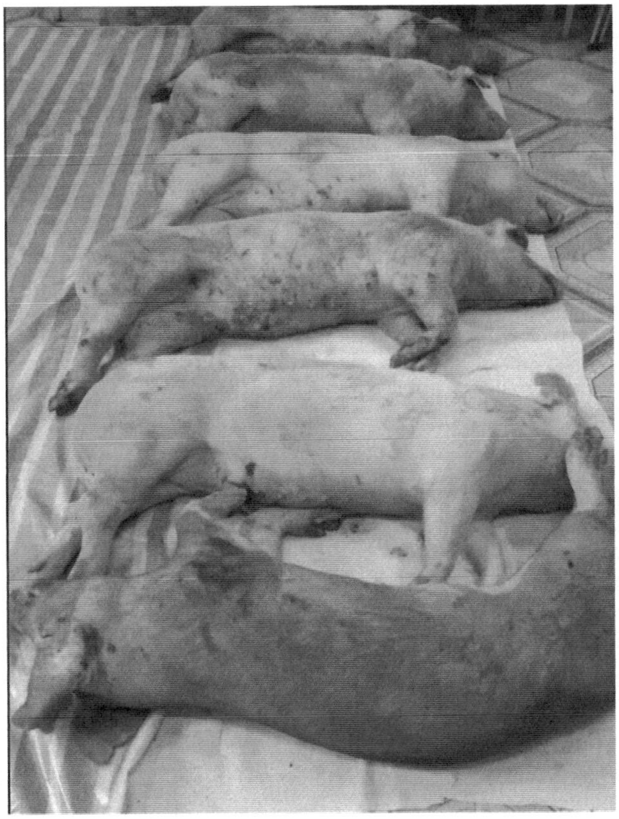

Figure 3: Compost piles for ASF study in Vietnam

Figure 4: Oklahoma Above Ground Burial research plots.

Above Ground Burial

In 2015, researchers in Virginia began evaluating the effectiveness of Above Ground Burial (AGB).[5] AGB is a hybrid of traditional deep burial and composting, designed to increase carcass decomposition and minimize environmental impacts. Since that time, the team has conducted projects in Virginia, North Carolina, Texas, Oklahoma (Figure 4), and Minnesota, looking at carcass decomposition, scavenger activity, nutrient migration, insect activity, and pathogen inactivation.

Results and discussion

Based on the research conducted to date, grinding prior to composting appears to have many advantages over whole carcass composting for the swine industry. While composting whole sow carcasses can take anywhere from 2 to 6 months, grinding carcasses prior to composting can reduce composting time to around 30 days. Grinding may also make the composting process more efficient, resulting in a 20 to 25 percent reduction in the amount of carbon material required. Finally, by thoroughly blending the carcass and carbon material, the compost piles heat more quickly and evenly, reducing the possibility of having cooler pockets in the compost pile where pathogens may survive. These benefits do come with a cost; the horizontal bed grinders used with this method are expensive, may not be universally available, and require a significant amount of time to clean and disinfect before moving to another site.

The work conducted in Vietnam has shown that composting can quickly and effectively inactivate ASFV in swine carcasses. Viable ASFV was not recovered in the compost piles constructed with rice hulls after 3 days, nor in the compost piles constructed with sawdust after 7 days.

Above Ground Burial also appears to be a viable carcass disposal option for the swine industry. When conducted in accordance with USDA's standard operating procedures,[6] AGB minimizes the vertical migration of contaminants, prevents scavenger activity, and enhances carcass decomposition. AGB has also been shown to inactivate a variety of pathogens, including the swine pox virus and Seneca Valley virus. Additionally, research is underway to evaluate the effectiveness of the method in inactivating ASFV.

References

1. Jones KE, Patel N, Levy M, et al. Global trends in emerging infectious diseases. Nature 2008; 451:990-94.

2. Centers for Disease Control and Prevention. (2020, August 20). Prioritizing and preventing deadly zoonotic diseases. Centers for Disease Control and Prevention. Retrieved November 6, 2021, from **https://www.cdc.gov/globalhealth/healthprotection/fieldupdates/winter-2017/prevent-zoonotic-diseases.html**

3. Centers for Disease Control and Prevention. (2020, July 29). CDC - rabies around the world - rabies. Centers for Disease Control and Prevention. Retrieved November 6, 2021, from **https://www.cdc.gov/rabies/location/world/index.html**

4. USEPA, 2018. Carcass Management During Avian Influenza Outbreaks [WWW Document]. United States Environ. Prot. Agency. Retrieved November 6, 2021, from **https://www.epa.gov/homeland-security-waste/carcass-management-during-avian-influenza-outbreaks#main-content**

5. Flory GA, Peer RW, Clark RA, Baccar MN, Le TT, Mbarek AB, Farsi S (2017) Aboveground burial for managing catastrophic losses of livestock, Int J One Health 2017;3:50-56.

6. Miller L.P., Flory G.A. . . . Peer RW. Guidelines for the Emergency Use of Above Ground Burial to Manage Catastrophic Livestock Mortality. United States Department of Agriculture Guidance. January 2021.

https://doi.org/10.54846/am2022/167

Resiliency debrief for triage and care: A proposed protocol and plan of action

Elizabeth B. Strand, PhD, LCSW
University of Tennessee, Knoxville, Colleges of Veterinary Medicine and Social Work

Veterinary and other animal-related professionals experience understandable psychological distress as a result of engaging in depopulation. This five-item resiliency debrief tool is designed to be used by veterinarians and other animal-related professionals who are preparing for, participating in, and recovering from animal-depopulation. These questions have been developed by an experienced veterinary social worker who has been implementing debriefs in the veterinary medical field for 20 years. For animal depopulation use, these questions have been additionally reviewed by three doctoral level traumatology experts as well as by several food animal veterinary medical professionals.

This tool is designed to provide credible and attuned social support and self-assessment. The use of peer support among professionals experiencing the same stressor has been shown to be an effective tool in protecting mental health.[1,2] The peer approach is important because it establishes credible social support. Peer support is "credible" because the individual offering support has lived experience with the stressor being experienced. Having an outside facilitator or mental health professional use traditional debrief tools (such as critical incident stress debriefing) has demonstrated iatrogenic effects.[3,4] Peer support can provide an alternative way to protect the mental health of professionals exposed to traumatic stress.[1]

Although credibility is important, that alone can also create iatrogenic effects if the peer connection only includes reliving a trauma, simple complaining about the experience, or a "pull yourself up by your bootstraps" type of advice-giving support. To help the social support avoid iatrogenesis, these questions are designed to guide animal professionals through a specific five-step socio-emotionally attuned traumatic growth process. Attunement means that the process invites professionals to share their experience in such a way that helps connect them with their own and others authentic experience and positive coping in managing animal depopulation. The steps are predictable, replicable, and can be implemented by animal related professionals themselves. The use of lay person public health approaches to help individuals facing a mental health challenge recognize it and get support accessing help is a growing approach to meet mental health needs.[5] Following is a description of each question in the five-step process including the keywords and rational behind the use of the question.

Debrief tool breakdown by question

Question #1 (keywords this question targets: intrusive thoughts, rumination, and social support)

Q1 Please share if there is anything you cannot shake right now? What is it? What do you remember about it (ie, what sounds, images, or sensations do you recall)? What emotions do you feel about it now?

Reasoning for the question: This question invites an animal professional to put into words any experience that is causing intrusive thoughts and rumination. Intrusive thoughts are unwanted images and/or memories of a stressful event. Rumination is the circular repetition of these unpleasant thoughts. These are typical experiences when someone is experiencing traumatic stress. An important skill in managing this is to "name" what the event is and connect that to emotions that occur in relation to the event. If answered in a dyad or group, this "storytelling" allows others to normalize the experience and feel more comfortable sharing their own experiences. Importantly, this question **invites** someone to share, but does not assume a professional has such an experience or wants to decide to share it. Animal professionals may or may not have an answer to this question and may or may not wish to answer it publicly. It is essential to respect animal professionals' self-determination on what and if to share answers to this question.

Question #2 (keywords this question targets: coping and self-efficacy)

Q2 Can you think of what you have DONE RIGHT in that situation-even the smallest thing counts (or any recent situation if there was no response to Q1)

Reasoning for the question: When someone is going through a stressful event, once it is acknowledged by Q1, inviting them to focus on how they are coping well is important. This question helps the animal professional recognize the ways in which they are demonstrating self-efficacy and positive intention in the midst of a difficult circumstance. Research indicates that attending to evidence of positive personally behaviors is helpful even if one thinks of a circumstance NOT related to the current stressor (as would be the case if there was no answer to Q1).[6]

Questions #3 and #4 (keywords these questions target: coping, self-efficacy, and problem-solving)

Q3 Is there anything you wish you had done differently?

Q4 Is there anything you have learned or need to adjust for tomorrow or next time?

Reasoning for the questions: When individuals can learn from their stressful experiences it promotes a sense of empowerment and self-efficacy. These questions also help a person engage active problem solving to avoid being simply overwhelmed by the stressor. They help professionals use their experience to change strategies and keep growing as an animal professional. Moreover, these questions invite practical problem-solving so that the individual may be more successful when they face the situation again.

Question #5: (keywords this question targets: gratitude, humor, and professional identity)

Q5 Is there anything you feel grateful for or made you laugh in this situation (or in your recent past)?

Reasoning for the question: Research in the field of positive psychology has demonstrated that attending to gratitude can have a positive impact on mental health.[7] Moreover, gratitude is also present for those with higher resiliency from traumatic stress.[8,9] Although the use of humor in times of tragedy is very dependent on individual preferences and traits, there is some evidence that it helps in cultivating an authentic professional identity.[10] This question allows professionals to choose which type of positive emotion to connect with that will be most personally helpful.

How to use the tool

Individually: This tool can be used individually by a professional who seeks to monitor his or her wellbeing throughout a depopulation experience. To do this an individual can use this as a writing prompt tool. Writing as a way to mitigate traumatic stress and promote post traumatic growth is evidenced in the research literature.[11–13]

Dyadically: This tool can be used as part of a dyadic "check-in friend" system that a professional sets up in preparation for a depopulation event. The "check-in friend" dyad would meet periodically, in person, by phone, or on a web meeting, and answer the questions together and share responses as a way to establish credible socio-emotionally attuned support before, during, and after the depopulation. The dyad can determine the frequency based on their own schedules and availability.

As a team: This tool can be used as a daily or weekly team debrief to establish and prioritize credible and socio-emotionally attuned support. Each member can be encouraged but not required to attend. If in attendance each member can share or not share. Each person should be allotted about 7 minutes to share responses to the questions. It would be helpful to print out the tool and have copies on hand.

Additional considerations

Resilience debrief as a psychological distress screener: If someone is consistently unable to identify and connect with positive feelings at the end of the 5-step process, it would be appropriate to encourage that individual to seek specialized mental health support from a licensed professional. It is important to respect self-determination of the individual who may or may not choose to seek such specialized support. These types of specialized mental health support resources can be offered but not required. Although the presence of positive feelings does not indicate the absence of psychological distress,[7] the presence of positive feelings does correlate with more resilience.[8,9] Therefore the prolonged absence of positive feelings would be an appropriate indicator that additional support could be helpful. Moreover, all individuals who are participating in depopulation should be informed of the mental health resources that area available to them. If the organization has access to the employee assistance benefits, alerting animal professionals of that resource in regular intervals before, during, and after the depopulation would be appropriate. Lastly there are mental health screeners that individuals can complete to help them monitor their psychological distress privately. For example these screeners from Mental Health America: **https://screening.mhanational.org/screening-tools/**

Summary review: Animal depopulation resiliency debrief tool

Prepare:

1. Locate employee behavioral health insurance benefits
2. Determine if company/farm has an Employee Assistance Program and what benefits are available
3. Provide mental health care information in advance
   a. Include free resources such as
      i. https://screening.mhanational.org/screening-tools/
      ii. https://suicidepreventionlifeline.org/
      iii. samhsa.gov/find-help/disaster-distress-helpline
4. Normalize consulting with mental health care as a part of sound self-care during depopulation

Five-step resiliency tool

Directions: These questions can be used 1) individually as a journaling tool, 2) as a discussion guide for "check-in friends," or 3) as a discussion guide for a daily or weekly team check-in debrief. Participation in the use of this tool should be voluntary. If someone decides to be present but quiet during a team debrief that is all right. If someone chooses not to participate at all, please ensure they have the 5-step process to do independently as well as information for accessing mental health care.

Question 1: *Please share if there is anything you cannot shake right now? What is it? What do you remember about it (ie, what sounds, images, or sensations do you recall)? What emotions do you feel about it now?*

Question 2: *Can you think of what you have DONE RIGHT in that situation-even the smallest thing counts (or any recent situation if there was no response to Q1)*

Question 3: *Is there anything you wish you had done differently?*

Question 4: *Is there anything you have learned or need to adjust for tomorrow or next time?*

Question 5: *Is there anything you feel grateful for or made you laugh in this situation (or in your recent past)*

References

1. Low, R., King, S. & Foster-Boucher, C. Learning to Bounce Back: A Scoping Review About Resiliency Education. *J. Nurs. Educ.* 58, 321–329 (2019).

2. Lane, M. A. *et al.* Supporting Clinicians After Adverse Events: Development of a Clinician Peer Support Program. *J. Patient Saf.* 14, e56–e60 (2018).

3. Regel, S. Post-trauma support in the workplace: the current status and practice of critical incident stress management (CISM) and psychological debriefing (PD) within organizations in the UK. *Occup. Med.* 57, 411–416 (2007).

4. Critical Incident Stress Debriefing: Helpful, Harmful, or Neither? *Journal of emergency nursing: JEN: official publication of the Emergency Department Nurses Association* vol. 45 611–612 (2019).

5. Synowski, J., Weiss, H. A., Velleman, R., Patel, V. & Nadkarni, A. A lay-counsellor delivered brief psychological treatment for men with comorbid Alcohol Use Disorder and depression in primary care: Secondary analysis of data from a randomized controlled trial. *Drug Alcohol Depend.* 227, 108961 (2021).

6. Creswell, J. D. *et al.* Affirmation of personal values buffers neuroendocrine and psychological stress responses. *Psychol. Sci.* 16, 846–851 (2005).

7. Jans-Beken, L. *et al.* Gratitude and health: An updated review. *J. Posit. Psychol.* 15, 743–782 (2020).

8. Zhou, X. & Wu, X. Longitudinal relationships between gratitude, deliberate rumination, and posttraumatic growth in adolescents following the Wenchuan earthquake in China. *Scand. J. Psychol.* 56, 567–572 (2015).

9. Israel-Cohen, Y., Uzefovsky, F., Kashy-Rosenbaum, G. & Kaplan, O. Gratitude and PTSD symptoms among Israeli youth exposed to missile attacks: examining the mediation of positive and negative affect and life satisfaction. *J. Posit. Psychol.* 10, 99–106 (2015).

10. Cain, C. L. Integrating Dark Humor and Compassion: Identities and Presentations of Self in the Front and Back Regions of Hospice. *J. Contemp. Ethnogr.* 41, 668–694 (2012).

11. Krupnick, J. L. *et al.* An Internet-based writing intervention for PTSD in veterans: A feasibility and pilot effectiveness trial. *Psychol. Trauma* 9, 461–470 (2017).

12. Canada, K. E., Brinkley, A., Peters, C. & Albright, D. L. Military veterans: Therapeutic journaling in a veterans treatment court. *Journal of Poetry Therapy* 28, 113–128 (2015).

13. Ullrich, P. M. & Lutgendorf, S. K. Journaling about stressful events: effects of cognitive processing and emotional expression. *Ann. Behav. Med.* 24, 244–250 (2002).

https://doi.org/10.54846/am2022/168

Best management practices for pen gestation: Opening Pandora's box

Thomas D. Parsons, VMD, PhD, DACAW
Swine Teaching and Research Center, School of Veterinary Medicine,
University of Pennsylvania, Kennett Square, Pennsylvania

Introduction

Gestation stalls have been the stalwart of housing pregnant sows world-wide since the 1980s. However, legislation first in the European Union and now in 10 states in the United States have banned the use of this type of housing. Market pressure in a variety of counties such Canada, South Africa, Brazil and Thailand also are moving sows out of stalls. Proposition 12, a new piece of California legislature, promises to further reduce the use of gestation stalls in the US. If Prop 12 survives its pending legal challenges, it will impact sows in states even without their own legislative appetite to ban gestation stalls. Taken together, the housing of pregnant sows in pens continues to grow and creates the opportunity for veterinarians to help implement best management practices for these new housing systems.

There are several viable alternatives to the gestation stall for sow housing and feeding, each with its unique strengths and weaknesses. Consulting veterinarians, educators, extension agents and other advisers must recognize the importance of identifying the option that best matches the needs and abilities of a particular producer. Another critical aspect to the success of these crate alternatives is the training and re-training of employees with requisite stockmanship skills to manage sows in a pen environment. While the discussion of gestation stall alternatives tends to revolve around differences in feed delivery systems, recognize that the gestation stall was a management system and all aspects of sow management need to be accounted for when implementing pen gestation. The goal of this paper is to familiarize you with the lexicon of these emerging technologies and help identify critical attributes of each crate-alternative technology that contribute to when it is an appropriate solution for you. These notions are re-enforced through a more specific discussion of one type of pen gestation system, electronic sow feeding (ESF).

Understanding pen gestation

The gestation crate provided several advantages that drove its ability to insure high productivity and widespread use. The gestation stall allowed for individual animal nutrition, individual animal care, and a way to manage the potential untoward effects of social hierarchy inherent in groups of pigs. When comparing crate alternatives, it is important to consider how any given approach addresses these basic needs for maintaining or improving herd productivity with group housed sows. Other important considerations are economic and include the cost of the technology and feed utilization.

Gestation stall alternatives

Competitive feeding systems

Floor feeding – Group housed sows are feed on the floor. In some systems, short stalls or stanchions are provided in an effort to reduce feed aggression. Animals are susceptible to disparities in body conditions associated with the absence of individual animal nutrition. Feeding discrepancies associated with social hierarchy issues can be lessened by feeding not every day or many times a day. Floor feeding may prove to be effective with small groups of sows (6 or less), but will require careful sorting of animals to insure that group has uniform nutritional requirements. Animals can be feed by hand or with standard feed drops. Space requirements, cost, and maintenance are similar to a crated facility, but the inability to feed individual animals may result in greater feed usage on per sow basis.

Trickle feeding – Feed is dispensed at a rate slower than the slowest eating sow can eat. Thus, there is no incentive/advantage for the boss sow to leave her feeding space. All sows in a group are fed the same amount. Animals must be grouped according to expected feed intake. Some success has been obtained with housing sows in small groups, but this can be difficult to manage in large herds. The slow rate of feed drop is achieved via a gear reduction on the final feed augers. The additional moving parts add cost to the system and require additional maintenance.

Non-competitive feeding systems

Cafeteria style – Sows are housed in group pens without provisions for feeding and moved once a day to a specialized area in the barn for feeding or so-called cafeteria. Typically, feed would be dropped in the feeding area, sows moved and locked into a crate until all sows in the group are finished eating. Opportunities for individual animal nutrition are limited, but it is possible to top dress additional feed to under conditioned animals. The advantage of this system is that it somewhat inexpensive to implement has you have only specialized feeding equipment for one pens' worth of sows. Disadvantages include that it is very labor intensive as all sows need to be moved, one pen at a time, to and from the feeding area. There is also some greater risk of injury or pregnancy loss given the daily movements.

Self-catching crates (free access stalls) – Animals can move freely between a loafing area and a freely accessible gestation crates. The crates are self-locking and unlocking to allow the animals to come and go at their own free will. As sows tend to re-use the same stalls there is some opportunity to meet nutritional needs of individual sows. This option typically requires additional space; up to 30 sq ft per sow. The crates themselves have lots of moving parts creating lots of opportunities for maintenance. The self-catching crate typically has been much more expensive than other alternatives to implement.

Conventional electronic sow feeding – Animals are uniquely identified with a microchip that allows for computer controlled individual feeding and management. Sows enter the feeding station one at a time and eat as individuals. This allows for less feed wastage and precise control of daily feed intake, and is likely superior to the gestation crate in terms of sow nutrition. Sows can be either automatically marked or selected for routine procedures (vaccination, preg-check, move to farrowing). Durable, reliable feed stations are required for the success of this technology. Some computer literacy is required and the management of an ESF facility likely embraces the largest deviation from a traditional crated barn. However, taken together, ESF offers tremendous opportunity to improve the feeding and management of sows. Our experience with pen gestation over the last decade suggests that electronic sow feeding (ESF) is the crate alternative with greatest upside potential for productivity and profitability. However, it also requires the most changes in standard operating procedures to fully realize its potential. Here I provide some specific examples regarding implementation of ESF.

Low-functioning electronic sow feeding – In recent years, a second type of electronic sow feeding has emerged based on a self-catching or free access stall. The sow enters the stall and feed is dispensed in a fashion similar to conventional ESF. However, when the sow is finished eating she must exit the crate by backing up and releasing the lock on the crate door. This approach is similar to the original designs of ESF from 30 years ago in Europe that ultimately was abandoned for the pass-through design now used in conventional ESF. In conventional ESF, the sow in part leaves the station due to the entry of another sow, which encourages her to leave the station through the front exit. With the low-functioning ESF stations, the sow actually controls when she leaves as she needs unlock the self-catching to exit. To encourage sows to leave the low functioning ESF stations, water is typically withheld from the sow in the station such that she will eventually get thirsty and leave. The absence of water in the feed station dramatically slows down the rate at which sows consume feed and thus reduces the throughput of the low functioning stations. The sow to feeder ratio with conventional ESF is often 60-80:1, but for the low functioning ESF that ratio drops to 15-25:1. As such, several more stations are needed to feed the same number of sows. The per station cost of the low functioning stations are less and often the per sow cost can be similar to that of conventional ESF. A potential strength of the low functioning ESF is that animals likely adapt to it with less training than conventional ESF. The low functioning ESF station also offers more flexibility in terms of group size (as few as 15 head) which can be convenient in retrofits and remodels. The disadvantage of the low functioning ESF is that it lacks much of the upside potential of conventional ESF (see below). Without a pass-through design, low functioning ESF cannot be used to sort animals. Furthermore, given the reduced sow to feeder ratio institution of things like spray marking, small quantity dispensers, or phase feeding from multiple diets will be more expensive as supplemental equipment costs may be as much as 4 times as much compared to conventional ESF.

Success with conventional electronic sow feeding

Electronic sow feeding (ESF) is the only alternative to gestation stalls that provides true individual animal nutrition. For producers with high production expectations, it is likely the best choice. Conventional ESF also promises the opportunity to further automate sow management (eg, spray marking of animals requiring vaccination or selection of animals needing to move to farrowing). We now have farms pushing above 30 pigs weaned per sow per year which for most US genetics is exemplary and thus the technology itself, if properly managed, is not a barrier to outstanding production. Our experience is drawn from the feeding of over 200,000 sows with ESF on 80 farms in Canada and the US. The farms range in size from 100 to 10,000 sows, utilize a variety of common genetic suppliers, and are either family-owned and operated or company-owned and run with hired labor. We do not face the same detailed regulations as found in the EU and thus have more flexibility in our solutions to pen gestation. Outlined below is a brief overview of a variety of topic we have found important.

Parity segregation

We routinely recommend that farms practice some degree of parity segregation when organizing the flow of their animal through a pen gestation barn. Gilts are one of the most valuable assets on the farm and the farms that have elected to flow the gilts separately from higher parity sows are afforded several opportunities. These animals can be feed differently (eg, different ration than the sows) and managed differently (perhaps attended to by best available labor). When mixed with higher parity sows, the smaller, still growing gilt can be at a disadvantage and suffer negative consequences from being at the bottom of the social order. It may also be possible to segregate smaller P1 animals with gilts or to further subdivide the older parity sows (eg, P2's from the rest) in an effort to reduce stress and competition in the pen environment.

Group structure

Dynamic – Group constituency is constantly changing. Essentially this is a continuous flow system and hence space utilization can be optimized. Dynamic flows are required on small farms where the weekly breeding group is less than that supported by a single ESF station. This will range between 50 and 75 sows depending upon the manufacturer of the ESF station. We have used dynamic flows on 700 sow farms, on 1400 sow farms that choose to flow gilts separately from the rest of the herd, and on 2800 sow farms in combination with static pens (see below). Dynamic flows work very well with big groups of sows as the turnover of animals would be disruptive to the social order if it were more strongly established. We have been able to take sows in and out of big dynamic pens essentially at will and with impunity. Dynamic flow also works well pre-implantation systems as the 21 day returns to heat can be simply rebred and left in the pen to join the appropriate farrowing group. One major challenge to the implementation of a dynamic system is the loss of the physical integrity of breeding group. To appropriately flow the big group dynamic system, the weekly breeding group is divided and spread across several different pens. Management practices need to be adapted to accommodate the distributed nature of individual breeding groups which often involves an increased reliance on management and ESF system software and its ability to color mark or sort out specific animals within a pen.

Static – Group is constituted once, social hierarchy stabilizes, and the group is left intact for the duration of gestation. The implementation of static groups is often attractive because the physical integrity of the breeding group is typically left intact and some semblance of a breeding snake can be organized in the gestation barn. However, variations in weekly breeding targets, and any unanticipated fallout from a group can lead to sub-optimal space utilization of the facility as stable groups function as an all-in-all out flow. We recommend static groups on large farms where a weekly breeding group is as large as or larger than the optimal number of sows supported by a single ESF station. A 1400-sow unit with a weekly breeding target of ~70 sows is the minimal sized system facility for implementation of static groups.

However, if one chooses to flow gilt separately from sows as we recommend, this reduces the number of sows available in week to constitute a static group to ~54 head and depending upon the ESF station maybe too small to fully utilize its capacity. We most often start the implementation of static groups with 2800 sow barns. Here we would have about 90 sows and 30 gilts comprising the weekly breeding group. We make one static group of ~75 sows per week and then run the gilts and remaining sows in separate dynamic flows. This captures the convenience of a static system, but helps maximize space and feeder utilization as expected from a dynamic flow. Static flows are simplest with 5600 or larger sow units. We more often see static flows coupled to post-implantation systems. On farms that are weaning and refilling farrowing rooms several days a week, they may constitute a static pen, but emptying it dynamically, eg, over a 3- or 4-day window. This can lead to some sub-optimal space utilization in gestation, but helps to maximize weaning age.

In sum, the decision about group structure largely depends on farm size. For farms practicing parity segregation of gilts in gestation, dynamic groups are used for herds of 1200 sows or less, static groups for herds of 5000 or more, and herds with size in between these can use some combination of static and dynamic groups optimize animal flow and productivity.

Time of group formation

Pre-implantation – Sows are crated after weaning and bred in the stalls. Groups are constituted as soon as animals are out of standing heat. The pre-implantation system also allows a sow earlier access to the ESF station during its production cycle which has the potential to provide a more sophisticated nutritional program than could be achieved in a gestation stall. The pre-implantation system minimizes the number of gestation stalls in the facility (eg, typically 2 weeks of production or less) and prioritizes the management of sows in pens. However, a pre-implantation system may be less forgiving to implement as there is a 3- to 5-day critical window for moving of animals into pens that must be respected to insure high farrowing rates and large litter sizes.

Post-implantation – Sows are crated after weaning and bred. Groups are constituted only after being confirmed pregnant at ~35 days. Implantation is complete before mixing sows, minimizing the possible reproductive negative impact. This approach is initially attractive as it leaves the basic reproductive management of the sow unaltered from a crated barn and physical integrity of breeding group can be left intact if farm size is large enough.

Our experience is that both pre- and post-implantation systems can support good production. Some of our best ESF herds are over 30 pigs weaned per sow per year and are using a pre-implantation system. Thus, pre-implantation group formation is not unto itself a barrier to a highly productive sow herd.

Design of pens

While ESF is an outstanding way to feed sows, it per se does not do much to help mitigate the untoward effects of animal-animal aggression common in groups of pigs. Successful implementation of ESF also requires the management of social hierarchy in pen gestation. There several important details in the design of the pen layout that helps to ensure success. Some factors to consider are:

Space allowance – We initially designed our barns for pens with stocking density of 18 to 20 sq ft (1.67 to 1.86 sq m) per head. As producers have become more accepting of alternatives to gestation stalls and are more open to questions of how best to implement these alternatives, we have been putting some barns in at 22 to 24 square feet (2.04 to 2.22 square meters), which is similar to the European standard for a pen with more than 40 sows. As stocking density decreases (or space allowances increase), the social dynamics for the sows are less intense, and management of the pens is more forgiving. On new construction, larger space allowances translate into increased construction cost, whereas on a retrofit this can result in loss of inventory.

Feeder capacity – We target the placement of ~75 animals per electronic sow feeder. However, feed capacity does vary between vendors in part based on different design feature of different feeders.

Pen size – For dynamic pens, we favor the use of pens with two or three feeders per pen or at least 150 sows per pen so that the social hierarchy is less well maintained. That makes it easier to introduce new animals. As the number of animals in the pen increases, it becomes harder for barn staff to provide quality individual animal care. That's why we typically do not use more than three feeders per pen, which results in a maximum number of animals in a pen up to approximately 230 head. It is unclear there are advantages to the sow with larger pens (eg, 300 sows), but it can make it harder on barn staff to manage. With static pens, we use a single feeder with 75 to 80 head per pen. In preimplantation barns, we typically place 80 to 85 head in static pens to anticipate the standard fallout of a few sows at 21 days postbreeding. These recycling animals are moved from the pens to the breed row to be inseminated again, and space utilization in the pen is optimized.

Pen shape – We most often use a rectangularshaped pen. A single feeder is placed on the fence line along one side of the pen adjacent to the alleyway. The bedrooms are placed along the back of the pen facing the feeder. A natural traffic zone develops between the bedrooms and the feeder/fence line and allows sows to move unimpeded to and from the feeder. The rectangular shape of the pen ensures there is sufficient flight distance for a sow to escape her aggressor and increases the amount of perimeter for a given square foot of pen, as sows like to lie along the perimeter. All animals are also directly visible from the alleyway. For large dynamic pens requiring more than one feeder, we have connected the rectangular pens end to end along the alleyway. These large multifeeder pens share all the attributes of singlefeeder pens and allow for automated sorting of animals from the pen directly into the alleyway.

The ESF technology is flexible in its implementation, so when necessary we have also utilized pens that are more square in their shape. In this case, the bedrooms are aligned facing at a right angle to the feeder. Appropriate spacing of the bedrooms allows for the development of similar traffic zones for sows to walk from their resting area to the feeder. However, sows are less visible from the alleyways in this layout. We have also applied a similar type of pen layout to large dynamic pens where multiple feeders are needed. Here the feeders are aggregated in a block at one end of the pen and may share a common sorting mechanism. The challenge with placing the feeders in close proximity is that a single boss sow has the potential to dominate multiple feeders. As such, we reduce the expected number of sows that can be feed by a single feeder in a day by about 10% to 15%, so this layout requires more ESF stations.

Pen dividers – We recommend the use of pen dividers along the back wall of the pen to create "bedrooms" for the sows. This further increases the perimeter of the pen which promotes orderly laying patterns and allows for the development of social sub-populations in the pen. We often observe the same animals lying in the same bedroom. We recommend that farmers label each of these laying areas as a convenient reference with which workers can find animals.

Solid laying areas - We have farms that utilize either solid flooring or slatted flooring in the bedrooms. We are not legislated in our country with regard to flooring and requirements for solid areas. However, the animals will prefer the solid area for laying compared to a slat and thus this helps to further order the pen.

Waterer placement - We prefer to place the waterers close to the entrance and the exit of the feeders. This discourages animals from sleeping there and creating congestion around the station.

People factor

Successful implementation of ESF depends critically on careful consideration of the design and flow of the barn discussed above. However correctly addressing these factors does not guarantee success because people have the biggest impact on the outcome of the transition away from gestation stalls. People working in the barn must be enthusiastic and committed to the project, as well as equipped with the necessary training and/or skills to succeed. In our experience, the implementation of ESF usually goes easier on owneroperator farms as the person who made the decision to implement ESF is usually working in the barn and their personal investment is often at stake. This is in contrast to the situation in the more integrated production companies where the individuals who are tasked with making ESF work were not involved in the decision process, and they rarely have a significant financial stake in the outcome of the decision. Regardless of the size or type of the farm, it is important to identify existing staff members who are looking for new challenges and not resistant to change. Less pig experience and an on-the-job workertraining program can in some cases be preferable for at least part of the staff rather than trying to reeducate experienced workers from a crated barn.

In addition to selecting the right people to staff an ESF barn, providing these workers with the appropriate training is also critical to success. It is highly recommended that as many staff and for as long as possible visit/work in an existing ESF facility before the startup. The staff must learn the procedural differences that exist between a crated barn and an ESF barn. Common challenges include successful management of tasks new to an ESF barn such as electronic identification tag management (sows cannot eat without an EID tag, and this needs to be monitored daily) or training of gilts to use the ESF station (poorly trained animals will struggle to get sufficient feed intake in an ESF barn).

Another task that is dramatically different is individual animal care. The larger the pen, usually the harder it is to track individual animals. Staff needs to become adept at recognizing individual animals in need of care and then being able to repeatedly locate them. The former requires the staff to better understand sow behavior and movement in order to identify treatable problems at the earliest possible stage. The latter necessitates a system to mark individual animals or at least a labeling schema to identify specific locations of animals within the pens.

Finally, visiting a working ESF barn allows staff new to this approach to get a vision of how well such a barn can run as there are more ways to make the system not work than work. Staff will benefit from having a clear view of their road to success.

Individual animal care

One of the biggest challenges I see during the transition to pen gestation is inability to maintain the same level of individual animal care as routinely achieved in the crated barn. The larger pen usually the larger the challenge as it is increasingly more difficult to use the physical location of the animal as the identifier of an animal and or the problem to be managed. Thus, barn staff needs to more skilled at identifying individual animals with the group as well as recognizing problems with in the pen. Several things can be done to facility the success management of these challenges:

1 - Uniquely label all physically identifiable locations within the barn. This would involve pens as well as demarcated sleeping areas. This allows barn staff to better communicate to management or fellow workers where problem sows are likely to be located.

2 - Assigning 300 to 400 individual sows per employee for them to monitor and observe daily in an effort make the barn staff more accountable for the well-being of specific animals.

3 - Use feed order through ESF station to identify animals that are at greatest risk of needing individual animal attention. Non-eaters would be the highest priority to investigate, then followed by the last 10% of the animals to eat. We recommend that you empirically determine how long it takes for 90% of your herd to eat and then daily printout a this of animals that have not eaten by this time of day.

4 - Not have too many sick pens in the barn as this forces the barn staff to identify problems early and treat the animals while they remain in the ESF pens.

5 - Organize staff's work schedule such that they spend much of their day either in or around the ESF pens. This increases the window of opportunity to observe animals in need of individual attention.

6 - Clearly mark animals identified as needing attention such that they can be readily found on subsequent days for continued treatment or observation.

Upside potential of conventional ESF

ESF offers several distinguishing features compared with other alternatives to the gestation stall. Our work has identified a variety of opportunities associated with ESF for the improvement of sow herd nutrition and management. This includes improved feed utilization by the reduction of feed needed to maintain individual animal body condition, better matching of feed delivered to changing nutritional needs of sows during gestation through the use of softwarecontrolled feeding curves, automated control of gilt estrus via the delivery of Matrix to selected individual animals feeding in the ESF station, and a practical solution to regulating the amount and timing of a feedback program in pen gestation to stimulate immunization against autogenous pathogens. The electronic identification of the sows via a radiofrequency identification tag also opens the door for additional digital management of the herd such as spray marking of animals requiring vaccination or selection of animals to move to farrowing. Based on these opportunities, we see ESF as perhaps the only alternative to the gestation stall that promises progress in how we will manage pregnant sows in the 21st century.

Conclusions

Our work has focused on the implementation of ESF on commercial farms in the US and Canada. We have employed a variety of approaches to ESF including static versus dynamic pens and pre versus postimplantation group formation. These choices were dictated by farm specific details, and none were found to preclude outstanding production. Successful implementation of ESF cannot be captured without some forethought on how the barn will be staffed and how staff will be trained. In the end people have the biggest impact on the outcome. We see ESF as an important and readily implementable alternative to the gestation stall that offers a tool for producers to advance the management of their sow herds.

https://doi.org/10.54846/am2022/169

California Prop 12: The slat-level experience

H. L. Frobose
JYGA Tech USA, Garnett, Kansas

Introduction

The passage of California Proposition 12 by voter referendum officially takes effect on January 1, 2022. This deadline has certainly captured the attention of the North American swine industry since its implementation directly impacts availability of pork for 40 million consumers. Attempts by farm and retailer organizations to overturn Prop 12 have failed thus far, so pork producers have begun making key investments to make compliant pork available for sale in CA in 2022. The key for sow farm compliance is the ability to meet the requirement that all female pigs kept for breeding have at least 24 square feet of useable floor space and be able to lie down, stand up, fully extend limbs, and turn around freely. This principally applies to gestating sows, as lactating sows nursing piglets are exempt, along with sows undergoing individual treatment and temporary animal husbandry purposes (< 6 hr per day). On paper, most can agree that additional space for sows is not inherently undesirable; however, the removal of gestation crates for breeding purposes adds significant challenges to animal management and increases likelihood of sow aggression. Perhaps equally significant, allowing 24 square feet per gestating sow space forces existing farms to recalibrate how to best use their limited space. To become compliant, most commonly farms must either reduce breeding stock inventory, cannibalize valuable farrowing or gilt development space, or add expensive new brick and mortar to keep farm output comparable to preexisting levels. The ensuing discussion focuses on these challenges and the various ways that producers are approaching them.

Sow housing

Although the US industry was already moving steadily towards group gestation housing, most farms had opted to continue to utilize gestation stalls during the period from weaning until pregnancy check, waiting to move sows into pens until 35 to 42 days post-mating. However, for Prop 12 compliance, gestation stalls cannot be used except for hospital sows and culls, or for temporary management purposes for less than 6 hours per day. While pen gestation after day 35 is most common in the US, it is also common to mix sows immediately post-breeding, which can yield similar reproductive performance but results in greater aggression at mixing.[1] For California Prop 12 compliance, producers must go one step further, by mixing sows into group environments immediately post-weaning. Mixing at weaning is the least understood, but the limited data available indicate that aggression, stress and injuries have the greatest implications on sows mixed at this time.[2]

US production systems have implemented a variety of feeding strategies to merge Prop 12 compliance with sow productivity and labor efficiency. These include electronic sow feeding systems (ESF), free-access stalls, shoulder stanchions, floor feeding, and others. A popular option chosen is to use free-access stalls (FAS) or modify existing gestation stalls for the weaned sows up until insemination, followed by a regrouping event where similar stage sows are mixed into electronic sow feeding pens where they can still be fed individually using RFID tags. Nonetheless, questions remain regarding the impact of multiple sow mixing events on hierarchy formation and performance, as well as the optimal way to utilize the available pen space. Most barn designs have attempted to concentrate labor tasks such as estrus detection, breeding, and pregnancy check in one section near the office. Existing barn layout and constraints such as slat locations, alley positions, and wall materials are critical factors in modifying a farm to Prop 12 compliant status

Compliance

Many producers have hesitated to invest in barn modifications for Prop 12 until the California Department of Food and Agriculture (CDFA) issued the final language to ensure compliance with the regulations. The proposed regulations were initially released May 28, 2021, and modifications to the text were made as recently as December 3, 2021, prior to the official implementation date of Jan 1, 2022.[3] Overall, the language has been consistent regarding the requirements for successful compliance. Some remaining areas that remain ambiguous include the type and fraction of gestating sow inventory in hospital areas, how biological variation between groups of sows will be interpreted, and the threshold for when replacement gilt officially need to be housed at 24 square feet. As implementation progresses into 2022, certifiers will provide additional clarification on how to handle these grey areas for overall success.

Observations

Many producers aptly point out that the CDFA acknowledges the standards set forth in Prop 12 are not based in specific peer-reviewed published scientific literature or accepted as standards within the scientific community. The economic analysis points to an increase in retail meat costs for CA consumers, increased cost of production for pork producers, and concedes minimal to no economic benefits because of implementation.[4] Nonetheless, farms representing at least 500,000 sow spaces have completed or in the process of converting facilities to Prop 12 compliant status, as a means to have access to the CA market.

In these farms, managing sow performance in a pen setting throughout the entire gestation period is the primary challenge. Optimizing barn design to minimize animal movements and worker tasks are critical. For some early adopters of Prop 12 standards, results have shown that housing sows in pens from weaning until farrowing can yield reproductive performance similar to sows in gestation stalls. Yet there are also increased opportunities for errors. Training workers to be good stockpersons and understand animal physiology is a challenge with today's slat-level workforce. For example, a pen designed to hold up to 50 sows with 24 ft^2 per animal does not mean that if you have 48 pregnant sows in a pre-preg check pen, you should add 2 additional sows to the group to maximize space utilization. Moreover, feeding sows in gestation pens creates challenges if the animals vary in body composition, parity, and stage of gestation. In Prop 12 facilities, the ability to individually feed a sow has increased importance because of the removal of the gestation stall as a management tool. Thus, many producers have re-evaluated ESF or similar feed systems that accommodate individualized sow feeding to combat the cost of body condition variation. Free-access stalls are also popular option to temporarily restrain the sow during breeding, but concerns over initial cost, maintenance, and compatibility with partially slatted flooring have resulted in wide variation of FAS use in Prop 12 sites.

Conclusions

The year 2022 will be a critical learning period for the US swine industry, as Prop 12 sites begin to be scrutinized by certifiers for compliance and a precedent is set for interpreting the remaining vague areas of the regulatory language. The industry has approached Prop 12 compliance in a variety of ways, and enough production data will start to be available to analyze which approaches yield the best sow performance and work environment. Although California only represents about 15% of per capita pork consumption, many producers are now viewing farms without pen gestation as a potential disadvantage in the future, and thus when existing equipment needs replacement, pen gestation after 35 days may be a wise investment. The question remains whether Prop 12 standards or equivalent will become required in other states or adopted by major retailers nationwide, and how soon. As an industry, we must also try to anticipate what new animal welfare requirements will be imposed in the future and be prepared to do so in the right way when necessary.

References

1. Stevens, B et al.: 2015 Appl. Anim. Behav. Sci. (165) 40-46.
2. Verdon, M et al.; 2015 J. Anim. Sci. (93) 1999-2017.
3. California Department of Food and Agriculture. Prop 12, Farm Animal confinement. 2021. Available at **https://www.cdfa.ca.gov/ahfss/Prop12.html** Accessed 5 December 2021.
4. California Department of Food and Agriculture. Prop 12, Proposed Regulations. 2021. Available at **https://www.cdfa.ca.gov/ahfss/pdfs/regulations/AnimalConfinement1stNoticePropReg_05252021.pdf** Accessed 5 December 2021.

https://doi.org/10.54846/am2022/170

The use of animal welfare as a tool to sustain public support for the use of animals in biomedical research: Lessons for the swine industry

J. O. Marx, DVM, PhD, DACLAM
Department of Pathobiology, School of Veterinary Medicine, and University Laboratory Animal Resources, University of Pennsylvania, Philadelphia, Pennsylvania

Introduction

The largest threat to the sustainability of the use of animals in biomedical research is an eroding public opinion of the use of animals in research.[1,2] The shifts in public opinion are primarily due to a well-organized attack of this process by animal rights organizations. While much of these organizations approaches rely on misinformation, the field of biomedical research must be cognizant of areas of potential improvement in our use of animals in research and then how we disseminate this information to the world. Ultimately, the goals and expectations are that improving animal welfare will not only improve public opinion of the use of animals in research, but also improve the quality of the scientific results produced.

The general public's concerns towards animal research revolve around: 1) that the animals are well cared for and 2) the information we get from using animals is directly applicable to human and animal health. Biomedical researchers are well aware of the link between the quality of the animals' welfare and the quality of the scientific results, which is very helpful in promoting the animals' welfare within our colonies. Even with that, financial costs of changes in husbandry and the potential impact of changes on the animal models are significant factors which need to be addressed with each proposed change. An important area of research the field is looking to enhance is the search to promote a positive affective state for the animals. Prevention of pain and distress has traditionally been the cornerstone of animal welfare in research, but more work has gone into providing for conditions which promote a positive perception of its life by the animal.[3] This is accomplished by multiple techniques, including giving the animals more control of their environment. When addressing our issues of sustainability, making changes is only half the story and disseminating this information is also critical to ensuring continues support for animal research.

Because of this, biomedical research has begun to promote transparency and to actively take the lead on promoting the discussion of the use of animals in research. For many years, researchers sat by quietly and felt safe relying on their important discoveries to guide the general public to appreciate the importance of their work. Over the past years however, as animal rights groups have become more organized, they have driven public opinion away from the acceptance of the importance of our work. Organizations have recently begun to "open up their doors" to the public, both literally and figuratively, using tours of their facilities, active promoting of the positive outcomes of their work, and the great lengths our field goes to in ensuring that animal welfare is preserved for our animals. Additionally, the field has started to use social media as a tool for dissemination of information.

Lastly, continued communication between other fields using animals will be imperative for continued improvement in the perception of the use of animals in research. All these fields can learn from the successes and failures of other groups in the battle for a positive public perception of their field.

References

1. Petetta et al: 2021, Addict Biol. 26(6).
2. Joffe et al: 2016, BMC Med Ethics. 17:17.
3. Jirkof et al: 2019, Animals (Basel). 9(12).

https://doi.org/10.54846/am2022/171

Statehouses and more: Legislation impacting animal agriculture

Elizabeth Rumley, JD, LLM (Agricultural Law)
Senior Staff Attorney, National Agricultural Law Center, Fayetteville, Arkansas

During sessions in 2021, state legislatures across the country have considered legislation that could, if passed, affect animal agriculture. Four of those types of legislation are discussed below.

Farm animal confinement

Laws affecting the housing requirements for farm animals have been proliferating for the past decade. They typically affect pregnant sows, veal calves and laying hens. In some states, they are more expansive and also require that products sold within the state come from animals living in those specific types of spaces (Figure 1). This leads to a primary effect on the living conditions of livestock raised on farms within the state boundaries, as well as a secondary effect on livestock raised on farms in other areas of the country, but whose producers would like to sell within that market.

In 2018, a ballot proposal known as "Proposition 12" was approved by the citizens of California. Sponsored by the Humane Society of the United States, it expanded previous state farm animal confinement requirements to include the following:

1. Eggs produced and/or sold in California must come from cage-free birds.
2. Pork and veal sold in California must come from farms without crates.
3. Breeding sows and gilts must be housed in enclosures with a minimum of 24 square feet per hog.

This proposal was challenged by agricultural organizations in two different cases, although the foundation of each case was similar. The cases were based on the "Commerce Clause" in the United States Constitution, where legislation affecting multiple states was reserved for Congress. Essentially, this provision prohibits states from creating barriers to trade by imposing obligations on out-of-state competitors in an effort to assist local businesses.

The first challenge to Prop 12 was led by the North American Meat Institute (NAMI). Each lower court that heard that case denied NAMI's request for an injunction that would stop the California Department of Food and Agriculture from enforcing the proposal. It was then appealed to the United States Supreme Court (USSC), which refused to consider it earlier this year. The second case was brought by the American Farm Bureau Federation and the National Pork Producers Council. The courts have not looked favorably on the AFBF/NPPC case either, with the District Court dismissing the original complaint and the appellate court upholding the dismissal. The plaintiffs AFBF/NPPC are currently requesting that the USSC agree to hear the case.

Unless the challenges are successful, Prop 12 will go into effect on January 1st of 2022 in the state of California. At that time, according to proposed regulations, sales of "whole pork meat" would be required to comply with the housing standards. Whole pork meat includes any uncooked cut of pork – including bacon, ham, chops, loin and ribs – that consists entirely of pork meat except for seasoning, curing agents, coloring, flavoring, preservatives or similar meat additives. Ground products such as sausage, ready-to-eat products and hot dogs are not covered under the proposed regulations.

Slaughter and processing

The processing of livestock- which includes animals such as cattle, sheep, swine, and goats - is governed on a national level by the Federal Meat Inspection Act. USDA-FSIS is given primary authority for oversight of meat products that will be offered for sale. One of the main components of that oversight is the requirement that the slaughter of livestock and processing of meat products be subject to continuous inspection by government inspectors.

That authority, however, may be designated to a state agency in those states that chose to apply for such authority, as long as the state requirements are "at least equal to" those enforced by USDA-FSIS. State inspection programs operate under a cooperative agreement with FSIS, and facilities in states with state inspection can choose between FSIS or state inspection. The difference between the two approaches is that state inspection programs only allow for meat processed in these facilities to be sold within the state - "intrastate"- while FSIS inspected facilities can export meat to other states, or "interstate."

There is a "in-between" level, so to speak. Plants inspected through programs such as Talmadge-Aiken and Cooperative Interstate Shipment allow products to be sold interstate while being overseen by inspectors who are state employees. However, all of those programs require both a state inspection program as well as additional cooperative agreements with FSIS.

Currently, 27 states have chosen to run a meat state inspection program, while 25 states have chosen to implement a poultry state inspection program. Within the past legislative session, several other states have considered beginning a state meat inspection program, and two of them- Arkansas and Oregon- have passed legislation to begin the process (Figure 2).

The slaughter and processing of livestock and poultry for the exclusive use of the owner, their household, guests or their employees, commonly called "custom exempt," are exceptions to the typical inspection requirements. In practice, producers may sell portions of an animal (eg, 1/4 steer, 1/2 hog) to several consumers while the animal is still alive. At that point, the consumers become co-owners of that animal, and once the animal is completely sold the producer acts as an agent to arrange transportation to the slaughter and processing facility. Each individual consumer/owner is then responsible for choosing how the animal should be processed, as well as paying both the producer (for the animal) and the processing facility (for the processing). Products that have been slaughtered and processed based on custom exempt guidelines may not be sold or donated.

Legislation has been proposed on both the state and federal level that would modify the current requirements for federal and state inspection. FSIS has been in contact with states who are considering this legislation, warning them that those modifications would be in violation of the Federal Meat Inspection Act, and would lead to suspension of the states' inspection programs.

Figure 1: States with statutes governing confinement and sales. April 2021.

■ Confinement statutes ■ Limit sales of eggs/pork/veal ■ Limit sales of eggs

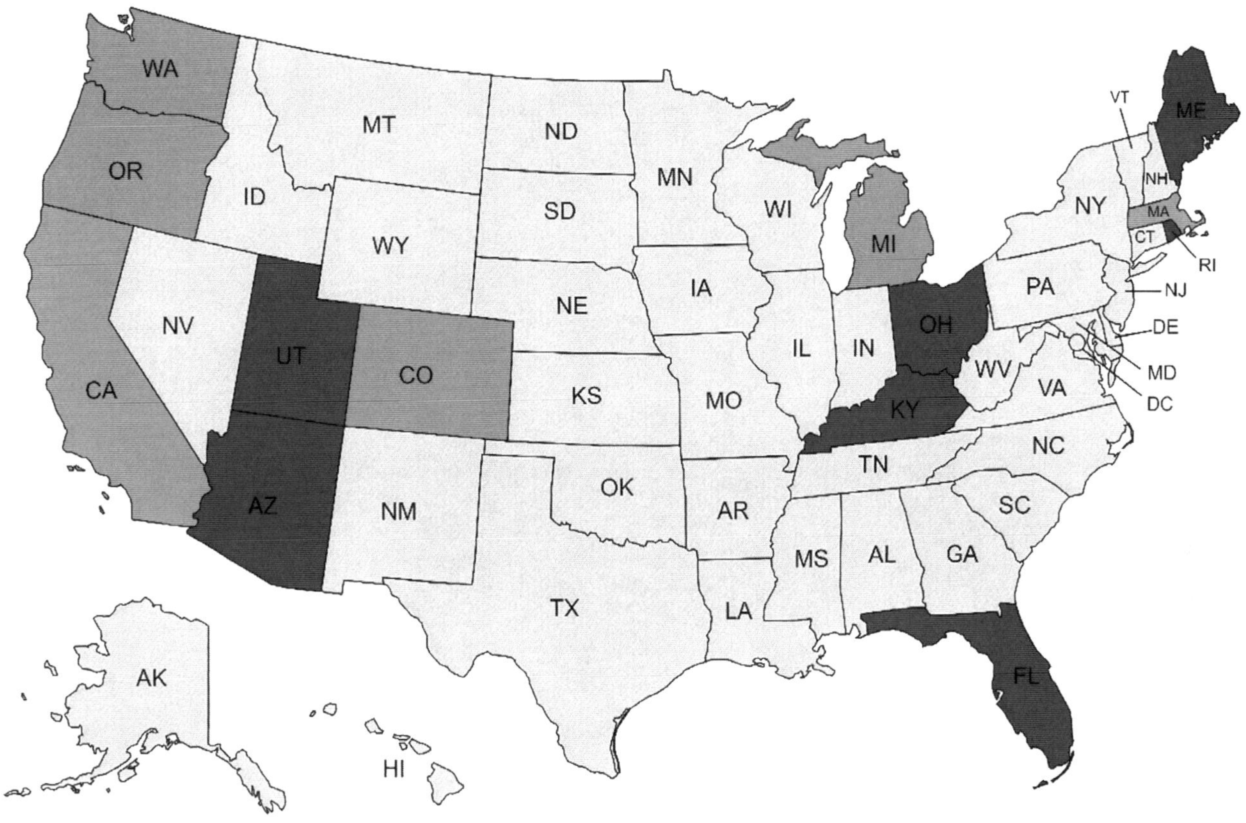

"Right to Farm" statutes

Legally, a nuisance is a substantial and ongoing interference with another's use and enjoyment of their property. In an agricultural context, this can include odor, lights, noise, dust, insects, or smoke. All fifty states have enacted right-to-farm laws that seek to protect qualifying farmers and ranchers from nuisance lawsuits filed by individuals who move into a rural area where normal farming operations exist, and who later use nuisance actions to attempt to stop those ongoing operations. This past legislative session, several states have proposed changes to their right to farm laws, although Florida is the only one that passed.

In part, the right-to-farm law included a definition for nuisance, which includes "any interference with reasonable use and enjoyment of land, including, but not limited to, noise, smoke, odors, dust, fumes, particle emissions, or vibration...." The term is further defined to include "all claims that meet the requirements of this definition, regardless of whether the plaintiff designates those claims as brought in nuisance, negligence, trespass, personal injury, strict liability, or other tort." These definitions limit the legal actions that can be brought by neighboring landowners. Under Florida's new law, plaintiffs may only bring lawsuits if the affected property is within a half mile of the alleged nuisance. For actions that are brought, however, the law requires that a plaintiff show by "clear and convincing evidence" that the alleged nuisance does not comply with state or federal environmental laws, regulations, or best management practices. If a plaintiff unsuccessfully brings a nuisance lawsuit against a compliant farming operation that has been in operation for over a year, that losing plaintiff will be liable for all costs, fees and expenses incurred during litigation.

In terms of damages, Florida lawmakers also limited compensatory damages to the reduction in fair market value of the property affected by the nuisance. Finally, a plaintiff cannot recover punitive damages unless the nuisance action has resulted in criminal enforcement or an environmental violation by a government agency within the last three years.

Ballot proposals

Ballot proposals are voted upon directly by citizens, rather than being placed in front of the state legislature. This type of legislative enactment has recently been important to animal agriculture because several farm animal confinement statutes, including California's, have been passed through this route. During the past legislative session, two states had ballot proposals that would have a significant impact on animal agriculture.

Colorado Initiative 16 would have prohibited the slaughter of livestock unless the animal has lived "one quarter of their natural lifespan." For pigs, that was defined as 15 years, thus forbidding the slaughter of hogs less than 45 months (3.75 years) old.

Figure 2: State meat inspection programs and proposals.

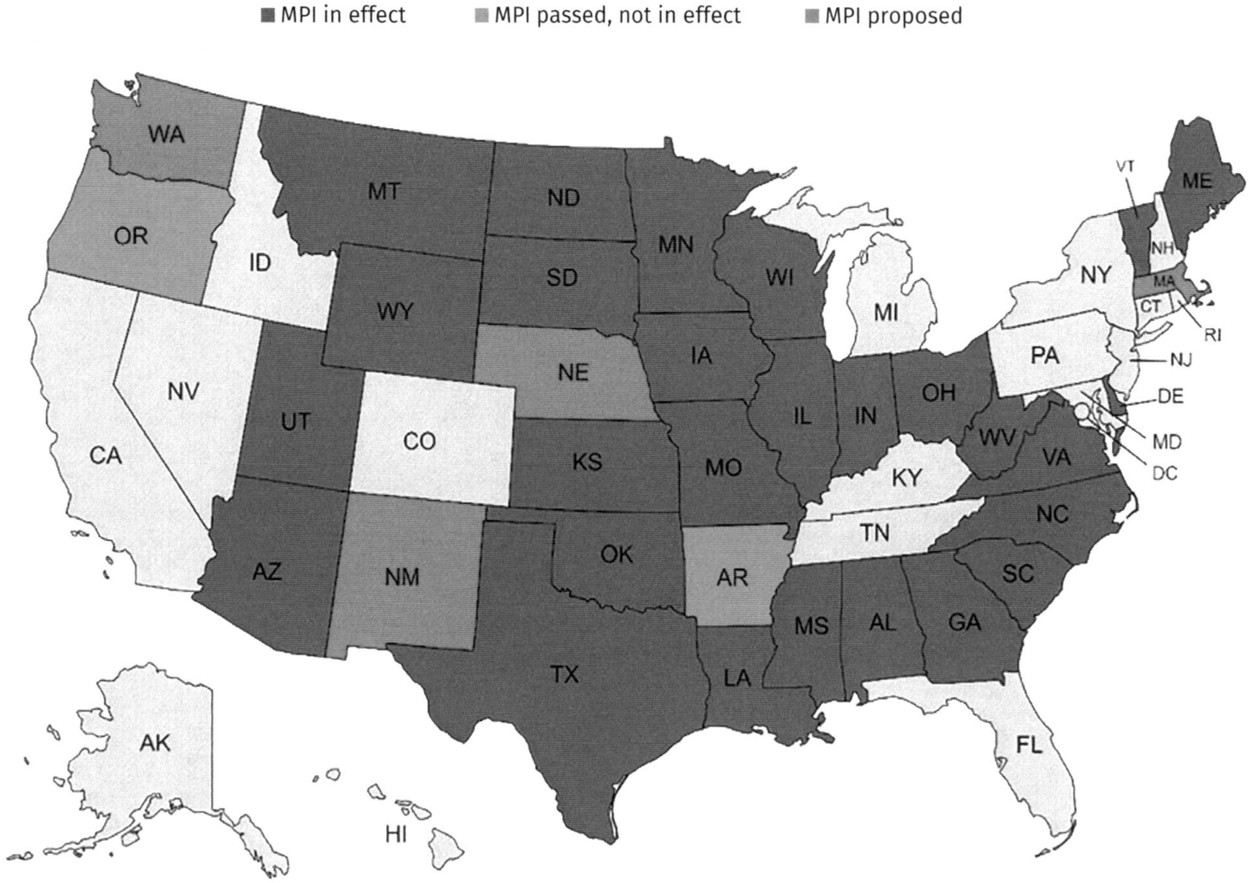

Further, it would have attached criminal penalties to some agricultural practices in breeding and animal husbandry by redefining a "sexual act with an animal" to include "any intrusion or penetration, however slight, with an object or part of a person's body into an animal's anus or genitals."

Initiative 16 had been approved for circulation before opponents to the proposal challenged it in the Colorado court system. The Colorado Supreme Court decided that it violated a state requirement that Initiatives only cover one subject, because the slaughter part of the initiative focused on livestock, while the definition of a sexual act included all animals. As a result of the ruling, it was not available for Colorado residents to vote on in November of 2021. However, proponents may regroup and resubmit similar proposals for future elections.

Oregon had a similar ballot proposal. Called Initiative Petition 13, it would have removed most current exemptions to animal husbandry practices. In other words, it would have banned the slaughter of livestock, along with dehorning, castration, branding and other practices. It would also have prohibited fishing, hunting, trapping, rodeo, animal exhibitions and animal research. Like Colorado's proposal, this initiative would have expanded the definition of sexual assault of an animal. In this case, it would become illegal to take actions for the purpose of 1) arousing or gratifying the sexual desire of either party; or 2) breeding domestic, livestock, and equine animals. Effectively, this would allow for embryo transfer by a veterinarian because there is a specific exemption for "animals subject to good veterinary practices" and IVF is specifically included as a veterinary practice under Oregon law. Another veterinary practice includes performing a "surgical... operation or procedure upon an animal". If that phrase could be re-interpreted to include artificial insemination, IVF, "flushing," and semen collection, those may be done as well, but only by a veterinarian.

This initiative made it through the initial part of the process, but proponents were unable to gather the 112,020 signatures that were required to place it on the ballot in 2021. However, it's still pending, and proponents are still able to continue gathering signatures in order to qualify it for the 2022 ballot.

Conclusion

While each state legislature is limited to passing laws affecting actions within their state boundaries, individual state laws can have significant effects on what happens across the country. For example, as evidenced by California Prop 12, when states with significant buying power pass laws about the products that may be sold in their states, they can move the market, so to speak, and force producers outside state boundaries to utilize preferred living conditions in order to capture that market-share. Secondly, opening the door by passing a law in one state makes it easier to pass in subsequent states. As a result, it is important to stay aware of pending legislation that may affect your practice.

https://doi.org/10.54846/am2022/172

Using technology to improve animal welfare

John Kolb, DVM, Dipl ABVP-SHM

Precision Livestock Farming (PLF) ... can offer an alternative philosophy to "farming by numbers" ... and ensure progress within the PLF field in favor of the livestock animals and their well-being.

Tomas Norton, 2019[1]

Introduction

Application of new technologies such as the internet of things (IoT), precision farming and sensors has been neglected in the animal farming segment. That may be changing, however, as increased interest in venture capital investment in agriculture generally and animal production specifically.[2] This brief review looks to summarize how different measures collected in production facilities relate to well-being and welfare of swine, what technologies are being developed to collect data directly or indirectly related to those measures and note the applications of these technologies by companies in the swine farming space.

Well-being versus welfare

These terms may seem to represent different concepts, and debate has centered on which is more appropriate for our use when dealing with food producing animals. In the context of this presentation, I will focus on well-being as a more encompassing term that relates not just to a particular set of recommendations or guidelines, but the overall ability of the animal to express both its biologic needs and potential for growth.[3]

Combined with economic models, companies may develop the "value proposition" to demonstrate both a biologic (Table 1) and financial return to their application; optimizing the well-being of the animal[4] and the financial return to the farmer at the same time. Other species have benefited from precision livestock concepts on a batch level.[5] Costs for installation and implementation for both groups of animals and an individual animal continue to decrease, and the analytic tools improve, making future application more cost effective. The increasing access to both 5G communications and improvements in rural internet access will also expand the ability of producers and veterinarians to evaluate and implement technologies in barn that may both improve the well-being of the animal and the profitability of the producer.[6]

Technology in precision livestock farming

Sensors and record systems utilize new technologies focus in one of two areas:

- Measurement of novel parameters, eg, activity levels in animals as an indication of health and welfare
- New technologies to improve upon existing data collection, eg, use of mobile applications to record data in near real time compared to paper-based systems
- Application of algorithms and other information systems to predict future performance on current behaviors

While it may not seem that mobile application data collection would be an improvement in animal well-being, the speed of notification of events such as treatments or mortality can assist in directly labor to areas of higher importance for the animal. Avoiding delayed treatment or expert observation can negatively impact the animal. In a weekly call-in system, for example, it may take more than 10 days from the occurrence of a morbidity or mortality event until it is both recorded in a central system and a report generated to management.

Tools may directly measure an event, ie, temperate, time at a drinker or feeder, video capture of an image.[7] Some of these measures are useful directly with no interpretation or modification - weight is weight! Other measures will need interpretation - how long at a drinker is really a drinking event?[6] What does this set of measures mean regarding estrus onset? The latter will require linking professional observations with indirect measures to develop and validate algorithms to interpret the data.

Discussion

Precision livestock farming offers great potential to improve both animal performance and well-being. High costs for devices or initial capital investment, limited connectivity, and failure to address producer needs has limited the uptake of PLF tools for swine. With declining hardware cost, improved access to broadband and 5G connections to upload data and utilize digital tools, PLF systems are becoming more cost effective to deploy. Matching the fit of digital technology with the appropriate value propositions may herald a new era of reconnecting the animal to the farmer in the swine production segment.

References

1. Norton T, et al (2019). *Review: Precision livestock farming: building "digital representations" to bring animals closer to the farmer.* Animal 13:12, pp 3009-3017. **doi:10.1017/S17517311199X**

2. Kimle K (2021). AgStartUp Engine, Iowa State University Research Park. Personal communication.

3. Ayre M, et al (2019). *Supporting and practicing digital innovation with advisors in smart farming.* Wageningen Journal of Life Sciences 90-91; **doi.org/10.1016/j.njas2019.05.001**

4. Gomez Y, et al (2021). *A systematic review on validated Precision Livestock Farming technologies for pig production and its potential to assess animal welfare.* Front Vet Sci 8:660565. **doi.3389/fvets.2021.660565**

5. Okinda C, et al (2020). *A review on computer vision systems in poultry: A welfare perspective.* Artificial Intelligence in Agriculture 4: 184-208. **doi.org/10.1016/j.aiia.2020.09.002**

6. Brown-Brandl T, et al (2019). *A review of passive radio frequency identification systems for animal monitoring in livestock facilities.* Applied Engineering in Agriculture 35(4): 579-591. **doi:10.13031/aea.12928**

7. Chen J, et al (2021). *Genetic parameters of drinking and feeding traits of wean-to-finish pigs under a polymicrobial natural disease challenge.* Journal of Animal Science and Biotechnology 12:105. **Doi.org/10.1186/s40104-021-00622-x**

8. Benjamin M and Yik S (2019). *Precision Livestock Farming in Swine Welfare: A review for swine practitioners.* Animals 9,133; **doi:10.3390/ani9040133**

Table 1: A partial list of technologies employed in swine precision farming to improve welfare (adapted from Gomez 2021)[4]

Technology	Measure	Welfare application	Companies in space
RFID enabled devices	Weigh scales	Weight gain/loss	Multiple
	Feeder & drinkers +/- weigh scales	Generation of eating and drinking events	BigDutchman Datamars Gestal/Jyga Hotreco Nedap Osborne PigEasy
Camera vision	Activity	Detect abnormal activity levels	Farm See MotionGrazer RoMain Serket
	Weight	Weight gain/loss	Asimetrix Fancom Lemberg Skov SmartAgriTech Wuggl
Force sensors	Weight Gait abnormality	Weight gain/loss Lameness, conformation	MotionGrazer
Thermal camera vision	Temperature	Environmental temp Body temperature (poultry)	
Audio	Cough Eating	Early detection of cough Eating event/activity	Sound Talks
Active monitoring tags	Location & movement, body temp		FarrPro PorkLogic Remote Insights
Connected sensors (temp, air quality, doors)	Environmental conditions, compliance to SOP, etc		BarnTools Distynct Intelligent Barn Realm Five
Cloud-based applications	Animal events, treatments, video records	Faster reporting of morbidity or mortality; treatment evaluation, action lists	AgriSyst Cloudfarm Farmera FarmHealth Gaurdian FarmTrace FarmWorx LeeO PTS PigCHAMP PigEasy Slide SwineTech

https://doi.org/10.54846/am2022/173

Utilizing vaccine to reduce the duration and impact of sow farm porcine epidemic diarrhea (PED) outbreaks

B. Sexton, DVM
The Maschhoffs, Carlyle, Illinois

Introduction

Since its emergence in the United States in 2013, porcine epidemic diarrhea virus (PEDV) has continued to negatively affect the swine industry. Naïve farms breaking with porcine epidemic diarrhea (PED) commonly experience pre-wean mortality (PWM) of 80 - 100% in suckling piglets. Shortly after the introduction of PEDV, vaccines became commercially available in the United States to help control the disease. There have been limited studies performed with PED vaccines, especially regarding impacts on elimination efforts. Commercially available PED vaccines have demonstrated an effective increase in antibody levels following natural exposure, as well as improved PWM in endemically infected farms.[1,2] However, pre-farrow vaccination of sows have not been found to effectively prevent the infection of piglets born from those sows.[3] Brown et al[4] showed no impact from the vaccine on the duration of viral shedding from gilts previously exposed to PED.

Across the industry, the use of PED vaccines is sporadic and system, or even vet, dependent. Like many swine production systems, upon diagnosis of PEDV, The Maschhoffs, LLC (TML) aim to eliminate the disease by utilizing controlled exposure and subsequent closure of the farm. In TML, PED vaccines have been used in specific instances at the discretion of the herd veterinarian to augment the elimination of the virus, especially in situations where closures have been prolonged. The objective of this project is to investigate the potential role of PED vaccine in reducing the duration and severity of PED outbreaks in commercial sow farms.

Case study

In February of 2021, four commercial sow farms representing approximately 12,000 sows within a five mile, were infected with PEDV over the span of 15 days. The farms were all PEDV naive, with no outbreaks the previous five years. Due to the perceived success of utilizing vaccines to help control PED in the past, the close proximity to several other naive sow farms, and the relatively high price of isowean piglets, the decision was made to implement a PED vaccine strategy to help control the disease.

Shortly after diagnosis, the piglets were early weaned to 12 days of age. All remaining pigs were euthanized. Newly farrowed piglets were subsequently euthanized for an additional 12 days. TML protocol is to euthanize for approximately 16-18 days, but the decision was made to attempt to save piglets starting at 12 days due to the high price of isowean piglets and the hypothesis that the vaccination would allow the successful saving of piglets at an earlier age. Sows were vaccinated pre-farrow with a commercially available PED vaccine, using the vaccine schedule in Table 1. Initially sows farrowed after receiving a single dose one week pre-farrow. At five weeks post-break, the farms farrowed three weeks of sows and gilts that had received vaccinations at four and one weeks pre-farrow. This was done to fill the farrowing house with three weeks of highly immunized dams that would hopefully create a "bubble" to help prevent the recirculation of PEDV within the farrowing house through susceptible piglets. At eight weeks post-break, the farms resumed farrowing animals vaccinated at one week pre-farrow and continued until the closure was completed. All four closures proceeded according to normal expectations, and the farms were declared PEDV negative following six consecutive weeks of 30 wean pig fecal swabs testing negative for PEDV via PCR testing when pooled in groups of five. This approach is summarized in Table 1.

Following the successful PEDV elimination at all 4 farms, a retrospective case-control study was performed utilizing production data in order to assess the perceived positive effects of the vaccine. Nine previous PED outbreaks with farms not utilizing vaccine were identified for inclusion as control farms, while the four vaccinated farms described earlier served as the case farms. All case and control farms: were either porcine reproduction and respiratory syndrome virus (PRRSV) Stage 2v or 3 according to AASV classification of PRRSV status, had the outbreak occur since 2016, and had no additional circumstances that had influenced the farm's ability to eliminate disease. Accurate production information was not available for three of the nine control farms, however closure length was available for all farms.

Data analysis

Three parameters were compared between case and control farms: closure length, pre-wean mortality (PWM), and pigs weaned per total sows weaned (PW/S). The closure length was measured in weeks from the time of the initial outbreak to the week of the sixth negative test. The lengths were compared using a standard mathematical mean comparison between the case and control groups. The production metrics were evaluated over the course of 25 weeks and broken down into four time frames. The time frames were: Pre-break (4 - 1 weeks pre-break), Acute Break (0 - 2 weeks post-break), Early Recovery (3 – 7 weeks post-break),

Table 1: PED vaccination schedule by farrow week post-break

Farrow week post-break	PED vaccine timing
0 – 2	No vaccine – piglets euthanized
3 – 4	1 week pre-farrow
5 – 7	4 & 1 week pre-farrow
8 – end of closure	1 week pre-farrow

and Late Recovery (8-20 weeks post-break). This time frame was chosen to allow the establishment of baseline production (Pre-break), and evaluate the production impacts over the course of the average length of the control closures. Pre-wean mortality (%) was transformed by the arcsine (square-root (%)) transformation prior to statistical analysis, to stabilize the variance and normalize data. Using the SAS Proc Mixed Procedure (SAS 9.4, Cary NC), transformed PWM and PW/S were analyzed with a model that considered the fixed effect of treatment, the random effects of farm and the residual error. Treatment Least Squares Means (LSMeans), for each variable separately, were calculated for each group and compared by the two-sided Student's t-test at the 5% level of significance. LSMeans for PWM were back transformed (geometric means) and presented.

It is important to note the limitations this data. One limitation of the test was that it is likely underpowered (we did not calculate a sample size) as the experimental unit was the farm and not the individual animal. When a study is underpowered we increase the risk of a Type II error (failing to observe a difference when the null hypothesis is actually false). The PWM is also a "quick-prewean" value, calculated as total piglets died during the week divided by total piglets born during the week. Because of this calculation method, mortality rates during the Acute Break period were able to exceed 100%. Finally, in order to assess PWM and PW/S along the same timeframe, it did not accommodate the lag between the identification of differences in the PWM and PW/S metrics.

Results

The comparison of average closure lengths between the case and control farms showed an improvement in closure length between the two groups. Vaccinated farms' closure lengths ranged from a duration of 12 to 16 weeks and had an average closure length of 14.75 weeks with a standard deviation of 1.64 weeks. The non-vaccinated farms' closure lengths ranged from 17 to 24 weeks in duration, and had an average closure length of 20.22 weeks with a standard deviation of 2.49 weeks. This represents a decreased closure length of greater than 5 weeks for farms utilizing the PED vaccine.

When evaluating PWM and PW/S, the data indicates an improvement in vaccinated groups, most notably in the Late Recovery period. Table 2 summarizes the results of PW/S based on treatment and timeframe. While there is a numerical difference in the Early Recovery period, this is likely due to management practices rather than the utilization of vaccine. The most important difference occurs in Late Recovery, with an advantage of almost 0.5 PW/S for vaccinated sows. Table 3 shows the geometric means of the pre-wean mortality of the case and control groups broken down by timeframe. There is a statistically significant difference in the pre-wean mortality of the Late Recovery period, showing a nearly 4% improvement for vaccinated farms.

Discussion

Case study

Although the data showed no difference between case and control farms during the Early Recovery period, observations from the farms indicated a substantial improvement in the ability to save pigs during the Early Recovery period for the vaccinated farms, as well as improved piglet survivability. Beyond the identified limits of the data, two primary explanations have been proposed to explain this discrepancy. The first is that coincidentally the case farms primarily broke late in the week, while the control farms primarily broke early in the week. This negated the five to six days the case farms were able to start saving pigs prior to the control farms, which could have shown differences in the Early Recovery period. The second potential factor is bias from the production team, as there was no blinding in this study.

A proposed mechanism of action

Overall, there is a relative lack of information on the clinical effect of utilizing vaccines to decrease the production impacts associated with PED outbreaks and duration of closures. A proposed mechanism for why pre-farrow vaccination could cause these improvements is based on the reduction of viral recirculation in the farrowing house. Utilizing the vaccine to bolster colostrum antibodies may improve the piglet's immunity to a point that infection can be prevented, as well as reduce the amount of virus produced by infected piglets. Theoretically, this would lead to fewer infected piglets over time and quicker elimination of the disease. Given that this mechanism of action has not been proven, it highlights the current gaps in our knowledge regarding PED vaccines. The impact of phase of production on immunity status, duration of viral shedding, degree of recirculation, and passive transfer of neutralizing antibodies via colostrum are all areas that need to be explored.

Table 2: Pigs weaned per sow differences between vaccinated and non-vaccinated groups

Timeframe	Treatment	Weeks observed	Mean weaned	Median weaned	Standard deviation	Minimum weaned	Maximum weaned
Pre-break	Non-Vx	24	11.86	11.56	0.92	10.57	13.92
Pre-break	Vx	16	11.97	11.94	0.30	11.58	12.50
Acute break	Non-Vx	18	3.06	0.00	4.73	0.00	12.96
Acute break	Vx	12	3.60	0.00	5.45	0.00	12.76
Early recovery	Non-Vx	30	6.87	9.74	5.08	0.00	12.38
Early recovery	Vx	20	5.62	5.43	5.52	0.00	12.31
Late recovery	Non-Vx	78	11.41	11.49	0.59	10.04	12.51
Late Rrecovery	Vx	52	11.85	11.87	0.50	10.41	14.33

Table 3: Pre-wean mortality differences between vaccinated and non-vaccinated groups

Timeframe	Geometric mean (Geo) vaccinate	Standard error (SE) vaccinated	Geo mean non-vaccinated	SE non-vaccinated
Pre-break	9.92	0.87	11.88	0.77
Early recovery	14.53	1.34	16.65	1.15
Late recovery	10.88[b]	1.01	14.61[a]	0.94

[a,b] Denote a statistical difference of $P < 0.05$

Conclusion

Based on clinical observations, as well as the results of the case study, there is a strong indication that the utilization of PED vaccine in sows pre-farrow can reduce the duration of closures and production impacts associated with PED outbreaks on sow farms. These improvements stand to have important financial impacts depending on the economic climate. While this study does not explicitly prove the benefits of PED vaccines, it does provide a basis for continued evaluation with the production system. Additionally, it highlights the need for continued efforts to grow our collective knowledge of PED.

References

1. DeDecker A, Coffey T, Rapp-Gabrielson V, Fredrickson D, Bandrick M, Sornsen S. Field efficacy of an experimental porcine epidemic diarrhea vaccine, killed virus, administered to pregnant sows. *Proc AASV*. Orlando, Florida. 2015:239-243.

2. Greiner L, Connor J, Graham A, Mellor J, Lowe J. Evaluation of a PED vaccine on piglet mortality and sow immunity. *Proc AASV*. Orlando, Florida. 2015:361.

3. Schwartz TJ, Rademacher CJ, Giménez-Lirola G, Sun Y, Zimmerman J. Evaluation of the effects of PEDV vaccine on PEDV naïve and previously PEDV exposed sows in a challenge model comparing immune response and preweaning mortality. *Proc AASV*. New Orleans, Louisiana. 2016:363-366.

4. Brown J, Rademacher C, Baker S, et al. Efficacy of a commercial porcine epidemic diarrhea virus vaccine at reducing duration of viral shedding in gilts. *J Swine Health Prod*. 2019;27(5):256-264.

https://doi.org/10.54846/am2022/174

The next frontier in disease elimination: Tackling the endemics *A suis*, *M hyorhinis* and *M hyosynoviae*

Maria J. Clavijo[1,2], DVM, PhD; Marisa Rotolo[1] DVM, PhD; Greg Ellis[1]; Joe Jobin[1]; Marney Jobin[1]; Marcelo Gottschalk[4], DVM, PhD; Jess Waddell[1], DVM, MS; Andrea Pitkin[1], DVM, MS; Deanne Hemker[1], DVM, MS; Thomas Riek[1], DVM; Jerome Geiger[1], DVM, MS; Wayne Cast[1], BS, MS; Kevin Eggers[5], DVM; Lucina Galina[5], DVM, PhD; Karen Harmon[2], PhD; Rodger Main[2], DVM, PhD; Bill Christianson[1] DVM, PhD; A. W.Tucker[6], DVM, PhD; Jean Paul Cano[3], DVM, PhD; Perry Harms[1], DVM, MS

[1]PIC®, Hendersonville, Tennessee; [2]Veterinary Diagnostic and Production Animal Medicine, Iowa State University, Ames, Iowa; [3]Pipestone Veterinary Service, Pipestone, Minnesota; [4]Department of Pathology and Microbiology, University of Montreal, Montreal, Quebec, Canada; [5]Zoetis Animal Health, Kalamazoo, New Jersey; [6]University of Cambridge, Cambridge, United Kingdom

Introduction

Actinobacillus suis (*A suis*), *Mycoplasma hyosynoviae* (*M hyosynoviae*) and *Mycoplasma hyorhinis* (*M hyorhinis*) are endemic pathogens that cause significant disease in late nursery and finishing phases. All three agents are considered normal inhabitands of the upper respiratory track of pigs, colonizing mainly the tonsils and nasal cavity. *A suis* infection is characterized by septicemia, polyserositis, sudden death, pleuropneumonia and arthritis typically in the grow-finish phase.[1] Likewise, MHS results in synovitis and arthritis in pigs mainly between 12 and 22 weeks of age.[2] Finally, *M hyorhinis* causes polyserositis and arthritis in nursery pigs.[3] A recent analysis of ISU VDL diagnostic data from the last 10 years showed an increase in the diagnosis of both *A suis* and *M hyorhinis* in polyserositis cases.[4] In contrast, detection and diagnosis of *M hyosynoviae* did not significantly increase over the last 10 years, however, *M hyosynoviae* continues to be the top cause of infectious lameness in grow-finish pigs as observed in the field.[5] Epidemiologically, all three agents have been traditionally classified as "late colonizers". These agents do not typically colonize a pig until late lactation providing an optimal opportunity for targeted pathogen elimination using early weaning techniques.[6-8]

The development of multi-site rearing systems in the 80s and 90s stemmed from the "Isowean principle" of separating suckling pigs from their sows to prevent vertical transmission of swine pathogens.[9] Since then, an assortment of isowean or medicated early weaning (MEW) programs have been developed for the elimination or control of various pathogens, such as, pseudorabies, *M hyopneumoniae*, PRRVs, toxigenic *P multocida*, *A pleuropneumoniae* and *A suis*, with varying degrees of success.[10-13] One of the main health principals for breeding stock companies is to stock farms with the healthiest population available. Thus, the objective of this work was to explore the feasibility of eliminating *A suis, M hyosynoviae* and *M hyowhinis* from piglets using a MEW program.

Materials and methods

The elimination program was based on a modified medicated early weaning protocol, and was divided into two phases.

Phase I consisted of a cross-sectional study carried out in a high health breeding site (Herd A) to estimate the prevalence of *A suis*, *M hyosynoviae* and *M hyorhinis* in different parity sows. A total of 60 tonsil scrapings were collected from parity 2, 3, and 4 sows, respectively, for a total of 180 sows sampled. In brief, sows were restrained using a snare and a oral speculum was utilized to open the oral cavity. A long-handled metal spoon was utilized to gently scrape the tonsil area to collect the mucosal excretion from the crypts. A rayon-tipped swab was utilized to collect the sample from the spoon and the spoon was disinfected between sows to avoid cross-contamination. The swabs were shipped refrigerated to the Veterinary Diagnostic Laboratory where they were tested for all three agents by PCR, using previously published protocols.[14,15]

In Phase II, 100 gestating parity 3+ sows from Herd A, were transported to an empty 1100 farrow-to-finish pen gestation sow herd (Herd B) located in KY. Prior to arrival of the sows, the entire site was thoroughly washed, cleaned, disinfected, dryed and audited. Internal clean and dirty lines were established which considered the gestation, office area and hallways as likely to be exposed with the targeted pathogens (Figure 1). Thus, a bench, visible signs, disposable coveralls and boots were placed at the entry of each farrowing, nursery and finishing room. Also, farm personnel disinfected their hands with a room-specific hand sanitizer and used a face mask prior to entering the room.

Sows were grouped into 8 cohorts by farrowing date. Prior to loading each cohort, sows were washed using a water-based detergent to reduce pathogen load. Once loaded into the farrowing rooms, sows were given Pulmotil® (tilmicosin) in feed and water during the farrowing phase, to achieve a dose of 15mg/kg. Farrowing rooms were managed under a strick McREBEL program. Piglets were given an injectable antibiotic (ceftiofur - Excede®, 5mg CE/kg) at birth, 7 and 14 days of age. Piglets were weaned at 10 days of age into on-site nursery rooms, where they were raised under an early wean intensive care program. Pigs were moved to the grow finish barns at 10 weeks of age through hallways that were washed, cleaned, disinfected and dried prior to the movement.

Tonsil scrapings were collected from sows at weaning. Oral fluids were collected from all rooms during the grow-finish stage. Serum and tonsil scrapings were collected from all pigs (n = 1220) between 22-26 weeks of age. At market age, the barrows and non-select gilts were sent for processing and the selected gilts were left on site for breeding. During mid-gestation, serum samples and tonsil scrapings were collected from all gilts on site (n = 465). All tonsil scrapings, except the swabs from the sows at weaning,

Figure 1: Herd B diagram showing the established clean (green) and dirty (red) areas of the farrow-to-finish site.

were pooled by 3 and all samples were tested by PCR for *A suis*, *M hyosynoviae* and *M hyorhinis* DNA. Sera was tested individually using an *A suis* ELISA, performed at University of Montreal.[16]

Results and discussion

The overall goal of Phase I was to identify the optimal parity population to use for the MEW program (Phase II), considering the genetic merit of the parity and the prevalence level for each agent. Table 1 shows the prevalence of each agent by parity group. The proportion of positive gilts for *A suis* and *M hyorhinis* between parity groups was not significantly different. In contrast, there was a significant difference in the proportion of *M hyosynoviae* positive sows between parities. Similar and contrasting literature has been previously published.[6,7,17,18] The decreasing trend observed in the study led to the decision to select parity 3 sows for Phase II, a population that would still have adecuate genetic merit but potentially a lower level of infection compared to younger females, increasing the chances of success.

In Phase II, 100% of sows at weaning tested *M hyosynoviae* and *M hyorhinis* negative by PCR, but 3% (3/100) tested PCR positive for *A suis*. While the timing of infection for these sows was unknown, the detection of *A suis* in P3 sows could reflect long term persistence of the bacterium in pigs. All oral fluids (n = 120) collected tested negative for all three agents. Similarly, all pools of tonsil scrapings (n = 406), collected from 1220 pigs, tested negative for all three agents by PCR, and all sera tested negative for the presence of *A suis* antibodies. Finally, all 465 gestating gilts tested negative by PCR for all agents and negative for *A suis* antibodies.

Geiger (2009) carried out a study comparing the effect of different piglet and/or sow treatment protocols for *A suis* elimination.[18] While *A suis* was not eliminated from the study groups, the prevalence of *A suis* in treated groups was significantly lower. In the present study, the use of an older parity sow, thorough cleaning and disifenction of the facilities, lower weaning age, a modified piglet medication program, an increased Pulmotil dose in sows, an intensive testing protocol, and the establishment of a clean and dirty line, likely contributed to the success of this project. Elimination efforts for *M hyosynoviae* and *M hyorhinis* based on MEW techniques have not been previously documented. Anderson et al,[19] 2017 showed the detection pattern of both agents in sows while under a herd closure program. In that study, a significant reduction in the prevalence (but not elimination) of *M hyosynoviae* and potential elimination of *M hyorhinis* in sows was observed following a 43-week herd closure.[19]

The establishment of an *A suis*, *M hyosynoviae* and *M hyorhinis*-negative piglet population from a positive P3 sow population was achieved by using a robust MEW program, applying a thorough biosecurity plan, and implementing comprehensive diagnostic testing.

Table 1: Proportion of positive sows by agent and parity group

	A suis	*M hyosynoviae*	*M hyorhinis*
P2	14/60 (23%)	9/60 (15%)	2/60 (3.3%)
P3	10/60 (17%)	6/60 (10%)	1/60 (1.6%)
P4	8/60 (13%)	0/60 (0%)	0/60 (0%)

Acknowledgments

The authors would like to acknowledge the invaluable work carried out by the farm employees at the Herd B site; Scott Norwood, Michael Pedigo and Todd Carter.

References

1. MacInnes J, Desrosiers R. 1999. Can J Vet Res 63:83–89.

2.Hagedorn-Olsen T, Nielsen NC, Friis NF, Nielsen J. Progression of *Mycoplasma hyosynoviae* infection in three pig herds. Development of tonsillar carrier state, arthritis and antibodies in serum and synovial fluid in pigs from birth to slaughter. Zentralblatt fur Veterinarmedizin. Reihe A. 1999 Nov 1;46(9):555-64.

3. Roberts, E. D., Awitzer, W. P. & Ramsey, F. K. (1963) Pathology of the visceral organs of swine inoculated with *Mycoplasma hyorhinis*. American Journal of Veterinary Research 24, 9–18.

4. Silva et al., 2021. Detection and diagnostic trends of five swine endemic bacterial pathogens (2010-2019). *AASV Proc*. 2021.

5. Canning P, Costello N, Mahan-Riggs E, et al. Retrospective study of lameness cases in growing pigs associated with joint and leg submissions to a veterinary diagnostic laboratory. *J Swine Health Prod* 2019;27(3):118-124.

6. Clavijo MJ, Murray D, Oliveira S, Rovira A. Infection dynamics of *Mycoplasma hyorhinis* in three commercial pig populations. Veterinary Record. 2017 Jul;181(3):68.

7. Roos LR, Surendran Nair M, Rendahl AK, Pieters M. *Mycoplasma hyorhinis* and *Mycoplasma hyosynoviae* dual detection patterns in dams and piglets. PloS one. 2019 Jan 3;14(1):e0209975.

8. Gottschalk M and Andre Broes, Actinobacillosis. In: Zimmerman JJ, Karriker LA, Ramirez A, Schwartz KJ, Stevenson GW, eds. Diseases of Swine. 11th ed. Ames, IA: Wiley-Blackwell; 2019.

9. Harris DL. Multi-site pig production. John Wiley & Sons; 2008 Feb 8.

10. Geiger JO, et al. 1991. Occurrence and elimination of Pseudorabies in an Isowean three-site production system, Proc AASP: 87–91.

11. Geiger JO, et al. 2002. PRRS Virus elimination in a 5000 sow multi-site system. Proc IPVS 1:280.

12. Baekbo P, 1999. Procedures to eliminate M hyo and produce M hyo free pigs, Proc AASP: 479–480.

13. Geiger JO et.al. 1992. Elimination of Toxigenic Pasteurella Utilizing Isowean Three-Site Production, Proc AASP: 45.

14. Kariyawasam S, Strait E, Jordan D, Kroll J. Development of a real-time polymerase chain reaction assay for detection of *Actinobacillus suis* in porcine lung. Journal of Veterinary Diagnostic Investigation. 2011 Sep;23(5):885-9.

15. Neto JC, Bower L, Erickson BZ, Wang C, Raymond M, Strait EL. Quantitative real-time polymerase chain reaction for detecting *Mycoplasma hyosynoviae* and *Mycoplasma hyorhinis* in pen-based oral, tonsillar, and nasal fluids. Journal of veterinary science. 2015 Jun 1;16(2):195-201.

16. Lapointe L, d'Allaire S, Lacouture S, Gottschalk M. Serologic profile of a cohort of pigs and antibody response to an autogenous vaccine for *Actinobacillus suis*. Veterinary research. 2001 Mar 1;32(2):175-83.

17. Geiger, J. 2009. Experiences with *A suis* – Part II. *AASV Proc.* 2009: 443-445.

18. Clavijo MJ, Davies P, Morrison R, Bruner L, Olson S, Rosey E, Rovira A. Temporal patterns of colonization and infection with *Mycoplasma hyorhinis* in two swine production systems in the USA. Veterinary microbiology. 2019 Jul 1;234:110-8.

19. Anderson A, et al., 2016. Effect of herd closure on the pattern of *Mycoplasma hyosynoviae* and *Mycoplasma hyorhinis* infection. AASV Proc. 2016: 92-93.

https://doi.org/10.54846/am2022/175

Rotavirus: Current experiences and thoughts with prevention and control

Jeremy S. Pittman, DVM, MS, DBAVP
Smithfield Hog Production – North Region, Waverly, Virginia

Introduction

Rotavirus (RV) continues to be a significant cause of piglet and post-weaning diarrhea resulting in increased mortality and reduced daily gain. In recent years, it appears that more practitioners have been frustrated in dealing with challenging cases of RV, particularly around RV type C (RVC), and has generated more "hallway" conversations around control strategies. It also seems that more focus has been brought to controlling farrowing house diarrhea after the experiences and lessons learned from the North American porcine epidemic diarrhea virus (PEDV) outbreaks of 2013-2015. This paper is an attempt to summarize the author's current thoughts and experiences with RV control and to shed light on the frustrations experienced by practitioners in a hope that more resources will be made available to practitioners. It should be noted, that while the author has presented previously on RV and control methods, he notes that often he still struggles with complete control, due to the challenges presented herein.[1,2] In fact, as this paper is being written, he is having discussion with a recent depopulated and repopulated farm on making a new on-farm RV "ice cubes" for use as feedback material. Finally, it must be stated that the work presented herein is based on the experiences shared by other colleagues, research conducted by scientists, practitioners and former students and the hard work of farm staff to implement the projects and protocols.

The basics

To start, we have to understand the virus, the piglet and immunity to the virus to understand how to develop control measures to address the issue of clinical disease. To borrow from a fellow practitioner; "the basic principles of diarrhea control and prevention are;

1) Decrease environmental exposure,

2) Increase passive immunity from sow to piglet and

3) Increase colostrum consumption to piglets."[3]

We will delve deeper into each of these aspects, particularly item 2, but fundamentally it comes back to these three principles and their interaction at the "crate-level".

Rotavirus

Rotaviruses are double stranded RNA viruses, with a genome of 11 segments, which allows for recombination and increased mutation rates.[4,5] There is no immunological cross-protection between serogroups RVA, RVB and RVC, meaning control must be independently considered and addressed. Within serogroup, there is further genetic diversity of the immunological epitopes; the outer capsid proteins VP7 (G serotyping) and VP4 (P serotyping), which induce neutralizing immunity.[5] Two other epitopes of interest are the more conserved inner capsid VP6, which defines the A/B/C serogroup and produces cross-reactive antibodies within group only, and the nonstructural glycoprotein NSP4, which acts as a viral enterotoxin.[6] More work is needed on the clinical effectiveness of immunity to VP6 and NSP4. In the author's opinion the genetic diversity and the influence on clinical immunity cannot be emphasized enough. A gilt, piglet, pig population or farm can have multiple RVs at the same or different times in their production life cycle, meaning clinical disease can reoccur with distinctly different isolates. RVs have been called the "influenza of the gut", and the level of immunological diversity appreciated with influenza viruses should also be bestowed to RVs.[4]

Environment

Rotaviruses are non-enveloped viruses and are very hardy in the environment, making sanitation an important component to control. In addition, infected piglets can shed high amounts of virus in both clinical and non-clinical infections. Facilities should be thoroughly cleaned of organic material, using a detergent and hot pressurized water, before using an effective disinfectant. Disinfectants in the classes of aldehydes and halogens (sodium hypochlorite) are most effective, followed by peroxygens and alkalis.[7] Finally, farrowing crates should be allowed to dry before being loaded, to allow desiccation of the virus and other pathogens. Often on commercial sow farms, this is a difficult series of events to achieve each turn of a farrowing room, which leads to increased risk of diarrheal disease.

Sanitation of the farm, outside of the farrowing crate, is also an important aspect of disease control. High levels of RVs in the breeding, gestation and hallway environment increase the risk of mechanical transmission to the farrowing environment. This risk was clearly experienced in the clean-up and elimination programs of PED. In cases of persistent uncontrolled piglet diarrhea it may be valuable to survey the farm environment to determine the level of RVs followed by a whole farm cleaning and disinfection event if the level is high. Greiner suggested a whole herd exposure to better control RV in endemic or epidemic herds, similar to the strategies utilized for PED elimination.[8] It would make sense that a whole herd sanitation measure following this approach would allow for increased homogenized immunity and reduced environmental load.

Internal biosecurity practices should be evaluated to reduce risk of transmission of RV to the farrowing environment and piglets; limited foot traffic, hand and equipment sanitation when handling litters (eg, day 1 care, processing, treating, vaccination), boot sanitation when entering farrowing, and other practices.

In addition to a sanitized environment, successful control of diarrhea in piglets requires and warm, draft-free micro-environment for piglets in the farrowing crate. This ensures that piglets are comfortable and have consistent suckling activity.

Immunity

Piglets are naïve to RV when born and are immediately and highly susceptible to infection. Incubation time for RV can be as short as 12 hours with virus being excreted in high quantities, leading to very rapid clinical disease and contamination of the environment.[9,10] This is mostly due to the fact that newborn intestines are lined with mature enterocytes, all of which are susceptible to rotavirus invasion and replication. Because of these factors, effective clinical immunity must come through passive immunity from the dam. Often as practitioners we are concerned and discuss the importance of colostrum quality and management, which is important for many reasons including initial control of RV, however control of clinical RV is more about ensuring lactogenic immunity through the entire sucking period.

Immunity to RV in piglets is mainly through neutralizing immunity from anti-VP7 and VP4 epitope antibodies in the colostrum and milk, which prevent attachment and entry to the enterocyte.[11] In recent years focus has primarily been on immunity to the VP7 epitope, but immunity to the VP4 epitope may increase clinical protection and reduce shedding, and protocols that addresses homologous immunity to both VP4 and VP7 epitopes may be more effective. In other species, immunity (luminal and systemic humoral and cellular) to VP6 and NSP4 may also be beneficial but this is less understood in pigs.[6,11]

Mucosal immunity and IgA antibodies are the most effective for neutralizing RVs in pigs. While IgG antibodies can also provide immunity, they appear less important for enteric pathogens. Colostrum has a higher IgG:IgA ratio, and this relationship flips with production of milk. This makes sense as IgG antibodies are important for systemic disease and pass through the intestinal tract into the piglet's body within the first hours of life, while IgA antibodies remain in the lumen, bathing and coating the intestinal tract throughout the lactation period. Fundamentally in RV control, this means that protocol focus should be not just on developing and ensuring good colostral immunity, but primarily generate a high quality and sustainable lactogenic immunity. In addition, factors that influence the pig's ability to access this high-quality immunity become important in managing RV infections. Activities that disrupt suckling behavior (manipulation of litters, cross-fostering, moving, sizing, sorting, vaccination, mass treatments, early weaning events), reduce suckling events (chilling, weakness, disease, competition) or produce poor lactation by the sow (anorexia, inadequate feed or water intake, mastitis, illness, etc) can result in disease or shedding by non-clinical piglets.

Once pigs are weaned, protection from RV is no longer present and the pig's own immune system must develop immunity to the different RVs to which it is exposed. In theory, RVs that piglets see in the lactation period may generate active immunity, however this immunity may not be of sufficient quality if lactogenic antibodies truncate active immune development. Also remember that the diversity of RVs (A/B/C serogroups and G/P serotypes) may allow a piglet population to develop immunity to different RVs preweaning, which does not cross-protect against post-weaning RVs.

Methods to influence maternal immunity

The primary focus of most RV control programs is to manipulate the immunity of the dam to provide optimal passive immunity. The common methods used by practitioners are 1) passive exposure, 2) intentional exposure (also known commonly as "feedback") and 3) vaccination. Each of these methods have advantages, disadvantages and variability in how they are accomplished.

Passive exposure

Passive exposure is simply the process by which replacement breeding stock are exposed to the farms' endemic pathogens through natural exposure to the environment and animals on the farm. This could be as simple as housing directly in a breeding or gestation barn, with close or direct contact to older breeding stock. It could be more involved, such as housing and mixing new replacement stock directly with to-be-culled or open animals. Exposure could be through housing in an isolation or acclimation barn or area that is not sanitized, allowing environmental contamination to build up and be a source of exposure. The advantage in this approach is that it doesn't take much extra labor, effort or cost above what is already required for entry of new breeding stock and is arguably a more natural method of exposure. The disadvantages are that shedding and exposure status is unknown and uncontrolled in both the new and old animals as well as the environment, and may result in destabilization of other diseases in the herd.

Passive exposure might also include a "do nothing" program, but in the author's experience "do nothing" programs result in "do something" programs later.

Intentional exposure

Intentional exposure covers a vast array of protocols with different terminology in the swine industry. Basically, an intentional exposure program means that there is an effort to collect some type of infectious material and present it to replacement or adult breeding stock in order to force exposure of the animal to the pathogen(s) of concern. This can be as simple as manure "feedback" or as complex as a system "ice cube" method. Other common terminology is oral controlled exposure (OCE) and natural planned exposure (NPE).

"Feedback"

The background of feedback has been well presented by others, as well as the risks and limitations.[1,12,13] Feedback as it is commonly termed in the US swine industry is the process of collecting manure, tissues or both that is assumed to contain desired infectious pathogens and then feeding this material back to adult animals in acclimation or gestation. Depending on the pathogens the practitioner or farm is attempting to control, the types of tissue may vary (enteric, systemic, respiratory, etc). For the purposes of this paper we will focus on RV (enteric) control and thus feedback herein will consist of a manure, intestinal tissues and contents or a combination.

The main issues with normal feedback programs are that the material's quality is relatively unknown at the time of exposure and it can create a "roller-coaster" effect of clinical disease and control and that.[3,13] These issues have been covered in more detail recently in these proceedings by the author.[14] These issues can be slightly mitigated with use of frozen batches and a "test and expose" process.[13]

"Ice cubes"

The ice cube method was developed by practitioners in the field to address the issues above with normal feedback.[1,3,15] It also addresses the issue surrounding the fastidious nature of RVB and RVC in cell culture, as colostrum deprived (CD) pigs are excellent replication vessels for RVs. This method clearly demonstrates the ingenuity of the swine practitioner. This process has been presented in detail elsewhere, but in short CD pigs are inoculated with known RV-positive infectious material (ie, PCR positive feces), which replicate high levels of virus over 18-24 hours of incubation. [1,2,9,10] The pigs are euthanized and intestinal material and contents are harvested, homogenized, diluted and aliquoted for storage. In the original protocols, the intestinal homogenate was aliquoted into ice cube trays and frozen, hence the terminology. Since then, others have stored aliquots in bags, tubes and other containers for storage.

This process can be conducted and implemented wholly at the farm, or a systematic approach can be taken for management within a pyramid or system.[1,2]

The advantages of this process are that pathogens within the inoculation and homogenate material can be tested for, controlled and quantified prior to use. Material can also be made continuously and in large volumes with similar quality for use over long periods of time, potentially indefinitely. The disadvantages are the labor and cost of the process (start-up and maintenance), the sacrifice of CD pigs and the oversight that must occur to ensure proper processes. Another disadvantage is that in some countries or jurisdictions, it is illegal to feed animal tissue to animals, and this process would qualify.

Vaccines for rotavirus

Another area of control used for RV is the use of vaccines. Currently there are 3 main types of RV vaccines available to swine practitioners in the United States, each with advantages and limitations. RV vaccines are normally used in a manner to elicit or booster immunity in gestating dams in mid to late gestation but could be used in replacement animal vaccine programs during acclimation. RV vaccines are less likely to be utilized in growing pigs, even though a commercial vaccine is labeled for baby and pre-weaned pigs.

Modified Live vaccine

A commercial RV modified live vaccine (MLV) is available through the ProSystem (Merck Animal Health, De Soto, KS) line of vaccines. This vaccine has been available for some time in the industry with reported good success. The main advantage of this vaccine is the modified live nature, allowing for a strong mucosal immunity to develop. The use of modified live virus allows for more broadly protecting immunity to be developed through exposure to all potentially protective epitopes; VP4, VP7, VP6 and possibly NSP4 enterotoxin, and thereby may provide more cross protection among the A serogroups, as compared to other vaccine options. The main disadvantage to the vaccine is it only contains RVA serogroup isolates and thus has no efficacy for RVB and RVC serogroup viruses. Another limitation is that the product contains RVAs with G5 and G9 serotypes, and thus may be less likely to protect against other RVA G serotypes. It should be noted here that the label for the ProSystem line of vaccines states; "Unique rotavirus protection includes two major serotypes G4 and G5 of Serogroup A", however based on recent sequence comparison, the isolates are better defined as G9 and G5, with the G9 possibly having cross-protection for G4 serotypes.[16,17]

Prescription vaccines

Prescription vaccines have been approved and used within the past 10 years [H3N2 influenza in 2012, RVC and PED in 2014 – Harrisvaccines, Ames, IA] in swine production in the United States and Canada. These are under a prescription platform until conditionally or fully licensed (eg, IAV, PED) and approved by the veterinarian of record. Currently there are two vaccine platforms available to swine veterinarians that use this technology: RNA particle (RP) vaccines through Sequivity (Merck Animal Health, De Soto, KS) and prescription platform vaccines (RxPP) through Medgene Labs (Brookings, SD). In general, for these vaccines, a gene(s) of interest is inserted into viral vector or backbone platform to produce and present a defined protective epitope or epitopes to the animal. What is unique about these platforms is that live antigen or pathogens do not need to be acquired for development of vaccine. This is possible because a sequence of the gene coding the epitope of interest is used to synthetically produce the gene that is then inserted into the virus vector or backbone. The main advantage of this is that pathogens that are fastidious or cannot be isolated can still have vaccines developed, such as with RVB and RVC. This also improves turn-around time for new vaccines or updates to new circulating isolates. The disadvantages to these vaccine platforms are that genes that code for protective epitopes need to be relatively well defined for the pathogen and only one or a few epitopes are normally included in each platform at a time. In addition, due to the limited and specific antigen presentation, immunity to these vaccines would not be as cross-protective as a MLV vaccine, due to the lack of presentation of other viral proteins (eg, VP6, VP4 if not included) and lack of multiple replication events in the host. In general, each antigen is prepared on its own vector or backbone. Multiple antigens can be combined after manufacturing to make poly-valent vaccines.

Due to the viral vector associated with RNA particle (RP) vaccines, the vaccine does enter the host cell, replicates and presents the antigen of interest, however continued replication does not occur. Thus, there is more stimulation of the immune system than a traditional killed vaccine. Currently RP vaccines are non-adjuvanted and RxPP vaccines are adjuvanted.

Autogenous vaccines

Traditional autogenous vaccines for RV are difficult to produce due to the need to isolate and replicate RVs to a high titer level to produce vaccine. To date, the author is not aware of any reported RVB or RVC autogenous vaccine and only a few RVA autogenous vaccines. It is unlikely that traditional autogenous vaccines will be a successful control strategy for RV until culture and propagation techniques are readily available and efficient.

Strategies of exposure

When applying the above feedback or vaccination methods there are a lot of considerations, options and variations in programs across farms. The author will attempt to address and share knowledge in some of those areas.

Historically timing of feedback in sow farms has been reported to be a 3 to 6-week pre-farrow (PF) time frame, usually with multiple exposures during this period. It is not clearly understood why this time frame is common among veterinarians but is similar to the time frame on common vaccine labels used pre-farrowing which allow for a time interval between the two doses and after the second dose to establish protective immunity. In addition, IgA producing plasma cells, B and T lymphocytes from the gut begin migration to the mammary tissue 1 to 3 weeks prior to farrowing.[8] Recently, Langel et al[18] demonstrated that mid-gestational exposure to PEDV resulted in a higher level of protection to the dam's piglets compared to early or late gestational exposure. This supports the concept that pregnancy is immunosuppressive, especially in the third trimester. It should be noted that dams in this were naïve prior to exposure, so the increased level of protection provided may also be an influence of timing of first exposure, initial timeline for immune development and level of viral shedding at the time of farrowing. It is not known if boostering previous immunity, with killed or live virus, would provide the same differences in protection given at these same time points.

One potential risk of late gestational feedback is an increased shedding of RV by dams at the time of loading into the farrowing environment and at the time of farrowing, thereby increasing the environmental load and risk to newly farrowed piglets. In some of our studies we have seen an increased level and percent of gilts shedding after exposure to NPE material, however this was in the weeks immediately following exposure and mostly resolved prior to farrowing, however a slight increase was seen post-farrowing. Others have reported similar findings of increased shedding by dams post-farrowing.[8] It is not understood if this is common due to stress of farrowing or secondary to increased exposure by NPE in late gestation.

Frequency of exposure to NPE, in addition to timing may also be an important factor of controlling RV. Anderson et al. demonstrated a linear and significant relationship to the number of NPE exposures gilts received during a 3 to 5-wk PF interval, with 3 weekly exposures resulting in 4% less PWM and 1.0 lb heaver wean pigs compared to wean pigs from non-exposed gilts.[19] Farkas et al[20] demonstrated a reduced number of pigs and litters and the number of litters shedding RV at 3 days of age when a second NPE was given at 1 wk PF, as compared to 3, 2 or 4 wks PF, after all animals received an initial 8 wk PF exposure.

Vaccination with a homologous RP vaccine 4 and 1 wk PF after a pre-breeding NPE showed significantly reduced mortality due to scours and increased adjusted wean weights when compared to pre-breeding NPE only.[21] Vaccination also numerically improved mortality and wean weights compared to NPE 5, 4 and 3 wks PF in this study. A similar study showed no difference between use of RP vaccine, NPE and control groups, however the authors cited lack of significant RV challenge in the piglets as a potential reason for the lack of difference observed.[22]

These studies may demonstrate an ideal situation of RV control where previous exposure to live virus, early in or prior to gestation, to more effectively prime the dam's (gut) immune system is followed by later parenteral boostering of that immunity with homologous vaccine to enhance the lactogenic immunity and thus increase piglet performance while reducing the risks associated with exposure to NPE late gestation (increased shedding, reproductive failure, etc). Interestingly, this methodology has been demonstrated and used in PED and TGE control or elimination protocols, as well in one controlled study for RV.[23-26]

Because of a lack of readily available immunological tests for RV, there is lack of information relative to the immune status and response to RV and exposure methods in gestating gilts and sows. When exposure is given (NPE or vaccination) there is no good way to measure immunological "take" by the animal or if and how much of a response the was by the immune system. In general, we do not know the level of RV needed to initiate or booter RV immunity in gilts and sows. We have no way to evaluate the animal's immune status or previous exposure status to RVs at the time of exposure, nor can we assess if the exposure stimulated the immune response and to what quality and quantity. In some cases, previous RV immunity may be enough to neutralize the current RV exposure, due to low virus load or strong immune protection. This has potentially been observed with control of other enteric pathogens.[23] This may be influenced by the individual animal, farm, parity, gilt entry group, exposure material, timing of previous exposure, RV serogroup or serotype, etc. Without a good way to measure the response we cannot effectively evaluate the control procedures and factors that may influence the response and are left blind to understanding how to improve the processes.

More work needs to be done to understand the optimum timing and sequence of exposures to NPE, vaccination or both in breeding age and gestating dams. Immunological assays are needed to better assess which RV control methods have the impact on piglet protection.

Sequencing

Sequencing of RV isolates has been reported on a limited based until recently and has usually been reserved to broad surveillance or research efforts. Aligned with the application of platform vaccines, where sequence of key genes is required, more sequencing of field isolates has been done to understand diversity as it relates to clinical disease, selection and updating of vaccine isolates. Currently the majority of sequences would be of the VP7, followed by the VP4 epitopes. Sequencing can be utilized to compare similarity and differences between two or more isolates. Most sequence databases are held at university diagnostic labs or vaccine companies that use platform methodologies. For RV, vaccine companies will manage a dashboard for clients with active, historical, vaccine and benchmark sequences (commercial vaccines, on-hand isolates and reference isolates).

It should be noted here that current sequence comparisons may be reported at the nucleotide or amino acid (AA) level. Practical comparisons should be made at the AA level, as that more likely dictates the unique folding of the protein when formed and is what the immune system is presented with. Advancements in identification of key antigenic sites within the protective epitopes for RV is needed, similar to work that has been done with the hemagglutinin gene key sites for some influenza A subtypes. It is very likely that nucleotide changes resulting in significant AA changes (eg, hydrophobic to hydrophilic, positive or negative charged) at key locations change the folding and presentation of the final epitope to the immune system are what result in important immunological and clinical differences. Therefore, care should be taken when calling isolates similar and different. Currently, for lack of a better threshold, recommendations are that sequences less than 95% similar on AA basis should be considered different and the closer to 100% similar is better for optimum isolate matching.

Utilization of sequences, especially as more refined methods of analysis are developed, is an important part of managing RV vaccines and "ice cubes" control strategies, both on-farm and at a system level. Understanding how and when clinical isolates change, move through a system (eg, genetic pyramids) and when current control isolates no longer match circulating isolates will be an important aspect of RV management, similar to IAV surveillance and control.

Examples of sequence comparison of RV VP7 genes have been provided in Tables 1 (RVA) and 2 (RVC). These comparisons show individual isolates from farms down the far-left column and the corresponding G serotype in the next column with reference isolates along the top row, the corresponding G serotypes in the next row with percent AA homology in the intersecting cells. This information can be utilized, along with farm history, clinical expression of disease, farm profile and date of isolation to make inferences about isolate selection and changes in isolate prevalence. Note that each additional vaccine isolate has a cost associated with it, and when used at a system level there has to be a significant cost:benefit evaluation of the number of isolates and the number of farms it will potentially provide clinical protection against their own RVs.

Table 1: VP7 amino acid homology of rotavirus type A isolates from a system of farms compared to reference vaccine isolates. Shaded cells represent > 95% homology.

			Reference strains							
			RA1	RA2	PSR1	PSR2	RA3	RA4	RA5	RA6
	Origin farm	Genotype	G5	G4	G9	G5	G9	G9	G3	G9
Rotavirus isolate	FARM A	G5	98.8	75.8	83.4	97.9	81.0	82.5	85.4	82.9
	FARM A	G9	82.9	78.6	95.4	82.5	94.2	94.8	85.8	96.0
	FARM B	G3	85.6	77.7	87.7	84.7	87.5	87.7	96.2	87.8
	FARM C	G9	83.5	78.3	97.9	83.1	94.8	96.6	86.1	95.7
	FARM C	G5	96.9	75.5	84.4	96.3	83.2	83.7	85.4	83.8
	FARM C	G5	96.9	75.5	84.4	96.3	82.6	83.7	85.4	83.8
	FARM E	G5	98.8	75.8	83.4	97.9	81.0	82.5	86.1	82.9
	FARM E	G4	76.5	98.8	78.2	74.9	78.0	78.8	78.8	77.7
	FARM F	G5	98.8	75.8	83.4	97.9	81.0	82.5	85.4	82.9
	FARM H	G5	98.2	73.7	81.9	97.2	79.0	80.8	84.3	81.1
	FARM I	G9	82.0	77.7	95.4	81.6	99.7	96.0	87.0	95.7
	FARM J	G9	82.0	77.7	95.4	81.6	99.7	96.0	87.0	95.7
	FARM K	G5	99.1	76.2	83.4	98.2	81.0	82.5	85.8	82.9
	FARM L	G5	99.4	75.6	83.1	98.4	80.5	82.1	86.0	82.4
	FARM P	G9	82.3	78.0	94.8	81.9	95.4	96.9	86.4	95.1

RA1 = System RNA particle vaccine isolate 1
RA2 = System RNA particle vaccine isolate 2
PSR1 = Merck ProSystem Rota vaccine isolate 1
PSR2 = Merck ProSystem Rota vaccine isolate 2
RA3-6 = Reference RNA particle vaccine isolates

Example 1 will utilize the RVA data in Table 1. The current RP vaccine for RVA contains a G5 [RP1] and G4 [RP2] isolate. Many of the G5 farm isolates have high similarity to both the RP1 and PSR2 vaccine isolates. There are several G9 farm isolates [FARMS A, C, I and J] and one G3 farm isolate [FARM B]. Given this information one might consider using ProSystem Rota vaccine [PSR] in the farms with a G9 isolate, developing a G9 RP vaccine or switching out a G9 for the current G4 RP isolate that is less prevalent [RA2]. It is not likely that the one farm with a G3 warrants a substantial change to the system vaccine at this time, unless there is some other epidemiological aspect (i.e., genetic farm). In this case, FARM B is a nursery and therefore does not, in the author's opinion, constitute a significant enough finding to change the vaccine. Currently in the system, farms with "one-off" isolates are managed through production of an on-farm "ice cube" specific to that isolate and not vaccine.

Example 2 will utilize the RVC data in Table 2. First you can see there are more isolates than Table 1, however the diversity appears to be similar. In this case there is only one RVC G6 isolate in the vaccine [RC1] and no ProSystem reference isolate (does not contain RVC). A majority, but not all, of the G6 farm isolates have high similarity to RC1 vaccine isolate. The remaining isolates are mostly a G1 or G9 serotype, more similar to the RC2 and RC3 vaccine isolates respectively. Note however that the G9 [RC3 similar] isolates are from the same farm, a repeat occurrence and not seen in other farms. It may or not be important to consider this as a viable vaccine candidate at a system-level, again depending on the epidemiological aspect of this farm. More important is the G1 isolate that is more RC2-like in its comparisons. This G1 is present in 3 different farms and depending on the characteristic of those farms, this may be a more meaningful isolate. In this case, one may elect to add RC2 to the vaccine for better coverage of the predominant RVC strains circulating. Note that there are four farms [FARMS J, L, M and N] that have distinctly different isolates with some potential similarity among them. Hopefully, one can see from evaluation of this dataset that it is more complicated to create a "one-size-fits-all" RVC vaccine for this system currently.

As a reminder, we are making assumptions of clinical cross-protection based on AA similarity and if we were able to compare isolates across smaller key sites, there may be less diversity between isolates and better prediction of which isolates would provide broader protection. In the author's opinion, this is not much different than the selection process for influenza isolate selection for autogenous vaccine use, we just don't have as much information on RVs as we do currently for influenzas (sequence database entries, key sites, surveillance, neutralization serology, etc).

Epidemiology

While there is a decent understanding of RV prevalence and distribution of RV serogroups and serotypes in the United States, Canada and the World, this is in constant flux.[5,27] One aspect

Table 2: VP7 amino acid homology of rotavirus type C isolates from a system of farms compared to reference vaccine isolates. Shaded cells represent > 95% homology.

			Reference strains							
			RC1	RC2	RC3	RC4	RC5	RC6	RC7	RC8
	Origin farm	Genotype	G6	G1	G9	G6	G6	G6	G6	G1
Rotavirus isolates	FARM A	G6	99.7	78.7	79.2	93.8	70.2	94.4	96.7	76.9
	FARM A	G9	78.4	83.5	95.2	78.1	69.3	77.5	79.0	83.8
	FARM A	G9	78.4	83.5	95.2	78.1	69.3	77.5	79.0	83.8
	FARM A	G9	78.1	83.2	94.9	77.8	69.0	77.2	78.7	83.5
	FARM C	G1	79.0	96.4	85.5	78.4	69.6	77.8	79.0	94.0
	FARM D	G6	100.0	78.7	79.2	94.1	70.2	94.6	97.0	76.9
	FARM D	G6	100.0	78.7	79.2	94.1	70.2	94.6	97.0	76.9
	FARM D	G1	79.0	96.4	85.5	78.4	69.9	77.8	79.0	94.0
	FARM D	G1	78.4	96.1	84.9	77.8	69.6	77.2	78.4	93.4
	FARM D	G1	79.0	96.4	85.5	78.4	69.9	77.8	79.0	94.0
	FARM E	G6	96.1	79.6	80.4	93.4	71.4	93.7	96.7	78.4
	FARM E	G6	99.4	78.4	79.2	93.5	69.9	94.1	96.4	76.6
	FARM E	G1	78.4	95.5	85.8	78.4	70.5	77.8	78.4	95.8
	FARM F	G6	100.0	78.7	79.2	94.1	70.2	94.6	97.0	76.9
	FARM F	G6	99.4	79.0	79.5	94.1	70.5	95.2	97.3	77.2
	FARM F	G6	94.7	78.4	78.9	94.1	70.5	97.6	94.4	76.9
	FARM G	G6	99.7	78.4	79.2	93.8	70.2	94.4	96.7	76.6
	FARM H	G6	99.7	79.0	79.5	93.8	69.9	94.4	96.7	77.2
	FARM I	G6	99.7	78.4	78.9	93.8	70.2	94.4	96.7	76.6
	FARM J	G3	70.9	70.9	71.4	71.5	96.1	70.3	70.3	70.3
	FARM K	G6	100.0	78.7	79.2	94.1	70.2	94.6	97.0	76.9
	FARM L	G6	71.2	71.2	71.1	71.8	95.8	70.6	70.6	70.6
	FARM M	G6	93.7	79.6	79.5	96.4	72.3	94.9	94.0	77.8
	FARM N	G6	95.0	79.6	78.9	96.4	71.4	94.9	95.8	77.5
	FARM O	G6	95.3	79.6	78.9	93.5	69.3	93.5	95.5	76.9

RC1 = System RNA particle vaccine isolate 1
RC2-8 = Reference RNA particle vaccine isolates

that is not understood well is the transmission of RV types across geographical regions as it relates to movement of pig populations. It is highly likely that new RV types arrive in distant geographical locations through the routine movement of RV positive wean and feeder pigs, as has been shown with other common diseases of swine (eg, PRRS, IAV). Another less reported, but equally important factor is the movement of replacement breeding stock from one geographical region to another, especially through significant genetic upgrades, and the linkage of genetic pyramids across geography. RVs can be commonly detected in growing and finishing aged animals, so developing replacement gilts would very likely also be exposed to and shedding RVs during the time in which they are selected and moved to distant sow farms, in addition to movement of wean age replacement gilts. In general, the author would assume that many practitioners, clinics and systems; 1) do not know the RV status of their incoming gilt sources, 2) do not test incoming gilts for RV, 3) do not have an active surveillance program for RV in their farms and 4) do not have enough sequence information to demonstrate these epidemiological links. In all fairness, the author does not either!

We also know that RVs are highly resistant in the environment and transmission between farms sites, likely via fomites (trucks, equipment, supplies, people), is possible if effective biosecurity measures are not in place. The swine industry is well attuned to the all too easy transmission of other pathogens (PED, DCoV, PRRS) through these same methods. We should consider that an increase of clinical RV diarrhea (especially if epidemic-like) may be due to the introduction of new RV strains into the farm through one of the above methods, rather than mutation or recombination of endemic isolates.

Other thoughts

In working with RV control programs there are a few additional questions and experiences that the author would like to pose and share to stimulate more discussion.

RV Co-infections as it relates to NPE

We understand that co-infections of RV serogroups are possible and actually common. CD pigs have been shown to be susceptible to combinations of RVA, RVB and RVC serogroups, with minor differences in onset and shedding duration.[9,10] What is not clear if and to what level co-infections within serogroups and among serotypes occur. It is very likely that a pig infected with both a RVA G5 and RVA G9 simultaneously would replicate both in its intestinal tract, but would the replication be balanced, resulting in relatively equal amounts of virus and equal levels of subsequent immunity. Vaccination with the commercial MLV would suggest this is the case, but the author is unaware of quantitative studies to confirm this. Is this dynamic the same for RVB and RVC serogroups? Also, do co-infections of RV serogroups result in a similar level of virus as mono-infections would? Do you get more virus if you make separate RVA and RVC NPE batches compared to a RVA/RVC co-infection batch? This information would be important if using co-infections in CD pigs to more efficiently propagate virus for NPE.

Method of NPE exposure

What method of exposure of manure, tissue or homogenate is most effective? In our system there are techniques used for NPE exposure to gilts or gestating sows. Most common is classic feedback with a 1:1 water:feed gruel that has NPE material mixed in (protocol is to thoroughly mix thawed NPE in water first, then add feed to optimize homogenization). Animals are provided a 4 oz volume of this material at a feeding event. Another common method would be use of a backpack or handheld garden pump sprayer. NPE material is thawed and diluted to a volume needed for the number of animals to be exposed. Using the sprayer, animals are exposed on their snout, mouth and feed. Some farms utilize dye to help visualize the material, which is especially helpful for groups of animals in pens. A less common method is to use the barn's water medication system to deliver the NPE material orally. The process is very similar to methodology in running commercial oral vaccines, where a defined amount of NPE is diluted in a quantity of water sufficient to "vaccinate" the group of animals in a 4-6 hour-period. Care should be taken to ensure the water sources for dilution are not treated with chemicals or disinfectants (eg, chlorines).[8] While this does risk the contamination of the waterlines, protocols encourage routine cleaning and sanitation of these water lines. This method has been successful in exposing naïve populations to other enteric vaccines or pathogens (PED, *E coli, Salmonella*). This technique is usually limited to acclimating gilt populations due to the complexities of running NPE to specific breeding groups in modern gestation barn water systems. Other novel exposure techniques have been described for electronic or group housed sows.[28,29]

NPE dose

The level of NPE needed to successfully expose animals is unknown, but most practitioners using this method would say get the highest amount of virus in the material that you can and expose sows multiple times to ensure exposure. Again, we do not have very good tests to determine or correlate exposure level or methodology to immune stimulation. Experience by the author and others suggests that NPE material PCR ct values need to be close to or less than 20 on final NPE material presented to the sow, based solely on post-exposure shedding, which admittedly is a poor measure of exposure in adult animals. When making NPE and calculating a "dose" the author's current standards and recipe are as follows: Intestine from 5 CD pigs are used to make a volume of 2 liters of tissue homogenate. This final diluted homogenate should have RV PCR ct values in the mid-teens (the lower the better), for all RV serogroups. The homogenate is frozen in 40 mL aliquots in 50 mL conical centrifuge tubes (our "ice cube") giving you approximately 50 tubes of homogenate. This 40 mL aliquot is used to expose 100 sows, mixed appropriately for whichever above method chosen. While this "recipe" has not been validated, it has been the standard recipe we have used for all rotavirus exposures and presented research using NPE to date. The background behind these calculations stem from the author contacting several swine practitioners across the US that were, at the time, using "ice cubes" and them graciously sharing their dose (sows per "ice cube"), estimated "ice cube" volume and PCR ct value ranges (if known).

Industry needs

In the author's opinion, the following are key industry needs for control of RV.

1. Production and economic impact of RV. Few have commented or demonstrated the impact of RV in commercial swine operations, but some reported values range from 3-20% pre-weaning mortality and 0.5 to 1.5 lbs reduced wean weight.[19,21,30] A better characterization of the production and economic impacts can allow practitioners and scientists to justify control programs and further research efforts.
2. Isolation techniques for RVB and RVC. Limitations in the ability to isolate and propagate RVB and RVC has stifled the development of diagnostic assays and vaccines that could be used to understand and control RV.
3. Modified live vaccines for RVB, RVC and RVA G serotypes not currently available. It appears that use of MLV vaccines for RVA can be effective measures to stimulate protective immunity in swine for homologous infections. A broader portfolio of RV MLVs would no doubt be advantageous in control.
4. Immunological assays for RVA, RVB and RVC. Methods to qualify and quantify an animal or population's immunological response and the correlation to clinical protection would greatly help practitioners optimize control strategies presented herein. It would also allow for more accurate troubleshooting when protocols result in ineffective or variable control.
5. Identification of key antigenic sites of the major immunological epitopes (VP7 and VP4). Understanding specifically where mutations or changes occur that influence epitope structure, presentation and recognition by the immune system can help focus in on what changes are immunologically and clinically important while making sequence comparisons less arbitrary.
6. Epidemiology. A better understanding of the epidemiology of RV in replacement breeding stock sources and pyramids can help practitioners understand if this is a key factor in RV disease control. Given what we know about control of other key swine pathogens (PRRS, IAV, *Mycoplasma hyopneumoniae, Strep suis*, etc), this would not be an unlikely risk factor.
7. Continued shared experiences. The experiences and information shared by the author are all based on implementation or

refinement of other's experiences and research (documented or "hallway" talks). Continued sharing of experiences by researchers and practitioners will help the industry better understand the strategies available that help control RV disease. The author encourages others to present this information with as much detail and qualification that can be given.

Summary

In summary, RV continues to be a frustrating pathogen for swine practitioners. The genetic diversity of RV isolates, the variability in generating consistent and quality immunity and the relative lack of diagnostic means to measure impact of different control measures make effective control of RV a complicated and often elusive process. Focus of RV control in farrowing still comes back to the 3 main principles of enteric disease control in piglets presented by Bruner; 1) Decrease environmental exposure, 2) Increase passive immunity from sow to piglet and 3) Increase colostrum consumption to piglets."[3] More tools are needs to help swine practitioners successfully control RV infections and more practitioners should quantify and share their experiences with others.

Acknowledgments

The author would like to thank all of the practitioners, diagnosticians, scientists, students and farm employees who have conducted, implemented, presented and shared their experiences with rotavirus and control measures.

References

1. Pittman JS. 2016. Field experiences with interventions for rotavirus control. ISU James D. McKean Swine Disease Conf. 30-36.

2. Pittman JS, Anderson A, Boyd W, Madigan J, Wolfe Z. 2019. Update on rotavirus control from the field. ISU James D. McKean Swine Disease Conference. 87-96.

3. Bruner L. 2011. Field experience with Rotavirus type C and *Clostridium spp.* Allen D. Leman Conference. 81-82.

4. Saif LJ, Vlasova A, Chattha K. 2011. The ABCs of swine rotaviruses. AASV. 435-440.

5. Marthaler D, Ciarlet M. 2014. Porcine Rotaviruses: What we've learned and what we're still missing. *Proc AASV.* 13-14.

6. Zhang M, Zeng C Q-Y, Morris AP, Estes MK. 2000. A Functional NSP4 Enterotoxin Peptide Secreted from Rotavirus-Infected Cells. *J Virology.* 74(24):11663-11670.

7. Antimicrobial Spectrum of Disinfectants. The Center for Food Security and Public Health. Iowa State University. Available at: **https://www.cfsph.iastate.edu/Disinfection/Assets/AntimicrobialSpectrumDisinfectants.pdf**

8. Greiner L. 2016. Practical Approaches to Managing Suckling Pig Rotavirus Infection. Allen D. Leman Conference. Available at: **https://www.youtube.com/watch?v=vf91nHbk8vs**

9. Madson D, Hoang H, Arruda P, Stevenson G, Yoon K-J. 2012. Rotavirus coinfection: diagnostic Overview. Iowa State Swine Disease Conference. 69-72.

10. Wolfe Z and Pittman JS. 2020. Dynamics of a rotavirus type A and C coinfection in colostrum deprived piglets. AASV. 283-284.

11. Franco MA, Angel J, Greenberg HB. 2006. Immunity and correlates of protection for rotavirus vaccines. Vaccine. 2718-2731.

12. Schwartz KJ. 2014. The science of feedback. ISU Swine Disease Conf. 51-66.

13. Robins RC, Byers EB. 2013. What do we really know about feedback to gestating dams? AASV. 537-540.

14. Pittman JS. 2022. Rotavirus: current experiences and thoughts on "feedback" diagnostics. AASV. *In process.*

15. Dufresne L. 2012. Rotavirus: Field challenges of prevention and control of a reemerging disease. AASV. 433-434.

16. Shepherd F. 2017. Longitudinal characterization of porcine RVA strains from the United States. Allen D. Leman Conference. Poster 51.

17. Shepherd FK, Dvorak CMT, Murtaugh MP and Marthaler DG. 2020. Leveraging a Validated *in silico* Approach to Elucidate Genotype-Specific VP7 Epitopes and Antigenic Relationships of Porcine Rotavirus A. Frontiers in Genetics. 11(Article 828):1-10.

18. Langel SN, Paim FC, Alhamo MA, Van Geelen A, Lager KM, Vlasova AN, Saif LJ. Stage of Gestation at Porcine Epidemic Diarrhea Virus Infection of Pregnant Swine Impacts Maternal Immunity and Lactogenic Immune Protection of Neonatal Suckling Piglets. Front Immunol. 10:727.

19. Anderson AV, Madigan J, Holtkamp D, Dominguez F, Pittman JS, Marthaler D. 2018. Evaluating natural planned exposure protocols on rotavirus shedding patterns in gilts and the impact on their suckling pigs. AASV. 64-66.

20. Farkas A, Ochoa L, Greiner L. 2020. Effect of different natural planned exposure (NPE) strategies on the shedding of rotavirus A, B, C, and pre-wean morbidity and mortality in an endemic sow farm. AASV. 357-358.

21. Boyd W, Pittman J, Wolfe A, Kumar D, Marthaler D. 2020. Comparison of prefarrow rotavirus vaccine and natural planned exposure on suckling pig performance. AASV. 228-229.

22. Anderson A, Pittman J, Marthaler D. 2019. The effect of vaccination of gilts with a pre-farrow rotavirus RNA particle vaccine following pre-breeding natural planned exposure on piglet growth and viral shedding. AASV. 270-271.

23. Saif LJ. 2014. Lactogenic immunity and vaccines for porcine epidemic diarrhea virus (PEDV): Revisiting transmissible gastroenteritis virus (TGEV) vaccine concepts. AASV. 403-406.

24. Park S, Sestak K, Hodgins DC, Shoup DI, Ward LA, Jackwood JD, Saif LJ. 1998. Immune response of sows vaccinated with attenuated transmissible gastroenteritis virus (TGEV) and recombinant TGEV spike protein vaccines and protection of their suckling pig against virulent TGEV challenge exposure. Am J Vet Res. 59:1002-1008.

25. Azevedo MP, Vlasova AN, Saif LJ. 2013. Human rotavirus virus-like particle vaccines evaluated in a neonatal gnotobiotic pig model of human rotavirus disease. Exper Rev Vaccines. 12:169-81.

26. DeDecker A, Coffey T, Rapp-Gabrielson V, Fredrickson D. Bandrick M, Sornsen S. 2015 Field efficacy of an experimental porcine epidemic diarrhea vaccine, killed virus, administered to pregnant sows. AASV. 239-243.

27. Marthaler D. 2016. Rotavirus Epidemiology. Allen D. Leman Conference. Available at: **https://www.youtube.com/watch?v=K5cHX2m8mYA**

28. Parsons TD, Beidel C, VanBlarcom J, Risser JC, Galina L. 2012. Automated feedback of autogenous pathogens in large pen gestation to reduce piglet diarrhea. AASV. 377-378.

29. John A, Pierdon M, Parsons T. 2015. A solution to controlled exposure of feedback material in group housed gestating sows. AASV. 125-126.

30. Groth D. 2014. Clinical management of rotavirus. AASV. 561-562.

https://doi.org/10.54846/am2022/176

Techniques to control/eliminate IAV-S in commercial sow farms

Cameron Schmitt, DVM, MS; Adam Schelkopf, DVM, MS; Joel Nerem, DVM

Currently most domestic US sow herds are considered IAV-S positive and are maintained in an endemic state where virus continually circulates in piglets in the farrowing house and/or in the gilt developer unit (GDU). We have previously proposed a nomenclature system for defining the negative/endemic/epidemic state of commercial sow farms.

When sow herds become acutely infected with a novel strain of IAV-S, the herd undergoes an epidemic with the introduction of this new strain. The clinical episode may be mild or severe clinically, but the farm is at an epidemic level while the adult population mounts an immune response and the virus is cleared from the animals. After the epidemic phase of infection is over, data and observations suggest most herds enter an endemic phase of infection where virus continues to circulate where naïve populations of animals exist – new piglets in the farrowing house and/or in gilt developer units (attached to the farm) where new gilts are periodically brought to the farm.

Control processes including vaccination, piglet and personnel movement restrictions, and others can be used to fully eliminate or control the virus from these populations of animals to reduce clinical impacts of the virus.

Vaccination is a critical element of a control/elimination effort. Vaccination with 100% homology autogenous products appears to be superior to near homology autogenous vaccine or commercial products. Two whole herd mass vaccinations at a 30-day interval will create a scenario where piglets receive enough colostral antibody protection to minimize circulation of the virus in the farrowing house.

After completion of the two whole herd mass vaccinations, specific internal biosecurity protocols in the farrowing house must be followed to prevent movement of the virus from older infected piglets to younger rooms of naïve piglets. Farrowing house staff is split into 3 types – front end (work with piglets from birth to processing), processors (work with piglets from processing to day 10) and late farrowing (work with piglets from day 10 until weaning). Personnel may move within their age bracket and older, but are required to change clothing, boots, and gloves to return to younger populations. No piglet movement from the air space where they were born is allowed until weaning. This process should continue for a period of ~12 weeks (targeting 3 turns of the farrowing barn).

Along with farrowing house restrictions, new gilts coming to the farm are likely positive coming from a multiplier farm. An extended quarantine of 70 days for these incoming gilts to allow them enough time to become immune and non-shedding before being in contact with other animals from the farm. Vaccination with a 100% homology autogenous vaccine at 7 and 21 days after placement will assist in reducing shedding and speeding up time for the groups to clear the virus. Entry into the GDU housing the extended quarantine animals may be done by Danish entry or shower-in/shower-out processes and may vary depending on the facility design/capabilities.

https://doi.org/10.54846/am2022/177

School of hard knocks: Disease prevention, control, and elimination

Paul Yeske, DVM, MS
Swine Vet Center, St Peter, Minnesota

Introduction

As a veterinarian, I think any time a disease can be prevented, this is the best form of control. Bioexclusion is focused on not allowing pathogens into the herd from an outside source and when works is by far the best form of disease control. In today's swine operations, we spend an increasing amount of time developing biosecurity programs and making sure that they are properly executed. We then spend time auditing and reauditing those programs, especially in the face of new health challenges. Although not always preventative, vaccination for various pathogens help to keep diseases under control and in many cases mitigate the effects of pathogens that may enter the herd. As practitioners our clients ask us to help them decide on the best course of medical management for their animals and herd. We understand that the recommendations are not going to be the same for every producer. The goals and risk tolerance of clients' needs to be considered. In capital intense projects need to make sure we understand the financial implications for the producer as this will impact the decision-making process and risk that these clients are willing to take. Sometimes we must convince the lenders more than the producer to move on these projects. The challenges are knowing when to push clients to make a more aggressive decision and when to take a more conservative approach.

The most important part of deciding on a direction to take the herd prevention, control or elimination is knowing the disease challenge and understanding the advantages and disadvantages if implementation. Then one can decide on what is best for the herd under the current economic conditions.

Materials and methods

When we talk about prevention most veterinarians are thinking of biosecurity. More specifically bio-exclusion, keeping the pathogen or agent from ever entering the herd. Clearly, this has been our best method of disease control. Certainly, all practitioners have been thinking more about this with the movement of African swine fever around the globe. Incorporating not only herd objectives but industry wide as well such as the Secure Pork Supply program.[1] We know that just having a program in place is not a complete solution, these programs must be dynamic and adapt and change over time. These programs are only successful if audited to ensure protocols are being trained and always followed.

However even with the best protocols and auditing, outbreaks still occur. Part of understanding how a new pathogen was introduced into a herd is completing a detailed outbreak investigation. These outbreak investigations are most successful if a uniform protocol is followed such as the procedures used by the emergency response corps veterinarians.[2,3]

Vaccination is part of a disease prevention program also. Vaccination is often geared towards pathogens already in the herd, but the vaccination and immunity are mitigating the effects of disease. Some good examples in today's industry include porcine circovirus Type 2, in which vaccination has been highly efficacious at controlling the disease. Ileitis vaccine is another example of a vaccine that when used has done well to prevent ileitis outbreaks within herds. Other vaccines may help to mitigate the disease but may not perform as well as we would like, and these are the greatest opportunities to look at elimination.

Perform a complete diagnostic workup to understand the disease or agent you want to control or eliminate. In the case for example of PRRSV understanding the sequence can be important to know whether the strain you are dealing with is likely to be a severe one or may just be a vaccine strain for example Understanding the herd immunity and level of spread within the population are very important before moving forward and deciding on a specific plans. In many cases there may be a period for stabilization before the herd is ready to enter an elimination program. Fully understanding the course of the disease in the herd may change your decision on what you will want to do. As an example, testing from every barn, every parity etc.

The owners' goals are also important, if selling breeding stock, the decision to depopulate and repopulate may be more important because there is shorter time to return to sales than if you do a control program followed by elimination. Essentially you can return to production as soon as the process has been completed.

Good records are essential for being able to make a decision regarding the economic impact that the disease is having. Is there a true cost, or is the pathogen or disease just an inconvenience? For herds in a system of production with the same genetics, nutrition, and management this can be a very good benchmark to compare against and decide what the real impact is in the herd. If a single herd using peers records may be beneficial or published databases as a reference may be help in deciding options to move forward.[2,3]

Partial budgets have long been used in the business world to do economic analysis of small changes in a business and projecting the outcomes of making these changes.[4-8] However, we must remember that these partial budgets or economic models are just an estimation of what the real outcome will look like. Just because it is in a nice spreadsheet model doesn't make it more correct. To quote Dr. Dennis DePietre "A spreadsheet can be precisely inaccurate."[9] These partial budgets are only as good as the data inputs.

Therefore, making sure that the data inputs are reasonable is very important. The best information that can be used is information within the farm or production system. If you are wanting to model the effects of a disease status on the production system, you need to know how will pigs perform without this disease problem. The best information is data from herds with pigs in the same management system with both disease statuses. Many times, this is not available especially if just one farm in the system. Then the next best thing that can be used is data from other systems or published data bases such as Meta Farms, PigCHAMP, etc, looking for herds or areas that are like the herd you are wanting to do analyses for.[10,11]

Once the inputs have been decided on then the model can be developed. The less variables changing at once the better and simpler the model. The more complex comparisons which are not uncommon, can be more difficult to construct and make sure that they yield reliable information. Comparing a limited number of items first can give the most reliable information. Understanding the interactions between multiple inputs can be difficult and must make sure that they are accounted for. The value of a spreadsheet model is that you can also do sensitivity analysis when results of outcomes are more questionable or if you want to look at best-case scenarios (producers) or worst-case scenarios (bankers).[12]

Disease elimination from herds has become a common practice. The most routine disease eliminations have been for PRRS, *Mycoplasma hyopneumoniae* and PEDV, all using loading, homogenizing, and closing procedures.

Some *Mycoplasma hyopneumoniae* eliminations have been done by just using medication and no herd closures; the odds of success are not as high, but results are faster and return to high production is when the medication program is completed.[13]

Depopulation and repopulation is the gold standard for disease elimination from a herd. This method has the advantage of being highly effective and can be used to eliminate multiple pathogens at once. This can also include upgrading genetics, parity distributions and the ability to fix multiple problems at once. The disadvantage is that it is the most expensive method. Especially if all the replacement breeding is done on site, because of the down time. This is the easiest and safest method from biosecurity standpoint since there is no additional site or movement risks. Offsite breeding projects have reduced the amount of downtime without producing pigs and allowed for minimal amount of down time but must be organized off site and carries the risk of that site and transport for reintroduction of the pathogens eliminated or possible new ones. These have historically been very successful if sites are in a good location and or filtered.

Results and discussion

Prevention is always the preferred method of control but even the best prevention programs can fail. Once this occurs, we must evaluate the options ahead of us as practitioners starting with diagnostics and understanding the severity or impact of that disease on-farm. With the use of filtration this allows for some options we didn't use to consider such as partial depopulation. Practitioners must move quickly and often make decisions with imperfect information, but we must keep an open mind to different possibilities, being creative with new unproven methods in these situations. Detailed outbreak investigations are important not to forget when caught up in managing the herd outbreaks and moving forward but valuable lessons can be learned from these mistakes.

Elimination is the next step if determined that the value is high enough. Depopulation and repopulation is always a possibility although the most expensive but looking back on many cases would have been a better decision in the long run. At times this just isn't possible due to many different constraints but many times the biggest constraint is not looking forward enough to really evaluate this. With farrow to wean flow and new antibiotics and new technologies we don't even know yet there will be other options available. As practitioners we owe it to our clients to think of new and creative methods to eliminate diseases. Certainly not all that I have tried have been successful and some of the best learnings have come from failures when we take the time to go back and re-evaluate what might have gone wrong, adapting, and improving with every effort.

References

1. Zaabel P., 2021 Secure Pork Supply plan website **securepork.org**

2. Holtkamp D., 2016 Rapid Response for Epidemiological Investigations, Proceedings James D. McKean Swine Disease Conference, pp 43-44.

3. Sundberg P., Rapid Response Core, SHIC website **swinehealth.org**

4. Boehlje MD, Eidman VR. *Farm Management.* New York, New York: Wiley and Sons. 1984.

5. Wittum TE, Dewey CE. Partial budget analysis of sow *Escherichia coli* vaccination. Swine Health Prod. 1996; 4:9-13.

6. Lane DM, Rendleman CM, Ott SL. Using partial budgets to analyze selected management practices associated with reduced preweaning mortality. Swine Health Prod. 1997; 3:95-102.

7. Drum S, Marsh W. How to use partial budgets to predict the impact of implementing segregated early weaning in a swine herd. Swine health and Prod. 1999; 13-18.

8. Alonso C, et.al. Financial implications of installing air filtration systems to prevent PRRS infection in large swine herds. Preventive Veterinary Medicine 2013; 268-277.

9. Dennis DePietre, personal communication.

10. MetaFarms 2021Quarterly Database

11. PigCHAMP 2021 Benchmark Spring edition

12. Yeske P. et.al., 2020, Old dog, same tricks: outbreak tools and budget using partial budgets. Proceedings AASV Preconvention session: The Pathogens That Ail Us pp3-6.

13. Yeske P., 2020, Eradication of *Mycoplasma hyopneumoniae* from pig herds, Mycoplasmas in Swine chapter 12 pp229-245.

https://doi.org/10.54846/am2022/178

Brachyspira in a production system

E. Noblett[1], DVM; R. Evelsizer[2], DVM; J. M. Hammer[2], DVM; M. Hensch[1], DVM; A. Maschhoff[1], DVM; M. Thompson[1], DVM
[1]The Maschhoffs LLC, Carlyle, Illinois; [2] Elanco Animal Health, Greenfield, Indiana

Introduction

Swine dysentery or bloody scours, caused by *Brachyspira hyodysenteriae* (*B hyo*) and *B hampsonii,* (*B hamp*) can lead to a reduction in growth performance and feed efficiency.[1] These beta hemolytic spirochete shaped bacteria are hardy in the environment.[1] Transmission can occur via pig to pig, soiled materials, rodents and/or fomites. Barn sanitation is a pivotal part of reducing the contaminant load that pigs may be exposed too. Lesions are typically found in the cecum, spiral colon and rectum. Clinical signs range from non-existent to mild lose stool to anorexia with severe mucohemorrhagic diarrhea.[1] In severe cases, morbidity can reach significant levels with mortality if the illness is left untreated.[2] *Brachyspira* spp can be treated with feed or water medications.[2] Due to a known positive case identified in one commercial sow farm, two case studies were completed within The Maschhoff system.

The objective of the case study was to conduct flow surveillance for investigation of bacterial burden in breeding stock, as historically these herds were not routinely monitored for these pathogens. Secondly, a feed medication protocol was developed for management of *B hampsonii* in a flow of growing pigs.

Case study and surveillance program

The farms included in this case study are six commercial sow farms and one multiplier located in central Oklahoma (OK), which wean approximately 10,000 pigs a week. Those wean pigs are then co-mingled and finished in Southern Illinois, Missouri, or Iowa. In December 2020, a wean to finish site in Illinois observed runny diarrhea and submitted diagnostics. Subsequent diagnostic testing revealed polymerase chain reaction (PCR) and culture positive for *B hamp. B hyo* had been previously eliminated in one OK sow farm in May 2019 via a sanitation and feed medication program. The remainder of the sow farms in the OK flow continued to test negative during and after this time frame and therefore did not complete an elimination program. The positive for *B hamp* in conjunction with the above history of *B hyo*, further testing was explored.

In December 2020 to February 2021, a total of 12 additional finishing sites housing OK flow pigs were tested with fecal samples. An additional three sites tested positive for *B hamp*, even after the elimination plan. During an additional 3-month surveillance period,four wean to finish sites tested positive for *B hamp.* Clinically, loose stools were observed in approximately 20-30% of the animals that did test positive for *B hamp*. Due to the results from the finishing flow, a surveillance program was put in place to investigate the potential pathogen burden at the sow farm level again. Thirty fecal samples per sow farm were submitted for a 4-week period. Sampling was targeted to younger parity animals (gilts entered in the past 30, 60, and 90 days) and younger parity animals under stressful conditions (about to farrow and heavy breeds). Samples were pooled in groups of 5 and tested via PCR, as well as culturefor *B hyo* and *B hamp.* Surveillance results (Table 1) returned with four sites testing positive either on PCR, culture, or both.

Feed medication protocol

Prior to the case study, pigs were started on 25g/ton Mecadox, followed by 50g/ton in the next phase of feed. After the OK sow farm surveillance was completed and the clinical dysentery observed in wean to finish pigs, Mecadox in the nursery rations was increased to 50g/ton in both feeding phases. The increase in Mecadox within the nursery ration resulted in a decrease of clinical dysentery in the finishing phase.

As part of the nursery ration case study, a more structure evaluated was conducted using feed grade medication pulses in place of Mecadox in the early nursery rations. Pigs from the OK flow were placed into a research barn in June 2021 and administered Denegard 35g/ton in the first feeding phase and CTC (450g/ton) with Denegard 35g/ton in the second feeding phase. The results of this feed medication trial are still pending.

Conclusions

Although a previous sow farm in the OK pod had completed a *B hyo* elimination program, finding evidence of *B hamp* in multiple sow and finishing farms prompted a wider system look. The system started precautionary surveillance at all multipliers, which is still being conducted today. Additional *B hamp* was found in a fecal sample at one multiplier, prompting a system wide surveillance program before the decision to pursue elimination programs. The response to the increase Mecadox dose (50g/ton) has been positive and decreased the presence of clinical signs in finishing age pigs aided in the discussions of treatment post weaning versus elimination at the breeding stock level. Furthermore, the in-consistency in positive and negative sites as well as lack of signs in the breeding stock herds, consideration needs to be given to other co-factors in *B hamp* disease expression. In addition, further investigation to understand alternative feed medication programs for pathogen control need to be conducted if elimination routes are not going to be pursued.

References

1. Burrough, Eric. Merck Veterinary Manual. 2021. Available at: **https://www.merckvetmanual.com** Accessed 1 Nov 2021.

2. Hammer, Mark. Personal interview. 29 January 2021

Table 1: Sow farm surveillance testing for *B hamp* and *B hyo*

	Week 6		Week 7		Week 8		Week 9	
Farm	**PCR**	**Culture**	**PCR**	**Culture**	**PCR**	**Culture**	**PCR**	**Culture**
1	Neg	*B murdochi/innocens*			Neg	Neg		
2	Neg	Neg			Neg	*B murdochi*	Neg	Neg
3	Neg	*B murdochi/innocens*			Neg	Neg	Neg	*B murdochi, B innocens*
4			**Hamp Pos 33.52**	Neg			Neg	Neg
5	**Hamp Pos 31.8**	***B hampsonii (3)***			**Hamp Pos 34.01**	Neg	**Hamp Pos (2)**	***B hampsonii***
6	Neg	***B hampsonii***	Neg	***B hampsonii, B murdochi***			Neg	Neg
7	Neg	*B murdochi*					Neg	***B murdochi, B innocens***
8	Not Submitted	Not Submitted			Neg	*B murdochi, B innocens*	***Hamp Pos 35.35***	***B murdochi, B innocens***

https://doi.org/10.54846/am2022/179

A cheat sheet for *Mycoplasma hyopneumoniae* surveillance

A. P. Poeta Silva[1], DVM, MS; M. Clavijo[1,2], DVM, PhD
[1]Veterinary Diagnostic and Population Animal Medicine, Iowa State University, Ames, Iowa;
[2]PIC, Hendersonville, Tennessee

Introduction

The need to improve the control and elimination of *Mycoplasma hyopneumoniae* (*M hyopneumoniae*) has intensified the search for effective diagnostic tools. Selecting an adequate diagnostic protocol hinges on several factors: objective of the sampling, sample size, labor/facility constraints, consequence of missing the agent, pathogen prevalence, test performance, and cost.

In the field, surveillance is typically founded on the two key characteristics of *M hyopneumoniae* epidemiology: 1) the slow rate of transmission and, 2) the long-term persistence within individual pigs and the population. Meyns et al[1] estimated a transmission rate of 1.28 pigs per week over a 4-week period of production. Pieters et al[2] detected *M hyopneumoniae* in bronchial swabs for 214 days post inoculation under experimental conditions. While, under field conditions, Takeuti et al[3] detected *M hyopneumoniae* in laryngeal swabs from replacement gilts over 150 days post placement, and Clavijo et al[4] detected *M hyopneumoniae* via tracheal sampling in a naturally exposed gilt for 284 days after first DNA detection. Thus, successful surveillance strategies are based on key *M hyopneumoniae* epidemiological and biological characteristics and the accuracy of available diagnostic tools.

In this session, we will synthesize recent findings related to *M hyopneumoniae* surveillance and provide recommendations depending on various scenarios of MHP infections.

M hyopneumoniae surveillance in negative populations

For naïve commercial herds, testing is typically triggered by the presence of clinical signs since timely detection of *M hyopneumoniae* is not warranted. For multipliers it is different. Timely detection is critical to avoid selling positive animals and infecting downstream farms. Thus, these herds usually implement monthly testing of serum or deep-tracheal samples. For *M hyopneumoniae*-naïve herds, surveillance resources should be focused on the incoming negative gilts during the quarantine period, either by collecting and testing a large set of samples (deep-tracheal samples) clearing the quarantine or by implementing longer quarantine periods, which increases the probability of detection and potentially minimizes testing cost.

Deep-tracheal samples

Given the site of *M hyopneumoniae* colonization (trachea and bronchi), in recently infected populations, the use of deep-tracheal samples provides the earliest detection[5]. Deep-tracheal samples have been shown to result in a higher diagnostic sensitivity and probability of detection, when compared to oral fluids PCR and serum ELISA within the first 28 days post pathogen introduction.[5] *M hyopneumoniae* detection probabilities of by sample type and day post exposure are given in Table 1. For instance, by collecting 30 samples at 28-day post exposure, the probability of detecting one *M hyopneumoniae*-infected pig using tracheal samples was 52% vs 13% using serum or 0% using oral fluids (Table 1), reflecting the less-than-ideal levels of detection with traditional sample sizes. However, if the sample is increased to 120 deep-tracheal samples, at the same days post-exposure, it would provide satisfactory levels of detection (~ 95%). While deep-tracheal samples are a highly accurate sample type, the cost (~$30 per PCR test) and labor involved, coupled with the need to collect a large number of tracheal samples to secure detection *M hyopneumoniae* limits their use for monitoring *M hyopneumoniae*-free herds.[5] In this sense, deep-tracheal sample pooling decreases testing costs and could improve herd sensitivity. Sponheim et al[6] demonstrated the use of pooling deep-tracheal samples, collected from pigs 113 days post exposure to detect *M hyopneumoniae* in a commercial sow farm. Furthermore, Mugabi et al[7] showed minimal loss of sensitivity with various pool sizes (3, 5, and 10), when the pools contain at least one positive PCR tracheal sample with a Ct $\leq$ 33. Still, the need for trained personnel to perform deep tracheal sampling may hinder wider adoption for scenarios with less risk associated with later detection of the agent.

Serum samples

Serum antibody testing by ELISA remains one of the most utilized tests to survey non-vaccinated, *M hyopneumoniae*-negative populations. As shown in Table 1, if a single pig is infected with *M hyopneumoniae*, 90 serum samples provide a ~50% probability of detection on DPI 35. However, diagnostic accuracy across commercially available ELISAs can vary. Poeta Silva et al[8] evaluated the performance of six commercial ELISAs under controlled and field conditions. Under controlled conditions, it was shown that three ELISAs resulted in no false positives and no cross-reactions with other mycoplasmas. When repeated under field conditions, these best performing ELISAs resulted in similar diagnostic sensitivity (Table 2). Therefore, either of these ELISAs (or a combination thereof) can also support *M hyopneumoniae* surveillance.

Given that the diagnostic specificity of these ELISAs is not perfect, unexpected positive serum samples can be observed during routine surveillance. Although early detection is paramount, false positives are highly disruptive to multiplier flows. Recently, the ISU VDL has onboarded the HIPRA ELISA platform (CIVTEST® SUIS MHYO) that can be used to troubleshoot unexpected positives generated with the traditional IDEXX ELISA platform (M hyo Ab Test). The performance of this novel test is still under validation. While implementing in-series testing can efficiently resolve most of unexpected positive cases, some will require additional sample collection, such as serum or tracheal samples for full resolution of the case.

Pen-based oral fluid and drinker samples

Oral fluid sampling is widely used in the US, including for *M hyopneumoniae*.[13,14] Oral fluids for PCR testing is convenient because it can be used to diagnose several agents in pig populations during

Table 1: Source: Clavijo et al[5] Probability of detecting one or five *M hyopneumoniae*-positive pigs in one pen over time by sample type (serum, SS; trachea sample, TS; oral fluids, OF) and sample size

Sample size	# of initially infected pigs	Sample type	Days post infection								
			0	7	14	21	28	35	42	49	56
15	1	TS	0.01	0.07	0.14	0.24	0.31	0.44	0.53	0.68	0.75
		SS	0.00	0.00	0.00	0.01	0.07	0.11	0.20	0.27	0.38
		OF	0.00	0.00	0.00	0.00	0.00	0.20	0.28	0.37	0.44
	5	TS	0.06	0.13	0.19	0.30	0.38	0.54	0.63	0.78	0.86
		SS	0.00	0.00	0.00	0.05	0.11	0.17	0.26	0.34	0.48
		OF	0.00	0.00	0.00	0.00	0.00	0.23	0.31	0.42	0.51
30	1	TS	0.03	0.15	0.27	0.42	0.52	0.67	0.76	0.87	0.92
		SS	0.00	0.00	0.00	0.02	0.13	0.22	0.36	0.48	0.61
		OF	0.00	0.00	0.00	0.00	0.00	0.39	0.49	0.61	0.70
	5	TS	0.11	0.24	0.36	0.51	0.62	0.77	0.84	0.93	0.96
		SS	0.00	0.00	0.00	0.10	0.21	0.31	0.46	0.56	0.71
		OF	0.00	0.00	0.00	0.00	0.00	0.42	0.53	0.66	0.75
60	1	TS	0.05	0.28	0.44	0.64	0.75	0.87	0.92	0.96	0.98
		SS	0.00	0.00	0.00	0.04	0.25	0.39	0.58	0.70	0.82
		OF	0.00	0.00	0.00	0.00	0.00	0.63	0.73	0.83	0.89
	5	TS	0.21	0.43	0.57	0.75	0.84	0.93	0.96	0.99	0.99
		SS	0.00	0.00	0.00	0.18	0.39	0.51	0.68	0.79	0.90
		OF	0.00	0.00	0.00	0.00	0.00	0.65	0.77	0.86	0.92
90	1	TS	0.06	0.38	0.57	0.77	0.86	0.94	0.97	0.99	0.99
		SS	0.00	0.00	0.00	0.05	0.34	0.51	0.71	0.81	0.91
		OF	0.00	0.00	0.00	0.00	0.00	0.77	0.86	0.91	0.95
	5	TS	0.31	0.57	0.71	0.86	0.92	0.97	0.99	1.00	1.00
		SS	0.00	0.00	0.00	0.26	0.51	0.66	0.82	0.89	0.95
		OF	0.00	0.00	0.00	0.00	0.00	0.78	0.87	0.94	0.97
120	1	TS	0.09	0.46	0.66	0.84	0.91	0.96	0.98	0.99	1.00
		SS	0.00	0.00	0.00	0.08	0.41	0.60	0.79	0.88	0.95
		OF	0.00	0.00	0.00	0.00	0.00	0.84	0.91	0.95	0.97
	5	TS	0.39	0.67	0.79	0.92	0.96	0.98	0.99	1.00	1.00
		SS	0.00	0.00	0.00	0.34	0.61	0.74	0.88	0.93	0.98
		OF	0.00	0.00	0.00	0.00	0.00	0.86	0.93	0.96	0.98

Table 2: Source: Poeta Silva et al[8] Diagnostic sensitivity and specificity of three *M hyopneumoniae* antibody ELISAs

ELISA	Sensitivity	Specificity	AUC
BioChek	47% (40, 55)	99% (96, 100)	80% (77, 84)
IDEXX	56% (49, 62)	99% (96, 100)	83% (79, 86)
HIPRA	62% (54, 69)	99% (96, 100)	85% (81, 87)

disease outbreak, such as *M hyopneumoniae*, PRRSV, IAV, and PEDV.[5,13-15] However, as previously mentioned, its use for timely detection of the agent in non-clinical populations is limited. As shown in Table 1, 30 or 60 oral fluids collected within 28 days post exposure would provide near 0% probability of detection. Thus, oral fluids are not adequate to early detection needed for MHP-negative multiplier herds.[5] Given that *M hyopneumoniae* is more likely to be found in the lower respiratory tract of pigs, coughing contributes to its presence in the oral cavity, increasing the probability of detection. Poeta Silva et al[12] described the value of cough as predictor for *M hyopneumoniae* DNA in aggregate specimens, such as novel pen-based drinker wipes and pen-based oral fluids. Analyses showed that the presence of coughing pigs in a pen increased the odds of detecting *M hyopneumoniae* in oral fluids by 14 times ($P < 0.001$, Odds ratio = 14.0; 95 CI% 3.7, 76.0), and by 6 times in drinker wipes ($P = 0.08$; Odds ratio = 6.0; 95 CI% 0.9, 67.3). That is, while oral fluids are not adequate to timely *M hyopneumoniae* detection in negative populations, in the presence of cough (*M hyopneumoniae*-positive populations), oral fluids can be utilized. Furthermore, diagnostic performance of oral fluids can be affected by the extraction and PCR protocol. Poeta Silva et al[16] used pen oral fluids daily collected (total of 322 samples) and side-by-side testing to compare four protocols: 1) ISU VDL extraction + ISU VDL *M hyopneumoniae* PCR, 2) ISU VDL extraction + ISU VDL *M hyopneumoniae* PCR + enzyme, 3) ISU VDL extraction + *M hyopneumoniae* PCR IDEXX platform, and 4) IDEXX extraction and *M hyopneumoniae* PCR IDEXX platform. A total of 173 oral fluid samples tested as *M hyopneumoniae* positive using Protocol 3 versus 148, 134, and 109 with Protocols 4, 2, and 1, respectively. The significant differences detected among protocols on *M hyopneumoniae* positivity rate in oral fluids suggest that further improvements in laboratory methods for the oral fluids matrix are warranted, which will favor their use to survey *M hyopneumoniae* in pig populations.

Alternative sample types

Novel sample types for *M hyopneumoniae* surveillance have been reported. Recent published research from UMN and ISU has shown the potential for detection of *M hyopneumoniae* DNA and antibodies in processing fluids. Vilalta et al [9], Canturri et al[10] described the detection of *M hyopneumoniae* DNA in processing fluids from an endemically infected sow herds. While Magtoto et al[11] demonstrated that an *M hyopneumoniae* serum ELISA (IDEXX ELISA platform) provided excellent discrimination between *M hyopneumoniae*-free and endemic sow herds, with high levels of diagnostic accuracy. These data suggest that processing fluids could be applied for routine surveillance in *M hyopneumoniae*-negative sow herds, especially since they are already being collected to test other pathogens. Still, more studies are needed to understand their usefulness for timely detection in negative populations.

More recently, Poeta Silva et al[12] showed that *M hyopneumoniae* DNA was detected in pen-based air samples (8 of 52 collected samples) in controlled animal study, with the first positive PCR samples taken at 10-days post inoculation. Thus, this lower detection observed in air samples may preclude its use to survey -negative populations. More studies (under field conditions) are also required to understand its usefulness in surveillance pig populations.

M hyopneumoniae surveillance in positive populations

The proper diagnosis of the primary agent is important to direct health interventions. Despite the observation of a dry, non-productive cough and bronchopneumonia-type lesions, the frequent occurrence of polymicrobial infections require diagnostic testing for a conclusive diagnosis of *M hyopneumoniae*. In endemically infected populations, *M hyopneumoniae* diagnostic testing is usually done to rule out the agent, given the presence of certain clinical signs, or to evaluate the effectivity of control and prevention programs, such as the use of vaccines or antimicrobials, and gilt acclimation efforts. For instance, clinical signs suggestive of *M hyopneumoniae* in recently introduced gilts or in downstream pigs may suggest a suboptimal gilt acclimation protocol, prompting testing of recently exposed gilts to verify the process (see next section). Likewise, testing of clinical populations, where the agent is detected on associated lung lesions, could provide insight into the need to modify timing of vaccine and antimicrobial use or check vaccine compliance.

Gilt acclimation

When exposure is applied, confirming successful infection is crucial when acclimating gilts for *M hyopneumoniae*, but *M hyopneumoniae* DNA and antibody detection can vary among pigs inoculated by different routes of exposure.[17] DNA in deep-tracheal samples was detected at 7 days post-exposure in *M hyopneumoniae*-exposed gilts using intratracheal catheters, mucosal atomization devices (intranasal), and cold foggers (aerosol). Whereas detecting DNA in pen-based oral fluids took 15 to 18 days post-exposure. In contrast, antibody was later detected with intratracheal and intranasal exposures at 21 days post-exposure, followed by aerosol exposure at 28 days post-exposure.

ELISA testing is more economical and can used through 5 to 6 weeks post-exposure, while pen-oral fluids are sensitive if cough is present in exposed gilt populations. Still, failures in exposure may occur and may not be detected or detected too late using these sample types. In contrast, deep-tracheal samples can be taken earlier as 14 to 21 days post-exposure. This approach is costlier but offers extra time to repeat the exposure if necessary. Sample sizes to estimate *M hyopneumoniae* exposure in gilt acclimation by GDU size and sample type are given in Table 3. For instance, if the site houses 500 gilts, at 21 days post-fogging (assuming 75% expectation of *M hyopneumoniae* infection rate), 21 gilts can be sample using deep-tracheal sampling to estimate exposure prevalence.

M hyopneumoniae elimination programs

For effective *M hyopneumoniae* clearance from a population, the herd is typically closed for 256 days but before this period begins, the whole herd must be homogenized.[2,5] That is, the breeding herd must be *M hyopneumoniae*-positive prior to the start of the closure period. Thus, verifying exposure of all female populations (gilt flows and breeding herd) that will be present at the re-opening of the herd is likely one of the most crucial steps for the success of an *M hyopneumoniae* elimination program. For this scenario a combination of serum (if no vaccine is used), deep-tracheal and oral fluid samples could be collected to establish day zero.[18] Close to the end of herd closure, the last group of exposed females should be targeted for testing to confirm lack of *M hyopneumoniae* shedding. However, it is important to realize that if still present, the *M hyopneumoniae* prevalence would likely be very low, requiring a very large sample size (Table 1). Testing negative incoming gilts should only be carried out after these have been present in the herd for at least 4 to 6 months.[5,19]

Considerations

As more swine production systems transition to either *M hyopneumoniae* negative or controlled statuses, it becomes imperative to rely on highly accurate diagnostic protocols that balance risk, cost, and labor. Recently, the *M hyopneumoniae* herd classification for breeding herds was proposed by Clavijo et al[19], involving all these aspects. Thus, selecting the appropriate sample type, sample size, and assay will depend on testing objectives, the expected prevalence, and economics. Overall, practitioners dealing with *M hyopneumoniae*-negative populations may target sample types and large sample sizes for early detection, while minimizing false alarms. For *M hyopneumoniae*-positive populations, practitioners may favor lower sample sizes, and sample types that are more convenient and cheaper but still allow them to evaluate and refine their current control programs. In addition to these scenarios, it was shown that there are opportunities to improve the diagnostic performance of *M hyopneumoniae* ELISAs and PCR platforms for certain sample types.

Table 3: Sampling[a] to estimate *M hyopneumoniae* exposure in gilt acclimation by GDU size and sample type (serum, SS; trachea sample, TS)

GDU size	Expected efficiency of exposure	Sample type	Gilts to be sampled
1,000	0.75	TS	21
		SS	78
	0.90	TS	12
		SS	80
750	0.75	TS	21
		SS	71
	0.90	TS	12
		SS	72
500	0.75	TS	21
		SS	68
	0.90	TS	12
		SS	69
200	0.75	TS	19
		SS	56
	0.90	TS	12
		SS	57

[a] Assuming 5% alpha, 97% diagnostic sensitivity and 100% specificity for TS, and 56% diagnostic sensitivity and 99% specificity for SS.[5,8]

References

1. Meyns T, et al. 2004. PVM. **doi: 10.1016/j.prevetmed.2004.10.001**
2. Pieters M, et al. 2009. VMJ. **doi: 10.1016/j.vetmic.2008.08.016**
3. Takeuti K, et al. 2017. VMJ. **doi: 10.1016/j.vetmic.2017.03.025**
4. Clavijo M, et al. 2018. Pig333. 13534
5. Clavijo M, et al. 2021. JCM. **doi: 10.1128/JCM.03051**
6. Sponheim A, et al. 2021. PVM. **doi: 10.1016/j.prevetmed.2021.105313**
7. Mugabi R, et al. 2021. AASV.
8. Poeta Silva APS, et al. 2020. JCM. **doi: 10.1128/JCM.00485-20**
9. Vilalta C, et al. 2019. VR. **doi: 10.1136/ vetrec-2019-105475**
10. Canturri et al. 2021. LEMAN.
11. Magtoto R, et al. 2022. AASV.
12. Poeta Silva APS, et al., 2022. AASV.
13. Trevisan G, et al. 2019. Plos One. **doi: 10.1371/journal.pone.0223544**
14. Gaurav R, et al. 2017. NHF.
15. Henao-Diaz A, et al. 2020. PHM. **doi: 10.1186/s40813-020-00168-w**
16. Poeta Silva APS, et al., 2020. IPVS.
17. Poeta Silva APS, et al. 2021. JVM. **doi: 10.1016/j.vetmic.2021.109118**
18. Yeske P. 2017. LEMAN.
19. Clavijo et al., 2021. JSHAP.

https://doi.org/10.54846/am2022/180

Addressing foreign animal diseases: It takes a village that cooperates

J. Kaisand, DVM
Iowa Department of Agriculture and Land Stewardship, Division of Food Safety and Animal Health, Animal Industry Bureau, Des Moines, Iowa

Addressing foreign animal diseases

It takes a village to address a foreign animal disease

The proverb that says "It takes a village to raise a child" applies to addressing foreign animal diseases. Whether it is in the preparedness phase, dealing with a foreign animal disease outbreak, or the recovery phase, it takes cooperation from all entities involved. The three general entities involved are the federal government, state government, and industry. The Iowa Department of Agriculture and Land Stewardship (IDALS) will describe these three entities being on each side of a triangle. This triangle is referred to as the "Triangle of Cooperation" (Figure 1). Each of these sides represents many other entities. Coordination of all entities on each side of the triangle is the goal, but everyone needs to realize that is not always the case. Industry starts with national organizations and then branches down to state level organizations. State level organizations may not always agree with other state level organizations, making it difficult to make decisions/recommendations at a national level. The same is true for state government. Each state government is bound by its code and rules, as well as representing its industry. This again can lead to the difficulties of regulations not being the same from state to state. The federal government is also challenged with coordinating many federal agencies as well as coordinating regulations that cross state lines. The goal of all sides of the Triangle of Cooperation is to coordinate as many things as possible before a disease outbreak occurs with the understanding that takes a lot of cooperation.

Figure 1: The triangle of cooperation.

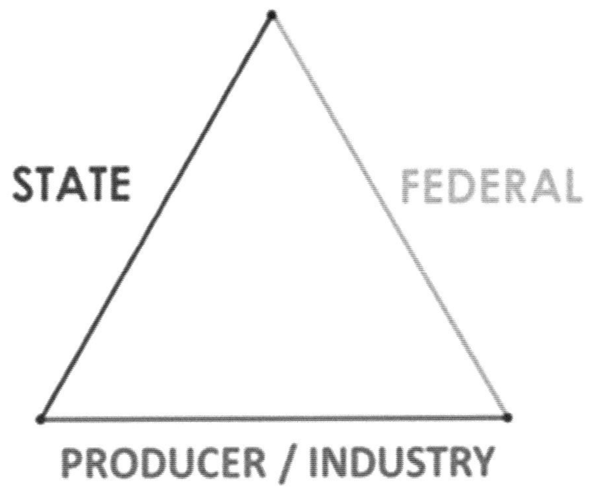

Prevention of a foreign animal disease

It is often mentioned that our best efforts lie in preventing a foreign animal disease from entering the United States because it will be easier to do that instead of dealing with a disease outbreak. Discussions of how to address a foreign animal disease outbreak often leads to very difficult issues that are complex. It is important to address both prevention and how to deal with an outbreak. Control what you can, and be aware of what you can't, is appropriate to consider at this point.

IDALS references the three levels of foreign animal disease prevention. They are to keep it in the foreign country where it is at, stopping it from crossing our border, and the last line of defense is keeping the foreign animal disease off a farm and infecting an animal.

Control what you can, and be aware of what you can't, is appropriate to think about in each of these levels. If you travel to a foreign country, receive things from a foreign country, or have any other connections to a foreign country, make sure you have a biosecurity plan in place. If you have no connections to a foreign country, at least be aware of prevention efforts. Understanding and supporting efforts at the border is critical.

The last line of defense at the farm level could be argued as the most critical. A foreign animal disease needs to infect an animal in the United States to establish itself here. A strong biosecurity plan to keep this from happening is paramount. Biosecurity is the protection from the unknowns, not the knowns. If you wait to put in, or change, your biosecurity plan based on what you know is happening, you are putting your farm at a great risk.

Dealing with a foreign animal disease outbreak

Dealing with a foreign animal disease outbreak is complex and requires a lot of coordination. A key reminder at this level is that you can't predict exactly how a disease outbreak will happen or progress. Having a plan for a foreign animal disease outbreak is important. It is also important to understand that the plan itself may not work or be relevant to the outbreak at hand. Components, or the building blocks, of the plan are the most important pieces of the plan. Having a plan is important, but planning is the core foundation to the disease response. It should also be remembered that during a disease outbreak, it becomes difficult to coordinate and incorporate all entities involved. Differing opinions from the "haves" and the "have nots" emerge during an outbreak. This means opinions become opposing from who has the foreign animal disease to who does not. This ranges from producers who have the foreign animal disease and ones that don't, to states that have the foreign animal disease and the ones that don't, and finally to countries that have the foreign animal disease and the ones that don't.

Critical components for veterinarians and producers to be prepared and should be included in a foreign animal disease response plan on a farm are:

1. **Biosecurity** - Have a biosecurity plan and be practicing it before an outbreak occurs. It is difficult to change peoples' behavior, especially in a short amount of time. If it is not being done today, it will not very likely be done tomorrow.
2. **Epidemiology** - Be prepared to quickly provide records to a state or federal animal health official during a disease outbreak. Records should include, at a minimum, all animal movements on and off a farm. Understanding epidemiology is important to understand where disease may be and/or where it may have come from. This is very important in stopping the spread of a foreign animal disease. The epidemiology is also important to show if a farm is connected or not connected to an infected farm. This becomes important in which farms receive quarantines, how quarantines are released, and affects a farms status in being able to move things.
3. **Quarantines** - Be prepared and understand what it means when a farm is quarantined. Is the farm quarantined because it is infected or is it quarantined because it is in a control area or another reason? Work with your state animal health official to understand how quarantines work in the state your farm or practice is in.
4. **Indemnity** - Be prepared and understand what indemnity is, how it works, and who is funding the indemnity.
5. **Mass depopulation** - Make sure that your farm has a plan for mass depopulation. Stamping out and eradicating a foreign animal disease will be dependent on how quickly it is found and how quickly it is dealt with.
6. **Disposal** - Make sure that your farm has a plan for disposal of the animals that are on your farm. Moving animals from an infected farm is problematic at best. Other producers that are in the category of a have not, are not very likely interested in another producer moving infected carcasses to another location.
7. **Permitting** - Understand what permitting is and how it works in a disease outbreak.

Understanding and being prepared for the components listed above are not the only things that producers and veterinarians should be prepared for but will go a long way in being prepared. Producers that take a proactive approach and have a plan, will be far better off than a producer that takes the approach that the government will take care of it.

Recovery from a foreign animal disease outbreak

Recovery from a foreign animal disease outbreak involves many things. Only a few will be mentioned here.

Recovery is dependent on how quickly the foreign animal disease is addressed. Stamping out and eradication that is done quickly means recovery happens quicker. Extra days and weeks taken at the beginning of a disease outbreak may mean extra months and years on the backside of the outbreak.

Records kept during an outbreak are important in the recovery phase. As foreign countries look at resuming trade with a country that has had a foreign disease outbreak, some of the things they look at are how the disease was handled, what records were kept in verifying infected farms, and what surveillance was, and is, being done to prove the foreign animal disease is no longer present.

Conclusion - it takes a village to address a foreign animal disease

The information in this paper is not intended to identify all things needed to address a foreign animal disease, but to identify that preparedness before, during, and after is dependent on all entities being engaged and prepared. It is truly a village that is needed to address a foreign animal disease.

https://doi.org/10.54846/am2022/181

Animal disease preparedness and response: The Indiana approach

Bret D. Marsh, DVM
Indiana State Board of Animal Health, Indianapolis, Indiana

Fundamentals of preparedness

The successful implementation of a disease prevention, preparedness and response plan requires an evaluation of several foundational pieces of information. Often, actions to address a disease threat shift immediately to the affected premises, but there are several key areas that must be addressed in advance to effectively manage a disease threat.

It cannot be overstated that the engagement of pork producers and practicing veterinarians is essential to a successful preparedness and response effort. Indiana producers and veterinarians have been very engaged in the Indiana State Board of Animal Health's (BOAH) five-step preparedness plan, and they have embraced the key elements of the initiative. Indiana producers are asked to have their premises validated, conduct a biosecurity assessment and training, engage in the certified swine sample collector program with their licensed and accredited veterinarian, practice submitting to BOAH an electronic movement permit and attend a meeting with the BOAH, either live or electronically. Hundreds of Indiana producers have completed all five steps of the program.

Fundamental to any animal disease preparedness initiative is having a thorough understanding of the legal authorities provided by state law. The ability to restrict movements of animals and animal products through quarantines is commonly available to State Animal Health Officials (SAHO), but the authority to establish emergency rules may not be provided by state law. Indiana law provides broad authority to the BOAH, an eleven-member board appointed by the Governor, to restrict movements of animals and animal products, establish rules by emergency order, mobilize personnel to address a specific disease issue, require certificates of veterinary inspection (either on paper or prepared electronically), require animal identification and, in Indiana's case, require premises identification for all individuals who buy, sell or exhibit livestock. Indiana law also addresses garbage feeding, feral swine, livestock marketing and compensation for losses associated with a disease incident. While the SAHOs of all states have authority over livestock, many do not have any authority over companion animals or wildlife, therefore, disease issues that affect these species, and that also may have an impact on livestock, can be very difficult to manage. Fortunately, Indiana law, under the BOAH, has broad authority over all species and a wide variety of animal products.

Traceability is critical to any disease preparedness effort, and Indiana has enjoyed substantial compliance with the Indiana traceability laws. Premises registration became an Indiana requirement in 2005, and the pork industry embraced this initiative. Additionally, the adoption of electronic certificates of veterinary inspection by veterinarians across the country has significantly improved the traceability of swine. The protocol for movement of swine within a production system has been widely adopted by producers across the country, and common and usual movements are now documented without the need for burdensome paperwork. Over 95% of the swine moving into and from Indiana are documented on electronic certificates of veterinary inspection. The rapid acceptance of this technology has greatly enhanced the national traceability objectives.

A robust database system is critical for the maintenance and management of massive quantities of data. Indiana has used the USAHERDS platform for over 15 years, and the system provides the sophisticated management of data needed for a successful preparedness and response plan. All HERDS states participate in ongoing discussions regarding enhancements to the database and any modifications to support one state are then enjoyed by all states using the system.

Disease surveillance is also a critical component of a preparedness and response plan. Indiana participates in the national classical swine fever (CSF)/African swine fever (ASF) surveillance program. Additionally, veterinary practitioners are encouraged to call the BOAH if they suspect a foreign animal disease (FAD). Licensed, accredited veterinarians are the cornerstone of Indiana's preparedness and response plan, and the BOAH does not function without their expertise. The BOAH's veterinary team has received additional training at Plum Island, New York, and the BOAH veterinarians are designated as Foreign Animal Disease Diagnosticians (FADD). Indiana's FADDs are frequently dispatched to farm locations to investigate suspected FADs in all species.

The BOAH relies heavily on the professional diagnostic services provided by the Indiana Animal Disease Diagnostic Laboratory (ADDL) located at Purdue University. As a member of the National Animal Health Laboratory Network (NAHLN), the ADDL is essential to the BOAH's prompt response to a disease incident. States have spent millions of dollars building, staffing and equipping laboratories to support the livestock sector, and the ability to get preliminary FAD laboratory results near the sites of production is critical.

Often overlooked, but essential to any incident, is the ability to effectively communicate to all stakeholders before, during and after an incident. The BOAH has a Director of Public Information who is critical to all preparedness and response planning. Speaking with one voice during an incident, based on the BOAH's experience with several disease events, is essential for success. Further, all communication must be coordinated with other state and federal agencies, which requires advanced outreach and preparedness.

Planning for the future

Although significant progress has been made over the last few years, more must be done to support the preparedness and planning efforts of the states. The swine industry is very dynamic and plans for protecting the industry must contemplate current industry practices while considering what changes the industry may undergo in the near future.

The diagnostic tools available for swine FADs have been improved, but it remains too cumbersome and laborious to dispatch FADDs, collect samples, ship the samples to an appropriate laboratory and then await the results. Point of care diagnostic tools must be officially recognized for preliminary results pending confirmation at a state or federal laboratory. Time is of the essence, and additional tools must be approved for use.

Additionally, there must be a program for recognizing lay personnel to collect samples at the farm level. The national Certified Swine Sample Collector program is essential to success. Trained under the auspices of a licensed, accredited veterinarian, these swine sample collectors will expand the sampling capability of the nation. The efforts of the National Pork Board and Iowa State University to launch this unique program are greatly appreciated.

The national experience with SARS-CoV2 taught us that a non-disease of swine can have significant consequences for the swine industry. Therefore, Indiana has redoubled its efforts to better prepare the industry for depopulation and disposal. Rather than relying on precious reserves of carbon dioxide (CO_2) in an incident, Indiana has continued to explore mass depopulation techniques that don't require CO_2. Electrocution and foaming are being explored as viable options for depopulation of large numbers of swine. Additionally, the Indiana disposal laws were recently changed to provide for above-ground burial giving producers another viable option for disposing of carcasses.

Indiana is an active participant in the Swine Health Improvement Plan (SHIP), and the BOAH has been designated as the Official State Agency (OSA) for Indiana. The SHIP provides a platform for all swine industry stakeholders to determine their own fate regarding FADs. Although the SHIP could later be used for endemic diseases, its focus is on CSF and ASF. The second meeting of the SHIP House of Delegates is scheduled for September 2022, and in the meantime several working groups are actively meeting to determine the standards for the program. As of August 2021, twenty-eight states have expressed interest in participating in the program, representing 98% of the nation's swine herd.

Consistency among the states regarding the movement of animal and animal products during a disease incident is of paramount importance to the pork industry. Since early 2019, a working group of SAHOs from the eighteen most populous pork states has met to streamline the process of moving pigs in interstate commerce during a disease incident. The discussions are continuing, and there has been considerable progress in developing a template for all states to use.

The collective efforts of stakeholders throughout the swine industry have substantially improved the preparedness and planning efforts of the nation. Continued collaborations are essential to find the best solutions for protecting the industry from the threats of an FAD. Gleaning from the experiences of countries across the globe that have had firsthand experience with ASF and CSF is very helpful to the planning efforts of the states, and their experiences have taught us the value of planning. Although sometimes tedious, planning efforts that recognize the fundamentals of preparedness and response are the most effective way to protect and preserve the agricultural assets of the country.

https://doi.org/10.54846/am2022/182

Animal disease preparedness and response

Beth S. Thompson, JD, DVM
MN State Veterinarian, Executive Director MN Board of Animal Health

When considering a response to foreign animal disease, the ultimate goal is to stop the disease while maintaining business continuity. Depending on the size of the outbreak, the disease itself, and the needs and resources for the response, there is a clear need for building partnerships and collaborations that are all-inclusive. Preparing for any foreign animal disease response is much like preparing to be outside on a Minnesota winter day, but that day is a week away. There is a large range of possibilities, and learning from past experience and taking advantage of new information as you get closer to that is helpful, but you won't know which coat to grab until the day arrives.

Historically, the highly pathogenic avian influenza (HPAI) outbreak in 2014 - 2015 lends good foundational information for preparedness in other livestock commodity groups. The extent of that outbreak was large; there were over 9 million birds on 109 Minnesota poultry farms that were affected, and the economic loss was estimated at $650 million. There were detections in 21 states, and indemnity paid by the federal government was at a cost of $1 billion dollars.[1] A key fact to keep in mind is that the response continued for a significant amount of time after the last positive case was detected and the birds were depopulated from that farm. Response to a foreign animal disease is much bigger and more involved than depopulation and disposal.

The following is an abbreviated list of needs and actions for a foreign animal disease response, specific to Minnesota and to swine.

Regulatory, farmer and veterinarian partnerships

There is no substitute for collaborative meetings with individuals and groups to discuss animal disease preparedness. Minnesota has built out an Emergency Disease Management Committee (EDMC) for both swine and poultry. The swine EDMC has a number of subcommittees to address every aspect of a response, from messaging to depopulation to regionalization goals. Livestock groups, farmers, veterinarians, regulators, University of Minnesota, and the Veterinary Diagnostic Laboratory staff are all involved in the discussions. Even with the establishment of EDMC, the question that needs to be asked at the end of every EDMC or subcommittee meeting: who else needs to be involved? The EDMC evolves as time passes, especially with new avenues for building out the response efforts.

Community involvement

During HPAI, I called a milk processor early in the response to see if there were milk tankers available for water transport; we needed water for depopulation foam used in the poultry barns. That processor wasn't able to assist, but gave a list of other possibilities, one of which was utilized. Later in the response, the involvement of the community helping with the "human" aspect of the response was important; churches, local community groups, University of Minnesota extension and other community groups were part of the support needed. We know that, as a whole, agriculture is very resilient, but we also know that individual farmers and responders may need assistance. In addition to alerting, educating, and utilizing local and community groups, the Minnesota Department of Agriculture has a department specifically targeted to rural and farm stress. Communities want to help, and in Minnesota, we continually seek out local resources.

Equipment

There have been a variety of prospects for building the equipment stockpile. Unfortunately, the COVID-19 slaughter plant and supply chain issues in the recent past have allowed us to better identify depopulation and disposal equipment and training. Additionally, programs such as the National Animal Disease Preparedness and Response program, other Farm Bill funding for the National Animal Health Laboratory Network and organizational opportunities, such as National Pork Board funding, have been put to good use in Minnesota and many other states. CO_2 depopulation equipment enhancement, Nitrogen foam exercises and making additions to state resource stockpiles are current projects.

External communication

While communication between all involved in the response is an overarching topic, communication to the general public needs to be agreed to upon prior to the event. There are key phrases that become part of the "talking points" for all parties involved. Additionally, any individual or group can also defer certain questions to another specific individual or group. Regulatory officials are in a position to talk about the regulatory process such as how federal and state officials are applying state or federal response plans; commodity-specific groups are best asked about the impact of a foreign animal disease on that commodity. Identifying key specific people who can speak to the response efforts is also important, as is availability of those people.

Movement of animals and product

The Secure Pork Supply (SPS) plan plays a key role in allowing farmers and veterinarians to plan ahead of any outbreak for movement requirements, while giving state animal health officials a guide to actions that may be taken in other states. Permitted movement based on the SPS plan utilizes risk assessments, because movement from, into or within a Control Area is not without risk. Risk mitigation via permitting and permit guidance will allow for animal and product movement. As tests, testing strategy and other mitigation components becomes available, Minnesota updates the permitting plan and the related response.

Biosecurity

While biosecurity is arguably part of the SPS planning process and other items listed above, it needs to be highlighted as imperative and central to a successful foreign animal disease response. From the perspective of a state animal health official, biosecurity is all encompassing once a foreign animal disease is found in the country. Keeping disease out of farms and preventing the spread of the disease drives the response. Also from the state animal health perspective, biosecurity is imperative for all responders in a disease event. Best practices for biosecurity should be in place now, and training for responders is an identified need.

Looking into the future as a regulatory official, it is part of my job to assume we will see an incursion of a foreign animal disease and to be prepared for it. While much effort will be placed on the site(s) with the positive diagnosis with resulting depopulation and disposal, there is also the work to be done on affected (but not infected) farms. Farms need to be able to continue to do business. Recognition of gaps in preparedness, addressing those gaps and other risks, and identifying all available assets is the backbone of a successful response.

Reference

1. Greene, J. L. (2015). *Update on the highly-pathogenic avian influenza outbreak of 2014-2015* (R44114). **https://fas.org/sgp/crs/misc/R44114.pdf** [Google Scholar]

https://doi.org/10.54846/am2022/183

African swine fever and plant-based feed ingredients: Canada's approach to risk management of imported feed products

Sharon Calvin[1]; Amy Snow[2]; Egan Brockhoff[3]
[1]Animal Health Risk Assessment and Intelligence Section, Canadian Food Inspection Agency, Ottawa, [2]Canada; Foreign Animal Disease Section, Canadian Food Inspection Agency, Ottawa, Canada; [3]Canadian Pork Council, Ottawa, Ontario

African swine fever (ASF) is a contagious viral disease that only affects pigs. While ASF does not cause disease in humans and is not a food safety risk, its impacts on industry and trade are significant. A single positive case in Canada would stop all hog and pork product exports as well as lead to an overwhelming market surplus resulting in the need for a welfare cull to reduce herd inventory.

In 2018, the emergence of ASF in Southeast Asia and extraordinary spread heightened concern about the potential risk to Canada's pork industry, should ASF be introduced to the country. This concern has resulted in unprecedented prevention and preparedness activities being collaboratively undertaken between government and industry.

One of the early activities that was conducted was the assessment of the possible pathways of ASF entry into Canada, including the identification of existing mitigation measures and gaps. Although details pertaining to the role of feed in disease transmission were still emerging, and this pathway of entry was considered to be a lower risk than other pathways, it was recognized that no risk mitigation measures were currently in place.

A scenario tree was developed to describe the introduction of ASF into the Canadian domestic or wild population via this pathway. For feed ingredients to be considered a hazard, a number of steps are required. First, virus excreted by infected animals must contaminate feed ingredients or ingredient containers in the country of origin. Containers may contaminate ingredients and vice versa. If the product undergoes processing in the country of origin, the virus must survive processing or contaminate the ingredient post-processing.

The ingredient is then shipped to its destination in Canada, and the virus must survive the environmental conditions for the length of the shipping process. Virus must then arrive in Canada in or on an ingredient or container. Some ingredients may then undergo processing in Canada, as well as manufacturing into a final feed product, and the virus must survive these steps. Alternatively, the virus contaminates the feed mill or other facility, where it then contaminates other feeds, such as those of Canadian origin. Final feed products in Canada may contaminate containers and vice versa.

Contaminated feed products or containers must then expose domestic pigs in Canada to the virus. This may occur directly if products are used as a swine feed, or indirectly via wild pigs who may become exposed by gaining access to feed used in an outdoor environment. Even if the feed product is not contaminated, virus could reach the farm environment in or on a contaminated container.

There are many potential points of contamination of ingredients in the country of origin (before, during and after processing). This can include the use of manure as fertilizer, contaminated irrigation sources, the movement of people or fomites between farms, contamination by wild pigs, and cross-contamination during processing, storage and transport. It was determined that the information required to analyse this probability is difficult to obtain without an on-site visit.

The assessment concluded that many aspects of product processing may affect the probability that viable virus survives, including the use of washing, solvents, changes in pH, drying and heating. Rigorous thermal processing is considered one of the best methods to inactivate foreign animal disease viruses.

Research has shown that ASF can maintain infectivity in various feed ingredients during the length (30 days) and environmental conditions simulating a Trans-Atlantic trip from Europe to the United States.[1] For a full assessment, further research is required on the survival of ASF in the ingredients of concern and under the environmental conditions and timing of a journey from affected countries to Canada. Since the virus survived on most ingredients tested, and in the stock virus control, it is assumed at this point in time that the virus will survive shipping equally in all ingredients imported.

There was a high level of uncertainty associated with many of the probabilities in this pathway, especially those related to contamination of ingredients in source countries, making it difficult to determine the actual probability of introduction. However, future updates in the available scientific evidence will result in reconsideration of the risk. Despite the uncertainties, it was concluded that plant-based feed ingredients produced in rural areas are more likely to be exposed to ASF than synthetically produced feed ingredients. For synthetic products, the lower risk is due to a lower likelihood of source contamination, extensive processing involving fermentation and/or high temperatures, greater oversight resulting in lower likelihood of cross-contamination, and lower inclusion rates in diets.[2-4] Synthetic products are more likely produced in facilities within industrial areas and facilities manufacturing synthetically produced feed ingredients are also more likely to be owned by large multinational companies that export widely and, as such, be subject to the oversight of various countries. This assumes that companies that produce vitamins and minerals for feed are not producing porcine blood products in the same facilities (eg,, blood meal, plasma, serum), which would present a risk for cross-contamination.

It was identified that the current gap in risk mitigation required a staged, multi-prong approach and it was also recognized that these measures strengthened Canada's overall level of protection against ASF but were not intended to reduce the level of risk to zero. In the very short term, CFIA used targeted communication with importers to facilitate compliance promotion. In the short to medium term, the CFIA used existing legislative authorities (the ability to declare secondary control zones to reduce the threat of disease introduction) to necessitate import requirements for select plant-based feed ingredients originating from countries of concern for ASF entering Canada through one of the six marine ports of entry that were identified. This approach can be adjusted as further information on the level of risk posed by these products emerges. A risk ranking of potential products was completed and the plant-based feed ingredients to be controlled included unprocessed grains, oilseeds and their associated meals, imported with the intent for use in livestock feed. These commodities were selected because, in the case of the unprocessed products, they undergo little or no processing or treatment prior to export to Canada, which leads to a greater likelihood of potential disease introduction. In the case of meals, the practice of cold-pressing or adding untreated material back in after processing also leads to a greater likelihood of potential disease introduction. The standard processing practices of the other commonly imported non-animal-origin feed ingredients (such as vitamins and minerals) were determined to be acceptable risk mitigation. It was determined that the new controls would also reduce the likelihood of cross-contamination from bulk packaging.

This approach to risk management contained an appropriate degree of flexibility and supports the OIE principle that risk assessment be 'flexible to deal with the complexity of real-life situations', including different 'types and amounts of data and information.[5] In the case of Canada, the import measures applied were seen to represent a measured approach to reduce an uncertain risk from comparatively higher risk commodities.

The role of plant-based feed in the transmission of ASF is not completely understood. However, even with an uncertain risk, countries such as Canada that would suffer severe consequences should ASF be introduced need to consider appropriate, risk-based mitigation measures. The measures implemented should balance the effect of reducing the impact of an uncertain risk, with the costs and risks associated with the implementation. In this example, it is believed that Canada was successful in achieving this balance. It should also be emphasized that the early engagement of relevant stakeholders was a major contributing factor to the success of this initiative.

References

1. Dee, S. A., Bauermann, F. V., Niederwerder, M. C., Singrey, A., Clement, T., de Lima, M., Long, C., Patterson, G., Sheahan, M. A., Stoian, A. M. M., Petrovan, V., Jones, C. K., De Jong, J., Spronk, J. J. G. D., Minion, L., Christopher-Hennings, J., Zimmerman, J. J., Rowland, R. R. R., Nelson, E., & Diel, D. G. (2018). Survival of viral pathogens in animal feed ingredients under transboundary shipping models. PLOS One, 13, e0194509. **https://doi.org/10.1371/journal.pone.0194509**

2. European Food Safety Authority (EFSA). (2011). Scientific Opinion on safety and efficacy of choline chloride as a feed additive for all animal species. EFSA Journal, 9(9), 2353. **https://doi.org/10.2903/j.efsa.2011.2353**

3. Jones, C., Woodworth, J., Dritz, S., Niederwerder, M. C., Tokach, M., Goodband, B., & DeRouchey, J. (2018). Frequently asked questions about ASFV and CSFV in feed. Kansas State University. **https://www.asi.k-state.edu/**

4. Swine Health Information Center (SHIC). (2019). African swine fever— Vitamin supply chain workshop notes. **https://www.swinehealth.org/wp-content/uploads/2019/05/4-30-19-ASF-Vit-Meeting-Summary-final002.pdf**

5. World Organisation for Animal Health (OIE). (2019). Terrestrial animal health code. **https://www.oie.int/index.php?id=169&L=0&htmfile=titre_1.2.htm**

https://doi.org/10.54846/am2022/184

US feed security response

Cassandra Jones, PhD
Kansas State University, Manhattan, Kansas

Ultimately, the United States animal health and pork production industries wish to reduce the risk of African swine fever virus (ASFV) entry, as well as be prepared for its potential transmission throughout the feed industry. When porcine epidemic diarrhea virus entered the US, our industry did not understand the potential role of feed-based transmission and, as such, were underprepared. This allowed for the disease to quickly spread to multiple sites and make large jumps to first enter farms and production systems through the feed supply chain. Both controlled and field research has demonstrated the ability for the feed supply chain to potentially play a role in ASFV transmission. We collectively understand that steps must be taken to reduce the risk of ASFV through feed and ingredients intended for consumption US pigs, but also that we must simultaneously prepare for rapid implementation of domestic feed biosecurity if the virus were to come within our borders, regardless of the means of entry.

The Feed Risk Task Force first met in June 2019 with representatives from US and Canadian regulatory agencies, US and Canadian feed industries, and the US pork, beef, milk, and poultry industries. This task force continues to provide a forum for dialogue between federal regulators, the feed industry, and the production animal industry regarding the risk of pathogen transmission in feed.

At the recent US Swine Health Improvement Plan (US SHIP) meeting held in August 2021, the House of Delegates requested "the commissioning of a working group to provide recommendations and next steps for the US SHIP program to address the risks associated with disease transmission in feed (including but not limited to regulated garbage feeding) for a broadly applicable feed biosecurity plan to be recognized nationally".

Simultaneously, a gateway program funded by National Pork Board completed its objectives to facilitate the development of a proposal for a regulatory gateway program for importing non animal-origin feed and ingredients from ASFV-endemic regions. Overwhelmingly, the working group strongly urged starting with a voluntary approach to such a gateway program, with potential regulatory additions only after sufficient success has been demonstrated by the swine industry. However, this working group also felt that the scope should be expanded to include feed and ingredients from classical swine fever virus- and foot-and-mouth disease virus-endemic regions.

These three projects have evolved independently to address the potential role of feed risk. However, it is now time for them to align and jointly address the importation expectations of non-animal-origin feed and ingredients from ASFV-endemic regions as one of the standards supported by US SHIP. A coordinated response from the entire agricultural sector is needed to continue to prevent foreign animal disease entry through feed, and would be paramount to control an outbreak if one were to occur in North America.

https://doi.org/10.54846/am2022/185

US Swine Health Improvement Plan pilot project, moving forward!

R. Main[1]; J. Roth[1]; J. Christopher-Hennings[2]; T. Holck[1]; J. Kraft[1]; J. Lowe[3]; C. Rademacher[1]; M. Torremorell[4]; J. Torrison[4]; G. Trevisan[1]; J. Zimmerman[1]
[1]Iowa State University, Ames, Iowa; [2]South Dakota State University, Brookings, South Dakota; [3]University of Illinois, Urbana, Illinois; [4]University of Minnesota, Saint Paul, Minnesota

The kick-off

The US Swine Health Improvement Plan (US SHIP) pilot project (Figure 1) is now moving forward in earnest. A formative congress of approximately 230 industry, state, and federal partners came together on August 23-24, 2021 in Des Moines Iowa to participate in the inaugural US SHIP House of Delegates (US SHIP HOD) meeting.

The inaugural forum served to:

1. Introduce and orientate interested US pork industry, state, and federal partners to this US SHIP pilot project.
2. Consider and finalize the initial (Year 1) program standards/requirements for conferring the ASF-CSF Monitored certification to participating production sites and slaughter facilities.
3. Provide feedback and direction on which additional items of high relevance (related to US swine health and foreign animal disease preparedness) are of interest to be explored further by working groups in the coming year, with findings and recommendations to be brought forward for subsequent consideration at the second US SHIP HOD meeting to be held in September 2022.

This inaugural US SHIP HOD was composed of US pork industry participants representing the interests of pork industry stakeholders across each of the 28 states that expressed interest in participating in this US SHIP pilot project. The 28 states expressing an interest in the US SHIP pilot house more than 99% of the domestic swine in the US.

Figure 1: US Swine Health Improvement Plan - ASF-CSF Monitored certification. Piloting a proven platform for safegarding, certifying, and bettering animal health. An industry, state, and federal partnership.

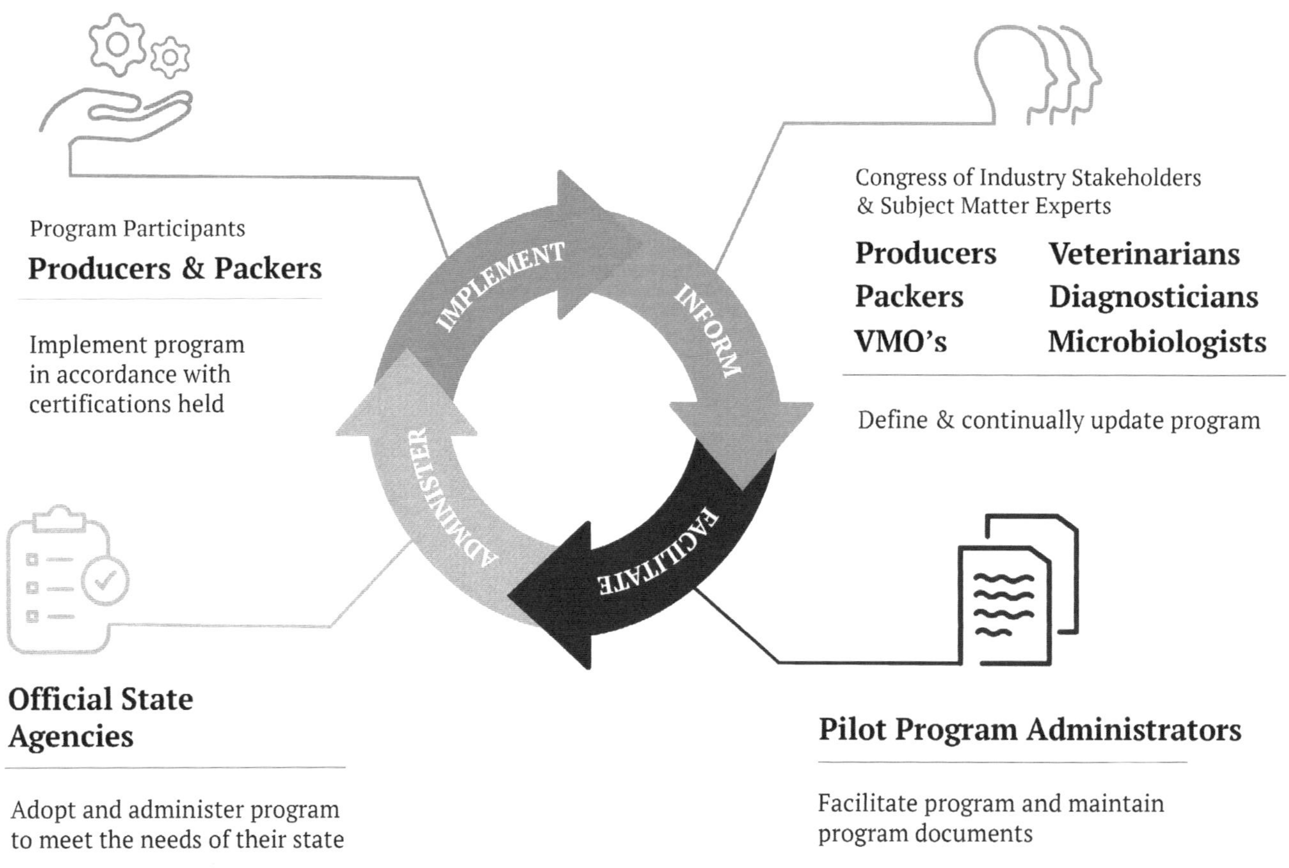

Each state expressing interest was allocated a specified number of voting delegates and the opportunity to invite two non-voting guests to attend this inaugural US SHIP HOD meeting. A formula was used to derive the number of voting delegates allocated to each state. The number of delegates included a baseline allocation to each state, as well as an allocation proportionate to the number of breeder and feeder swine (respectively) located within each participating state.

State pork producer associations selected volunteers from their respective states to serve as voting delegates or non-voting guests in this inaugural US SHIP HOD. Each participating state's voting delegation was inclusive of the State Animal Health Official or their designee. State level participation in this US SHIP pilot project is at the discretion of the State Animal Health Official.

US SHIP pilot project investigators, staff, and technical committees (involving the contributions of more than 90 subject matter experts/industry participants from across the US) contributed to the development of a draft set of program standards for Year 1 of this US SHIP pilot project. The investigators, staff, and technical committees worked earnestly to ensure the Year 1 program standards being set forth for consideration represented a logical starting point and a tangible step forward in preparedness that can be achieved within the first-12 months of this start-up venture across the full-breadth of interested industry participants and states. These proposed program standards/requirements for certification were brought forth for consideration, discussion, amending, and a formal vote at the US SHIP HOD meeting.

The US SHIP technical committees (Biosecurity, Traceability, and Sampling & Testing) also worked to identify several key items of high relevance that were beyond the scope of the program standards to be considered for Year 1. These additional items of high relevance were discussed more in-depth at a series of breakout sessions at the US SHIP HOD meeting.

Objectives

The primary objectives of this US SHIP pilot project endeavor are to develop and implement an US SHIP "ASF-CSF Monitored" certification of US pork production operations modelled after the basic tenets of the National Poultry Improvements Plan's "H5/H7 Avian Influenza Monitored" certification held by greater than 99% of the US Commercial Poultry operations.

This US SHIP pilot project aims to establish a "national playbook" of common technical standards and definitions to be recognized across all program participants and participating states. The pilot project intends to proactively establish an active and working system of certification that centers on prevention of disease and demonstrating evidence of freedom of disease outside of ASF-CSF control areas.

Working and seamless systems of traceability, and implementing uniform and effective systems for early detection and demonstrating evidence of freedom of disease (across supply chains, areas, & regions) are each foundational elements needed to support ongoing interstate and international commerce over the course of a response and recovery period.

Purpose

The overarching purpose of this US SHIP pilot are to:

1. Enhance all three aspects (prevention, response, & recovery) of trade impacting disease (TID) preparedness amongst participating producers, slaughter facilities, and states through proactively establishing an industry-informed and working system of operations and certification built upon well-defined program requirements for biosecurity, traceability, and disease surveillance.

2. Reduce the impact of recurring endemic diseases of high consequence through the sustainable advancement of sanitary standards and practices that mitigate disease spread into and between farms.

3. Provide US pork industry participants a first-hand experience in developing and participating in an "National Poultry Improvement Plan - like" program customized to meet the needs of the US pork industry.

Upon the conclusion of this US SHIP pilot project (should there be interest), the experiences gained and operations established through the pilot could be transitioned into a more formal and ongoing platform (ie, US Swine Health Improvement Plan or US SHIP) for safeguarding, certifying, and bettering the health of US swine and longer-term competitiveness of the US pork industry.

While advancing practices that mitigate risks of disease introduction into the country is the top priority, proactively developing and implementing an industry-informed and functional system prior to an ASF-CSF incursion will also enable participants and states to readily scale up the necessary testing to demonstrate evidence of freedom of disease across specified supply chains, areas, regions, and market segments throughout a response and recovery phase.

Advancing biosecurity practices that mitigate widespread indirect transmission of disease through unknowingly contaminated fomites; working and seamless systems of traceability; and well-understood, practical, and effective disease surveillance are each critical aspects of FAD preparedness and response.

This US SHIP ASF-CSF Monitored certification program aims to play a primary role in helping support the responsible movement of swine and continuity of business and trade outside of ASF-CSF control areas.

Operating structure, roles and responsibilities

This US SHIP pilot project is utilizing the same basic operational structure (Figure 1) as the US poultry and egg industries' long-standing National Poultry Improvement Plan (NPIP). This approach towards establishing a cooperative industry, state, and federal partnership presents a proven platform for safeguarding, certifying, and bettering animal health.

Participation is voluntary. Program participants include US pork producers, packers (slaughter facilities), swine interest veterinarians, and both state and federal veterinary medical officials. Certificate holders will be the US pork producers and packing (slaughter) facilities electing to participate. Producer and slaughter facility participation is at the level of a given premises (geographic location). Due the start-up nature of this endeavor, slaughter facilities will not be required to have 100% of their supply chain originating from ASF-CSF Monitored certified farm sites to participate in this pilot project.

Requirements for certification include enrollment with the US SHIP Official State Agency (US SHIP OSA), maintaining a valid veterinary-client-patient relationship with a licensed and accredited veterinarian, and meeting or exceeding the program standards. US SHIP program requirements center on biosecurity, traceability, and disease surveillance. An abbreviated summary of the program standards for conferring this US SHIP ASF-CSF Monitored certification as passed at the inaugural US SHIP HOD have been included at the end of this proceedings paper for your reference. A more complete description can be found on the US SHIP website (**https://usswinehealthimprovementplan.com**).

Each participating state will be responsible for designating an entity (US SHIP Official State Agency, US SHIP OSA) to administer the program, verify participant status, and maintain a current list of the certified participants in their respective states.

Participant responsibilities are to enroll, meet or exceed requirements for certification, and maintain the ability to verify compliance on an as needed basis.

US SHIP program administrators (ie, team of investigators and program staff) serve to facilitate central program development, technical committees and working groups, and the HOD; provide support to the US SHIP OSA's; keep the official US SHIP program documents up to date; and produce aggregate level program participation statistics (by state).

US SHIP technical committees (Biosecurity, Traceability, and Sampling & Testing) are composed of subject matter experts (volunteers) appointed by the US SHIP program administrators. The technical committees serve to draft and review proposed updates to the program standards to be brought forth for consideration by the US SHIP HOD.

The US SHIP HOD is a democratic forum of US pork industry participants representing each participating state that serves to determine the programs content, direction, and requirements for certification by majority vote.

The National Swine Disease Council (NSDC) is serving in an advisory capacity to the US SHIP program administrators.

USDA Veterinary Services works closely in providing input to US SHIP program administrators; participates in the various US SHIP technical committees and the US SHIP HOD; and provides a final review and approval of the program standards derived from the US SHIP HOD.

Funding

The USDA funding received to support this pilot project ($495,000, over 2-years) aims to provide support for the human resources, management systems, and outreach necessary to facilitate the initiation and central coordination of this pilot project.

Each state electing to participate is responsible for funding the operations of the US SHIP OSA within their respective state. Funding of the NPIP OSA operations varies significantly among states, but commonly comes from multiple sources (eg, NPIP participant fees, state departments of agriculture, state poultry and egg industry organizations, and USDA APHIS cooperative agreements).

Producer and packer participants will be responsible for the costs incurred associated with meeting or exceeding the requirements of certification.

Next steps

Each state electing to participate in the pilot project are in the process of determining the entity that will administer (house) the US SHIP OSA, and working to form and begin to stand-up their US SHIP OSA in Q4 2021, in preparation for initiating enrollment of participants in Q1 2022.

Initial participant sign-ups and the associated certification process are anticipated to move ahead in earnest in Q1 2022.

A number of concurrent US SHIP related projects and working groups are also currently moving forward in accordance per the recommendations set forth by seven resolutions that were passed at the inaugural US SHIP HOD in August 2021. The resolutions center on a series of efforts that collectively aim to further investigate a number of higher-order items related preparedness. Working group findings and any associated recommendations or proposed amendments to the program standards stemming from such study will be brought forth for review and consideration at the second US SHIP House of Delegates meeting planned for September 2022.

An abbreviated summary of the seven resolutions passed at the inaugural US SHIP HOD have been included at the end of this proceedings paper for your reference. A more complete description of these resolutions passed at the US SHIP HOD can be found on the US SHIP website (**https://usswinehealthimprovementplan.com**).

Eye on the long game

As you consider your engagement in this US SHIP pilot project endeavor, I would ask each of you to keep the long view in mind. Thinking of both the current needs, as well as the future systems of health assurance that will be needed by the next generation of US pork industry participants (producers and slaughter facilities) in the years ahead.

This US SHIP pilot project is anticipated to be a learning and growing experience for all parties involved. The pace of change; urgency for continuous improvement; need for adapting to changing domestic and international consumer demands; and the risks and realities of the intercontinental movement of diseases of high consequence are not likely to lessen or slow-down anytime soon. Twenty-first century approaches and solutions are needed to address the ever-more complex and consequential swine health challenges (and opportunities) that extend beyond immediate influence the individual producer's or packer's farm gate.

This US SHIP pilot project endeavor presents as an opportunity for pulling industry, state, and federal partners together to aid in the US pork industry's ongoing efforts to forge a sustainable path to a better, more robust, and less vulnerable US pork industry of tomorrow. Continuous improvement, asset preservation, and protecting the opportunities and way of life for the current and future US pork industry participants and associated local communities come to mind.

Summary of program standards (Year 1) as passed at inaugural US SHIP HOD

Enrollment

- Participating premises are to be enrolled with the US SHIP Official State Agency (US SHIP OSA) in the state in which the premises is located.

Veterinary service provider

- Producers are to maintain a valid veterinary client-patient relationship with a licensed and federally accredited veterinarian.

Traceability

Premises level information:

- Premises level demographic information for each participating premises is to be complete, accurate, current, and on-file with the US SHIP Official State Agency in which the premises is located.

Swine movement information:

- Participants are to maintain records of the intrastate and interstate movements of live swine into and out of each participating premises.
- Participants must demonstrate competency in providing at least 30 days of movement information electronically in a common format (eg, a prescribed CSV file) to the US SHIP Official State Agency in a timely manner (eg, < 72 hours).

For participants with multiple participating premises within a given state, such competency can be demonstrated on a site-by-site basis or en-masse.

Semen movement information

- Boar stud premises participants are to maintain records of the intrastate and interstate movements of semen distributed out of each participating premises.
- Boar stud premises participants must demonstrate competency in providing at least 30 days of semen movement information electronically in a common format (eg, a prescribed CSV file) to the US SHIP Official State Agency in a timely manner (eg, < 72 hours).

For participants with multiple participating premises within a given state, such competency can be demonstrated on a site-by-site basis or en-masse.

Animal identification

- Certified ASF-CSF monitored participants must comply with existing state and federal laws regarding animal/group/lot identification.

Biosecurity

Feed supply:

- The feeding of swill, garbage, or table waste that have the potential to include meat products is strictly prohibited.

Personnel:

- Permissioned individuals that have recently been exposed to livestock, feral/wild pigs or slaughter facilities in ASF/CSF/FMD positive regions or countries abroad should only visit farms or slaughter facilities in the US after observing a 5-day downtime since arriving in the US, and donning PPE (boots/coveralls, etc) provided by farm site or slaughter facility being visited.

Enrollment survey (biosecurity practices)

- At enrollment, participating premises will complete a survey to provide a simplistic categorization of some of the high-level biosecurity practices being implemented at the premises. Information from this survey is to provide quantitative data to assess current standards of practice across a broad spectrum of program participants. Results will help provide insight towards consideration of additional biosecurity related program standards in the future.

Sampling and testing (disease surveillance)

- Initial 12-month research period: No sampling and testing requirements of participants

In the absence of an introduction of ASF/CSF, there will be no additional ASF/CSF sampling and testing requirements of participants beyond the current and/ongoing systems foreign animal disease (FAD) surveillance taking place across the US.

The first 12-months of the testing related activities will serve to develop informational and training materials, further modeling of disease spread and sensitivity of detection across herds and regions, and to conduct an expanded negative-cohort study of commercially available ASF-CSF PCR assays.

- Maintain compliance with ASF-CSF sampling and testing requirements

US SHIP sampling and testing requirements will vary by Production Site Type and the ASF-CSF status of the US, State, or Region (Figures 2, 3, and 4).

The program is based on targeted testing of animals of poor or sub-standard health. Targeted sampling enhances both the efficiency of detection and the simplicity of sample collection across the spectrum of commercial and non-commercial farms in the US.

The frequency of on-site sampling is a function of time and is independent of the timing of pig movement, thereby providing for a uniform and continuous system of disease monitoring across production sites, areas, and regions.

US SHIP ASF-CSF tests are to be used for screening purposes only. Non-negative results would result in the testing laboratory (USDA NAHLN lab certified to conduct ASF-CSF testing) contacting the appropriate state and federal animal health officials to initiate a foreign animal disease investigation (FADI) for the collection of additional samples for official ASF-CSF testing (confirmatory) purposes.

Summary of resolutions as passed at the inaugural US SHIP HOD

Resolution # 1: Traceability case study

The US SHIP House of Delegates requests the conduct of a case study of traceability standards of practice and systems used among other export centered pork producing countries from which future technical standards may be developed and implemented for US SHIP.

Figure 2: Sampling and testing requirements for ASF-CSF risk level 1.

ASF/CSF Status = Level 1, US Negative (Peace Time)

			Sampling & Testing Requirements (Alternative Options)			
				Option 1 Individual Only		Option 2 Aggregate Only (Group or Pen)
Production Site Type	**Specimen Type(s)**	**I or A**[1]	**Frequency / Timing of Sampling**	**# of Individuals**	**# of Pools (Groups of up to 5)**	**# of Samples**
Boar Stud Mature Boars, Distributing Semen, ± On-Site Isolation	Oral Swab Blood Swab Oral Fluids	I I I	Research Period: No Sampling and Testing Requirements			
Breeding Herd Breed to Wean, Breeding/ Gestation/ or Farrow Only, ± On-Site GDU or Isolation	Oral Swab Blood Swab Oral Fluids	I I I	Research Period: No Sampling and Testing Requirements			
Growing Pig Nursery, Grower, Finisher, Isolation	Oral Swab Blood Swab Oral Fluids	I I A	Research Period: No Sampling and Testing Requirements			
Farrow to Feeder **Farrow to Finish**	Requirements of Breeding Herd + Growing Pig In Numbers, and Growing Pig Only in Frequency					
Small Holding ≥ 100 or < 1,000 Breeder or Feeder Swine	Oral Swab Blood Swab Oral Fluids	I I I or A	Research Period: No Sampling and Testing Requirements			
Non-Commercial < 100 Breeder or Feeder Swine	Oral Swab Blood Swab Oral Fluids	I I A	Research Period: No Sampling and Testing Requirements			

[1] I = Individual Sample, A = Aggregate (Group or Pen) Sample

Figure 3: Sampling and testing requirements for ASF-CSF risk level 2.

ASF/CSF Status = Level 2,
US Positive, Operations Normalizing, and State or Region Negative (All US SHIP Testing is outside of Control Areas)

				Sampling & Testing Requirements (Alternative Options)		
				Option 1 Individual Only		Option 2 Aggregate Only (Group or Pen)
Production Site Type	**Specimen Type(s)**	**I or A**[1]	**Frequency / Timing of Sampling**	**# of Individuals**	**# of Pools (Groups of up to 5)**	**# of Samples**
Boar Stud Mature Boars, Distributing Semen, ± On-Site Isolation	Oral Swab Blood Swab Oral Fluids	I I I	2X per month	10	2	-
Breeding Herd Breed to Wean, Breeding/ Gestation/ or Farrow Only, ± On-Site GDU or Isolation	Oral Swab Blood Swab Oral Fluids	I I I	Monthly	10	2	-
Growing Pig Nursery, Grower, Finisher, Isolation	Oral Swab Blood Swab Oral Fluids	I I A	Monthly	10	2	2
Farrow to Feeder Farrow to Finish	Requirements of Breeding Herd + Growing Pig In Numbers, and Growing Pig Only in Frequency					
Small Holding ≥ 100 or < 1,000 Breeder or Feeder Swine	Oral Swab Blood Swab Oral Fluids	I I I or A	Monthly	5	1	1 per 500, or 2 if > 500 pigs
Non-Commercial < 100 Breeder or Feeder Swine	Oral Swab Blood Swab Oral Fluids	I I A	Quarterly	5	1	1

[1] I = Individual Sample, A = Aggregate (Group or Pen) Sample

Figure 4: Sampling and testing requirements for ASF-CSF risk level 3.

ASF/CSF Status = Level 3,
US Positive, Immediately After Incursion, or if State or Region Positive. (All US SHIP Testing is Outside of Control Areas)

Sampling & Testing Requirements (Alternative Options)

				Option 1 Individual Only		Option 2 Aggregate Only (Group or Pen)
Production Site Type	**Specimen Type(s)**	**I or A[1]**	**Frequency / Timing of Sampling**	**# of Individuals**	**# of Pools (Groups of up to 5)**	**# of Samples**
Boar Stud Mature Boars, Distributing Semen, ± On-Site Isolation	Oral Swab Blood Swab Oral Fluids	I I I	Weekly	10	2	-
Breeding Herd Breed to Wean, Breeding/ Gestation/ or Farrow Only, ± On-Site GDU or Isolation	Oral Swab Blood Swab Oral Fluids	I I I	2X per month	10	2	-
Growing Pig Nursery, Grower, Finisher, Isolation	Oral Swab Blood Swab Oral Fluids	I I A	Monthly	20	4	1 per 500 pigs with maximum of 8 per site
Farrow to Feeder **Farrow to Finish**	Requirements of Breeding Herd + Growing Pig In Numbers, and Growing Pig Only in Frequency					
Small Holding ≥ 100 or < 1,000 Breeder or Feeder Swine	Oral Swab Blood Swab Oral Fluids	I I I or A	Monthly	10	2	1 per 500, or 2 if > 500 pigs
Non-Commercial < 100 Breeder or Feeder Swine	Oral Swab Blood Swab Oral Fluids	I I A	Monthly	5	1	1

[1] I = Individual Sample, A = Aggregate (Group or Pen) Sample

Resolution # 2: Pilot demonstration of a more comprehensive approach and system of traceability in the US pork industry (ie, similar to PigTRACE™ Canada)

The US SHIP House of Delegates requests:

Completion of an in-depth gap analysis assessing the feasibility of implementing a system of traceability standards patterned after the basic tenets of PigTRACE™ Canada.

Pursue the completion of a smaller-scale demonstration project of a more comprehensive approach and system of traceability within a subset of states and US SHIP pork industry participants (ie, 1 or 2 states with a subset of interested supply chains tracking all inter-premises movements of pigs through slaughter, with opportunity for use of AgView software).

Resolution # 3: Feed biosafety

The US SHIP House of Delegates requests the commissioning of a working group to provide recommendations and next steps for the US SHIP program to address the risks associated with disease transmission in feed (including but not limited to regulated garbage feeding) for a broadly applicable feed biosafety plan to be recognized nationally.

Resolution # 4: Biosecurity site plans

The US SHIP House of Delegates requests the commissioning of a working group to integrate the Secure Pork Supply plan and provide recommendations and next steps for the US SHIP program for a broadly applicable biosecurity site plan to be recognized nationally.

Resolution # 5: Sanitary standards of transportation to/from terminal markets

The US SHIP House of Delegates requests the commissioning of a working group centering on:

1. Quantifiably understanding the status-quo of the standards of practice and existing infrastructure for cleaning and disinfecting trailers returning from terminal points concentration.
2. Identifying existing production systems and supply chains routinely cleaning and disinfecting all trucks leaving terminal points of concentration prior to returning to farm sites to gauge best practices, costs, and infrastructure requirements.
3. Obtaining stakeholder sentiment from a broad range of producers and slaughter facilities across the country.
4. Preparing a summary of findings and formulating a set of recommendations to be presented and considered at the US SHIP HOD in 2022.

Resolution # 6: Live animal marketing channels

The US SHIP House of Delegates requests the commissioning of a working group to leverage the previous NASAHO efforts and recommendations. This working group would provide recommended next steps for the US SHIP program as it relates to additional research or recommended program standards to better mitigate the risk and impact of disease transmission in and from all live animal marketing channels. Any such recommendations related to a US SHIP program standard would then be considered further at the US SHIP 2022 HOD.

The marketing segment will have an opportunity for voting representation beginning with the 2022 US SHIP HOD.

Resolution # 7: Sampling & testing, 12-month research period, plan of work

The US SHIP House of Delegates provides its support for the following US SHIP sampling & testing related items to proceed forward as part of the plan of work to be initiated during this initial 12-month Research Period.

1. Further explore and prepare a summary document with options, considerations, and recommendations for US SHIP ASF-CSF Risk Level 1 (Peace Time) sampling and testing requirements.
2. Complete further modeling of disease spread and sensitivities of detection achieved via US SHIP ASF-CSF Monitored participatory sampling and testing requirements.
3. Work closely with USDA, SAHO, and industry partners to complete a study (ASF/CSF PCR Negative Cohort Study) that aims to build upon a number of USDA sponsored efforts looking to expand the number of ASF/CSF PCR assays and sample types approved for use to support ASF/CSF diagnostic efforts. This study would make a substantive contribution towards a much larger collective effort focused on creating a step-change in the ASF/CSF PCR diagnostic testing capacity and state of readiness across the NAHLN.
4. Pursuant of applied research evaluating the presence/ absence of ASF/CSF in processing fluids of infected piglets. This course of study would aim provide the foundational information needed to understand the potential fitness for using processing fluids as a sample type for ASF/CSF disease monitoring purposes in otherwise asymptomatic breeding herds.
5. Clarify process for review and approval of screening assays to be used for US SHIP ASF-CSF Monitored certification purposes.

https://doi.org/10.54846/am2022/186

Validation of extended storage protocols

S. Dee[1], DVM, MS, PhD, Dipl, ACVM; K. Havas[1], DVM, PhD, MBA, Dipl, ACVPM; A. Shah[2], MBA; M. Niederwerder[3], DVM, PhD; E. Nelson[4], PhD
[1]Pipestone Applied Research; [2]SAM Nutrition; [3]Kansas State University; [4]South Dakota State University

Introduction

It has been shown that animal feed can harbor pathogens of swine for extended periods.[1] One component of the Responsible Imports protocol is climate-controlled storage of feed to allow time for viral degradation.[2] While this concept has been evaluated under laboratory conditions,[3] it has not been tested under conditions representative of commercial feed production. In addition, the specific temperature required to effectively inactive viral contaminants in feed ingredients under conditions of extended storage has not been determined. Therefore, the objective of this study is to determine the necessary ambient temperature required to inactivate Seneca virus A and PRRSV 144 L1C in 1-ton totes of conventional soybean meal during a 30-day extended storage period.

Materials and methods

A 30-day storage period will be used throughout the study and during the 30-day storage period, totes will be stored at controlled temperatures using refrigerated trailers. The animal component of the study will be conducted at the Pipestone Research BSL-2 facility consisting of 6 rooms, each with its own feed bin. Each airspace is self-contained, allowing for individual treatments to be assigned per room without the risk of cross-contamination. Based on its documented protective effect,[1] conventional (high protein/low fat) soybean meal (SBM) will be used in the study. To represent commercial feeding conditions, SBM will be stored in 1-ton totes. To produce a "hot spot" of contamination, totes will be spiked with a 10 mL ice cube containing Seneca virus A (SVA) and PRRSV Lineage 1C (PRRSV) at 5 logs $TCID_{50}$ total dose. SVA was chosen for this study as it can be used under BSL-2 conditions, is an effective surrogate for foot and mouth disease virus and survives well in conventional soybean meal.[1] For this study, we will evaluate virus survival at 3 different temperatures in conjunction with a 30-day storage period. Temperatures (treatments) used to store totes will be 75°, 60°, or 50°F. We will use 6 totes per treatment with the 1-ton tote serving as experimental unit. A sample size of 6 totes per treatment yields a 95% confidence and 80% power to detect a 75% difference in the proportion of totes that cause disease. To evaluate whether SVA and PRRSV 144 L1C are still viable at the end of each storage period, the entire 1-ton tote of SBM will be fed to cull finishing pigs for a 14-day period at the Pipestone BSL-2 (natural feeding behavior challenge). Thirty finishing pigs, capable of consuming 5 lbs./head/day will be housed in each room. Following consumption of the SBM ton, pigs will be fed complete feed to supplement potential calcium and phosphorus shortages secondary to the SBM feeding for an additional 14 days. During this 30-day period, oral fluid samples will be collected at day 0, 14, and 28 post challenge and tested by PCR for evidence of SVA RNA and PRRSV RNA. In addition, individual pigs demonstrating clinical signs of SVA (lameness, vesicular lesions, poor condition, etc.) and PRRSV (dyspnea, anorexia, pyrexia, etc.) will be necropsied and tissues will be tested for the presence of SVA and PRRSV. The proportion of totes that cause disease will be evaluated using the Fisher Exact test or Z-test. If a correction is needed due to a perfect difference between the tote storage groups, a Yates Correction will be applied. Appropriate binomial confidence intervals will also be calculated and compared and presented at the AASV meeting.

Results and discussion

As of this writing, the project is underway. One replicate has been completed and SVA and PRRSV were inactivated across all totes stored at 75°F. Results from the viability assessment at 60°F and 50°F will be presented at the AASV meeting.

References

1. Dee SA et al: 2018, PLOS ONE, 13(3): e0194509.
2. Patterson G et al: 2019, JAVMA,254,7: 1-2.
3. Dee SA et al: 2021, J Swine Health Prod, 29(3):124-128.

https://doi.org/10.54846/am2022/187

Comprehensive procedure of responsible animal feed importation from ASFV positive countries

A. Shah; S. Shah; S. Dee, DVM, MS, PhD, Dipl ACVM; R. Cochrane, PhD; A. Wu, PhD; T. Swenson; A. Yang

Introduction

The Responsible Importer Guidelines and Policies (RIG) was designed for the safety and integrity of animal feed ingredients, and biosecure import methods from ASFV positive countries. All imports follow a chain of command to which they are purchased from an approved supplier or manufacturer, selected from a rigorous onboarding process; imported via shipping cargo containers under strict guidelines into the United States, or an ASFV negative country; and quarantined in a heated, controlled environment greater than 70 degrees Fahrenheit for a minimum of thirty-five days. After undergoing the import process, imported goods are handled and shipped in accordance with stringent standard operating procedures in efforts to reduce any potential or known domestic risks and hazards, in accordance SAM Nutrition's internal policies and regulations established by the US Department of Transportation, US Department of Agriculture, and Food and Drug Administration, as well as other applicable regulatory agencies.

Materials and methods

As defined by Kansas State University, the "African swine fever virus (ASFV) is a contagious, rapidly spreading, transboundary animal disease and a major threat to pork production".[1] It is a unique double-stranded DNA virus that causes lethal hemorrhagic disease of domestic swine. Due to its high mortality rate, effected countries and regions experience significant negative economic impact. An effective vaccine has not yet been developed.

Phase one: Safe loading and operating requirements at/in the country of origin, in an ASFV positive country

The production process of high-risk feed ingredients consists of (or is similar to) the batching of ingredients, heat treatment and fermentation, filtration, distillation and crystallization, and spray drying conducted at high temperatures between 90 and 120 degrees Celsius. At such heat, the likeliness of positive ASFV in the actual consumable product is near zero - warranting product integrity and high quality. Therefore, the only form of possible transmission is in supply chain and transportation, residing on the outer packages or materials used to transport the actual goods, if any at all. Pallets are to be heat treated and fumigated prior to use, and all final products - palletized, shrink-wrapped cargo - must be sprayed with a disinfectant prior to loading in a certified clean container.

While on standby for international shipment, all final products are safely stored an average of two weeks in a secured warehouse with limited access and is thoroughly covered in plastic sheets in protection to any potential hazards. Warehouses operate in respect to product for quality control and assurance purposes; no two products are warehoused together. Clean overalls and booties are required to be always worn when working in a secured warehouse.

The shipment loading of goods first requires completion of the Certified Clean Container (CCC) policy established and specifically designed to guarantee the cleanliness of the mode of transportation. A general purpose 20-foot or 40-foot container that is certified clean must undergo 120 hours of heat and fumigation treatment of Aluminum Phosphide at $4g/m^3$, or other similar cleaning agent and concentration, compliant with US FDA regulations for the transportation of food-grade products, composing a shipment method that is near free of all potential cross contamination. Post-treatment, containers are to remain closed and sealed until product loading at the production plant and re-sealed immediately thereafter, in accordance with US import laws and the export regulations of the country of origin, in preparation for vessel loading and voyage (see figure 1). Container seals may only be stripped at final destination by an authorized, trained staff; or at any location on route by the ocean carrier, US Customs and Border Protection or other applicable regulatory agencies for import clearance and inspection purposes only, accompanied with official paperwork and proper notification. All documentation associated with inland movement in the country of origin, loading at the production plant, and cargo importation are carefully recorded and archived, in compliance with the Foreign Supplier Verification Program (FSVP) under jurisdiction of the US FDA.

Phase two: Product arrival, unloading and handling at final destination in an ASFV negative country

The arrival of shipping containers is strictly managed through numerous layers of documentation to ensure a strong traceability system for quality control. All container and seal numbers must be cross referenced with the declared import documents to ensure the integrity of the method of shipment. Declared import documents consists of commercial invoices, packing lists, bills of lading, telex releases, certificates of origin, certificates of insurance, certificates of clean container fumigation, and all associated certificates of analysis with corresponding lot numbers and quantity breakdown. The average overall transit time with transshipment stops and rail transfers are not a risk to biosecurity and quality control if, and only if, containers arrive with the original seal, sealed at the loading origin, intact and unimpaired. Shipments with discrepancies may or may not be offloaded following a thorough investigation to identify its nonconformity (see figure 2).

Approved and verified containers are to be offloaded only at the arranged inbound loading dock under operating biosecurity guidelines, in preparation to undergo the quarantine process, to which the container numbers are accurately documented, the number of off-loaded products by pallet count, the product type, brand or manufacturer, package size, and the associated lot numbers of each pallet and its quantity. All preceding documents are stored securely on hard drives as well as cloud-based platforms to ensure that they remain available as long as needed, accessible to authorized staff, under US CBP and FSVP consumable goods traceability guidelines.

All shipments that consist of transcontinental shipping through, produced or packaged in a country identified with the presence of positive ASFV must undergo SAM Nutrition's Transit + 35-days

Figure 1: Product inland travel at country of origin, factory to cargo vessel.

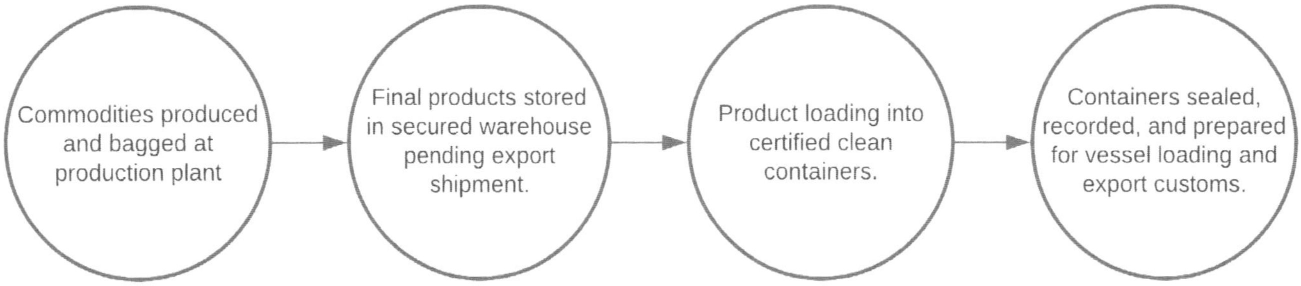

Figure 2: Overall product handling and import process.

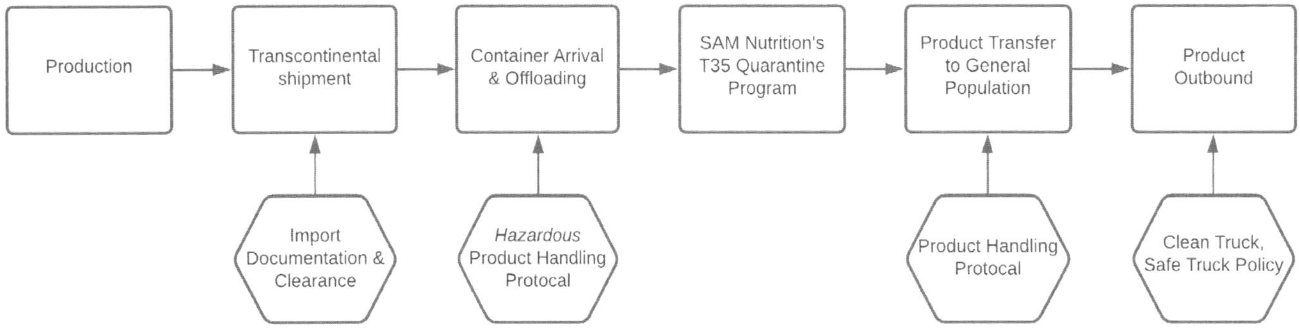

(T35) quarantine program. Prior to the offloading of any products subjected to the quarantine program, additional high-risk product handling protocol is practiced as precautionary measures of cross-contamination between new and unquarantined import goods, and the general warehouse space consisting of non-high-risk materials. All trained and authorized staff operating in the controlled environment during the offloading of high-risk materials must wear a disposable overall, gloves, shoe covers, hair net and goggles with any long or loose hair tied back. Upon completion of the required work, SAM personnel are to carefully remove the protective wear and discard it in the designated enclosed bin in the Sanitize Control Room (see figure 3).

All equipment such as forklifts, pallets jacks, disposal bins, brooms, and mops are to be used only in the space they are assigned to; for example, disposal bins in the warehouse are red in color, and those in the general warehouse space are green for efficient identification. Pallets entering quarantine are sprayed with Concentrated HDX Germicidal Bleach, a disinfectant, non-food contact sanitizer and then arranged in such a way to best utilize the space, while staying a minimum of two feet from each product. The additional high-risk product handling protocols are designed only for the operations in the controlled environment and includes guidelines on training, proper attire and use of equipment.

Phase three: True quarantine program of thirty-five days

SAM's ASFV T35 Quarantine Program is a specific comprehensive structure of additional biosecurity measures for quality assurance and was established under the guidelines provided by professionals from the Swine Health Information Center, the National Pork Board, the National Pork Producers Council, and the American Association of Swine Veterinarians, and is a 'true quarantine' due to system cycles prohibiting new arrivals to be placed next to products that are already in quarantine and never to those that have been moved from the quarantine control room to the general population.

Quarantine products are prohibited to be released from the control zone without the appropriate written authorization from executive management and prior to the completion of their 35-day quarantine program against the date to which the controlled zone has been sealed. The quarantine loading dock(s) is strictly prohibited for open to the offloading of new products until the quarantine-controlled zone has been cleared of and in no possession of previous ASFV subjected products. The number of quarantine days, of 35 days was selected in guidance from the mean holding time for 99.99% degradation of ASFV in Lysine at 13 days at 59 degrees Fahrenheit, to which we have nearly tripled in addition to an average transcontinental travel time of 35 to 42 days to ensure minimum contamination and nonviable existence of the virus (see Table 1).[2]

As additional biosecurity measures, staff visitors, and personnel who were in close contact with any and all swine creatures in the United States and other areas, region, and countries with the absence of ASFV, are not permitted to enter the warehouse space or its premises for 36 hours, and those who visited ASFV positive locations are prohibited from entering for a minimum of seven days. Those who traveled directly to farms or locations in possession of swine (livestock or meat, frozen or unfrozen) of infected areas are prohibited for a minimum of fourteen days. All shoes, attire and wear associated with the visit must be thoroughly washed.

Figure 3: Warehouse layout.

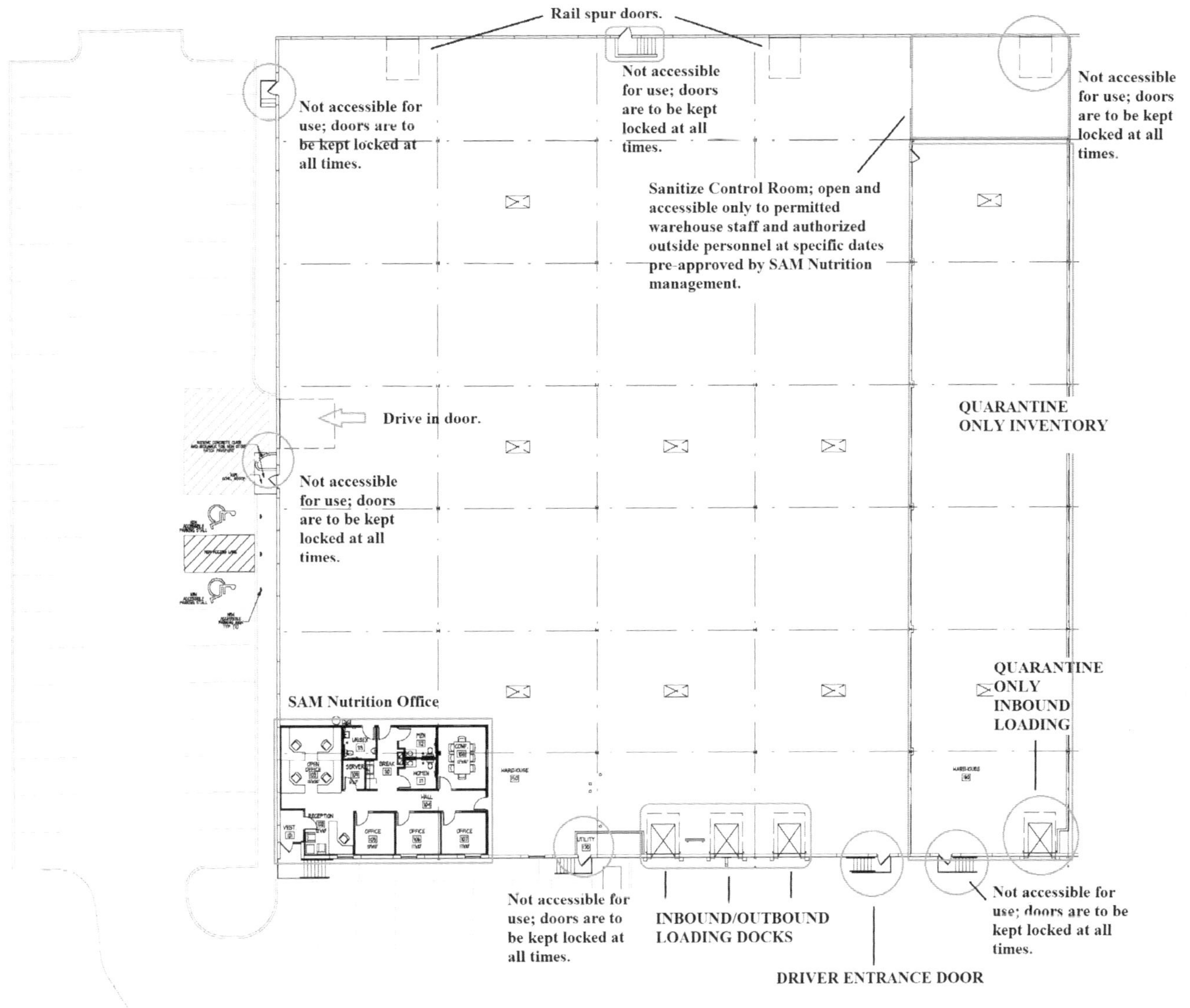

Phase four: Safe truck, clean truck outbound and additional warehouse standard operating procedures

To maintain high-quality products and a safe operating facility, all carriers of SAM Nutrition must abide with the Safe Truck, Clean Truck policy designed to proactively avert cross-contamination from known or potential hazards. All trucks operating for, on behalf of, or providing services SAM Nutrition must clean and free of substances, objects, or items that can and/or may compromise product integrity.

All drivers are required to sign in at the driver front desk to which they are a asked a number of questions including if they and/or their trailer has recently visited, delivered or entered the premises of a feed mill, farm or come into close contact with livestock. Drivers and trailers who answer yes are required to provide a valid washout ticket to be loaded. All wash certificates must be washed out at a minimum of 180 degrees Fahrenheit hot water or with a "Quat" Sanitizer/Disinfectant compound (Ammonium Acid base), compliant with FDA regulations on the transportation of feed and food grade products. No animal-based proteins are permitted to be carried in trucks that deliver for SAM Nutrition. In addition, any trailer containing any cattle origin must be thoroughly washed to reduce the spread of bovine *spongiform encephalopathy* in the United States. All carriers must comply with Federal Code 21 CFR §589.2000 and Federal Code 21 CFR §589.2001. SAM Nutrition reserves the right to refuse truck loading for whatsoever reason.

Additional measures of warehouse standard operating procedures include a strict visitor policy and robust security plan. All non-associates on company property must be issued a pass by the appropriate administrative associate or receptionist at the check-in point. Identification cards and all other applicable licenses or badges must be present for access and visitors must be accompanied by a SAM Nutrition staff at all times.

Table 1: The number of suggested holding days for 99.99% ASFV degradation based on controlled temperatures

Mean holding time for 99.99% degradation	Number of days at 4°C (36.9°F)	Number of days at 15° (59.0°F)	Number of days at 30° (86.0°F)
Conventional soybean meal	143	52	26
DDGS	494	182	26
Vitamin D	39	26	26
Lysine	78	13	13

SAM Nutrition's warehouse also has a state-of-the-art security plan consisting of sensors at all entry points, camera footages at all loading docks, entrance access codes with varies settings based on staff access needs and requirements, and a full emergency response and alarm system.

Results and discussion

All the described measures, approaches and regulations established by SAM Nutrition was precisely developed for upkeeping product integrity and its distribution methods from ASFV and any known and potential hazards, both domestic and international, to protect the nation and its feed industry as a whole. This process includes long-term partnership with verified and reputable animal feed manufacturers, a comprehensive import supply chain including food-grade certified clean containers, and successful completion of the T35 Quarantine Program combined with additional biosecurity protocols to successfully achieve an ASFV degradation rate of 100 percent.

References

1. Niederwerder, M. C., Stoian, A., Rowland, R., Dritz, S. S., Petrovan, V., Constance, L. A....Hefley, T. J. (2019). Infectious Dose of African Swine Fever Virus When Consumed Naturally in Liquid or Feed. Emerging Infectious Diseases, 25(5), 891-897. **https://doi.org/10.3201/eid2505.181495**

2. Dee SA, Bauermann FV, Niederwerder MC, Singrey A, Clement T, de Lima M, et al. (2018) Survival of viral pathogens in animal feed ingredients under transboundary shipping models. PLoS ONE 13(3): e0194509. **https://doi.org/10.1371/journal.pone.0194509**

https://doi.org/10.54846/am2022/188

African swine fever: A practitioner's perspective

Scott Dee, DVM, MS, PhD, DACVM; Gordon Spronk, DVM; Joseph Yaros, DVM
Pipestone Applied Research, Pipestone, Minnesota

African swine fever virus (ASFV) was officially reported by the Ministry of Agriculture and Rural Affairs (MARA) in the domestic swine herd in China on August 3, 2018.[1] Pipestone staff have been visiting swine herds throughout China and Asia for over 20 years as requested by Chinese producers, vendors to the Chinese industry, governmental agencies, and non-governmental agencies (NGOs). These visits have given our staff extensive field observations of clinical disease in swine herds in Asia both prior to and after the official reporting of ASFV. It may be of interest that swine disease presentation in Asia in the field typically results in the same field observations of the same pathogens as in North American herds and the rest of the world. In the majority of cases, the pathogen presentation is as widely described and published. In a small subset of field observations however, the presentation may be different and vary widely due to co-infections, pathogen isolate differences, operational differences, and implemented interventions or medications. Additionally, several pathogens described as foreign animal diseases, (FADs) including foot-and-mouth disease virus (FMDV), classical swine fever virus (CSFV), pseudorabies virus (PRV) that have been eliminated from swine herds in North America and Europe have been commonly observed in China swine herds. These field observations of isolated pathogens (FADs) or with co-infections may provide opportunity to educate future global veterinarians and health officials allowing improvement in the health of the global swine herd. This paper will focus on the field observations of ASFV in commercial operations in Asia and actions practitioners in the US can take to prevent spread and movement of the virus upon possible entry into the US.

ASFV in the world swine herd

Prior to official reports of ASFV in the largest swine herd in the world, ASFV was commonly known to have spread widely over the past nearly 100 years from Africa to the Iberian Peninsula, to eastern Europe, across Russia, and finally into Asia. Movement of the virus is commonly understood to be the result of "people or human" movement of the pathogen; the virus movement speculated to be by movement of either infected pigs, or meat by producers or staff.

ASFV is known and reported to be one of the deadliest viruses (if not the deadliest as measured by mortality) to infect swine of all ages. In small stakeholder operations in Asia, the virus may result in death loss of 100% of the pigs in the herd (note that any surviving pigs are commonly sold prior to death resulting in under reporting of true mortality). In larger operations, a small percentage of pigs have been observed and reported to survive an ASFV episode.

Clinical presentation of ASFV infection is widely published and commonly understood to include a hemorrhagic fever, slow spread in a population, and high mortality.

In China, from August 2018 until August 2019, there were official reports of 160 cases in 32 of the 33 provinces in China. The virus has spread throughout 15 countries in Asia, affecting pigs of all ages and genetics. Additionally, all types of production systems have been infected and observed under field conditions including smaller traditional farms (less than 500 head) to modern western style barns (over 8,000 sows) of all designs and operational practices. Since its identification in herds in China in August 2018 to January 2022, producers have learned to live the virus in the country and restored the number of sows in operation back to July 2018 numbers to an estimated 42 million sows.[5]

Reports of vaccines for ASFV and their availability in China are widely described. The first tool that veterinarians and producers traditionally implement to control any pathogen is a vaccine which successfully reduces mortality in an affected population and stops the spread of the pathogen to naïve animals. One additional feature of a good vaccine is the ability to discern the difference between the vaccinated pig and a pig exposed and infected with field virus. Researchers have struggled for decades to develop an effective ASFV vaccine and as of this writing, there is no officially approved vaccine for ASFV to prevent infection of naïve pigs.[2] ASFV vaccines used in China reportedly revert to virulence after 6 months post vaccination, resulting in clinical signs identical to wild type virus infection. The volume of ASFV shed from these infections is low, limiting timely diagnosis and allowing these attenuated viruses to spread from one operation to another.

On July 28, 2021 ASFV was diagnosed in the Dominican Republic and in Haiti on September 20, 2021.[6] This marks the first time ASFV has been diagnosed in the Western Hemisphere since it was last identified in Haiti in 1983 and brings the virus within 70 miles of Puerto Rico. In the US, all producers have benefited from ASFV not being identified within the country. To maintain this status, producers need to be aware of the clinical signs of the virus and correct sampling procedures for diagnosis. The optimal diagnostic samples are collected from pigs exhibiting symptoms of concern and include whole blood, spleen, and superficial lymph nodes including notifying the state veterinarian located in the associated state. Only if ASFV is diagnosed rapidly before spreading to multiple sites is there a chance at elimination from the national herd.

ASFV epidemiology

The epidemiology of ASFV is widely published under controlled research conditions: the virus is shed from all porcine bodily fluids, with blood containing the highest virus concentrations. Pigs can be infected via nose to nose contact, semen, milk, fecal/oral, injections under the skin containing viral particles, or ingestion of tissue from infected pigs. The incubation time of the virus is 3-14 days, with clinical signs including lethargy, erythema, piling of pigs, febricity, bloody scours, and sudden death.

More epidemiological research and publication is needed as field reports from China in large commercial operations with potential co-infections dispute some of these reported results. Some of the variation in field reports may be explained by differences in virus isolate commonly reported as high, moderate and low virulence. These ASFV isolates also may have different field presentations and pathogenesis described as peracute, acute, subacute and chronic. Field observations support these reported differences in clinical presentation.

Differential diagnosis for ASFV include, but are not limited to, classical swine fever, porcine reproductive and respiratory syndrome (PRRS), pseudorabies virus, Salmonella, and post-weaning multisystem wasting syndrome and other bacterial pathogens including Clostridium and Streptococcus.

ASFV is highly pathogenic, but not reported to be highly transmissible. Some authors describe this as "contagiosity" (there may be high mortality with low infection rate when compared to other pathogens like FMDV).[3] In clinical field cases in large populations of commercial pigs where the virus has been left uncontrolled and epidemiological data has been monitored, total mortality can more than double in the matter of days until the majority of the population is deceased.

Investigations into the source of the infections into a population in Asia have been conducted to understand the routes of entry which may allow for future prevention of movement of the virus to other area farms. In general, field observations in Asia support rapid area spread with speculated routes of movement of the virus attributed to movement of infected pigs, movement of infected meat, contaminated fomites, rats, flies, mosquitoes, staff, feed, and lack of implementation of proper biosecurity protocols. The role of contaminated feed cannot be overlooked in the transmission dynamics. Improper disposal of the carcasses and movement of infected pigs are high on the list of methods that may have moved this virus across all of China. Pigs that have died from ASFV require proper disposal and enhanced biosecurity practices to limit the spread of ASFV to naïve populations since carcasses are shedding high levels of virus.

Site depopulation and cleanup protocols have been implemented and observed under field conditions and vary widely. It is critical that following depopulation of a site, a cleaning and disinfection protocol that removes all organic material and viral particles is completed to prevent viral transmission to the next group of pigs and for the site to remain healthy and free from ASFV. Commonly accepted protocols for other viruses (including but not limited to porcine epidemic diarrhea virus or PRRS) have been successfully used in the field. Attempts to eliminate ASF virus from sow populations using test and removal techniques are reported in China.[4] This elimination technique using herd closure, partial depopulation and whole herd sampling methods may hold promise and needs further research and documentation to understand if effective for this pathogen.

To ensure that ASFV does not enter other sites, a full biosecurity evaluation should be conducted to ensure that protocols regarding personnel entry, trucking, loading, feed, downtime, and other possible routes of entry are properly written, staff are educated on the topics, and standard operating procedures are properly implemented. A critical step is to be sure daily monitoring is also completed to maintain daily review and correction of any deficiencies in implementation and education by staff on the farm.

Additional capital expenditure projects to improve biosecurity are often necessary to prevent the infection and loss of more sites and farms. The only country that has successfully eliminated ASFV is Spain, in 1995, after years of unsuccessful attempts to eliminate ASFV from their national herd.[2] All across the world it is the responsibility of the individuals with the greatest expertise in the swine industry to provide assistance and knowledge on preventing the spread and circulation of ASFV. In the United States, with leaders in the swine industry leading the charge to keep ASFV out of the country, it is hopeful that government authorities will support and follow.

References

1. 2019. ASF Situation in Asia Update. *FAO ASF Situation Update - African Swine Fever (ASF) - FAO Emergency Prevention System for Animal Health (EMPRES-AH).* Food and Agriculture Organization of the United Nations.

2. Sanchez-Vizcaino, Jose M., et al. 2019. African swine fever virus. *Diseases of Swine. 11:443-452.*

3. Depner, K. 5 Aug. 2018. The World of ASF A World Apart. 7th ECPHM Resident Workshop. 7th ECPHM Resident Workshop.

4. Yan Jason 2019. Why New Generation Testing and Removal is Winning ASF Control in China. Proceedings, 2019 Leman China University of Minnesota. Zhengzhou, China. October, 2019.

5. Beijing Boyar Communication co., Ltd. Sep 2021. 2022 China Swine Industry Outlook Report.

6. United States Department of Agriculture. USDA Statement on Confirmation of African Swine Fever in Haiti. 21 Sep. 2021.